Business Calculus

Revised Second Edition

Maijian Qian and Adam Glesser

Mathematics Department

California State University, Fullerton

Cover design courtesy of Kendall Hunt Publishing

Chapter 6 figures created by Maijian Qian and Adam Glesser

www.kendallhunt.com
Send all inquiries to:
4050 Westmark Drive
Dubuque, IA 52004-1840

Copyright © 2010, 2017 by Maijian Qian and Adam Glesser

PAK ISBN 978-1-5249-6128-2
TEXT ISBN 978-1-5249-6129-4

Kendall Hunt Publishing Company has the exclusive rights to reproduce this work,
to prepare derivative works from this work, to publicly distribute this work,
to publicly perform this work and to publicly display this work.

All rights reserved. No part of this publication may be reproduced,
stored in a retrieval system, or transmitted, in any form or by any
means, electronic, mechanical, photocopying, recording, or otherwise,
without the prior written permission of the copyright owner.

Published in the United States of America

Errata

Items from the Second Edition corrected for Revised Second Edition

Page and line numbers refer to the Second Edition.

Page	Line or Section	Correction
v	-5	Replaced "author's" with "authors"
18	8	Replaced "thousands of dollars" with "thousand dollars"
84	-1	Replaced "$\lim_{x \to 1^+} f(x) = 1$" with "$\lim_{x \to 1^+} f(x) = 2$"
88	2	Replaced "Tables 2.2.1 and 2.2.2" with "Tables 2.2.3 and 2.2.4"
96	12	Replaced "$30,00$" with "$3,000$"
129–130	Example 1(b),(c)	Fixed $f(x)$ to equal $\sqrt{x}$ rather than $\sqrt{x^2+1}$
132	3	Replaced "composition" with "composite"
156	3–4	Replaced "radiuses" with "radii"
158	6	Replaced $(0.5)(0.02) = 0.01$ to $(0.4)(0.2) = 0.008$
178	15	Replaced "increasing" to "decreasing"
186	-3	Added a hyphen between "First" and "Derivative"
189	4	Added a hyphen between "Second" and "Derivative"
213	-14	Added a parenthesis in the numerator of the third expression, after $-x^2 + 9 + 4x^2$
215	-5	Removed a superfluous parenthesis
239	Example 6	Changed the language to make it clear that the storage cost per couch does not depend on whether the couch is currently being stored
239	-14	Replaced "Therefor" with "Therefore"
251	Reality Check	Cleaned up the language
254	6	Added a missing parenthesis to the second expression
264	3, 5	Replaced "if they exist" with "if it exists"
288	9	Corrected the Constant Multiple Property to exclude $k = 0$
288	-13	Replaced "provide" with "provided that"

Page	Line or Section	Correction
327	Exercise Set 4.5	Deleted repeats of problems 31–34
367	7	Replaced "500,00" with "50,000"
369	2	Replaced "2,2903" with "22,903"
386	Problem 32(a)	Replaced "an uniform" with "a uniform"
441	-15	Eliminated unneeded reference to another example
441	-4	Replaced "six" with "five"

Contents

	Preface	vi
	Preface (1st Edition)	viii
1	**Functions and Mathematical Models**	**1**
1.1	Functions and Graphs	2
1.2	Slopes and Linear Functions	8
1.3	Application of Functions to Economics	16
1.4	Nonlinear Functions and Algebraic Equations	26
1.5	Exponential Functions and Financial Models	40
1.6	Logarithmic Functions and Exponential Equations	48
	Chapter 1 Summary	60
	Chapter 1 Review Exercises	64
	Chapter 1 Test	68
	Chapter 1 Projects	71
2	**Differentiation**	**72**
2.1	Limits: Approached Numerically and Graphically	73
2.2	Limits and Continuity	84
2.3	Average Rates of Change	94
2.4	Differentiation Using Limits of Difference Quotients	103
2.5	Basic Derivative Formulas and Properties	111
2.6	Differentiation Techniques: Product and Quotient Rules	122
2.7	Extended Power and Natural Exponential Rules	128
2.8	Chain Rule and Derivatives of Log Functions	133
	Chapter 2 Summary	142
	Chapter 2 Review Exercises	149
	Chapter 2 Test	152
	Chapter 2 Projects	155
3	**Applications of Differentiation**	**156**
3.1	Differentials and Linear Approximation	157
3.2	Marginal Analysis	165
3.3	How Derivatives Affect the Shape of a Graph	178
3.4	Derivative Tests and Relative Extrema	189
3.5	Graph Sketching: Using the Derivatives	202
3.6	Graph Sketching: Asymptotes and Rational Functions	212
3.7	Absolute Maximum and Minimum	226
3.8	Optimization Problems	237
3.9	Elasticity of Demand	250
	Chapter 3 Summary	258
	Chapter 3 Review Exercises	264
	Chapter 3 Test	267
	Chapter 3 Project	273

4 Integration — **274**

- 4.1 The Area under a Graph … 275
- 4.2 Areas and Antiderivatives … 288
- 4.3 The Fundamental Theorem of Calculus … 301
- 4.4 Areas and Definite Integrals … 314
- 4.5 Integration by Substitution or Algebraic Manipulation … 326
- 4.6 Integration by Tables … 335
  - Chapter 4 Summary … 340
  - Chapter 4 Review Exercises … 345
  - Chapter 4 Test … 348
  - Chapter 4 Project … 352

5 Applications of Integration — **353**

- 5.1 Consumers' and Producers' Surpluses … 354
- 5.2 Definite Integrals in Finance … 369
- 5.3 Improper Integrals … 380
- 5.4 Probability Distributions and Density Functions (I) … 386
- 5.5 Probability Distributions and Density Functions (II) … 395
- 5.6 Differential Equations … 407
  - Chapter 5 Summary … 415
  - Chapter 5 Review Exercises … 424
  - Chapter 5 Test … 427
  - Chapter 5 Projects … 431

6 Functions of Two Variables — **433**

- 6.1 Functions of Two Variables and Partial Derivatives … 434
- 6.2 Maximum–Minimum Problems … 444
  - Chapter 6 Summary … 453
  - Chapter 6 Review Exercises … 457
  - Chapter 6 Test … 458
- Chapter 1 Exercises … 460
- Chapter 2 Exercises … 465
- Chapter 3 Exercises … 470
- Chapter 4 Exercises … 490
- Chapter 5 Exercises … 497
- Chapter 6 Exercises … 501

7 TABLES — **505**

- Table 1: Integration Formulas … 506
- Table 2: Areas of a Standard Normal Distribution … 507

Index — **509**

Preface (Second Edition)

For the Student

Why calculus? It doesn't make sense. Ask your favorite uncle. He worked at Raytheon for 25 years, the last fifteen as a Product Control Specialist, and never used calculus once. Of course, you will notice that he spent the last *fifteen* years in the same position. He lost vertical employment mobility when it became clear that, despite being a hard worker who was well-liked by his peers, he was unable to communicate technically with the engineers. Furthermore, in the last five years, most of the upper-management positions were assumed by former engineers. The engineers don't need colleagues who are technical masters, but they do need colleagues who are technically proficient, and who can at least parse their language. That language is calculus.

This book is meant to help you understand enough calculus to be able to work in just about any business setting, and to even see how your uncle might have used calculus—if he had only known what it was. Pay special attention to the red Warnings. Avoiding the common mistakes that students make is a sure fire way to make your instructor happy. And a happy instructor gives out happy grades. True Story. Also be sure to look through the green Reality Checks. These give you what may be your most important mathematical tool in the real world—the ability to recognize when you've done something wrong.

For Whomever Is Still Reading

In the debate over content versus style, mathematicians come down firmly on the side of content. Find a mathematician who routinely and knowingly divides by zero, and you have also found someone who probably likes to kick puppies. Find a mathematician who writes in a dull, colorless style with little interest in helping their reader understand via an innovative presentation, and you have found a typical mathematician. Focusing primarily on content is not a problem when you write to an audience of mathematicians. However, when your audience consists of college students taking their terminal (and it can't end soon enough) math course, you better bring a little more to the table than just Greek letters and ridiculously clever proofs.

This new edition is an attempt bring stylistic sensibilities to devastatingly interesting mathematical content. The exposition remains direct and intuitive, with even more helpful diagrammatic solution aids. The book, however, is now made much more usable by a graphical layout that better helps the student to compartmentalize the examples, results, mnemonics, and warnings. The text has been streamlined somewhat by the removal of a few sections that the authors found were frequently skipped. For instructors and students still wishing to cover these sections, the authors have made the sections freely available online.

Acknowledgements

Both of us wish to express our appreciation for all of our colleagues in the Mathematics Department of California State University, Fullerton, who used the previous edition of the textbook and provided extremely helpful feedback. We would also like to thank Linda Chapman and Lara Sanders at Kendall Hunt for their supervision of the project.

Dr. Glesser would like to thank Dr. Qian for inviting him to join this wonderful project. He would also like to thank his wife for supporting him through countless nights when he stayed up to make revisions and optimize LaTeX code.

October 15, 2016

Revised Second Edition

This Revised 2nd Edition corrects several errors found in the original 2nd edition while making a few minor cosmetic changes that, while unlikely to have bothered students, certainly offended the aesthetic sensibilities of the authors. The authors would like to thank all of the instructors who used the textbook this past year and sent in corrections and feedback. In particular, we thank Kim Norman for her exceptional attention to detail in reading through both the main text as well as the exercises. There is a online homework supplement to the textbook, and we thank Fatemeh Khatami and Paul Farnham for helping us to track down so many of the "lost in translation" errors that inevitably occur when transferring from one medium to another.

May 31, 2018

Preface (First Edition)

Business Calculus is suitable for use in a one-semester introductory calculus course for students majoring in business and economics. The prerequisite for this course is a course in college algebra. However, many important mathematical models and concepts, like linear and exponential functions, are revisited in the first chapter.

The calculus text is presented in a direct, intuitive manner with plenty of examples. To help students solve problems and understand concepts, examples are given in detail so that students will not be distracted by missing algebraic steps. Topics that are difficult for students to understand are written with extra care. Each chapter ends with a summary, a set of exercises, and a chapter test so that students can check how they have learned the material.

Applications related to the fields of business and economics are supplied throughout the book. The realistic applications illustrate the use of calculus in these disciplines. Often a new concept is motivated by an application, and many applications are related to real life. Therefore, students who use this book will learn valuable mathematical skills that can be applied to real-life financial decisions.

The text is written for student comprehension. Particular cases are usually discussed before concepts and results are introduced. The interplay among graphical, numeric, and algebraic concepts and properties is carried out throughout the text. The content focuses on computational skills, ideas, and problem-solving rather than mathematical theory. Most derivations and proofs are omitted. Boxes are used to highlight important definitions, results, and step-by-step procedures. Theorems are both numbered within the section and named for easy reference and clarification. Many important concepts and results are followed by detailed notes for further explanations. New terms are introduced with boldface type.

Exercise problems are carefully selected. Each exercise set is designed so that every student will experience success both in formula manipulations and problem-solving in applications. Solutions to the odd-numbered exercise problems are in the back of the book. Most of these problems have corresponding examples in the content to follow. The numbers of the reference examples are listed along with the solutions. Even-numbered exercise problems have similar odd-numbered problems to match, so that students can practice on them and check their answers with the solutions.

This textbook can be used without access to a graphing calculator. However, it is likely that many students will want to make use of a graphing calculator or mathematical software. Many examples are designed with this possibility in consideration. Often students are encouraged to use graphs to explore some important characteristics of functions.

Content

This text begins with the introduction of several important functions and mathematical models in business, economics, and finance. These functions and mathematical models are utilized throughout the text. The main topics in Chapter 1 include linear, exponential, and piecewise defined functions, along with their graphs; supply, demand, and equilibrium point; cost, revenue, and profit; and savings accounts and compound interest. Many phenomena in finance, such as compound interest, can be modeled as exponential functions. Therefore, exponential and logarithmic functions and models are introduced in the first chapter, rather than waiting until differential calculus and its applications on algebraic functions have been studied. Techniques of solving various types of algebraic equations are covered in Section 1.4. Depending on the syllabus for the course and the backgrounds of the students, selected portions of Section 1.4 may be referred to as needed later in a course, instead of being covered at the beginning of the course.

The material in Chapters 2 and 3 consists of differential calculus and its applications. The material in Chapters 4 and 5 consists of integral calculus and its applications. Chapter 6 deals with multivariable differential calculus and its applications. Some sections can be given brief treatments or omitted without affecting the remaining text. These sections are pointed out in the following discussions.

Chapter 2 introduces limits and derivatives. The first section introduces an intuitive approach (through numerical tables and graphs) to the concept of limits (including infinity limits and limits at infinity). The second section covers limit properties and quickly turns to the concept of continuity. The concept of limits is essential to understanding the definition of the derivative of a function, which is introduced in Section 2.4, after detailed discussions on the average rate of change in Section 2.3. Sections 2.5 to 2.8 cover various techniques of differentiation. The derivatives of power and natural exponential functions and basic rules on coefficients and sums/differences, as well as higher order derivatives, are introduced early on in Section 2.5, followed by the product and quotient rules introduced in Section 2.6. The extended rules for powers and exponentials are introduced in Section 2.7, before the general chain rule is introduced in Section 2.8. The derivative of the natural logarithmic function is introduced in Section 2.8, as an application of the chain rule. Since all exponential and logarithmic functions with general bases can be turned into composite natural exponential and logarithmic functions using the change of base formulas, their derivatives are briefly discussed in the end of Section 2.8.

The material in Chapter 3 consists of various topics related to differentiation. The first section introduces differentials and linear approximation. The first few examples in that section use real-life situations to enhance the understanding of meanings of the derivative at a given point, and how the derivatives can be used to estimate the function values around a given point. These techniques are then applied in Section 3.2 to analyze productivity. In that section, a particular type of optimization, maximization of profit, is introduced.

To solve many problems raised in the business world, mathematical models are used. Such mathematical models involve either mathematical equations or the study of certain properties of functions. The graph of a function often supplies very important information about the behavior of the function. The importance of calculus in graph sketching does not disappear with the broad use of graphing devices such as graphing calculators, since calculus helps us to grasp the major features of a function its graph should include. Therefore, a significant portion of Chapter 3 is devoted to the study of graph sketching using calculus. Section 3.3 connects the shape of the graph of a function to the first and second derivatives of the function. Section 3.4 shows how to apply the derivative tests to determine relative extrema at a critical number. Sections 3.5 and 3.6 cover graphing strategies, with and without a graphing devise. Section 3.5 focuses on nonrational functions, and Section 3.6 focuses on rational functions. The strategies of finding absolute maximum and/or absolute minimum values of a function on various types of intervals is covered in Section 3.7, optimization is covered in Section 3.8, elasticity is introduced in Section 3.9.

Chapter 4 introduces integration. Throughout the first four sections, the interplay among graphical, numeric, and algebraic properties is extensively used to emphasize the linkages between areas under curves, definite integrals, and indefinite integrals (antiderivatives). A summary of the these linkages is presented in Section 4.3 and in the chapter summary. The chapter opens with some explorations of areas under graphs of positive functions, followed by the study, in Section 4.2, of the link between a function and the area under its graph, thus naturally leading to the topic of antiderivatives. In Section 4.3, the Fundamental Theorem of Calculus is introduced. The Fundamental Theorem of Calculus is unquestionably the most important theorem in calculus. It ranks as one of the great accomplishments of the human mind. Therefore, it deserves to be presented in its original format, where the definite integral is defined by the limit of Riemann sums, although the name "Riemann sums" is not explicitly mentioned in the book. Section 4.4 covers other properties of definite integrals, and the applications such as areas between two curves and average values of functions.

The second half of Chapter 4 is devoted to the properties and techniques of integrating. Section 4.5 covers integration by substitution or algebraic manipulation and Section 4.6 covers integration by integration tables. Given the fact that the integration table presented in Section 4.6 includes all the formulas used in this course which may be derived by the technique of integration by parts, that integration technique is not included in this book.

The material in Chapter 5 consists of topics related to applications of integration in economics, finance, and other related fields, including consumers' and producers' surpluses (Section 5.1), financial models with continuous money flow (Section 5.2), probability and important distributions (Sections 5.4 and 5.5), and differential equations (Section 5.6). Improper integrals are introduced in Section 5.3. The instructor may choose to omit some of the sections. However, Sections 5.4 and 5.5 do use improper integrals, therefore, Section 5.3 should be included if Sections 5.4 and 5.5 are not omitted.

Chapter 6 extends differential calculus to functions that involve two independent variables. Partial derivatives are covered in Section 6.1. Unconstrained maximizing/minimizing techniques are presented in Section 6.2.

Chapter 1

Functions and Mathematical Models

Introduction

In this chapter, we study several important functions and mathematical models in business, economics, and finance. These functions and mathematical models will be considered throughout the text. The main topics in this chapter include linear, exponential, and piecewise defined functions, along with their graphs; supply, demand, and equilibrium point; cost, revenue, and profit; and savings accounts and compound interest.

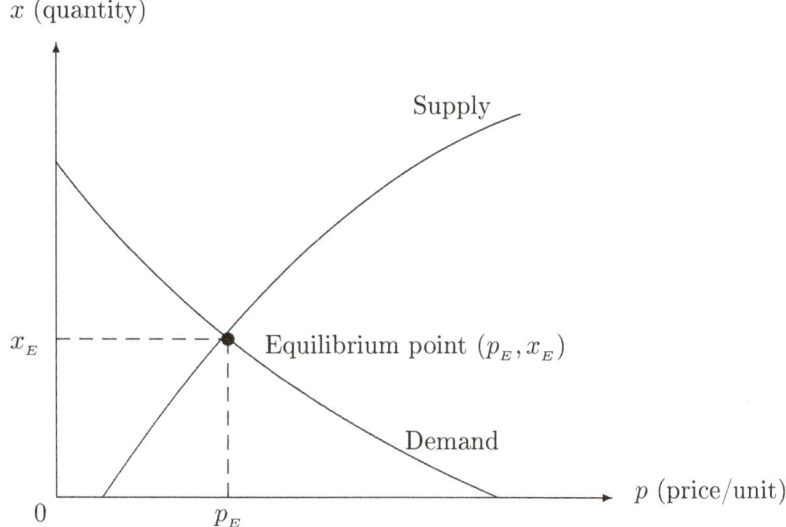

Study Objectives

1. Understand the formulary and graphical notions of a function.

2. Determine the outputs of a function by either its formula or its graph.

3. Set up functions in economical and financial applications.

4. Interpret slopes in economical and financial applications.

1.1 Functions and Graphs

Calculus is the study of change of functions — how they change and how quickly they change. The idea of a **function** is one of the most important concepts in mathematics. So *what is a function*? Simply put, a function is a special relationship between two sets. Let us start with a situation which affects nearly all of us: A driver realizes that he needs to add gas to his car. He pulls into a gas station and notices that the price of a gallon of gas is $3.90. How much does it cost him to purchase gas? You probably will reply that it *depends* on the quantity he purchases. We can assume that the car's gas tank can hold at most 20 gallons of gas. Now, the relationship between the cost and the amount of gas he purchases can be represented in four ways:

1. With words: The cost of purchasing x gallons of gas, $C(x)$, is the price of $3.90 per gallon multiplied by x gallons, where x is at least 0 gallons and at most 20 gallons.

2. With a numerical table (Table 1.1.1):

Gas (gal)	Cost of gas ($)
0	0
1	3.90
1.5	5.85
5	19.50
10	39
20	78

Table 1.1.1: Cost for a given amount of gas

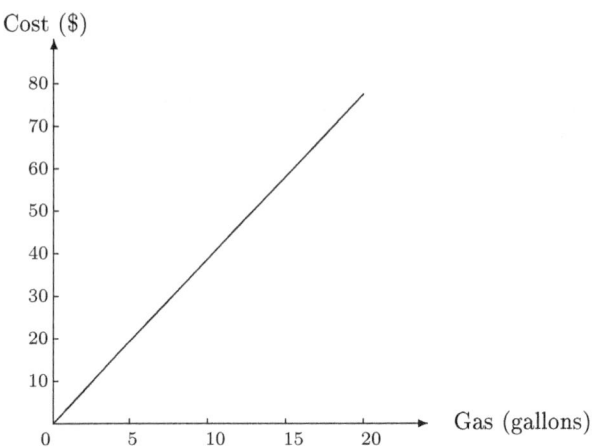

Figure 1.1.1: Cost vs. amount purchased

3. With a graph as presented in Figure 1.1.1.

4. With an equation: $C(x) = 3.9x$, $0 \leq x \leq 20$, where x is in gallons and $C(x)$ is in dollars.

Most of the functions studied in this text can be viewed from each of the four representations: verbal, numerical, graphical, and algebraic. Each of these representations helps us to understand the function from different perspectives. However, among these four representations, only the graph and the algebraic equation allow us to apply calculus concepts to study the change behaviors of the function.

━ Function Notations

As mentioned before, a function is a special relationship between two sets. In the example of the gas purchase, the two sets are the amounts of gas purchased (in gallons), and the costs (in dollars). It is clear that the cost **depends** on the amount of gas purchased. Hence, the amount of gas purchased is called the **independent variable**, or the **input**, while the cost is called the **dependent variable**, or the **output**. The first set is called the **domain** and the second set is called the **range**. Therefore, a function f can be thought as a process (or a rule, or a correspondence) in which an independent value x in the domain is mapped to exactly one dependent value in the range (Figure 1.1.2).

1.1. FUNCTIONS AND GRAPHS

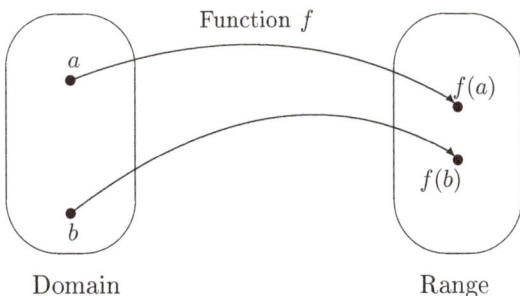

Figure 1.1.2: Domain and range

> **Definition** *function, domain, range*
>
> A **function** is a rule that assigns exactly one element in the range to each element in the domain. The **domain** is the set of all possible independent variable values. The **range** is the set of all possible dependent variable values.

Let us return to the gas-purchasing example. For any amount of gas purchased, there is one and only one cost, calculated by multiplying the price of $3.90 per gallon by the amount of gallons. Therefore, the cost is a function of the amount of gas purchased. Since the amount of the purchase is at least 0 gallons and at most 20 gallons, the domain of this function is the interval $[0, 20]$, and the range is the interval $[0, 78]$.

Example 1: Given the set $\{1, 2, 3, 4, 5\}$ and the rule as "assign each number in this set its square": Is this rule a function? If it is a function, what is the domain? What is the range?

Solution Since each number has only one square, this rule is a function. The domain is the set $\{1, 2, 3, 4, 5\}$ since each of the element can be assigned an output. The range is the set $\{1, 4, 9, 16, 25\}$.

Example 2: Find the domain for the following functions: (a) $y = f(x) = \dfrac{2}{x-3}$, (b) $y = g(x) = \sqrt{x-2}$.

Solution (a) For $y = f(x) = \dfrac{2}{x-3}$, if $x = 3$, then the denominator becomes zero. Therefore, 3 is not in the domain. For all other values, the formula is valid. Hence, the domain is all real numbers except $x = 3$, or all real numbers x such that $x \neq 3$, or simply $x \neq 3$.

(b) For $y = g(x) = \sqrt{x-2}$, since only non-negative numbers have square roots, we need that $x - 2 \geq 0$, or $x \geq 2$. Therefore, the domain is all real numbers x such that $x \geq 2$, or simply $x \geq 2$. We can also write the domain as the interval $[2, \infty)$.

■ Evaluating Functions

Most functions in this text are represented by an algebraic equation as in $C(x) = 2.89x$. To evaluate a function at a given value of x, we "plug in" the given value to replace x and carry on the algebraic operations. For example, to find the cost of purchasing 8 gallons of gas, we write $C(8) = 2.89 \times 8 = \$23.12$.

Example 3: The function f is given by the equation $f(x) = 2x^2 - 3x$. Find $f(2)$, $f(5a)$, $f(x+h)$, and $f(\sqrt[3]{t})$.

Solution Notice that this time the variable x occurs in three places. By replacing every x in the equation with the given values, we obtain:

$$f(2) = 2(2)^2 - 3(2) = 2(4) - 6 = 8 - 6 = 2.$$

$$f(5a) = 2(5a)^2 - 3(5a) = 2(25)a^2 - 15a = 50a^2 - 15a.$$

$$f(x+h) = 2(x+h)^2 - 3(x+h) = 2\left(x^2 + 2xh + h^2\right) - 3(x+h) = 2x^2 + 4xh + 2h^2 - 3x - 3h.$$

$$f\left(\sqrt[3]{t}\right) = 2\left(\sqrt[3]{t}\right)^2 - 3\sqrt[3]{t} = 2\sqrt[3]{t^2} - 3\sqrt[3]{t}.$$

Example 4: For $f(x) = 2x - 5x^2$, find $f(x+h)$ and $\dfrac{f(x+h) - f(x)}{h}$ for $h \neq 0$.

Solution By replacing every x in the equation with the given $x + h$, we obtain:

$$f(x+h) = 2(x+h) - 5(x+h)^2 = 2(x+h) - 5\left(x^2 + 2xh + h^2\right)$$

$$= 2x + 2h - 5x^2 - 10xh - 5h^2, \quad \text{and}$$

$$\frac{f(x+h) - f(x)}{h} = \frac{2x + 2h - 5x^2 - 10xh - 5h^2 - \left(2x - 5x^2\right)}{h}$$

$$= \frac{2x + 2h - 5x^2 - 10xh - 5h^2 - 2x + 5x^2}{h}$$

$$= \frac{2h - 10xh - 5h^2}{h} = \frac{h(2 - 10x - 5h)}{h}$$

$$= 2 - 10x - 5h, \quad \text{for } h \neq 0.$$

Graphs of Functions: Identifying and Evaluating

The most common method for visualizing a function is its graph. The study of graphs is a very important aspect of calculus. If f is a function, then the graph of f consists of all points (x, y) in the coordinate plane such that x is in the domain and $y = f(x)$.

Example 5: Draw the graph of the function given in Example 1.

Solution
The graph of the function in Example 1 consists of the following five points: $(1,1)$, $(2,4)$, $(3,9)$, $(4,16)$, and $(5,25)$. Plotting these five points in the coordinate plane, we obtain the graph (Figure 1.1.3).

Example 6: The function s is given by $s(x) = 1/x$. Find its domain and draw its graph.

Solution Since the only value of x that does not have a correspondence is $x = 0$, which makes the denominator equal zero, the domain of the function is {All real numbers except $x = 0$}, or simply $x \neq 0$. To draw the graph, we sample some numerical values (Table 1.1.2). Plotting these points in the coordinate plane and linking the dots we obtain the graph (Figure 1.1.4). Notice that the graph is not connected around $x = 0$.

1.1. FUNCTIONS AND GRAPHS

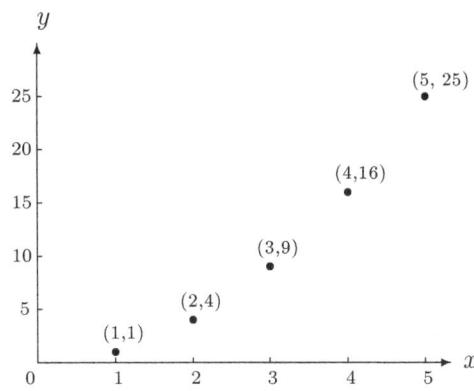

Figure 1.1.3: Plot of the function given in Example 1

x	$s(x)$
-3	-0.33
-1	-1
-0.5	-2
-0.3	-3.3
0	—
0.25	4
0.5	2
1	1
2	0.5
4	0.25

Table 1.1.2: Values of $s(x)$

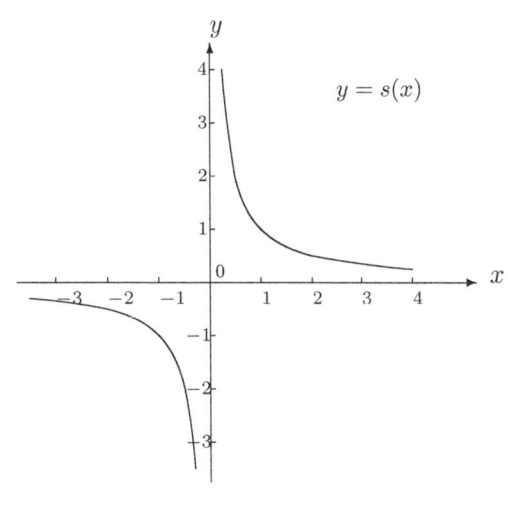

Figure 1.1.4: Graph of $s(x)$

Reality Check   *Graphing*

Making the graph of a function by sampling points can be tedious as well as inaccurate—we may miss some very important features of the graph. Even using a graphing calculator to plot the graph needs some guides and caution. In chapter 3 we study how calculus can help us to capture the important features of a function to plot its graph appropriately.

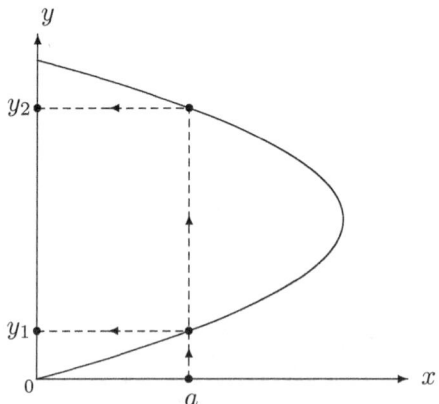

Figure 1.1.5: A graph that does not represent a function of $f(x)$

It should be pointed out that not every given graph in a coordinate plane represents the graph of a function of x. Consider the graph shown in Figure 1.1.5. Note that for the input a, there are two outputs y_1 and y_2. Since a function assigns only *one* output to each input, this graph does not represent a function. This example also suggests the following method of visual assessment known as the **vertical line test**.

Reality Check   *Vertical line test*

A graph does not represent a function if you can draw a vertical line that crosses the graph in two or more places.

The following example illustrates how to evaluate a function by its graph.

Example 7: The function G is represented by the graph in Figure 1.1.6. Find $G(-2)$, $G(-1)$, and $G(1)$.

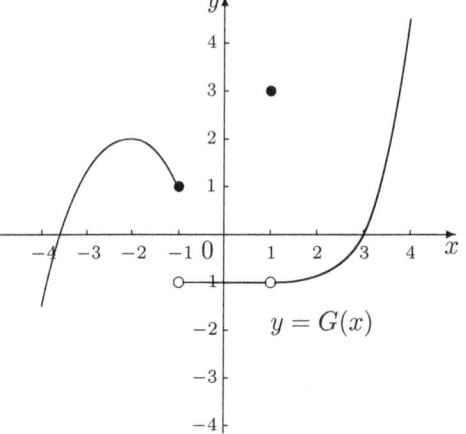

Figure 1.1.6: Graph of G

Solution To evaluate a function on a graph, first make a vertical line passing the given input value. At the location where the vertical line crosses the graph, the y-coordinate is the output $G(x)$.

At $x = -2$, the vertical line crosses the graph at $(-2, 2)$. Therefore $G(-2) = 2$.

At $x = -1$, the vertical line crosses the graph at $(-1, 1)$ (the solid dot). Therefore $G(-1) = 1$.

Similarly, at $x = 1$, the vertical line crosses the graph at $(1, 3)$. Therefore $G(1) = 3$.

Exercise Set 1.1

1. The function f is given by the equation
$$f(x) = 2x + 3.$$
   (a) Find the domain of this function.
   (b) Find $f(-1)$, $f(0)$, and $f(2)$.
   (c) Find $f(3a)$ and $f(t^2)$.
   (d) Find $f(x+h)$.
   (e) Find $\dfrac{f(x+h) - f(x)}{h}$ for $h \neq 0$.

2. The function S is given by the equation
$$S(x) = 4x - 1.$$
   (a) Find the domain of this function.
   (b) Find $S(-1)$, $S(0)$, and $S(2)$.
   (c) Find $S(3a)$ and $S(t^2)$.
   (d) Find $S(x+h)$.
   (e) Find $\dfrac{S(x+h) - S(x)}{h}$ for $h \neq 0$.

3. For the function f in Exercise #1, use the results in part (b) to make a numerical table, then use the table to plot a graph of the function.

4. For the function S in Exercise #2, use the results in part (b) to make a numerical table, then use the table to plot a graph of the function.

5. The function F is given by the equation
$$F(x) = 2x^2 - 2.$$
   (a) Find the domain of this function.
   (b) Find $F(-2)$, $F(-1)$, $F(0)$, $F(1)$, and $F(2)$.
   (c) Find $F(5a)$ and $F(\sqrt{a})$.
   (d) Find $F(x+h)$.
   (e) Find $\dfrac{F(x+h) - F(x)}{h}$ for $h \neq 0$.

6. The function H is given by the equation
$$H(x) = x^2 + 1.$$
   (a) Find the domain of this function.
   (b) Find $H(-2)$, $H(-1)$, $H(0)$, $H(1)$, and $H(2)$.
   (c) Find $H(a^2)$ and $H(\sqrt[4]{a})$.
   (d) Find $H(x+h)$.
   (e) Find $\dfrac{H(x+h) - H(x)}{h}$ for $h \neq 0$.

7. For the function F in Exercise #5, use the results in part (b) to make a numerical table, then use the table to plot a graph of the function.

8. For the function H in Exercise #6, use the results in part (b) to make a numerical table, then use the table to plot a graph of the function.

9. The function G is given by the equation
$$G(x) = 2/(x+1).$$
   (a) Find the domain of this function.
   (b) Find $G(-2)$, $G(-1.5)$, $G(-1.1)$, $G(-0.9)$, $G(0)$, and $G(1)$.
   (c) Use the results in part (b) to make a numerical table, then use the table to plot a graph of the function.

10. The function T is given by the equation
$$T(x) = 1/(x-1).$$
    (a) Find the domain of this function.
    (b) Find $T(-1)$, $T(0)$, $T(0.5)$, $T(0.9)$, $T(1.1)$, and $T(3)$.
    (c) Use the results in part (b) to make a numerical table, then use the table to plot a graph of the function.

11. The function f is represented by the following graph. Use the graph to find the following function values:
    (a) $f(-3)$ (b) $f(-2)$
    (c) $f(-1)$ (d) $f(0)$
    (e) $f(1)$ (f) $f(3)$

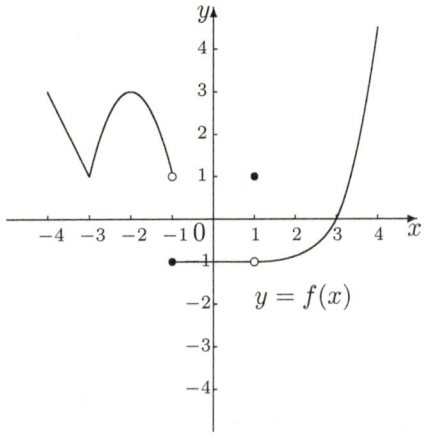

(a) $g(-2)$ (b) $g(-1)$
(c) $g(0)$ (d) $g(1)$
(e) $g(2)$ (f) $g(3)$

12. The function g is represented by the following graph. Use the graph to find the following function values:

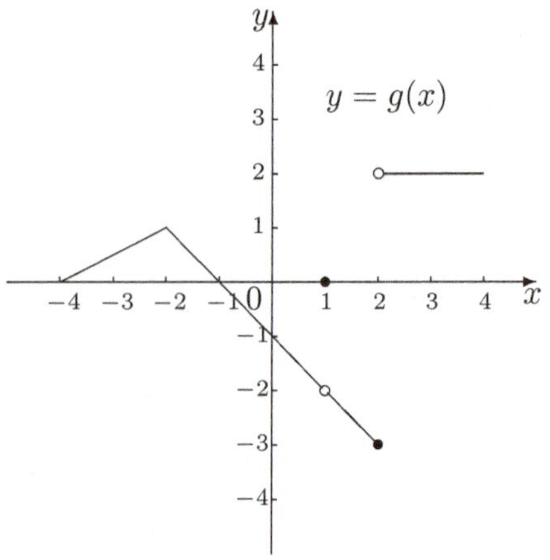

1.2 Slopes and Linear Functions

The first type of function that we study is the **linear function**, that is, any function whose graph is a straight line. Linear functions are probably the most commonly used functions. The gas purchasing function described in the previous section is a linear function. Linear functions have the character that the *change* in the function value is proportional to the *change* of the variable. In other words, the **slope** of a linear function is a constant. Later, we realize that the concept of the slope plays an extremely important role in calculus. We start by exploring the relation of lines and their slopes.

▬ Lines and Slopes

We first take a look at two special types of lines: **vertical lines** and **horizontal lines**. The graphs of vertical lines as illustrated in Figure 1.2.1 link to the equations of the type of $x = a$. The graph of a vertical line does not represent a function since it fails the vertical line test.

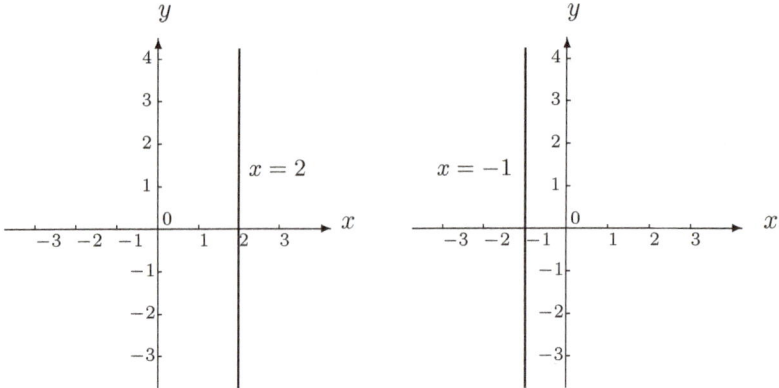

Figure 1.2.1: $x = 2$ and $x = -1$

The graphs of horizontal lines illustrated in Figures 1.2.2 and 1.2.3 link to the equations of the type of $y = b$. Each of the graphs represents a function since it passes the vertical line test.

1.2. SLOPES AND LINEAR FUNCTIONS

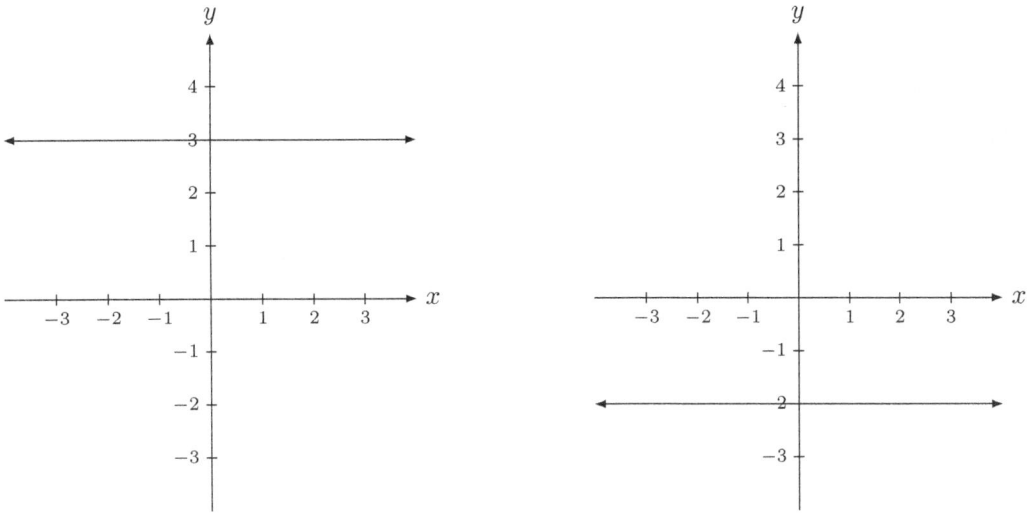

Figure 1.2.2: $y = 3$ **Figure 1.2.3:** $y = -2$

Figure 1.2.4 illustrates some lines that are neither horizontal nor vertical. Such a line can either go from southwest to northeast (the first graph), or go from northwest to southeast (the second graph). Both lines pass the vertical line test; therefore, each line represents a function. The first line represents an **increasing** function and the second line represents a **decreasing** function.

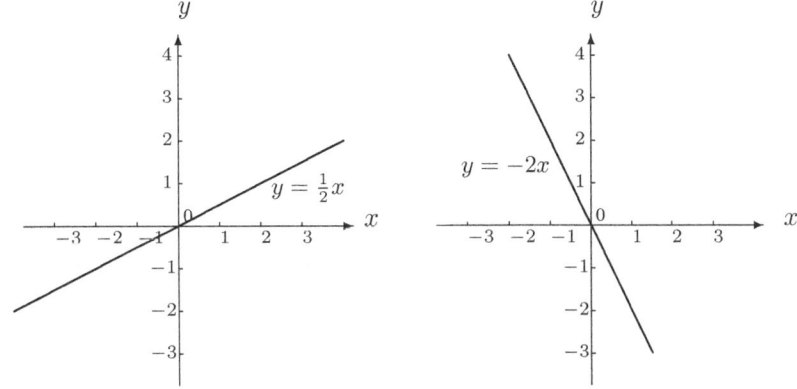

Figure 1.2.4: Increasing and decreasing lines

The **slope** of a line, denoted by m, is a measurement of the steepness as well as the direction of the line. Given two points on a line, (x_1, y_1) and (x_2, y_2), the slope of the line is computed by the equation

$$m = \frac{y_2 - y_1}{x_2 - x_1}.$$

Figure 1.2.5 illustrates two types of slopes. A line going from southwest to northeast (i.e., an increasing function) has positive slope (the first graph), and a line going from northwest to southeast (i.e., a decreasing function) has negative slope (the second graph). A horizontal line has slope equal to zero, and the slope of a vertical line is undefined, since any two points on that line have the same x-coordinate, and the denominator of the slope becomes zero.

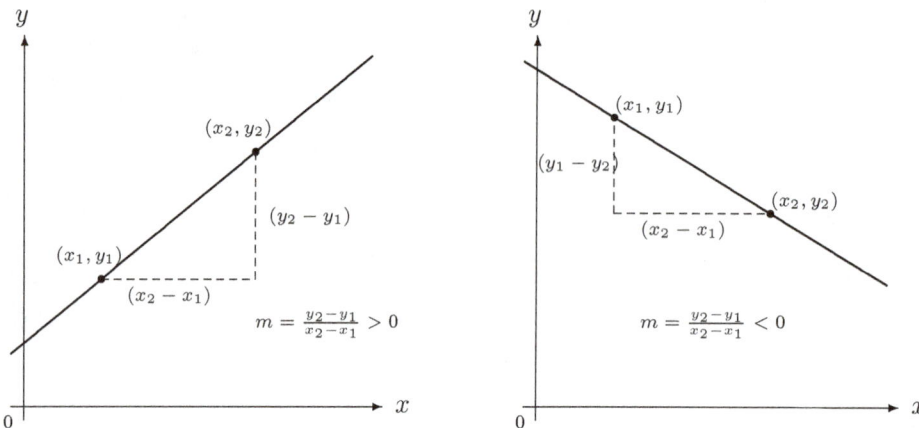

Figure 1.2.5: Positive and negative slopes

Linear Functions

We now formally define a linear function.

> **Definition** *linear function, slope, y-intercept, slope-intercept form, slope-point form*
>
> A **linear function** has the graph of a straight line and has the form
>
> $$y = f(x) = mx + b$$
>
> where m is the **slope** of the line and b is a constant. The point $(0, b)$ is called the **y-intercept**, or the **vertical intercept**. The above equation $y = mx + b$ is called the **slope-intercept equation** of the line. Another important form used for describing a line is the **slope-point equation**
>
> $$y - y_1 = m(x - x_1)$$
>
> where m is the slope of the line, and (x_1, y_1) is any point on the line.

A line can be described in other forms of equations, such as the general equation $Ay + Bx + C = 0$. The slope-point equation will be used to derive the **linear approximation** in Chapter 3.

Example 1: A line with slope $m = -2$ contains the point $(2, 1)$.

(a) Find the slope-point equation of the line;

(b) Find the slope-intercept equation of the line;

(c) Plot its graph.

Solution (a) Since both the slope and one point are given, we obtain the slope-point equation by directly plugging in $m = -2$, $x_1 = 2$, and $y_1 = 1$. Hence, the slope-point equation of the line is:

$$y - 1 = (-2)(x - 2).$$

1.2. SLOPES AND LINEAR FUNCTIONS

(b) We first set the slope-intercept equation with $m = -2$:

$$y = -2x + b.$$

Since the point $(2, 1)$ is on the line, $x = 2$ and $y = 1$ satisfy the equation:

$$1 = -2(2) + b.$$

Solving for b, we obtain that $b = 5$. Hence, the slope-intercept equation of the line is:

$$y = -2x + 5.$$

Note that we can also obtain this equation by working on the slope-point equation: From

$$y - 1 = (-2)(x - 2)$$

we obtain

$$y = (-2)(x - 2) + 1 = (-2)x - (-2)(2) + 1 = -2x + 4 + 1 = -2x + 5.$$

(c) The most convenient way to plot the graph of a line is to plot two points which are on the line, and then link these two points. Since both $(2, 1)$ and $(0, 5)$ are on the line, we obtain the plot as shown in Figure 1.2.6:

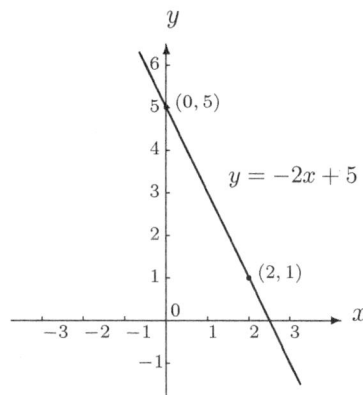

Figure 1.2.6: Graph of $y = -2x + 5$

Example 2: Find the slope of the line

(a) Containing both points $(1, 3)$ and $(-2, 5)$;

(b) Containing both points $(1, 3)$ and $(0, 3)$;

(c) Containing both points $(1, 3)$ and $(1, 5)$.

Solution (a) We obtain the slope using the formula $m = \dfrac{y_2 - y_1}{x_2 - x_1}$ with $x_1 = 1$, $y_1 = 3$, $x_2 = -2$, and $y_2 = 5$. Hence,

$$m = \frac{5 - 3}{(-2) - 1} = -\frac{2}{3}.$$

(b) Similar to part (a), with $x_1 = 1$, $y_1 = 3$, $x_2 = 0$, and $y_2 = 3$, we obtain the slope:

$$m = \frac{3-3}{0-1} = \frac{0}{-1} = 0.$$

(c) Note that, in this case, we have $x_1 = x_2 = 1$, therefore, the denominator of the slope equation becomes zero. Hence, the slope is undefined. Indeed, the line is vertical with the equation $x = 1$.

Example 3: A line contains both points $(1, 1)$ and $(5, 3)$.

(a) Find the slope-point equation of the line;

(b) Find the slope-intercept equation of the line.

Solution (a) We first obtain the slope using the formula $m = \dfrac{y_2 - y_1}{x_2 - x_1}$ with $x_1 = 1$, $y_1 = 1$, $x_2 = 5$, and $y_2 = 3$. Hence

$$m = \frac{3-1}{5-1} = \frac{1}{2}.$$

We then obtain the slope-point equation by plugging in $m = \frac{1}{2}$, $x_1 = 1$, and $y_1 = 1$. Hence, the slope-point equation of the line is:

$$y - 1 = \frac{1}{2}(x - 1).$$

(b) We set the slope-interception equation with $m = \frac{1}{2}$ obtained in part (a):

$$y = \frac{1}{2}x + b.$$

Since the point $(1, 1)$ is on the line, $x = 1$ and $y = 1$ satisfy the equation:

$$1 = \frac{1}{2}(1) + b.$$

Solving for b, we obtain that $b = \frac{1}{2}$. Hence, the slope-intercept equation of the line is:

$$y = \frac{1}{2}x + \frac{1}{2}.$$

Applications: Linear Models

When a real-world situation is described in mathematical language, it is called a **mathematical model**. If a linear equation is derived from a real-world situation, we say that we have a **linear model**.

Example 4: Mark has been working for a cloth manufacturing firm for 3 weeks and his wage is $618 per week. Six weeks later (that is, his total working history is 9 weeks), his weekly wage is $654. Assuming that Mark's weekly wage W is a linear function of the number of weeks, t, he works for the firm:

(a) Find a formula for Mark's weekly wage W as a function of the number of weeks, t, of working for the firm.

(b) What is Mark's weekly wage when he has worked for the firm for 20 weeks?

(c) What is the slope of the line? Explain what the value of the slope means in the context of this problem.

(d) What is the vertical intercept of the line? Explain what the value of the vertical intercept means in the context of this problem.

Solution (a) Assuming W is a linear function of t, we can find the slope using the formula $m = \dfrac{W_2 - W_1}{t_2 - t_1}$. Let $t_1 = 3$, $t_2 = 9$, $W_1 = 618$, and $W_2 = 654$, we obtain the slope

$$m = \frac{654 - 618}{9 - 3} = 6.$$

We now set the linear function $W(t) = mt + b$ with $m = 6$:

$$W(t) = 6t + b.$$

Since the point $(3, 618)$ is on the line, $t = 3$ and $W = 618$ satisfy the equation:

$$618 = 6(3) + b.$$

Solving for b, we obtain that $b = 600$. Hence, the linear function is:

$$W(t) = 6t + 600.$$

(b) Substituting $t = 20$ into the function $W(t)$, we obtain that $W(20) = 6(20) + 600 = \$720$. Hence, Mark's weekly wage is \$720 when he has worked for the firm for 20 weeks.

(c) The slope of the line is $m = 6$. The slope represents the increased rate of the weekly wage per week. The value $m = 6$ means that if the number of weeks that Mark has worked for the firm goes up by 1, then his weekly wage goes up by \$6. Therefore, each additional week raises the weekly wage by 6 dollars.

(d) The vertical intercept is $b = \$600$. This means that when Mark started working for the firm, his starting weekly wage was \$600.

Exercise Set 1.2

1. The linear function f is given by the equation $f(x) = 2x - 4$.

   (a) Find the slope.

   (b) Plot the graph of this function.

2. The linear function f is given by the equation $f(x) = -\frac{1}{3}x + 6$.

   (a) Find the slope.

   (b) Plot the graph of this function.

3. A line is represented by the slope-point equation $y + 2 = 5(x + 1)$.

   (a) Find the slope-intercept equation for the line.

   (b) Plot the graph of this line.

4. A line is represented by the slope-point equation $y - 3 = -4(x + 2)$.

   (a) Find the slope-intercept equation for the line.

   (b) Plot the graph of this line.

5. Use the graph of the line to find the following:

   (a) The slope;

   (b) The vertical intercept;

   (c) The slope-intercept equation of this line.

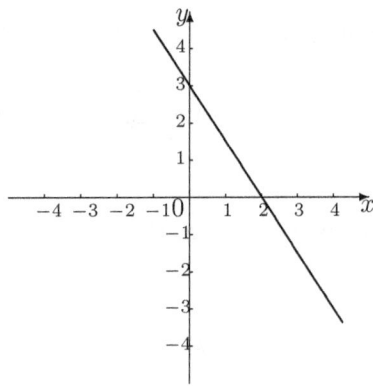

6. Use the graph of the line to find the following:

(a) The slope;

(b) The vertical intercept;

(c) The slope-intercept equation of this line.

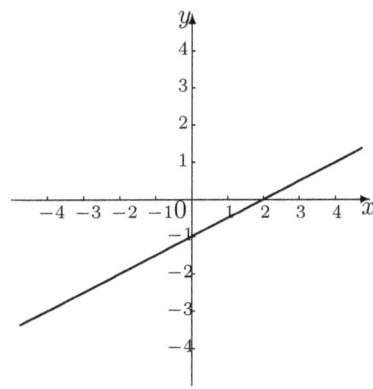

For Exercises #7 to #18, find both the slope-point equation and the slope-intercept equation of the line:

7. With slope 2, containing point $(2, 0)$.

8. With slope 3, containing point $(-1, 0)$.

9. With slope $-\frac{3}{5}$, containing point $(3, 1)$.

10. With slope $-\frac{4}{7}$, containing point $(-1, 1)$.

11. With slope -2, containing point $(0, 1)$.

12. With slope -3, containing point $(0, 4)$.

13. With slope 0, containing point $(3, 7)$.

14. With slope 0, containing point $(2, -4)$.

15. Containing both points $(1, 2)$ and $(3, -4)$.

16. Containing both points $(-2, 2)$ and $(1, 5)$.

17. Containing both points $(-1, 2)$ and $(0, 2)$.

18. Containing both points $(3, 4)$ and $(5, 4)$.

For Exercises #19 to #32, find the slope of the line containing the given pair of points. If the slope is undefined, state the fact.

19. $(\frac{1}{2}, 0)$ and $(2, 5)$

20. $(\frac{1}{3}, -1)$ and $(0, 1)$

21. $(1, -1)$ and $(3, 2)$

22. $(-3, -2)$ and $(-2, 0)$

23. $(3, -1)$ and $(-1, -1)$

24. $(1, 2)$ and $(-2, 2)$

25. $(-2, 1)$ and $(-2, 2)$

26. $(1, -2)$ and $(1, 1)$

27. $(x, 2x)$ and $(x + h, 2(x + h))$

28. $(x, 3x)$ and $(x + h, 3(x + h))$

29. $(x, 2x + 4)$ and $(x + h, 2(x + h) + 4)$

30. $(x, 3x - 1)$ and $(x + h, 3(x + h) - 1)$

31. $(x, -4x)$ and $(x + h, -4x - 4h)$

32. $(x, -1)$ and $(x + h, -1)$

APPLICATIONS

33. A teller of a bank has worked for 5 weeks and his weekly wage is $700 per week. Five weeks later (i.e., his total working history is 10 weeks), the weekly wage is $720. Assuming that the weekly wage S is a linear function of the number of weeks, x, the teller works for the bank:

(a) Find a formula for the weekly wage S as a function of the number of weeks, x, of working for the bank.

(b) What is the teller's weekly wage when he has worked for the bank for 23 weeks?

(c) What is the slope of the line? Explain what the value of the slope means in the context of this problem.

(d) What is the vertical intercept of the line? Explain what the value of the vertical intercept means in the context of this problem.

34. A technician of an IT firm has a starting weekly wage of $1,500 per week. Five weeks later, the weekly wage is $1,650. Assuming that the weekly wage S is a linear function of the number of weeks, x, the technician works for the firm:

(a) Find a formula for the weekly wage S as a function of the number of weeks, x, of working for the firm.

(b) What is the technician's weekly wage when she has worked for the firm for 15 weeks?

(c) What is the slope of the line? Explain what the value of the slope means in the context of this problem.

35. The resale value of a used car is $25,000 when the car is 2 years old. The resale value drops by $6,000 from the previous given value when the car is 4 years old. Assuming that the resale value of that car, V, is a linear function of the car's age, t:

(a) Find a formula for the resale value of the car, V, as a function of the number of years, t, of the age of the car.

(b) What is the car's resale value when the car is 7 years old?

(c) What is the slope of the line? Explain what the value of the slope means in the context of this problem.

(d) What is the vertical intercept of the line? Explain what the value of the vertical intercept means in the context of this problem.

(e) Find the age of the car when it has essentially no value.

36. The resale value of a used car is $45,000 when the car is 3 year old. The resale value drops by $8,000 from the previous given value when the car is 5 years old. Assuming that the resale value of that car, V, is a linear function of the car's age, t:

(a) Find a formula for the resale value of the car, V, as a function of the number of years, t, of the age of the car.

(b) What is the car's resale value when the car is 7 years old?

(c) What is the slope of the line? Explain what the value of the slope means in the context of this problem.

(d) What is the vertical intercept of the line? Explain what the value of the vertical intercept means in the context of this problem.

(e) Find the age of the car when it has essentially no value.

37. The relationship between the Fahrenheit (F) and Celsius (C) temperature scales is given by the linear function $F = \frac{9}{5}C + 32$.

(a) What is the slope of the line? Explain what the value of the slope means in the context of this problem.

(b) What is the vertical intercept of the line? Explain what the value of the vertical intercept means in the context of this problem.

38. A car owner found that her monthly cost of driving her car depends on the number of miles she drives. She calculates that if she drives 320 miles per month, her monthly cost is $270. If she drives 400 miles per month, her monthly cost is $290. Assuming that the monthly cost, C, of driving the car is a linear function of the number of miles, d, she drives in a month:

(a) Find a formula for the monthly cost, C, of driving the car as a function of the number of miles, d, she drives in a month.

(b) What is the her monthly cost when she drives 520 miles per month?

(c) What is the slope of the line? Explain what the value of the slope means in the context of this problem.

(d) What is the vertical intercept of the line? Explain what the value of the vertical intercept means in the context of this problem.

1.3 Application of Functions to Economics

In this section, we explore some functions that occur in the economics field. In economics, the basic activities are production and sales. Therefore, we start by studying how a firm calculates its cost, revenue, and profit for producing and selling certain quantities of goods.

▬ Total Costs, Total Revenue, and Total Profit

> **Definition**   *total cost function*
>
> The **total cost function**, $C(x)$, gives the total cost of producing a quantity x of some good.

> **Reality Check**   *Fixed costs and variable costs*
>
> Since the more goods that are made, the higher the total cost, we expect that $C(x)$ is an increasing function. Cost of production can be separated into two parts: the **fixed costs**, such as rent, maintenance, etc., that occur even when nothing is produced; and the **variable costs**, that depend on how many units are produced. Many cost functions can be modeled as linear functions.

Example 1: The owner of a dairy firm determined that the total cost, in thousands of dollars, of producing x thousand gallons of milk each month can be modeled by

$$C(x) = 20 + 0.3x, \quad 0 \leq x \leq 30$$

(a) Find the fixed costs and the variable costs;

(b) Plot the graph of the total cost function.

(c) What is the total cost of producing 10,000 gallons of milk? 20,000 gallons of milk?

(d) What does the slope of the equation represent?

Solution (a) The fixed costs are 20 thousands of dollars, or \$20,000. The vertical intercept of the equation is 20. The variable costs are $0.3x$ thousands of dollars, or $300x$.

(b) The graph is shown in Figure 1.3.1. Notice that the graph corresponds to the domain of the function which is the interval $[0, 30]$.

(c) Since $C(10) = 20 + 0.3(10) = 23$ thousands of dollars, it costs \$23,000 to produce 10,000 gallons of milk. Similarly, it costs $C(20) = 20 + 0.3(20) = 26$ thousands of dollars, or \$26,000, to produce 20,000 gallons of milk.

(d) The slope of this function is $m = 0.3$ thousands of dollars per thousand gallons, or dollars per gallon. That represents the variable cost per gallon. That is, to produce each additional gallon of milk, the total cost will increase by \$0.3.

1.3. APPLICATION OF FUNCTIONS TO ECONOMICS

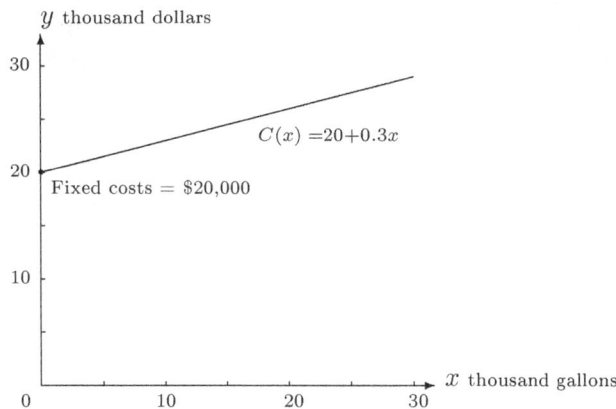

Figure 1.3.1: Graph of the milk-production cost function, C

> **Reality Check**   *Units in applications*
>
> In this example, the unit for the input is thousand gallons and the unit for the output is thousand of dollars. In application problems, choosing appropriate units for the variables makes both the equation and the graph of the function to be balanced.

Example 2: A coffeehouse's daily costs are $252 when they sell 80 coffee drinks and $278 when they sell 120 coffee drinks. Assume that the daily cost C is a linear function of the number of coffee drinks, x.

(a) Find a formula for the daily cost C as a function of the number of coffee drinks, x.

(b) What is the slope of the line? Explain what the value of the slope means in the context of this problem.

(c) What is the vertical intercept of the line? Explain what the value of the vertical intercept means in the context of this problem.

Solution (a) Assuming C is a linear function of x, we can find the slope using the formula $m = \dfrac{C_2 - C_1}{x_2 - x_1}$. Letting $x_1 = 80$, $x_2 = 120$, $C_1 = 252$, and $C_2 = 278$, we obtain the slope

$$m = \frac{278 - 252}{120 - 80} = 0.65.$$

We now set the linear function $C(x) = mx + b$ with $m = 0.65$:

$$C(x) = 0.65x + b.$$

Since the point (80,252) is on the line, $x = 80$ and $C = 252$ satisfy the equation:

$$252 = 0.65(80) + b.$$

Solving for b we obtain that $b = 200$. Hence, the linear function is:

$$C(x) = 0.65x + 200.$$

(b) The slope of the line is $m = \$0.65$ per coffee drink. The slope represents the variable cost per coffee drink. The value $m = 0.65$ means that if the number of coffee drinks sold goes up by 1, then the costs go up by \$0.65. Therefore, each additional coffee drink sold costs the coffeehouse 65 cents. This would correspond to the cost of the coffee beans and other variable costs such as electricity, water, and so on, used to make each drink.

(c) The vertical intercept is $b = \$200$. This means that even if the coffeehouse sells no drinks, it still has daily costs of \$200. These would correspond to fixed costs such as rent, utilities, and employee wages.

When a company sells its product, it receives revenue.

Definition   *total revenue function*

The **total revenue function**, $R(x)$, gives the total revenue received by selling a quantity x of some good.

Reality Check   *Revenue*

The revenue is always calculated as the price times the quantity sold. If the price is fixed, we can use a constant p for the price, and write $R(x) = p \cdot x$.

Example 3: The dairy firm in Example 1 can sell its milk at the price of \$2.00 per gallon.

(a) Find the total revenue function for the firm, in thousands of dollars, from selling x thousand gallons of milk each month;

(b) Plot the graph of the revenue function.

(c) What is the revenue from selling 10,000 gallons of milk? 20,000 gallons of milk?

(d) What does the slope of the equation represent?

Solution (a) The total revenue function, in thousands of dollars, for the firm from selling x thousand gallons of milk each month is

$$R(x) = 2x, \quad 0 \leq x \leq 30.$$

(b) The graph is shown in Figure 1.3.2. Notice that graph has vertical intercept $(0,0)$, since if the firm does not sell any milk, then it does not receive any revenue.

(c) Since $R(10) = 2(10) = 20$, the revenue from selling 10,000 gallons of milk is \$20,000. Similarly, the revenue from selling 20,000 gallons of milk is $R(20) = 2(20) = 40$ thousand dollars, or \$40,000.

(d) The slope of this function is $m = \$2/\text{gallon}$, or dollars per gallon. That represents the price of selling one gallon of the milk.

1.3. APPLICATION OF FUNCTIONS TO ECONOMICS

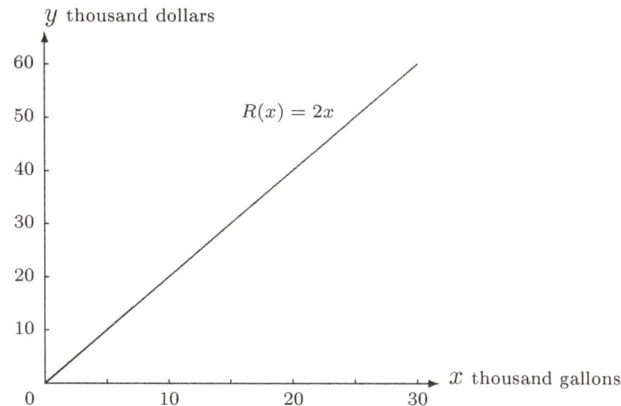

Figure 1.3.2: Graph of the revenue function, R, for milk sales

When a company produces and sells x units of its product, it makes profit if the total revenue is greater than the total cost. We introduce the **total profit function**.

> **Definition**   *total profit function*
>
> The **total profit function**, $P(x)$, is the difference of the total revenue and the total cost, i.e., $P(x) = R(x) - C(x)$, of producing and selling a quantity x of some good.

> **Reality Check**   *Profit and break-even point*
>
> Notice that most total cost functions include the fixed costs, while the revenue functions all start from zero. Hence, when x is small, the total profit function may take negative values, representing a loss. A company will **break even** at the value of x for which $P(x) = 0$. This x value is often called the **break-even point** since on the graph, it corresponds to the point where the graph of the total cost function meets the total revenue function.

Example 4: Consider the dairy firm in Examples 1 and 3, where the total cost function is given by

$$C(x) = 20 + 0.3x, \quad 0 \le x \le 30$$

and the total revenue function is given by

$$R(x) = 2x, \quad 0 \le x \le 30.$$

(a) Find the total profit function for the firm, in thousands of dollars, from producing and selling x thousand gallons of milk each month;

(b) Plot the graph of the profit function.

(c) What is the profit from producing and selling 10,000 gallons of milk? 20,000 gallons of milk?

(d) Find the break-even point for the firm.

(e) Plot the break-even point with the graphs of the total cost function and the total revenue function.

Solution (a) The total profit function, in thousands of dollars, for the firm from producing and selling x thousand gallons of milk each month is

$$P(x) = R(x) - C(x) = 2x - (20 + 0.3x) = 1.7x - 20, \quad 0 \leq x \leq 30.$$

(b) The graph is shown in Figure 1.3.3.

(c) Since $P(10) = 1.7(10) - 20 = -3$, the firm will have a loss of \$3,000 from producing and selling 10,000 gallons of milk. The profit from producing and selling 20,000 gallons of milk is $P(20) = 1.7(20) - 20 = 14$ thousands of dollars, or \$14,000.

(d) We solve for x in the equation $P(x) = 0$:

$$1.7x - 20 = 0,$$

hence, $x = 20/1.7 \approx 11.765$. That is, the firm needs to produce and sell 11,765 gallons of milk to break even.

(e) The graph is shown in Figure 1.3.4.

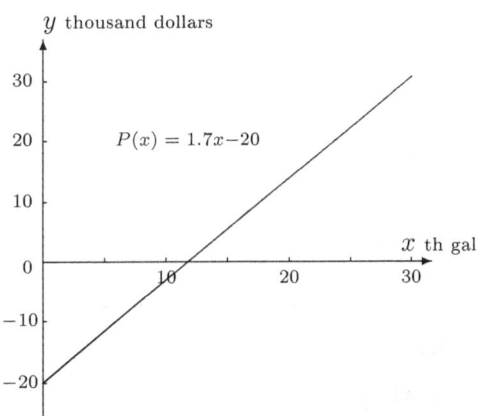

Figure 1.3.3: The profit function, P

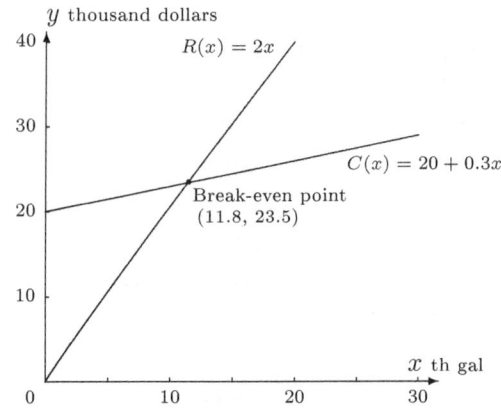

Figure 1.3.4: The break-even point, where $C(x) = R(x)$

Example 5: The coffeehouse in Example 2 sells its coffee drinks at the average price of \$3.00 per drink.

(a) Find the daily revenue function for the coffeehouse, in dollars, for selling x coffee drinks;

(b) Find the daily profit function for the coffeehouse, in dollars, for making and selling x coffee drinks;

(c) How many coffee drinks must the coffeehouse make and sell daily to break even?

Solution (a) The daily revenue function for selling x coffee drinks is

$$R(x) = 3x.$$

(b) In Example 2, we obtained the daily cost $C(x) = 0.65x + 200$; hence the daily profit function is

$$P(x) = R(x) - C(x) = 3x - (0.65x + 200) = 2.35x - 200.$$

(c) We solve $P(x) = 0$ to find the break-even point:

$$2.35x - 200 = 0,$$

hence $x = 200/2.35 \approx 85$ drinks. That is, the coffeehouse must make and sell 85 coffee drinks daily to break even. Or, in other words, to make a profit, the coffeehouse must make and sell at least 85 coffee drinks daily.

Supply and Demand Functions

In a free market, for a certain product, the price determines the quantity that consumers are willing to purchase, as well as the quantity that manufacturers are willing to supply. In general, we assume that if the price increases, then the manufacturers are willing to supply more of the product, while the consumers' demand will decrease. We now introduce the **supply function** and **demand function**.

Definition *supply and demand functions*

The **supply function**, $x = S(p)$, gives the quantity, x, of the product that manufacturers are willing to make per unit time under the price, p, for which the product can be sold.

The **demand function**, $x = D(p)$, gives the quantity, x, of the product that consumers are willing to purchase per unit time under the price, p.

Figure 1.3.5 illustrates some typical supply and demand curves.

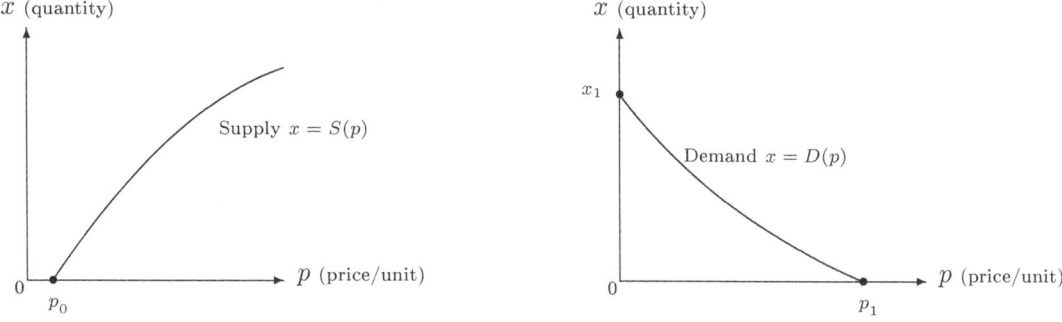

Figure 1.3.5: Typical supply and demand curves

Reality Check *Demand and supply*

If the price increases, then the manufacturers are willing to supply more of the product, hence the supply function is an **increasing** function. Since the consumer demands will decrease if the price increases, the demand function is a **decreasing** function.

> **Warning**   *Economists switch axes*
>
> In most economics textbooks, the price is placed on the vertical (y-) axis while the quantity supplied (or demanded) is placed on the horizontal (x-) axis. This convention dates back to Alfred Marshall's influential *Principles of Economics* (1890). Since we usually think of the quantity supplied (or demanded) as depending on the price, Marshall's convention runs counter to the modern standard of placing the independent variable on the horizontal axis and the dependent variable on the vertical axis. In order to continue stressing the modern standard, we will keep price on the horizontal axis. The good news for when you take economics classes is that, even with this switch, supply curves will still generally upward sloping, and demand curves will still generally be downward sloping; it is only the axis labels that will change.

Example 6: Explain the economic meanings of the prices p_0 and p_1, and the quantity x_1 in the graphs shown in Figure 1.3.5.

Solution The point $(p_0, 0)$ on the supply curve indicates that below or at that price, the manufacturers are not willing to supply anything. Hence, p_0 is the minimum price at which the manufacturers are willing to start producing. The point $(p_1, 0)$ on the demand curve indicates that at or above that price, the consumers are not willing to purchase anything. Hence, p_1 is the maximum price at which the consumers are willing to start purchasing. The point $(x_1, 0)$ on the demand curve indicates that even if the good is free, x_1 units of the quantity will be taken. Hence, x_1 represents the market capacity.

If we plot the supply and demand curves on the same axes, we can study the market behavior. At the price p_3, the demand is higher than the supply, that is, there are not enough goods in the market. When the supply is short, the price will be driven up due to the demand. That promotes the manufacturers to increase the production. Conversely, at the price p_4, the supply is higher than the demand. That is, there are more goods than the consumers will purchase. The over supply will drive the price to go down (Figure 1.3.6).

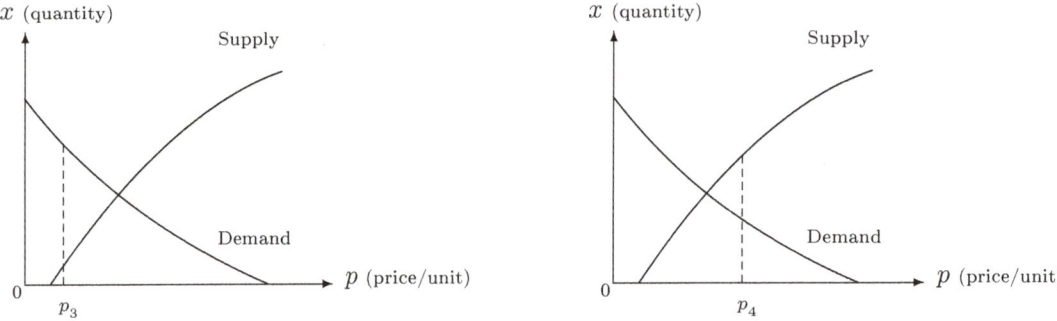

Figure 1.3.6: Short supply and over supply

The point of interception of the demand and supply curves is called the **equilibrium point**. At the price p_E, the manufacturers are willing to produce x_E quantity of a good, while the consumers are willing to purchase x_E quantity of the good. It is assumed that the market naturally settles to this equilibrium point, as presented in Figure 1.3.7.

1.3. APPLICATION OF FUNCTIONS TO ECONOMICS

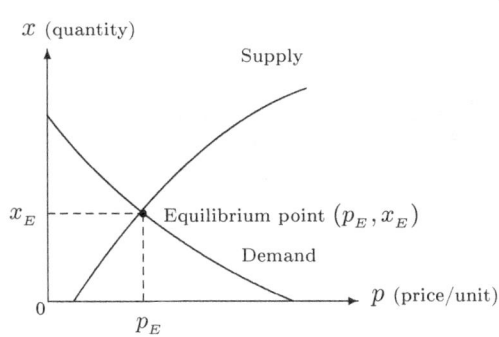

Figure 1.3.7: Equilibrium point

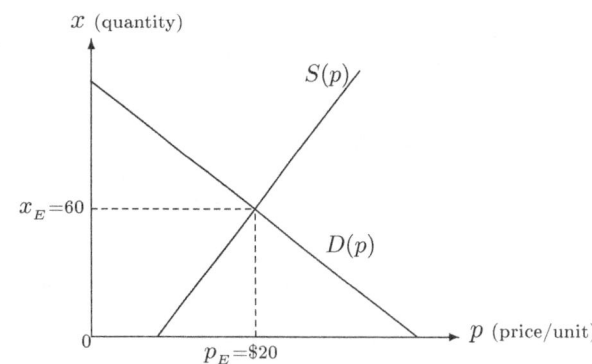

Figure 1.3.8: Equilibrium point for Example 7

Example 7: Find the equilibrium price and quantity if the supply function is $x = S(p) = 5p - 40$ and the demand function is $x = D(p) = 120 - 3p$, where p is in dollars and x is in items.

Solution At the equilibrium price, the quantities of supply and demand must be the same. To find the equilibrium price, we set $S(p) = D(p)$ and solve for p:

$5p - 40 = 120 - 3p$

$5p + 3p = 120 + 40$

$8p = 160$

$p = 20$

Hence, the equilibrium price is $p_E = \$20$. To find the equilibrium quantity, we substitute $p = 20$ into either the supply or the demand function to obtain

$x_E = S(20) = 5(20) - 40 = 60$, or

$x_E = D(20) = 120 - 3(20) = 60.$

Therefore, the equilibrium price is $20, and the equilibrium quantity is 60 items, as presented in Figure 1.3.8.

Exercise Set 1.3

1. The owner of a dairy firm determined that the total cost, in dollars, of producing x gallons of milk each day can be modeled by

   $C(x) = 600 + 0.3x, \quad 0 \leq x \leq 1{,}000.$

   (a) Find the fixed costs and the variable costs;

   (b) Plot the graph of the total cost function.

   (c) What is the total cost of producing 200 gallons of milk? 800 gallons of milk?

   (d) What does the slope of the equation represent?

2. The manager of a furniture manufacturing firm determined that the total cost, in dollars, of producing x chairs each week can be modeled by

   $C(x) = 6{,}000 + 15x, \quad 0 \leq x \leq 500.$

   (a) Find the fixed costs and the variable costs;

   (b) Plot the graph of the total cost function.

   (c) What is the total cost of producing 100 chairs? 300 chairs?

   (d) What does the slope of the equation represent?

3. The dairy firm in Exercise #1 sells the milk at $1.20 per gallon.

   (a) Find the total revenue function $R(x)$ for selling x gallons of milk;

   (b) Plot the graph of the total revenue function.

   (c) What is the total revenue from selling 200 gallons of milk? 800 gallons of milk?

   (d) What does the slope of the equation represent?

4. The furniture manufacturing firm in Exercise #2 sells the chairs for $55 each.

   (a) Find the total revenue function $R(x)$ from selling x chairs.

   (b) Plot the graph of the total revenue function.

   (c) What is the total revenue from selling 100 chairs? 300 chairs?

   (d) What does the slope of the equation represent?

5. For the dairy firm in Exercises #1 and #3, do the following:

   (a) Find the total profit function $P(x)$ for producing and selling x gallons of milk;

   (b) Plot the graph of the total profit function.

   (c) What is the total profit from producing and selling 200 gallons of milk? 800 gallons of milk?

   (d) Find the break-even point for the firm by solving the equation $P(x) = 0$.

   (e) Plot the break-even point with the graphs of the total cost function and the total revenue function.

6. For the furniture manufacturing firm in Exercises #2 and #4, do the following:

   (a) Find the total profit function $P(x)$ for producing and selling x chairs.

   (b) Plot the graph of the total profit function.

   (c) What is the total profit from producing and selling 100 chairs? 300 chairs?

   (d) Find the break-even point for the firm by solving the equation $P(x) = 0$.

   (e) Plot the break-even point with the graphs of the total cost function and the total revenue function.

7. A radio manufacturing firm has fixed costs of $25,000 and variable costs of $8 per radio. The company sells the radios for $16 each.

   (a) Find the total cost function $C(x)$ for producing x radios.

   (b) Find the total revenue function $R(x)$ for selling x radios.

   (c) Find the total profit function $P(x)$ for producing and selling x radios.

   (d) Find the break-even point for the firm by solving the equation $P(x) = 0$.

   (e) Plot the break-even point with the graphs of the total cost function and the total revenue function.

8. A sports shoes manufacturing firm has fixed costs of $45,000 and variable costs of $25 per pair of shoes. Each pair of shoes sells for $85.

   (a) Find the total cost function $C(x)$ for producing x pairs of shoes.

   (b) Find the total revenue function $R(x)$ for selling x pairs of shoes.

   (c) Find the total profit function $P(x)$ for producing and selling x pairs of shoes.

   (d) Find the break-even point for the firm by solving the equation $P(x) = 0$.

   (e) Plot the break-even point with the graphs of the total cost function and the total revenue function.

9. A coffeehouse's daily costs are $184 when they sell 70 coffee drinks and $208 when they sell 100 coffee drinks. Assume the daily cost C is a linear function of the number of coffee drinks, x.

   (a) Find a formula for the daily cost C as a function of the number of coffee drinks, x.

   (b) What is the slope of the line? Explain what the value of the slope means in the context of this problem.

   (c) What is the vertical intercept of the line? Explain what the value of the vertical intercept means in the context of this problem.

10. An ice cream store's daily costs are $180 when they sell 100 scoops of ice cream and $240 when they sell 200 scoops of ice cream. Assume the daily cost C is a linear function of the number of scoops of ice cream, x.

    (a) Find a formula for the daily cost C as a function of the number of scoops of ice cream, x.

    (b) What is the slope of the line? Explain what the value of the slope means in the context of this problem.

    (c) What is the vertical intercept of the line? Explain what the value of the vertical intercept means in the context of this problem.

11. The coffeehouse in Exercise #9 sells its coffee drinks at the average price of $2.50 per drink.

    (a) Find the daily revenue function for the coffeehouse, in dollars, for selling x coffee drinks;

    (b) Find the daily profit function for the coffeehouse, in dollars, for making and selling x coffee drinks;

    (c) How many drinks must the coffeehouse make and sell daily to break even?

12. The ice cream store in Exercise #10 sells its ice cream at the average price of $2.00 per scoop.

    (a) Find the daily revenue function for the ice cream store, in dollars, for selling x scoops of ice cream;

    (b) Find the daily profit function for the ice cream store, in dollars, for making and selling x scoops of ice cream;

    (c) How many scoops of ice cream must the ice cream store make and sell daily to break even?

13. A demand curve is given by

    $$45p + 50x = 3{,}000,$$

    where p is the price of the product, in dollars, and x is the quantity demanded at that price. Find the p-intercept and x-intercept of the demand curve and interpret them in terms of consumer demand.

14. A demand curve is given by

    $$x = D(p) = 1{,}500 - 20p,$$

    where p is the price of the product, in dollars, and x is the quantity demanded at that price. Find the p-intercept and x-intercept of the demand curve and interpret them in terms of consumer demand.

15. A city tour company finds that if the price p, charged for a 2-hour bus tour, is $30, the average number of passengers per week, x, is 200. When the price is reduced to $25, the average number of passengers per week increases to 250. Assuming that the demand curve $x = D(p)$ is linear, find its formula.

16. A boat tour company finds that if the price p, charged for a 1-hour harbor tour, is $20, the average number of passengers per week, x, is 300. When the price is reduced to $18, the average number of passengers per week increases to 360. Assuming that the demand curve $x = D(p)$ is linear, find its formula.

17. Find the equilibrium price and quantity if the supply function is $x = S(p) = 6p - 48$ and the demand function is $x = D(p) = 120 - 5p$, where p is in dollars and x is in items.

18. Find the equilibrium price and quantity if the supply function is $x = S(p) = 3p - 60$ and the demand function is $x = D(p) = 140 - 2p$, where p is in dollars and x is in items.

19. The demand and supply curves of a certain brand of sunglasses are given by $x = D(p) = 2{,}000 - 10p$ and $x = S(p) = 20p - 400$, where p is the price in dollars and x is the quantity.

    (a) At a price of $50, how many pairs of the sunglasses would consumers purchase?

    (b) At a price of $50, how many pairs of the sunglasses would manufacturers supply?

    (c) Will the market price be higher or lower than $50?

    (d) What would the price have to be for consumers to buy at least 1,000 pairs of the sunglasses?

    (e) What would the price have to be for the manufacturers to supply at least 600 pairs of the sunglasses?

    (f) Find the equilibrium price and quantity.

(g) Represent the equilibrium price and quantity on a graph.

20. The demand and supply curves of a certain brand of running shoes are given by $x = D(p) = 2{,}800 - 25p$ and $x = S(p) = 15p - 800$, where p is the price in dollars and x is the quantity.

   (a) At a price of $80, how many pairs of the running shoes would consumers purchase?

   (b) At a price of $80, how many pairs of the running shoes would manufacturers supply?

   (c) Will the market price be higher or lower than $80?

   (d) What would the price have to be for consumers to buy at least 200 pairs of the running shoes?

   (e) What would the price have to be for the manufacturers to supply at least 250 pairs of the running shoes?

   (f) Find the equilibrium price and quantity.

   (g) Represent the equilibrium price and quantity on a graph.

1.4 Nonlinear Functions and Algebraic Equations

In addition to the linear functions, there are other functions that frequently occur in this course. We also often need to solve equations involving those functions. Those functions include polynomials, rational functions, power functions with rational components, exponential and logarithmic functions, and piecewise defined functions. We study the exponential and logarithmic functions in the next two sections. In this section, we explore some basic properties and particular graphs, as well as the techniques required in solving equations, of each of the remaining groups of functions.

▬ Polynomial Functions

A **polynomial function** takes the form of

$$f(x) = a_n x^n + a_{n-1} x^{n-1} + \ldots + a_2 x^2 + a_1 x + a_0$$

where n is a nonnegative integer and $a_n, a_{n-1}, \ldots, a_2, a_1, a_0$ are real numbers called **coefficients**. The number n indicates the highest power term of a polynomial, and is called the **degree** of the polynomial. The following are a few examples of polynomial functions:

$f(x) = 4$ (A constant function)

$f(x) = 2x - 5$ (A linear function)

$f(x) = 3x^2 - 4x + 1$ (A quadratic function)

$f(x) = x^3 - 3x^2 + 1$ (A cubic, or third degree, polynomial)

$f(x) = 4x^7 - 2x^3 + 1$ (A seventh degree polynomial)

The domain of a polynomial function is the set of all real numbers, denoted by the interval $(-\infty, \infty)$.

We have studied the linear functions (which include the constant functions) in Section 1.2. The graph of a linear function is a line and is easy to plot. For a general polynomial function with degree higher than two, it is difficult to plot its graph without the help of calculus or a graphing device. Figure 1.4.1 shows the graph of the cubic polynomial $f(x) = x^3 - 3x^2 + 1$.

1.4. NONLINEAR FUNCTIONS AND ALGEBRAIC EQUATIONS

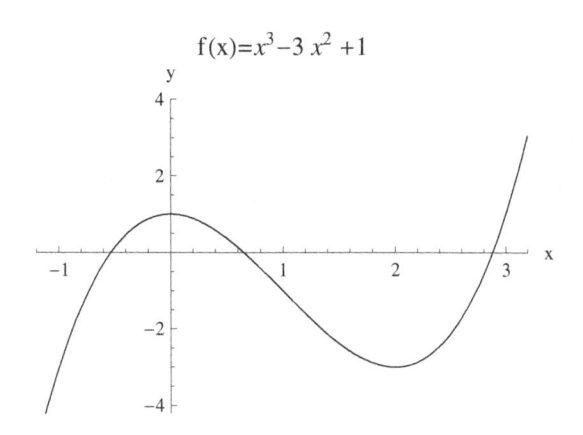

Figure 1.4.1: Graph of $f(x) = x^3 - 3x^2 + 1$

However, other than the linear functions, there is another group of polynomials whose graphs have special shapes. These are the **quadratic functions**. The graph of a quadratic function is a **parabola**. A parabola has two main features: The opening direction and the location of its **vertex**, the turning point. A quadratic function takes the form of

$$f(x) = ax^2 + bx + c$$

where $a \neq 0$ (If $a = 0$ then the function becomes linear). If $a > 0$, then its graph opens upward, while if $a < 0$, its graph opens downward. The first coordinate of its vertex is

$$x = -\frac{b}{2a}.$$

Example 1: Plot the graphs of the following quadratic functions:

(a) $f(x) = 2x^2 - 3x + 1$;

(b) $f(x) = -x^2 + 2x$;

(c) $f(x) = x^2 - 2x + 2$.

Solution (a) We have $a = 2$, $b = -3$, and $c = 1$. The parabola opens upward, and the first coordinate of its vertex is

$$x = -\frac{b}{2a} = -\frac{-3}{2(2)} = \frac{3}{4}.$$

Substituting $x = 3/4$ into the function, we obtain the second coordinate of the vertex:

$$f\left(\frac{3}{4}\right) = 2\left(\frac{3}{4}\right)^2 - 3\left(\frac{3}{4}\right) + 1 = 2\left(\frac{9}{16}\right) - \frac{9}{4} + 1 = \frac{9}{8} - \frac{18}{8} + \frac{8}{8} = -\frac{1}{8}.$$

Hence, the vertex point is at $(3/4, -1/8)$. We sample a few additional numerical values (Table 1.4.1). Plotting these points in the coordinate plane and linking the dots we obtain the graph as shown in Figure 1.4.2.

x	$f(x)$
$-\frac{1}{2}$	3
0	1
$\frac{1}{2}$	0
$\frac{3}{4}$	$-\frac{1}{8}$
1	0
2	3

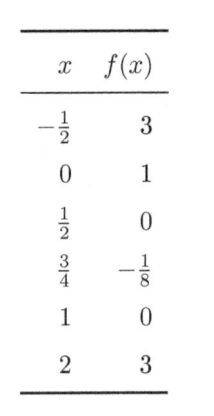

Table 1.4.1: Values of $f(x)$

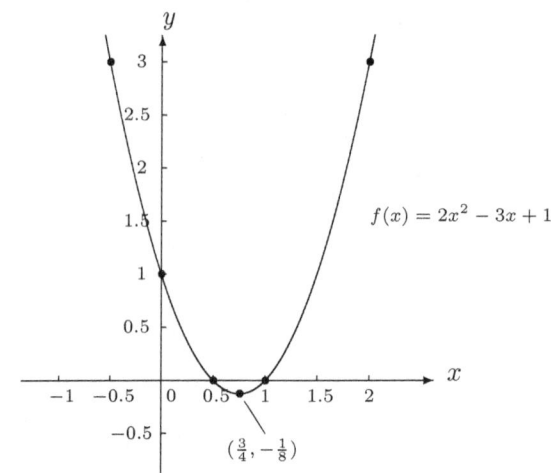

Figure 1.4.2: Graph of $f(x) = 2x^2 - 3x + 1$

(b) We have $a = -1$, $b = 2$, and $c = 0$. The parabola opens downward, and the first coordinate of its vertex is

$$x = -\frac{b}{2a} = -\frac{2}{2(-1)} = -\frac{2}{-2} = 1.$$

Substituting $x = 1$ into the function, we obtain the second coordinate of the vertex:

$$f(1) = -(1)^2 + 2(1) = -1 + 2 = 1.$$

Hence, the vertex point is at $(1, 1)$. We sample a few additional numerical values (Table 1.4.2). Plotting these points in the coordinate plane and linking the dots we obtain the graph as shown in Figure 1.4.3.

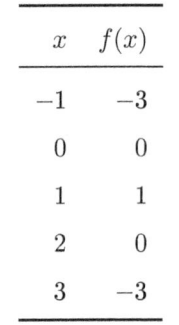

x	$f(x)$
-1	-3
0	0
1	1
2	0
3	-3

Table 1.4.2: Values of $f(x)$

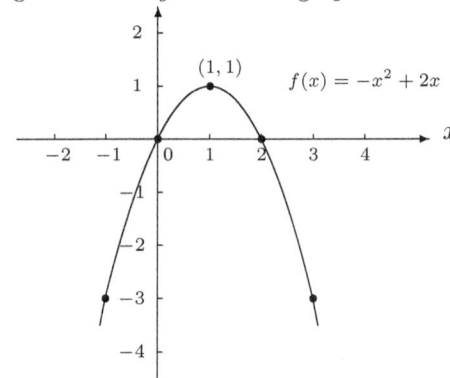

Figure 1.4.3: Graph of $f(x) = -x^2 + 2x$

(c) We have $a = 1$, $b = -2$, and $c = 2$. The parabola opens upward, and the first coordinate of its vertex is

$$x = -\frac{b}{2a} = -\frac{-2}{2(1)} = \frac{2}{2} = 1.$$

Substituting $x = 1$ into the function, we obtain the second coordinate of the vertex:

$$f(1) = (1)^2 - 2(1) + 2 = 1 - 2 + 2 = 1.$$

Hence, the vertex point is at $(1, 1)$. We sample a few additional numerical values (Table 1.4.3). Plotting these points in the coordinate plane and linking the dots we obtain the graph as shown in Figure 1.4.4.

1.4. NONLINEAR FUNCTIONS AND ALGEBRAIC EQUATIONS

x	$f(x)$
-1	5
0	2
1	1
2	2
3	5

Table 1.4.3: Values of $f(x)$

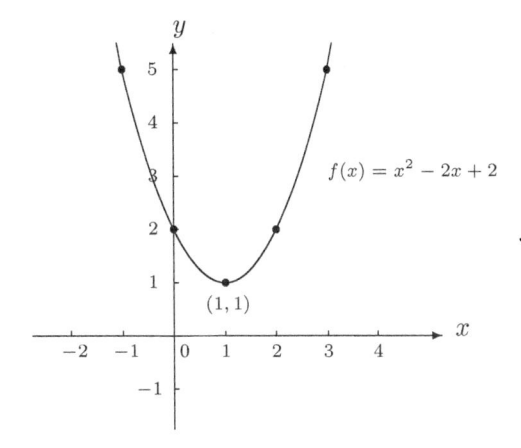

Figure 1.4.4: Graph of $f(x) = x^2 - 2x + 2$

▬ Solving Polynomial Equations

In Chapter 3, we often solve equations such as $x^3 - 2x^2 - 3x = 0$. In general, unlike solving linear equations, the approach is to factor the polynomial.

Example 2: Solve the following equations:

(a) $x^3 - 4x = 0$; (b) $2x^2 - 3x + 1 = 0$; (c) $x^2 + 4x + 4 = 0$.

Solution (a) We can factor the polynomial as:

$$x^3 - 4x = x(x^2 - 4) = x(x-2)(x+2),$$

then the equation becomes

$$x(x-2)(x+2) = 0.$$

Hence, the equation has three solutions: $x = 0$, $x = 2$, and $x = -2$.

(b) We can factor the polynomial as: $2x^2 - 3x + 1 = (2x-1)(x-1)$, then the equation becomes

$$(2x-1)(x-1) = 0.$$

Hence, the equation has two solutions: $x = 1/2$, and $x = 1$.

(c) We can factor the polynomial as:

$$x^2 + 4x + 4 = (x+2)^2,$$

then the equation becomes

$$(x+2)^2 = 0.$$

Hence, the equation has only one solution: $x = -2$. This is called a **double root** of the function $f(x) = x^2 + 4x + 4$.

Notice that in Example 2, part (b), the quadratic function $f(x) = 2x^2 - 3x + 1$ is the same function as that in Example 1, part (a). From the graph of this function we realize that the solutions to the equation $f(x) = 0$ are the first coordinates of the x-**intercepts**, or **horizontal intercepts**.

Factoring a high-degree polynomial is often a difficult task. Even factors of a quadratic function are sometimes not obvious. In addition, many polynomials cannot be factored. When solving a **quadratic equation**, as in Example 2, parts (b) and (c), in addition to factorization, we can make use of the **quadratic formula**:

Theorem 1   *Quadratic Formula*

For the equation $ax^2 + bx + c = 0$, where $a \neq 0$: Let $S = b^2 - 4ac$. Then:

(a) If $S < 0$, then the equation has no real solution.

(b) If $S = 0$, then the equation has one real solution $x = \dfrac{-b}{2a}$.

(c) If $S > 0$, then the equation has two real solutions:

$$x_1 = \frac{-b + \sqrt{S}}{2a} \quad \text{and} \quad x_2 = \frac{-b - \sqrt{S}}{2a}.$$

Example 3: Solve the following quadratic equations using the quadratic formula:

(a) $2x^2 - 3x + 1 = 0$; (b) $x^2 + 4x + 4 = 0$; (c) $x^2 - 2x + 2 = 0$.

Solution (a) We have $a = 2$, $b = -3$, and $c = 1$. Hence, $S = b^2 - 4ac = (-3)^2 - 4(2)(1) = 9 - 8 = 1$. Since $S > 0$, this equation has two solutions:

$$x_1 = \frac{-b + \sqrt{S}}{2a} = \frac{-(-3) + \sqrt{1}}{2(2)} = \frac{3 + 1}{4} = 1;$$

and

$$x_2 = \frac{-b - \sqrt{S}}{2a} = \frac{-(-3) - \sqrt{1}}{2(2)} = \frac{3 - 1}{4} = \frac{1}{2}.$$

(b) We have $a = 1$, $b = 4$, and $c = 4$. Hence, $S = b^2 - 4ac = (4)^2 - 4(1)(4) = 16 - 16 = 0$. Since $S = 0$, the equation has one solution $x = \dfrac{-b}{2a} = \dfrac{-4}{2(1)} = 2$.

(c) We have $a = 1$, $b = -2$, and $c = 2$. Hence, $S = b^2 - 4ac = (-2)^2 - 4(1)(2) = 4 - 8 = -4$. Since $S < 0$, the equation has no real solution.

Notice that the graph of the function $f(x) = x^2 - 2x + 2$ is illustrated by Example 1 (c). The graph has no horizontal intercept.

— APPLICATIONS

Example 4: A rock is thrown upward from the top of a cliff that is 196 m above the ground with initial velocity 29.4 m/s. The height of the rock at time t seconds follows the function

$$H(t) = -4.9t^2 + 29.4t + 196.$$

1.4. NONLINEAR FUNCTIONS AND ALGEBRAIC EQUATIONS

(a) Find the time the rock reaches its maximum height.

(b) Find the time the rock hits the ground.

(c) Plot the graph of $H(t)$.

Solution (a) The height function is quadratic with $a = -4.9$, $b = 29.4$, and $c = 196$. The graph of this quadratic function is a parabola which opens downward. It's vertex point represents the highest position of the rock. By the vertex formula, we have

$$t = -\frac{b}{2a} = -\frac{29.4}{2(-4.9)} = \frac{29.4}{9.8} = 3.$$

Hence, the rock reaches its maximum height at 3 seconds.

(b) The rock hits the ground when $H(t) = 0$. Hence, we need to solve the quadratic equation

$$-4.9t^2 + 29.4t + 196 = 0.$$

We can factor this equation to obtain

$$-4.9t^2 + 29.4t + 196 = -4.9(t^2 - 6t - 40) = -4.9(t+4)(t-10).$$

Hence, the equation has two roots: $t = -4$ and $t = 10$. Since t has to be nonnegative, we drop the negative root. Hence, the rock hits the ground at 10 seconds.

(c) The height function is meaningful only when $0 \leq t \leq 10$. Hence, the graph of the height function is part of the parabola where both the first and the second coordinates are nonnegative. From the discussion in part (a), we know that the parabola opens downward, and the first coordinate of its vertex is $t = 3$. Substituting $t = 3$ into the function, we obtain the second coordinate of the vertex:

$$H(3) = -4.9(3)^2 + 29.4(3) + 196 = 240.1.$$

Hence, the vertex point is at $(3, 240.1)$. We sample a few additional numerical values (Table 1.4.4). Plotting these points in the coordinate plane and linking the dots we obtain the graph as shown in Figure 1.4.5.

t	$H(t)$
0	196
3	240.1
6	196
10	0

Table 1.4.4: Values of $H(t)$

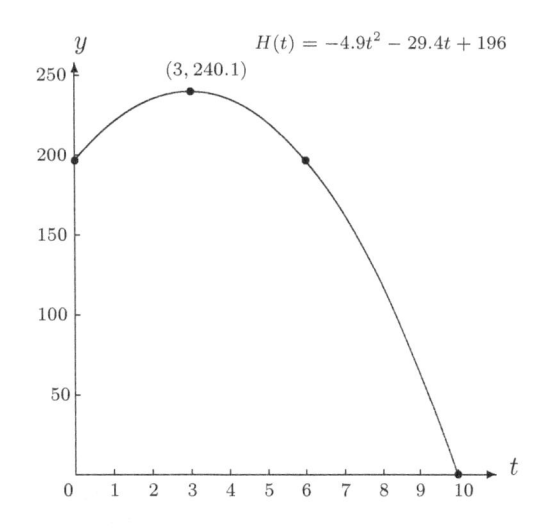

Figure 1.4.5: Graph of $H(t) = -4.9t^2 - 29.4t + 196$

Example 5: Economics: Equilibrium Point Find the equilibrium point if the demand and supply functions are given by

$$D(p) = \frac{10}{p+1}, \quad \text{and} \quad S(p) = 2p + 3.$$

Solution To find the first coordinate of the equilibrium point, we set $D(p) = S(p)$:

$$\frac{10}{p+1} = 2p + 3.$$

Multiply both sides by $p + 1$:

$$10 = (2p+3)(p+1) = 2p^2 + 5p + 3.$$

Subtracting 10 from both sides we obtain the quadratic equation:

$$2p^2 + 5p - 7 = 0.$$

To solve this equation, we can apply the quadratic formula: We have $a = 2$, $b = 5$, and $c = -7$. Hence $S = b^2 - 4ac = (5)^2 - 4(2)(-7) = 25 + 56 = 81$. Since $S > 0$, this equation has two solutions:

$$p_1 = \frac{-b + \sqrt{S}}{2a} = \frac{-(5) + \sqrt{81}}{2(2)} = \frac{-5 + 9}{4} = 1;$$

and

$$p_2 = \frac{-b - \sqrt{S}}{2a} = \frac{-(5) - \sqrt{81}}{2(2)} = \frac{-5 - 9}{4} = \frac{-7}{2}.$$

Since p must be nonnegative, we drop p_2. Therefore $p_E = 1$. Substituting into the supply function we obtain $x_E = S(1) = 2(1) + 3 = 5$. Hence, the equilibrium point is $(p_E, x_E) = (1, 5)$.

Rational Functions and Equations

A **rational function** is defined as the ratio of two polynomials:

$$R(x) = \frac{P(x)}{Q(x)} = \frac{a_n x^n + a_{n-1} x^{n-1} + \ldots + a_1 x + a_0}{b_m x^m + b_{m-1} x^{m-1} + \ldots + b_1 x + b_0}, \quad (b_m \neq 0).$$

The following are a few examples of rational functions:

$$s(x) = \frac{1}{x}; \quad f(x) = \frac{x^3 + 1}{x^2 - 1}; \quad g(x) = \frac{x+2}{x^2 + 1}.$$

The domain of a rational function is the set of all real numbers such that the denominator is not zero. The function $s(x) = 1/x$ is studied in Example 6 in Section 1.1, with its graph. The domain of s is all real numbers except 0 (since $x = 0$ would result in a denominator of zero). For the function f, since the equation for the denominator equals zero is

$$x^2 - 1 = (x+1)(x-1) = 0$$

and it results in two solutions $x = 1$ and $x = -1$, the domain of f is all real numbers except -1 and 1. For the function g, since its denominator is always positive, the domain of g is all real numbers.

For a general rational function, it is again difficult to plot its graph without the help of calculus or a graphing device. We study the graphing techniques in Chapter 3.

1.4. NONLINEAR FUNCTIONS AND ALGEBRAIC EQUATIONS

In this course we need to solve equations involving a rational function,

$$R(x) = \frac{P(x)}{Q(x)} = 0.$$

To solve such equations, we look for the roots of the numerator,

$$P(x) = 0,$$

and check that the roots are in the domain of $R(x)$.

Example 6: Consider the function

$$f(x) = \frac{x^2 + 2x - 3}{x - 1}:$$

(a) Find the domain of this function.

(b) Solve the equation $f(x) = 0$.

(c) Draw the graph of f.

Solution (a) The domain of the function is the set of all real numbers except 1, since $x = 1$ is the only number resulting in zero in the denominator.

(b) To solve the equation

$$f(x) = \frac{x^2 + 2x - 3}{x - 1} = 0,$$

we look for the solution to the equation

$$x^2 + 2x - 3 = 0.$$

We may use factorization for this equation:

$$x^2 + 2x - 3 = (x - 1)(x + 3).$$

Hence, the equation has two solutions: $x = 1$ and $x = -3$. However, notice that $x = 1$ is not in the domain. Hence, the equation $f(x) = 0$ only has one solution $x = -3$.

(c) To draw the graph, we realize that both the numerator and the denominator of the function have the factor $x - 1$. When $x \neq 1$, we can cancel the common factor $x - 1$. Therefore, this function can be simplified as

$$f(x) = \frac{x^2 + 2x - 3}{x - 1} = \frac{(x - 1)(x + 3)}{x - 1} = x + 3, \quad x \neq 1.$$

Hence, the graph of the function is formed by a line except for the part when $x = 1$, which corresponds an open circle at $(1, 4)$. Since a line is uniquely determined by two points, we only need to sample two numerical values for the graph (Table 1.4.5). Plotting these points in the coordinate plane and linking the dots we obtain the graph as shown in Figure 1.4.6.

x	$f(x)$
-3	0
0	3

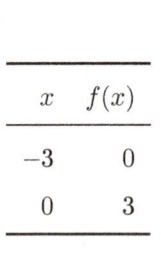

Table 1.4.5: Values of $f(x)$

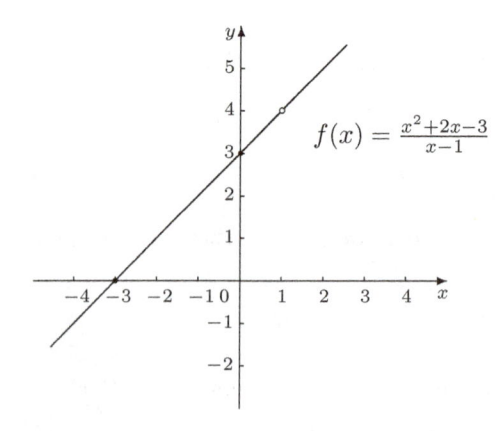

Figure 1.4.6: Graph of $f(x) = \dfrac{x^2 + 2x - 3}{x - 1}$

Power Functions with Rational Exponents

A **power function** has the format $f(x) = x^p$. If p is a positive integer, as in $f(x) = x^3$, then it can be viewed as a particular polynomial with only one term. If p is a negative integer, as in $f(x) = x^{-1} = 1/x$, then it can be viewed as a particular rational function. In this course, we also deal with the case where p is a **rational number**, as in

$$f(x) = x^{\frac{1}{2}}, \quad \text{or in general,} \quad f(x) = x^{\frac{n}{m}},$$

where n and m are integers. Such functions can also be expressed using **radical** notations, as in

$$f(x) = x^{\frac{1}{2}} = \sqrt{x}.$$

> **Reality Check**   *No even roots of negative numbers*
>
> If a function has a formula involving a rational exponent with an even root, then inside the root the expression cannot be negative.

We summarize the laws of operations with exponents and radicals:

> **Theorem 2**   *Laws of Operations with Exponents and Radicals*
>
> For any nonzero real numbers x and y, and any integers n and m:
>
> $$x^n \cdot x^m = x^{n+m} \qquad (x^n)^m = x^{n \cdot m} \qquad (xy)^n = x^n \cdot y^n$$
>
> $$\dfrac{x^n}{x^m} = x^{n-m} \qquad x^{-n} = \dfrac{1}{x^n} \qquad \left(\dfrac{x}{y}\right)^n = \dfrac{x^n}{y^n}$$
>
> $$x^{\frac{1}{n}} = \sqrt[n]{x} \qquad x^{\frac{m}{n}} = \sqrt[n]{x^m} = \left(\sqrt[n]{x}\right)^m \qquad \sqrt[n]{xy} = \sqrt[n]{x} \cdot \sqrt[n]{y}$$

Example 7: Consider the function $f(x) = \sqrt{x}$:

(a) Find the domain of this function.

(b) Draw the graph of f.

1.4. NONLINEAR FUNCTIONS AND ALGEBRAIC EQUATIONS

Solution (a) The domain of the function is the set of all nonnegative real numbers, denoted by the interval $[0, \infty)$.

(b) To draw the graph, we sample some numerical values for the graph (Table 1.4.6). Plotting these points in the coordinate plane and linking the dots we obtain the graph as shown in Figure 1.4.7.

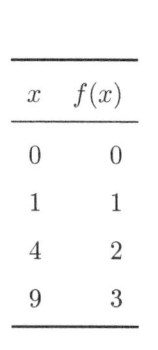

x	$f(x)$
0	0
1	1
4	2
9	3

Table 1.4.6: Values of $f(x)$

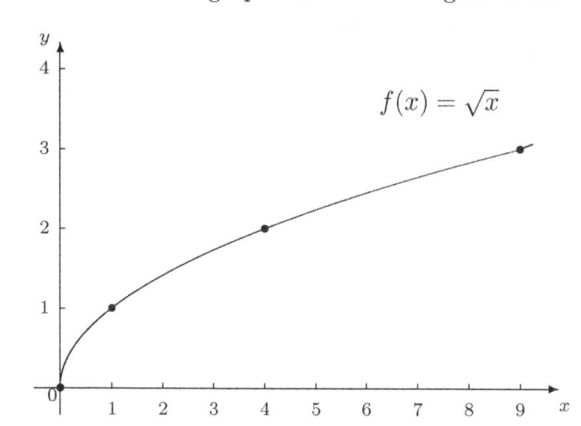

Figure 1.4.7: Graph of $f(x) = \sqrt{x}$

In this course we need to solve equations involving radicals. To solve such equations, we need to isolate the radical, then take the appropriate power for both sides of the equation to cancel the radical. After solving the revised equation, we also need to check that the solutions are valid for the original radical.

Example 8: Consider the function $f(x) = \sqrt{5x - 6} - x$:

(a) Find the domain of this function.

(b) Solve the equation $f(x) = 0$.

Solution (a) For $f(x)$ to be a real number, we need the expression inside the square root to be nonnegative. That is, $5x - 6 \geq 0$. By adding 6 to both sides and then dividing by 5, we obtain $x \geq 6/5$. Hence, the domain of the function is the set of all real numbers greater or equal to 6/5, or the interval $[6/5, \infty)$.

(b) To solve the equation $f(x) = \sqrt{5x - 6} - x = 0$, we first add x to both sides:

$$\sqrt{5x - 6} = x.$$

We then square both sides:

$$\left(\sqrt{5x - 6}\right)^2 = x^2, \text{ or } 5x - 6 = x^2.$$

We now have a quadratic equation:

$$x^2 - 5x + 6 = 0.$$

By factoring

$$x^2 - 5x + 6 = (x - 2)(x - 3),$$

we obtain two solutions to the **quadratic** equation: $x = 2$ and $x = 3$. Since both $x = 2$ and $x = 3$ are in the domain of f, and $f(2) = f(3) = 0$, they are both solutions to the equation $f(x) = 0$.

Piecewise Defined Functions

A function may be defined by different formulas in different parts of its domain. Such functions are called **piecewise defined functions**. The following three examples illustrate how to graph and evaluate such functions.

Example 9: Consider the function $f(x) = |x|$.

(a) Rewrite f as a piecewise defined function.

(b) Find $f(-2)$, $f(0)$, and $f(1)$.

(c) Draw the graph of f.

Solution (a) The absolute-value function is defined such that, if x takes negative values, then the output is the negative of x, while if x takes positive values or zero, then the output is the same as x. Hence f can be re-written as:

$$f(x) = \begin{cases} -x & \text{if } x < 0, \\ x & \text{if } x \geq 0. \end{cases}$$

(b) Since $-2 < 0$, we have $f(-2) = -(-2) = 2$. Since both 0 and 1 are located in $x \geq 0$, we have $f(0) = 0$ and $f(1) = 1$.

(c) To draw the graph, we realize that the function is formed by two linear functions in different sections. Since a line is uniquely determined by two points, we only need to sample two numerical values for each section (Table 1.4.7). We then obtain the graph of f as shown in Figure 1.4.8.

x	$f(x)$
-3	3
-1	1
0	0
1	1

Table 1.4.7: Values of $f(x)$

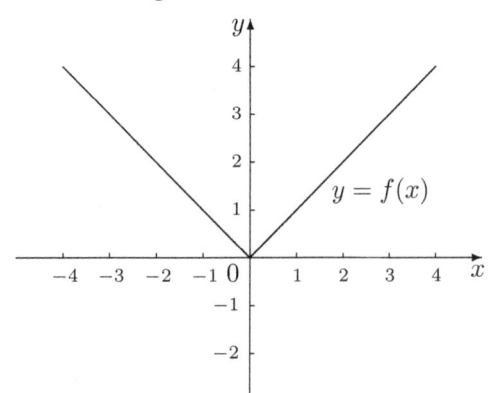

Figure 1.4.8: Graph of $f(x) = |x|$

Example 10: The function g is given by $g(x) = \begin{cases} x + 2 & \text{if } x \leq 1, \\ 2 & \text{if } x > 1. \end{cases}$

(a) Find $g(0)$, $g(1)$, and $g(3)$. (b) Draw the graph of g.

Solution (a) Since $0 < 1$, we have $g(0) = (0) + 2 = 2$. For $x = 1$, since the first formula is assigned to all x such that $x \leq 1$, we have $g(1) = (1) + 2 = 3$. Since $3 > 1$, we have $g(3) = 2$.

(b) To draw the graph, we again realize that this function is formed by two linear functions in different sections. We sample some numerical values (Table 1.4.8), then obtain the graph of g as shown in Figure 1.4.9. Notice that an open dot is put at the position $(1, 2)$.

1.4. NONLINEAR FUNCTIONS AND ALGEBRAIC EQUATIONS

x	$g(x)$
-3	-1
1	3
1.5	2
3	2

Table 1.4.8: Values of $g(x)$

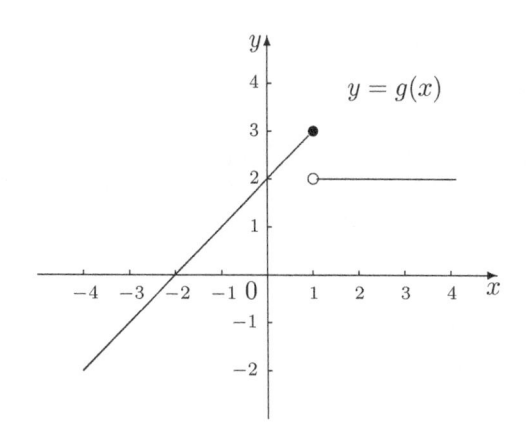

Figure 1.4.9: Graph of $g(x)$

Example 11: The function h is given by $h(x) = \begin{cases} x+2 & \text{if } x \neq 1, \\ 2 & \text{if } x = 1. \end{cases}$

(a) Find $f(0)$, $f(1)$, and $f(3)$. (b) Draw the graph of f.

Solution (a) The function is defined as $h(1) = 2$, and for all other x values, $h(x) = x + 2$. Hence, $h(0) = (0) + 2 = 2$, $h(1) = 2$, and $h(3) = (3) + 2 = 5$.

(b) To draw the graph, note that the point $(1, 2)$ is on the graph. When $x \neq 1$, the function represents a line. We then sample some numerical values for $x \neq 1$ (Table 1.4.9) to obtain the graph of h as shown in Figure 1.4.10. Notice that an open dot is put at the spot $(1, 3)$.

x	$h(x)$
-3	-1
2	4

Table 1.4.9: Values of $h(x)$

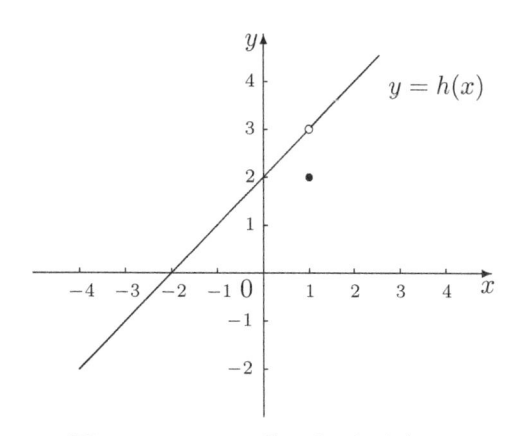

Figure 1.4.10: Graph of $h(x)$

Exercise Set 1.4

For each of the following quadratic functions, find the parabola's vertex (listed as a pair of coordinates), then draw the parabola.

1. $f(x) = 2x^2 - 1$
2. $f(x) = -x^2 + 1$
3. $f(x) = x^2 - 2x$
4. $f(x) = -x^2 + 3x$
5. $f(x) = 2x^2 + 6x - 1$
6. $f(x) = x^2 + 2x + 3$
7. $f(x) = -2x^2 + 6x + 1$
8. $f(x) = x^2 + 3x - 1$

Solve the polynomial equations using factorization

9. $x^3 - 4x = 0$
10. $x^4 - 4x^3 + 4x^2 = 0$
11. $x^4 + x^2 = 2$
12. $x^4 - 5x^2 = -4$
13. $x^2 - 4x = -3$
14. $x^2 - 3x = -2$

Solve the following equations using the quadratic formula. If a real solution does not exist, state the fact.

15. $x^2 - 4x = 2$
16. $2x^2 - 3x + 1 = 0$
17. $2x^2 - 3x + 2 = 0$
18. $x^2 - 5x + 1 = 0$
19. $4x^2 - 4x + 1 = 0$
20. $x^2 - 6x + 9 = 0$

For the functions in Exercises #21 to #26: (a) Find the domain; (b) Graph.

21. $f(x) = -\sqrt{x-1}$
22. $f(x) = \sqrt{x+1}$
23. $f(x) = \dfrac{x^2 - 9}{x - 3}$
24. $f(x) = \dfrac{x^2 - 4}{x + 2}$
25. $f(x) = \dfrac{x^2 + 3x + 2}{x + 1}$
26. $f(x) = \dfrac{x^2 - 4x + 4}{x - 2}$

For the functions in Exercises #27 to #32: (a) Find the domain; (b) Solve the equation $f(x) = 0$. If a real solution does not exist, state the fact.

27. $f(x) = \sqrt{x^2 - 1} + 2x$
28. $f(x) = \sqrt{2x + 1} - 2x$
29. $f(x) = \dfrac{x^2 - 9}{x^2 - 4x + 3}$
30. $f(x) = \dfrac{x^2 - 4}{x^2 - x - 6}$
31. $f(x) = \dfrac{x^2 - 3x + 2}{x - 2}$
32. $f(x) = \dfrac{x^2 + 4x + 4}{x + 2}$

33. Consider the function given by

$$f(x) = \begin{cases} x + 2 & \text{if } x \leq 1, \\ -2x + 5 & \text{if } x > 1. \end{cases}$$

(a) Find $f(-1)$, $f(0)$, $f(1)$, $f(2)$, and $f(3)$.

(b) Sample a few function values for $x \leq 1$ and make a numerical table.

(c) Sample a few function values for $x > 1$ and make a numerical table.

(d) Use the results in parts (b) and (c) to plot a graph of the function.

34. Consider the function given by

$$g(x) = \begin{cases} 3 - x & \text{if } x \leq 0, \\ 2x + 3 & \text{if } x > 0. \end{cases}$$

(a) Find $g(-1)$, $g(0)$, $g(1)$, and $g(2)$.

(b) Sample a few function values for $x \leq 0$ and make a numerical table.

1.4. NONLINEAR FUNCTIONS AND ALGEBRAIC EQUATIONS

(c) Sample a few function values for $x > 0$ and make a numerical table.

(d) Use the results in parts (b) and (c) to plot a graph of the function.

35. For the function given by $F(x) = |x+2|$:

(a) Use piecewise formulas to represent this function.

(b) Find $F(-3)$, $F(-2)$, $F(-1)$, and $F(2)$.

(c) Sample a few function values for $x < -2$ and make a numerical table.

(d) Sample a few function values for $x \geq 2$ and make a numerical table.

(e) Use the results in parts (c) and (d) to plot a graph of the function.

36. For the function given by $G(x) = |x-1|$:

(a) Use piecewise formulas to represent this function.

(b) Find $G(-1)$, $G(0)$, $G(1)$, and $G(2)$.

(c) Sample a few function values for $x < 1$ and make a numerical table.

(d) Sample a few function values for $x \geq 1$ and make a numerical table.

(e) Use the results in parts (c) and (d) to plot a graph of the function.

37. Consider the function given by

$$f(x) = \begin{cases} x+2 & \text{if } x \leq 1, \\ x^2 - 3 & \text{if } x > 1. \end{cases}$$

(a) Find $f(-1)$, $f(0)$, $f(1)$, $f(2)$, and $f(3)$.

(b) Sample a few function values for $x \leq 1$ and make a numerical table.

(c) Sample a few function values for $x > 1$ and make a numerical table.

(d) Use the results in parts (b) and (c) to plot a graph of the function.

38. Consider the function given by

$$g(x) = \begin{cases} 3 - x^2 & \text{if } x \leq 0, \\ 2x + 3 & \text{if } x > 0. \end{cases}$$

(a) Find $g(-1)$, $g(0)$, $g(1)$, and $g(2)$.

(b) Sample a few function values for $x \leq 0$ and make a numerical table.

(c) Sample a few function values for $x > 0$ and make a numerical table.

(d) Use the results in parts (b) and (c) to plot a graph of the function.

39. Consider the function given by

$$f(x) = \begin{cases} x+1 & \text{if } x \neq 1, \\ 3 & \text{if } x = 1. \end{cases}$$

(a) Find $f(-1)$, $f(0)$, $f(1)$, $f(2)$, and $f(3)$.

(b) Sample a few function values and make a numerical table. Then use the table to plot a graph of the function.

40. Consider the function given by

$$g(x) = \begin{cases} 3 - x & \text{if } x \neq 2, \\ 4 & \text{if } x = 2. \end{cases}$$

(a) Find $g(-1)$, $g(0)$, $g(2)$, and $g(3)$.

(b) Sample a few function values and make a numerical table. Then use the table to plot a graph of the function.

APPLICATIONS

41. A rock is thrown upward from the top of a tower that is 98 m above the ground with initial velocity 19.6 m/s. The height of the rock at time t seconds follows the function

$$H(t) = -4.9t^2 + 19.6t + 98.$$

(a) Find the time the rock reaches its maximum height.

(b) Find the time the rock hits the ground.

(c) Plot the graph of $H(t)$.

42. A rock is thrown upward from the top of a building that is 120 m above the ground with initial velocity 24.5 m/s. The height of the rock at time t seconds follows the function

$$H(t) = -4.9t^2 + 24.5t + 120.$$

(a) Find the time the rock reaches its maximum height.

(b) Find the time the rock hits the ground.

(c) Plot the graph of $H(t)$.

Economics *Find the equilibrium point for the demand and supply functions in Exercises #43 to #46*

43. $D(p) = \dfrac{21}{p+3}$; $S(p) = 3p - 2$.

44. $D(p) = \dfrac{30}{p+2}$; $S(p) = 2p - 5$.

45. $D(p) = 20 - 2\sqrt{p+36}$; $S(p) = p - 2$.

46. $D(p) = 10 - \sqrt{p+1}$; $S(p) = 2p - 3$.

1.5 Exponential Functions and Financial Models

In this section, we study exponential functions. Many real life phenomena, such as population growth, can be modeled as exponential functions. We start this section by exploring the financial application, **compound interest**.

We deposit $1,000 dollars in a bank paying interest at a rate of 6% a year, for t years. How much is in the account after t years? It depends on the method the bank uses to calculate the interest.

Simple interest: The method of simple interest is that, after the interest is calculated each year, it is set aside. Hence, the principal stays the same as it started. Since the interest is at the rate of 6% per year, each year there is the interest of $1,000 × 6% = $1,000 × 0.06 = $60. Therefore, at end of tth year, the amount in the account is the initial principal, $1,000, plus the interest of $60 × t. We can express the amount in the account as $P(t) = 1,000 + 60t$. This is a *linear* function.

Interest compounded annually: This time, by the end of the first year, the interest is still computed by $1,000 × 0.06 = $60. However, it is then added to the principal. Let us denote the amount in the account after 1 year as $P(1)$, then

$P(1) = 1,000 + 60 = 1,000 + 1,000 \times 0.06 = 1,000(1 + 0.06)$.

Therefore, by the end of the second year, the interest is $1,060 × 0.06 = $63.60, and the amount in the account is $1,060 + $63.60 = $1,123.60. We can express the amount as

$P(2) = P(1) + P(1) \times 0.06 = P(1)(1 + 0.06) = 1,000(1 + 0.06)(1 + 0.06) = 1,000(1 + 0.06)^2$.

Similarly, $P(3) = 1,000(1 + 0.06)^3$, and after t years, the amount in the account is

$P(t) = 1,000(1 + 0.06)^t = 1,000(1 + 0.06)^t$.

Notice that the variable t occurs at the exponent. Hence, the amount of money in the account is an **exponential** function of t.

Definition *exponential functions*

An **exponential function** takes the form of

$$f(x) = P_0 a^x$$

where a is a positive real number other than 1, called the **base**, and P_0 is a constant, called the **initial value**.

The function $P(t) = 1,000(1 + 0.06)^t$ is an exponential function of t with base 1.06 and initial value $P_0 = 1,000$. In general, if an amount of P_0 dollars is deposited into a savings account with annual interest rate r and compounded annually, then the amount in the account after t years is an exponential function

$$P(t) = P_0(1 + r)^t.$$

1.5. EXPONENTIAL FUNCTIONS AND FINANCIAL MODELS

The following are a few more examples of exponential functions:

$$f(x) = 2^x \qquad g(x) = \left(\frac{1}{2}\right)^x \qquad h(x) = 5 \cdot \pi^x$$

The domain of an exponential function is the set of all real numbers, denoted by the interval $(-\infty, \infty)$.

We now explore the graphs of the exponential functions $f(x)$ and $g(x)$ in the above examples. Sampling a few numerical values (Table 1.5.1), plotting these points in the coordinate plane, and linking the dots we obtain the graphs as shown in Figures 1.5.1 and 1.5.2:

x	2^x	$\left(\frac{1}{2}\right)^x$
-3	$\frac{1}{8}$	8
-2	$\frac{1}{4}$	4
-1	$\frac{1}{2}$	2
0	1	1
1	2	$\frac{1}{2}$
2	4	$\frac{1}{4}$
3	8	$\frac{1}{8}$

Table 1.5.1: Values of 2^x, $\left(\frac{1}{2}\right)^x$

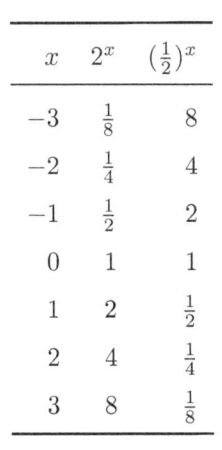

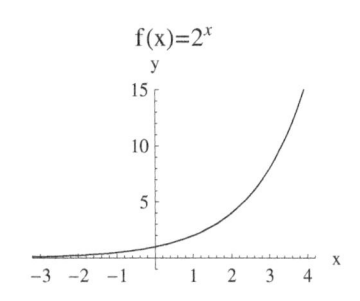

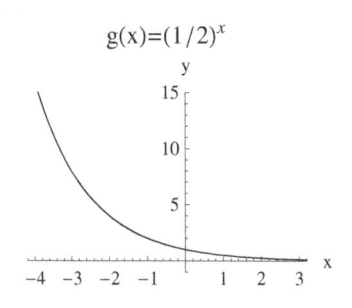

Figure 1.5.1: Graph of $y = 2^x$

Figure 1.5.2: Graph of $y = \left(\frac{1}{2}\right)^x$

The graph of an exponential function $y = a^x$ is similar to one of the above graphs. Since $a^0 = 1$ for any positive real number a, the graph must pass the point $(0, 1)$. If $a > 1$, then the graph is increasing and resembles that of $y = 2^x$. If $0 < a < 1$, then the graph is decreasing and resembles that of $y = \left(\frac{1}{2}\right)^x$. Since by the laws of exponential operations illustrated in Section 1.4,

$$\left(\frac{1}{2}\right)^x = \frac{1}{2^x} = 2^{-x},$$

the two graphs are the reflections along the y-axis.

Reality Check   *Exponential vs. power functions*

When we look at the graphs of $y = x^2$ and $y = 2^x$, for the positive x values, the two graphs look similar: both are increasing. However, an exponential function increases much, much, faster than a power function when x is positive and large.

Example 1: Let $f(x) = x^6$ and $g(x) = 2^x$. Compare:

(a) $f(2)$ and $g(2)$ (b) $f(30)$ and $g(30)$ (c) $f(100)$ and $g(100)$

Solution (a) We have $f(2) = 2^6 = 64$ and $g(2) = 2^2 = 4$. Hence, $f(2)$ is larger.

(b) We have $f(30) = (30)^6 \approx 7.3 \times 10^8$ and $g(30) = 2^{30} \approx 1.07 \times 10^9$. Hence, $g(30)$ is larger.

(c) We have $f(100) = (100)^6 = 10^{12}$ and $g(100) = 2^{100} \approx 1.3 \times 10^{30}$. Hence, $g(100)$ is much larger: $g(100) \approx 130{,}000{,}000{,}000{,}000{,}000 \times f(100)$.

Compound Interests and the Base e

A special exponential function takes the base

$$e = 2.718281828459\ldots$$

The importance of this particular base is indicated by the fact that most calculators have an e^x button. The special features of the function $y = e^x$ is illustrated in Chapter 2. Here, we link this particular base with the compound interest models.

Suppose that $1 is invested in an account with an annual interest rate of 100%. We calculate the amount in the bank by the end of 1 year.

1. If the interest is compounded once a year, then by the end of 1 year, the amount in the account is $P(1) = 1(1 + 1) = \$2$.

2. If the interest is compounded two times a year, then for each period, the interest rate is at 50% per half year. Hence, by the end of half of a year, the amount in the account is $1(1+0.5) = \$1.50$. By the end of 1 year, the amount in the account is $P(1) = 1.5(1+0.5) = \$2.25$, or

$$P(1) = 1(1+0.5)^2 = 1\left(1+\frac{1}{2}\right)^2 = \$2.25.$$

3. If the interest is compounded monthly, that is, 12 times a year, then for each period, the interest rate is 100/12% per month, or 1/12 per 1/12 of a year. By the end of the year, the amount in the account is

$$P(1) = 1\left(1+\frac{1}{12}\right)^{12} \approx \$2.61304.$$

4. Similarly, if the interest is compounded daily, hourly, every minute, every second, the amounts in the account by the end of the year are:

Compounded daily: $P(1) = 1\left(1+\dfrac{1}{365}\right)^{365} \approx \$2.7145748202.$

Compounded hourly: $P(1) = 1\left(1+\dfrac{1}{365 \times 24}\right)^{365 \times 24} \approx \$2.7181266906.$

Compounded every minute: $P(1) = 1\left(1+\dfrac{1}{365 \times 24 \times 60}\right)^{365 \times 24 \times 60} \approx \$2.7182792154.$

Compounded every second: $P(1) = 1\left(1+\dfrac{1}{365 \times 24 \times 360}\right)^{365 \times 24 \times 360} \approx \$2.7182812274.$

Comparing with the number $e = 2.718281828459\ldots$, we realize that the more frequently the interest is compounded, the closer the amount is to $\$e$. Indeed, if the interest is compounded **continuously**, then by the end of the year, the amount is indeed $\$e$. We extend the above discussion to more general situations.

1.5. EXPONENTIAL FUNCTIONS AND FINANCIAL MODELS

> **Theorem 1**   *Compounded Interests*
>
> Suppose that $\$P_0$ is invested at the annual interest rate of r for t years. If the interest is compounded n times per year, then by the end of tth year, the balance is
>
> $$P(t) = P_0 \left(1 + \frac{r}{n}\right)^{nt}.$$
>
> If the interest is compounded continuously, then by the end of tth year, the balance is
>
> $$P(t) = P_0 e^{rt}.$$

Example 2: The grandparents of a 1-year-old girl invest $5,000 as her college education fund. Suppose the annual interest rate is at 5%. Find the amount in the account and the total interest made 17 years later, if the interest is compounded

(a) Four times a year;

(b) Monthly;

(c) Continuously.

Solution (a) We have $P_0 = 5,000$, $r = 0.05$, $n = 4$, and $t = 17$. Hence, the amount in the account is

$$P(17) = 5,000\left(1 + \frac{0.05}{4}\right)^{4 \times 17} \approx \$11,636.76.$$

Since the account starts with $5,000 and ends with $11,636.76, the difference $6,636.76 is the total interest made.

(b) We have $P_0 = 5,000$, $r = 0.05$, $n = 12$, and $t = 17$. Hence, the amount in the account is

$$P(17) = 5,000\left(1 + \frac{0.05}{12}\right)^{12 \times 17} \approx \$11,677.59.$$

The total interest made is $11,677.59 − $5,000 = $6,677.59.

(c) We have $P_0 = 5,000$, $r = 0.05$, and $t = 17$. Hence, the amount in the account is

$$P(17) = 5,000 e^{0.05 \times 17} = 5,000 e^{0.85} \approx \$11,698.23.$$

The total interest made is $11,698.23 − $5,000 = $6,698.23.

> **Reality Check**   *Continuous vs. discrete frequency*
>
> Observe that the frequency of compounding only slightly affect the future amount of the account. Indeed, this feature allows us to use the continues model to approximate the discrete model later.

▬ Present and Future Values

Jack has won a lottery. He has two options to receive the award: He can either take it in three installments of $100,000 each, at the end of each year, for the next 3 years, for a total of $300,000; or he can take a lump–sum of $270,000 now. Which option should he take?

Although not many people win large amounts in lotteries, in many business or daily life situations we encounter payments in the future, such as car loans and house mortgages. The $1,000 your friend borrowed from you is worth less if it is paid back five years later, due both to the inflation and to the interest you could have earned if you had invested it instead.

The effect of inflation can be conveniently combined with the effect of interest, by reducing the promised interest. For example, suppose that the interest in an investment is at an annual interest rate of 6% compounded annually. Suppose that during that year inflation is at the rate of 3%. Then by the end of the year, the $1,000 earns $60 to become $1,060. However, due to inflation, the same goods which cost $1,000 at the beginning of the year now cost $1,030. Hence, the purchasing capacity of the $1,060 is reduced to $1,060/1,030 = \$1,029.13$. Hence, the "true" annual interest rate is about 2.91%. Due to the above analysis, we consider only the effect of interest.

Suppose that $1,000 is invested at the annual interest rate of 6% compounded annually. By the end of 1 year, the amount in the account is $1,060. We call the $1,060 the **future value** of the $1,000. We can also call the $1,000 the **present value** of the $1,060. The relationship between a present value, P_0, and a future value, P, depends on the annual interest rate, r, the time of the investment, t, and the way the interest is compounded.

Theorem 2   *Present and Future Values*

Suppose that P_0 is the present value of P, and P is the future value of P_0. Suppose that the annual interest rate is r and the time between the present and the future is t years.

(a) If the interest is compounded n times a year, then

$$P = P_0\left(1 + \frac{r}{n}\right)^{nt} \quad \text{and} \quad P_0 = \frac{P}{\left(1 + \frac{r}{n}\right)^{nt}} = P\left(1 + \frac{r}{n}\right)^{-nt}.$$

(b) If the interest is compounded continuously, then

$$P = P_0 e^{rt} \quad \text{and} \quad P_0 = \frac{P}{e^{rt}} = Pe^{-rt}.$$

Reality Check   *Present value vs. future value*

Notice that the two future value equations in Theorem 2 are identical as the equations in Theorem 1. Indeed, the two present value equations can be obtained from the corresponding future value equations easily. Therefore, given a financial case, as long as the correct equation provided by Theorem 1 is picked, and all known information is filled in, finding the remaining unknown becomes very simple.

Example 3: Jack has won a lottery. He was offered two options to receive the award: He can either take it in three installments of $100,000 each, at the end of each year, for the next 3 years; or he can take a lump–sum of $270,000 now. Assuming (a) interest at an annual rate of 5% compounded annually, or (b) interest at an annual rate of 5.5% compounded continuously, which option should he choose under the consideration of cash value only?

Solution (a) We compute the present value for each installment under the assumption that $r = 0.05$ and $n = 1$. The first installment occurs one year later, and has the present value

$$P_{01} = P_1\left(1 + \frac{r}{n}\right)^{-nt} = 100{,}000(1 + 0.05)^{-1} \approx \$95{,}238;$$

1.5. EXPONENTIAL FUNCTIONS AND FINANCIAL MODELS

The second installment occurs two years later, and has the present value

$$P_{02} = 100{,}000\,(1+0.05)^{-2} \approx \$90{,}703;$$

The third installment occurs three years later, and has the present value

$$P_{03} = 100{,}000(1+0.05)^{-3} \approx \$86{,}384.$$

Hence, the total present value is

$$P_0 = P_{01} + P_{02} + P_{03} = 95{,}238 + 90{,}703 + 86{,}384 = \$272{,}325.$$

Since the present value is larger than the lump-sum of $270{,}000$, Jack should choose the option of three installments.

(b) We compute the present value for each installment under the assumption that $r = 0.055$ and interest is compounded continuously. The first installment occurs one year later, and has the present value

$$P_{01} = P_1 e^{-rt} = 100{,}000 e^{-0.055} \approx \$94{,}649;$$

The second installment occurs two years later, and has the present value

$$P_{02} = 100{,}000 e^{-0.055(2)} \approx \$89{,}584;$$

The third installment occurs three years later, and has the present value

$$P_{03} = 100{,}000 e^{-0.055(3)} \approx \$84{,}789.$$

Hence, the total present value is

$$P_0 = P_{01} + P_{02} + P_{03} = 94{,}649 + 89{,}584 + 84{,}789 = \$269{,}022.$$

Since the present value is less than the lump–sum of $270{,}000$, Jack should choose the option of the lump–sum.

Example 4: The grandparents of a 1-year-old girl plan to invest for her college education. Their target is that when she is 18 years old, the fund should have the amount of $60{,}000$. Find the initial amount needed to set up the fund now, assuming the following annual interest rates and compounding methods:

(a) At the annual rate of 5%, compounded monthly;

(b) At the annual rate of 7%, compounded continuously;

(c) At the annual rate of 9%, compounded continuously.

Solution (a) We compute the present value of the $60{,}000$ under the assumption that $r = 0.05$, $n = 12$, and $t = 17$. The present value is

$$P_0 = P\left(1+\frac{r}{n}\right)^{-nt} = 60{,}000\left(1+\frac{0.05}{12}\right)^{-12(17)} \approx \$25{,}690.$$

Hence, the grandparents need to set up the fund with $25{,}690$.

(b) We compute the present value of the $60{,}000$ under the assumption that $r = 0.07$, compounded continuously, and $t = 17$. The present value is

$$P_0 = P e^{-rt} = 60{,}000 e^{-0.07(17)} \approx \$18{,}253.$$

Hence, the grandparents need to set up the fund with $18,253.

(c) Similar to Part (b) with $r = 0.09$, the present value is

$$P_0 = Pe^{-rt} = 60{,}000e^{-0.09(17)} \approx \$12{,}992.$$

Hence, the grandparents need to set up the fund with $12,992.

> **Reality Check** *Effect of rate*
>
> Observe that, in the long run, the interest rate has a significant effect on the yield of an investment with compounded interest.

Exercise Set 1.5

For Exercises #1 to #8, do the following:

(a) Make a table of function values for $x = -4$ to $x = 4$.

(b) Graph the function.

1. $f(x) = 3^x$
2. $f(x) = (\frac{3}{2})^x$
3. $f(x) = (\frac{1}{3})^x$
4. $f(x) = (\frac{2}{3})^x$
5. $f(x) = e^{0.2x}$
6. $f(x) = e^{0.5x}$
7. $f(x) = e^{-0.2x}$
8. $f(x) = e^{-0.5x}$

9. If Joe deposits $800 into an account that has a 3% annual interest rate, how much will be in the account after 5 years under the simple interest formula? How much interest will he make?

10. If Mark deposits $1,000 into an account that has a 1.6% annual interest rate, how much will be in the account after 5 years under the simple interest formula? How much interest will he make?

11. If Joe deposits $800 into an account that has a 3% annual interest rate, how much will be in the account after 5 years if the interest is compounded four times a year? How much interest will he make?

12. If Mark deposits $1,000 into an account that has a 1.6% annual interest rate, how much will be in the account after 5 years if the interest is compounded quarterly (four times a year)? How much interest will he make?

13. If Lucy deposits $3,000 into an account that has a 2% annual interest rate, how much will be in the account after 6 years if the interest is compounded monthly (12 times a year)? How much interest will she make?

14. If Elizabeth deposits $2,000 into an account that has a 1.5% annual interest rate, how much will be in the account after 4 years if the interest is compounded monthly (12 times a year)? How much interest will she make?

15. If the Lopez family deposits $12,000 into an account that has a 2.1% annual interest rate, how much will be in the account after 3 years if the interest is compounded quarterly (4 times a year)? How much interest will they make?

16. If the Nguyen family deposits $15,000 into an account that has a 1.9% annual interest rate, how much will be in the account after 5 years if the interest is compounded weekly (52 times a year)? How much interest will they make?

17. Jack deposits $5,000 into an account that has a 2.5% annual interest rate. Find the amount in the account after 10 years if the interest is compounded

1.5. EXPONENTIAL FUNCTIONS AND FINANCIAL MODELS

(a) Annually   (b) Monthly

(c) Weekly   (d) Daily

(e) Continuously

18. John deposits $8,000 into an account that has a 2.4% annual interest rate. Find the amount in the account after 8 years if the interest is compounded

    (a) Annually   (b) Monthly

    (c) Weekly   (d) Daily

    (e) Continuously

19. Joan deposits $6,000 into an account that has a 2.4% annual interest rate. Find the amount in the account after 9 years if the interest is compounded

    (a) Annually   (b) Monthly

    (c) Weekly   (d) Daily

    (e) Continuously

20. Jan deposits $7,000 into an account that has a 3.2% annual interest rate. Find the amount in the account after 11 years if the interest is compounded

    (a) Annually   (b) Monthly

    (c) Weekly   (d) Daily

    (e) Continuously

For Exercises #21 to #30, do the following:

(a) Use an appropriate formula to evaluate $P(5)$;

(b) Find how much interest is made at the end of the 5-year period.

21. $5,500 is invested at an annual interest rate of 3% compounded monthly.

22. $9,000 is invested at an annual interest rate of 2.5% compounded daily.

23. $6,000 is invested at an annual interest rate of 2.5% compounded annually.

24. $7,500 is invested at an annual interest rate of 2.4% compounded weekly.

25. $8,000 is invested at an annual interest rate of 5% compounded continuously.

26. $9,000 is invested at an annual interest rate of 4.1% compounded continuously.

27. $3,000 is invested at an annual interest rate of 3.5% compounded quarterly.

28. $4,500 is invested at an annual interest rate of 4.3% compounded quarterly.

29. $7,000 is invested at an annual interest rate of 4% compounded daily.

30. $8,500 is invested at an annual interest rate of 5% compounded continuously.

For Exercises #31 to #40: Use an appropriate formula to evaluate either the future value or the present value.

31. The future value of an investment of $5,000 for 3 years at an annual interest rate of 3% compounded monthly.

32. The future value of an investment of $6,000 for 6 years at an annual interest rate of 2.5% compounded monthly.

33. The present value of an investment of $5,000 for 3 years at an annual interest rate of 3% compounded monthly.

34. The present value of an investment of $6,000 for 6 years at an annual interest rate of 2.5% compounded monthly.

35. The future value of an investment of $20,000 for 4 years at an annual interest rate of 2.5% compounded continuously.

36. The future value of an investment of $15,000 for 5 years at an annual interest rate of 2.2% compounded continuously.

37. The present value of an investment of $20,000 for 4 years at an annual interest rate of 2.5% compounded continuously.

38. The present value of an investment of $15,000 for 5 years at an annual interest rate of 2.2% compounded continuously.

39. The present value of an investment of $7,000 for 7 years at an annual interest rate of 8% compounded annually.

40. The present value of an investment of $8,000 for 8 years at an annual interest rate of 9% compounded annually.

41. The parents of a 1-year-old boy plan to invest for his college education. Their target is that when he is 18 years old, the fund should have the amount of $80,000. Find the initial amount needed to set up the fund now, assuming the following annual interest rates and compounding methods:

    (a) At the annual rate of 4.5%, compounded monthly;

    (b) At the annual rate of 6.8%, compounded continuously;

    (c) At the annual rate of 9.2%, compounded continuously.

42. The parents of a 1-year-old boy plan to invest for his college education. Their target is that when he is 18 years old, the fund should have the amount of $75,000. Find the initial amount needed to set up the fund now, assuming the following annual interest rates and compounding methods:

    (a) At the annual rate of 4.6%, compounded monthly;

    (b) At the annual rate of 6.7%, compounded monthly;

    (c) At the annual rate of 9.3%, compounded continuously.

43. Dean has won a lottery. He was offered two options to receive the award: he can either take it in four installments of $50,000 each, at the end of each year, for the next 4 years; or he can take a lump–sum of $175,000 now. Assuming (a) interest at an annual rate of 5% compounded annually, or (b) interest at an annual rate of 5.5% compounded continuously, which option should he choose under the consideration of cash value only?

44. David has won a lottery. He was offered two options to receive the award: he can either take it in three installments of $80,000 each, at the end of each year, for the next 3 years; or he can take a lump–sum of $218,000 now. Assuming (a) interest at an annual rate of 4.5% compounded annually, or (b) interest at an annual rate of 5% compounded continuously, which option should he choose under the consideration of cash value only?

45. Sharon has won a lottery. She was offered two options to receive the award: she can either take it in four installments of $100,000 annually, starting from now; or she can take a lump–sum of $370,000 now. Assuming (a) interest at an annual rate of 5.2% compounded annually, or (b) interest at an annual rate of 5.8% compounded continuously, which option should she choose under the consideration of cash value only?

46. Ana has won a lottery. She was offered two options to receive the award: she can either take it in five installments of $60,000 annually, starting from now; or she can take a lump–sum of $271,000 now. Assuming (a) interest at an annual rate of 5% compounded annually, or (b) interest at an annual rate of 6% compounded continuously, which option should she choose under the consideration of cash value only?

1.6 Logarithmic Functions and Exponential Equations

Logarithmic Functions

Katherine invests $5,000 in an account with an annual interest rate of 7% compounded continuously. How long does it take for her investment to reach $8,000? From Section 1.5, we know that the amount in the account after t years can be calculated by $P(t) = 5{,}000e^{0.07t}$. Hence, we need to find t such that

$$8{,}000 = 5{,}000e^{0.07t}.$$

This is an **exponential equation**, since the unknown variable t is in the exponent. To solve such equations, we have to first study the **logarithmic functions**.

1.6. LOGARITHMIC FUNCTIONS AND EXPONENTIAL EQUATIONS

> **Definition**   *logarithmic function*
>
> The **logarithmic function with base $a > 0$ and $a \neq 1$**, denoted by $f(x) = \log_a(x)$, is defined as:
>
> $y = \log_a(x)$ if and only if $x = a^y$.

This definition tells us, for example, that

$$2^3 = 8 \quad \text{means} \quad \log_2(8) = 3,$$

and

$$5^{-2} = 0.04 \quad \text{means} \quad \log_5(0.04) = -2.$$

Although any positive real number a can be a base of a logarithmic function, there are two special bases that are most widely used. They are base 10 and base e.

> **Definition**   *two special logarithmic functions*
>
> The **common logarithmic** is $y = \log_{10}(x)$, denoted by $y = \log(x)$. Therefore
>
> $y = \log(x)$ if and only if $x = 10^y$.
>
> The **natural logarithmic** is $y = \log_e(x)$, denoted by $y = \ln(x)$. Therefore
>
> $y = \ln(x)$ if and only if $x = e^y$.

We call the logarithmic function $g(x) = \log_a(x)$ the **inverse function** of $f(x) = a^x$. Conversely, if we switch x and y in the definition, we can define the exponential function a^x by the logarithmic function $\log_a(x)$. Therefore, the two functions are inverse functions to each other. We say that $g(x) = \log_a(x)$ and $f(x) = a^x$ are inverse functions. In particular, $g(x) = \ln(x)$ and $f(x) = e^x$ are inverse functions.

Since an exponential expression $x = a^y$ is only valid when x is positive, the domain of a logarithmic function is the set of all positive real numbers, denoted by the interval $(0, \infty)$.

We now explore the graphs of the logarithmic functions. Let $f(x) = e^x$ and $g(x) = \ln(x)$. We first sample a few numerical values for $f(x) = e^x$. Since $y = g(x) = \ln(x)$ means $x = e^y = f(y)$, we can switch the x and y values in the numerical table of $f(x)$ to fill the numerical table of $g(x)$ (Table 1.6.1). Plotting these points in the coordinate plane and linking the dots we obtain the graphs of $y = e^x$ and $y = \ln(x)$ as shown in Figures 1.6.1 and 1.6.2.

x	$f(x)$	x	$g(x)$
-2	0.136	0.136	-2
-1	0.38	0.38	-1
0	1	1	0
0.25	1.28	1.28	0.25
0.5	1.65	1.65	0.5
1	2.72	2.72	1
2	7.39	7.39	2
4	54.6	54.6	4

Table 1.6.1: Values of $f(x)$ and $g(x)$

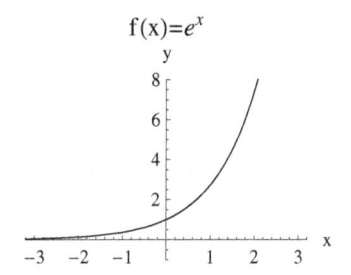

Figure 1.6.1: Graph of $f(x) = e^x$

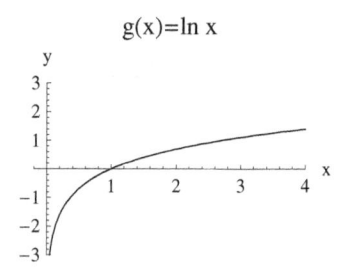

Figure 1.6.2: Graph of $g(x) = \ln(x)$

Observe that the two graphs are the reflections along the line $y = x$ (Figure 1.6.3):

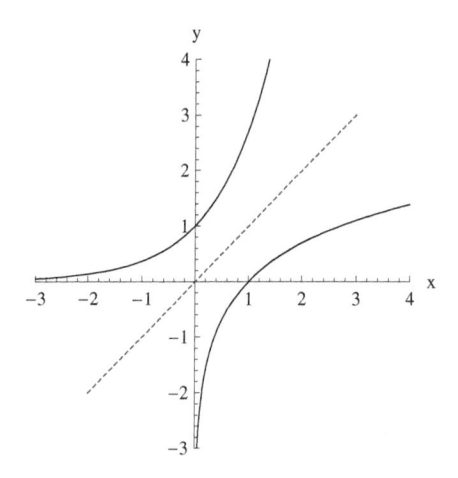

Figure 1.6.3: Graph of $\ln(x)$ obtained by reflecting the graph of e^x across the line $y = x$

The properties of logarithms are much like those of exponents. We summarize the laws of operations with logarithms:

1.6. LOGARITHMIC FUNCTIONS AND EXPONENTIAL EQUATIONS

Theorem 1   *Laws of Operations with Logarithms*

For any positive real numbers a, x, and y, and any real number p :

$$\log_a(xy) = \log_a(x) + \log_a(y) \qquad \log_a(x^p) = p \cdot \log_a(x)$$

$$\log_a\left(\frac{x}{y}\right) = \log_a(x) - \log_a(y) \qquad \log_a(x) = \log_a(y) \text{ if and only if } x = y$$

$$\log_a(1) = 0 \qquad\qquad\qquad \log_a(a) = 1$$

$$x = a^{\log_a(x)} \qquad\qquad\qquad x = \log_a(a^x)$$

Since the natural logarithm is the most often used logarithm in this course, we summarize the laws of operations with natural logarithms:

Theorem 2   *Laws of Operations with Natural Logarithms*

For any positive real numbers x and y, and any real number p :

$$\ln(xy) = \ln(x) + \ln(y) \qquad \ln(x^p) = p \cdot \ln(x)$$

$$\ln\left(\frac{x}{y}\right) = \ln(x) - \ln(y) \qquad \ln(x) = \ln(y) \text{ if and only if } x = y$$

$$\ln(1) = 0 \qquad\qquad\qquad \ln(e) = 1$$

$$x = e^{\ln(x)} \qquad\qquad\qquad x = \ln(e^x)$$

Example 1: Use the properties of logarithms to rewrite the following as sums and/or differences, where a and b are positive real numbers.

(a) $\ln(\sqrt{a \cdot b})$ (b) $\ln\left(\dfrac{a^2}{b^3}\right)$ (c) $\log_2\left(\dfrac{\sqrt[5]{a}}{b^7}\right)$

Solution (a) $\ln(\sqrt{a \cdot b}) = \ln(a \cdot b)^{\frac{1}{2}} = \frac{1}{2}\ln(a \cdot b) = \frac{1}{2}(\ln(a) + \ln(b))$.

(b) $\ln\left(\dfrac{a^2}{b^3}\right) = \ln(a^2) - \ln(b^3) = 2\ln(a) - 3\ln(b)$.

(c) $\log_2\left(\dfrac{\sqrt[5]{a}}{b^7}\right) = \log_2(\sqrt[5]{a}) - \log_2(b^7) = \dfrac{1}{5}\log_2(a) - 7\log_2(b)$.

Example 2: Use the properties of logarithms to simplify the following:

(a) $\ln(\sqrt[3]{e})$ (b) $e^{3\ln(2)}$ (c) $\log_2\left(\dfrac{8}{\sqrt{2}}\right)$

Solution (a) $\ln(\sqrt[3]{e}) = \ln e^{\frac{1}{3}} = \dfrac{1}{3}\ln e = \dfrac{1}{3}\cdot 1 = \dfrac{1}{3}.$

(b) $e^{3\ln(2)} = e^{\ln 2^3} = 2^3 = 8.$

(c) $\log_2\left(\dfrac{8}{\sqrt{2}}\right) = \log_2(8) - \log_2(\sqrt{2}) = \log_2(2^3) - \log_2\left(2^{\frac{1}{2}}\right) = 3\log_2(2) - \dfrac{1}{2}\log_2(2) = 3\cdot 1 - \dfrac{1}{2}\cdot 1 = 2\dfrac{1}{2}.$

Using the property $a = e^{\ln(a)}$, we can change any exponential function $y = a^x$ to the exponential function with base e:

$$y = a^x = (e^{\ln(a)})^x = e^{(\ln(a))x}.$$

On the other hand, for any logarithmic function $y = \log_a(x)$, we have

$$x = a^y = e^{(\ln(a))y}, \quad \text{therefore} \quad (\ln a)y = \ln(x), \quad \text{or} \quad y = \dfrac{\ln(x)}{\ln(a)}.$$

We summarize these results:

Theorem 3 **Change of Bases**

For any positive real numbers a, x, and any real number t :

$$a^t = e^{(\ln(a))t} \quad \text{and} \quad \log_a(x) = \dfrac{\ln(x)}{\ln(a)}.$$

One of the important applications of the logarithms occurs in solving equations that involve exponential unknowns.

▬ Exponential Equations and Doubling Time ▬

We can now solve an exponential equation by using logarithms.

Example 3: Use logarithms to solve the following exponential equations for t:

(a) $e^t = 10$ (b) $e^{-0.03t} = 0.08$ (c) $5^t = 8$ (d) $5{,}000 e^{0.07t} = 8{,}000$

Solution (a) We can use the property that $e^y = x$ if and only if $y = \ln(x)$. Hence,

$$t = \ln(10) \approx 2.3.$$

(b) This time, we try another approach by taking the natural logarithm on both sides:

$$\ln\left(e^{-0.03t}\right) = \ln(0.08).$$

Using the simplification

$$\ln\left(e^{-0.03t}\right) = -0.03t \ln(e) = -0.03t,$$

we have

$$-0.03t = \ln(0.08), \quad \text{hence} \quad t = \dfrac{\ln(0.08)}{-0.03} \approx \dfrac{-2.52573}{-0.03} \approx 84.2.$$

(c) Taking the natural logarithm on both sides we have:

$$\ln\left(5^t\right) = \ln(8), \quad \text{or} \quad t\ln(5) = \ln(8).$$

Hence, $\quad t = \dfrac{\ln(8)}{\ln(5)} \approx \dfrac{2.07944}{1.60944} \approx 1.29.$

(d) We need to revise the equation by dividing by 5,000 on both sides:

$$e^{0.07t} = \dfrac{8{,}000}{5{,}000} = 1.6.$$

We then have

$$0.07t = \ln(1.6), \quad \text{hence} \quad t = \dfrac{\ln(1.6)}{0.07} \approx \dfrac{0.47}{0.07} \approx 6.7.$$

Recall that this equation occurred at the beginning of this section for Katherine's investment. Hence, it takes about 6.7 years for her investment to reach $8,000.

Example 4: The grandparents of a 1-year-old girl invest $6,000 for her college education fund.

(a) At what annual interest rate, compounded continuously, will the fund have $20,000 when the girl is 18 years old?

(b) Suppose that the fund earns interest at an annual rate of 6%, compounded continuously. How long does it take for the fund to double itself?

Solution (a) Assume that the annual interest rate is r. From Section 1.5, we know that the amount in the fund after t years can be calculated by $P(t) = 6{,}000 e^{rt}$. When the girl is 18 years old, $t = 17$. Hence, we are looking for the value r such that $P(17) = 20{,}000$, or

$$6{,}000 e^{r(17)} = 20{,}000.$$

Dividing by 6,000 on both sides, we have

$$e^{r(17)} = \dfrac{20{,}000}{6{,}000} = \dfrac{10}{3}.$$

We then have

$$17r = \ln\left(\dfrac{10}{3}\right), \quad \text{hence} \quad r = \dfrac{\ln(10/3)}{17} \approx \dfrac{1.204}{17} \approx 0.0708 = 7.08\%.$$

Therefore, the annual interest rate needs to be at 7.08%.

(b) Suppose that the fund earns interest at an annual rate of 6%, compounded continuously. Then the amount in the fund after t years is calculated by

$$P(t) = 6{,}000 e^{0.06t}.$$

To find the number of years needed for the fund to double itself, we solve the equation

$$12{,}000 = 6{,}000 e^{0.06t}, \quad \text{or} \quad 2 = e^{0.06t}.$$

Therefore,

$$0.06t = \ln(2), \quad \text{or} \quad t = \dfrac{\ln(2)}{0.06} \approx 11.55.$$

That is, it takes about 11 years and 7 months for the fund to double itself.

From part (b) of Example 4, we observe that the time T it takes for an investment to double itself does not depend on the initial investment P_0. The number T is called the **doubling time**, and it has a simple relationship with the annual interest rate r under continuous compounding. We summarize this relation as follows for a general exponential growth model:

Theorem 4 *Doubling Time and Growth Rate*

For the exponential growth model $P(t) = P_0 e^{rt}$ ($r > 0$), the **growth rate** r and the **doubling time** T are related by

$$T = \frac{\ln(2)}{r} \quad \text{and} \quad r = \frac{\ln(2)}{T}.$$

A simple rule to compute the doubling time in finance is called the **Rule of 70**. Since $\ln(2) = 0.693147\cdots \approx 0.7$, we have

$$T = \frac{\ln(2)}{r} \approx \frac{0.7}{r} = \frac{70}{100r}.$$

For example, if the annual interest rate is at 5%, then it takes about $70/5 = 14$ years for an investment to double itself. If the annual interest rate is at 10%, then it only takes about $70/10 = 7$ years for an investment to double itself. For an investment of \$10,000, at the annual interest rate of 5%, it will grow to \$40,000 after 28 years (doubled twice). However, at the annual interest rate of 10%, it will grow to \$160,000 after 28 years (doubled four times)!

General Exponential Growth/Decay Models

There are many applications of the exponential model other than an investment. For example, an uninhibited population can be described by the equation

$$P(t) = P_0 e^{rt}, \; r > 0,$$

and the mass of a radioactive element can be described by the equation

$$P(t) = P_0 e^{-rt}, \; r > 0.$$

It is natural to call the first model the **exponential growth** and the second model the **exponential decay**. For an investment, the future value is an exponential growth model, while the present value is an exponential decay model.

Example 5: A bacteria culture initially contains 200 cells. After 2 hours, the population has increased to 360 cells.

(a) Assuming the growth is exponential, find an expression for the number of bacteria after t hours.

(b) Find the number of bacterial after 5 hours.

(c) How long does it take for the population to double?

(d) When will the population reach 5,000?

Solution (a) The number of bacteria after t hours, $P(t)$, satisfies the exponential growth model

$$P(t) = P_0 e^{rt}.$$

Since the initial population is 200 cells, $P_0 = 200$. Now we are looking for the growth rate r. Since the population is 360 after 2 hours, we have

$$P(2) = 360, \quad \text{or} \quad 200e^{r(2)} = 360.$$

By dividing by 200 on both sides we have

$$e^{r(2)} = \frac{360}{200} = 1.8.$$

We then have

$$2r = \ln(1.8), \quad \text{hence} \quad r = \frac{\ln(1.8)}{2}.$$

Therefore, the number of bacteria after t hours is

$$P(t) = 200e^{\left(\frac{\ln(1.8)}{2}\right)t}.$$

(b) The number of bacterial after 5 hours is

$$P(5) = 200e^{\left(\frac{\ln(1.8)}{2}\right)5} \approx 869 \text{ cells}.$$

(c) Applying the doubling time formula with $r = \frac{\ln(1.8)}{2}$, we have

$$T = \frac{\ln 2}{r} = \frac{\ln 2}{(\ln(1.8))/2} \approx 2.36 \text{ hours}.$$

Hence, it takes about 2.36 hours for the population to double itself.

(d) To find the time it needs for the population to reach 5,000, we set the exponential equation

$$5,000 = 200e^{\left(\frac{\ln(1.8)}{2}\right)t}.$$

By dividing by 200 on both sides we have

$$e^{\left(\frac{\ln(1.8)}{2}\right)t} = \frac{5,000}{200} = 25.$$

We then have

$$\left(\frac{\ln(1.8)}{2}\right)t = \ln(25), \quad \text{hence} \quad t = \frac{\ln(25)}{(\ln(1.8))/2} \approx 11 \text{ hours}.$$

Therefore, the number of bacteria will reach 5,000 after about 11 hours.

Newton's Law of Cooling

One of the interesting daily life applications of an exponential decay model is called **Newton's law of cooling**. It is stated that the temperature of a cooling object follows the function

$$T(t) = (T_0 - C)e^{-rt} + C,$$

where r is a constant, T_0 is the initial temperature of the object, and C is the constant temperature of the surrounding medium.

Example 6: A cup of freshly brewed coffee with temperature at 130°F is put on a table in a room with temperature at 70°F. After 1.5 minutes, the coffee cools to 125°F.

(a) Find the temperature of the coffee, $T(t)$, after t minutes.

(b) After the coffee has been put on the table, how many minutes does it take for the coffee to cool down to 112°F?

Solution (a) According to Newton's law of cooling, the temperature of the coffee t minutes after being put on the table, $T(t)$, satisfies the equation

$$T(t) = (T_0 - C)e^{-rt} + C.$$

We have $C = 70°F$ and $T_0 = 130°F$. Hence

$$T(t) = 60e^{-rt} + 70.$$

Now we are looking for the cooling rate r. Since the temperature of the coffee is 125°F after 1.5 minutes, we have

$$T(1.5) = 125, \quad \text{or} \quad 60e^{-r(1.5)} + 70 = 125.$$

By subtracting 70 from both sides we have

$$60e^{-r(1.5)} = 55.$$

By dividing by 60 on both sides we have

$$e^{-r(1.5)} = \frac{55}{60} = \frac{11}{12}.$$

We then have

$$-r(1.5) = \ln\left(\frac{11}{12}\right), \quad \text{hence} \quad r = \frac{\ln(11/12)}{-1.5} \approx 0.058.$$

Therefore, the temperature of the coffee after t minutes is

$$T(t) = 60e^{-0.058t} + 70.$$

(b) To find the time it needs for the coffee to cool down to 112°F, we set the exponential equation

$$112 = 60e^{-0.058t} + 70.$$

By subtracting 70 and then dividing by 60 on both sides we have

$$e^{-0.058t} = \frac{42}{60} = 0.7.$$

We then have

$$-0.058t = \ln(0.7), \quad \text{hence} \quad t = \frac{\ln(0.7)}{-0.058} \approx 6.15 \text{ minutes}.$$

Therefore, it takes about 6.15 minutes for the coffee to cool down to 112°F.

Exercise Set 1.6

For Exercises #1 to #6, write an equivalent logarithmic equation

1. $10^3 = 1,000$
2. $27^{1/3} = 3$
3. $e^k = b$
4. $e^{-2} = M$
5. $a^h = J$
6. $10^{-2} = 0.01$

For Exercises #7 to #12, write an equivalent exponential equation

7. $\log_3(9) = 2$
8. $\log_{16}(2) = \frac{1}{4}$
9. $\ln(10) = k$
10. $\ln(h) = -3$
11. $\log_a(H) = G$
12. $-\log_a(K) = h$

13. (a) Make a numerical table for the exponential function $f(x) = 2^x$ and graph the function;

(b). Use the table from part (a) to make a numerical table for the logarithmic function $g(x) = \log_2(x)$ and graph the function;

(c) Graph both $f(x) = 2^x$ and $g(x) = \log_2(x)$ in the same coordinate plane.

14. (a) Make a numerical table for the exponential function $f(x) = (3/2)^x$ and graph the function;

(b). Use the table from part (a) to make a numerical table for the logarithmic function $g(x) = \log_{3/2}(x)$ and graph the function;

(c) Graph both $f(x) = (3/2)^x$ and $g(x) = \log_{3/2}(x)$ in the same coordinate plane.

For Exercises #15 to #26, solve for t using natural logarithms.

15. $e^t = 10$
16. $e^t = 20$
17. $5^t = 9$
18. $120 = 10^t$
19. $a^t = M$
20. $3^{-t} = 0.1$
21. $e^{-3t} = 0.1$
22. $e^{-0.05t} = 0.2$
23. $3e^{4t} = 9$
24. $100e^{-0.05t} = 30$
25. $60e^{-0.02t} + 40 = 90$
26. $7e^{0.2t} - 10 = 25$

For Exercises #27 to #32, rewrite the exponential expressions using base e

27. 10^b
28. $2^{1/3}$
29. 3^{-x}
30. 5^{6x}
31. a^h
32. $10^{-2.5}$

For Exercises #33 to #38, rewrite the logarithms using natural logarithms

33. $\log_3(K)$
34. $\log_5(10)$
35. $\log_2(7)$
36. $\log_6(G)$
37. $\log_a(H)$
38. $-\log_a(32)$

For Exercises #39 to #42, simplify the natural logarithms

39. $\ln(e^3)$
40. $\ln\left(\sqrt[5]{e}\right)$
41. $\ln[(2e)^{-3}] + 3\ln(2)$
42. $\ln[(3e)^{-4}] + 4\ln 3$
43. $\ln\left(\sqrt[3]{ae}\right) - \frac{1}{3}\ln(a)$
44. $\ln\left(\sqrt[4]{ae}\right) - \frac{1}{4}\ln(a)$

45. If Mark deposits $1,000 into an account that pays interest at an annual rate of 2.6% compounded continuously, how long does it take for the investment to reach $1,500?

46. If Joe deposits $8,000 into an account that

pays interest at an annual rate of 5% compounded continuously, how long does it take for the investment to reach $10,000?

47. If Mark deposits $1,000 into an account that pays interest at an annual rate of 7.6% compounded continuously, how long does it take for the investment to reach $1,500?

48. If Lucy deposits $3,000 into an account that pays interest at an annual rate of 6% compounded continuously, how long does it take for the investment to reach $5,500?

49. Elizabeth has $8,000 to invest. She wants the investment to grow to $10,000 in 4 years. What should the annual interest rate be, compounded continuously?

50. The Lopez family has $12,000 to invest. What should the annual interest rate be, compounded continuously, for the investment to grow to $20,000 in 7 years?

51. How long does it take for an investment to double, at an annual interest rate of 5.6%, compounded continuously? Do not use the Rule of 70.

52. How long does it take for an investment to double, at an annual interest rate of 3.8%, compounded continuously? Do not use the Rule of 70.

53. Find the annual interest rate, compounded continuously, for an investment to double in 5 years. Do not use the Rule of 70.

54. Find the annual interest rate, compounded continuously, for an investment to double in 7 years. Do not use the Rule of 70.

55. Use the Rule of 70 to predict the doubling time of an investment which is earning interest at an annual interest rate of 6%, compounded continuously. Compare your answer with the exact doubling time.

56. Use the Rule of 70 to predict the doubling time of an investment which is earning interest at an annual interest rate of 5.5%, compounded continuously. Compare your answer with the exact doubling time.

57. Jack deposits $5,000 into an account that pays interest at an annual rate of 4.5%, compounded continuously.

(a) Find an expression for the amount in the account after t years.

(b) Find the amount in the account after 5 years.

(c) How long does it take for the investment to double?

(d) When will the investment reach $8,000?

58. John deposits $8,000 into an account that pays interest at an annual rate of 5.1%, compounded continuously.

(a) Find an expression for the amount in the account after t years.

(b) Find the amount in the account after 5 years.

(c) How long does it take for the investment to double?

(d) When will the investment reach $14,000?

59. The parents of a 1-year-old boy plan to invest for his college education. Their target is that when he is 18 years old, the fund should have the amount of $80,000.

(a) Find the initial amount needed to set up the fund now, assuming an annual rate of 5.8%, compounded continuously.

(b) If they set up the fund with $30,000, at what annual interest rate, compounded continuously, can the fund reach the target?

60. The parents of a 1-year-old boy plan to invest for his college education. Their target is that when he is 18 years old, the fund should have the amount of $75,000.

(a) Find the initial amount needed to set up the fund now, assuming the annual rate of 5.8%, compounded continuously.

(b) If they set up the fund with $30,000, at what annual interest rate, compounded continuously, can the fund reach the target?

61. Dean has won a lottery. He was offered two options to receive the award: either (a) $15,000 now and $30,000 one year later; or (b) $19,000 now and $25,000 one year later. Assume that the interest is compounded continuously. What annual interest rate $r\%$ would lead Dean to switch choices?

1.6. LOGARITHMIC FUNCTIONS AND EXPONENTIAL EQUATIONS

62. David has won a lottery. He was offered two options to receive the award: either (a) $25,000 now and $20,000 one year later; or (b) $29,000 now and $15,000 one year later. Assume that the interest is compounded continuously. What annual interest rate $r\%$ would lead David to switch choices?

63. Sharon is to be paid $3,000 for work done over a year. She was offered two options for the payment: $2,900 paid now; or $1,000 paid now and $2,000 paid one year later. Using only financial reasons to make her decision, under what annual interest rate should Sharon switch her decision? Assume that the interest is compounded continuously.

64. A business associate who owes Ana $4,000 offered to pay her $3,800 now or to pay her $2,000 now and $2,000 two years later. Using only financial reasons to make her decision, under what annual interest rate should Ana switch her decision? Assume that the interest is compounded continuously.

65. A bacteria culture initially contains 600 cells. After 1.5 hours the population has increased to 900 cells.

(a) Assuming the growth is exponential, find an expression for the number of bacteria after t hours.

(b) Find the number of bacteria after 5 hours.

(c) How long does it take for the population to double?

(d) When will the population reach 8,000?

66. After 30 minutes, a bacteria culture contains 600 cells. After 2 hours the population has increased to 1,500 cells.

(a) Assuming the growth is exponential, find an expression for the number of bacteria after t hours.

(b) Find the number of bacterial after 4 hours.

(c) How long does it take for the population to double?

(d) When will the population reach 10,000?

67. The total world marine catch in 1950 was 17 million tons and in 1995 was 91 million tons. If the marine catch is increasing exponentially, find an expression for the total world marine catch since 1950, and use it to predict the total world marine catch in the year 2016.

68. In 1950, the U.S. Federal budget was $39.4 billion. In 2006, the federal budget was $2153.9 billion.

(a) Assuming the growth is exponential, find an expression for the federal budget since 1950.

(b) Use the expression to predict the federal budget in 2015.

(c) When will the federal budget exceed $12 trillion, that is, $12,000 billion?

69. A cup of freshly brewed coffee with temperature at 130°F is put on a table in a room with temperature at 75°F. Two minutes later the coffee cools to 120°F.

(a) Find the temperature of the coffee, $T(t)$, t minutes after it is put on the table.

(b) After the coffee has been put on the table, how many minutes does it take for the coffee to cool down to 115°F?

70. A bottle of soda pop at room temperature of 72°F is put into a refrigerator where the temperature is 42°F. Twenty minutes later the soda pop has cooled to 60°F.

(a) Find the temperature of the soda pop, $T(t)$, t minutes after it is put into the refrigerator.

(b) After the soda pop has been put in the refrigerator, how many minutes does it take for it to cool down to 50°F?

71. A roast turkey is taken from an oven when its temperature has reached 185°F. It is put on a table in a room where the temperature is 75°F. Thirty minutes later the turkey has cooled to 140°F.

(a) Find the temperature of the turkey, $T(t)$, t minutes after it is put on the table.

(b) After the turkey has been put on the table, how many minutes does it take for it to cool down to 100°F?

72. During a winter storm, a house in Wisconsin loses electrical power at 1:00 p.m. The heater does not work without electricity. The living room temperature is 70°F when the power goes out. At 8:00 p.m. the room temperature is 60°F. The outside temperature is 10°F.

(a) Find the temperature of the living room, $T(t)$, t hours after 1:00 p.m.

(b) If the power does not come back at 8:00 a.m. the next morning, should the house owner worry about the water pipes freezing?

Chapter 1 Summary

This chapter prepares us for the study of calculus. It starts with a review of the concept of a function and its graph.

A **function** is a special relationship between two sets. It is a rule that assigns exactly one element in the range to each element in the domain. The **domain** is the set of all possible independent variable values. The **range** is the set of all possible dependent variable values.

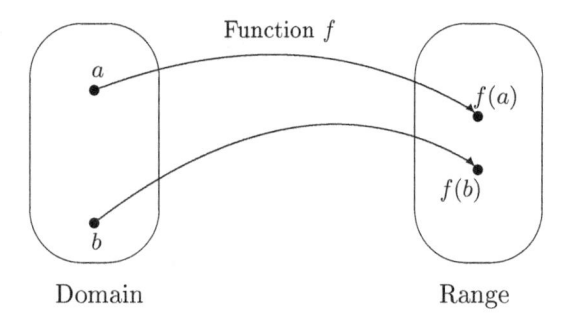

Most functions in this text are represented by an algebraic equation, such as $f(x) = 2x^2 - 3x$. To **evaluate** a function at a particular independent variable value, we replace every x in the equation with that particular value. For example,

$$f(2) = 2(2)^2 - 3(2) = 2(4) - 6 = 8 - 6 = 2.$$

Sometimes the particular value of the independent variable is a letter or a formula. It is treated the same way as a number. For example,

$$f(x+h) = 2(x+h)^2 - 3(x+h) = 2(x^2 + 2xh + h^2) - 3(x+h) = 2x^2 + 4xh + 2h^2 - 3x - 3h.$$

Some functions have more than one formula for different independent variable values. Such a function is called a **piecewise defined** function. For example, the function g is given by

$$g(x) = \begin{cases} x + 2 & \text{if } x \leq 1, \\ 2 & \text{if } x > 1. \end{cases}$$

To evaluate $g(0)$, since $0 < 1$, we have $g(0) = (0) + 2 = 2$. To evaluate $g(3)$, since $3 > 1$, we have $g(3) = 2$. More examples can be found in Section 1.4.

The **graph** of a function $y = f(x)$ consists of all points (x, y) in the coordinate plane such that x is in the domain and $y = f(x)$. A graph that represents a function must pass the **vertical line test**. That is, you cannot draw a vertical line that crosses the graph in two or more places.

To evaluate a function on a graph, say $f(a)$, first make a vertical line passing through the given input value, a. At the location where the vertical line crosses the graph, the y-coordinate is the output $f(a)$.

CHAPTER 1 SUMMARY

Important Graphs and Functions

The graph of a constant function $f(x) = b$ is a horizontal line. The graph of an equation of the form $x = a$ is a vertical line. It is not a graph of a function. The graph of a **linear function** $f(x) = mx + b$ is a line with **slope** m and y-intercept b.

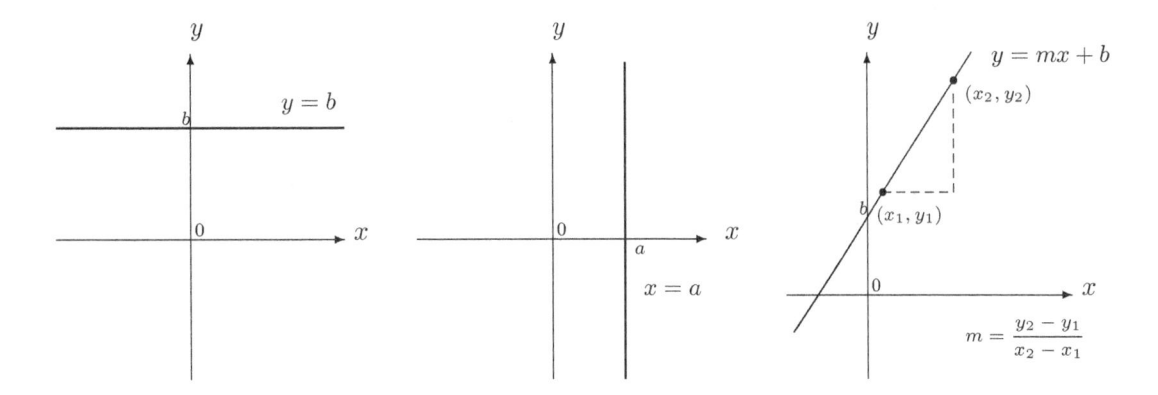

The slope, m, of a linear function $f(x) = mx + b$, is a measurement of the steepness as well as the direction of the line. Given two points on a line, (x_1, y_1) and (x_2, y_2), the slope of the line is computed by

$$m = \frac{y_2 - y_1}{x_2 - x_1}$$

Positive slope indicates that f is increasing, while negative slope indicates that f is decreasing. It also measures the rate of change of a linear function.

The graph of a **quadratic function** $f(x) = ax^2 + bx + c$ ($a \neq 0$) is a **parabola**. A parabola has two main features: The opening direction and the location of its **vertex**, the turning point. If $a > 0$, then its graph opens upward, while if $a < 0$, its graph opens downward. The first coordinate of its vertex is $x = -b/(2a)$.

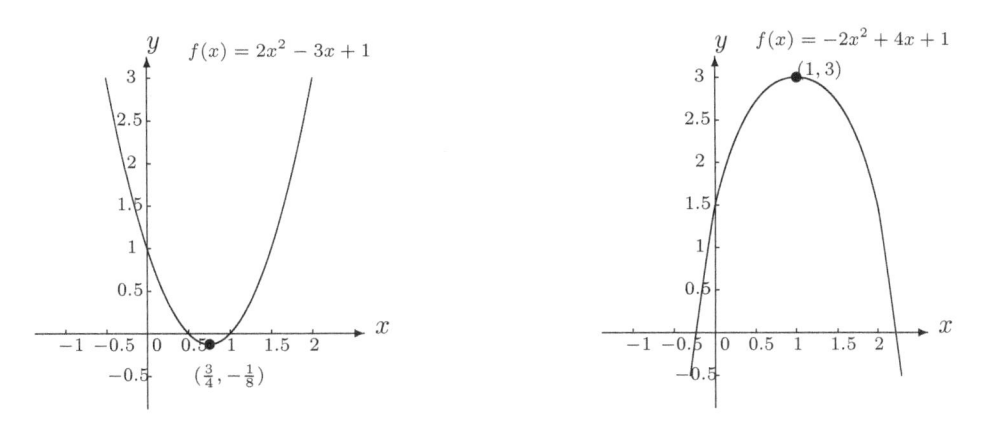

The quadratic functions are special cases of **polynomial functions**. A polynomial function takes the form of

$$f(x) = a_n x^n + a_{n-1} x^{n-1} + \ldots + a_2 x^2 + a_1 x + a_0$$

where n is a nonnegative integer and $a_n, a_{n-1}, \ldots, a_2, a_1, a_0$ are real numbers called **coefficients**. The number n indicates the highest power term of a polynomial, and is called the **degree** of the polynomial.

Other important classes of functions include **rational functions, power functions with rational exponents, exponential functions**, and **logarithmic functions**. A rational function is defined as the ratio of two polynomials:

$$R(x) = \frac{P(x)}{Q(x)} = \frac{a_n x^n + a_{n-1} x^{n-1} + \ldots + a_1 x + a_0}{b_m x^m + b_{m-1} x^{m-1} + \ldots + b_1 x + b_0}, \quad (b_m \neq 0).$$

A power function takes the form

$$f(x) = ax^p, \quad \text{where } p \text{ is a real number.}$$

When p is a positive integer, it is a particular polynomial. When p is a negative integer, it is a particular rational function. When p is a rational number $p = m/n$ where n and m are integers, the power function $f(x) = x^{\frac{n}{m}}$ can also be expressed using **radical** notation, as in $f(x) = x^{\frac{m}{n}} = \sqrt[n]{x^m}$. The laws of operations with exponents and radicals are summarized in Section 1.4.

An exponential function takes the form of

$$f(x) = P_0 a^x$$

where a is a positive real number other than 1, called the **base**, and P_0 is a constant, called the **initial value**. A special exponential function takes the base

$$e = 2.718281828459\ldots$$

The function $y = P_0 e^x$ will be studied throughout the course.

The logarithmic function with base $a > 0$, denoted by $f(x) = \log_a(x)$, is defined as:

$$y = \log_a(x) \quad \text{if and only if} \quad x = a^y.$$

Two special logarithmic functions are the **common logarithmic** $y = \log_{10}(x)$, denoted by $y = \log(x)$, and the **natural logarithmic** $y = \log_e(x)$, denoted by $y = \ln(x)$. The laws of operations with natural logarithms are summarized in Section 1.6.

We call the logarithmic function $g(x) = \log_a(x)$ the **inverse function** of $f(x) = a^x$. Conversely, if we switch x and y in the definition, we can define the exponential function a^x by the logarithmic function $\log_a(x)$. Therefore, the two functions are inverse functions to each other. We say that $g(x) = \log_a(x)$ and $f(x) = a^x$ are inverse functions. In particular, $g(x) = \ln(x)$ and $f(x) = e^x$ are inverse functions.

We can change any exponential function $y = a^t$ to the exponential function with base e, and change any logarithmic function $y = \log_a(x)$ to the natural logarithmic:

$$a^t = e^{(\ln(a))t} \quad \text{and} \quad \log_a(x) = \frac{\ln(x)}{\ln(a)}.$$

▬ Solving Algebraic and Exponential Equations

Throughout the course we will be solving various types of equations. The simplest equation to solve is a **linear equation** $ax + b = 0$, with the solution $x = -b/a$. Next is a **quadratic equation** $ax^2 + bx + c = 0$, where $a \neq 0$. We can apply the **quadratic formula**: let $S = b^2 - 4ac$.

(a) If $S < 0$, then the equation has no real solution.

(b) If $S = 0$, then the equation has one real solution $x = \dfrac{-b}{2a}$.

(c) If $S > 0$, then the equation has two real solutions:
$$x_1 = \frac{-b + \sqrt{S}}{2a} \quad \text{and} \quad x_2 = \frac{-b - \sqrt{S}}{2a}.$$

For solving a **polynomial equation** $p(x) = 0$ with degree of p higher than 2, the approach is to factor the polynomial. For example, to solve $x^3 - 4x = 0$, we can factor the polynomial as:
$$x^3 - 4x = x(x^2 - 4) = x(x-2)(x+2),$$
then
$$x(x-2)(x+2) = 0.$$

Hence, the equation has three solutions: $x = 0$, $x = 2$, and $x = -2$.

For solving a **rational equation** $r(x) = p(x)/q(x) = 0$, we solve the polynomial equation of the numerator, $p(x) = 0$. Example 6 in Section 1.4 illustrates the detailed process.

To solve **equations involving radicals**, we need to isolate the radical, then take the appropriate power for both sides of the equation to cancel the radical. After solving the revised equation, we also need to check that the solutions are valid for the original radical. Example 8 in Section 1.4 illustrates the detailed process.

To solve an **exponential equation**, we need to isolate the exponential term, then take the appropriate logarithmic for both sides of the equation. Several examples in Section 1.6 illustrate the detailed process.

Mathematical Models in Economics, Finance, and Other Fields

When a real-world situation is described in mathematical language, it is called a **mathematical model**. If a linear equation is derived from a real-world situation, we say that we have a **linear model**. Similarly, if an exponential equation is derived from a real-world situation, we say that we have a **exponential model**.

The **cost, revenue, and profit** functions, as well as the **supply and demand** functions and the **equilibrium point** are introduced in Section 1.3, with many examples. The formulas for **future and present** values of a savings account with **compounded interest** are introduced in Sections 1.5 and 1.6. Additional models of exponential growth or decay, including **Newton's law of cooling**, are discussed in Section 1.6. Many application examples in these three sections illustrate the steps of solving problems involved with these models.

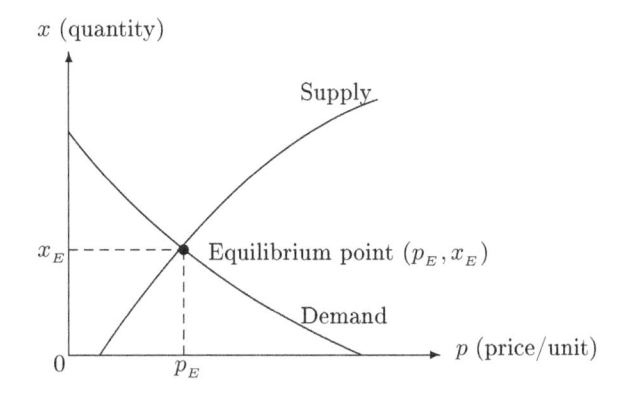

Chapter 1 Review Exercises

1. The function f is represented by the following graph. Use the graph to find the following function values:

   (a) $f(-3)$ (b) $f(-2)$

   (c) $f(-1)$ (d) $f(0)$

   (e) $f(1)$ (f) $f(3)$

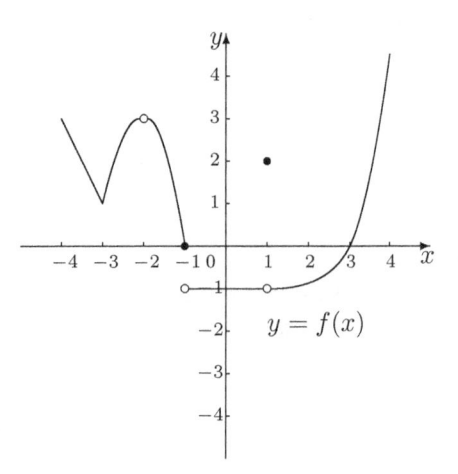

2. The function F is given by the equation
   $$F(x) = 2x^2 - 2.$$

   (a) Find the domain of this function.

   (b) Find $F(-2)$, $F(-1)$, $F(0)$, $F(1)$, and $F(2)$.

   (c) Find $F(5a)$ and $F(\sqrt{a})$.

   (d) Find $F(x+h)$.

   (e) Find $\dfrac{F(x+h) - F(x)}{h}$ for $h \neq 0$.

   (f) Use the results in part (b) to make a numerical table, then use the table to plot a graph of the function.

3. The linear function f is given by the equation $f(x) = -\tfrac{1}{3}x + 6$.

   (a) Find the slope.

   (b) Plot the graph of this function.

For Exercises #4 and #5: Find both the slope-point equation and the slope-intercept equation of the line.

4. With slope 3, containing point $(-1, 0)$.

5. Containing both points $(1, 2)$ and $(3, -4)$.

6. The resale value of a used car is $45,000 when the car is 3 year old. The resale value drops by $8,000 from the previous given value when the car is 5 years old. Assuming that the resale value of that car, V, is a linear function of the car's age, t:

   (a) Find a formula for the resale value of the car, V, as a function of the number of years, t, of the age of the car.

   (b) What is the car's resale value when the car is 7 years old?

   (c) What is the slope of the line? Explain what the value of the slope means in the context of this problem.

   (d) What is the vertical intercept of the line? Explain what the value of the vertical intercept means in the context of this problem.

   (e) Find the age of the car when it has essentially no value.

7. The manager of a furniture manufacturing firm determined that the total cost, in dollars, of producing x chairs each week can be modeled by
   $$C(x) = 6{,}000 + 15x, \quad 0 \leq x \leq 1{,}000.$$
   The firm sells the chairs for $25 each.

   (a) Find the fixed costs.

   (b) Find the variable costs for producing x chairs.

   (c) Find the total revenue function $R(x)$ for selling x chairs.

   (d) Find the total profit function $P(x)$ for producing and selling x chairs.

   (e) Find the total cost, total revenue, and total profit for producing and selling 500 chairs.

   (f) Find the total cost, total revenue, and total profit for producing and selling 800 chairs.

   (g) Find the break-even point for the firm by solving the equation $P(x) = 0$.

   (h) Plot the break-even point with the graphs of the total cost function and the total revenue function.

CHAPTER 1 REVIEW EXERCISES

(i) What does the slope in the the function $P(x)$ represent?

8. A coffeehouse's daily costs are $184 when they sell 70 coffee drinks and $208 when they sell 100 coffee drinks. Assume the daily cost C is a linear function of the number of coffee drinks, x.

   (a) Find a formula for the daily cost C as a function of the number of coffee drinks, x.

   (b) What is the slope of the line? Explain what the value of the slope means in the context of this problem.

   (c) What is the vertical intercept of the line? Explain what the value of the vertical intercept means in the context of this problem.

9. For each of the following quadratic functions, find the parabola's vertex (listed as a pair of coordinates), then draw the parabola.

   (a) $f(x) = 2x^2 + 6x - 1$

   (b) $f(x) = -2x^2 + 6x + 1$

10. Solve the polynomial equations using factorization

    (a) $x^3 - 4x = 0$

    (b) $x^4 - 5x^2 = -4$

11. Solve the following equations using the quadratic formula. If a real solution does not exist, state the fact.

    (a) $x^2 - 4x = 2$

    (b) $x^2 - 6x + 9 = 0$

For the functions in Exercises #12 to #15: (a) Find the domain; (b) Solve the equation $f(x) = 0$. If a real solution does not exist, state the fact.

12. $f(x) = \dfrac{x^2 - 9}{x^2 - 4x + 3}$

13. $f(x) = \dfrac{x^2 - 4}{x^2 - x - 6}$

14. $f(x) = \dfrac{x^2 + 4x + 4}{x + 2}$

15. $f(x) = \sqrt{x^2 - 1} + 2x$

16. Consider the function given by

$$f(x) = \begin{cases} x + 2 & \text{if } x \leq 1, \\ x^2 - 3 & \text{if } x > 1. \end{cases}$$

   (a) Find $f(-1)$, $f(0)$, $f(1)$, $f(2)$, and $f(3)$.

   (b) Sample a few function values for $x \leq 1$ and make a numerical table.

   (c) Sample a few function values for $x > 1$ and make a numerical table.

   (d) Use the results in parts (b) and (c) to plot a graph of the function.

17. If Jack deposits $5,000 into an account that has a 4.5% annual interest rate, how much will be in the account after 10 years if the interest is compounded

   (a) Annually?   (b) Monthly?
   (c) Weekly?   (d) Daily?
   (e) Continuously?

For Exercises #18 to #20: Use an appropriate formula to evaluate either the future value or the present value.

18. The present value of an investment of $6,000 for 6 years at an annual interest rate of 4.3% compounded monthly.

19. The future value of an investment of $20,000 for 4 years at an annual interest rate of 6% compounded continuously.

20. The present value of an investment of $20,000 for 4 years at an annual interest rate of 5.2% compounded continuously.

21. The parents of a 1-year-old boy plan to invest for his college education. Their target is that when he is 18 years old, the fund should have the amount of $80,000. Find the initial amount needed to set up the fund now, assuming the following annual interest rates and compounding methods:

   (a) at the annual rate of 4.5%, compounded monthly;

   (b) at the annual rate of 6.8%, compounded continuously;

   (c) at the annual rate of 9.2%, compounded continuously.

22. John has won a lottery. He was offered two options to receive the award: he can either take it in four installments of $50,000 each, at the end of each year, for the next 4 years; or he can take a lump-sum of $175,000 now. Assuming (a) interest at an annual rate of 5% compounded annually, or (b) interest at an annual rate of 5.5% compounded continuously, which option should he choose under the consideration of cash value only?

23. The demand and supply curves of a certain brand of sunglasses are given by

    $x = D(p) = 2{,}000 - 10p$

    and $x = S(p) = 20p - 400$

    where p is the price in dollars and x is the quantity.

    (a) At a price of $50, how many pairs of the sunglasses would consumers purchase?

    (b) At a price of $50, how many pairs of the sunglasses would manufacturers supply?

    (c) Will the market price be higher or lower than $50?

    (d) What would the price have to be for consumers to buy at least 1,000 pairs of the sunglasses?

    (e) What would the price have to be for the manufacturers to supply at least 600 pairs of the sunglasses?

    (f) Find the equilibrium price and quantity.

    (g) Represent the equilibrium price and quantity on a graph.

24. Solve for t using natural logarithms:

    (a) $e^t = 100$ (b) $120 = 10^t$

    (c) $e^{-3t} = 0.01$ (d) $60e^{-0.02t} + 40 = 90$

25. Rewrite the exponential expressions using base e:

    (a) 5^{6x} (b) $2^{1/3}$ (c) a^h (d) $10^{-2.5}$

26. If Mark deposits $1,000 into an account that pays interest at an annual rate of 3.6%, compounded continuously, how long does it take for the investment to reach $1,500?

27. Elizabeth has $8,000 to invest. She wants the investment to grow to $10,000 in four years. What should the annual interest rate be, compounded continuously?

28. Jack deposits $5,000 into an account that pays interest at an annual rate of 4.5%, compounded continuously.

    (a) Find an expression for the amount in the account after t years.

    (b) Find the amount in the account after five years.

    (c) How long does it take for the investment to double?

    (d) When will the investment reach $8,000?

29. The parents of a 1-year-old boy plan to invest for his college education. Their target is that when he is 18 years old, the fund should have the amount of $80,000.

    (a) Find the initial amount needed to set up the fund now, assuming the annual rate of 5.8%, compounded continuously.

    (b) If they set up the fund with $30,000, at what interest rate, compounded continuously, can the fund reach the target?

30. Dean has won a lottery. He was offered two options to receive the award: either (a) $15,000 now and $30,000 one year later; or (b) $19,000 now and $25,000 one year later. Assume that the interest is compounded continuously. What annual interest rate r% would lead Dean to switch choices?

31. A bacteria culture initially contains 500 cells. After 1.5 hours the population has increased to 900 cells.

    (a) Assuming the growth is exponential, find an expression for the number of bacteria after t hours.

    (b) Find the number of bacteria after 5 hours.

    (c) How long does it take for the population to double?

    (d) When will the population reach 8,000?

32. A cup of freshly brewed coffee with temperature at 130°F is put on a table in a room with temperature at 72°F. Two minutes later the coffee cools to 120°F.

(a) Find the temperature of the coffee, $T(t)$, after t minutes.

(b) After the coffee has been put on the table, how many minutes does it take for the coffee to cool down to 115°F?

Chapter 1 Test (100 points)

SCORE _____

NAME _____

ANSWERS

1. (10pts) The function F is given by the equation $F(x) = x^2 - 2x$.

    (a) Find $F(-1)$ and $F(2)$. 1. (a) _____

    (b) Find $F(5a)$ and $F(\sqrt{a})$. (b) _____

    (c) Find $F(x+h)$. (c) _____

    (d) Find $\dfrac{F(x+h) - F(x)}{h}$ for $h \neq 0$. (d) _____

2. (6pts) Find the slope-intercept equation of the line with slope -2 and containing point $(4, 0)$.

    2. _____

3. (6pts) Solve the polynomial equation $x^3 - 9x = 0$ using factorization.

    3. _____

4. (6pts) Solve the equation $x^2 - 4x = 1$ using the quadratic formula.

    4. _____

5. (6pts) Solve the exponential equation $60e^{-0.02t} + 40 = 90$.

    5. _____

6. (18pts) The manager of a furniture manufacturing firm determined that the total cost, in dollars, of producing x chairs each week can be modeled by $C(x) = 5{,}000 + 20x$, $0 \leq x \leq 1{,}000$. The firm sells the chairs for $60 each.

    (a) Find the total revenue function $R(x)$ for selling x chairs. 6. (a) _____

    (b) Find the total profit function $P(x)$ for producing and selling x chairs. (b) _____

(c) Find the break-even point for the firm.

(c) _____

(d) What does the slope in the the function $P(x)$ represent?

(d) _____

7. (18pts) The demand and supply curves of a certain brand of sunglasses are given by

$$x = D(p) = 3{,}000 - 10p \quad \text{and} \quad x = S(p) = 20p - 600$$

where p is the price in dollars and x is the quantity.

(a) At a price of $80, how many pairs of the sunglasses would consumers purchase?

7. (a) _____

(b) At a price of $80, how many pairs of the sunglasses would manufacturers supply?

(b) _____

(c) Will the market price be higher or lower than $80?

(c) _____

(d) What would the price have to be for consumers to buy at least 1,000 pairs of the sunglasses?

(d) _____

(e) What would the price have to be for the manufacturers to supply at least 600 pairs of the sunglasses?

(e) _____

(f) Find the equilibrium price and quantity.

(f) _____

(g) Represent the equilibrium price and quantity on a graph.

(g) _____

8. (8pts) The parents of a 1-year-old boy plan to invest for his college education. Their target is that when he is 18 years old, the fund should have the amount of $60,000. Find the initial amount needed to set up the fund now, assuming the following annual interest rates and compounding methods:

(a) At the annual rate of 4.5%, compounded monthly;   8. (a) _____

(b) At the annual rate of 6.8%, compounded continuously.   (b) _____

9. (6pts) If Mark deposits $2,000 into an account that pays interest at an annual rate of 4.6%, compounded continuously, how long does it take for the investment to reach $4,200?

9. _____

10. (6pts) Elizabeth has $6,000 to invest. She wants the investment to grow to $8,000 in four years. What should the annual interest rate be, compounded continuously?

10. _____

11. (10pts) A cup of freshly brewed coffee with temperature at 130°F is put on a table in a room with temperature at 76°F. After 1.5 minutes, the coffee cools to 122°F.

(a) Find the temperature of the coffee, $T(t)$, after t minutes.

11. (a) _____

(b) Find the temperature of the coffee after 3 minutes.

(b) _____

(c) After the coffee has been put on the table, how many minutes does it take for the coffee to cool down to 110°F?

(c) _____

Chapter 1 Projects

Project 1

Life is exciting for Mara those days! She graduated from college last summer as a business major and has found a nice job. As a graduation gift, her grandmother gave her $5,000 as seed money for the down payment on a condominium Mara plans to purchase in 3 years.

Mara wants to invest her $5,000 in a savings account. She wants to find a bank in the neighborhood that will give her best return. She found the following information:

Bank A has an interest rate of 6.03%, compounded every 3 months.

Bank B has an interest rate of 6.05%, but Mara forgets about how often the interest is compounded.

Bank C has an interest rate of 6%, compounded continuously.

Mara wants to invest her money in whichever one of the three banks that will give her the best return on her money. She plans to go to Bank B once more to ask about their compounding frequency, but she wants to make the right choice as soon as she knows the compounding frequency.

Let us help Mara choose the right bank! Show how you compare those banks. Your report must include formulas (with explanation), Excel spreadsheets, and Excel graphs showing the differences.

Project 2

The **Rule of 70**, introduced in Section 1.6, can be used to estimate the doubling time of an investment. How accurately does this rule work?

(1) Construct a table of doubling times for interest rates from 2% to 12%, in increments of 0.2%, when interest is compounded annually, monthly, weekly, daily, and continuously.

(2) Construct a table of doubling times for interest rates from 2% to 12%, in increments of 0.2%, using the Rule of 70. Devise similar rules for 69, 71, and 72. Then construct tables for those rules.

(3) For each interest rate listed in the first table, compute the errors when using each of the rules listed in part (2). Report the errors by percent, that is,

$$\text{Percent error} = \frac{\text{Estimate} - \text{True value}}{\text{True value}} 100\%.$$

(4) For each interest rate listed in the first table, determine the "best" approximating rule, and determine whether the rule overestimates or underestimates the true doubling time. Comment on your preference.

Chapter 2

Differentiation

Introduction

In this chapter, we study the first of the two main building blocks of calculus: *differentiation*. We l first learn a new concept called the **limit**, and link this concept to **continuity**. The concept of the limit is used to introduce the **derivative** of a function. A derivative represents an instantaneous rate of change, as well as the slope of a tangent line at a specific point. We learn various techniques for finding derivatives.

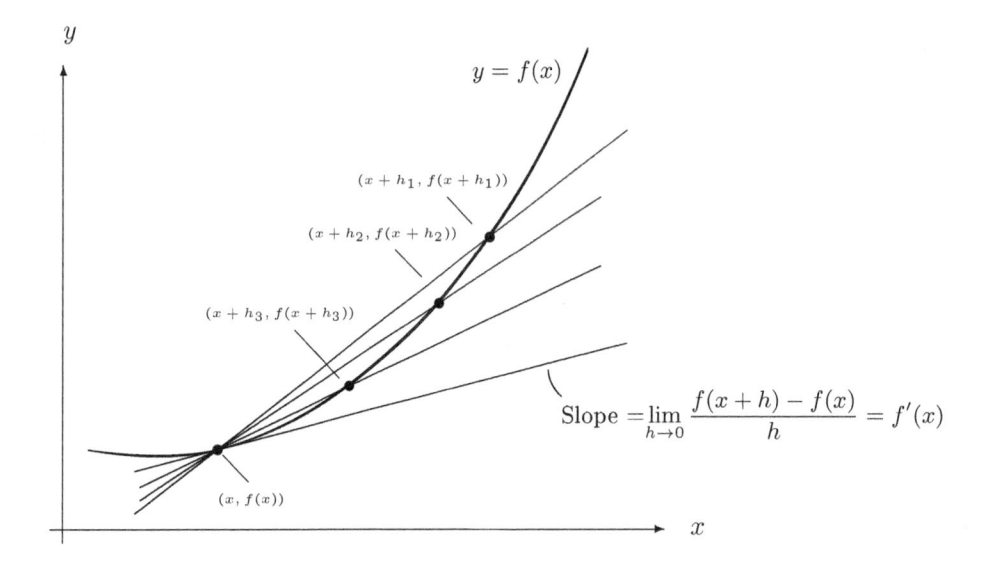

Study Objectives

1. Understand the numerical, algebraic, and graphical notions of a limit.
2. Determine where a function is continuous.
3. Compute the average rate of change of a function.
4. Compute the derivative of a function by it's definition.
5. Compute the derivative(s) of a function by the properties and rules.
6. Determine the equation of the tangent line of the graph of a function at a given point.

2.1 Limits: Approached Numerically and Graphically

The concepts which form the two main branches of calculus, differential calculus and integral calculus, are all based on the concept of limits. In this section we use intuitive approaches, through numerical and graphical examination, to explore the concept of a limit.

Limits

We introduce the concept of a limit through an example, which will lead to an intuitive definition of the concept.

Example 1: Let $f(x) = 2x + 1$. Discuss the behavior of the values of $f(x)$ when x is close to 2.

Solution We begin by making two tables to study the numerical behavior of $f(x)$ when x is close to 2. Since x can be either less than or greater than 2, we study both cases (Tables 2.1.1 and 2.1.2).

x ($x < 2$)	$f(x)$
1	3
1.5	4
1.9	4.8
1.99	4.98
1.999	4.998

x ($x > 2$)	$f(x)$
3	7
2.5	6
2.1	5.2
2.01	5.02
2.001	5.002

Table 2.1.1: Limit as x approaches 2 from the left

Table 2.1.2: Limit as x approaches 2 from the right

We observe that the closer x is to 2 (from either side of 2), the closer $f(x)$ is to 5. Next, we draw a graph of f that excludes the value of f when $x = 2$ (Figure 2.1.1). The graph confirms what we have observed from Tables 2.1.1 and 2.1.2: As x approaches 2 from either side (notice that x *does not* take value 2, either in the tables or on the graph), $f(x)$ approaches 5. We write this behavior as $\lim_{x \to 2} f(x) = 5$. This equation is read as "The limit of $f(x)$ as x approaches 2 is 5."

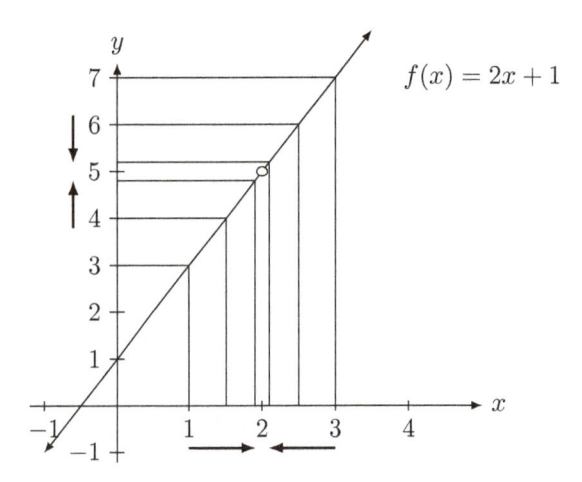

Figure 2.1.1: As x approaches (but does not reach) 2, $f(x)$ approaches 5

What follows is an intuitive definition of the important concept of a limit. This definition is not precise enough to prove deep mathematical results, but it is good enough for practical purposes that we forgo a more rigorous definition.

Definition   *limit, two-sided limit*

We write

$$\lim_{x \to a} f(x) = L, \quad \text{or} \quad f(x) \to L \text{ as } x \to a,$$

and say "when x approaches a, the **limit** of the function is L," if the function value $f(x)$ is close to L whenever x is close, but not equal, to a (on either side of a). We also call such a limit a **two-sided limit**.

Definition   *left-hand limit, right-hand limit, one-sided limit*

We write

$$\lim_{x \to a^-} f(x) = L$$

if the function value $f(x)$ is close to L whenever x is close, but not equal, to a **from the left** (i.e., $x < a$) and call it the **left-hand limit**. We write

$$\lim_{x \to a^+} f(x) = L$$

if the function value $f(x)$ is close to L whenever x is close, but not equal, to a **from the right** (i.e., $x > a$) and call it the **right-hand limit**. More generally, we refer to left-hand limits and right-hand limits as **one-sided limits**.

The word *close* is the source of mathematical difficulty in the definition. Colloquially, *close* is a heavily context dependent notion. An asteroid that comes within 25,000 miles of Earth would be considered close. For some, close means getting within 50 yards of their favorite singer. For others, close only counts in horseshoes and hand grenades. A mathematician must use a context-neutral definition. In this course, you can think of close as meaning *to any degree of accuracy*.

For a function that is defined on an interval, the definition of limit is problematic at the endpoints of the interval because the function is only defined on one side of the endpoints. To describe the behavior

2.1. LIMITS: APPROACHED NUMERICALLY AND GRAPHICALLY

of a function when x approaches a only from one side, we employ the terminology and notations in the second definition.

In order for a (two-sided) limit to exist, both the left-hand limit and the right-hand limit must exist and be the same. Conversely, if the limit does exist, then the one-sided limits from both sides exist and equal the limit. We state the relationship as the following theorem.

Theorem 1   *The Existence of a Limit*

Let f be a function defined on an interval containing a. As x approaches a, a (two-sided) limit of f exists if and only if the the left-hand limit of f at a and the right-hand limit of f at a exist and are equal. That is,

$$\lim_{x \to a} f(x) = L \text{ if and only if } \lim_{x \to a^-} f(x) = \lim_{x \to a^+} f(x) = L.$$

Example 2:   The function g is given by $g(x) = \begin{cases} x+2 & \text{if } x < 2, \\ 2x+1 & \text{if } x \geq 2. \end{cases}$ Find each of the following limits, if they exist. When necessary, state that the limit does not exist.

(a) $\lim\limits_{x \to 2^-} g(x)$;   (b) $\lim\limits_{x \to 2^+} g(x)$;   (c) $\lim\limits_{x \to 2} g(x)$;   (d) $\lim\limits_{x \to 1} g(x)$.

Solution   We inspect the limits both numerically and graphically. For the limits in parts (a)–(c), from Tables 2.1.3 and 2.1.4 and the graph as shown in Figure 2.1.2, we conclude that

(a) $\lim\limits_{x \to 2^-} g(x) = 4$,

(b) $\lim\limits_{x \to 2^+} g(x) = 5$,

(c) $\lim\limits_{x \to 2} g(x)$ does not exist since $\lim\limits_{x \to 2^-} g(x) \neq \lim\limits_{x \to 2^+} g(x)$.

$x \to 2^- (x < 2)$	$g(x)$
1	3
1.5	3.5
1.9	3.9
1.99	3.99
1.999	3.998

Table 2.1.3: Limit of $g(x)$ as x approaches 2 from the left

$x \to 2^+ (x > 2)$	$g(x)$
3	7
2.5	6
2.1	5.2
2.01	5.02
2.001	5.002

Table 2.1.4: Limit of $g(x)$ as x approaches 2 from the right

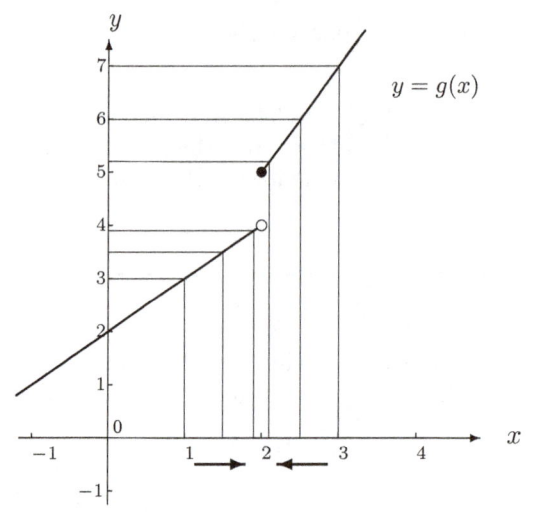

Figure 2.1.2: The graph of g near $x = 2$

> **Warning** *One-sided limits don't tell you about the value at the point*
>
> The limit from one side at a point may exist even when it does not equal the value of the function at that point. For example, $\lim_{x \to 2^-} g(x) = 4$, even though $g(2) = 5 \neq 4$. In fact, the function may not be defined at a point and the (one- or two-sided) limit may still exist at that point.

Now we inspect the limit in part (d) both numerically and graphically. From Tables 2.1.5 and 2.1.6 the graph as shown in Figure 2.1.3, we conclude that

(d) $\lim_{x \to 1} g(x) = 3$.

$x \to 1^- (x < 1)$	$g(x)$
0	2
0.5	2.5
0.9	2.9
0.99	2.99
0.999	2.999

Table 2.1.5: Limit of $g(x)$ as x approaches 1 from the left

$x \to 1^+ (x > 1)$	$g(x)$
1.9	3.9
1.5	3.5
1.1	3.1
1.01	3.01
1.001	3.001

Table 2.1.6: Limit of $g(x)$ as x approaches 1 from the left

2.1. LIMITS: APPROACHED NUMERICALLY AND GRAPHICALLY

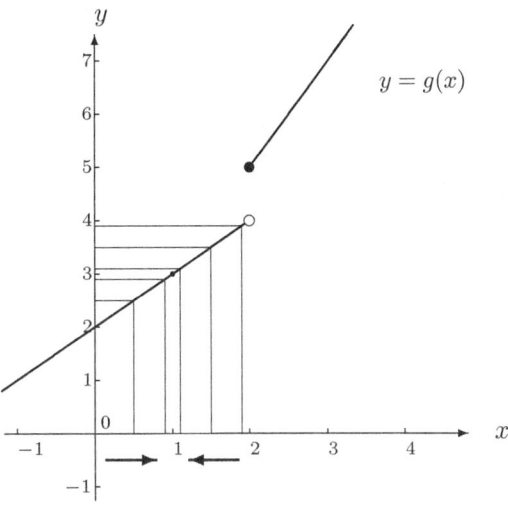

Figure 2.1.3: Graph of $g(x)$ near $x = 1$

Reality Check   *Graphical inspection of limits*

In Example 2, when inspecting the left-hand limit at $a = 2$ on the graph, we can block the graph on the right side of $a = 2$ with a piece of paper. We then follow the curve from left to right using a pencil until it hits the edge of the paper. The y value at that location is the left-hand limit, $\lim\limits_{x \to 2^-} g(x)$. Perform the same process for the right-hand limit, but with the paper covering the left side of $a = 2$. If the left-hand and right-hand limits are the same, then the limit exists, as in the case of $a = 1$. In this case, since the two side limits are not the same, we conclude that $\lim\limits_{x \to 2} g(x)$ does not exist.

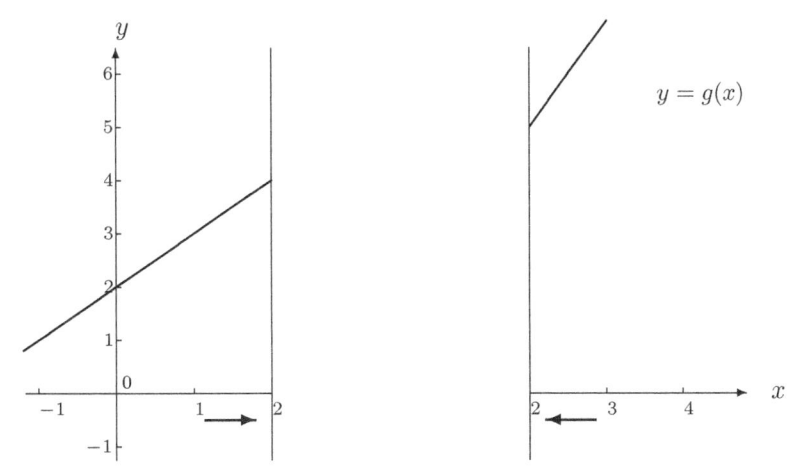

Figure 2.1.4: Inspect the limit of $g(x)$ as x approaches 2 from left and right

Example 3: The function f is given by $f(x) = \begin{cases} x + 2 & \text{if } x \neq 2, \\ 5 & \text{if } x = 2. \end{cases}$ Find each of the following limits, if they exist. When necessary, state that the limit does not exist.

(a) $\lim\limits_{x \to 2^-} f(x)$;   (b) $\lim\limits_{x \to 2^+} f(x)$;   (c) $\lim\limits_{x \to 2} f(x)$.

Solution We inspect the limits both numerically and graphically. From Tables 2.1.7 and 2.1.8 and the graph as shown in Figure 2.1.5, we conclude that

(a) $\lim_{x \to 2^-} f(x) = 4,$ (b) $\lim_{x \to 2^+} f(x) = 4,$ (c) $\lim_{x \to 2} f(x) = 4.$

$x \to 2^- (x < 2)$	$f(x)$
1	3
1.5	3.5
1.9	3.9
1.99	3.99
1.999	3.999

$x \to 2^+ (x > 2)$	$f(x)$
3	5
2.5	4.5
2.1	4.1
2.01	4.01
2.001	4.001

Table 2.1.7: Limit of $f(x)$ as x approaches 2 from the left

Table 2.1.8: Limit of $f(x)$ as x approaches 2 from the right

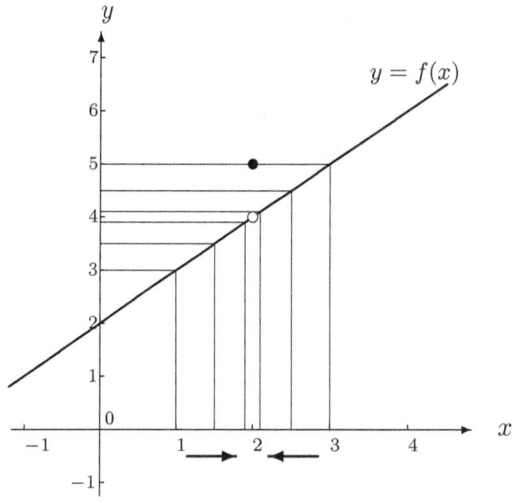

Figure 2.1.5: Graph of $f(x)$ near $x = 2$

Observe that, in Example 3, $\lim_{x \to 2} f(x) = 4$, even though $f(2) = 5 \neq 4$.

Example 4: The function h is given by $h(x) = \dfrac{x^2 - 4}{x - 2}$. Find each of the following limits, if they exist. When necessary, state that the limit does not exist.

(a) $\lim_{x \to 2^-} h(x);$ (b) $\lim_{x \to 2^+} h(x);$ (c) $\lim_{x \to 2} h(x).$

Solution Recall that, when computing the limit as x approaches 2 (from any direction), the value of the function at $x = 2$ is completely irrelevant. When $x \neq 2$, we have

$$h(x) = \frac{x^2 - 4}{x - 2} = \frac{(x+2)(x-2)}{x - 2} = x + 2, \quad x \neq 2.$$

Hence, when $x \neq 2$, h outputs the same values as the function f in Example 3, while h is not defined at

2. Using the tables and the graph in Example 3, but deleting the dot at $(2,5)$, we conclude that

(a) $\lim_{x \to 2^-} h(x) = 4$,   (b) $\lim_{x \to 2^+} h(x) = 4$,   (c) $\lim_{x \to 2} h(x) = 4$.

Observe that, in Example 4, $\lim_{x \to 2} h(x) = 4$, even though h is not defined at $x = 2$.

> **Warning**   *Evaluating a limit by just plugging the number in is wrong*
>
> A common mistake when evaluating the limits in the previous example is to write:
>
> $$\lim_{x \to 2} \frac{x^2 - 4}{x - 2} = \frac{2^2 - 4}{2 - 2} = \frac{0}{0} \quad \text{does not exist}.$$
>
> This is poor form! The limit is not determined by the value of the function at $x = 2$, but rather by the values of the function near $x = 2$. It may happen that the limit of a function at a point equals the value of the function at that point, but this is a special case covered in the next section.

— Limits Involving Infinity

In this section, we examine limits where the magnitude of the input or the output of the function grows without bound. The former will be of particular importance in later sections on *perpetual accumulated present value* and *probability density functions*.

> **Mnemonic**
>
> Keeping the following in mind will be useful. Be sure you believe that the statements are true before proceeding.
>
> $$\frac{1}{\text{tiny}} = \text{big} \qquad \frac{1}{\text{big}} = \text{tiny}$$
>
> A word of caution. The aforementioned mnemonic does not take into account whether the number is positive or negative. *Tiny* and *big* refer to the number's distance from 0, that is, their absolute value.

Example 5: The function s is given by $s(x) = \dfrac{1}{x - 1} + 2$. Find each of the following limits, if they exist. When necessary, state that the limit does not exist.

(a) $\lim_{x \to 1^-} s(x)$;   (b) $\lim_{x \to 1^+} s(x)$;   (c) $\lim_{x \to 1} s(x)$.

Solution We inspect the limits both numerically (Tables 2.1.9 and 2.1.10) and graphically (Figure 2.1.6). As x approaches 1 from the left, the function values become more and more negative without bound. We denote such behavior as

$$\lim_{x \to 1^-} s(x) = -\infty$$

and say that the left-hand limit is negative infinity. As x approaches 1 from the right, the function values become more and more positive without bound. We denote such behavior as

$$\lim_{x \to 1^+} s(x) = \infty$$

and say that the right-hand limit is infinity. Since the the left-hand limit and the right-hand limit differ, we conclude that the limit does not exist. Hence,

(a) $\lim_{x \to 1^-} s(x) = -\infty$,   (b) $\lim_{x \to 1^+} s(x) = \infty$,

(c) $\lim_{x \to 1} s(x)$ does not exist since $\lim_{x \to 1^-} s(x)$ differs from $\lim_{x \to 1^+} s(x)$.

$x \to 1^- \ (x < 1)$	$s(x)$
0	1
0.5	0
0.9	−8
0.99	−98
0.999	−998

Table 2.1.9: Limit as x approaches 1 from the left.

$x \to 1^+ \ (x > 1)$	$s(x)$
2	3
1.5	4
1.1	12
1.01	102
1.001	1002

Table 2.1.10: Limit as x approaches 1

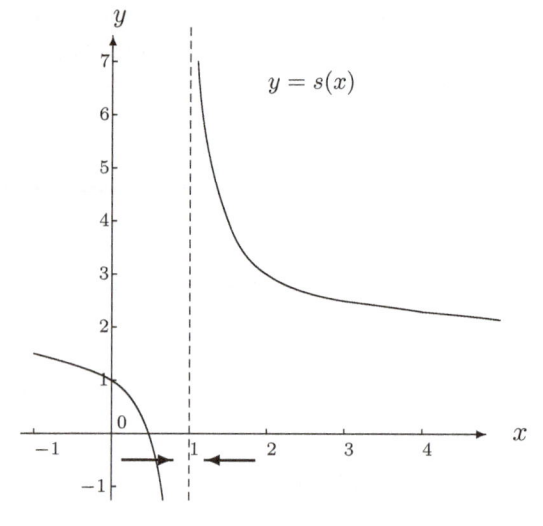

Figure 2.1.6: Graph of $s(x)$ near $x = 1$

Definition   infinite (output) limit

We say that a function f has an *infinite (output) limit* as x approaches a from the left if, as x approaches a from the left, the values of $f(x)$ grow without bound as positive or negative numbers. In the former case, we write

$$\lim_{x \to a^-} f(x) = \infty,$$

while in the latter case we write

$$\lim_{x \to a^-} f(x) = -\infty.$$

We define infinite limits from the right in a nearly identical way (replacing left with right).

Warning   ∞ is not a number

The symbols ∞ and $-\infty$ are not real numbers. They are used to describe the numbers increasing (for ∞) or decreasing (for $-\infty$) without bound, as occurred in the interval notation $(-\infty, \infty)$.

2.1. LIMITS: APPROACHED NUMERICALLY AND GRAPHICALLY

Example 6: The function H is given by $H(x) = \dfrac{1}{(x-1)^2}$. Find each of the following limits, if they exist. When necessary, state that the limit does not exist.

(a) $\lim\limits_{x \to 1^-} H(x)$; (b) $\lim\limits_{x \to 1^+} H(x)$; (c) $\lim\limits_{x \to 1} H(x)$.

Solution We inspect the limits numerically (Tables 2.1.11 and 2.1.12).

$x \to 1^- (x < 1)$	$H(x)$
0	1
0.5	4
0.9	100
0.99	10,000
0.999	1,000,000

Table 2.1.11: Limit as x approaches 1 from the left

$x \to 1^+ (x > 1)$	$H(x)$
2	1
1.5	4
1.1	100
1.01	10,000
1.001	1,000,000

Table 2.1.12: Limit as x approaches 1 from the right

As x approaches 1 from both sides, the function values become more and more positive without bound. Hence, we conclude that

(a) $\lim\limits_{x \to 1^-} H(x) = \infty$, (b) $\lim\limits_{x \to 1^+} H(x) = \infty$, (c) $\lim\limits_{x \to 1} H(x) = \infty$.

Sometimes we need to explore the behavior of a function, say $y = f(x)$, when the input x grows larger and larger without bound. If the function's values approach a certain real number L, then we say that the limit of $f(x)$, as x approaches infinity, is L, and denote this by

$$\lim_{x \to \infty} f(x) = L.$$

Example 7: Consider the function s in Example 4 given by $s(x) = \dfrac{1}{x-1} + 2$. Find $\lim\limits_{x \to \infty} s(x)$.

Solution

$x \to \infty$	$s(x)$
10	2.1111
100	2.0101
1,000	2.001
10,000	2.0001

Table 2.1.13: Limit as $x \to \infty$

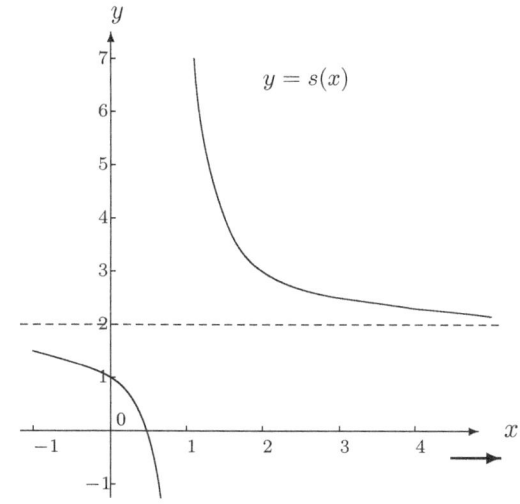

Figure 2.1.7: Graph of $s(x)$

We inspect the limits both numerically and graphically. From Table 2.1.13 and the graph as shown in Figure 2.1.7 we conclude that

$$\lim_{x \to \infty} s(x) = 2.$$

> **Vertical and Horizontal Asymptotes Preview**
>
> The graph of the function $s(x) = \dfrac{1}{x-1} + 2$ in Figure 2.1.6 contains a vertical dotted line, and the graph in Figure 2.1.7 contains a horizontal dotted line. The function curve approaches the vertical dotted line when the variable value approaches 1 from either side, and the function curve approaches the horizontal dotted line when the variable value approaches infinity. These dotted lines are called a **vertical asymptote** and a **horizontal asymptote**, respectively. They are formally defined in Chapter 3.

Exercise Set 2.1

For Exercises 1–10, use the graph of G to find each limit. If the limit does not exist, state that fact.

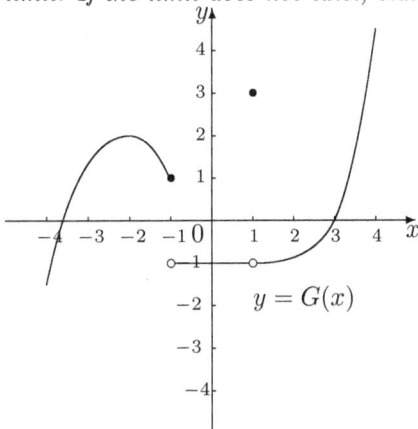

1. Find $\lim\limits_{x \to -1^-} G(x)$
2. Find $\lim\limits_{x \to -1^+} G(x)$
3. Find $\lim\limits_{x \to -1} G(x)$
4. Find $\lim\limits_{x \to 0^-} G(x)$
5. Find $\lim\limits_{x \to 0^+} G(x)$
6. Find $\lim\limits_{x \to 0} G(x)$
7. Find $\lim\limits_{x \to 1^-} G(x)$
8. Find $\lim\limits_{x \to 1^+} G(x)$
9. Find $\lim\limits_{x \to 1} G(x)$
10. Find $\lim\limits_{x \to 3} G(x)$

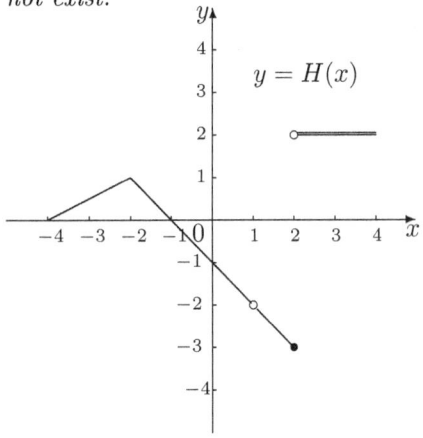

11. Find $\lim\limits_{x \to -2^-} H(x)$
12. Find $\lim\limits_{x \to -2^+} H(x)$
13. Find $\lim\limits_{x \to -2} H(x)$
14. Find $\lim\limits_{x \to 2^-} H(x)$
15. Find $\lim\limits_{x \to 2^+} H(x)$
16. Find $\lim\limits_{x \to 2} H(x)$
17. Find $\lim\limits_{x \to 1^-} H(x)$
18. Find $\lim\limits_{x \to 1^+} H(x)$
19. Find $\lim\limits_{x \to 1} H(x)$
20. Find $\lim\limits_{x \to 3} H(x)$

For Exercises 11–20, use the graph of H to find each limit. If necessary, state that the limit does not exist.

For Exercises 21–30, use the graph of T to find each limit. If necessary, state that the limit does not exist.

21. Find $\lim\limits_{x \to -3^-} T(x)$
22. Find $\lim\limits_{x \to -3^+} T(x)$
23. Find $\lim\limits_{x \to -3} T(x)$
24. Find $\lim\limits_{x \to 0^-} T(x)$
25. Find $\lim\limits_{x \to 0^+} T(x)$
26. Find $\lim\limits_{x \to 0} T(x)$
27. Find $\lim\limits_{x \to -1^-} T(x)$
28. Find $\lim\limits_{x \to -1^+} T(x)$
29. Find $\lim\limits_{x \to -1} T(x)$
30. Find $\lim\limits_{x \to \infty} T(x)$

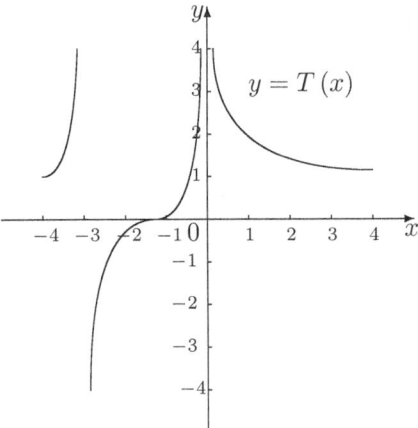

2.1. LIMITS: APPROACHED NUMERICALLY AND GRAPHICALLY

[Graph of $y = T(x)$ shown]

For Exercises 31–40, use both numerical tables and graphs to find the specified limits. If the limit does not exist, state the fact.

31. $f(x) = 2x + 3$; find $\lim_{x \to -2} f(x)$ and $\lim_{x \to 1} f(x)$.

32. $f(x) = x^2 + 1$; find $\lim_{x \to -1} f(x)$ and $\lim_{x \to 0} f(x)$.

33. $g(x) = \dfrac{1}{x-2}$; find $\lim_{x \to -2} g(x)$ and $\lim_{x \to 2} g(x)$.

34. $g(x) = \dfrac{1}{x} + 3$; find $\lim_{x \to 0} g(x)$ and $\lim_{x \to 1} g(x)$.

35. $s(x) = \dfrac{1}{(x-1)^2} + 4$;
    find $\lim_{x \to 1} s(x)$ and $\lim_{x \to \infty} s(x)$.

36. $s(x) = \dfrac{1}{x+3} - 5$;
    find $\lim_{x \to -3} s(x)$ and $\lim_{x \to \infty} s(x)$.

37. $F(x) = \begin{cases} x + 5 & \text{if } x \leq 1, \\ x - 6 & \text{if } x > 1. \end{cases}$
    Find $\lim_{x \to 1^-} F(x)$, $\lim_{x \to 1^+} F(x)$, and $\lim_{x \to 1} F(x)$.

38. $G(x) = \begin{cases} 2x + 3 & \text{if } x \leq 2, \\ x - 1 & \text{if } x > 2. \end{cases}$
    Find $\lim_{x \to 2^-} G(x)$, $\lim_{x \to 2^+} G(x)$, and $\lim_{x \to 2} G(x)$.

39. $S(x) = \begin{cases} 3x & \text{if } x < 0, \\ -2x & \text{if } x > 0. \end{cases}$
    Find $\lim_{x \to 0^-} S(x)$, $\lim_{x \to 0^+} S(x)$, and $\lim_{x \to 0} S(x)$.

40. $H(x) = \begin{cases} -x + 2 & \text{if } x \leq -1, \\ 3 & \text{if } x > -1. \end{cases}$

    Find $\lim_{x \to -1^-} H(x)$, $\lim_{x \to -1^+} H(x)$,
    and $\lim_{x \to -1} H(x)$.

For Exercises 41–50, graph each function and then find the specified limits. If the limit does not exist, state the fact.

41. $f(x) = 2x^2$; find $\lim_{x \to 0} f(x)$ and $\lim_{x \to 2} f(x)$.

42. $f(x) = |x|$; find $\lim_{x \to -1} f(x)$ and $\lim_{x \to 0} f(x)$.

43. $g(x) = \dfrac{x^2 - 4}{(x+2)^2}$;
    find $\lim_{x \to -2} g(x)$ and $\lim_{x \to 0} g(x)$.

44. $g(x) = \dfrac{x^2 - 1}{x - 1}$; find $\lim_{x \to 0} g(x)$ and $\lim_{x \to 1} g(x)$.

45. $s(x) = 2e^{-x}$;
    find $\lim_{x \to 0} s(x)$ and $\lim_{x \to \infty} s(x)$.

46. $s(x) = \ln x$;
    find $\lim_{x \to 1} s(x)$ and $\lim_{x \to 0^+} s(x)$.

47. $F(x) = \begin{cases} 4 & \text{if } x = 1, \\ 2x & \text{if } x \neq 1. \end{cases}$
    Find $\lim_{x \to 1^-} F(x)$, $\lim_{x \to 1^+} F(x)$, and $\lim_{x \to 1} F(x)$.

48. $G(x) = \begin{cases} 5 & \text{if } x = 2, \\ x + 1 & \text{if } x \neq 2. \end{cases}$
    Find $\lim_{x \to 2^-} G(x)$, $\lim_{x \to 2^+} G(x)$, and $\lim_{x \to 2} G(x)$.

49. $S(x) = \begin{cases} 4x - 1 & \text{if } x < 0, \\ x & \text{if } x > 0. \end{cases}$
    Find $\lim_{x \to 0^-} S(x)$, $\lim_{x \to 0^+} S(x)$, and $\lim_{x \to 0} S(x)$.

50. $H(x) = \begin{cases} 2 & \text{if } x \leq -1, \\ 3x & \text{if } x > -1. \end{cases}$
    Find $\lim_{x \to -1^-} H(x)$, $\lim_{x \to -1^+} H(x)$,
    and $\lim_{x \to -1} H(x)$.

For Exercises 51–60, use numerical tables to find the specified limits. If the limit does not exist, state the fact.

51. $f(x) = \sqrt{x}$; find $\lim_{x \to 4} f(x)$ and $\lim_{x \to 1} f(x)$.

52. $f(x) = \sqrt[3]{x}$; find $\lim_{x \to -1} f(x)$ and $\lim_{x \to 0} f(x)$.

53. $g(x) = \dfrac{x^2 + x - 6}{x - 2}$; find $\lim_{x \to -2} g(x)$.

54. $g(x) = \dfrac{x^2 + 3x + 2}{x + 1}$; find $\lim_{x \to -1} g(x)$.

55. $s(x) = \dfrac{1}{(x-1)^2} - 1$;

    find $\lim_{x \to 1} s(x)$ and $\lim_{x \to \infty} s(x)$.

56. $s(x) = \dfrac{1}{x + 1}$;

    find $\lim_{x \to -1} s(x)$ and $\lim_{x \to \infty} s(x)$.

57. $F(x) = \begin{cases} x^2 & \text{if } x \leq 2, \\ x + 2 & \text{if } x > 2. \end{cases}$

    Find $\lim_{x \to 2^-} F(x)$, $\lim_{x \to 2^+} F(x)$, and $\lim_{x \to 2} F(x)$.

58. $G(x) = \begin{cases} e^x & \text{if } x \leq 0, \\ x^3 + 1 & \text{if } x > 0. \end{cases}$

    Find $\lim_{x \to 0^-} G(x)$, $\lim_{x \to 0^+} G(x)$, and $\lim_{x \to 0} G(x)$.

59. $S(x) = \begin{cases} 3x - 3 & \text{if } x < 1, \\ \ln x & \text{if } x > 1. \end{cases}$

    Find $\lim_{x \to 1^-} S(x)$, $\lim_{x \to 1^+} S(x)$, and $\lim_{x \to 1} S(x)$.

60. $H(x) = \begin{cases} 1/(x+1) & \text{if } x < -1, \\ \ln(x+1) & \text{if } x > -1. \end{cases}$

    Find $\lim_{x \to -1^-} H(x)$, $\lim_{x \to -1^+} H(x)$,

    and $\lim_{x \to -1} H(x)$.

2.2 Limits and Continuity

In this section, we introduce the methods that can be applied to evaluate limits directly by knowing the rule that define a function, without having to look at a graph or a numerical table. We then use limits to study **continuity**, a concept of great importance in calculus. After establishing some continuity results, we illustrate how the property of continuity can be used to find limits for a wide variety of functions.

▬ Limit Properties

We first introduce some important properties of limits. These properties are summarized in Theorem 1. The next two examples illustrate how to apply the properties to find a limit.

Example 1: Find $\lim_{x \to 2}(x^2 + 3x - 5)$.

Solution By Properties **L1**, and **L2**,

$$\lim_{x \to 2} x = 2 \quad \text{and} \quad \lim_{x \to 2} 5 = 5.$$

By Properties **L3**, and **L4**,

$$\lim_{x \to 2} 3x = 3 \cdot 2 = 6 \quad \text{and} \quad \lim_{x \to 2} x^2 = 2^2 = 4.$$

Now by Property **L6**,

$$\lim_{x \to 2}(x^2 + 3x - 5) = 4 + 6 - 5 = 5.$$

Note that the previously illustrated steps can be combined into

$$\lim_{x \to 2}(x^2 + 3x - 5) = (2)^2 + 3(2) - 5 = 4 + 6 - 5 = 5.$$

2.2. LIMITS AND CONTINUITY

Example 2: Find $\lim_{x \to 1} \sqrt{x^2 - 2x + 5}$.

Solution By the limit properties listed in Example 1 and Property **L5**,

$$\lim_{x \to 1} \sqrt{x^2 - 2x + 5} = \sqrt{(1)^2 - 2(1) + 5} = \sqrt{4} = 2.$$

Theorem 1 *Properties of Limits*

L1 For a constant number, c, $\lim_{x \to a} c = c$, that is, the limits of a constant function is the constant.

L2 $\lim_{x \to a} x = a$.

If $\lim_{x \to a} f(x) = L$, then

L3 For a constant number c, $\lim_{x \to a} cf(x) = cL$, that is, the limit of a constant times a function is the constant times the limit of the function.

L4 For any positive integer n, $\lim_{x \to a} [f(x)]^n = [\lim_{x \to a} f(x)]^n = L^n$, that is, the limit of a power is the power of the limit.

L5 For any positive integer n, $\lim_{x \to a} \sqrt[n]{f(x)} = \sqrt[n]{\lim_{x \to a} f(x)} = \sqrt[n]{L}$, that is, the limit of a root is the root of the limit. When n is even, we assume that $L \geq 0$.

If $\lim_{x \to a} g(x) = M$, then we have the following.

L6 $\lim_{x \to a} [f(x) \pm g(x)] = L \pm M$, that is, the limit of a sum or a difference is the sum or the difference of the limits.

L7 $\lim_{x \to a} [f(x) \cdot g(x)] = [\lim_{x \to a} f(x)] \cdot [\lim_{x \to a} g(x)] = L \cdot M$, that is, the limit of a product is the product of the limits.

L8 If $M \neq 0$, then $\lim_{x \to a} \dfrac{f(x)}{g(x)} = \dfrac{\lim_{x \to a} f(x)}{\lim_{x \to a} g(x)} = \dfrac{L}{M}$, that is, the limit of a quotient is the quotient of the limits, provided that the limit of the denominator is not zero.

Warning *Illegal use of limit properties*

Properties **L3**–**L8** of Theorem 1 require the user to already know that a certain limit exists. For example, by **L1** we have

$$0 = \lim_{x \to 0} 0 = \lim_{x \to 0} \left(\frac{1}{x} - \frac{1}{x} \right).$$

If we use **L6** to split the aforementioned limit as

$$\left(\lim_{x \to 0} \frac{1}{x} \right) - \left(\lim_{x \to 0} \frac{1}{x} \right),$$

we encounter a problem: neither of the aforementioned limits exist! Do not conclude that the original limit does not exist. Conclude, instead, that one may not apply **L6** in this context. In practice, this means that one must use a bit of foresight when using the limit properties.

Continuity

We now turn our focus to the concept of **continuity**. If the graph of a function is given, then it is easy to judge whether the function is continuous by the "**pencil test**."

> **Mnemonic**   Pencil test for continuity
>
> A function is continuous, if its graph can be traced without lifting the pencil from the paper. That is, the graph is connected, meaning that it does not have "holes" or "jumps."

The functions graphed in Figure 2.2.1 are continuous, whereas the functions graphed in Figure 2.2.2 are not continuous.

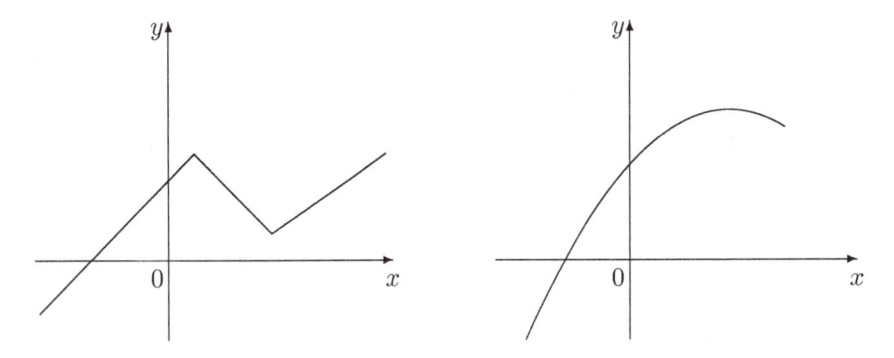

Figure 2.2.1: Graphs of continuous functions

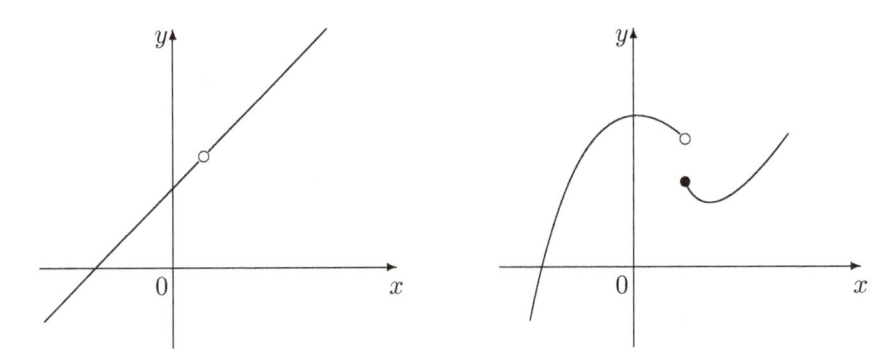

Figure 2.2.2: Graphs of discontinuous functions

Observe that when the graph of $f(x)$ is connected at an x value a, the limit $\lim_{x \to a} f(x)$ exists and equals the function value $f(a)$. We now give the formal definition of continuity:

> **Definition**   *continuous at, continuous over an interval*
>
> A function f is **continuous at** $x = a$ if all three of the following are true:
> a) $f(a)$ exists. (f is defined at $x = a$.)
>
> b) $\lim_{x \to a} f(x)$ exists. (The limit of the function exists when x approaches a.)
>
> c) $\lim_{x \to a} f(x) = f(a)$. (The limit is the same as the function value.)
>
> A function is **continuous over an interval** I if it is continuous at each point in I.

2.2. LIMITS AND CONTINUITY

Example 3: Is the function given by $f(x) = x^2 + 3x - 5$ continuous at $x = 2$?

Solution First,
$$f(2) = (2)^2 + 3(2) - 5 = 5,$$
hence $f(2)$ exists. Second, by Example 1,
$$\lim_{x \to 2} f(x) = 5,$$
hence the limit exists. Now we have
$$\lim_{x \to 2} f(x) = f(2).$$
Therefore, f is continuous at $x = 2$.

Observe that if a function is not continuous at $x = a$, then

(a) $f(a)$ does not exist, that is, f is not defined at $x = a$, or

(b) $\lim_{x \to a} f(x)$ does not exists, that is, the limit of the function does not exist when x approaches a, or

(c) Both $f(a)$ and $\lim_{x \to a} f(x)$ exist, and $\lim_{x \to a} f(x) \neq f(a)$, that is, the limit is not the same as the function value.

Example 4: Is the function given by $f(x) = \dfrac{x^2 - 4}{x - 2}$ continuous at $x = 2$? Why or why not?

Solution Since at $x = 2$, the function is not defined, $f(2)$ does not exist. Therefore, f is not continuous at $x = 2$.

Example 5: Consider the function given by

$$f(x) = \begin{cases} x + 2 & \text{if } x \leq 1, \\ 2 & \text{if } x > 1. \end{cases}$$

Is the function continuous at $x = 1$? Why or why not?

Solution From the graph (Figure 2.1.3) we obtain that
$$\lim_{x \to 1^-} f(x) = 3 \quad \text{and} \quad \lim_{x \to 1^+} f(x) = 2.$$
Since $\lim_{x \to 1^-} f(x) \neq \lim_{x \to 1^+} f(x)$, we conclude that $\lim_{x \to 1} f(x)$ does not exist. Therefore, the function is not continuous at $x = 1$.

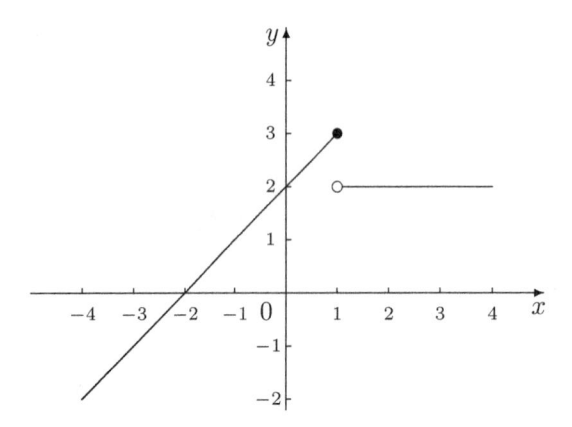

Figure 2.2.3: Graph of $f(x)$

Example 6: Consider the function given by

$$f(x) = \begin{cases} x+2 & \text{if } x \neq 1, \\ 2 & \text{if } x = 1. \end{cases}$$

Is the function continuous at $x = 1$? Why or why not?

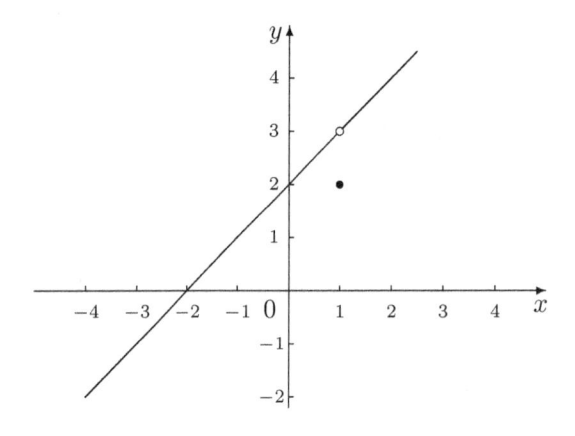

Figure 2.2.4: Graph of $f(x)$

Solution We have $f(1) = 2$. From the graph (Figure 2.2.4) we obtain that

$$\lim_{x \to 1} f(x) = 3.$$

Hence $\lim_{x \to 1} f(x) \neq f(1)$. Therefore, the function is not continuous at $x = 1$.

Continuity and Limits

Observing the definition of continuity, we notice that, if a function f is *known* to be continuous, then we can apply the definition to obtain its limit, since in such a case, we have

$$\lim_{x \to a} f(x) = f(a).$$

2.2. LIMITS AND CONTINUITY

Indeed, most of the functions studied in this course *are* continuous.

Theorem 2   *A List of Continuous Functions*

The following types of functions are continuous at every number in their domains:
(a) Polynomials

(b) Rational functions

(c) Root functions

(d) Exponential functions

(e) Logarithmic functions

(f) Sums, differences, products, and quotients of the aforementioned functions

---

Applying the theorem, we can obtain many of the limits by simply obtaining the function value.

Example 7: Find $\lim\limits_{x \to 0} \dfrac{6e^{2x} + x^2}{\sqrt{2x + 9}}$.

Solution Since $f(x) = \dfrac{6e^{2x} + x^2}{\sqrt{2x + 9}}$ is continuous at $x = 0$, we have

$$\lim_{x \to 0} \frac{6e^{2x} + x^2}{\sqrt{2x + 9}} = \frac{6e^{2(0)} + (0)^2}{\sqrt{2(0) + 9}} = \frac{6}{3} = 2.$$

Example 8: Find $\lim\limits_{h \to 0}(6x^2 - 4xh + h^2)$.

Solution Since, in the limit, x does not change its values, we treat x as a constant. Note that the function is continuous at $h = 0$. We have

$$\lim_{h \to 0}(6x^2 - 4xh + h^2) = 6x^2 - 4x(0) + 0^2 = 6x^2.$$

Warning   *Continuity Carelessness*

Two common misapplications of Theorem 2 occur when
(a) The limit involves a quotient with a denominator approaches zero, or

(b) The limit involves a (piecewise-defined) function formed by two or more formulas, with the variable approaching a "joint," that is, the function changes formulas around the variable value.

For example, $x + 2$ and $x - 4$ are continuous functions, and, therefore, $\dfrac{x + 2}{x - 4}$ is continuous *except* when the denominator is 0, that is, when $x = 4$. Also, the function given by the rule

$$\begin{cases} x + 2 & x \leq 0 \\ x - 4 & x > 0, \end{cases}$$

is continuous *except* at the "joint," that is, when $x = 0$. There are instances where one has continuity even at a joint. For example, replace $x - 4$ with $3x - 4$ in the aforementioned example.

We illustrate the two cases and the strategies for finding the limits in the following examples.

Example 9: Find $\lim_{x \to -1} \dfrac{x^2 - 1}{x + 1}$.

Solution Note that when x approaches -1, both the numerator and the denominator approach zero. Since in taking the limit, $x \neq -1$, we can simplify the quotient before taking the limit:

$$\lim_{x \to -1} \frac{x^2 - 1}{x + 1} = \lim_{x \to -1} \frac{(x-1)(x+1)}{x+1} = \lim_{x \to -1} (x - 1) = (-1) - 1 = -2.$$

Example 10: Find $\lim_{h \to 0} \dfrac{\frac{2}{3+h} - \frac{2}{3}}{h}$.

Solution Note that when h approaches 0, both the numerator and the denominator approach zero. Since, in taking the limit, $h \neq 0$, we can simplify the quotient before taking the limit. We first simplify the numerator:

$$\frac{2}{3+h} - \frac{2}{3} = \frac{2}{3+h} \cdot \frac{3}{3} - \frac{2}{3} \cdot \frac{3+h}{3+h} = \frac{6 - (6 + 2h)}{3(3+h)} = \frac{6 - 6 - 2h}{3(3+h)} = \frac{-2h}{3(3+h)}.$$

Thus, (note that dividing by h is equivalent to multiplying by $\frac{1}{h}$)

$$\frac{\frac{2}{3+h} - \frac{2}{3}}{h} = \frac{-2h}{3(3+h)} \cdot \frac{1}{h} = \frac{-2}{3(3+h)}, \text{ for } h \neq 0.$$

Therefore,

$$\lim_{h \to 0} \frac{\frac{2}{3+h} - \frac{2}{3}}{h} = \lim_{h \to 0} \frac{-2}{3(3+h)} = -\frac{2}{3(3+0)} = -\frac{2}{9}.$$

Example 11: Find $\lim_{h \to 0} \dfrac{(x+h)^2 - x^2}{h}$.

Solution As in Example 8, we treat x as a constant. Note that when h approaches 0, both the numerator and the denominator approach zero. Since, in taking the limit, $h \neq 0$, we can simplify the quotient before taking the limit:

$$\lim_{h \to 0} \frac{(x+h)^2 - x^2}{h} = \lim_{h \to 0} \frac{x^2 + 2xh + h^2 - x^2}{h}$$
$$= \lim_{h \to 0} \frac{h(2x + h)}{h}$$
$$= \lim_{h \to 0} (2x + h)$$
$$= 2x.$$

Example 12: Find $\lim_{x \to -1} \dfrac{x - 1}{x + 1}$.

Solution Note that when x approaches -1, the denominator approaches zero and the numerator approaches -2. In this case, taking the limit numerically (Tables 2.2.1 and 2.2.2) is more convenient. We observe that when x approaches -1^-, the limit is ∞, whereas when $x \to -1^+$, the limit is $-\infty$. Hence, $\lim_{x \to -1} \dfrac{x - 1}{x + 1}$ does not exist.

2.2. LIMITS AND CONTINUITY

$x \to -1^-$ $(x<-1)$	$f(x)$
-1.1	21
-1.01	201
-1.001	$2{,}001$
-1.0001	$20{,}001$

Table 2.2.1: Limit of $f(x)$ as x approaches -1 from the left

$x \to -1^+$ $(x>-1)$	$f(x)$
-0.9	-19
-0.99	-199
-0.999	$-1{,}999$
-0.9999	$-19{,}999$

Table 2.2.2: Limit of $f(x)$ as x approaches -1 from the right

Example 13: Find $\displaystyle\lim_{x \to 1} \frac{x}{(x-1)^2}$.

Solution Note that when x approaches 1, the denominator approaches zero and the numerator approaches 1. Again we take the limit numerically (Tables 2.2.3 and 2.2.4). We observe that the limit is ∞. Hence $\displaystyle\lim_{x \to 1} \frac{x}{(x-1)^2} = \infty$.

$x \to 1^-$ $(x<1)$	$f(x)$
0.9	90
0.99	$9{,}900$
0.999	$999{,}000$

Table 2.2.3: Limit of $f(x)$ as x approaches 1 from the left

$x \to 1^+$ $(x>1)$	$f(x)$
1.1	110
1.01	$10{,}100$
1.001	$1{,}001{,}000$

Table 2.2.4: Limit of $f(x)$ as x approaches 1 from the right

Example 14: The function f is given by
$$f(x) = \begin{cases} 3x^2 + 2 & \text{if } x \le 3, \\ 2 - 4x & \text{if } x > 3. \end{cases}$$
Find (a) $\displaystyle\lim_{x \to 2} f(x)$, and (b) $\displaystyle\lim_{x \to 3} f(x)$.

Solution (a) Since $f(x) = 3x^2 + 2$ for x near 2, and $3x^2 + 2$ is continuous, we can take the function value as the limit:
$$\lim_{x \to 2} f(x) = \lim_{x \to 2} 3x^2 + 2 = 3(2)^2 + 2 = 14.$$

(b) Since $x = 3$ is a "joint," we must take the left-hand limit and the right-hand limit using different formulas:

$$\lim_{x \to 3^-} f(x) = \lim_{x \to 3^-} (3x^2 + 2) = 3(3)^2 + 2 = 29;\text{ and}$$
$$\lim_{x \to 3^+} f(x) = \lim_{x \to 3^+} (2 - 4x) = 2 - 4(3) = -10.$$

Since $\displaystyle\lim_{x \to 3^-} f(x) \ne \lim_{x \to 3^+} f(x)$, we conclude that $\displaystyle\lim_{x \to 3} f(x)$ does not exist.

Example 15: The function f is given by

$$f(x) = \begin{cases} x^2 - 3x & \text{if } x \leq 2, \\ 4x - 10 & \text{if } x > 2. \end{cases}$$

(a) Find $\lim_{x \to 2} f(x)$;   (b) Is f continuous at $x = 2$? Why or why not?

Solution (a) Since $x = 2$ is a "joint," we must take the left-hand limit and the right-hand limit using different formulas:

$$\lim_{x \to 2^-} f(x) = \lim_{x \to 2^-} (x^2 - 3x) = (2)^2 - 3(2) = -2; \text{ and}$$
$$\lim_{x \to 2^+} f(x) = \lim_{x \to 2^+} (4x - 10) = 4(2) - 10 = -2.$$

Since $\lim_{x \to 2^-} f(x) = \lim_{x \to 2^+} f(x) = -2$, we conclude that $\lim_{x \to 2} f(x) = -2$.
(b) Since $f(2) = -2$ and $\lim_{x \to 2} f(x) = -2$, all three conditions in the definition of continuity are satisfied. Hence f is continuous at $x = 2$.

Exercise Set 2.2

Use the continuity of the functions to find the following limits.

1. $\lim_{x \to 2} (2x - 5)$
2. $\lim_{x \to -1} (6x + 3)$
3. $\lim_{x \to 1} (5x^2 - 3x + 1)$
4. $\lim_{x \to -3} (x^2 - 9)$
5. $\lim_{x \to 1} \dfrac{x^2 - 4}{x - 2}$
6. $\lim_{x \to 2} \dfrac{x - 5}{x^2 - 1}$
7. $\lim_{x \to 1} (e^{3x} + \ln x)$
8. $\lim_{x \to -1} (2e^{-x} + 5x^2)$
9. $\lim_{h \to 0} (3x^2 + 3xh + h^2)$
10. $\lim_{h \to 0} \dfrac{-1}{x(x + h)}$

Determine whether each of the functions shown in Exercises 11–14 is continuous over the interval $(-4, 4)$.

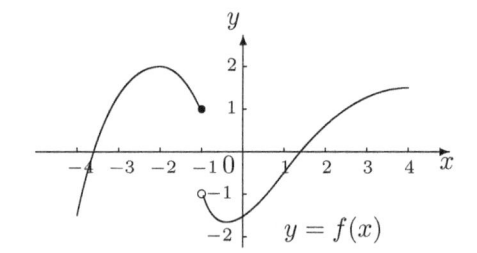

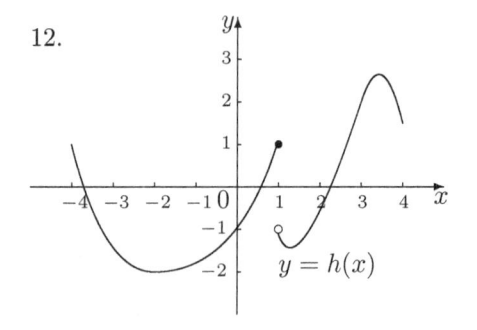

11.

$y = f(x)$

12.

$y = h(x)$

2.2. LIMITS AND CONTINUITY

13.

[Graph of $y = g(x)$]

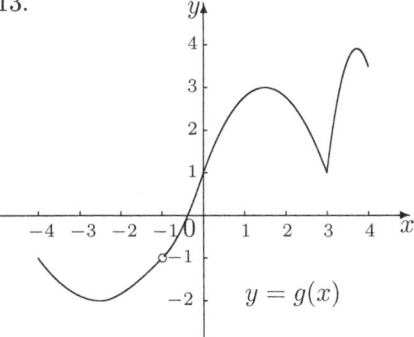

14.

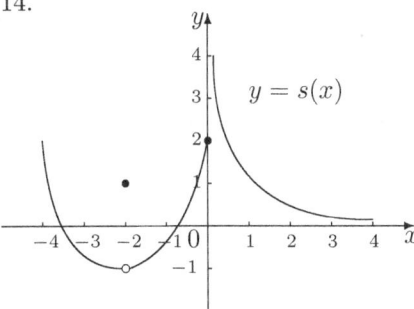

[Graph of $y = s(x)$]

Use the graphs in Exercises 11–14 to answer the following questions:

15. (a) Find $f(-1)$.

   (b) Find $\lim_{x \to -1^-} f(x)$, $\lim_{x \to -1^+} f(x)$, and $\lim_{x \to -1} f(x)$.

   (c) Is f continuous at $x = -1$? Why or why not?

   (d) Find $f(-2)$.

   (e) Find $\lim_{x \to -2^-} f(x)$, $\lim_{x \to -2^+} f(x)$, and $\lim_{x \to -2} f(x)$.

   (f) Is f continuous at $x = -2$? Why or why not?

16. (a) Find $h(1)$.

   (b) Find $\lim_{x \to 1^-} h(x)$, $\lim_{x \to 1^+} h(x)$, and $\lim_{x \to 1} h(x)$.

   (c) Is h continuous at $x = 1$? Why or why not?

   (d) Find $h(3)$.

   (e) Find $\lim_{x \to 3^-} h(x)$, $\lim_{x \to 3^+} h(x)$, and $\lim_{x \to 3} h(x)$.

   (f) Is h continuous at $x = 3$? Why or why not?

17. (a) Find $g(-1)$.

   (b) Find $\lim_{x \to -1^-} g(x)$, $\lim_{x \to -1^+} g(x)$, and $\lim_{x \to -1} g(x)$.

   (c) Is g continuous at $x = -1$? Why or why not?

   (d) Find $g(3)$.

   (e) Find $\lim_{x \to 3^-} g(x)$, $\lim_{x \to 3^+} g(x)$, and $\lim_{x \to 3} g(x)$.

   (f) Is g continuous at $x = 3$? Why or why not?

18. (a) Find $s(-2)$.

   (b) Find $\lim_{x \to -2^-} s(x)$, $\lim_{x \to -2^+} s(x)$, and $\lim_{x \to -2} s(x)$.

   (c) Is s continuous at $x = -2$? Why or why not?

   (d) Find $s(0)$.

   (e) Find $\lim_{x \to 0^-} s(x)$, $\lim_{x \to 0^+} s(x)$, and $\lim_{x \to 0} s(x)$.

   (f) Is s continuous at $x = 0$? Why or why not?

19. Consider the function given by
$$f(x) = \begin{cases} x + 2 & \text{if } x \leq 1, \\ x^2 - 3 & \text{if } x > 1. \end{cases}$$

   (a) Find $f(1)$.

   (b) Find $\lim_{x \to 1^-} f(x)$, $\lim_{x \to 1^+} f(x)$, and $\lim_{x \to 1} f(x)$.

   (c) Is f continuous at $x = 1$? Why or why not?

20. Consider the function given by
$$f(x) = \begin{cases} x - 2 & \text{if } x \leq 1, \\ x^2 & \text{if } x > 1. \end{cases}$$

   (a) Find $f(1)$.

   (b) Find $\lim_{x \to 1^-} f(x)$, $\lim_{x \to 1^+} f(x)$, and $\lim_{x \to 1} f(x)$.

   (c) Is f continuous at $x = 1$? Why or why not?

21. Consider the function given by
$$f(x) = \begin{cases} x^2 - 4 & \text{if } x \leq 1, \\ 3x - 6 & \text{if } x > 1. \end{cases}$$

   (a) Find $f(1)$.

   (b) Find $\lim_{x \to 1^-} f(x)$, $\lim_{x \to 1^+} f(x)$, and $\lim_{x \to 1} f(x)$.

   (c) Is f continuous at $x = 1$? Why or why not?

22. Consider the function given by
$$f(x) = \begin{cases} x + 5 & \text{if } x \leq 1, \\ 4x + 2 & \text{if } x > 1. \end{cases}$$

   (a) Find $f(1)$.

   (b) Find $\lim_{x \to 1^-} f(x)$, $\lim_{x \to 1^+} f(x)$, and $\lim_{x \to 1} f(x)$.

   (c) Is f continuous at $x = 1$? Why or why not?

Use algebraic simplifications to find the following limits.

23. $\lim_{x \to 2} \dfrac{x^2 - 4}{x - 2}$

24. $\lim_{x \to -2} \dfrac{x^2 - 4}{x + 2}$

25. $\lim_{x \to 1} \dfrac{x^2 + x - 2}{x - 1}$

26. $\lim_{x \to -3} \dfrac{x^2 + x - 6}{x + 3}$

27. $\lim_{x \to 1} \dfrac{x - 1}{x^2 - 1}$

28. $\lim_{x \to 2} \dfrac{x - 2}{x^2 - 4}$

29. $\lim_{h \to 0} \dfrac{(2 + h)^2 - 4}{h}$

30. $\lim_{h \to 0} \dfrac{(1 + h)^2 - 1}{h}$

31. $\lim_{h \to 0} \dfrac{(1 + h)^3 - 1}{h}$

32. $\lim_{h \to 0} \dfrac{(2 + h)^3 - 8}{h}$

33. $\lim_{h \to 0} \dfrac{(x + h)^2 - x^2}{h}$

34. $\lim_{h \to 0} \dfrac{3(x + h)^2 - 3x^2}{h}$

35. $\lim_{h \to 0} \dfrac{\frac{1}{1+h} - 1}{h}$

36. $\lim_{h \to 0} \dfrac{\frac{3}{4+h} - \frac{3}{4}}{h}$

37. $\lim_{h \to 0} \dfrac{\frac{1}{x+h} - \frac{1}{x}}{h}$

38. $\lim_{h \to 0} \dfrac{\frac{3}{x+h} - \frac{3}{x}}{h}$

Find the following limits. When necessary, state that the limit does not exist.

39. $\lim_{x \to 2} \dfrac{x^2}{x - 2}$

40. $\lim_{x \to 1} \dfrac{-4}{x - 1}$

41. $\lim_{x \to 2} \dfrac{3}{(x - 2)^2}$

42. $\lim_{x \to 3} \dfrac{x + 1}{(x - 3)^2}$

43. $\lim_{x \to 0} \dfrac{2}{x}$

44. $\lim_{x \to 0} \dfrac{x + 3}{x}$

45. $\lim_{x \to 0} \dfrac{x + 1}{x^2}$

46. $\lim_{x \to 0} \dfrac{x - 2}{x^2}$

2.3 Average Rates of Change

What is the difference between when we say "My average speed in the last five minutes of driving is 45 mph" and when we say "In this instant my speedometer says that I am driving at 65 mph?" The first statement talks about the **average rate of change** of the car's position, while the second statement talks about the **instantaneous rate of change** of the car's position. The instantaneous rate of change is exactly the concept of the derivative of a function. To understand the instantaneous rate of change, we first need to fully understand the average rate of change in different settings.

Example 1: A car is moving along a straight road between 4 p.m. and 6 p.m.. The following graph (Figure 2.3.1) shows the mileage covered in the 2 hours:

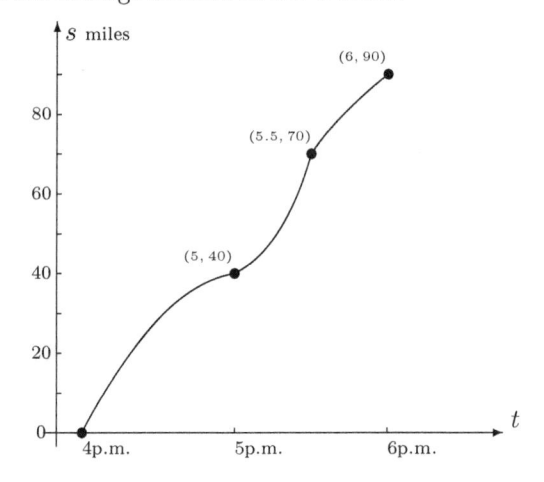

Figure 2.3.1: Graph of mileage vs. time

(a) What is the average speed of the car in mph between 4 p.m. and 5 p.m.?

2.3. AVERAGE RATES OF CHANGE

(b) What is the average speed of the car in mph between 5 p.m. and 5:30 p.m.?

Solution (a) From the graph we have

$$\frac{40 \text{ miles } - 0 \text{ miles}}{5 \text{ p.m. } - 4 \text{ PM}} = \frac{40 \text{ miles}}{1 \text{ hour}} = 40 \text{ mph}.$$

(b) From the graph we have

$$\frac{70 \text{ miles } - 40 \text{ miles}}{5.5 \text{ PM } - 5 \text{ PM}} = \frac{30 \text{ miles}}{0.5 \text{ hour}} = 60 \text{ mph}.$$

Now, let us consider a function $y = f(x)$. We are interested in the ratio of the change of the output, y, with respect to the change of the input, x. There are several ways to express such a ratio. The first set of notations, the most intuitive, denotes the ratio of changes when the input changes from a particular value x_1 to another value x_2. Then, the output values corresponding to the two input values are $y_1 = f(x_1)$ and $y_2 = f(x_2)$, hence the **average rate of change of y with respect to x** is the ratio

$$\frac{y_2 - y_1}{x_2 - x_1} = \frac{f(x_2) - f(x_1)}{x_2 - x_1} \quad \text{(where } x_2 \neq x_1\text{)}.$$

On a graph, this ratio represents the **slope of the secant line** going through the two points $(x_1, f(x_1))$ and $(x_2, f(x_2))$ (Figure 2.3.2).

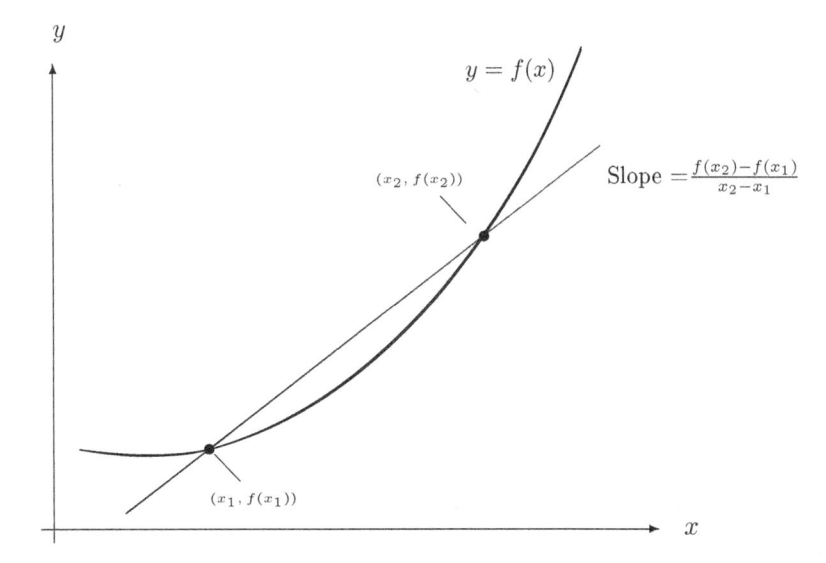

Figure 2.3.2: Average rate of change as the slope of a secant line

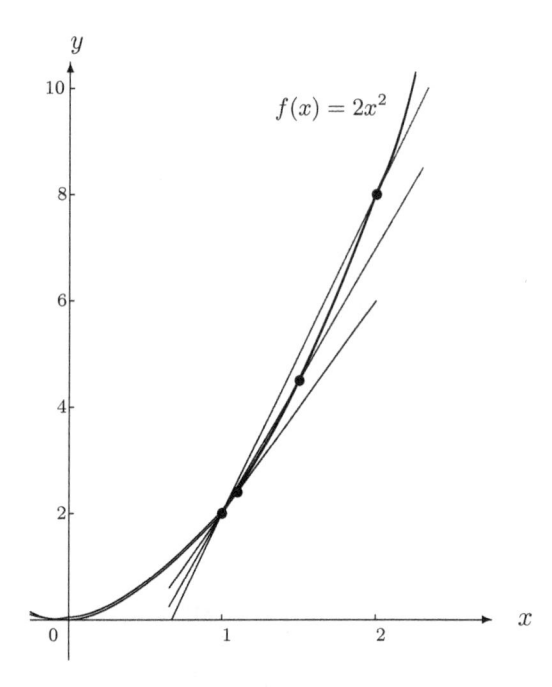

Figure 2.3.3: Graph of $f(x) = 2x^2$ with 3 secant lines

Example 2: For $y = f(x) = 2x^2$, use the formula $\dfrac{f(x_2) - f(x_1)}{x_2 - x_1}$ to find the average rate of change as
(a) x changes from 1 to 2.
(b) x changes from 1 to 1.5.
(c) x changes from 1 to 1.1.

Solution The three secant lines whose slopes are being computed are illustrated in Figure 2.3.3.

(a) We substitute $x_1 = 1$ and $x_2 = 2$ into the formula to obtain

$$\frac{f(2) - f(1)}{2 - 1} = \frac{f(2) - f(1)}{1} = \frac{2(2)^2 - 2(1)^2}{1} = \frac{8 - 2}{1} = 6.$$

(b) We substitute $x_1 = 1$ and $x_2 = 1.5$ into the formula to obtain

$$\frac{f(1.5) - f(1)}{1.5 - 1} = \frac{f(1.5) - f(1)}{0.5} = \frac{2(1.5)^2 - 2(1)^2}{0.5} = \frac{4.5 - 2}{0.5} = 5.$$

(c) We substitute $x = 1$ and $x_2 = 1.1$ into the formula to obtain

$$\frac{f(1.1) - f(1)}{1.1 - 1} = \frac{f(1.1) - f(1)}{0.1} = \frac{2(1.1)^2 - 2(1)^2}{0.1} = \frac{2.42 - 2}{0.1} = 4.2.$$

To study the **instantaneous rate of change,** we need to study the behavior of the average rate of change when the second input, x_2, is moving closer and closer to the first input, x_1. That is, we are more interested in the *change* in the input. We now introduce the new notation to represent this behavior. We replace x_1 by x and x_2 by $x + h$, where h denotes the change. Hence the change in the input, $x_2 - x_1$, is now simply h. Correspondingly, the output values are now represented by $f(x)$ and $f(x + h)$. Hence, now the average rate of change is the ratio (where $x_2 \neq x_1$, hence $h \neq 0$)

$$\frac{y_2 - y_1}{x_2 - x_1} = \frac{f(x_2) - f(x_1)}{x_2 - x_1} = \frac{f(x + h) - f(x)}{(x + h) - x} = \frac{f(x + h) - f(x)}{h}.$$

2.3. AVERAGE RATES OF CHANGE

The last ratio is also called the **difference quotient**. On a graph, again, the difference quotient represents the slope of the secant line passing through the two points $(x, f(x))$ and $(x+h, f(x+h))$ (Figure 2.3.4).

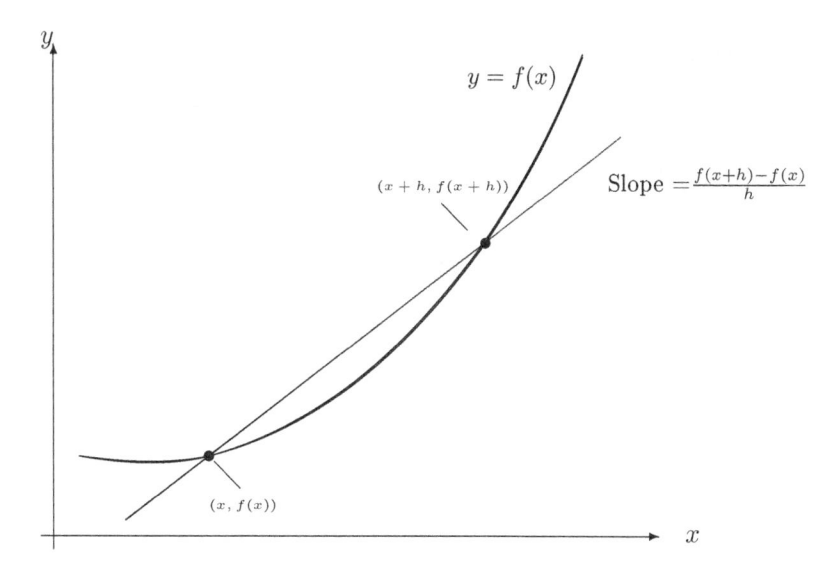

Figure 2.3.4: Difference quotient as the slope of a secant line

Example 3: For $y = f(x) = 2x^2$, use the difference quotient $\dfrac{f(x+h) - f(x)}{h}$ to find the average rate of change as

(a) x changes from 1 to 2.

(b) x changes from 1 to 1.5.

(c) x changes from 1 to 1.1.

Solution Note that the questions are exactly the same as those in Example 2, hence we expect the same results.

(a) We substitute $x = 1$ and $h = 2 - 1 = 1$ into the difference quotient to obtain

$$\frac{f(1+1) - f(1)}{1} = \frac{f(2) - f(1)}{1} = \frac{2(2)^2 - 2(1)^2}{1} = \frac{8 - 2}{1} = 6.$$

(b) We substitute $x = 1$ and $h = 1.5 - 1 = 0.5$ into the difference quotient to obtain

$$\frac{f(1+0.5) - f(1)}{0.5} = \frac{f(1.5) - f(1)}{0.5} = \frac{2(1.5)^2 - 2(1)^2}{0.5} = \frac{4.5 - 2}{0.5} = 5.$$

(c) We substitute $x = 1$ and $h = 1.1 - 1 = 0.1$ into the difference quotient to obtain

$$\frac{f(1+0.1) - f(1)}{0.1} = \frac{f(1.1) - f(1)}{0.1} = \frac{2(1.1)^2 - 2(1)^2}{0.1} = \frac{2.42 - 2}{0.1} = 4.2.$$

Let us observe the computations of the three average rates of change in the aforementioned two examples. It seems that the computations are rather cumbersome and repetitive. Indeed, the formula can be simplified before the values are substituted.

Example 4: For $y = f(x) = 2x^2$:

(a) Simplify the difference quotient $\dfrac{f(x+h) - f(x)}{h}$;

(b) Use the simplified formula to find the average rate of change at $x = 1$ as $h =$ (i) 1 (ii) 0.5 (iii) 0.1.

Solution (a) We have $f(x) = 2x^2$ so
$$f(x+h) = 2(x+h)^2 = 2\left(x^2 + 2xh + h^2\right) = 2x^2 + 4xh + 2h^2.$$

Therefore,
$$f(x+h) - f(x) = (2x^2 + 4xh + 2h^2) - 2x^2 = 4xh + 2h^2 = h(4x + 2h).$$

Then,
$$\frac{f(x+h) - f(x)}{h} = \frac{h(4x + 2h)}{h} = 4x + 2h, \quad h \neq 0.$$

(b(i)) We now substitute $x = 1$ and $h = 1$ into the simplified formula to obtain
$$\frac{f(x+h) - f(x)}{h} = 4x + 2h = 4(1) + 2(1) = 4 + 2 = 6.$$

(b(ii)) We substitute $x = 1$ and $h = 0.5$ into the simplified formula to obtain
$$4(1) + 2(0.5) = 4 + 1 = 5.$$

(b(iii)) We substitute $x = 1$ and $h = 0.1$ into the simplified formula to obtain
$$4(1) + 2(0.1) = 4 + 0.2 = 4.2.$$

It is obvious that computations are much easier when a simplified form of the difference quotient is used. Furthermore, the behavior of the average rate of change as h is approaching zero is much clearer. It is worth emphasizing that the simplification is valid only for nonzero values of h.

Example 5: For $y = f(x) = x^3$, find a simplified form of the difference quotient $\dfrac{f(x+h) - f(x)}{h}$.

Solution For $f(x) = x^3$, we have
$$f(x+h) = (x+h)^3 = x^3 + 3x^2h + 3xh^2 + h^3.$$

Therefore,
$$f(x+h) - f(x) = (x^3 + 3x^2h + 3xh^2 + h^3) - x^3$$
$$= 3x^2h + 3xh^2 + h^3 = h(3x^2 + 3xh + h^2).$$

Then,
$$\frac{f(x+h) - f(x)}{h} = \frac{h(3x^2 + 3xh + h^2)}{h} = 3x^2 + 3xh + h^2, \quad h \neq 0.$$

2.3. AVERAGE RATES OF CHANGE

Example 6: For $y = f(x) = 4x + 3$, find a simplified form of the difference quotient $\dfrac{f(x+h) - f(x)}{h}$.

Solution For $f(x) = 4x + 3$, we have

$$f(x+h) = 4(x+h) + 3 = 4x + 4h + 3.$$

Therefore,

$$f(x+h) - f(x) = (4x + 4h + 3) - (4x + 3) = 4x + 4h + 3 - 4x - 3 = 4h.$$

Then,

$$\frac{f(x+h) - f(x)}{h} = \frac{4h}{h} = 4, \quad h \neq 0.$$

Note that for a linear function, the average rate of change is a constant. On the graph, it is obvious that for a line, all the secant lines occur as the same as the line itself, hence they share the same slope.

Example 7: For $y = f(x) = \frac{1}{x}$, find a simplified form of the difference quotient $\dfrac{f(x+h) - f(x)}{h}$.

Solution For $f(x) = \frac{1}{x}$, we have $f(x+h) = \dfrac{1}{x+h}$. Therefore,

$$\begin{aligned}
f(x+h) - f(x) &= \frac{1}{x+h} - \frac{1}{x} = \frac{1}{x+h} \cdot \frac{x}{x} - \frac{1}{x} \cdot \frac{x+h}{x+h} \\
&= \frac{x - (x+h)}{x(x+h)} = \frac{x - x - h}{x(x+h)} = \frac{-h}{x(x+h)}.
\end{aligned}$$

Then, (note that dividing by h is equivalent to multiplying by $\frac{1}{h}$)

$$\frac{f(x+h) - f(x)}{h} = \frac{-h}{x(x+h)} \cdot \frac{1}{h} = \frac{-1}{x(x+h)}, \quad h \neq 0.$$

Before we leave this section, let us review the means of the average rate of change in some applications.

Example 8: A car starts traveling at 8 a.m. with the odometer reading as 23,822 miles. It stops at 10 a.m. with the odometer reading as 23,914 miles. Let $t = 0$ correspond to 8 a.m. and $t = 2$ correspond 10 a.m. Then the odometer reading is a function of t. We have $s(0) = 23,822$ and $s(2) = 23,914$.

(a) Find $s(2) - s(0)$. What does this quantity represent?

(b) Find $\dfrac{s(2) - s(0)}{2 - 0}$. What does this quantity represent?

Solution (a) We have $s(2) - s(0) = 23,914 - 23,822 = 92$ miles. It represents the distance the car travelled during the time from 8 a.m. to 10 a.m.

(b) We have $\dfrac{s(2) - s(0)}{2 - 0} = \dfrac{92}{2} = 46$ mph. This average rate of change represents the average speed of the car traveling from 8 a.m. to 10 a.m.

Example 9: Mary invests \$3,000 in a savings account that pays interest at an annual rate of 5%, compounded quarterly. Her balance after t years is given by

$$A(t) = 3{,}000(1 + 0.05/4)^{4t} = 3{,}000(1.0125)^{4t}.$$

(a) Find $A(2) - A(0)$. What does this quantity represent?

(b) Find $\dfrac{A(2) - A(0)}{2 - 0}$. What does this quantity represent?

(c) Find $A(4) - A(2)$. What does this quantity represent?

(d) Find $\dfrac{A(4) - A(2)}{4 - 2}$. What does this quantity represent?

Solution (a) We have $A(2) - A(0) = 3{,}000(1.0125)^{4(2)} - 3{,}000(1.0125)^{4(0)} = 3{,}000(1.0125)^8 - 3{,}000 \approx 3{,}313.46 - 3{,}000 = \$313.46$. It represents the increase in, or the interest earned on, Mary's investment during the first two years.

(b) We have $\dfrac{A(2) - A(0)}{2 - 0} \approx \dfrac{313.46}{2} = 156.73$ \$/year. This average rate of change represents the average annual interest earned by Mary's investment during the first two years.

(c) We have $A(4) - A(2) = 3{,}000(1.0125)^{4(4)} - 3{,}000(1.0125)^{4(2)} = 3{,}000(1.0125)^{16} - 3{,}000(1.0125)^8 \approx 3{,}659.67 - 3{,}313.46 = \$346.21$. It represents the increase in, or the interest earned on, Mary's investment during the third and forth years.

(d) We have $\dfrac{A(4) - A(2)}{4 - 2} \approx \dfrac{346.21}{2} = 173.105$ \$/year. This average rate of change represents the average annual interest earned by Mary's investment during the third and fourth years.

Example 10: If a rock is dropped from the top of a 144-ft-tall building, then the height of the rock, in feet, at t seconds can be modeled by

$$s(t) = 144 - 16t^2 \quad 0 \leq t \leq 3.$$

(a) Find $s(2) - s(0)$. What does this quantity represent?

(b) Find $\dfrac{s(2) - s(0)}{2 - 0}$. What does this quantity represent?

Solution (a) We have $s(0) = 144 - 16(0)^2 = 144 - 0 = 144$ ft, and $s(2) = 144 - 16(2)^2 = 80$ ft. Hence, $s(2) - s(0) = 80 - 144 = -64$ ft. It represents the **displacement** of the rock during the first 2 seconds. That is, during the first 2 seconds after the rock is dropped, it falls by 64 ft.

(b) We have $\dfrac{s(2) - s(0)}{2 - 0} = \dfrac{-64}{2} = -32$ ft/sec. This average rate of change represents the **average velocity** of the rock during the first 2 seconds. That is, during the first 2 seconds after the rock is dropped, the average speed at which the rock is falling is 32 ft/sec.

Exercise Set 2.3

For the functions in each of Exercises 1–18, do the following:

(a) Find a simplified form of the difference quotient $\frac{f(x+h)-f(x)}{h}$.

(b) Use the simplified form to compute the difference quotient for the given x and h.

1. $f(x) = 3x^2$, $x = 3$,
   $h =$ (i) 1 (ii) 0.2 (iii) 0.1 (iv) 0.01.

2. $f(x) = -2x^2$, $x = 2$,
   $h =$ (i) 1 (ii) 0.2 (iii) 0.1 (iv) 0.01.

3. $f(x) = x^2 - x$, $x = 1$,
   $h =$ (i) 1 (ii) 0.2 (iii) 0.1 (iv) 0.01.

4. $f(x) = 2x^2 + 3x$, $x = 1$,
   $h =$ (i) 1 (ii) 0.2 (iii) 0.1 (iv) 0.01.

5. $f(x) = 5x + 2$, $x = 3$,
   $h =$ (i) 1 (ii) 0.2 (iii) 0.1 (iv) 0.01.

6. $f(x) = 2x - 5$, $x = 1$,
   $h =$ (i) 1 (ii) 0.2 (iii) 0.1 (iv) 0.01.

7. $f(x) = \frac{3}{x}$, $x = 1$,
   $h =$ (i) 0.5 (ii) 0.1 (iii) 0.01 (iv) 0.001.

8. $f(x) = \frac{5}{x}$, $x = 2$,
   $h =$ (i) 0.5 (ii) 0.1 (iii) 0.01 (iv) 0.001.

9. $f(x) = \frac{2}{x^2}$, $x = 2$,
   $h =$ (i) 0.5 (ii) 0.1 (iii) 0.01 (iv) 0.001.

10. $f(x) = \frac{1}{x^2}$, $x = 2$,
    $h =$ (i) 0.5 (ii) 0.1 (iii) 0.01 (iv) 0.001.

11. $f(x) = 4$, $x = 2$,
    $h =$ (i) 1 (ii) 0.2 (iii) 0.1 (iv) 0.01.

12. $f(x) = -2$, $x = 1$,
    $h =$ (i) 1 (ii) 0.2 (iii) 0.1 (iv) 0.01.

13. $f(x) = 2 - 2x^2$, $x = 1$,
    $h =$ (i) 0.5 (ii) 0.1 (iii) 0.01 (iv) 0.001.

14. $f(x) = 4 - x^2$, $x = 2$,
    $h =$ (i) 0.5 (ii) 0.1 (iii) 0.01 (iv) 0.001.

15. $f(x) = -x^3$, $x = 2$,
    $h =$ (i) 1 (ii) 0.2 (iii) 0.1 (iv) 0.01.

16. $f(x) = 2x^3$, $x = 1$,
    $h =$ (i) 1 (ii) 0.2 (iii) 0.1 (iv) 0.01.

17. $f(x) = x^2 - 2x + 3$, $x = 5$,
    $h =$ (i) 1 (ii) 0.2 (iii) 0.1 (iv) 0.01.

18. $f(x) = x^2 + 3x - 1$, $x = 3$,
    $h =$ (i) 1 (ii) 0.2 (iii) 0.1 (iv) 0.01.

APPLICATIONS

19. If a rock is dropped from the top of a 256-ft-tall building, then the height of the rock, in feet, at t seconds can be modeled by
    $$s(t) = 256 - 16t^2 \quad 0 \leq t \leq 4.$$
    (a) Find $s(3) - s(1)$.
    What does this quantity represent?
    (b) Find $\frac{s(3) - s(1)}{3 - 1}$.
    What does this quantity represent?

20. If a rock is dropped from the top of a 196-ft-tall building, then the height of the rock, in feet, at t seconds can be modeled by
    $$s(t) = 196 - 16t^2 \quad 0 \leq t \leq 3.5.$$
    (a) Find $s(2) - s(1)$.
    What does this quantity represent?
    (b) Find $\frac{s(2) - s(1)}{2 - 1}$.
    What does this quantity represent?

21. If a rock is thrown from the ground with an initial velocity of 80 ft/sec, then the height of the rock, in feet, at t seconds can be modeled by
    $$s(t) = -16t^2 + 80t \quad 0 \leq t \leq 5.$$
    (a) Find $s(2) - s(0)$.
    What does this quantity represent?
    (b) Find $\frac{s(2) - s(0)}{2 - 0}$.
    What does this quantity represent?

22. John invests $2,000 in a savings account that pays interest at an annual rate of 4.5%, compounded monthly. His balance after t years is given by
$$A(t) = 2{,}000(1.00375)^{12t}.$$
   (a) Find $A(1) - A(0)$.
       What does this quantity represent?
   (b) Find $\dfrac{A(1) - A(0)}{1 - 0}$.
       What does this quantity represent?
   (c) Find $A(2) - A(1)$.
       What does this quantity represent?
   (d) Find $\dfrac{A(2) - A(1)}{2 - 1}$.
       What does this quantity represent?
   (e) Compare the answers obtained in parts (b) and (d). Which is greater? Why?

23. Rich invests $4,000 in a savings account that pays interest at an annual rate of 4%, compounded continuously. His balance after t years is given by
$$A(t) = 4{,}000 e^{0.04t}.$$
   (a) Find $A(1) - A(0)$.
       What does this quantity represent?
   (b) Find $\dfrac{A(1) - A(0)}{1 - 0}$.
       What does this quantity represent?
   (c) Find $A(2) - A(0)$.
       What does this quantity represent?
   (d) Find $\dfrac{A(2) - A(0)}{2 - 0}$.
       What does this quantity represent?
   (e) Compare the answers obtained in parts (b) and (d). Which is greater? Why?

24. Jan has a $2,000 balance on her credit card that charges interest at an annual rate of 16%, compounded quarterly. Her monthly payment is waived for 3 years. Her balance after t years is given by
$$A(t) = 2{,}000(1.04)^{4t} \quad 0 \le t \le 3.$$
   (a) Find $A(3) - A(0)$.
       What does this quantity represent?
   (b) Find $\dfrac{A(3) - A(0)}{3 - 0}$.
       What does this quantity represent?
   (c) Find $A(3) - A(2)$.
       What does this quantity represent?
   (d) Find $\dfrac{A(3) - A(2)}{3 - 2}$.
       What does this quantity represent?
   (e) Compare the answers obtained in parts (b) and (d). Which is greater? Why?

25. The number of unbreakable sunglasses sold at a specialty store during a sales promotion can be modeled by
$$g(x) = -2x^2 + 60x \quad 0 \le x \le 10.$$
where x represents the number of weeks since the promotion began and $g(x)$ represents the number of sunglasses sold.
   (a) Find $g(2) - g(0)$.
       What does this quantity represent?
   (b) Find $\dfrac{g(2) - g(0)}{2 - 0}$.
       What does this quantity represent?
   (c) Find $g(10) - g(8)$.
       What does this quantity represent?
   (d) Find $\dfrac{g(10) - g(8)}{10 - 8}$.
       What does this quantity represent?
   (e) Compare the answers obtained in parts (b) and (d). Which is greater? Why?

26. The number of a certain brand of skateboard sold at a sporting goods store during a sales promotion can be modeled by
$$S(x) = -x^2 + 60x \quad 0 \le x \le 30.$$
where x represents the number of days since the promotion began and $S(x)$ represents the number of skateboards sold.
   (a) Find $S(10) - S(0)$.
       What does this quantity represent?
   (b) Find $\dfrac{S(10) - S(0)}{10 - 0}$.
       What does this quantity represent?
   (c) Find $S(30) - g(20)$.
       What does this quantity represent?
   (d) Find $\dfrac{S(20) - S(10)}{20 - 10}$.
       What does this quantity represent?
   (e) Compare the answers obtained in parts (b) and (d). Which is greater? Why?

27. Suppose that a shoe manufacturer determines that the daily cost, in dollars, of producing x pairs of a certain brand of shoes is given by

$$C(x) = -0.02x^2 + 40x + 200 \quad 0 \leq x \leq 250.$$

Find $\dfrac{C(101) - C(100)}{101 - 100}$, and explain to the company what this quantity represents.

28. Suppose that a refrigerator manufacturer determines that its weekly cost, in dollars, of producing x refrigerators is given by

$$C(x) = 2x^2 + 15x + 1500 \quad 0 \leq x \leq 200.$$

Find $\dfrac{C(101) - C(100)}{101 - 100}$, and explain to the company what this quantity represents.

29. Suppose that a specialty store owner determines that the revenue, in dollars, from selling x pairs of its unbreakable sunglasses is given by

$$R(x) = -0.1x^2 + 100x \quad 0 \leq x \leq 200.$$

Find $\dfrac{R(101) - R(100)}{101 - 100}$, and explain to the store owner what this quantity represents.

30. Suppose that a sporting goods store manager determines that the revenue, in dollars, from selling x of a certain brand of skateboard is given by

$$R(x) = -0.02x^2 + 100x \quad 0 \leq x \leq 100.$$

Find $\dfrac{R(51) - R(50)}{51 - 50}$, and explain to the store manager what this quantity represents.

2.4 Differentiation Using Limits of Difference Quotients

The formal definition of the **derivative** of a function is introduced in this section. We will use the definition to calculate the derivatives, and explore the meanings of the derivative at a certain variable value.

Definition of the Derivative

In Section 2.3, we studied the average rate of change of a function $y = f(x)$ in the form of a difference quotient

$$\frac{f(x+h) - f(x)}{h}.$$

From the examples in that section, we observe that when h approaches zero, the simplified form of the difference quotient often has a limit. We now formally define such a limit as the **derivative** of the function at x, and denote it as $f'(x)$, read as "f prime of x." Note that the derivative itself is a function of x.

> **Definition**
>
> For a function $y = f(x)$, its **derivative** at x, denoted as $f'(x)$, is defined by
>
> $$f'(x) = \lim_{h \to 0} \frac{f(x+h) - f(x)}{h},$$
>
> if such a limit exists. If $f'(x)$ exists, then we say that f is **differentiable** at x. We call f' the **derivative function**.

Recall from the examples in the previous section, the limit is much easier to calculate when the difference quotient is simplified first. Therefore we perform the following three steps when calculating the derivative of a given function $y = f(x)$:

Three steps in calculating a derivative

1. Write down the difference quotient, $\dfrac{f(x+h) - f(x)}{h}$.

2. Simplify the difference quotient so that the denominator no longer approaches 0 when h approaches 0.

3. Find the limit as h approaches 0.

Example 1: For $y = f(x) = 2x^2$,

(a) find $f'(x)$; then

(b) find $f'(1)$, and

(c) find $f'(-2)$.

Solution (a) In Example 4 of Section 2.3, we showed that the difference quotient can be simplified to $4x + 2h$. We now restate the process in the fashion of the three steps:

Step 1: $\dfrac{f(x+h) - f(x)}{h} = \dfrac{2(x+h)^2 - 2x^2}{h}$.

Step 2: $\dfrac{f(x+h) - f(x)}{h} = \dfrac{(2x^2 + 4xh + 2h^2) - 2x^2}{h}$

$= \dfrac{4xh + 2h^2}{h} = \dfrac{h(4x + 2h)}{h}$

$= 4x + 2h, \quad h \neq 0.$

Step 3: Now we take the limit of the simplified difference quotient as h approaches 0:

$f'(x) = \lim\limits_{h \to 0} \dfrac{f(x+h) - f(x)}{h} = \lim\limits_{h \to 0} (4x + 2h) = 4x.$

Therefore, $f'(x) = 4x$.

(b) We now substitute $x = 1$ into $f'(x) = 4x$ to obtain $f'(1) = 4(1) = 4$.

(c) Similarly, $f'(-2) = 4(-2) = -8$.

Example 2: For $y = f(x) = x^3$,

(a) find $f'(x)$; then

(b) find $f'(-0.5)$, and

(c) find $f'(0.2)$.

Solution (a) In Example 5 of Section 2.3, we showed that the difference quotient can be simplified to $3x^2 + 3xh + h^2$. We now proceed to **Step 3** directly:

2.4. DIFFERENTIATION USING LIMITS OF DIFFERENCE QUOTIENTS

Step 3: We take the limit of the simplified difference quotient as h approaches 0:

$$f'(x) = \lim_{h \to 0} \frac{f(x+h) - f(x)}{h} = \lim_{h \to 0} (3x^2 + 3xh + h^2) = 3x^2.$$

Therefore, $f'(x) = 3x^2$.

(b) We now substitute $x = -0.5$ into $f'(x) = 3x^2$ to obtain $f'(-0.5) = 3(-0.5)^2 = 3(0.25) = 0.75$.

(c) Similarly, $f'(0.2) = 3(0.2)^2 = 3(0.04) = 0.12$.

Example 3: For $y = f(x) = 4x + 3$, find $f'(x)$ and $f'(5)$.

Solution In Example 6 of Section 2.3, we showed that the difference quotient can be simplified to 4, a constant. Hence,

$$f'(x) = \lim_{h \to 0} \frac{f(x+h) - f(x)}{h} = \lim_{h \to 0} 4 = 4.$$

Therefore, $f'(x) = 4$, and $f'(5) = 4$.

Example 4: For $y = f(x) = \dfrac{1}{x}$:

(a) Find a simplified form of the difference quotient $\dfrac{f(x+h) - f(x)}{h}$.

(b) Find $f'(x)$ by $\lim_{h \to 0} \dfrac{f(x+h) - f(x)}{h}$ using the simplified form obtained in (a).

(c) Compute $f'(a)$ for $a =$ (i) 3, (ii) -1, (iii) -2.

Solution (a) In Example 7 of Section 2.3, we showed that the difference quotient can be simplified to

$$\frac{f(x+h) - f(x)}{h} = \frac{-1}{x(x+h)}.$$

(b) We take the limit of the simplified difference quotient as h approaches 0:

$$f'(x) = \lim_{h \to 0} \frac{f(x+h) - f(x)}{h} = \lim_{h \to 0} \frac{-1}{x(x+h)} = \frac{-1}{x \cdot x} = \frac{-1}{x^2}.$$

Therefore, $f'(x) = \dfrac{-1}{x^2}$.

(c(i)) We now substitute $x = 3$ into $f'(x) = \dfrac{-1}{x^2}$ to obtain

$$f'(3) = \frac{-1}{(3)^2} = \frac{-1}{9} = -\frac{1}{9}.$$

(c(ii)) Similarly, $f'(-1) = \dfrac{-1}{(-1)^2} = \dfrac{-1}{1} = -1.$

(c(iii)) Similarly, $f'(-2) = \dfrac{-1}{(-2)^2} = \dfrac{-1}{4} = -\dfrac{1}{4}.$

So far we have learned how to calculate a derivative by its definition. We realize that the process is quite tedious. In the next four sections, we learn how to calculate derivatives using various rules. In the remaining part of this section we focus on understanding what a derivative represents, and when a derivative does not exist.

Tangent Lines and Meanings of Derivatives

Recall that on the graph of $f(x)$, the difference quotient represents the slope of the secant line passing through the two points $(x, f(x))$ and $(x + h, f(x + h))$. When h approaches 0, the second point moves toward the first, and the secant lines are approaching the **tangent line**. Therefore, the derivative $f'(x)$, as the limit of the slope of the secant lines, represents the slope of the tangent line passing through $(x, f(x))$ (Figure 2.4.1).

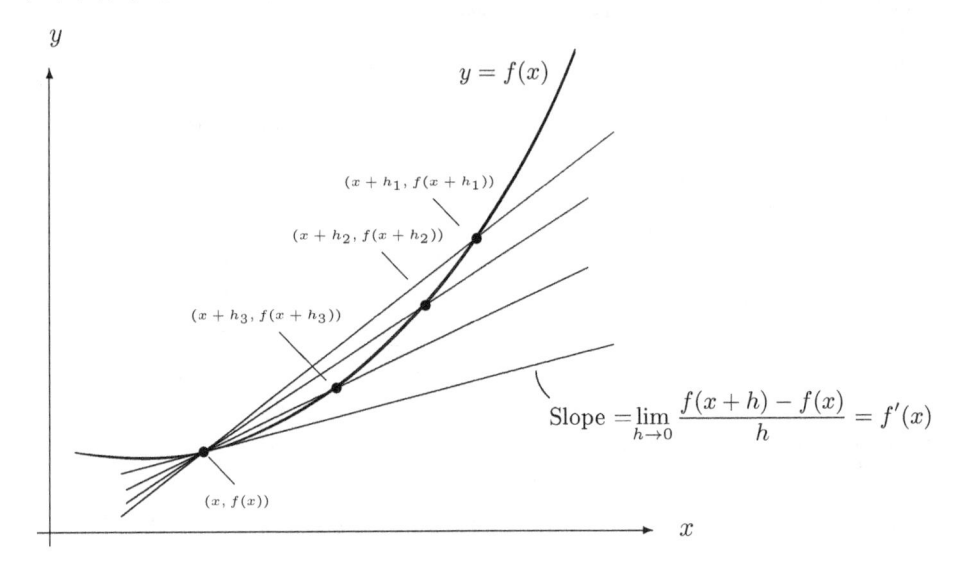

Figure 2.4.1: Derivative as the slope of a tangent line

Example 5: Find the equation of the tangent line to the graph of $f(x) = 2x^2$ at $x = 1$.

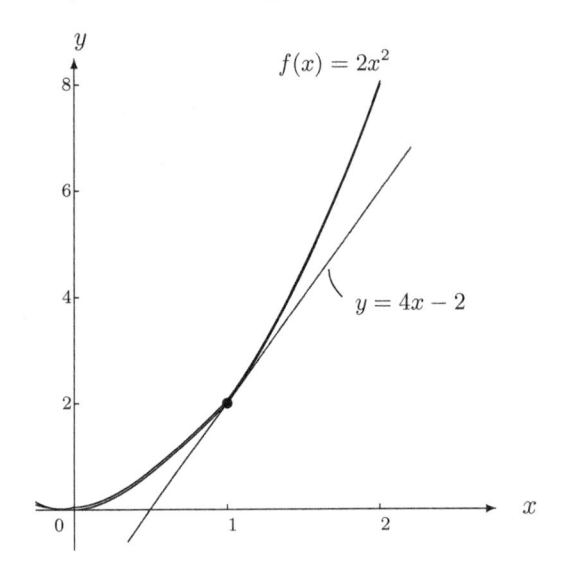

Figure 2.4.2: Graph of $f(x) = 2x^2$ with tangent line at $x = 1$

Solution In Example 1, we already obtained that $f'(1) = 4$. When $x = 1$, we have $f(1) = 2$. Hence the tangent line has slope 4 and passes the point $(1, 2)$. Using the slope-point equation, we have

$$y - 2 = 4(x - 1).$$

2.4. DIFFERENTIATION USING LIMITS OF DIFFERENCE QUOTIENTS

Rewriting this equation in the slope-intercept form we obtain $y = 4x - 2$ (Figure 2.4.2).

Recall that the difference quotient represents the average rate of change. Hence, by taking the limit of the difference quotient as h, the change in the input variable, approaches 0, the result (i.e., the derivative) represents the **instantaneous rate of change**.

Recall that in Section 2.3, Example 5, for the height function s, the difference quotient represents the average velocity of the rock. Therefore, the derivative represents the **instantaneous velocity**, or simply the velocity, that is, $v(t) = s'(t)$.

Example 6: If a rock is dropped from the top of a 144-ft-tall building, then the height of the rock, in feet, at t seconds can be modeled by

$$s(t) = 144 - 16t^2 \quad 0 \leq t \leq 3.$$

(a) Find $v(t)$.

(b) Find $v(2)$. What does this quantity represent?

Solution (a) We apply the three steps to find $v(t) = s'(t)$:

Step 1: $\dfrac{s(t+h) - s(t)}{h} = \dfrac{(144 - 16(t+h)^2) - (144 - 16t^2)}{h}.$

Step 2: $\dfrac{s(t+h) - s(t)}{h} = \dfrac{[144 - 16(t^2 + 2th + h^2)] - [144 - 16t^2]}{h}$

$= \dfrac{[144 - 16t^2 - 32th - 16h^2] - [144 - 16t^2]}{h}$

$= \dfrac{144 - 16t^2 - 32th - 16h^2 - 144 + 16t^2}{h}$

$= \dfrac{-32th - 16h^2}{h} = \dfrac{h(-32t - 16h)}{h}$

$= -32t - 16h, \quad h \neq 0.$

Step 3: Now, we take the limit of the simplified difference quotient as h approaches 0:

$$s'(t) = \lim_{h \to 0} \dfrac{s(t+h) - s(t)}{h} = \lim_{h \to 0} (-32t - 16h) = -32t.$$

Therefore, $v(t) = s'(t) = -32t$.

(b) We have $v(2) = -64$ ft/sec. This quantity represents the velocity of the rock at 2 seconds. That is, at 2 seconds after the rock is dropped, it is falling with a speed of 64 ft/sec.

▬ Nonexistence of the Derivative ▬

Recall that in the definition of a derivative, we say that f is differentiable at x if $f'(x)$ exists. Here we study the situations at which f is *not* differentiable, that is, the derivative does not exist.

By the definition of the derivative, the existence of a derivative at $x = a$ depends on the existence of the limit of the difference quotient at $x = a$. That is,

$$f'(a) = \lim_{h \to 0} \dfrac{f(a+h) - f(a)}{h}.$$

If the limit does not exist, then we say that f is **nondifferentiable** at $x = a$, or $f'(a)$ does not exist.

Since in the difference quotient, the denominator approaches 0 when h approaches 0, we conclude that for the limit to exist, the numerator must approach 0, too. Notice that,

$$\lim_{h \to 0} [f(a+h) - f(a)] = 0$$

is equivalent to
$$\lim_{h \to 0} f(a+h) = f(a).$$
Recall that in Section 2.2 this equation means that f is continuous at $x = a$. Therefore, we conclude: For the derivative to exist, the function must be continuous at $x = a$, or equivalently, if the function is not continuous at $x = a$, then the function is nondifferentiable at $x = a$.

Even when the function is continuous at $x = a$, the derivative may still not exist. One such example is for the function $f(x) = |x|$ at $x = 0$. Observe that the graph of $f(x) = |x|$ has a **corner** at $x = 0$ (Figure 2.4.3).

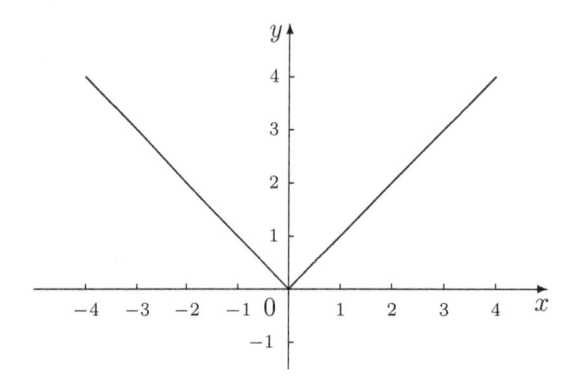

Figure 2.4.3: $y = |x|$

Example 7: Show that $f(x) = |x|$ is nondifferentiable at $x = 0$.

Solution We inspect the limit
$$\lim_{h \to 0} \frac{f(0+h) - f(0)}{h}.$$

Note that the absolute value function can be rewritten as a piecewise function:
$$f(x) = \begin{cases} -x & \text{if } x < 0, \\ x & \text{if } x \geq 0. \end{cases}$$

Therefore, $f(0) = 0$ and $f(h) = -h$ if $h < 0$; and $f(h) = h$ if $h > 0$. Now for the difference quotient of f we have
$$\frac{f(h) - f(0)}{h} = \begin{cases} \dfrac{-h}{h} = -1 & \text{if } h < 0, \\ \dfrac{h}{h} = 1 & \text{if } h > 0. \end{cases}$$

Since $h = 0$ is a "joint," we must take the left-hand limit and the right-hand limit using different formulas:
$$\lim_{h \to 0^-} \frac{f(h) - f(0)}{h} = \lim_{h \to 0^-} \frac{-h}{h} = \lim_{h \to 0^-} -1 = -1,$$

and
$$\lim_{h \to 0^+} \frac{f(h) - f(0)}{h} = \lim_{h \to 0^+} \frac{h}{h} = \lim_{h \to 0^+} 1 = 1.$$

Since the left-hand limit and the right-hand limit differ, we conclude that the derivative

$$f'(0) = \lim_{h \to 0} \frac{f(0+h) - f(0)}{h}$$

does not exist, or f is nondifferentiable at $x = 0$.

Recall that on the graph of a function, the derivative $f'(a)$ represents the slope of the tangent line at $x = a$. If the tangent line is **vertical**, then its slope does not exist. Hence, the function is nondifferentiable at $x = a$ (Figure 2.4.4, case IV).

We now summarize the conditions under which f is nondifferentiable at $x = a$ (Figure 2.4.4).

A function is nondifferentiable at $x = a$ if

1. The function is not continuous at $x = a$, that is, the graph of the function has a break at $x = a$;

2. The graph of the function has a corner at $x = a$;

3. The graph of the function has a cusp at $x = a$;

4. The tangent line of the function at $x = a$ is vertical.

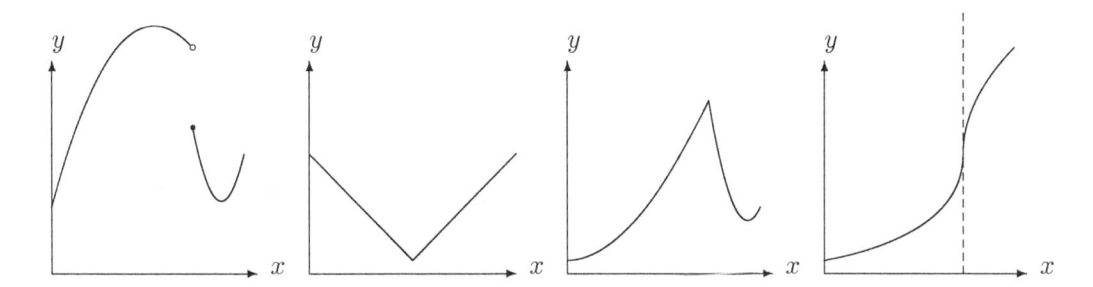

Figure 2.4.4: Four cases at which the derivative does not exist

Exercise Set 2.4

For the functions in each of the Exercises 1–8, do the following:

(a) *Find a simplified form of the difference quotient $\frac{f(x+h)-f(x)}{h}$.*

(b) *Find $f'(x)$ by $\lim_{h \to 0} \frac{f(x+h)-f(x)}{h}$ using the simplified form obtained in (a).*

(c) *Compute $f'(a)$ for the given values of a.*

1. $f(x) = 3x^2$

    $a = $ (i) -1 (ii) 0 (iii) 1 (iv) 2

2. $f(x) = -2x^2$,

    $a = $ (i) -1 (ii) 0 (iii) 1 (iv) 2

3. $f(x) = x^2 - x$

    $a = $ (i) -2 (ii) 0 (iii) 0.5 (iv) 1

4. $f(x) = 2x^2 + 3x$

    $a = $ (i) -2 (ii) 0 (iii) 0.5 (iv) 1

5. $f(x) = 5x + 2$

    $a = $ (i) -1 (ii) 0 (iii) 3

6. $f(x) = 2x - 5$

    $a = $ (i) -1 (ii) 0 (iii) 3

7. $f(x) = \dfrac{3}{x}$

    $a = $ (i) -1 (ii) 1 (iii) 3

8. $f(x) = \dfrac{5}{x}$

   $a =$ (i) -1 (ii) 1 (iii) 5

For the functions in each of the Exercises 9–18, do the following:

(a) *Find $f'(x)$ by $\lim\limits_{h \to 0} \dfrac{f(x+h)-f(x)}{h}$.*

(b) *Compute $f'(a)$ for the given values of a.*

(c) *Find the equation of the tangent line at points $(a, f(a))$ for the given values of a.*

(d) *Graph the function and draw tangent lines at these points to confirm that the slopes match the derivatives obtained in part (b).*

9. $f(x) = \dfrac{-2}{x}$

   $a =$ (i) -1 (ii) 2

10. $f(x) = \dfrac{1}{x}$

    $a =$ (i) -1 (ii) 2

11. $f(x) = -x^2$,

    $a =$ (i) -2 (ii) 0 (iii) 1

12. $f(x) = \dfrac{1}{2}x^2$

    $a =$ (i) -1 (ii) 0 (iii) 2

13. $f(x) = 2 - 2x^2$

    $a =$ (i) -1 (ii) 0 (iii) 2

14. $f(x) = 4 - x^2$

    $a =$ (i) -2 (ii) 0 (iii) 1

15. $f(x) = -x^3$

    $a =$ (i) -1 (ii) 0 (iii) 1

16. $f(x) = 2x^3$

    $a =$ (i) -1 (ii) 0 (iii) 1

17. $f(x) = x^2 - 2x + 3$

    $a =$ (i) -2 (ii) 0 (iii) 1

18. $f(x) = x^2 + 3x - 1$

    $a =$ (i) -1 (ii) 0 (iii) 1

For Exercises 19 and 20, list the points in the graph at which each function is not differentiable, and classify each such point as discontinuity, corner, or vertical tangent line.

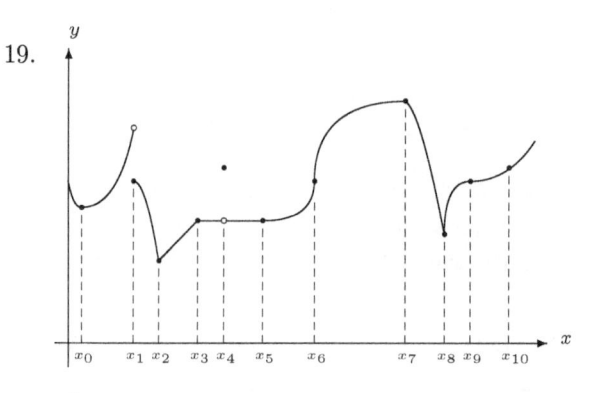

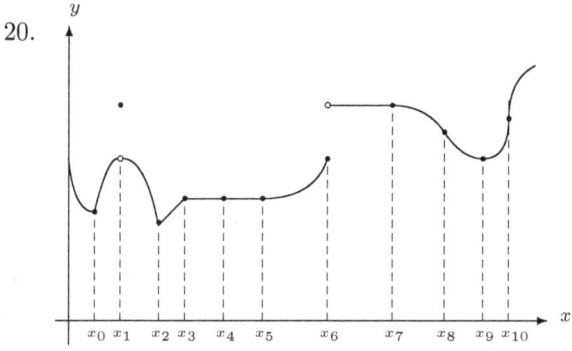

19.

20.

APPLICATIONS

21. If a rock is dropped from the top of a 256-ft-tall building, then the height of the rock, in feet, at t seconds can be modeled by

    $$s(t) = 256 - 16t^2 \quad 0 \le t \le 4.$$

    (a) Find $v(t)$.

    (b) Find $v(3)$. What does this quantity represent?

22. If a rock is dropped from the top of a 196-ft-tall building, then the height of the rock, in feet, at t seconds can be modeled by

    $$s(t) = 196 - 16t^2 \quad 0 \le t \le 3.5.$$

    (a) Find $v(t)$.

    (b) Find $v(2)$. What does this quantity represent?

23. If a rock is thrown from the ground with an initial velocity of 80 ft/sec, then the height of the rock, in feet, at t seconds can be modeled by

    $$s(t) = -16t^2 + 80t \quad 0 \le t \le 5.$$

    (a) Find $v(t)$.

    (b) Find $v(3)$. What does this quantity represent?

(c) At $t = 3$ seconds, is the rock rising up or falling down?

24. If a rock is thrown from the ground with an initial velocity of 120 ft/sec, then the height of the rock, in feet, at t seconds can be modeled by

$$s(t) = -16t^2 + 120t \quad 0 \le t \le 7.5.$$

(a) Find $v(t)$.

(b) Find $v(3)$. What does this quantity represent?

(c) At $t = 3$ seconds, is the rock rising up or falling down?

2.5 Basic Derivative Formulas and Properties

Some basic rules and properties of differentiation are introduced in this section. With the help of these rules and properties, we will be able to find the derivatives of a broad group of function directly. We start this section by learning some new notations for the derivatives.

▬ Notations for Derivatives

In computing derivatives of functions, several versions of notations, in addition to $f'(x)$, are often used. We introduce these notations here.

When y is a function of x, that is, $y = f(x)$, its derivative, $f'(x)$, can be denoted as

$$\frac{dy}{dx}.$$

This notation of a derivative is called the **Leibniz notation**, since it was invented by the German mathematician Gottfried Leibniz. It is read as "the derivative of y with respect to x."

Note that it does not mean dy divided by dx. However, recalling that a derivative does represent a ratio or a slope, this notation indeed reflects the features of a derivative.

If $y = x^2$, then we can denote its derivative by

$$f'(x) = \frac{dy}{dx} = 2x.$$

We can also write

$$\frac{d}{dx} f(x) = \frac{d}{dx} x^2 = 2x.$$

The value of the derivative at $x = 3$ can then be denoted as either

$$f'(3) \quad \text{or} \quad \left.\frac{dy}{dx}\right|_{x=3}.$$

Hence if $\frac{dy}{dx} = 2x$, then

$$\left.\frac{dy}{dx}\right|_{x=3} = 2x \Big| x = 3 = 2 \cdot 3 = 6.$$

Often, instead of using $f'(x)$ or $\frac{dy}{dx}$, we can use y' (read as "y prime"). For example, if $y = x^2$, then

$$y' = 2x.$$

Yet another notation, $D_x y$, is sometimes used for derivatives. Each of these symbols for the derivatives has its particular advantage in certain situation. These symbols will become familiar after some experience.

> **Warning**   *Derivative notation*
>
> While
> $$y',\ f'(x),\ \frac{dy}{dx},\ \frac{d}{dx}f(x),\ \frac{df}{dx},\ D_x f,\ \text{and}\ D_x y$$
> are all acceptable notations for the derivative, the notation
> $$\frac{dy}{dx}f(x)$$
> is not. Strictly speaking, this last notation means the derivative of the function times the function. So, for example,
> $$\frac{d}{dx}x^2 = 2x$$
> whereas
> $$\frac{dy}{dx}x^2 \neq 2x.$$

Now we are ready to learn some basic differentiation rules and properties. We start with the derivative of a constant function.

The Derivative of a Constant Function

If $y = f(x) = C$, where C is a constant, then we have $f(x+h) = C$ and $f(x+h) - f(x) = C - C = 0$. Therefore, the difference quotient has the numerator equal 0 for any $h \neq 0$, and the limit of the difference quotient, as h approaches 0, is also 0 by Limit Property L1. That is,

$$f'(x) = \lim_{h \to 0} \frac{f(x+h) - f(x)}{h} = \lim_{h \to 0} \frac{C - C}{h} = \lim_{h \to 0} \frac{0}{h} = \lim_{h \to 0} 0 = 0.$$

Therefore, $\dfrac{d}{dx} C = 0$.

It is also obvious from the graph of the function (Figure 2.5.1): The graph is a horizontal line, hence the slope is 0 for all the x values.

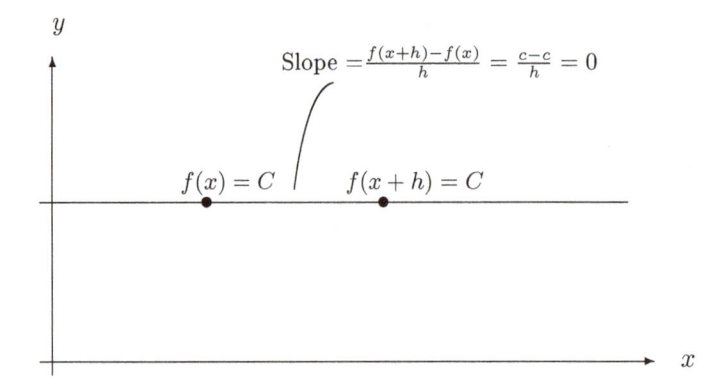

Figure 2.5.1: Slope of a horizontal line

This property is stated in the following theorem.

2.5. BASIC DERIVATIVE FORMULAS AND PROPERTIES

> **Theorem 1** *The Derivative of a Constant Function*
>
> The derivative of a constant function is 0. That is,
>
> $$\frac{d}{dx}C = 0,$$
>
> or equivalently, if $y = f(x) = C$, then
>
> $$y' = f'(x) = \frac{dy}{dx} = 0.$$

Example 1: Differentiate each of the following:

(a) $y = -4$;   (b) $y = \frac{1}{3}$;   (c) $y = \sqrt{\pi}$;   (d) $y = e^{\frac{5}{3}}$.

Solution Since each of the four functions is a constant function, the derivative is 0, that is,

(a) $\dfrac{d}{dx}(-4) = 0$ (b) $\dfrac{d}{dx}\left(\dfrac{1}{3}\right) = 0$

(c) $\dfrac{d}{dx}\left(\sqrt{\pi}\right) = 0$ (d) $\dfrac{d}{dx}\left(e^{\frac{5}{3}}\right) = 0$

The Power Rule

In Section 2.4, we obtained the derivatives of a few functions. Note that most of those functions are power functions. We summarize these results in Table 2.5.1 and look for a pattern. The table illustrates the pattern of the derivative of a power function $y = f(x) = x^k$, where k=−1, −2, 1, 2, and 3: The power of the function becomes the coefficient of its derivative, while the power of the derivative is one lower than the power of the function. This same pattern works for any real power k, and the rule is called the **Power Rule**. This rule is stated in the Theorem 2.

Function	Derivative
x^3	$3x^2$
x^2	$2x^1 = 2x$
$x = x^1$	$1 \cdot x^0 = 1$
$\dfrac{1}{x} = x^{-1}$	$-1 \cdot x^{-2} = \dfrac{-1}{x^2}$
$\dfrac{1}{x^2} = x^{-2}$	$-2 \cdot x^{-3} = \dfrac{-2}{x^3}$

Table 2.5.1: Derivatives of a few power functions

> **Theorem 2**   **The Power Rule**
>
> For any real number k,
>
> $$\frac{d}{dx} x^k = k \cdot x^{k-1},$$
>
> or equivalently, if $y = f(x) = x^k$, then
>
> $$y' = f'(x) = \frac{dy}{dx} = k \cdot x^{k-1}.$$

Note that the power, k, can be any real number. To differentiate a power function x^k, we write the power k as the coefficient of the derivative, then write x with the power 1 less than k.

Example 2: Differentiate each of the following:

(a) $y = x^4$;   (b) $y = x^{-3}$;   (c) $y = x^{1/2}$;   (d) $y = x^\pi$.

Solution   We apply the Power Rule to each of the functions:

(a) $\dfrac{d}{dx} x^4 = 4 \cdot x^{4-1} = 4x^3$.

(b) $\dfrac{d}{dx} x^{-3} = -3 \cdot x^{-3-1} = -3x^{-4} = \dfrac{-3}{x^4}$.

(c) $\dfrac{d}{dx} x^{\frac{1}{2}} = \dfrac{1}{2} \cdot x^{\frac{1}{2}-1} = \dfrac{1}{2} x^{-\frac{1}{2}}$.

(d) $\dfrac{d}{dx} x^\pi = \pi \cdot x^{\pi-1}$.

Example 3: Find each of the following derivatives:

(a) $\dfrac{d}{dx} \dfrac{1}{x^4}$;   (b) $\dfrac{d}{dx} \dfrac{1}{\sqrt{x}}$;   (c) $\dfrac{d}{dx} \sqrt[3]{x^5}$.

Solution   We first express each function as a power form x^k, then apply the Power Rule:

(a) $\dfrac{d}{dx} \dfrac{1}{x^4} = \dfrac{d}{dx} x^{-4} = -4 \cdot x^{-4-1} = -4x^{-5}$,   or $\dfrac{-4}{x^5}$.

(b) $\dfrac{d}{dx} \dfrac{1}{\sqrt{x}} = \dfrac{d}{dx} x^{-\frac{1}{2}} = -\dfrac{1}{2} \cdot x^{-\frac{1}{2}-1} = -\dfrac{1}{2} x^{-\frac{3}{2}}$,   or $\dfrac{-1}{2\sqrt{x^3}}$.

(c) $\dfrac{d}{dx} \sqrt[3]{x^5} = \dfrac{d}{dx} x^{\frac{5}{3}} = \dfrac{5}{3} \cdot x^{\frac{5}{3}-1} = \dfrac{5}{3} x^{\frac{2}{3}}$,   or $\dfrac{5}{3} \sqrt[3]{x^2}$.

The Derivative of Natural Exponential Function $y = e^x$

In the process of finding the derivative of $y = f(x) = e^x$, we will use (without proof) the fact that

$$\lim_{h \to 0} \frac{e^h - 1}{h} = 1.$$

We illustrate this limit by Tables 2.5.2 and 2.5.3:

$h \to 0^-$ ($h < 0$)	$\frac{e^h - 1}{h}$
-0.1	0.9516
-0.01	0.9950
-0.001	0.9995

$h \to 0^+$ ($h > 0$)	$\frac{e^h - 1}{h}$
0.1	1.0517
0.01	1.0050
0.001	1.0005

Table 2.5.2: Limit of $\frac{e^h-1}{h}$ as $h \to 0^-$

Table 2.5.3: Limit of $\frac{e^h-1}{h}$ as $h \to 0^+$

We now apply the definition introduced in Section 2.4 to find the derivative of $f(x) = e^x$. We first reorganize the difference quotient. Note that $e^{x+h} = e^x \cdot e^h$:

$$\frac{f(x+h) - f(x)}{h} = \frac{e^{x+h} - e^x}{h} = \frac{e^x \cdot e^h - e^x}{h}$$
$$= \frac{e^x(e^h - 1)}{h} = e^x \cdot \frac{e^h - 1}{h}, \quad h \neq 0.$$

By definition, the derivative $f'(x)$ is the limit of the difference quotient as h approaches 0. Using Limit Property L3 and the fact that $\lim_{h \to 0} \frac{e^h - 1}{h} = 1$, we obtain

$$\begin{aligned} f'(x) &= \lim_{h \to 0} \frac{f(x+h) - f(x)}{h} = \lim_{h \to 0} e^x \cdot \frac{e^h - 1}{h} \\ &= e^x \lim_{h \to 0} \frac{e^h - 1}{h} = e^x \cdot 1 = e^x. \end{aligned}$$

Therefore, $\frac{d}{dx} e^x = e^x$, that is, the derivative of the natural exponential function is itself. In fact, the natural exponential function is the only function having such a character.

Theorem 3   *The Natural Exponential Rule*

The derivative of the natural exponential function e^x is e^x. That is,

$$\frac{d}{dx} e^x = e^x,$$

or equivalently, if $y = f(x) = e^x$, then

$$y' = f'(x) = \frac{dy}{dx} = e^x.$$

So far we have learned how to find the derivative of a function that is a constant, a power, or a natural exponential. We now consider functions with more general forms, such as polynomials. A polynomial can be expressed as a combination of the power functions. First we introduce some basic properties of differentiation, which will empower us to differentiate a broader class of functions.

The Constant Multiple Property

Let $F(x) = k\dot{f}(x)$, where $f'(x)$ exists. We now apply the definition introduced in Section 2.4 to find the derivative of $F(x)$. We first reorganize the difference quotient. Note that $F(x+h) = k \cdot f(x+h)$:

$$\frac{F(x+h) - F(x)}{h} = \frac{k \cdot f(x+h) - k \cdot f(x)}{h} = \frac{k \cdot [f(x+h) - f(x)]}{h}$$

$$= k \cdot \frac{f(x+h) - f(x)}{h}, \quad h \neq 0.$$

By definition, the derivative $F'(x)$ is the limit of the difference quotient as h approaches 0. Using the Limit Property L3 and the definition for the derivative of $f(x)$:

$$F'(x) = \lim_{h \to 0} \frac{F(x+h) - F(x)}{h} = \lim_{h \to 0} k \frac{f(x+h) - f(x)}{h}$$

$$= k \lim_{h \to 0} \frac{f(x+h) - f(x)}{h} = k \cdot f'(x).$$

Therefore,

$$\frac{d}{dx}[k \cdot f(x)] = k \cdot \frac{d}{dx} f(x).$$

Theorem 4 *Constant Multiple Property*

If $y = F(x) = k \cdot f(x)$, then

$$y' = F'(x) = k \cdot f'(x),$$

or equivalently,

$$\frac{d}{dx}[k \cdot f(x)] = k \cdot \frac{d}{dx} f(x).$$

Example 4: Find each of the following derivatives:

(a) $\dfrac{d}{dx} \dfrac{5x^3}{2}$; (b) $\dfrac{d}{dx} \dfrac{3}{x}$; (c) $\dfrac{d}{dx} 7e^x$; (d) $\dfrac{d}{dx} \dfrac{-1}{4\sqrt[3]{x}}$.

Solution We apply the rules introduced in Theorem 1 to Theorem 3, with the Constant Multiple Property:

(a) $\dfrac{d}{dx} \dfrac{5x^3}{2} = \dfrac{5}{2} \dfrac{d}{dx} x^3 = \dfrac{5}{2} \cdot 3 \cdot x^2 = \dfrac{15}{2} x^2.$

(b) $\dfrac{d}{dx} \dfrac{3}{x} = 3 \dfrac{d}{dx} x^{-1} = 3 \cdot -1 \cdot x^{-1-1} = -3x^{-2}$, or $\dfrac{-3}{x^2}.$

2.5. BASIC DERIVATIVE FORMULAS AND PROPERTIES

(c) $\dfrac{d}{dx} 7e^x = 7 \dfrac{d}{dx} e^x = 7e^x$.

(d) $\dfrac{d}{dx} \dfrac{-1}{4\sqrt[3]{x}} = \dfrac{d}{dx}\left(-\dfrac{1}{4} x^{-\frac{1}{3}}\right) = -\dfrac{1}{4} \dfrac{d}{dx} x^{-\frac{1}{3}} = -\dfrac{1}{4} \cdot -\dfrac{1}{3} \cdot x^{-\frac{1}{3}-1} = \dfrac{1}{12} x^{-\frac{4}{3}}$, or $\dfrac{1}{12\sqrt[3]{x^4}}$.

The Sum and Difference Properties

Let $F(x) = f(x) + g(x)$, where $f'(x)$ and $g'(x)$ exist. We now apply the definition introduced in Section 2.4 to find the derivative of $F(x)$. We first reorganize the difference quotient:

$$\begin{aligned}
\dfrac{F(x+h) - F(x)}{h} &= \dfrac{[f(x+h) + g(x+h)] - [f(x) + g(x)]}{h} \\
&= \dfrac{[f(x+h) - f(x)] + [g(x+h) - g(x)]}{h} \\
&= \dfrac{f(x+h) - f(x)}{h} + \dfrac{g(x+h) - g(x)}{h}, \quad h \neq 0.
\end{aligned}$$

By definition, the derivative $F'(x)$ is the limit of the difference quotient as h approaches 0. Using the Limit Property L6 and the definition for the derivative of $f(x)$ and $g(x)$:

$$\begin{aligned}
F'(x) &= \lim_{h \to 0} \dfrac{F(x+h) - F(x)}{h} \\
&= \lim_{h \to 0} \left[\dfrac{f(x+h) - f(x)}{h} + \dfrac{g(x+h) - g(x)}{h}\right] \\
&= \lim_{h \to 0} \dfrac{f(x+h) - f(x)}{h} + \lim_{h \to 0} \dfrac{g(x+h) - g(x)}{h} \\
&= f'(x) + g'(x).
\end{aligned}$$

Therefore,

$$\dfrac{d}{dx}[f(x) + g(x)] = \dfrac{d}{dx} f(x) + \dfrac{d}{dx} g(x).$$

Similarly,

$$\dfrac{d}{dx}[f(x) - g(x)] = \dfrac{d}{dx} f(x) - \dfrac{d}{dx} g(x).$$

Theorem 5 *The Sum and Difference Properties*

If $y = F(x) = f(x) + g(x)$, then

$$y' = F'(x) = f'(x) + g'(x),$$

or equivalently,

$$\dfrac{d}{dx}[f(x) + g(x)] = \dfrac{d}{dx} f(x) + \dfrac{d}{dx} g(x).$$

If $y = F(x) = f(x) - g(x)$, then

$$y' = F'(x) = f'(x) - g'(x),$$

or equivalently,

$$\dfrac{d}{dx}[f(x) - g(x)] = \dfrac{d}{dx} f(x) - \dfrac{d}{dx} g(x).$$

Example 5: Find each of the following derivatives:

(a) $\dfrac{d}{dx}\left(3x^2 - 5x + 8e^x - 7\right);$ (b) $\dfrac{d}{dx}\left(\dfrac{3}{x^2} - 4\sqrt[5]{x^3}\right).$

Solution We apply the rules introduced in Theorem 1 to Theorem 3, with the Constant Multiple and Sum and Difference Properties:

(a) $\dfrac{d}{dx}\left(3x^2 - 5x + 8e^x - 7\right) = \dfrac{d}{dx}\left(3x^2\right) - \dfrac{d}{dx}(5x) + \dfrac{d}{dx}(8e^x) - \dfrac{d}{dx}7$

$= 3\dfrac{d}{dx}x^2 - 5\dfrac{d}{dx}x + 8\dfrac{d}{dx}e^x - \dfrac{d}{dx}7 = 3\cdot 2x - 5\cdot 1 + 8e^x - 0 = 6x - 5 + 8e^x.$

(b) $\dfrac{d}{dx}\left(\dfrac{3}{x^2} - 4\sqrt[5]{x^3}\right) = \dfrac{d}{dx}\left(\dfrac{3}{x^2}\right) - \dfrac{d}{dx}\left(4\sqrt[5]{x^3}\right) = 3\dfrac{d}{dx}x^{-2} - 4\dfrac{d}{dx}x^{\frac{3}{5}}$

$= 3\cdot -2\cdot x^{-2-1} - 4\cdot \dfrac{3}{5}\cdot x^{\frac{3}{5}-1} = -6x^{-3} - \dfrac{12}{5}x^{-\frac{2}{5}} = -\dfrac{6}{x^3} - \dfrac{12}{5\sqrt[5]{x^2}}.$

▬ Slopes of Tangent Lines ▬

We now return to the topic of tangent lines to the graph of a given function $y = f(x)$. Recall that, at a point $(x_0, f(x_0))$ on the graph of the function, the slope of the tangent line is exactly $f'(x_0)$, the derivative value at x_0. With the rules and properties of differentiation introduced in the theorems, the task of finding the equation of a tangent line becomes much simpler.

Example 6: Find the equation of the tangent line to the graph of $f(x) = x^3 - 3x^2 + 1$ at $x = 1$.

Solution We first find $f'(x)$:

$\dfrac{d}{dx}\left(x^3 - 3x^2 + 1\right) = \dfrac{d}{dx}x^3 - \dfrac{d}{dx}\left(3x^2\right) + \dfrac{d}{dx}1$

$= \dfrac{d}{dx}x^3 - 3\dfrac{d}{dx}x^2 + \dfrac{d}{dx}1 = 3\cdot x^{3-1} - 3\cdot 2\cdot x^{2-1} + 0 = 3x^2 - 6x.$

Therefore, $f'(1) = 3(1)^2 - 6(1) = -3$. That is the slope of the tangent line. Now we need to find a point on the tangent line. When $x = 1$, we have $f(1) = -1$. Hence, the tangent line has slope $m = -3$ and passes the point $(1, -1)$. Using the slope-point equation, we have

$$y - (-1) = -3(x - 1).$$

Rewriting this equation in the slope-intercept form we obtain

$$y = -3x + 2.$$

This equation represents the line tangent to the graph of the function at the point $(1, -1)$.

In Chapter 3 we often need to determine points at which the tangent to the curve has a certain slope. Since the derivative represents the slope of the tangent lines, we are looking for the points at which the derivative has a particular value.

Example 7: Find the points on the graph of $f(x) = x^3 - 3x^2 + 1$ at which the tangent line is horizontal.

2.5. BASIC DERIVATIVE FORMULAS AND PROPERTIES

Solution We are looking for the points $(a, f(a))$ such that $f'(a) = 0$. From Example 6 we have $f'(x) = 3x^2 - 6x$. Hence, we need to solve the equation

$$3x^2 - 6x = 0.$$

We factor and solve:

$$\begin{aligned} 3x(x-2) &= 0 \\ 3x = 0 \quad &\text{or} \quad x - 2 = 0 \\ x = 0 \quad &\text{or} \quad x = 2. \end{aligned}$$

To find the points on the graph, we need to find the function values at $x = 0$ and $x = 2$:

$$f(0) = (0)^3 - 3(0)^2 + 1 = 1.$$

$$f(2) = (2)^3 - 3(2)^2 + 1 = 8 - 12 + 1 = -3.$$

Hence, the points at which the tangent line is horizontal are $(0, 1)$ and $(2, -3)$ (Figure 2.5.2).

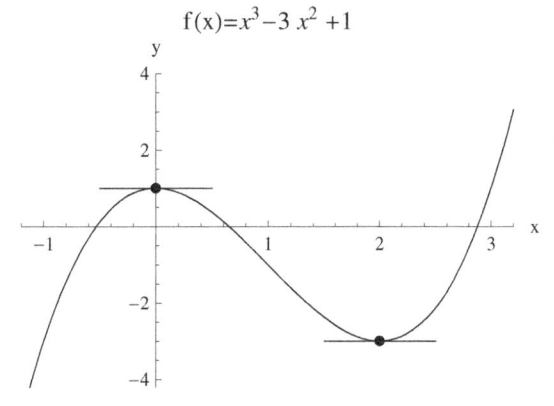

Figure 2.5.2: Graph of $f(x) = x^3 - 3x^2 + 1$ with horizontal tangent lines

■ Higher Order Derivatives

Consider the function $y = f(x) = x^4$. Its derivative is $y' = f'(x) = 4x^3$. The derivative function can also be differentiated. We call the derivative of a derivative function the **second order of derivative**, or the **second derivative of** $f(x)$ and denote it as

$$y'' = f''(x) = \frac{d^2 y}{dx^2} = \frac{d^2}{dx^2} f(x).$$

Hence for $y = f(x) = x^4$ we have

$$y'' = f''(x) = \frac{d^2 y}{dx} = \frac{d^2}{dx} x^4 = \frac{d}{dx}\left(\frac{d}{dx} x^4\right) = \frac{d}{dx} 4x^3 = 12x^2.$$

Now notice that the second derivative function is again differentiable, thus we can take the **third order of derivative**

$$y''' = f'''(x) = \frac{d^3 y}{dx^3} = \frac{d^3}{dx^3} x^4 = 24x.$$

In general, the notations used for the nth order derivative for $n \geq 4$ are

$$y^n = f^{(n)}(x) = \frac{d^n y}{dx^n} = \frac{d^n}{dx^n} f(x).$$

Example 8: Find each of the following derivatives:

(a) $\dfrac{d^3}{dx^3}\left(3x^2 - 5x + 8e^x - 7\right)$; (b) $\dfrac{d^2}{dx^2}\left(\dfrac{3}{x^2} - 4\sqrt[5]{x^3}\right)$.

Solution (a) From Example 5 (a) we have $\dfrac{d}{dx}\left(3x^2 - 5x + 8e^x - 7\right) = 6x - 5 + 8e^x$. Hence

$$\begin{aligned}
\dfrac{d^2}{dx^2}\left(3x^2 - 5x + 8e^x - 7\right) &= \dfrac{d}{dx}(6x - 5 + 8e^x) \\
&= 6\dfrac{d}{dx}x - \dfrac{d}{dx}5 + 8\dfrac{d}{dx}e^x \\
&= 6\cdot 1 - 0 + 8e^x \\
&= 6 + 8e^x, \quad \text{and}
\end{aligned}$$

$$\begin{aligned}
\dfrac{d^3}{dx^3}\left(3x^2 - 5x + 8e^x - 7\right) &= \dfrac{d}{dx}(6 + 8e^x) \\
&= 0 + 8e^x = 8e^x.
\end{aligned}$$

(b) From Example 5(b) we have $\dfrac{d}{dx}\left(\dfrac{3}{x^2} - 4\sqrt[5]{x^3}\right) = -6x^{-3} - \dfrac{12}{5}x^{-\frac{2}{5}}$. Hence

$$\begin{aligned}
\dfrac{d^2}{dx^2}\left(\dfrac{3}{x^2} - 4\sqrt[5]{x^3}\right) &= \dfrac{d}{dx}\left(-6x^{-3} - \dfrac{12}{5}x^{-\frac{2}{5}}\right) \\
&= -6\dfrac{d}{dx}x^{-3} - \dfrac{12}{5}\dfrac{d}{dx}x^{-\frac{2}{5}} \\
&= -6\cdot -3\cdot x^{-3-1} - \dfrac{12}{5}\cdot\dfrac{-2}{5}\cdot x^{-\frac{2}{5}-1} \\
&= 18x^{-4} + \dfrac{24}{25}x^{-\frac{7}{5}} = \dfrac{18}{x^4} + \dfrac{24}{25\sqrt[5]{x^7}}.
\end{aligned}$$

Recall that in Section 2.4, Example 6, for the height function s, the derivative represents the velocity, that is, $v(t) = s'(t)$. The second derivative of s represents the **acceleration**, which measures how fast the velocity is changing:

$$a(t) = v'(t) = s''(t).$$

Example 9: If a rock is dropped from the top of a 144-ft-tall building, then the height of the rock, in feet, at t seconds can be modeled by

$$s(t) = 144 - 16t^2 \quad 0 \le t \le 3.$$

(a) Find $v(t)$ and $a(t)$.

(b) Find $a(2)$. What does this quantity represent?

Solution (a) From Section 2.4 Example 5 we have $v(t) = s'(t) = -32t$, hence

$$a(t) = v'(t) = -32.$$

(b) We have $a(2) = -32$ ft/sec^2. This quantity represents the acceleration of the rock at 2 seconds. That is, at 2 seconds after the rock is dropped, its speed of falling is increasing by 32 ft/sec per second.

Exercise Set 2.5

Find $f'(x)$

1. $f(x) = 2x^7$
2. $f(x) = -4x^5$
3. $f(x) = 3\sqrt{x}$
4. $f(x) = 5\sqrt[3]{x}$
5. $f(x) = -3\sqrt[5]{x^7}$
6. $f(x) = 2\sqrt{x^5}$
7. $f(x) = -3\sqrt[5]{x^3}$
8. $f(x) = 2\sqrt[3]{x^2}$
9. $f(x) = \dfrac{2}{x^2}$
10. $f(x) = \dfrac{-3}{x^4}$
11. $f(x) = \dfrac{2}{\sqrt{x}}$
12. $f(x) = \dfrac{3}{\sqrt[3]{x^2}}$

Find $\dfrac{dy}{dx}$

13. $y = \dfrac{5x}{7}$
14. $y = \dfrac{3x^2}{2}$
15. $y = \dfrac{7e^x}{2}$
16. $y = \dfrac{2e^x}{3}$
17. $y = \dfrac{-3}{2\sqrt[4]{x}}$
18. $y = \dfrac{-2}{3\sqrt[5]{x}}$
19. $y = -2.4x^{-0.1}$
20. $y = -1.7x^{0.4}$
21. $y = 2x^3 - 10x^2 + 6e^x - 3$
22. $y = 5x^4 - 3x - 7e^x + 8$
23. $y = 3\sqrt{x} - 3e^x + \dfrac{2}{x}$
24. $y = 2\sqrt[3]{x} - 5e^x + \dfrac{5}{x^3}$

Find y'

25. $y = \dfrac{5x}{7} + 3\sqrt{x} - \dfrac{5}{\sqrt[3]{x^2}}$
26. $y = \dfrac{-3x}{4} + 5\sqrt{x^3} - \dfrac{2}{\sqrt[3]{x}}$

27. $y = \dfrac{x^2}{3} + \dfrac{3}{x^2}$
28. $y = \dfrac{x^3}{2} + \dfrac{2}{x^3}$
29. $y = 5e^x - x^e$
30. $y = 3x^e + 2e^x$

Differentiate:

31. $\dfrac{d}{dx}(8x^2 - 6x + 4)$
32. $\dfrac{d}{dx}(3x^2 - 5x + 10)$
33. $\dfrac{d^2}{dx^2}(8x^2 - 6x + 4)$
34. $\dfrac{d^2}{dx^2}(3x^2 - 5x + 10)$
35. $\dfrac{d}{dx}\left(5\sqrt{x} - \dfrac{3}{\sqrt{x}}\right)$
36. $\dfrac{d}{dx}\left(2\sqrt[3]{x} - \dfrac{7}{\sqrt[3]{x}}\right)$
37. $\dfrac{d^5}{dx^5} x^7$
38. $\dfrac{d^4}{dx^4}(3x^4)$
39. $\dfrac{d^7}{dx^7}(3e^x)$
40. $\dfrac{d^8}{dx^8}(4e^x)$

41. For $f(x) = \sqrt{x^3}$, find $f'(4)$ and $f''(4)$.
42. For $f(x) = \dfrac{2}{x^2}$, find $f'(2)$ and $f''(2)$.
43. Find an equation of the tangent line to the graph of the function $f(x) = 2x^3 - 3x^2 + x - 1$ at
    (a) $x = -1$; (b) $x = 0$; (c) $x = 1$.
44. Find an equation of the tangent line to the graph of the function $f(x) = x^2 - \dfrac{1}{x}$ at
    (a) $x = -1$; (b) $x = 1$; (c) $x = 2$.

For Exercises 45 to 62, find the points on the curve where the tangent line is horizontal. If such a point does not exist, state the fact.

45. $f(x) = \dfrac{1}{2}x^2 - 1$
46. $f(x) = -2x^2 + 3$

47. $f(x) = \frac{1}{3}x^3 - 2$

48. $f(x) = -5x^3 + 1$

49. $f(x) = 10x - x^2$

50. $f(x) = 6x - 3x^2$

51. $f(x) = 0.3x^2 - 1.2x + 3$

52. $f(x) = 0.5x^2 + 2x - 1$

53. $f(x) = 2x^3 + 3x^2 - 12x + 1$

54. $f(x) = 2x^3 - 7x^2 + 4x + 3$

55. $f(x) = \frac{1}{3}x^3 - 2x^2$

56. $f(x) = 2x^3 - 3x^2 + 5$

57. $f(x) = x^3 - 12x + 3$

58. $f(x) = x^3 - 3x - 2$

59. $f(x) = 7$ 60. $f(x) = -5$

61. $f(x) = 3x - 1$ 62. $f(x) = 4x + 5$

APPLICATIONS

63. If a rock is dropped from the top of a 256-ft-tall building, then the height of the rock, in feet, at t seconds can be modeled by

$$s(t) = 256 - 16t^2 \quad 0 \le t \le 4.$$

(a) Find $v(t)$ and $a(t)$.

(b) Find $s(3)$, $v(3)$, and $a(3)$. What does each quantity represent?

64. If a rock is dropped from the top of a 196-ft-tall building, then the height of the rock, in feet, at t seconds can be modeled by

$$s(t) = 196 - 16t^2 \quad 0 \le t \le 3.5.$$

(a) Find $v(t)$ and $a(t)$.

(b) Find $s(2)$, $v(2)$, and $a(2)$. What does each quantity represent?

65. If a rock is thrown from the ground with an initial velocity of 80 ft/sec, then the height of the rock, in feet, at t seconds can be modeled by

$$s(t) = -16t^2 + 80t \quad 0 \le t \le 5.$$

(a) Find $v(t)$ and $a(t)$.

(b) At what time does the rock reach the highest height?

(c) What is the highest height the rock reaches?

(d) What is the acceleration of the rock when it reaches the highest height?

66. If a rock is thrown from the ground with an initial velocity of 120 ft/sec, then the height of the rock, in feet, at t seconds can be modeled by

$$s(t) = -16t^2 + 120t \quad 0 \le t \le 7.5.$$

(a) Find $v(t)$ and $a(t)$.

(b) At what time does the rock reach the highest height?

(c) What is the highest height the rock reaches?

(d) What is the acceleration of the rock when it reaches the highest height?

2.6   Differentiation Techniques: Product and Quotient Rules

In Section 2.5, we learned how to find the derivative of functions such as polynomials. In this section, we study the rules which will help us to find the derivative of a rational function, or a function such as $y = x^2 e^x$, or as $y = \dfrac{x^2 + 1}{e^x}$. These functions are the **product** or the **quotient** of other functions.

The Product Rule

Consider two differentiable functions $f(x)$ and $g(x)$ and a constant c. We know that

$$\frac{d}{dx}[f(x) + g(x)] = \frac{d}{dx}f(x) + \frac{d}{dx}g(x) \quad \text{and} \quad \frac{d}{dx}[c \cdot f(x)] = c \cdot \frac{d}{dx}f(x),$$

That is, the derivative of a sum is the sum of the derivatives, and the derivative of a constant times a function is the constant times the derivative of the function. However, the derivative of the product $f(x) \cdot g(x)$ is not the product of the derivatives. For example, if $f(x) = x^2$ and $g(x) = x^3$, then

$$f(x) \cdot g(x) = x^2 \cdot x^3 = x^5,$$

therefore,

$$\frac{d}{dx}[f(x) \cdot g(x)] = \frac{d}{dx}x^5 = 5x^4.$$

Meanwhile, since $f'(x) = 2x$ and $g'(x) = 3x^2$, $f'(x) \cdot g'(x) = 2x \cdot 3x^2 = 6x^3$, which does not equal $5x^4$, the derivative of $f(x) \cdot g(x)$.

We now introduce the rule of finding the derivative of a product. This rule is stated in the following theorem:

Theorem 1 *The Product Rule*

If $f(x)$ and $g(x)$ are two differentiable functions, and $y = F(x) = f(x) \cdot g(x)$, then

$$y' = F'(x) = f(x) \cdot g'(x) + f'(x) \cdot g(x),$$

or equivalently,

$$\frac{d}{dx}[f(x) \cdot g(x)] = f(x) \cdot \left[\frac{d}{dx}g(x)\right] + \left[\frac{d}{dx}f(x)\right] \cdot g(x).$$

This rule can be efficiently expressed as an "sum of cross-products":

$$\begin{array}{ccc} f(x) & & g(x) \\ & \times & \\ f'(x) & & g'(x) \end{array}$$

$$f'(x) \cdot g(x) \quad + \quad f(x) \cdot g'(x)$$

Note: Since both sum and product follow the communicative law, the orders do not matter.

We verify this rule with the example $f(x) = x^2$ and $g(x) = x^3$:

$$\begin{array}{ccc} f(x) & & g(x) \\ & \times & \\ f'(x) & & g'(x) \end{array}$$

$$f'(x) \cdot g(x) \quad + \quad f(x) \cdot g'(x)$$

$$= \begin{array}{ccc} x^2 & & x^3 \\ & \times & \\ 2x & & 3x^2 \end{array}$$

$$2x \cdot x^3 \quad + \quad x^2 \cdot 3x^2 \quad = 2x^4 + 3x^4 = 5x^4.$$

Hence, $\dfrac{d}{dx}(x^2 \cdot x^3) = 5x^4$.

Example 1: Differentiate each of the following:

(a) $y = (x^3 - 2x^2 + 1)(x^4 - 2)$; (b) $y = x^{-3}e^x$.

Solution (a) We apply the "sum of cross-products" process with $f(x) = x^3 - 2x^2 + 1$ and $g(x) = x^4 - 2$:

$$\begin{array}{ccc} f(x) & & g(x) \\ & \times & \\ f'(x) & & g'(x) \end{array}$$

$$f'(x) \cdot g(x) \quad + \quad f(x) \cdot g'(x)$$

$$= \begin{array}{ccc} x^3 - 2x^2 + 1 & & x^4 - 2 \\ & \times & \\ 3x^2 - 4x & & 4x^3 \end{array}$$

$$(3x^2 - 4x) \cdot (x^4 - 2) \quad + \quad (x^3 - 2x^2 + 1) \cdot 4x^3.$$

Hence, $\dfrac{d}{dx}(x^3 - 2x^2 + 1)(x^4 - 2) = (3x^2 - 4x) \cdot (x^4 - 2) + (x^3 - 2x^2 + 1) \cdot 4x^3.$

Note: When it becomes too tedious to simplify the resulting derivative, we may leave the product forms as is.

(b) We apply the "sum of cross-products" process with $f(x) = x^{-3}$ and $g(x) = e^x$:

$$\begin{array}{ccc} f(x) & & g(x) \\ & \times & \\ f'(x) & & g'(x) \end{array}$$

$$f'(x) \cdot g(x) \quad + \quad f(x) \cdot g'(x)$$

2.6. DIFFERENTIATION TECHNIQUES: PRODUCT AND QUOTIENT RULES

$$
\begin{array}{ccc}
 & x^{-3} & e^x \\
 & \times & \\
= & -3x^{-4} & e^x \\
\\
 & -3x^{-4} \cdot e^x & + \quad x^{-3} \cdot e^x.
\end{array}
$$

Hence, $\dfrac{d}{dx} x^{-3} e^x = -3x^{-4} \cdot e^x + x^{-3} \cdot e^x = (x-3)x^{-4}e^x$, or $\dfrac{(x-3)e^x}{x^4}$.

The Quotient Rule

We now introduce the rule of finding the derivative of a quotient. This rule is stated in Theorem 2. We verify this rule with the example $f(x) = x^2$ and $g(x) = x^3$. Let $F(x) = \dfrac{f(x)}{g(x)} = \dfrac{x^2}{x^3} = \dfrac{1}{x}$. Then $F'(x) = \dfrac{-1}{x^2}$. We now apply the Quotient Rule:

$$F'(x) = \frac{f'(x) \cdot g(x) - f(x) \cdot g'(x)}{[g(x)]^2} = \frac{2x \cdot x^3 - x^2 \cdot 3x^2}{[x^3]^2}$$

$$= \frac{2x^4 - 3x^4}{x^6} = \frac{-x^4}{x^6} = \frac{-1}{x^2}.$$

Hence, $\dfrac{d}{dx}\left(\dfrac{x^2}{x^3}\right) = \dfrac{-1}{x^2}$.

Theorem 2   *The Quotient Rule*

If $f(x)$ and $g(x)$ are two differentiable functions, and $y = F(x) = \dfrac{f(x)}{g(x)}$, then

$$y' = F'(x) = \frac{f'(x) \cdot g(x) - f(x) \cdot g'(x)}{[g(x)]^2},$$

or equivalently,

$$\frac{d}{dx}\left[\frac{f(x)}{g(x)}\right] = \frac{[\frac{d}{dx} f(x)] \cdot g(x) - f(x) \cdot [\frac{d}{dx} g(x)]}{[g(x)]^2}.$$

Note: In the Quotient Rule, the orders in the numerator of the resulting formula **do** matter. The term associated with the derivative of the denominator function, $g(x)$, has a negative sign.

Example 2: Differentiate each of the following:

(a) $y = \dfrac{x^3 - 2x^2 + 1}{x^4 - 2}$;   (b) $y = \dfrac{e^x}{x^3}$.

Solution (a) We apply the Quotient Rule with $f(x) = x^3 - 2x^2 + 1$ and $g(x) = x^4 - 2$:

$$\begin{aligned}\frac{dy}{dx} &= \frac{f'(x)\cdot g(x) - f(x)\cdot g'(x)}{[g(x)]^2}\\ &= \frac{(3x^2 - 4x)\cdot (x^4 - 2) - (x^3 - 2x^2 + 1)\cdot 4x^3}{[x^4 - 2]^2}\\ &= \frac{(3x^6 - 4x^5 - 6x^2 + 8x) - (4x^6 - 8x^5 + 4x^3)}{[x^4 - 2]^2}\\ &= \frac{-x^6 + 4x^5 - 4x^3 - 6x^2 + 8x}{[x^4 - 2]^2}.\end{aligned}$$

Hence, $\dfrac{d}{dx}\dfrac{x^3 - 2x^2 + 1}{x^4 - 2} = \dfrac{-x^6 + 4x^5 - 4x^3 - 6x^2 + 8x}{(x^4 - 2)^2}$.

(b) We apply the the Quotient Rule with $f(x) = e^x$ and $g(x) = x^3$:

$$\begin{aligned}\frac{dy}{dx} &= \frac{f'(x)\cdot g(x) - f(x)\cdot g'(x)}{[g(x)]^2}\\ &= \frac{e^x \cdot x^3 - e^x \cdot 3x^2}{(x^3)^2}\\ &= \frac{e^x x^2 (x - 3)}{x^6} = \frac{e^x(x - 3)}{x^4}.\end{aligned}$$

Hence, $\dfrac{d}{dx}\dfrac{e^x}{x^3} = \dfrac{e^x(x - 3)}{x^4}$.

Note: The function in Example 2 (b), $y = \dfrac{e^x}{x^3}$, is identical to the function in Example 1 (b), $y = x^{-3}e^x$, where the Product Rule is applied. The two derivatives obtained by different rules are identical, too. This example illustrates an interesting fact: A quotient can be treated as a product with the denominator function carrying a negative power. After the Extended Power Rule is introduced in the next section, we may apply the Product Rule with the Extended Power Rule to find the derivative of a quotient.

Example 3: Differentiate each of the following:

(a) $y = \dfrac{x^3 - 2x^2 + 1}{x^2}$;   (b) $y = \dfrac{xe^x}{x - 1}$.

Solution (a) We can apply the Quotient Rule with $f(x) = x^3 - 2x^2 + 1$ and $g(x) = x^2$. However, we also can divide the expression and then apply the Power Rule:

$$y = \frac{x^3 - 2x^2 + 1}{x^2} = \frac{x^3}{x^2} - \frac{2x^2}{x^2} + \frac{1}{x^2} = x - 2 + x^{-2}.$$

Hence,

$$\frac{dy}{dx} = 1 - 2x^{-3} = 1 - \frac{2}{x^3}.$$

(b) We apply the Quotient Rule with $f(x) = xe^x$ and $g(x) = x - 1$. Note that we need to apply the Product Rule for $f'(x)$:

$$\frac{dy}{dx} = \frac{f'(x) \cdot g(x) - f(x) \cdot g'(x)}{[g(x)]^2}$$

$$= \frac{(e^x + xe^x) \cdot (x-1) - xe^x \cdot 1}{(x-1)^2}$$

$$= \frac{xe^x + x^2 e^x - e^x - xe^x - xe^x}{(x-1)^2} = \frac{e^x(x^2 - x - 1)}{(x-1)^2}.$$

Hence, $\dfrac{d}{dx} \dfrac{xe^x}{x-1} = \dfrac{e^x(x^2 - x - 1)}{(x-1)^2}.$

Exercise Set 2.6

Find $f'(x)$ in two ways: First, by multiplying the expressions before differentiating; Second, by applying the Product Rule. Compare your results.

1. $f(x) = x^7 \cdot x^2$
2. $f(x) = x^5 \cdot x^3$
3. $f(x) = 3x^3 (2x^2 + x)$
4. $f(x) = x (5x^4 - 3x^2)$
5. $f(x) = \sqrt{x} \cdot \sqrt[3]{x}$
6. $f(x) = \sqrt[3]{x} \cdot \sqrt[4]{x}$

Find $f'(x)$ in two ways: First, by dividing the expressions before differentiating; Second, by applying the Quotient Rule. Compare your results.

7. $f(x) = \dfrac{x^6}{x^2}$
8. $f(x) = \dfrac{x^7}{x^4}$
9. $f(x) = \dfrac{x^2 - 2x^8}{x^5}$
10. $f(x) = \dfrac{x^3 + 3x}{x^2}$
11. $f(x) = \dfrac{x^2 - 1}{x + 1}$
12. $f(x) = \dfrac{x^4 - 4}{x^2 + 2}$

Find $f'(x)$

13. $f(x) = (2x^7 + 5x^4 - 3)(7x^2 - 6x + 1)$
14. $f(x) = (3x^5 - 2x^4 + 6)(5x^3 + 2x^2 - 2)$
15. $f(x) = \left(3\sqrt{x} + \dfrac{8}{x}\right)(e^x + x^2 + 4)$
16. $f(x) = (2e^x + 5\sqrt[3]{x} - 8)\left(3x^4 + \dfrac{6}{x^2} + 3\right)$

17. $f(x) = \dfrac{2x - 1}{x^3 + 5}$
18. $f(x) = \dfrac{3x^2 + 5}{2x - 1}$
19. $f(x) = \dfrac{2e^x}{x^2}$
20. $f(x) = \dfrac{-3e^x}{x^4}$
21. $f(x) = \dfrac{2x}{e^x}$
22. $f(x) = \dfrac{3x}{e^x}$

Find $\dfrac{dy}{dx}$

23. $y = \dfrac{5x + 3}{7}$
24. $y = \dfrac{3x^2 - 5}{2}$
25. $y = \dfrac{7e^x}{2x + 1}$
26. $y = \dfrac{2e^x}{3x - 4}$
27. $y = \dfrac{(x^2 - 3x + 2)}{2\sqrt[4]{x} + 3}$
28. $y = \dfrac{x^3 - 2x^2 + 1}{3\sqrt[5]{x} + 5}$
29. $y = 2x + x^3 e^x$
30. $y = -2x^4 + 7xe^x$
31. $y = (2x^3 - 10x^2 + 6e^x - 3) e^x$
32. $y = (5x^4 - 3x - 7e^x + 8) e^x$
33. $y = 3\sqrt{x} e^x + \dfrac{2}{x}$
34. $y = 2\sqrt[3]{x} e^x + \dfrac{5}{x^3}$

Find y'

35. $y = \dfrac{5x}{7x^2 + 3} + (3\sqrt{x} - 2)(x + 3)$
36. $y = \dfrac{3x}{4x^3 - 5} + (5\sqrt{x^3} + 3)(2x - 1)$
37. $y = \dfrac{x^2}{3e^x + 2} + \dfrac{3e^x + 2}{x^2}$

38. $y = \dfrac{x^3}{2e^x + 3} + \dfrac{2e^x + 3}{x^3}$

Differentiate:

39. $\dfrac{d}{dx}\left(\dfrac{3xe^x}{x^2 + 4}\right)$ 40. $\dfrac{d}{dx}\left(\dfrac{x^2 e^x}{x + 3}\right)$

41. $\dfrac{d^2}{dx^2}(3x + 1)(x^2 - 5x + 2)$

42. $\dfrac{d^2}{dx^2}(4x + 3)(2x^2 + 7x - 1)$

43. $\dfrac{d^2}{dx^2}(x^2 e^x)$ 44. $\dfrac{d^2}{dx^2}(3xe^x)$

45. $\dfrac{d^2}{dx^2}\dfrac{3}{3x + 1}$ 46. $\dfrac{d^2}{dx^2}\dfrac{1}{x^2 + 4}$

47. Find an equation of the tangent line to the graph of the function $f(x) = (3x + 2)e^x$ at $(0, 2)$.

48. Find an equation of the tangent line to the graph of the function $f(x) = (2x^3 - 5x)e^x$ at $x = 0$.

49. Find an equation of the tangent line to the graph of the function $f(x) = \dfrac{4}{x^2 + 2}$ at

    (a) $x = -1$;   (b) $x = 0$;   (c) $x = 1$.

50. Find an equation of the tangent line to the graph of the function $f(x) = \dfrac{x}{x^2 + 4}$ at

    (a) $x = -1$;   (b) $x = 0$;   (c) $x = 1$.

2.7 Differentiation Techniques: The Extended Power and Natural Exponential Rules

Many functions that arise in economic or business models are more complicated than simple power or exponential functions. For example, if a savings account starts with an initial deposit of $500 and pays interest at an annual rate of 4%, compounded continuously, then its balance after t years can be modeled as

$$S(t) = 500 e^{0.04t}.$$

What is the derivative of $S(t)$? Can we just apply the Natural Exponential Rule? Let us explore with the function

$$y = \left(1 + x^3\right)^2,$$

which can be extended as

$$y = 1 + 2x^3 + x^6,$$

and we know that

$$\dfrac{dy}{dx} = 6x^2 + 6x^5.$$

However, if we apply the Power Rule to the power $p = 2$ only, we obtain $2(1 + x^3)$, and that is not the correct derivative of y.

In this section, we study how to find the derivative of the power of a function, like $f(x) = \left(1 + x^2\right)^4$, or a function with the form as $S(t) = 500 e^{0.04t}$.

2.7. EXTENDED POWER AND NATURAL EXPONENTIAL RULES

■ The Extended Power Rule

Consider the functions

$$y = \left(1 + x^3\right)^2$$

again. We note that y is the composition of two functions:

$$y = f(u) = u^2 \quad \text{and} \quad u = g(x) = 1 + x^3.$$

We also note that if we only apply the Power Rule to the outer power 2, we obtain $2(1+x^3)$, while the true derivative $\frac{dy}{dx}$ is

$$\frac{dy}{dx} = 6x^2 + 6x^5 = 2(1+x^3) \cdot 3x^2.$$

Also notice that the derivative of $u = g(x) = 1 + x^3$ is $\frac{du}{dx} = 3x^2$. Hence it seems that, to find the derivative of a function with the form $y = [g(x)]^k$, in addition to the Power Rule, we need to multiply the derivative of the "inside" function $g(x)$, provided that $g'(x)$ exists. Therefore,

$$\frac{dy}{dx} = k[g(x)]^{k-1} \cdot \frac{d}{dx} g(x).$$

We call this rule as the Extended Power Rule. This rule is stated in the following theorem:

Theorem 1 *The Extended Power Rule*

Suppose that $g(x)$ is a differentiable function. Then for any real number k, if $y = F(x) = [g(x)]^k$, then

$$y' = F'(x) = k[g(x)]^{k-1} \cdot g'(x),$$

or equivalently,

$$\frac{d}{dx}[g(x)]^k = k[g(x)]^{k-1} \cdot \frac{d}{dx} g(x).$$

Example 1: Differentiate each of the following:

(a) $y = \left(x^3 - 4x^2 + 5\right)^5$; (b) $y = \dfrac{1}{(2x + e^x)^3}$; (c) $y = \sqrt{1 + x^3}$.

Solution (a) We apply the Extended Power Rule with $k = 5$ and $g(x) = x^3 - 4x^2 + 5$. Since $g'(x) = 3x^2 - 8x$, we have

$$\frac{d}{dx}\left(x^3 - 4x^2 + 5\right)^5 = 5\left(x^3 - 4x^2 + 5\right)^4 \cdot \left(3x^2 - 8x\right).$$

(b) Note that the function can be written as $y = (2x + e^x)^{-3}$. We apply the Extended Power Rule with $k = -3$ and $g(x) = 2x + e^x$. Since $g'(x) = 2 + e^x$, we have

$$\frac{d}{dx}\frac{1}{(2x+e^x)^3} = \frac{d}{dx}(2x+e^x)^{-3} = (-3)(2x+e^x)^{-4} \cdot (2+e^x) = \frac{-3(2+e^x)}{(2x+e^x)^4}.$$

(c) Note that the function can be written as $y = \left(1 + x^3\right)^{1/2}$. We apply the Extended Power Rule with

$k = 1/2$ and $g(x) = 1 + x^3$. Since $g'(x) = 3x^2$, we have

$$\frac{d}{dx}\sqrt{1+x^3} = \frac{d}{dx}(1+x^3)^{1/2} = \frac{1}{2}(1+x^3)^{-1/2} \cdot 3x^2 = \frac{3x^2}{2\sqrt{1+x^3}}.$$

▬ The Extended Natural Exponential Rule ▬

We now introduce the rule of finding the derivative of a function with the form $y = e^{g(x)}$ where $g(x)$ is a differentiable function. This rule is stated in the following theorem:

> **Theorem 2**   *The Extended Natural Exponential Rule*
>
> Suppose that $g(x)$ is a differentiable function. Then if $y = F(x) = e^{g(x)}$, then
>
> $$y' = F'(x) = e^{g(x)} \cdot g'(x),$$
>
> or equivalently,
>
> $$\frac{d}{dx} e^{g(x)} = e^{g(x)} \cdot \frac{d}{dx} g(x).$$

Example 2: Differentiate each of the following:

(a) $y = e^{-0.04x}$;   (b) $y = e^{x^3+2x}$;   (c) $y = e^{\sqrt{x}}$.

Solution (a) We apply the Extended Natural Exponential Rule with $g(x) = -0.04x$. Since $g'(x) = -0.04$, we have

$$\frac{d}{dx} e^{-0.04x} = e^{-0.04x} \cdot (-0.04) = -0.04 e^{-0.04x}.$$

(b) We apply the Extended Natural Exponential Rule with $g(x) = x^3 + 2x$. Since $g'(x) = 3x^2 + 2$, we have

$$\frac{d}{dx} e^{x^3+2x} = e^{x^3+2x} \cdot (3x^2+2), \quad \text{or} \quad (3x^2+2) e^{x^3+2x}.$$

(c) We apply the Extended Natural Exponential Rule with $g(x) = \sqrt{x}$. Since $g'(x) = \dfrac{1}{2\sqrt{x}}$, we have

$$\frac{d}{dx} e^{\sqrt{x}} = e^{\sqrt{x}} \cdot \frac{1}{2\sqrt{x}} = \frac{e^{\sqrt{x}}}{2\sqrt{x}}.$$

Example 3: Differentiate $y = (x^3 - 2x^2 + 1)^3 e^{3x^2}$.

Solution Notice that y is a product. Hence, the Product Rule is called for. However, when computing the derivative of each function, an extended rule is needed. We apply the "sum of cross-products" process with $f(x) = (x^3 - 2x^2 + 1)^3$ and $g(x) = e^{3x^2}$. We apply the Extended Power Rule to obtain $f'(x) = 3(x^3 - 2x^2 + 1)^2 (3x^2 - 4x)$. We apply the Extended Natural Exponential Rule to obtain $g'(x) = 6x e^{3x^2}$. Hence, the following diagram holds:

2.7. EXTENDED POWER AND NATURAL EXPONENTIAL RULES

$$
\begin{array}{ccc}
f(x) & & g(x) \\
& \times & \\
f'(x) & & g'(x)
\end{array}
$$

$$f'(x) \cdot g(x) \quad + \quad f(x) \cdot g'(x)$$

$$
= \quad
\begin{array}{ccc}
(x^3 - 2x^2 + 1)^3 & & e^{3x^2} \\
& \times & \\
3(x^3 - 2x^2 + 1)^2 (3x^2 - 4x) & & 6xe^{3x^2}
\end{array}
$$

$$3(x^3 - 2x^2 + 1)^2 (3x^2 - 4x) \cdot e^{3x^2} \quad + \quad (x^3 - 2x^2 + 1)^3 \cdot 6xe^{3x^2}.$$

Hence,

$$\frac{d}{dx}(x^3 - 2x^2 + 1)^3 e^{3x^2}$$

$$= 3(x^3 - 2x^2 + 1)^2 (3x^2 - 4x) \cdot e^{3x^2} + (x^3 - 2x^2 + 1)^3 \cdot 6xe^{3x^2}$$

$$= [3(3x^2 - 4x) + 6x(x^3 - 2x^2 + 1)](x^3 - 2x^2 + 1)^2 e^{3x^2}$$

$$= 3(2x^4 - 4x^3 + 3x^2 - 2x)(x^3 - 2x^2 + 1)^2 e^{3x^2}.$$

Note: When it becomes too tedious to simplify the resulting derivative, we may leave the product forms as is.

Example 4: Differentiate $y = \dfrac{x^2 - 1}{e^{5x}}$.

Solution We can apply the Quotient Rule. However, we now can also rewrite the function as $y = (x^2 - 1)(e^{5x})^{-1} = (x^2 - 1)e^{-5x}$ and apply the Product Rule. We illustrate both approaches:

(a) Apply the Quotient Rule:

$$\frac{d}{dx}\frac{x^2 - 1}{e^{5x}} = \frac{\left(\frac{d}{dx}(x^2 - 1)\right) \cdot e^{5x} - (x^2 - 1) \cdot \frac{d}{dx} e^{5x}}{(e^{5x})^2}$$

$$= \frac{2x \cdot e^{5x} - (x^2 - 1) \cdot 5e^{5x}}{(e^{5x})^2} = \frac{(-5x^2 + 2x + 5)e^{5x}}{(e^{5x})^2}$$

$$= \frac{-5x^2 + 2x + 5}{e^{5x}}.$$

(b) Apply the Product Rule with $f(x) = x^2 - 1$ and $g(x) = e^{-5x}$:

$$\begin{array}{ccc} f(x) & & g(x) \\ & \times & \\ f'(x) & & g'(x) \end{array}$$

$$f'(x) \cdot g(x) \quad + \quad f(x) \cdot g'(x)$$

$$= \begin{array}{ccc} x^2 - 1 & & e^{-5x} \\ & \times & \\ 2x & & -5e^{5x} \end{array}$$

$$2x \cdot e^{-5x} \quad + \quad (x^2 - 1) \cdot (-5)e^{-5x}.$$

Hence,

$$\frac{d}{dx} \frac{x^2 - 1}{e^{5x}} = \frac{d}{dx} (x^2 - 1) e^{-5x}$$

$$= 2x \cdot e^{-5x} + (x^2 - 1) \cdot (-5)e^{-5x}$$

$$= (2x + (x^2 - 1)(-5)) e^{-5x}$$

$$= (-5x^2 + 2x + 5) e^{-5x}.$$

Exercise Set 2.7

Find $f'(x)$

1. $f(x) = (2x^3 + 5x^2 - 3)^7$
2. $f(x) = (3x^5 - 2x^4 + 6)^5$
3. $f(x) = 3\sqrt{2x^3 + 2e^x}$
4. $f(x) = 5\sqrt{6x^3 + 2e^x}$
5. $f(x) = 5e^{0.06x}$ 6. $f(x) = 7e^{0.2x}$

7. $f(x) = 2e^{3x^2 + x}$ 8. $f(x) = 3e^{4x^3 - 5x}$

9. $f(x) = 2e^{-0.02x}$ 10. $f(x) = 5e^{\frac{x^2}{2}}$

11. $f(x) = (x+1)^7 \cdot (3x-2)^2$

12. $f(x) = (2x+3)^5 \cdot (x-2)^4$

13. $f(x) = 4x^5 \cdot e^{3x^2}$

14. $f(x) = 2(x^2 + 1)^3 \cdot e^x$

15. $f(x) = x \cdot \sqrt{x^2 + 1}$

16. $f(x) = x \cdot \sqrt{3x^2 + 5}$

Find $f'(x)$ in two ways: First, by rewriting the quotient as a product and applying the Product Rule; Second, by applying the Quotient Rule. Compare your results.

17. $f(x) = \dfrac{x^3 + 2x}{x^2 + 1}$

18. $f(x) = \dfrac{x^2 - 3x}{2x + 2}$

19. $f(x) = \dfrac{x^2 - 2x}{e^x}$

20. $f(x) = \dfrac{x^3 + 3x}{e^x}$

Find $\dfrac{dy}{dx}$

21. $y = \left(\dfrac{5x + 3}{7x - 2}\right)^3$

22. $y = \left(\dfrac{3x^2 - 5}{2x + 1}\right)^4$

23. $y = \sqrt{\dfrac{2x + 3}{3x - 5}}$

24. $y = \sqrt{\dfrac{5x + 1}{7x - 3}}$

25. $y = \dfrac{7}{(2x + 1)^3}$

26. $y = \dfrac{2}{(3x - 4)^5}$

27. $y = (3x^2 + e^x)^5$

28. $y = (4x^3 - 2e^x)^4$

29. $y = 2x^2 + x^3 e^{4x+1}$

30. $y = -2x^4 + 7xe^{3x-2}$

31. $y = (10x^2 + 6x - 3)^2 \cdot e^{2x+5}$

32. $y = (5x^4 - 3x + 8)^3 \cdot e^{3x+1}$

33. $y = 3\sqrt{2x + 1} \cdot e^x$

34. $y = 2\sqrt{3x + 5} \cdot e^x$

Find y'

35. $y = \sqrt[3]{x^5 - 3x^3 + 2}$

36. $y = \sqrt[3]{x^4 - 2x^3 - 3}$

37. $y = \dfrac{(x^2 + 1)^4}{3e^x + 2}$

38. $y = \dfrac{(x^3 + 2)^2}{2e^x + 3}$

Differentiate:

39. $\dfrac{d}{dx}\left(3e^{x^2} + (2x + 1)^5\right)$

40. $\dfrac{d}{dx}\left(5e^{x^3} + (3x - 2)^4\right)$

41. $\dfrac{d^2}{dx^2} e^{x^2}$

42. $\dfrac{d^2}{dx^2} e^{2x^3}$

43. $\dfrac{d^2}{dx^2} 3(x^2 + 1)^4$

44. $\dfrac{d^2}{dx^2} 2(x^2 + 4)^3$

45. Find an equation of the tangent line to the graph of the function $f(x) = e^{x^2 + 2x}$ at $(0, 1)$.

46. Find an equation of the tangent line to the graph of the function $f(x) = 2e^{2x^2 + 3x}$ at $x = 0$.

47. Find an equation of the tangent line to the graph of the function $f(x) = \dfrac{4}{(x^2+1)^2}$ at

   (a) $x = -1$; (b) $x = 0$; (c) $x = 1$.

48. Find an equation of the tangent line to the graph of the function $f(x) = \dfrac{3}{(x^2 + 4)^2}$ at

   (a) $x = -1$; (b) $x = 0$; (c) $x = 1$.

2.8 The Chain Rule and The Derivatives of Logarithmic Functions

The Extended Power Rules and the Extended Natural Exponential Rules are special cases of a more general rule called the **Chain Rule**. The Chain Rule deals with the derivative of a **composite function**. In this section we study how to find the derivatives of composite functions, including the functions studied in Section 2.7, like $f(x) = (1 + x^2)^4$, or $f(x) = 3e^{-0.4x}$. Later, by applying the Chain Rule, we can derive the formula of the derivative of the natural logarithmic function $f(x) = \ln x$, as well as the composition of the natural logarithmic function and other functions.

Composite Functions

Many functions used in economic or business models are composite functions. Take the example mentioned in Section 2.7. If a savings account starts with an initial deposit of $500 and pays interest at an annual rate of 4%, compounded continuously, then its balance after t years can be modeled as

$$S(t) = 500e^{0.04t}.$$

If we write $u = g(t) = 0.04t$, and $f(u) = 500e^u$, then $S(t)$ can be written as

$$S(t) = f(g(t)), \quad \text{or} \quad S = f(u) \text{ and } u = g(t).$$

We say that S is the **composition** of f and g.

> **Mnemonic   The Chain Train**
>
> We may visualize the composition aforementioned as in the following diagram. 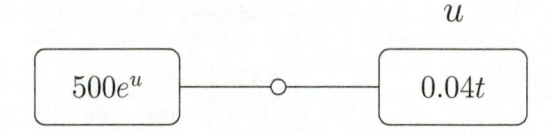
>
> $$\boxed{500e^u} \;-\!\!\circ\!\!-\; \overset{u}{\boxed{0.04t}}$$
>
> A more complicated function like $e^{\sqrt{x^2+1}}$ is diagrammed as follows.
>
> $$\boxed{e^u} \;-\!\!\circ\!\!-\; \overset{u}{\boxed{\sqrt{v}}} \;-\!\!\circ\!\!-\; \overset{v}{\boxed{x^2+1}}$$

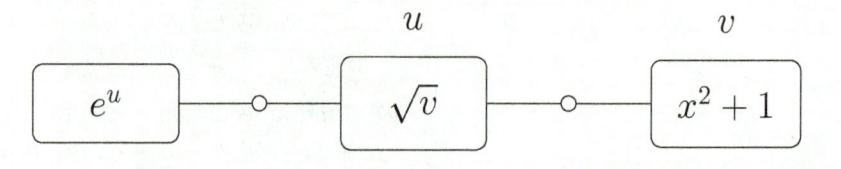

Example 1: Given $f(x) = \sqrt{x}$ and $g(x) = x^3 - x$, find

(a) $f(g(x))$;   (b) $g(f(x))$;   (c) $f(f(x))$;   (d) $g(g(x))$.

Solution (a) We replace the variable x in the "outside" function f by the formula for the "inside" function g and then simplify:

$$f(g(x)) = f(x^3 - x) = \sqrt{x^3 - x}.$$

Pictorially, we write:

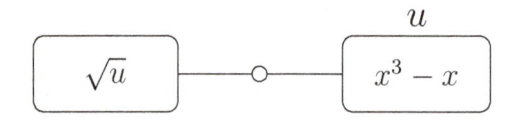

$$\boxed{\sqrt{u}} \;-\!\!\circ\!\!-\; \overset{u}{\boxed{x^3 - x}}$$

(b) We replace the variable x in the "outside" function g by the formula for the "inside" function f and then simplify:

$$g(f(x)) = g(\sqrt{x}) = (\sqrt{x})^3 - (\sqrt{x}) = \sqrt{x}\left(\sqrt{x}^2 - 1\right) = \sqrt{x}(x - 1).$$

Pictorially, we write:

2.8. CHAIN RULE AND DERIVATIVES OF LOG FUNCTIONS

$$\boxed{u^3 - u} \!-\!\!\circ\!\!-\! \boxed{\overset{u}{\sqrt{x}}}$$

(c) We replace the variable x in the "outside" function f by the formula for the "inside" function f and then simplify:

$$f(f(x)) = f\left(\sqrt{x}\right) = \sqrt{\sqrt{x}} = \sqrt[4]{x}$$

The graphical representation is

$$\boxed{\sqrt{u}} \!-\!\!\circ\!\!-\! \boxed{\overset{u}{\sqrt{x}}}$$

(d) We replace the variable x in the "outside" function g by the formula for the "inside" function g and then simplify:

$$g(g(x)) = g\left(x^3 - x\right) = \left(x^3 - x\right)^3 - \left(x^3 - x\right)$$
$$= \left(x^9 - 3x^7 + 3x^5 - x^3\right) - \left(x^3 - x\right) = x^9 - 3x^7 + 3x^5 - 2x^3 + x.$$

The graphical representation is

$$\boxed{u^3 - u} \!-\!\!\circ\!\!-\! \boxed{\overset{u}{x^3 - x}}$$

The Chain Rule

We now introduce the rule for finding the derivative of a function with the form $y = f(g(x))$ where both $f(x)$ and $g(x)$ are differentiable functions. Another often-used expression for $y = f(g(x))$ is to write that

$$y = f(u) \quad \text{and} \quad u = g(x).$$

Pictorially, we have

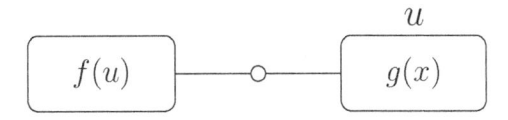

$$\boxed{f(u)} \!-\!\!\circ\!\!-\! \boxed{\overset{u}{g(x)}}$$

Recall that in the Extended Power Rule, $y = [g(x)]^k$ is actually the composition of

$$y = f(u) = u^k \quad \text{and} \quad u = g(x),$$

which we can visualize as

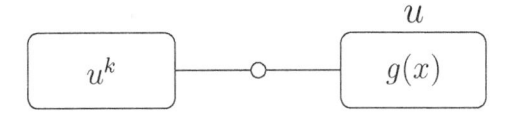

$$\boxed{u^k} \!-\!\!\circ\!\!-\! \boxed{\overset{u}{g(x)}}$$

The Extended Power Rule states that the derivative of such a composition is obtained by taking the derivative of the "outside" function, the power function, and then multiplying by the derivative of the "inside" function g. Applying the derivative notation $\frac{dy}{dx}$, $\frac{dy}{dx}$, and $\frac{dy}{dx}$, we can re-state the Extended Power Rule as:

If $y = u^k$ and $u = g(x)$, then $\dfrac{dy}{dx} = \dfrac{dy}{du} \cdot \dfrac{du}{dx}$.

Graphically, we organize this rule as follows.

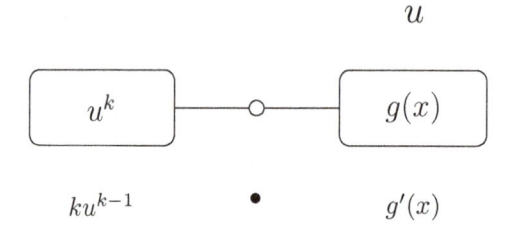

$$u$$

$$\boxed{u^k} \longrightarrow \boxed{g(x)}$$

$$ku^{k-1} \qquad \bullet \qquad g'(x)$$

Similarly, in the Extended Natural Exponential Rule, $y = e^{g(x)}$ is actually the composition of

$$y = f(u) = e^u \quad \text{and} \quad u = g(x).$$

The Extended Natural Exponential Rule states that the derivative of such a composition is obtained by taking the derivative of the "outside" function, the natural exponential function, and then multiplying by the derivative of the "inside" function g. Applying the derivative notation $\frac{dy}{dx}$, $\frac{dy}{dx}$, and $\frac{dy}{dx}$, we can re-state the Extended Natural Exponential Rule as:

If $y = e^u$ and $u = g(x)$, then $\dfrac{dy}{dx} = \dfrac{dy}{du} \cdot \dfrac{du}{dx}$.

We indicate this with the following picture.

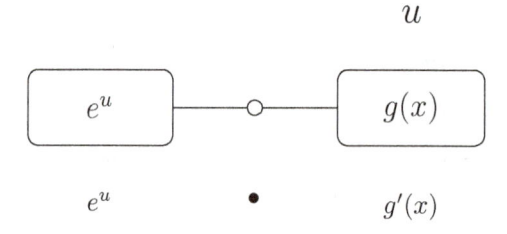

$$u$$

$$\boxed{e^u} \longrightarrow \boxed{g(x)}$$

$$e^u \qquad \bullet \qquad g'(x)$$

The rule for finding the derivative of a function with the form $y = f(g(x)) = (f \circ g)(x)$ takes the exact format as the extended rules, with f as a general differentiable function. It is called the **Chain Rule** and is stated in the following theorem:

Theorem 1 *The Chain Rule*

Suppose that $f(x)$ and $g(x)$ are differentiable functions. If $y = F(x) = f(g(x))$, then

$$y' = F'(x) = f'(g(x)) \cdot g'(x),$$

or equivalently, if $y = f(u)$ and $u = g(x)$, then

$$\frac{dy}{dx} = \frac{dy}{du} \cdot \frac{du}{dx}.$$

Graphically, we represent the chain rule by

2.8. CHAIN RULE AND DERIVATIVES OF LOG FUNCTIONS

$$u$$

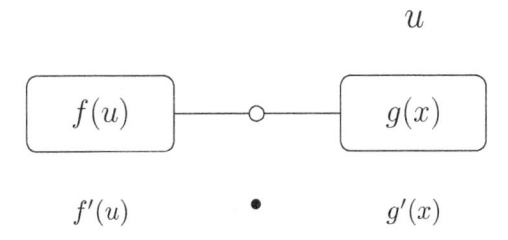

$$\boxed{f(u)} \longrightarrow \boxed{g(x)}$$

$$f'(u) \qquad \bullet \qquad g'(x)$$

The process for using the Chain Rule employing the notation of $\dfrac{dy}{dx} = \dfrac{dy}{du} \cdot \dfrac{du}{dx}$ is as follows:

1. Given a composite function $y = f(g(x))$, rewrite the function as $y = f(u)$ and $u = g(x)$.

2. Find $\dfrac{dy}{du} = f'(u)$ and $\dfrac{du}{dx} = g'(x)$ by applying the appropriate differentiation rules.

3. Write $\dfrac{dy}{dx}$ by multiplying the two derivatives obtained in step 2, and replacing the variable u in the first derivative by the function $g(x)$.

Example 2: Differentiate each of the following using the $\dfrac{dy}{dx} = \dfrac{dy}{du} \cdot \dfrac{du}{dx}$ notation:

(a) $y = \sqrt{e^x + x^2}$;   (b) $y = e^{x^3 + 2x}$.

Solution (a) We rewrite this function as the composition of $y = \sqrt{u}$ and $u = e^x + x^2$. Then we have

$$\frac{dy}{du} = \frac{d}{du} u^{\frac{1}{2}} = \frac{1}{2} u^{-\frac{1}{2}} = \frac{1}{2\sqrt{u}},$$

and

$$\frac{du}{dx} = \frac{d}{dx}\left(e^x + x^2\right) = e^x + 2x.$$

Hence,

$$\frac{dy}{dx} = \frac{dy}{du} \cdot \frac{du}{dx} = \frac{1}{2\sqrt{u}} \cdot (e^x + 2x) = \frac{e^x + 2x}{2\sqrt{e^x + x^2}}.$$

Graphically, this application of the chain rule appears as follows.

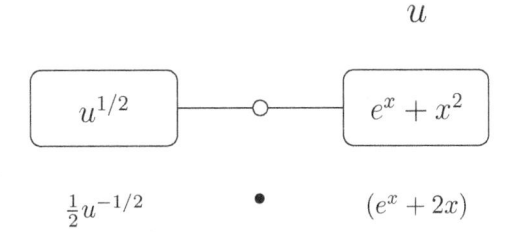

$$u$$

$$\boxed{u^{1/2}} \longrightarrow \boxed{e^x + x^2}$$

$$\tfrac{1}{2} u^{-1/2} \qquad \bullet \qquad (e^x + 2x)$$

(b) We rewrite this function as the composition of $y = e^u$ and $u = x^3 + 2x$. Then we have

$$\frac{dy}{du} = \frac{d}{du} e^u = e^u,$$

and

$$\frac{du}{dx} = \frac{d}{dx}\left(x^3 + 2x\right) = 3x^2 + 2.$$

Hence,

$$\frac{dy}{dx} = \frac{dy}{du} \cdot \frac{du}{dx} = e^u \cdot (3x^2 + 2) = e^{x^3+2x} \cdot (3x^2 + 2) \quad \text{or} \quad (3x^2 + 2)\, e^{x^3+2x}.$$

The graphical representation is as follows.

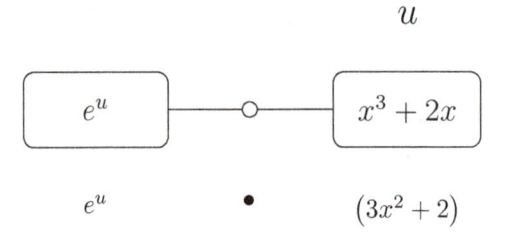

$$\boxed{e^u} \longleftarrow \overset{u}{\circ} \longleftarrow \boxed{x^3 + 2x}$$

$$e^u \qquad \cdot \qquad (3x^2 + 2)$$

Example 3: If $y = 3u^2$ and $u = 7x^5 + 2e^{x^2}$, find the following:

(a) $\dfrac{dy}{du}$;   (b) $\dfrac{du}{dx}$;   (c) $\dfrac{dy}{dx}$.

Solution (a) $\dfrac{dy}{du} = \dfrac{d}{du} 3u^2 = 6u.$

(b) $\dfrac{du}{dx} = \dfrac{d}{dx}\left(7x^5 + 2e^{x^2}\right) = 35x^4 + 2e^{x^2} \cdot 2x = 35x^4 + 4xe^{x^2}.$

(c) $\dfrac{dy}{dx} = \dfrac{dy}{du} \cdot \dfrac{du}{dx} = 6u \cdot \left(35x^4 + 4xe^{x^2}\right) = 6\left(7x^5 + 2e^{x^2}\right) \cdot \left(35x^4 + 4xe^{x^2}\right).$

The Derivative of The Natural Logarithmic Function $y = \ln x$

We are now ready to find the derivative of $f(x) = \ln x$. Recall that by the definition of $f(x) = \ln x$, we have

$$y = \ln x \quad \text{if and only if} \quad x = e^y.$$

Replacing y by $f(x)$ we have the exponential equation

$$e^{f(x)} = x. \tag{2.1}$$

We now differentiate on both sides of the equation:

$$\frac{d}{dx} e^{f(x)} = \frac{d}{dx} x.$$

On the right side of the equation, we have that $\frac{d}{dx} x = 1$. We now apply the Extended Exponential Rule to the left side of the equation:

$$\frac{d}{dx} e^{f(x)} = e^{f(x)} \cdot f'(x).$$

Notice that $e^{f(x)} = x$, hence on the left side we now have

$$\frac{d}{dx} e^{f(x)} = e^{f(x)} \cdot f'(x) = x \cdot f'(x).$$

Hence, after differentiating (2.1), we obtain

$$x \cdot f'(x) = 1,$$

or equivalently,

$$f'(x) = \frac{1}{x}.$$

> **Theorem 2**   **The Natural Logarithmic Rule**
>
> The derivative of the natural logarithmic function $\ln x$ is $\frac{1}{x}$. That is, for any positive number x,
>
> $$\frac{d}{dx} \ln x = \frac{1}{x},$$
>
> or equivalently, if $y = f(x) = \ln x$, then
>
> $$y' = f'(x) = \frac{dy}{dx} = \frac{1}{x}.$$

Example 4: Differentiate each of the following:

(a) $y = \sqrt{x} \ln x$;   (b) $y = (\ln x)^2$.

Solution (a) Notice that y is a product. We apply the "sum of cross-products" process with $f(x) = \sqrt{x}$ and $g(x) = \ln x$.

$$\begin{array}{cc} f(x) & g(x) \\ & \times \\ f'(x) & g'(x) \\ \\ f'(x) \cdot g(x) & + \quad f(x) \cdot g'(x) \end{array} \quad = \quad \begin{array}{cc} \sqrt{x} & \ln x \\ & \times \\ \frac{1}{2\sqrt{x}} & \frac{1}{x} \\ \\ \frac{1}{2\sqrt{x}} \cdot \ln x & + \quad \sqrt{x} \cdot \frac{1}{x}. \end{array}$$

Hence, $\dfrac{d}{dx}\left(\sqrt{x} \ln x\right) = \dfrac{1}{2\sqrt{x}} \cdot \ln x + \sqrt{x} \cdot \dfrac{1}{x}$

$$= \frac{\ln x}{2\sqrt{x}} + \frac{1}{\sqrt{x}} = \frac{\ln x + 2}{2\sqrt{x}}.$$

(b) We apply the Extended Power Rule with $g(x) = \ln x$ to obtain

$$\frac{d}{dx}(\ln x)^2 = 2(\ln x) \cdot \frac{1}{x} = \frac{2 \ln x}{x}.$$

We can apply the Chain Rule to the function $y = \ln g(x)$ to find its derivative:

> **Theorem 3**   *The Extended Natural Logarithmic Rule*
>
> The derivative of the natural logarithm of a differentiable function is its derivative divided by the function. That is,
> $$\frac{d}{dx}\ln g(x) = \frac{g'(x)}{g(x)},$$
> or equivalently, if $y = \ln u$ and $u = g(x)$, then
> $$\frac{dy}{dx} = \frac{1}{u}\frac{du}{dx}.$$

Example 5: Differentiate each of the following:

(a) $y = \ln\left(e^x + x^2\right)$;   (b) $y = \ln(\ln x)$;   (c) $y = \ln(3x)$.

Solution (a) We apply the Extended Natural Logarithmic Rule with $g(x) = e^x + x^2$ to obtain
$$\frac{d}{dx}\ln\left(e^x + x^2\right) = \frac{e^x + 2x}{e^x + x^2}.$$

(b) We apply the Extended Natural Logarithmic Rule with $g(x) = \ln x$ to obtain
$$\frac{d}{dx}\ln(\ln x) = \frac{1/x}{\ln x} = \frac{1}{x\ln x}.$$

(c) We apply the Extended Natural Logarithmic Rule with $g(x) = 3x$ to obtain
$$\frac{d}{dx}\ln(3x) = \frac{3}{3x} = \frac{1}{x}.$$

Example 6: Differentiate each of the following:

(a) $y = \ln\left(\dfrac{x^3 + 5}{2x + 1}\right)$;   (b) $y = \ln\left(\sqrt[3]{x^4}\right)$.

Solution (a) Since $\ln\dfrac{M}{N} = \ln M - \ln N$, we can rewrite the function as
$$y = \ln\left(x^3 + 5\right) - \ln(2x + 1).$$

Hence,
$$\frac{d}{dx}\ln\left(\frac{x^3 + 5}{2x + 1}\right) = \frac{d}{dx}\left(\ln\left(x^3 + 5\right) - \ln(2x + 1)\right) = \frac{3x^2}{x^3 + 5} - \frac{2}{2x + 1}.$$

(b) Since $\ln(a)^k = k\ln a$, we can rewrite the function as
$$y = \ln x^{4/3} = \frac{4}{3}\ln x.$$

Hence,
$$\frac{d}{dx}\ln\left(\sqrt[3]{x^4}\right) = \frac{d}{dx}\frac{4}{3}\ln x = \frac{4}{3}\cdot\frac{1}{x} = \frac{4}{3x}.$$

The Derivatives of $y = a^x$ and $y = \log_a x$

We end this section by exploring how the Extended Natural Exponential and Logarithmic Rules can be used to find the derivatives of the exponential and logarithmic functions with an arbitrary base $a > 0$. Using the identity $a = e^{\ln a}$ we can write the function $y = a^x$ as

$$y = \left(e^{\ln a}\right)^x = e^{(\ln a)x}.$$

Hence, the Extended Natural Exponential Rule can be applied to obtain

$$\frac{dy}{dx} = (\ln a) e^{(\ln a)x} = (\ln a) a^x.$$

Therefore,

$$\frac{d}{dx} a^x = (\ln a) a^x.$$

Similarly, for $y = \log_a x$, we have

$$x = a^y = e^{(\ln a)y},$$

hence $(\ln a)y = \ln x$, or $y = \dfrac{\ln x}{\ln a}$. Therefore,

$$\frac{d}{dx} \log_a x = \frac{dy}{dx} = \frac{1}{(\ln a)x}.$$

Exercise Set 2.8

Find $f(g(x))$ and $g(f(x))$ for the following functions:

1. $f(x) = x^2$, $g(x) = 3x + 1$

2. $f(x) = x^3$, $g(x) = 2x + 4$

3. $f(x) = \sqrt{x}$, $g(x) = x^2 - 1$

4. $f(x) = \sqrt[3]{x}$, $g(x) = x^2 + 3$

5. $f(x) = 3x^2 + 2x + 1$, $g(x) = e^x$

6. $f(x) = x^2 - 5x$, $g(x) = 2e^x$

Find $f(u)$ and $g(x)$ so that $y = f(u)$ and $u = g(x)$. Answers may vary.

7. $y = (2x + 3)^5$

8. $y = \left(4x^2 + 5x + 1\right)^4$

9. $y = \sqrt{e^x + 2}$

10. $y = \dfrac{1}{\sqrt{5x+4}}$

11. $y = e^{2x^2 + x}$

12. $y = \ln\left(3x^2 + 1\right)$

Find $\dfrac{dy}{du}$, $\dfrac{du}{dx}$, and $\dfrac{dy}{dx}$

13. $y = u^3$ and $u = 3e^x + 2$

14. $y = \sqrt{u}$ and $u = x^2 + 2$

15. $y = e^u$ and $u = 3x^2 + 4$

16. $y = u^2 - 2u$ and $u = \sqrt{x}$

17. $y = \ln u$ and $u = x^4 + 1$

18. $y = u^2$ and $u = \ln x$

Find $f'(x)$

19. $f(x) = 2x + 3\ln x + e^x$

20. $f(x) = x^3 - 4\ln x + 3e^x$

21. $f(x) = 5\ln(2x^2 + 3)$

22. $f(x) = 4\ln(3x^4 + x^2 + 5)$

Find $\dfrac{dy}{dx}$

23. $y = \ln\left(\dfrac{5x+3}{4x+1}\right)$ 24. $y = \ln\left(\dfrac{2x+1}{3x-2}\right)$

25. $y = \ln(4x)$ 26. $y = \ln(5x)$

27. $y = \ln\left(\sqrt[4]{x}\right)$ 28. $y = \ln\left(\sqrt[5]{x^3}\right)$

29. $y = x\ln x$ 30. $y = x^2 \ln x$

Find y'

31. $y = (\ln x + 4e^x - 3x^2)^6$

32. $y = (5x^4 - 4\ln x + 3)^5$

33. $y = 3\sqrt{x}\ln 6x$

34. $y = 2\sqrt[3]{x}\ln 7x$

Differentiate:

35. $\dfrac{d}{dx}\left((\ln x)^3 + \ln 4\right)$

36. $\dfrac{d}{dx}\left(\sqrt{(\ln x)} + \ln 5\right)$

37. $\dfrac{d^2}{dx^2}\ln(3x+1)$

38. $\dfrac{d^2}{dx^2}\ln(2x-2)$

39. $\dfrac{d^2}{dx^2}\left(x^2\ln x\right)$

40. $\dfrac{d^2}{dx^2}\left(3x\ln x\right)$

41. Find an equation of the tangent line to the graph of the function $f(x) = (3x+2)\ln x$ at $(1, 0)$.

42. Find an equation of the tangent line to the graph of the function $f(x) = e^x \ln(x^2 + 2)$ at $x = 0$.

43. Find an equation of the tangent line to the graph of the function $f(x) = \ln(x+4)$ at

   (a) $x = -1$;   (b) $x = 0$;   (c) $x = 1$.

44. Find an equation of the tangent line to the graph of the function $f(x) = \ln(2x+5)$ at

   (a) $x = -1$;   (b) $x = 0$;   (c) $x = 1$.

Chapter 2 Summary

This chapter starts our discussion of the first of the two main building blocks of calculus: *differentiation*. The first new concept introduced is the **limit**. This concept is the foundation of both branches of calculus — differentiation and integration. The concept of the limit is defined in an intuitive manner: We say that "when x approaches a, the limit of the function is L" if the function value $f(x)$ is close to L whenever x is close, but not equal, to a (on either side of a). We denote the limit as

$$\lim_{x \to a} f(x) = L, \quad \text{or} \quad f(x) \to L \text{ as } x \to a.$$

The left-limit and right-limit are defined similarly. A limit exists if and only if both side limits exist and are equal.

The limit can be computed algebraically after a group of limit laws is introduced in Section 2.2. However, using a numerical and/or graphical approach is not only very helpful in understanding the concept of the limit, but also in computing limits in the following cases: (1) when $f(x)$ is defined piecewise and $x = a$ is where f changes its formula; (2) when $f(a)$ is not defined; and (3) when $x = a$ makes the denominator, but not the numerator, of f equal zero (in such cases infinity is involved in the limit).

CHAPTER 2 SUMMARY

Example 1: The function g is given by

$$g(x) = \begin{cases} x + 2 & \text{if } x < 2, \\ 2x + 1 & \text{if } x \geq 2. \end{cases}$$

Find $\lim_{x \to 2} g(x)$.

Solution We inspect the limits both numerically and graphically. From the tables and the graph we conclude that

$$\lim_{x \to 2^-} g(x) = 4 \quad \text{and} \quad \lim_{x \to 2^+} g(x) = 5,$$

Hence, $\lim_{x \to 2} g(x)$ does not exist since $\lim_{x \to 2^-} g(x) \neq \lim_{x \to 2^+} g(x)$.

$x \to 2^- \ (x < 2)$	$g(x)$
1	3
1.5	3.5
1.9	3.9
1.99	3.99

Limit of $g(x)$ as $x \to 2^-$

$x \to 2^+ \ (x > 2)$	$g(x)$
3	7
2.5	6
2.1	5.2
2.001	5.002

Limit of $g(x)$ as $x \to 2^+$

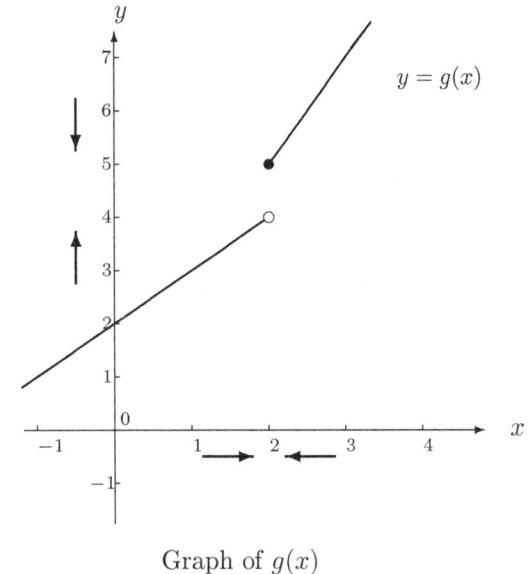

Graph of $g(x)$

Example 2: The function H is given by

$$H(x) = \frac{1}{(x-1)^2}.$$

Find $\lim_{x \to 1} H(x)$.

Solution We inspect the limits numerically.

$x \to 1^-(x<1)$	$H(x)$
0	1
0.5	4
0.9	100
0.99	10,000
0.999	1,000,000

Limit of $H(x)$ as $x \to 1^-$

$x \to 1^+(x>1)$	$H(x)$
2	1
1.5	4
1.1	100
1.01	10,000
1.001	1,000,000

Limit of $H(x)$ as $x \to 1^+$

As x approaches 1 from both sides, the function values become more and more positive without bound. Hence, we conclude that

$$\lim_{x \to 1} H(x) = \infty.$$

For a function $f(x)$, if $f(a)$ is defined and $\lim_{x \to a} f(x) = f(a)$, then we say that f is **continuous** at $x = a$. Since most of the functions studied in this course are continuous, we can apply the definition to obtain its limit, since in such cases, we have

$$\lim_{x \to a} f(x) = f(a).$$

However, if the limit involves a quotient with a denominator that approaches zero when $x \to a$, or the limit involves a piecewise defined function that changes formulas at $x = a$, then the "direct plug-in" approach does not work. Many examples can be found in Section 2.2 in dealing with such cases. One particularly important example is the following:

Example 3: Find $\lim_{h \to 0} \dfrac{(x+h)^2 - x^2}{h}$.

Solution Since in the limit only h approaches 0, while x keeps the same value, we treat x as a constant. Note that when $h \to 0$, both the numerator and the denominator approach zero. Since in taking the limit $h \neq 0$, we can simplify the quotient before taking the limit:

$$\frac{(x+h)^2 - x^2}{h} = \frac{x^2 + 2xh + h^2 - x^2}{h}$$

$$= \frac{h(2x+h)}{h} = (2x+h), \quad (h \neq 0).$$

Hence,

$$\lim_{h \to 0} \frac{(x+h)^2 - x^2}{h} = \lim_{h \to 0} (2x+h) = 2x.$$

The expression $\left[(x+h)^2 - x^2\right]/h$ is a special case (where $f(x) = x^2$) of the **difference quotient**

$$\frac{f(x+h) - f(x)}{h}.$$

CHAPTER 2 SUMMARY

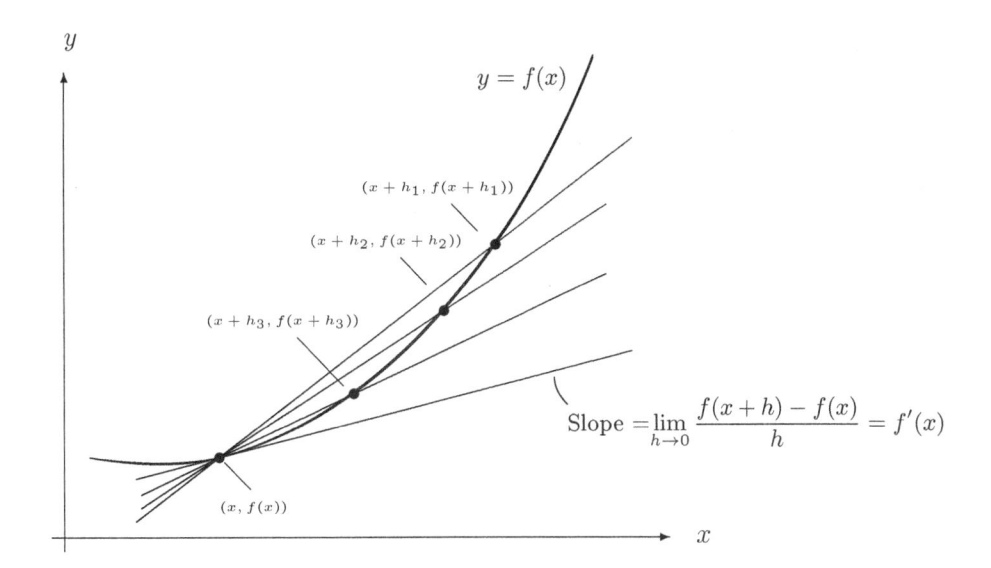

On a graph, the difference quotient represents the **slope of the secant line** passing through the two points $(x, f(x))$ and $(x+h, f(x+h))$. It also represents the **average rate of change** of the function when the input variable changes from x to $x+h$. Applications of the average rate of change can be found in Section 2.3.

When h approaches zero, the average rate of change approaches the **instantaneous rate of change** of the function at x. On a graph, the slope of the secant line approaches the **slope of the tangent line** at $(x, f(x))$. This number is referred to as the **derivative** of f at x.

$$f'(x) = \lim_{h \to 0} \frac{f(x+h) - f(x)}{h}.$$

Example 4: For $y = f(x) = 2x^2 - x$, find (a) $f'(x)$; (b) $f'(1)$; (c) the equation of the tangent line to the graph of $f(x) = 2x^2 - x$ at $x = 1$.

Solution (a) We simplify the difference quotient first:

$$\frac{f(x+h) - f(x)}{h} = \frac{[2(x+h)^2 - (x+h)] - [2x^2 - x]}{h}$$

$$= \frac{[2x^2 + 4xh + 2h^2 - x - h] - [2x^2 - x]}{h} = \frac{2x^2 + 4xh + 2h^2 - x - h - 2x^2 + x}{h}$$

$$= \frac{4xh + 2h^2 - h}{h} = \frac{h(4x + 2h - 1)}{h} = 4x + 2h - 1, \quad h \neq 0.$$

We then take the limit of the simplified difference quotient as h approaches 0:

$$f'(x) = \lim_{h \to 0} \frac{f(x+h) - f(x)}{h} = \lim_{h \to 0} 4x + 2h - 1 = 4x - 1.$$

Therefore, $f'(x) = 4x - 1$.

(b) We now substitute $x = 1$ into $f'(x) = 4x - 1$ to obtain $f'(1) = 4(1) - 1 = 3$.

(c) Since the derivative $f'(1)$ is the slope of the tangent line to the graph of $f(x) = 2x^2 - x$ at $x = 1$, and the line passes through the point $(1, f(1)) = (1, 1)$, the equation is

$$y - 1 = 3(x - 1), \quad \text{or} \quad y = 3x - 2.$$

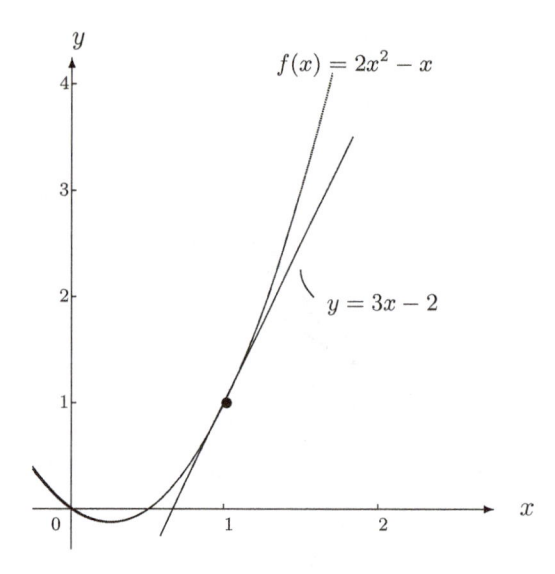

There are several other notations used for the derivative. If $y = f(x)$, then the derivative can be denoted as

$$f'(x) \quad \text{or} \quad y' \quad \text{or} \quad \frac{dy}{dx} \quad \text{or} \quad \frac{d}{dx} f(x).$$

The derivative itself is a function of x, hence higher order derivatives can be defined in a similar manner. The notations for the **second derivative** are as follows:

$$f''(x) \quad \text{or} \quad y'' \quad \text{or} \quad \frac{d^2 y}{dx^2} \quad \text{or} \quad \frac{d^2}{dx^2} f(x).$$

Differentiation Properties and Rules

Several differentiation properties and rules enable us to take short cuts in computing the derivative of many functions.

- **Derivative of Constants** $\dfrac{d}{dx} c = 0$.

Example (a) $\dfrac{d}{dx} 3 = 0$; (b) $\dfrac{d}{dx} e^{-0.2} = 0$.

- **Power Rule** $\dfrac{d}{dx} x^k = k x^{k-1}$.

Example (a) $\dfrac{d}{dx} x^3 = 3x^2$; (b) $\dfrac{d}{dx} \dfrac{1}{x^3} = \dfrac{d}{dx} x^{-3} = -3x^{-4} = -\dfrac{3}{x^4}$;
(c) $\dfrac{d}{dx} \sqrt[3]{x} = \dfrac{d}{dx} x^{1/3} = \dfrac{1}{3} x^{-2/3} = \dfrac{1}{3\sqrt[3]{x^2}}$.

- **Natural Exponential Rule** $\dfrac{d}{dx} e^x = e^x$.

- **Natural Logarithmic Rule** $\dfrac{d}{dx} \ln x = \dfrac{1}{x}$.

- **Basic Differential Properties** (a) $\dfrac{d}{dx}[k \cdot f(x)] = k \cdot f'(x)$; (b) $\dfrac{d}{dx}[f(x) \pm g(x)] = f'(x) \pm g'(x)$.

Example $\dfrac{d}{dx}\left(2e^x - 3\ln x + x^2 - 5\right) = 2e^x - \dfrac{3}{x} + 2x$.

- **Product Rule** $\dfrac{d}{dx}[f(x) \cdot g(x)] = f'(x) \cdot g(x) + f(x) \cdot g'(x)$.

Example $\dfrac{d}{dx}\left(x^4 e^x\right) = 4x^3 e^x + x^4 e^x = (4 + x)x^3 e^x$.

- **Quotient Rule** $\dfrac{d}{dx}\dfrac{f(x)}{g(x)} = \dfrac{f'(x) \cdot g(x) - f(x) \cdot g'(x)}{[g(x)]^2}$.

Example $\dfrac{d}{dx}\dfrac{x^2}{e^x} = \dfrac{2xe^x - x^2 e^x}{(e^x)^2} = \dfrac{(2-x)xe^x}{(e^x)^2} = \dfrac{(2-x)x}{e^x}$.

- **Chain Rule** $\dfrac{d}{dx} f(g(x)) = f'(g(x)) \cdot g'(x)$, or $\dfrac{dy}{dx} = \dfrac{dy}{du} \cdot \dfrac{du}{dx}$.

Example 5: If $y = 3e^u + u^4$ and $u = 5x^3 + 4x^2$, find $\dfrac{dy}{dx}$.

Solution We have $\dfrac{dy}{du} = 3e^u + 4u^3$ and $\dfrac{du}{dx} = 15x^2 + 8x$. Hence,

$$\dfrac{dy}{dx} = \dfrac{dy}{du} \cdot \dfrac{du}{dx} = \left(3e^u + 4u^3\right)\left(15x^2 + 8x\right)$$

$$= \left[3e^{5x^3 + 4x^2} + 4\left(5x^3 + 4x^2\right)^3\right]\left(15x^2 + 8x\right).$$

Special Cases of Chain Rule

- **Extended Power Rule** $\dfrac{d}{dx}[g(x)]^k = k[g(x)]^{k-1} \cdot g'(x)$.

Example $\dfrac{d}{dx}(2 + 3e^x)^{10} = 10(2 + 3e^x)^9 (3e^x) = 30e^x (2 + 3e^x)^9$.

- **Extended Natural Exponential Rule** $\dfrac{d}{dx} e^{g(x)} = g'(x) \cdot e^{g(x)}$.

Example $\dfrac{d}{dx} e^{3x^2} = 6xe^{3x^2}$.

- **Extended Natural Logarithmic Rule** $\dfrac{d}{dx} \ln[g(x)] = \dfrac{g'(x)}{g(x)}$.

Example $\dfrac{d}{dx} \ln\left(3x^5 + 2e^x\right) = \dfrac{15x^4 + 2e^x}{3x^5 + 2e^x}.$

Applications

One of the most common uses of derivatives is in physics, where velocity is the derivative of position with respect to time. The derivative of velocity, with respect to time, is acceleration. Hence the acceleration is the second derivative of position with respect to time. Other applications of derivatives include finding rates of change in finance and productivity settings.

Example 6: If a rock is dropped from the top of a 144-ft-tall building, then the height of the rock, in feet, at t seconds can be modeled by

$$s(t) = 144 - 16t^2 \quad 0 \le t \le 3.$$

(a) Find the velocity $v(t)$ and the acceleration $a(t)$ of the rock.

(b) Find $v(2)$ and $a(2)$. What do these quantities represent?

Solution (a) We have

$$v(t) = s'(t) = -32t, \quad \text{and} \quad a(t) = v'(t) = s''(t) = -32.$$

(b) We have $v(2) = -32(2) = -64$ ft/sec and $a(2) = -32$ ft/sec^2. These quantities represent the velocity and acceleration of the rock after 2 seconds. That is, at 2 seconds after the rock is dropped, it is falling at a speed of 64 ft/sec, and its speed of falling is increasing by 32 ft/sec per second.

Example 7: Suppose that \$1,000 is invested in a savings account paying interest at an annual rate of 5%, compounded continuously.

(a) Find the balance of the investment after 3 years.

(b) Find the rate at which the balance is changing after 3 years.

Solution (a) We have $k = 0.05$ and

$$P(t) = 1{,}000 e^{0.05t}.$$

Hence the balance in the account after 3 years is

$$P(3) = 1{,}000 e^{0.05(3)} = 1{,}000 e^{0.15} \approx \$1{,}161.83$$

(b) We apply the Extended Natural Exponential Rule to find the derivative of $P(t)$:

$$P'(t) = 1{,}000(0.05) e^{0.05t} = 50 e^{0.05t}.$$

Hence after 3 years, the rate at which the balance is changing is

$$P'(3) = 50 e^{0.05(3)} = 50 e^{0.15} \approx \$58.09/\text{year}.$$

That is, at the end of the third year, the balance of the investment is growing at the rate of \$58.09 per year.

Chapter 2 Review Exercises

1. The graph of the function H is represented here. Use the graph to find each limit. If necessary, state that the limit does not exist.

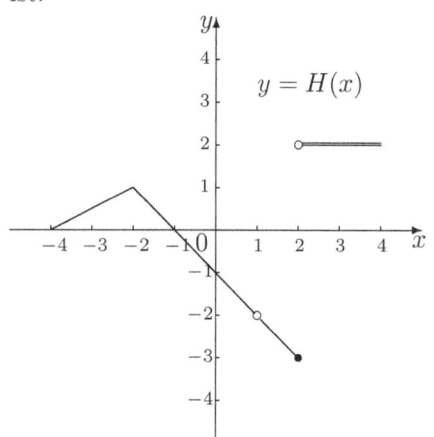

   (a) $\lim_{x \to -2^-} H(x)$ (b) $\lim_{x \to -2^+} H(x)$

   (c) $\lim_{x \to -2} H(x)$ (d) $\lim_{x \to 2^-} H(x)$

   (e) $\lim_{x \to 2^+} H(x)$ (f) $\lim_{x \to 2} H(x)$

   (g) $\lim_{x \to 1^-} H(x)$ (h) $\lim_{x \to 1^+} H(x)$

   (i) $\lim_{x \to 1} H(x)$ (j) $\lim_{x \to 3} H(x)$

2. Use numerical tables to find the specified limits. If the limit does not exist, state the fact.

   (a) $g(x) = \dfrac{x^2 + x - 6}{x - 2}$; find $\lim_{x \to -2} g(x)$.

   (b) $s(x) = \dfrac{1}{(x-1)^2} - 1$;

   find $\lim_{x \to 1} s(x)$ and $\lim_{x \to \infty} s(x)$.

   (c) $f(x) = \dfrac{1}{x+1}$;

   find $\lim_{x \to -1} f(x)$ and $\lim_{x \to \infty} f(x)$.

3. Use the continuity of the functions to find the following limits.

   (a) $\lim_{x \to 1} \dfrac{x^2 - 4}{x - 2}$;

   (b) $\lim_{h \to 0} \dfrac{-4}{x(x+h)}$.

4. The graph of the function f is represented here. Use the graph to answer the following questions.

   (a) Is f continuous over the interval (-4, 4)?

   (b) Find $\lim_{x \to -1^-} f(x)$, $\lim_{x \to -1^+} f(x)$, $\lim_{x \to -1} f(x)$, and $f(-1)$.

   (c) Is f continuous at $x = -1$? Why or why not?

   (d) Find $\lim_{x \to -2^-} f(x)$, $\lim_{x \to -2^+} f(x)$, $\lim_{x \to -2} f(x)$, and $f(-2)$.

   (e) Is f continuous at $x = -2$? Why or why not?

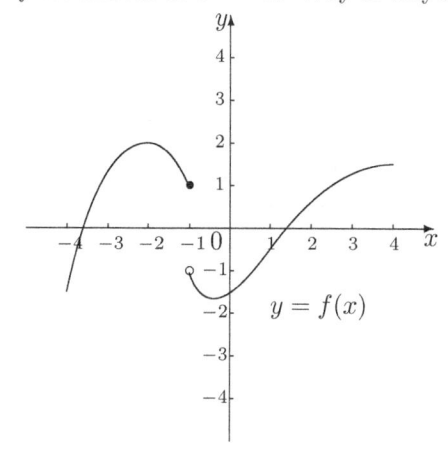

5. Consider the function given by
   $$f(x) = \begin{cases} x - 2 & \text{if } x \leq 1, \\ x^2 & \text{if } x > 1. \end{cases}$$

   (a) Find $f(1)$.

   (b) Find $\lim_{x \to 1^-} f(x)$, $\lim_{x \to 1^+} f(x)$, and $\lim_{x \to 1} f(x)$.

   (c) Is f continuous at $x = 1$? Why or why not?

6. Use algebraic simplifications to find the following limits.

   (a) $\lim_{x \to 1} \dfrac{x^2 + x - 2}{x - 1}$

   (b) $\lim_{x \to -1} \dfrac{x + 1}{x^2 - 1}$

   (c) $\lim_{h \to 0} \dfrac{(2+h)^3 - 8}{h}$

   (d) $\lim_{h \to 0} \dfrac{5(x+h)^2 - 5x^2}{h}$

(e) $\lim_{h \to 0} \dfrac{\frac{2}{x+h} - \frac{2}{x}}{h}$

7. If a rock is dropped from the top of a 256-ft-tall building, then the height of the rock, in feet, at t seconds can be modeled by
$$s(t) = 256 - 16t^2 \quad 0 \le t \le 4.$$

(a) Find $s(3) - s(1)$.

What does this quantity represent?

(b) Find $\dfrac{s(3) - s(1)}{3 - 1}$.

What does this quantity represent?

(c) Find $v(t)$.

(d) Find $v(2)$. What does this quantity represent?

8. Jan has a $2,000 balance on her credit card that charges a 16% annual interest rate, compounded quarterly. Her monthly payment is waived for three years. Her balance after t years is given by
$$A(t) = 2{,}000(1.04)^{4t} \quad 0 \le t \le 3.$$

(a) Find $A(3) - A(0)$.

What does this quantity represent?

(b) Find $\dfrac{A(3) - A(0)}{3 - 0}$.

What does this quantity represent?

9. Find an equation of the tangent line to the graph of the function $f(x) = 2x^3 - 3x^2 + x - 1$ at

(a) $x = -1$; (b) $x = 0$; (c) $x = 1$.

10. For $f(x) = 2x^3 - 7x^2 + 4x + 3$, find the points on the curve where the tangent line is horizontal.

For Exercises 11–23: Find the derivative of each function.

11. $f(x) = 5x^4 - 3x - 7e^x - 2\ln x + 8$

12. $f(x) = \left(3\sqrt{x} + \dfrac{8}{x}\right)(e^x + x^2 + 4)$

13. $f(x) = \dfrac{5x}{7} + 3\sqrt{x} - \dfrac{5}{\sqrt[3]{x^2}}$

14. $f(x) = \dfrac{x^2 + x + 4}{x - 2}$

15. $f(x) = 5e^x - x^e$

16. $f(x) = 5\ln(x^2 + 1) - 7e^{-0.04x}$

17. $f(x) = (2x^3 + 5x^2 - 3)^7$

18. $f(x) = 5e^{\frac{x^2}{2}}$

19. $f(x) = (x^2 + 1)^7 \cdot (3x - 2)^2$

20. $f(x) = \dfrac{x^3 + 3x}{e^x}$

21. $f(x) = \sqrt{\dfrac{5x + 1}{7x - 3}}$

22. $f(x) = (10x^2 + 6x - 3)^2 \cdot e^{2x+5}$

23. $f(x) = 3\sqrt{2x + 1} \cdot \ln x$

24. Determine $\dfrac{d^4}{dx^4} x^7$.

25. For $f(x) = \sqrt{x^3}$, find $f'(4)$ and $f''(4)$.

26. Find $\dfrac{dy}{du}, \dfrac{du}{dx},$ and $\dfrac{dy}{dx}$:

(a) $y = u^3 + e^u$ and $u = \sqrt{x}$.

(b) $y = \sqrt{u}$ and $u = x^2 + \ln x$.

27. If a rock is thrown from the ground with an initial velocity of 80 ft/sec, then the height of the rock, in feet, at t seconds can be modeled by
$$s(t) = -16t^2 + 80t \quad 0 \le t \le 5.$$

(a) Find $v(t)$ and $a(t)$.

(b) At what time does the rock reach the highest height?

(c) What is the highest height the rock reaches?

(d) What is the acceleration of the rock when it reaches the highest height?

28. John invests $2,000 in a savings account paying interest at an annual rate of 4.5%, compounded monthly. His balance after t years is given by
$$A(t) = 2{,}000(1.00375)^{12t}.$$

(a) Find $A'(1)$.

What does this quantity represent?

(b) Find $A'(5)$.

What does this quantity represent?

(c) Compare the answers obtained in parts (a) and (b). Which is greater? Why?

(d) Which of the following quantities change with time?

(i) The balance

(ii) The annual interest rate

(iii) The rate at which the balance is changing.

Chapter 2 Test (100 points)

SCORE _____

NAME _____

ANSWERS

1. (10pts) For the function g graphed here, answer the following.

    (a) Find $\lim_{x \to -1} g(x)$.

    (b) Find $g(-1)$.

    (c) Is g continuous at x=-1?

    (d) Find $\lim_{x \to 3} g(x)$.

    (e) Find $g(3)$.

    (f) Is g continuous at x=3?

    (g) Is g continuous at x=2?

    (h) Find $\lim_{x \to 0} g(x)$.

1. (a) _____

    (b) _____

    (c) _____

    (d) _____

    (e) _____

    (f) _____

    (g) _____

    (h) _____

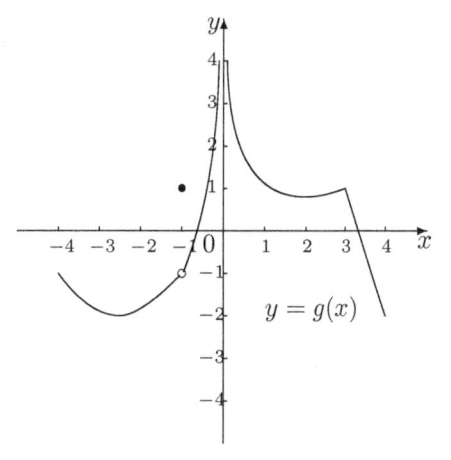

2. (4pts) Find $\lim_{x \to -5} \dfrac{x+5}{x^2 - 25}$.

    2. _____

3. (4pts) Find $\lim_{x \to 0} \dfrac{6}{x}$.

    3. _____

4. (4pts) Find $\lim_{x \to 0} \dfrac{x+6}{x^2 + 3x - 18}$.

    4. _____

5. (8pts) Find a simplified difference quotient for $f(x) = 2x^2 + 4x$.

    5. _____

6. (3pts) Find the limit of the simplified difference quotient in Problem 5 when $h \to 0$.

    6. _____

CHAPTER 2 TEST

7. (8pts) Find an equation of the tangent line to the graph
of $y = x + (4/x)$ at the point $(2, 4)$.

7. _____

8. (6pts) Find the points on the graph of $y = x^2 + 6x$ at which the tangent line is horizontal.

8. _____

9. (6pts) Find $\dfrac{dy}{dx}$ for $y = \dfrac{1}{6}x^6 - \dfrac{2}{x^2} + 2\sqrt[4]{x} + 6e^x - 3\ln x$.

9. _____

10. (6pts) Find $f'(x)$ for $f(x) = (3x^3 - 1)^{-4}$.

10. _____

11. (6pts) Find $f'(x)$ for $f(x) = (4x - 7)^{1/3} e^{x^2}$.

11. _____

12. (6pts) Find $f'(x)$ for $f(x) = \dfrac{x^2 + 4}{3 - x}$.

12. _____

13. (6pts) For $y = x^6 - 2x^4 + 3x^2$, find $\dfrac{d^3 y}{dx^3}$.

13. _____

14. (8pts) Given $s(t) = 2t^3 + 2t$, where s is measured in feet and t is in seconds:

    (a) Find $s(5) - s(3)$ and explain its meaning. Include the unit.

    14. (a) _____

    (b) Find the average velocity as t changes from $t_1 = 3$ to $t_2 = 5$. Include the unit.
    Show your work:

    (b) _____

    (c) Find the velocity when $t = 3$ second. Include the unit.
    Show your work:

    (c) _____

(d) On the **graph** of $s(t)$, the answer in part (c) represents the

_____.

15. (8pts) Find (a) $\dfrac{dy}{du}$, (b) $\dfrac{du}{dx}$, and (c) $\dfrac{dy}{dx}$ given that $y = \dfrac{3}{u^2} + \ln u$ and $u = 2x + 1$.

15. (a) _____

(b) _____

(c) _____

16. (8pts) Rich invests $4,000 in a savings account paying interest at an annual rate of 4%, compounded continuously. His balance after t years is given by
$$A(t) = 4{,}000 e^{0.04t}.$$

(a) Find $A'(3)$.

16. (a) _____

(b) What does $A'(3)$ represent?

(b) _____

(c) Which of the following quantities change with time?

(i) The balance

(ii) The annual interest rate

(iii) The rate at which the balance is changing.

(c) _____

Chapter 2 Projects

Project 1

Jerry is helping his elderly mother to organize her financial investments. One of her investment funds started 7 years ago, with the initial deposit unknown. The interest rate did not change.

(1) Draw a possible graph of the balance A against time t (years) since the fund started. Explain any interesting feature of the graph.

(2) Jerry found that 2 years after the fund was set up, the balance was $60,000 and was increasing at the (instantaneous) rate of $2,400 per year. Using this information, plus what you know about the shape of the A graph, estimate the balance of the fund at $t = 8$.

(3) Using all the data given so far, derive a function $A(t)$. You may use a continuous compounding model. Explain to Jerry why such a model works, even though the interest may not be compounded continuously.

(4) Use the function obtained in part (3) to check how accurate the estimate in part (2) is.

(5) Use the function obtained in part (3) to find how long will it take from now for the fund to reach the balance of $100,000.

Project 2

The following graph represents your velocity, v (in miles per hour) with respect to time t (in hours), on a bicycle trip along a straight road. Suppose that you started out 25 miles west of your home, and positive velocities take you toward home (i.e., eastbound) and negative velocities take you away from home. Answer each of the questions; explain clearly how you reached each answer.

(a) At what time(s) do you change directions?

(b) At $t = 2$ hour, are you riding toward home or away from home?

(c) At what time are you peddling most rapidly away from home?

(d) At what time are you peddling most rapidly toward home?

(e) During what time period(s) are you speeding up?

(f) At what instant are you speeding up most rapidly?

(g) During what time period(s) are you slowing down?

(h) At what instant are you slowing down most rapidly?

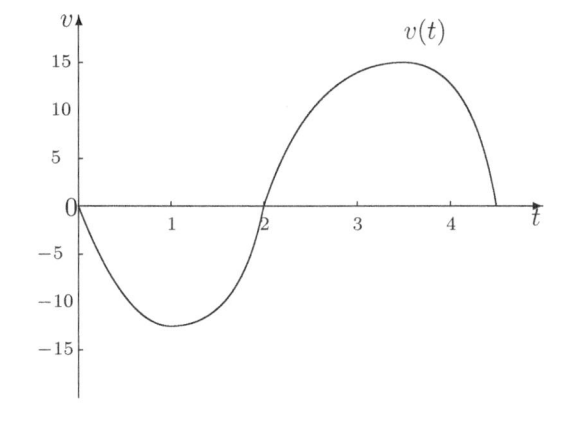

Chapter 3

Applications of Differentiation

Introduction

In Chapter 2, we learned that a derivative represents an instantaneous rate of change, as well as the slope of a tangent line at a specific point. We also learned various techniques for finding derivatives. In this chapter, we explore how the concept of rate of change (i.e., the derivatives) help us to study the behavior of a function. We first study how the derivatives can help us to estimate the function values around a given point. We then apply these techniques to analyze productivity, which is called the marginal analysis in economics.

Mathematical models can be used to solve many problems raised in the business world. Such mathematical models involve either mathematical equations or the study of certain properties of functions. The graph of a function often supplies very important information about the behavior of the function. The importance of calculus in graph sketching does not disappear with the broad use of graphing devices such as graphing calculators, since calculus helps us to grasp the major features of a function its graph should include. Therefore, a significant portion of this chapter is devoted to the study of graph sketching using calculus.

In the remainder of the chapter, we study several mathematical models arising from economics, such as how to maximize revenue or profit, or to minimize costs. We illustrate how calculus can help us to find solutions to such problems.

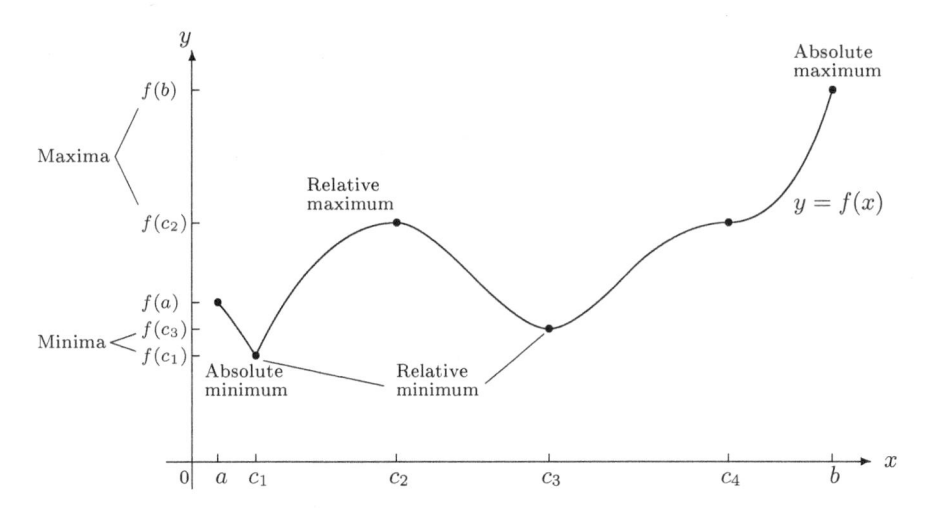

> **Study Objectives**
> 1. Use derivatives to approximate function values.
> 2. Use derivatives to determine where a function is increasing or decreasing, concave up or concave down.
> 3. Apply the derivative tests to determine relative and absolute extrema of a function.
> 4. Sketch the graph of a function using a graphing device or by hand.
> 5. Solve optimization problems.
> 6. Apply the aforementioned skills to solve business related problems, such as profit maximization, cost minimization, elasticity analysis, etc.

3.1 Differentials and Linear Approximation

In this section, we study how knowing the derivative at a point can help us to estimate the values of the function at nearby points. We start with an example.

Example 1: The number of bottles of a new brand of vitamin pills, y, sold in a week is a function of the amount, x, in dollars, spent on advertising in that week. So $y = f(x)$.

(a) What does $f'(150) = 3$ mean?

(b) If $f(150) = 160$, use the information given in part (a) to estimate $f(151)$ and $f(155)$. Which estimate is more reliable?

Solution (a) Since the derivative represents the rate of change, $f'(150) = 3$ tells us that the function is increasing at the rate of three bottles per dollar. That is, if the amount spent on advertising is $150 and increases by $1, then the sales will go up by about three bottles.

(b) $f(150) = 160$ means that if $150 is spent on advertising, there are 160 bottles of vitamin sold. By part (a), if $151 is spent on advertising, then the sales will increase by about three bottles, hence, the total sales will be about 163 bottles. That is,

$$f(151) \approx f(150) + f'(150)(1) = 160 + 3(1) = 163.$$

Similarly, if $155 is spent on advertising, then the amount spent on advertising is increased from $150 by $5, hence, the sales will increase by about $3(5) = 15$ bottles, hence, the total sales will be about 175 bottles. That is,

$$f(155) \approx f(150) + f'(150)(5) = 160 + 3(5) = 175.$$

In the second case, we have to assume that the rate of three additional sales for each additional dollar continues from 150 up to 155; therefore, the estimate of $f(151)$ is more reliable.

Example 1 illustrates how the derivative value at a particular variable value can be used to approximate the function values at nearby variable values. This type of approximation is called **linear approximation**. We now represent the linear approximation in formal mathematical notation. We first introduce the **delta** notation. We use the symbol Δx, read "delta x," to represent the *change in variable x*. That is,

$$\Delta x = x_2 - x_1, \quad \text{or if} \quad x_1 = x \text{ and } x_2 = x+h, \quad \text{then } \Delta x = (x+h) - x = h.$$

For a function $y = f(x)$, when the variable x makes a change by Δx, it causes a change in the function value. We use the symbol Δy to represent the *change in function*. That is,

$$\Delta y = y_2 - y_1 = f(x_2) - f(x_1), \quad \text{or if} \quad x_1 = x \text{ and } x_2 = x+h, \quad \text{then } \Delta y = f(x+h) - f(x).$$

The relationships are illustrated in the following graph (Figure 3.1.1).

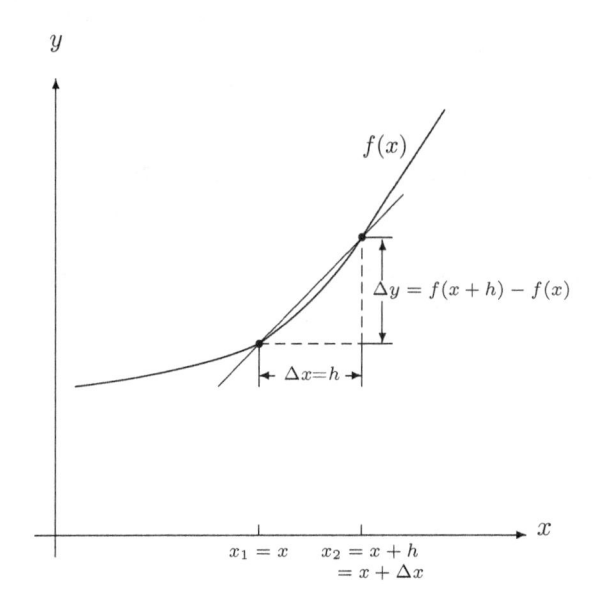

Figure 3.1.1: Δx and Δy

Recall that in Section 2.3, the difference quotient of $y = f(x)$ is defined by

$$\frac{f(x+h) - f(x)}{h},$$

and in Section 2.4, the derivative of $y = f(x)$ is defined by

$$\frac{dy}{dx} = f'(x) = \lim_{h \to 0} \frac{f(x+h) - f(x)}{h}.$$

Using the delta notation, the difference quotient becomes

$$\frac{f(x+h) - f(x)}{h} = \frac{\Delta y}{\Delta x},$$

and the derivative becomes

$$\frac{dy}{dx} = f'(x) = \lim_{\Delta x \to 0} \frac{\Delta y}{\Delta x}.$$

When Δx is close to zero, we have the approximation

$$\frac{dy}{dx} \approx \frac{\Delta y}{\Delta x}, \quad \text{or} \quad f'(x) \approx \frac{\Delta y}{\Delta x}.$$

3.1. DIFFERENTIALS AND LINEAR APPROXIMATION

Multiplying both sides of the second equation by Δx, we obtain the approximation

$$\Delta y \approx f'(x)\Delta x.$$

Since $\Delta y = f(x + \Delta x) - f(x)$, an equivalent approximation is

$$f(x + \Delta x) - f(x) \approx f'(x)\Delta x, \quad \text{or } f(x + \Delta x) \approx f(x) + f'(x)\Delta x.$$

The relationships are illustrated in the following graph (Figure 3.1.2). Note that the approximating point is on the tangent line. That is how the name "linear approximation" is given.

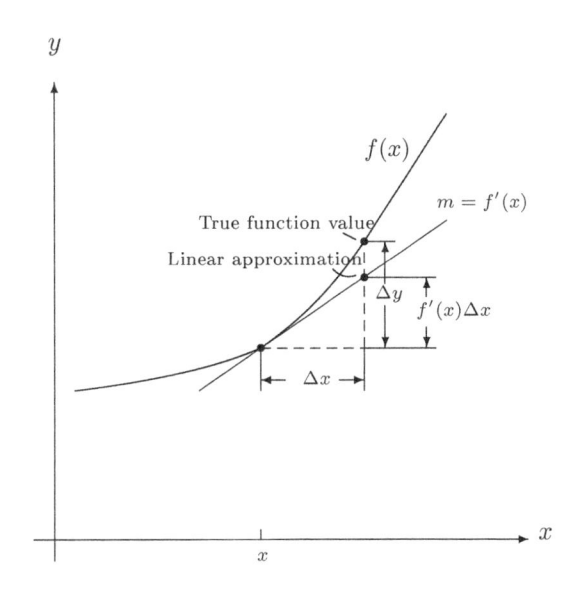

Figure 3.1.2: Linear approximation using Δx

We summarize the approximations as the following:

Linear Approximation

Suppose that $y = f(x)$ is a continuous and differentiable function, then for small Δx,

(a) $f'(x) \approx \dfrac{\Delta y}{\Delta x}$. That is, the difference quotient is an approximation of the derivative.

(b) $\Delta y \approx f'(x)\Delta x$. That is, the vertical change along the tangent line is an approximation of the change in the function.

(c) $f(x+\Delta x) \approx f(x)+f'(x)\Delta x$. That is, the vertical value on the tangent line is an approximation of the function value.

Example 2: Consider the function $y = f(x) = 3x^2 - 2x$:

(a) Find Δx and Δy for $x_1 = 2$ and $x_2 = 2.5$.

(b) Find Δx and Δy for $x_1 = 2$ and $x_2 = 2 + h$.

(c) Find Δy for $x = 2$ and $\Delta x = 0.1$, and compare with $f'(2)\Delta x$.

(d) Use the results in part (c) to find the linear approximation of $f(2.1)$ and compare it with the true value $f(2.1)$.

Solution (a) According to the definition,

$$\Delta x = x_2 - x_1 = 2.5 - 2 = 0.5, \quad \text{and}$$

$$\Delta y = f(x_2) - f(x_1) = f(2.5) - f(2)$$
$$= [3(2.5)^2 - 2(2.5)] - [3(2)^2 - 2(2)] = 13.75 - 8 = 5.75.$$

(b) According to the definition,

$$\Delta x = x_2 - x_1 = (2 + h) - 2 = h, \quad \text{and}$$

$$\Delta y = f(x_2) - f(x_1) = f(2+h) - f(2) = \left[3(2+h)^2 - 2(2+h)\right] - \left[3(2)^2 - 2(2)\right]$$
$$= \left[3(4 + 4h + h^2) - 4 - 2h\right] - 8 = \left[8 + 10h + 3h^2\right] - 8 = 10h + 3h^2.$$

(c) According to the definition,

$$\Delta y = f(x + \Delta x) - f(x) = f(2 + 0.1) - f(2) = f(2.1) - f(2)$$
$$= \left[3(2.1)^2 - 2(2.1)\right] - \left[3(2)^2 - 2(2)\right] = 9.03 - 8 = 1.03.$$

Since $f'(x) = 6x - 2$, we have $f'(2) = 6(2) - 2 = 10$. Therefore, $f'(2)\Delta x = 10(0.1) = 1$. It is a fairly accurate approximation of $\Delta y = 1.03$.

(d) We apply the linear approximation formula with $x = 2$ and $\Delta x = 0.1$. From part (c), we know that $f'(2)\Delta x = 1$:

$$f(2.1) = f(2 + 0.1) \approx f(2) + f'(2)(0.1) = 8 + 1 = 9.$$

From part (c), the true value $f(2.1)$ is 9.03. Hence, the linear approximation is fairly accurate.

Example 3: Estimate $\sqrt{18}$ using linear approximation.

Solution Let $f(x) = \sqrt{x}$. Since a perfect square near 18 is 16, we let $x = 16$. Then $\Delta x = 18 - 16 = 2$. The linear approximation formula gives

$$f(18) = f(16 + 2) \approx f(16) + f'(16)(2).$$

We have $f(16) = \sqrt{16} = 4$. Since

$$f'(x) = \frac{1}{2\sqrt{x}},$$

we have $f'(16) = \dfrac{1}{2\sqrt{16}} = \dfrac{1}{8}$. Therefore

$$f(18) \approx f(16) + f'(16)(2) = 4 + \frac{1}{8}(2) = 4 + 0.25 = 4.25.$$

Note that $\sqrt{18} = 4.24264\ldots$. Hence, the linear approximation is fairly accurate.

Example 4: The owner of a small painting company wins a bid to paint 20 decorative balls for the city for holiday street decorations. The balls are 6 ft in diameter, and the paint coat should be 0.005 in.

thick. The owner remembers that the volume formula for a sphere is $V(r) = \left(\dfrac{4\pi}{3}\right)r^3$. How should he use linear approximation to estimate the amount of paint he needs to purchase for the job?

Solution Since the balls have 6 ft diameters, their radius is $3\,\text{ft} = 36\,\text{in}$. After the painting, the radius increases to $36 + 0.005 = 36.005\,\text{in}$. Hence, the paint needed for each ball has volume $\Delta V = V(36.005) - V(36)$. The amount of paint needed for each ball is approximately

$$\Delta V \approx V'(36)\Delta r = V'(36)(0.005).$$

Since $V(r) = \left(\dfrac{4\pi}{3}\right)r^3$, we have $V'(r) = \dfrac{4\pi}{3}\cdot 3r^2 = 4\pi r^2$. Hence, $V'(36) = 4\pi(36)^2 = 5{,}184\pi$. Therefore

$$\Delta V \approx V'(36)(0.005) = 5{,}184\pi(0.005) \approx 81.4\,\text{in.}^3$$

The total volume of paint for 20 balls is about $81.4 \times 20 = 1{,}628\,\text{in.}^3$ Since 1 gallon equals $231\,\text{in.}^3$, the owner needs to purchase about 7 gallons of paint for the job.

Reality Check   *Volume and surface area*

Notice that the derivative of the volume is the surface area. The linear approximation in Example 4 indicates that, to estimate the amount of paint needed for painting a ball (which is the additional volume of the ball due to the paint), we can multiply the surface area of the unpainted ball by the thickness of the paint.

Reality Check   *Significance of linear approximations*

Linear approximations were very valuable before computers and graphing calculators became common tools for computations. Today, the use of linear approximation as a computation method has become much less significant. However, it still remains an important approach in the applied mathematics fields.

The Differentials

Let us observe the approximation relationship

$$\frac{dy}{dx} \approx \frac{\Delta y}{\Delta x}.$$

On the left side is the derivative using Leibniz notation. It looks like a fraction but is actually a single symbol. On the right hand there is a true fraction, the ratio of change in y to the change in x. Each side is an approximation of the other. Also, recall the Chain Rule in Leibniz notation:

$$\frac{dy}{dx} = \frac{dy}{du} \cdot \frac{du}{dx}.$$

Again they act like fractions. Indeed, we are now ready to define dy and dx as two separate symbols. They are called **differentials**.

> **Definition** *differentials*
>
> Let $y = f(x)$ be a differentiable function. We define the **differential of** x, denoted by dx, as
>
> $$dx = \Delta x.$$
>
> We define the **differential of** y, denoted by dy, as
>
> $$dy = f'(x)dx.$$

The differentials dx and dy are illustrated in the following graph (Figure 3.1.3):

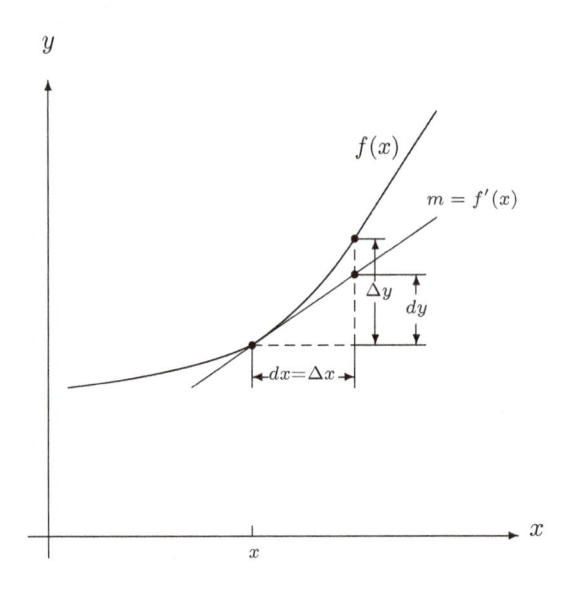

Figure 3.1.3: dx and dy vs. Δx and Δy

> **Reality Check** *Graphical meaning of differentials*
>
> Note that on graph, dy represents the change of y along the tangent line. Therefore, the linear approximation can be denoted as
>
> $$\Delta y \approx dy.$$
>
> From the definition of dy and dx, the slope of the tangent line is indeed the ratio of dy over dx. The differentials as symbols play important roles in integrals and differential equations in later chapters.

Example 5: For $y = \ln(x^2 + 4)$:

(a) Find dy.

(b). Find dy when $x = 1$ and $dx = 0.02$.

Solution (a) We first find the derivative $\frac{dy}{dx}$. Applying the Extended Logarithmic Rule, we obtain

$$\frac{dy}{dx} = \frac{2x}{x^2+4}.$$

Hence

$$dy = \frac{dy}{dx} \cdot dx = \left[\frac{2x}{x^2+4}\right] dx.$$

(b) When $x = 1$ and $dx = 0.02$, we have

$$dy = \left[\frac{2(1)}{(1)^2+4}\right](0.02) = (0.4)(0.02) = 0.008.$$

Exercise Set 3.1

1. The daily heating cost in the winter for a house in Wisconsin is a function of the outside temperature. Hence $C = f(T)$, where T is in degrees Fahrenheit and C is in dollars.

   (a) Using units, explain the meaning of $f(36) = 4.50$.

   (b) Using units, explain the meaning of $f'(36) = -0.25$.

   (c) Using the information in parts (a) and (b), estimate the daily heating cost when the outside temperature is 38°F.

   (d) Using the information in parts (a) and (b), estimate the daily heating cost when the outside temperature is 33°F.

2. The waste collection cost for a restaurant is a function of the weight of the waste collected. Hence $C = f(w)$, where w is in kilograms and C is in dollars.

   (a) Using units, explain the meaning of $f(100) = 30$.

   (b) Using units, explain the meaning of $f'(100) = 0.15$.

   (c) Using the information in parts (a) and (b), estimate the cost of collecting 105 kg of waste.

   (d) Using the information in parts (a) and (b), estimate the cost of collecting 97 kg of waste.

3. After investing $1,000 in a mutual fund, your balance is a function of the time passed. Hence $B = f(t)$, where t is in months and B is in dollars.

   (a) Using units, explain the meaning of $f(12) = 1,095$.

   (b) Using units, explain the meaning of $f'(12) = 8$.

   (c) Using the information in parts (a) and (b), estimate and interpret $f(12.5)$.

4. The total amount required to pay off a 5 year car loan is a function of the annual interest rate charged. Hence $C = f(r)$, where r is a percentage and C is in dollars.

   (a) Using units, explain the meaning of $f(8) = 300$.

   (b) Using units, explain the meaning of $f'(8) = 3,800$.

   (c) Using the information in parts (a) and (b), estimate and interpret $f(8.4)$.

5. The quantity of a special brand of skateboard sold annually in a sporting goods store is a function of the price charged. Hence $q = f(p)$, where p is in dollars per skateboard and q is the number of skateboards sold.

   (a) Using units, explain the meaning of $f(120) = 500$.

   (b) Using units, explain the meaning of $f'(120) = -10$.

(c) Using the information in parts (a) and (b), estimate and interpret $f(124)$.

6. The revenue of a car dealer from car sales is a function of the advertising expenditure. Hence $R = f(a)$, where both of a and R are in thousands of dollars.

   (a) Using units, explain the meaning of $f(10) = 50$.

   (b) Using units, explain the meaning of $f'(10) = 2$.

   (c) Using the information in parts (a) and (b), estimate and interpret $f(10.8)$.

7. A chicken is taken out of the refrigerator and placed in a hot oven to cook. The temperature of the middle of the chicken is a function of the time passed. Hence $H = f(t)$, where t is in minutes and H is in degrees Fahrenheit.

   (a) Using units, explain the meaning of $f(25) = 95$.

   (b) Using units, explain the meaning of $f'(25) = 5$.

   (c) Using the information in parts (a) and (b), estimate the temperature of the chicken 28 minutes after it is put in the oven.

   (d) Using the information in parts (a) and (b), estimate the temperature of the chicken 32 minutes after it is put in the oven.

   (e) Which estimate is more accurate? Why?

8. A cup of freshly made coffee is put on a table. The temperature of the coffee is a function of the time passed. Hence $H = f(t)$, where t is in minutes and H is in degrees Fahrenheit.

   (a) Using units, explain the meaning of $f(3) = 120$.

   (b) Using units, explain the meaning of $f'(3) = -10$.

   (c) Using the information in parts (a) and (b), estimate the temperature of the coffee 3.5 minutes after it is put on the table.

   (d) Using the information in parts (a) and (b), estimate the temperature of the coffee 4 minutes after it is put on the table.

   (e) Which estimate is more accurate? Why?

9. Consider the function $y = f(x) = x^2 + 2x$:

   (a) Find Δx and Δy for $x_1 = 2$ and $x_2 = 2.5$.

   (b) Find Δx and Δy for $x_1 = 2$ and $x_2 = 2 + h$.

   (c) Find Δy for $x = 2$ and $\Delta x = 0.1$, and compare with $f'(2)\Delta x$.

   (d) Use the results in part (c) to find the linear approximation of $f(2.1)$ and compare with the true value $f(2.1)$.

10. Consider the function $y = f(x) = -4x^2 + x$:

    (a) Find Δx and Δy for $x_1 = 2$ and $x_2 = 2.5$.

    (b) Find Δx and Δy for $x_1 = 2$ and $x_2 = 2 + h$.

    (c) Find Δy for $x = 2$ and $\Delta x = 0.01$, and compare with $f'(2)\Delta x$.

    (d) Use the results in part (c) to find the linear approximation of $f(2.01)$ and compare with the true value $f(2.01)$.

For Exercises #11 to #22: Find Δy and $f'(x)\Delta x$. Round to three decimal places.

11. For $y = f(x) = 3x + 1$, $x = 3$, and $\Delta x = 0.5$.

12. For $y = f(x) = -4x + 3$, $x = 5$, and $\Delta x = 0.2$.

13. For $y = f(x) = 1/x$, $x = 1$, and $\Delta x = 0.3$.

14. For $y = f(x) = -2/x$, $x = 2$, and $\Delta x = 0.2$.

15. For $y = f(x) = 1/x^2$, $x = 1$, and $\Delta x = 0.04$.

16. For $y = f(x) = 3/x^2$, $x = 1$, and $\Delta x = 0.03$.

17. For $y = f(x) = 2e^{-0.5x}$, $x = 0$, and $\Delta x = 0.01$.

18. For $y = f(x) = e^{0.4x}$, $x = 0$, and $\Delta x = 0.01$.

19. For $y = f(x) = \ln(x + 1)$, $x = 0$, and $\Delta x = 0.1$.

20. For $y = f(x) = \ln(3x + 1)$, $x = 0$, and $\Delta x = 0.1$.

21. For $y = f(x) = x^3$, $x = 1$, and $\Delta x = 0.04$.

22. For $y = f(x) = x - x^3$, $x = 1$, and $\Delta x = 0.02$.

23. Use linear approximation to estimate $\ln 1.02$.

24. Use linear approximation to estimate $e^{0.05}$.

For Exercises #25 to #36: Use linear approximation to find a decimal approximation of each radical expression. Round to three decimal places.

25. $\sqrt{8.8}$ 26. $\sqrt{15.6}$

27. $\sqrt[3]{29}$ 28. $\sqrt[3]{28}$

29. $\sqrt{5}$ 30. $\sqrt{4.5}$

31. $\sqrt[3]{25.5}$ 32. $\sqrt[3]{26.2}$

33. $\dfrac{2}{\sqrt{50}}$ 34. $\dfrac{2}{\sqrt{17}}$

35. $\dfrac{1}{\sqrt[4]{15.5}}$ 36. $\dfrac{1}{\sqrt[3]{7.5}}$

37. For $y = (2x+1)^3$, find dy when $x = 2$ and $dx = 0.01$.

38. For $y = (3x+1)^2$, find dy when $x = 1$ and $dx = 0.02$.

39. For $y = 3x^3 + 2x + 1$, find Δy and dy when $x = 1$ and $\Delta x = dx = 0.01$.

40. For $y = (x^2 - 3)^3$, find Δy and dy when $x = 2$ and $\Delta x = dx = 0.01$.

41. For $f(x) = x^3 - 2x - 2$, use a differential to approximate $f(2.1)$.

42. For $f(x) = xe^{x-5}$, use a differential to approximate $f(5.1)$.

3.2 Marginal Analysis

Many economic decisions are based on analysis of cost and revenue "at the margin." For example, a trucking company owner needs to decide whether to add another truck to a route. Based purely on financial grounds, if the added truck will make money for the company, then it should be added. Hence, the owner has to consider the cost and revenue involved. Since the choice is between adding this truck and leaving things the way they are, the crucial question is whether the **additional** costs incurred are greater or smaller than the **additional revenue** generated by the truck. These additional costs and revenues are called **marginal costs** and **marginal revenue**.

Suppose that $C(x)$ and $R(x)$ are the cost and revenue functions of operating x trucks, and the trucking company is currently running 20 trucks. With the additional truck, the additional cost and revenue will be $C(21) - C(20)$ and $R(21) - R(20)$, respectively. However, from Section 3.1, we have

$$C(21) - C(20) = \Delta C \approx C'(20)(21 - 20) = C'(20) \text{ and}$$

$$R(21) - R(20) = \Delta R \approx R'(20)(21 - 20) = R'(20).$$

Often it is easier to compute $C'(20)$ and $R'(20)$ than to compute $C(21) - C(20)$ and $R(21) - R(20)$. Hence, economists often define the marginal cost and marginal revenue by the derivatives of the cost and revenue functions. We summarize the definitions as the following:

> **Marginal Cost, Revenue, and Profit**
>
> Suppose that $C(x)$, $R(x)$, and $P(x)$ represent, respectively, the total cost, revenue, and profit of producing and selling x items.
>
> The **marginal cost, revenue, and profit** at x are given by $MC(x) = C'(x)$, $MR(x) = R'(x)$, and $MP(x) = P'(x)$, respectively. They are the approximations of the cost, revenue, and profit, respectively, of producing and selling the $(x+1)$th item.
>
> $$MC(x) \approx C(x+1) - C(x), \quad MR(x) \approx R(x+1) - R(x), \quad \text{and} \quad MP(x) \approx P(x+1) - P(x).$$

Notice that the relationship among the marginal cost, revenue, and profit follow the same relationship among the total cost, revenue, and profit:

$$MP(x) = MR(x) - MC(x), \quad \text{or} \quad P'(x) = R'(x) - C'(x).$$

Example 1: A salsa company has cost function (in dollars)

$$C(x) = 0.01x^3 - 0.5x^2 + 12x + 1{,}500$$

where x is the number of cases of salsa produced. The revenue function (in dollars) for selling x cases of salsa is given by

$$R(x) = 120x^{0.98}.$$

(a) Find the total profit when x cases of salsa are produced and sold.

(b) Find the total cost, revenue, and profit when 100 cases of salsa are produced and sold.

(c) Find the marginal cost, revenue, and profit when 100 cases of salsa are produced and sold.

(d) Find the exact cost, revenue, and profit for producing and selling the 101th case of salsa. Compare the results with those obtained in part (c).

Solution (a) According to the definition, the total profit for producing and selling x cases of salsa is

$$P(x) = R(x) - C(x) = 120x^{0.98} - 0.01x^3 + 0.5x^2 - 12x - 1{,}500.$$

(b) Setting $x = 100$, we obtain

$$C(100) = 0.01(100)^3 - 0.5(100)^2 + 12(100) + 1{,}500 = 7{,}700,$$

$$R(100) = 120(100)^{0.98} = 10{,}944.13,$$

and

$$P(100) = R(100) - C(100) = 10{,}944.13 - 7{,}700 = 3{,}244.13.$$

Hence, if 100 cases of salsa are produced and sold, the total cost is $7,700, the total revenue is $10,944.13, and the total profit is $3,244.13.

(c) We find the derivatives for the functions $C(x)$ and $R(x)$:

$$C'(x) = 0.03x^2 - x + 12,$$

3.2. MARGINAL ANALYSIS

$$R'(x) = 120(0.98)x^{-0.02} = 117.6x^{-0.02}.$$

Hence, at $x = 100$,

$$C'(100) = 0.03(100)^2 - (100) + 12 = 300 - 100 + 12 = 212,$$

and

$$R'(100) = 117.6(100)^{-0.02} = 107.25.$$

Then

$$P'(100) = R'(100) - C'(100) = 107.25 - 212 = -104.75.$$

Therefore, when 100 cases of salsa are produced and sold, the marginal cost is $212 per case, the marginal revenue is $107.25 per case, and the marginal profit is $-\$104.75$ per case, that is, the company will lose about $104.75 for producing and selling the 101th case of salsa.

(d) The exact cost, revenue, and profit for producing and selling the 101th case of salsa are $C(101) - C(100)$, $R(101) - R(100)$, and $P(101) - P(100)$, respectively. From part (b) we already obtained $C(100) = 7{,}700$, $R(100) = 10{,}944.13$, and $P(100) = 3{,}244.13$. Now that

$$C(101) = 0.01(101)^3 - 0.5(101)^2 + 12(101) + 1{,}500 = 7{,}914.51,$$

$$R(101) = 120(101)^{0.98} = 11{,}051.40, \quad \text{and} \quad P(101) = R(101) - C(101) = 3{,}136.87,$$

we have

$$C(101) - C(100) = 7{,}914.51 - 7{,}700 = 214.51,$$

$$R(101) - R(100) = 11{,}051.40 - 10{,}944.13 = 107.24, \quad \text{and} \quad P(101) - P(100) = 107.24 - 214.51 = -107.27.$$

Therefore, for producing and selling the 101th case of salsa, the exact cost is $214.51, the exact revenue is $107.24, and the exact loss is $107.27.

Now we compare those numbers with the marginal cost, revenue, and profit obtained in part (c): For the marginal cost, the relative error is

$$\frac{|212 - 214.51|}{214.51} \approx 0.012 = 1.2\%.$$

That is, the approximations is within 1.2% of the actual value. Similarly, the relative errors in the marginal revenue and profit are 0.01% and 2.3%, respectively.

Reality Check *Marginal vs. actual*

The last part of Example 1 illustrates that the marginal cost, revenue, and profit are very good approximations of the exact cost, revenue, and profit for producing and selling the additional item. In addition, the computations of the marginal values are relatively simpler compared with the computations of the actual ones.

In Section 3.1, we studied how to use the function and derivative values at a particular location to approximate the nearby function values. This computation is especially helpful in business, since often formulas for $C(x)$, $R(x)$, and $P(x)$ may not be readily available.

Example 2: The owner of a clothing manufacturer finds that the total cost of producing 200 of a certain brand of suits is $8,000, and the marginal cost when 200 suits are produced is $35 per suit. Estimate the total cost of producing 204 suits.

Solution Since $C(200) = 8,000$ and $C'(200) = 35$, we apply the linear approximation formula to obtain

$$C(204) \approx C(200) + C'(200)(204 - 200) = 8,000 + 35(4) = 8,140.$$

Hence, the total cost of producing 204 suits is about $8,140.

Example 3: A shoe manufacturer has the cost function (in dollars)

$$C(x) = 1,000 + 15x + 0.1x^2 + 200\ln(x+1)$$

where x is the number of pairs of shoes produced each week. The revenue function (in dollars) for selling x pairs of shoes is given by

$$R(x) = 80x.$$

(a) Find the total profit when x pairs of shoes are produced and sold each week.

(b) Find the total cost, revenue, and profit when 50 pairs of shoes are produced and sold each week.

(c) Find the marginal cost, revenue, and profit when 50 pairs of shoes are produced and sold each week.

(d) Use the information in parts (b) and (c) to estimate the total cost, revenue, and profit when 52 pairs of shoes are produced and sold each week.

Solution (a) According to the definition, the total profit for producing and selling x pairs of shoes is

$$P(x) = R(x) - C(x) = 80x - 1,000 - 15x - 0.1x^2 - 200\ln(x+1).$$

(b) Setting $x = 50$ we obtain

$$C(50) = 1,000 + 15(50) + 0.1(50)^2 + 200\ln(50+1) = 2,786.37,$$

$$R(50) = 80(50) = 4,000,$$

and

$$P(50) = R(50) - C(50) = 4,000 - 2,786.37 = 1,213.63.$$

Hence, if 50 pairs of shoes are produced and sold each week, the total cost is $2,786.37, the total revenue is $4,000, and the total profit is $1,213.63.

(c) We find the derivatives for the functions $C(x)$ and $R(x)$:

$$C'(x) = 15 + 0.2x + \frac{200}{x+1}, \quad \text{and} \quad R'(x) = 80.$$

Hence, at $x = 50$,

$$C'(50) = 15 + 0.2(50) + \frac{200}{51} \approx 28.92,$$

and

$$R'(50) = 80.$$

3.2. MARGINAL ANALYSIS

Then
$$P'(50) = R'(50) - C'(50) = 80 - 28.92 = 51.08.$$

Therefore, when 50 pairs of shoes are produced and sold each week, the marginal cost is \$28.92 per pair, the marginal revenue is \$80 per pair, and the marginal profit is \$51.08 per pair.

(d) We apply the linear approximation formula to obtain
$$C(52) \approx C(50) + C'(50)(52 - 50) = 2{,}786.37 + 28.92(2) = 2{,}844.21.$$

Hence, the total cost of producing 52 pairs of shoes is about \$2,844.21. Since $R(x)$ itself is a linear function,
$$R(52) = R(50) + R'(50)(52 - 50) = 4{,}000 + 80(2) = 4{,}160.$$

Hence, the total revenue from selling 52 pairs of shoes is \$5,600. The total profit from producing and selling 52 pairs of shoes is about \$4,160 − \$2,844.21 = \$1,315.79.

Average Cost, Revenue, and Profit

In addition to the marginal cost, revenue, and profit, it is often in a business's interest to calculate the **average** cost, revenue, and profit, when x items of the product are produced and sold. These average functions are defined as the following.

Average Cost, Revenue, and Profit

Suppose that $C(x)$, $R(x)$, and $P(x)$ represent, respectively, the total cost, revenue, and profit of producing and selling x items.

The **average cost, revenue, and profit** at x are given by $A_C(x) = \dfrac{C(x)}{x}$, $A_R(x) = \dfrac{R(x)}{x}$, and $A_P(x) = \dfrac{P(x)}{x}$, respectively.

Example 4: Refer to Example 1, the salsa company has cost and revenue functions
$$C(x) = 0.01x^3 - 0.5x^2 + 12x + 1{,}500 \quad \text{and} \quad R(x) = 120x^{0.98}.$$

(a) Find the average cost, revenue, and profit when x cases of salsa are produced and sold each week.

(b) Find the average cost, revenue, and profit when 100 cases of salsa are produced and sold each week.

(c) Find the rate at which the average profit is changing when 100 cases of salsa are produced and sold each week.

Solution (a) According to the definition, we have
$$A_C(x) = \frac{C(x)}{x} = \frac{0.01x^3 - 0.5x^2 + 12x + 1{,}500}{x} = 0.01x^2 - 0.5x + 12 + \frac{1{,}500}{x},$$
$$A_R(x) = \frac{R(x)}{x} = \frac{120x^{0.98}}{x} = 120x^{-0.02},$$
$$A_P(x) = \frac{P(x)}{x} = A_R(x) - A_C(x) = 120x^{-0.02} - 0.01x^2 + 0.5x - 12 - \frac{1{,}500}{x}.$$

(b) Setting $x = 100$ we obtain $A_C(100) = 77$, $A_R(100) = 109.44$, and $A_P(100) = 32.44$. Hence, if 100 cases of salsa are produced and sold each week, the average cost is \$77 per case, the average revenue is \$109.44 per case, and the average profit is \$32.44 per case.

(c) We find the derivative for the function $A_P(x)$. The Power Rule is used for each term:

$$A_P'(x) = \frac{d}{dx}\left(120x^{-0.02} - 0.01x^2 + 0.5x - 12 - \frac{1{,}500}{x}\right)$$

$$= 120 \cdot (-0.02)x^{-1.02} - 0.01 \cdot (2)x^1 + 0.5 - \frac{-1{,}500}{x^2}$$

$$= -2.4x^{-1.02} - 0.02x + \frac{1{,}500}{x^2}.$$

At $x = 100$, the rate of change is $A_P'(100) = -1.87$. That is, when 100 cases of salsa are produced and sold each week, the average profit is decreasing at a rate of \$1.87/case per case.

Reality Check *Marginal vs. average*

It is worth mentioning that marginal cost, revenue, and profit are different from the average cost, revenue, and profit. For example, in Examples 1 and 4, when 100 cases of salsa are produced the marginal cost is \$212 per case, while the average cost of producing 100 cases of salsa is \$77 per case. Indeed, the marginal cost is an approximation of the cost of producing the 101th case of salsa, while the average cost is $C(100)/100$, that is, the total cost of producing 100 cases of salsa divided by 100.

Maximum Profit

We now return to the example of the trucking company owner who needs to decide whether to add another truck to a route. Based purely on financial grounds, if the added truck will make money for the company, then it should be added. Applying the marginal analysis, if the marginal revenue is larger than the marginal cost, then the marginal profit is positive. That is, by adding a truck to the route the total profit will increase. Conversely, if the marginal revenue is less than the marginal cost, then the marginal profit is negative. That is, by adding a truck to the route the total profit will decrease. Therefore, the total profit reaches its maximum when the marginal cost equals the marginal revenue. The result is summarized in Theorem 1.

From the graphs of the cost function and the revenue function shown in Figure 3.2.1, we see that the maximum profit occurs at the x value at which the two curves have parallel tangent lines. Meanwhile, from the graph of the profit function shown in Figure 3.2.2, we see that the maximum profit occurs at the x value at which the curve has a horizontal tangent line.

3.2. MARGINAL ANALYSIS

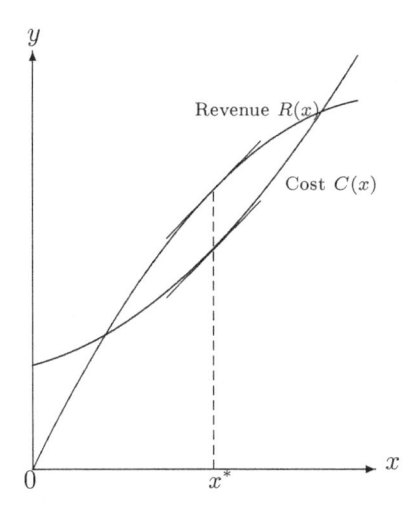

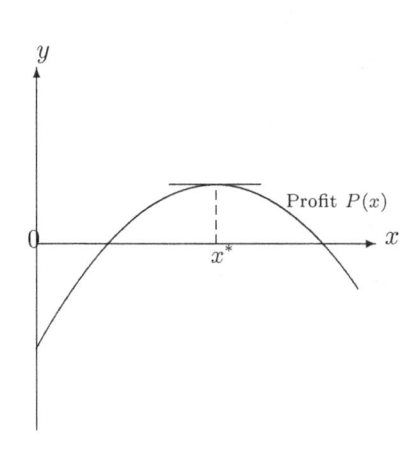

Figure 3.2.1: Maximum profit vs. MC and MR

Figure 3.2.2: Maximum profit vs. MP

Theorem 1   *The Marginal Test for Maximum Profit*

Given the cost function $C(x)$ and the revenue function $R(x)$, maximum profit occurs at x^* if

$MC(x^*) = MR(x^*)$ and

$MC(x) < MR(x)$ for nearby $x < x^*$, $MC(x) > MR(x)$ for nearby $x > x^*$;

Or equivalently, if

$MP(x^*) = 0$ and

$MP(x) > 0$ for nearby $x < x^*$, $MP(x) < 0$ for nearby $x > x^*$.

Example 5: A salsa company has the cost function (in dollars)

$$C(x) = 0.01x^3 - 0.5x^2 + 12x + 1{,}500$$

where x is the number of cases of salsa produced each week. The revenue function (in dollars) for selling x cases of salsa is given by

$$R(x) = 100x.$$

(a) Find the number of cases of salsa that should be produced and sold each week for the company to make the maximum profit.

(b) Find the maximum profit the company can make each week.

Solution (a) We first find the marginal cost and revenue:

$$MC(x) = C'(x) = 0.03x^2 - x + 12 \quad \text{and} \quad MR(x) = R'(x) = 100.$$

We set the equation

$$MC(x) = MR(x)$$

and solve for x:

$$0.03x^2 - x + 12 = 100 \quad \text{or} \quad 0.03x^2 - x - 88 = 0.$$

This is a quadratic equation. Applying the quadratic formula we have

$$x = \frac{1 \pm \sqrt{1 - 4(0.03)(-88)}}{2(0.03)} = \frac{1 \pm \sqrt{11.56}}{0.06} = \frac{1 \pm 3.4}{0.06}.$$

We take the positive root $x^* \approx 73$. We now check the other conditions listed in Theorem 1. We have

$$MP(x) = P'(x) = R'(x) - C'(x) = -0.03x^2 + x + 88.$$

Hence, $MP(72) = 4.48$ and $MP(74) = -2.28$. Therefore,

$$MP(x) > 0 \text{ for nearby } x < 73 \quad \text{and} \quad MP(x) < 0 \text{ for nearby } x > 73.$$

According to Theorem 1, the profit function reaches its maximum at $x = 73$. That means, to achieve the maximum profit, 73 cases of salsa should be produced and sold each week.

(b) At $x^* = 73$, the maximum profit is

$$P(73) = R(73) - C(73) = 3{,}698.33$$

Hence, the maximum weekly profit the company can make is \$3,698.33.

Example 6: A shoe manufacturer has the cost function (in dollars)

$$C(x) = 1{,}000 + 15x + 0.1x^2 + 200\ln(x+1)$$

where x is the number of pairs of shoes produced each week. The revenue function (in dollars) for selling x pairs of shoes is given by

$$R(x) = 80x$$

(a) Find the number of pairs of shoes that should be produced and sold each week for the company to make the maximum profit.

(b) Find the maximum profit the company can make each week.

Solution (a) We first find the marginal cost and revenue:

$$MC(x) = C'(x) = 15 + 0.2x + \frac{200}{x+1} \quad \text{and} \quad MR(x) = R'(x) = 80.$$

We set the equation

$$MC(x) = MR(x)$$

and solve for x:

$$15 + 0.2x + \frac{200}{x+1} = 80 \quad \text{or} \quad 0.2x + \frac{200}{x+1} - 65 = 0.$$

By multiplying $x+1$ on both side of the equation, we obtain

$$0.2x(x+1) + 200 - 65(x+1) = 0 \quad \text{or} \quad 0.2x^2 - 64.8x + 135 = 0.$$

This is a quadratic equation. Applying the quadratic formula, we have

$$x = \frac{64.8 \pm \sqrt{64.8^2 - 4(0.2)(135)}}{2(0.2)} = \frac{64.8 \pm \sqrt{4091}}{0.4} = \frac{64.98 \pm 63.96}{0.4}.$$

3.2. MARGINAL ANALYSIS

Both roots are positive: $x_1 \approx 2$ and $x_1 \approx 322$. We now check the other conditions. We have

$$MC(1) = 115.2 > MR(1) = 80.$$

Hence, the first root does not satisfy the condition listed in Theorem 1, which requires that $MC(x) < MR(x)$ for $x < x^*$. It can be verified that

$$MC(x) < MR(x) \text{ for nearby } x < 322, \quad \text{and} \quad MC(x) > MR(x) \text{ for nearby } x > 322.$$

According to Theorem 1, the profit function reaches its maximum at $x = 322$. That means, to achieve the maximum profit, 322 pairs of shoes should be produced and sold each week.

(b) At $x^* = 322$, the maximum profit is

$$P(322) = R(322) - C(322) = 8,406.07$$

Hence, the maximum weekly profit the company can make is $8,406.07.

In Section 1.3, we defined the revenue as the price times the quantity sold. In the market, the price, p, is related to the quantity, x, by the demand function $x = D(p)$. Therefore, the revenue can be expressed either as a function of the quantity x, as in the production study, and later in Section 3.8; or as a function of the price p, as later in Section 3.9. The following example illustrates how the revenue can be formed as a function of x using the given marketing information.

Example 7: An appliance manufacturer has the cost function (in dollars)

$$C(x) = 4,000 + 33x + 0.25x^2$$

where x is the number of a certain brand of microwave ovens produced each week. In the market, the demand function for that brand of microwave oven is found to be

$$x = D(p) = 120 - 0.2p$$

where p, in dollars, represents the price of each microwave oven.

(a) Rewrite the demand function to express p as a function of x.

(b) Find the revenue function $R(x)$.

(c) Find the number of microwave ovens that should be produced and sold each week for the company to make the maximum profit.

Solution (a) From

$$x = 120 - 0.2p,$$

we solve for p to obtain

$$p = \frac{120 - x}{0.2} = 600 - 5x.$$

(b) Since the revenue is the price times the quantity sold, or $R = p \cdot x$, and from part (a) we have $p = 600 - 5x$, the revenue function is

$$R(x) = p \cdot x = (600 - 5x) \cdot x = 600x - 5x^2.$$

(c) We first find the marginal cost and revenue:

$$MC(x) = C'(x) = 33 + 0.5x \quad \text{and} \quad MR(x) = R'(x) = 600 - 10x.$$

We set the equation $MC(x) = MR(x)$ and solve for x:

$$33 + 0.5x = 600 - 10x \quad \text{or} \quad 10.5x = 567 \quad \text{or} \quad x = 54.$$

We now apply Theorem 1 to verify. We have

$$MP(x) = 567 - 10.5x. \quad \text{Hence,} \quad MP(53) = 10.5 \text{ and } MP(55) = -10.5.$$

According to Theorem 1, the profit function reaches its maximum at $x = 54$. Therefore, to achieve the maximum profit, 54 microwave ovens should be produced and sold each week.

Example 8: At a price of $60 for a 3-hour city tour, a bus-tour company attracts 120 customers. For every $2 decrease in price, an additional four customers are attracted. The cost, in dollars, for carrying x customers in the tour is

$$C(x) = 1{,}500 + 12x + 0.1x^2.$$

(a) Find the demand function; express the price, p, as a function of x, the number of customers.

(b) Find the price for the 3-hour city tour the company should charge to maximize its profit.

Solution (a) The demand function follows a linear model. Hence, we assume

$$p = mx + b.$$

The slope is $m = -2/4 = -0.5$. Using $x = 120$ and $p = 60$ we have

$$60 = -0.5(120) + b,$$

and we obtain $b = 120$. Hence, the demand function is

$$p = -0.5x + 120.$$

(b) We first need to find the revenue function $R(x)$. Since the revenue is the price times the quantity sold, or $R = p \cdot x$, and from part (a) we have $p = -0.5x + 120$, the revenue function is

$$R(x) = p \cdot x = (-0.5x + 120) \cdot x = -0.5x^2 + 120x.$$

We then find the marginal cost and revenue:

$$MC(x) = C'(x) = 12 + 0.2x \quad \text{and} \quad MR(x) = R'(x) = -x + 120.$$

We set the equation $MC(x) = MR(x)$ and solve for x:

$$12 + 0.2x = -x + 120 \quad \text{or} \quad 1.2x = 108 \quad \text{or} \quad x = 90.$$

We now apply Theorem 1 to verify. We have

$$MP(x) = 108 - 1.2x. \quad \text{Hence,} \quad MP(89) = 1.2 \text{ and } MP(91) = -1.2.$$

According to Theorem 1, the profit function reaches its maximum at $x = 90$. According to part (a), when $x = 90$, the price is

$$p = -0.5(90) + 120 = 75.$$

Therefore, to achieve its maximum profit, the company should charge $75 for the tour.

Exercise Set 3.2

1. The cost function, in dollars, for producing x items of a certain brand of skateboards is given by $C(x) = 1{,}200 + 53x^{4/5}$.

   (a) Find $MC(x) = C'(x)$.

   (b) Find $MC(77)$ and explain what this quantity represents.

   (c) Calculate the exact cost of producing the 78th skateboard, and compare with the result in part (b).

   (d) Use $C(77)$ and $MC(77)$ to estimate the total cost of producing 79 skateboards.

2. The cost function, in dollars, for producing x items of a certain brand of barstool is given by $C(x) = 0.01x^3 - 0.6x^2 + 13x + 200$, $0 \leq x \leq 80$.

   (a) Find $MC(x) = C'(x)$.

   (b) Find $MC(50)$ and explain what this quantity represents.

   (c) Calculate the exact cost of producing the 51th barstool, and compare with the result in part (b).

   (d) Use $C(50)$ and $MC(50)$ to estimate the total cost of producing 53 barstools.

3. The manager of a furniture manufacturer finds that the total cost of producing 150 of a certain brand of chairs is $1,400, and the marginal cost when 150 chairs are produced is $8 per chair. Estimate the total cost of producing 153 chairs.

4. The owner of a clothing manufacturer finds that the total cost of producing 300 of a certain brand of shirts is $2,000, and the marginal cost when 300 shirts are produced is $15 per shirt. Estimate the total cost of producing 302 shirts.

5. A furniture manufacturer has the cost function

   $$C(x) = 3{,}000 + 125x + 0.05x^2 - 300e^{-0.02x}$$

   (in dollars), where x is the number of dining sets produced each month. The revenue function for selling x dining sets is given by

   $R(x) = 400x$ (in dollars).

   (a) Find the total profit when x dining sets are produced and sold each month.

   (b) Find the total cost, revenue, and profit when 100 dining sets are produced and sold each month.

   (c) Find the marginal cost, revenue, and profit when 100 dining sets are produced and sold each month.

   (d) Use the information in parts (b) and (c) to estimate the total cost, revenue, and profit when 102 dining sets are produced and sold each month.

6. An appliance manufacturer has the cost function

   $$C(x) = 4{,}000 + 36x + 0.5x^2 \text{ (in dollars)},$$

   where x is the number of a certain brand of microwave ovens produced each week. The revenue function for selling x microwave ovens is given by

   $$R(x) = 500x - 4x^2 \text{ (in dollars)}.$$

   (a) Find the total profit when x microwave ovens are produced and sold each week.

   (b) Find the total cost, revenue, and profit when 40 microwave ovens are produced and sold each week.

   (c) Find the marginal cost, revenue, and profit when 40 microwave ovens are produced and sold each week.

   (d) Use the information in parts (b) and (c) to estimate the total cost, revenue, and profit when 41 microwave ovens are produced and sold each week.

7. A shoe manufacturer has the cost function $C(x) = 1{,}500 + 20x + 0.2x^2 + 100\ln(x+5)$ (in dollars), where x is the number of pairs of shoes produced each week. The revenue function for selling x pairs of shoes is given by $R(x) = 90x$ (in dollars).

   (a) Find the total profit when x pairs of shoes are produced and sold each week.

(b) Find the total cost, revenue, and profit when 60 pairs of shoes are produced and sold each week.

(c) Find the marginal cost, revenue, and profit when 60 pairs of shoes are produced and sold each week.

(d) Use the information in parts (b) and (c) to estimate the total cost, revenue, and profit when 63 pairs of shoes are produced and sold each week.

8. A clothing manufacturer has the cost function
$$C(x) = 1{,}200 + 18x + 0.1x^2 \text{ (in dollars)},$$
where x is the number of suits produced each week. The revenue function for selling x suits is given by $R(x) = 120x$ (in dollars).

(a) Find the total profit when x suits are produced and sold each week.

(b) Find the total cost, revenue, and profit when 70 suits are produced and sold each week.

(c) Find the marginal cost, revenue, and profit when 70 suits are produced and sold each week.

(d) Use the information in parts (b) and (c) to estimate the total cost, revenue, and profit when 74 suits are produced and sold each week.

9. The cost function, in dollars, for producing x items of a certain brand of skateboard is given by
$$C(x) = 1{,}200 + 53x^{4/5}.$$

Find the rate at which the average cost is changing when 77 skateboards have been produced.

10. The cost function, in dollars, for producing x pairs of a certain brand of shoes is given by
$$C(x) = 800 + 22\ln(x+5).$$

Find the rate at which the average cost is changing when 92 pairs of shoes have been produced.

11. The revenue function, in dollars, for selling x items of the skateboards listed in Exercise 11 is given by
$$R(x) = 120\sqrt[3]{x^2}.$$

Find the rate at which the average revenue is changing when 77 skateboards have been sold.

12. The revenue function, in dollars, for selling x pairs of shoes listed in Exercise 12 is given by
$$R(x) = 80x + 40xe^{-0.04x}.$$

Find the rate at which the average revenue is changing when 92 pairs of shoes have been sold.

13. Use the information in Exercises 9 and 11 to find the rate at which the average profit is changing when 77 skateboards have been produced and sold.

14. Use the information in Exercises 10 and 12 to find the rate at which the average profit is changing when 92 pairs of shoes have been produced and sold.

15. Suppose that a refrigerator manufacturer determines that its cost, in dollars, of producing x refrigerators is given by
$$C(x) = 0.02x^2 + 15x + 4{,}500$$
and its revenue, in dollars, of selling x refrigerators is given by
$$R(x) = 200x.$$

Find the rate at which the average profit is changing when 120 refrigerators have been produced and sold.

16. Suppose that a fruit juice manufacturer determines that its cost, in dollars, of producing x bottles of apple juice is given by
$$C(x) = 285 + 0.55\sqrt{x},$$
and its revenue, in dollars, of selling x bottles of apple juice is given by
$$R(x) = 5.5x^{0.6}.$$

Find the rate at which the average profit is changing when 500 bottles of apple juice have been produced and sold.

17. A furniture manufacturer has the cost function

$$C(x) = 25{,}000 + 125x + 0.1x^2$$

(in dollars), where x is the number of dining sets produced each year. The revenue function for selling x dining sets is given by

$$R(x) = 350x \text{ (in dollars)}.$$

(a) Find the number of dining sets that should be produced and sold each year for the company to make the maximum profit.

(b) Find the maximum profit the company can make each year.

18. For the small appliance manufacturer in Exercise #6,

(a) Find the number of microwave ovens that should be produced and sold each week for the company to make the maximum profit.

(b) Find the maximum profit the company can make each week.

19. A shoe manufacturer has the cost function

$$C(x) = 1{,}500 + 20x + 0.2x^2 \text{ (in dollars)},$$

where x is the number of pairs of shoes produced each week. The revenue function for selling x pairs of shoes is given by

$$R(x) = 90x \text{ (in dollars)}.$$

(a) Find the number of pairs of shoes that should be produced and sold each week for the company to make the maximum profit.

(b) Find the maximum profit the company can make each week.

20. A clothing manufacturer has the cost function

$$C(x) = 1{,}000 + 18x + 0.1x^2$$

(in dollars), where x is the number of suits produced each week.

The revenue function for selling x suits is given by

$$R(x) = 120x \text{ (in dollars)}.$$

(a) Find the number of suits that should be produced and sold each week for the company to make the maximum profit.

(b) Find the maximum profit the company can make each week.

21. When the production is 1,500 units, the marginal revenue is \$5 per unit and the marginal cost is \$4.7 per unit. Do you expect the maximum profit to occur at a production level above or below 1,500 units? Explain.

22. When the production is 1,000 units, the marginal revenue is \$5 per unit and the marginal cost is \$5.2 per unit. Do you expect the maximum profit to occur at a production level above or below 1,000 units? Explain.

23. Suppose that a refrigerator manufacturer determines that its cost, in dollars, of producing x refrigerators each week is given by $C(x) = 2x^2 + 15x + 1{,}500$. In the market, the demand function for that brand of refrigerator is found to be $x = D(p) = 420 - 0.5p$, where p is in dollars.

(a) Rewrite the demand function to express p as a function of x.

(b) Find the revenue function $R(x)$.

(c) Find the number of refrigerator that should be produced and sold each week for the company to make the maximum profit.

24. Suppose that a fruit juice manufacturer determines that its weekly cost, in dollars, of producing x thousands of bottles of apple juice is given by

$$C(x) = 285 + 0.5x + 0.01x^2.$$

In the market, the demand function for that brand of apple juice is found to be

$$x = D(p) = 2{,}000 - 400p,$$

where p is in dollars.

(a) Rewrite the demand function to express p as a function of x.

(b) Find the revenue function $R(x)$.

(c) Find the number of bottles of apple juice that should be produced and sold each week for the company to make the maximum profit.

25. At a price of $40 for a 2-hour harbor tour, a water-tour company attracts 200 customers. For every $1 decrease in price, an additional five customers are attracted. The cost of carrying x customers in the tour is
$$C(x) = 3{,}000 + 5x + 0.01x^2.$$

  (a) Find the demand function; express the price, p, as a function of x, the number of customers.

  (b) Find the revenue function $R(x)$.

  (c) Find the number of customers the company should attract to maximize its profit.

  (d) Find the price for the 2-hour harbor tour the company should charge to maximize its profit.

26. At a price of $50 for a 2-hour city tour, a bus-tour company attracts 100 customers. For every $2 decrease in price, an additional six customers are attracted. The cost for carrying x customers in the tour is
$$C(x) = 1{,}000 + 10x + 0.1x^2.$$

  (a) Find the demand function; express the price, p, as a function of x, the number of customers.

  (b) Find the revenue function $R(x)$.

  (c) Find the number of customers the company should attract to maximize its profit.

  (d) Find the price for the 2-hour city tour the company should charge to maximize its profit.

3.3 How Derivatives Affect the Shape of a Graph

In this section, we examine more deeply the relationship that exists between the graph of a function and its derivatives. The study will reveal how the sign of the first derivative on an interval indicates whether the function is increasing or decreasing, and how the signs of the second derivative on an interval indicate the **concavity** of a function. Further, we see how the derivatives can help us to locate the **relative extrema** of a function.

Intervals of Increase and Decrease for Functions

A graph of any function can be partitioned into portions that rise from left to right and that drop from left to right. If the graph rises from left to right on a certain interval I, we say that the function is **increasing** on I. If the graph drops from left to right on a certain interval I, we say that the function is **decreasing** on I. In Figure 3.3.1, the function is increasing on (a, b) and (c, d) and is decreasing on (b, c). We now describe the increasing and decreasing of a function using mathematical formulas.

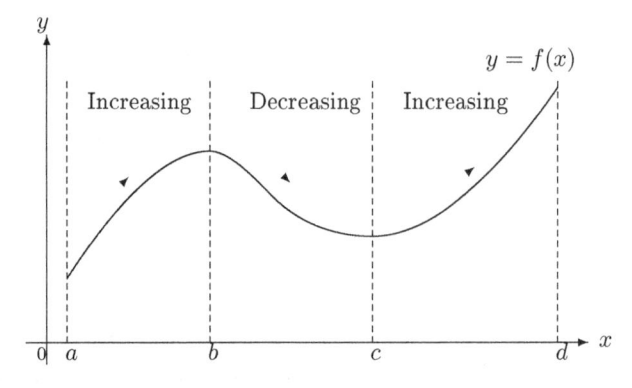

Figure 3.3.1: Intervals of increasing and decreasing of f

3.3. HOW DERIVATIVES AFFECT THE SHAPE OF A GRAPH

> **Definition** *increasing and decreasing*
>
> A function f is **increasing** on an interval (a,b) if for every x_1 and x_2 in (a,b),
>
> if $x_1 < x_2$, then $f(x_1) < f(x_2)$.
>
> A function f is **decreasing** on an interval (a,b) if for every x_1 and x_2 in (a,b),
>
> if $x_1 < x_2$, then $f(x_1) > f(x_2)$.
>
> If a function has a constant value on an interval (a,b), then it is neither increasing nor decreasing on (a,b).

From the graph of a function, it is obvious that, when the tangent lines of the function have *positive* slopes, the function is increasing; and when the tangent lines of the function have *negative* slopes, it is decreasing. Also, if the function is constant on an interval, then its slopes are zero on that interval. Recall that the derivative of a function represents the slopes of its tangent lines. Hence, the increasing and decreasing of a function can be indicated by the signs of its first derivative (Figures 3.3.2-3.3.4).

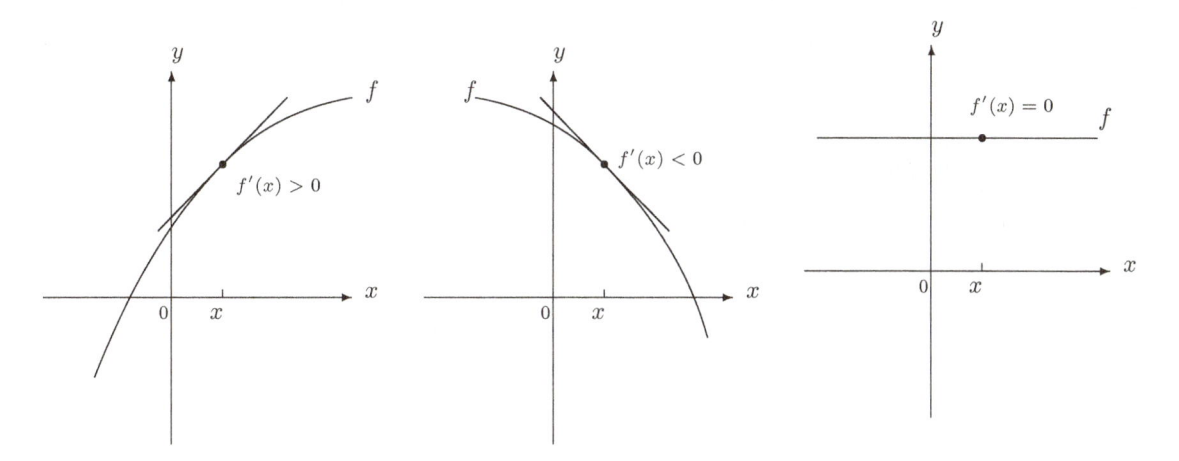

Figure 3.3.2: f vs. f' **Figure 3.3.3:** f vs. f' **Figure 3.3.4:** f vs. f'

We introduce the following theorem that indicates how to use the first derivative to determine whether a function is increasing, decreasing, or is a constant on an interval.

> **Theorem 1** *First-Derivatives and Increasing, Decreasing*
>
> Suppose that a function f is continuous and differentiable on an open interval (a,b):
> 1. If $f'(x) > 0$ for all x in (a,b), then f is increasing on (a,b).
>
> 2. If $f'(x) < 0$ for all x in (a,b), then f is decreasing on (a,b).
>
> 3. If $f'(x) = 0$ for all x in (a,b), then f is constant on (a,b).

Example 1: The graphs of $f(x) = x^3 - 3x + 1$ is shown in Figure 3.3.5. Use the graph to determine where:

(a) f is increasing;   (b) $f' \geq 0$;   (c) f is decreasing;   (d) $f' \leq 0$.

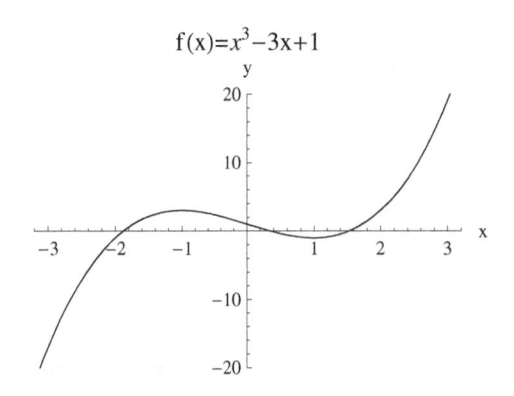

Figure 3.3.5: Graph of $f(x) = x^3 - 3x + 1$

Solution (a) f is increasing on the intervals $(-\infty, -1)$ and $(1, \infty)$.

(b) When a function is increasing it has nonnegative derivative. Hence, $f' \geq 0$ on the intervals $(-\infty, -1)$ and $(1, \infty)$.

(c) f is decreasing on the interval $(-1, 1)$.

(d) When a function is increasing it has nonnegative derivative. Hence, $f' \leq 0$ on the interval $(-1, 1)$.

***Example* 2:** The graphs of the derivative of $f(x) = x^3 - 3x + 1$, that is, $f'(x) = 3x^2 - 3$, is shown in Figure 3.3.6. Use the graph to determine where:

(a) $f' > 0$;   (b) f is increasing;   (c) $f' < 0$;   (d) f is decreasing.

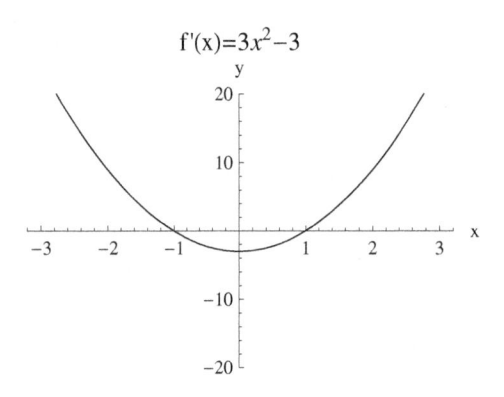

Figure 3.3.6: Graph of $f'(x) = 3x^2 - 3$

Solution (a) Recall that a function is greater than zero whenever its graph is *above* the x-axis. Hence, $f' > 0$ on the intervals $(-\infty, -1)$ and $(1, \infty)$.

(b) By Theorem 1 and part (a), f is increasing on the intervals $(-\infty, -1)$ and $(1, \infty)$.

(c) Recall that a function is less than zero whenever its graph is *below* the x-axis. Hence, $f' < 0$ on the interval $(-1, 1)$.

(d) By Theorem 1 and part (c), f is decreasing on the interval $(-1, 1)$.

Relative and Absolute Extrema

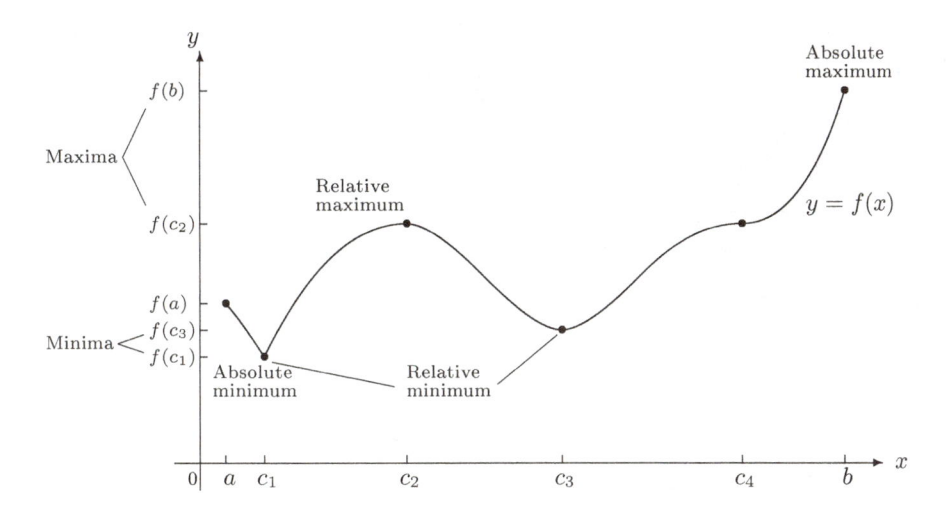

Figure 3.3.7: Maxima and minima of f on interval $[a, b]$

Observe the graph in Figure 3.3.7 and notice that the graph has "valleys" and "hills" at the *interior* points c_1, c_2, and c_3. We call the function value $f(c_2)$ a **relative** (or **local**) **maximum** (plural: **maxima**). We say that "f has a relative maximum at $(c_2, f(c_2))$," or "f has a relative maximum $f(c_2)$ at $x = c_2$." For example, if $c_2 = 5$ and $f(5) = 20$, then it can be stated as "$f(5) = 20$ is a relative maximum," or "f has a relative maximum (that equals) 20 at $x = 5$," or "f has a relative maximum at $(5, 20)$," or sometimes simply "f has a relative maximum at $x = 5$" if the location of a relative maximum is of interest. Similar manners can be applied to the minima. Each of $f(c_1)$ and $f(c_3)$ is called a **relative** (or **local**) **minimum** (plural: **minima**). We now describe a relative maximum and a relative minimum of a function using mathematical formulas.

Definition   *relative maximum and relative minimum*

Assume that an open interval (d, e) is contained in the domain of a function f, and c is contained in the interval (d, e). Then

$f(c)$ is a **relative maximum** if $f(c) \geq f(x)$ for all x in the interval.

$f(c)$ is a **relative minimum** if $f(c) \leq f(x)$ for all x in the interval.

Reality Check   *Relative and absolute extrema*

A relative maximum may or may not be the greatest function value over the whole interval, and a relative minimum may or may not be the least value over the whole interval, as indicated in Figure 3.3.7. The highest point there occurs at $(b, f(b))$, which is a "boundary point," and the lowest point occurs at $(c_1, f(c_1))$, at which f happens to have a relative minimum. The greatest function value over the whole interval is called the **absolute** (or **global**) **maximum**, and the least function value over the whole interval is called the **absolute** (or **global**) **minimum**. In Section 3.7, we study the techniques of using derivatives to find the absolute extrema (if either exists) of a given function. Currently we focus on finding the relative extrema.

> **Reality Check** *Relative and local*
>
> A relative maximum or minimum is "local," hence, the open interval (d, e) in the definition can be arbitrary small. For example, in Figure 3.3.7, around $x = c_2$, the largest such open interval is (a, c_4). However, one can choose d to be any number between a and c_2, and choose e to be any number between c_2 and c_4.

> **Reality Check** *Relative extrema*
>
> All the relative maxima and relative minima are referred to as **relative extrema** (single: **extremum**).

> **Reality Check** *Relative extrema and constant functions*
>
> If a function has a constant value on an interval (d, e), then by the definitions, $f(c)$ is both a relative maximum and a relative minimum for every number c in the interval (d, e).

Concavity and Inflection Points

Observe the following graphs and notice that the curves in Figure 3.3.8 are "bending up" and the curves in Figure 3.3.9 are "bending down." We call the graphs in Figure 3.3.8 **concave up** and the graphs in Figure 3.3.9 **concave down**. Another way to observe the concavity is that a concave up graph can "hold water" and a concave down graph makes a drop of water running away. Yet one more character of the concavity on a graph is that when any two points on the graph are connected by a line, then the whole line segment stays *above* the curve in a concave up graph, while the whole line segment stays *below* the curve in a concave down graph.

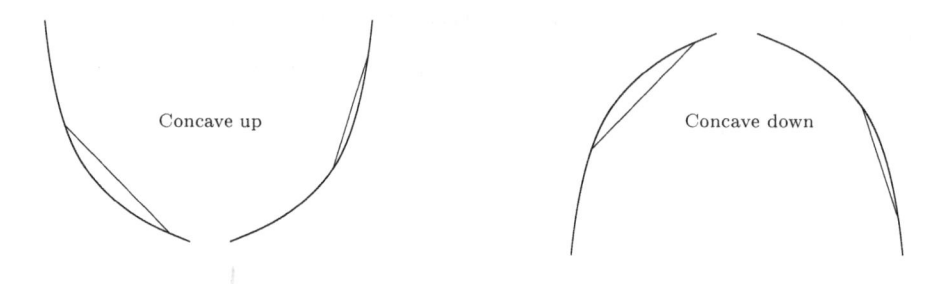

Figure 3.3.8: Concave up curves   **Figure 3.3.9:** Concave down curves

We now link the concavity of a function with its derivatives. Notice that if a function is decreasing and concave up, when the variable moves from left to right, the tangent lines become flatter. That is, the slope of its tangent lines (i.e., the first derivative) becomes less negative. If a function is increasing and concave up, when the variable moves from left to right, the tangent lines become steeper. That is, the slope of its tangent lines (i.e., the first derivative) becomes more positive. Both cases indicate that the first derivative is *increasing* if the function is concave up. Similarly, we can conclude that the first derivative is *decreasing* if the function is concave down. We use this feature to define the concavity of a function.

3.3. HOW DERIVATIVES AFFECT THE SHAPE OF A GRAPH

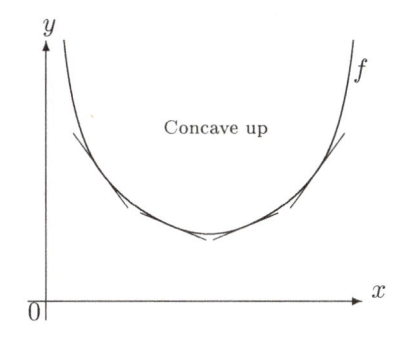

Figure 3.3.10: Concavity vs. f'

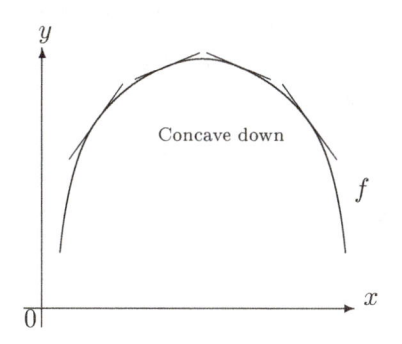

Figure 3.3.11: Concavity vs. f'

Reality Check *Lines and concavity*

A linear function is neither concave up nor concave down.

On the graph of a concave up function, all the tangent lines are below the curve, and on the graph of a concave down function, all the tangent lines are above the curve.

Definition *concave up, concave down, and inflection point*

Assume that an open interval (a,b) is contained in the domain of a function f, and the derivative f' exists at every number x in the interval (a,b). Then

f is **concave up** on (a,b) if f' is increasing on (a,b).

f is **concave down** on (a,b) if f' is decreasing on (a,b).

A point $(p, f(p))$ is called an **inflection point** if the function changes its concavity at this point.

Reality Check *Relative extrema and concavity*

If a function has a relative minimum at a point $(c, f(c))$ and the derivative $f'(c)$ exists, then the function must be concave up at this point (observe Figure 3.3.10). Similarly, if a function has a relative maximum at a point $(c, f(c))$ and the derivative $f'(c)$ exists, then the function must be concave down at this point (observe Figure 3.3.11).

Reality Check *Second derivatives and concavity*

Since the second derivative is the derivative of the first derivative, then by Theorem 1, positive second derivative indicates that the first derivative is increasing, therefore, the function is concave up. Similarly, negative second derivative indicates that the first derivative is decreasing, therefore, the function is concave down. We state this relationship in Theorem 2.

> **Theorem 2 Second-Derivatives and Concavity**
>
> Suppose that the second derivative f'' of a function f exists on an open interval (a,b).
>
> If $f''(x) > 0$ on (a,b), then f is concave up on (a,b).
>
> If $f''(x) < 0$ on (a,b), then f is concave down on (a,b).

We now illustrate how to locate an inflection point. Since the concavity changes at an inflection point, the sign of f'' changes there: it is positive on one side of the inflection point and negative on the other side (Figures 3.3.12-3.3.13). Therefore, at the inflection point, f'' is either zero or undefined.

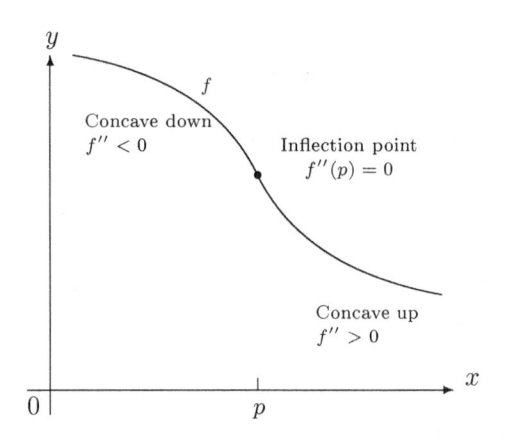

Figure 3.3.12: An inflection point

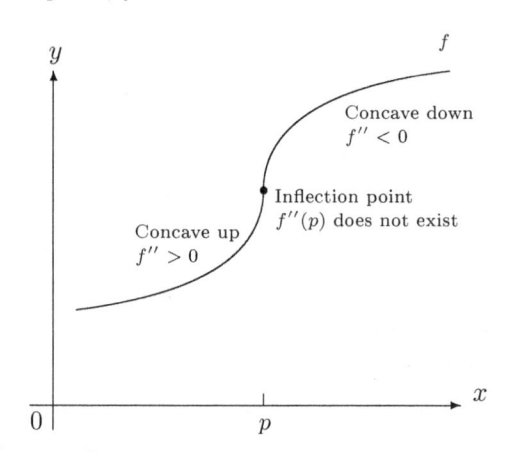

Figure 3.3.13: An inflection point

Example 3: Determine the inflection points for each function:

(a) $f(x) = x^4 - 6x^2$ (b) $g(x) = xe^{2x}$

Solution (a) We first take the derivatives. We have $f'(x) = 4x^3 - 12x$ and $f''(x) = 12x^2 - 12$. Since f'' exists everywhere, the first coordinate p of all inflection points satisfies $f''(p) = 0$:

$$12x^2 - 12 = 0$$

$$12(x-1)(x+1) = 0$$

$$x - 1 = 0 \quad \text{or} \quad x + 1 = 0$$

$$x = 1 \quad \text{or} \quad x = -1.$$

We now check the signs of f'' around these two numbers. We have

$$f''(-2) = 12(-2)^2 - 12 = 48 - 12 = 36 > 0,$$

$$f''(0) = -12 < 0,$$

$$f''(2) = 12(2)^2 - 12 = 48 - 12 = 36 > 0,$$

Hence, f'' does change signs at the two numbers. We have

$$f(1) = (1)^4 - 6(1)^2 = 1 - 6 = -5, \quad \text{and}$$

$$f(-1) = (-1)^4 - 6(-1)^2 = 1 - 6 = -5.$$

Therefore, the inflection points are $(-1, -5)$ and $(1, -5)$.

(b) We first take the derivatives. Applying the Product Rule and the Chain Rule, we obtain

$$g'(x) = e^{2x} + 2xe^{2x} = (1+2x)e^{2x}.$$

Applying the Product Rule and the Chain Rule again, we obtain

$$g''(x) = 2e^{2x} + 2(1+2x)e^{2x} = 4(1+x)e^{2x}.$$

Since g'' exists everywhere, the first coordinate p of all inflection points satisfies $g''(p) = 0$. Since an exponential function has all positive values, the only solution is from $1 + x = 0$, or $x = -1$. We now check the signs of g'' around $x = -1$. We have

$$g''(-3) = 4(-2)e^{-6} < 0,$$

$$g''(0) = 4(1)e^0 > 0.$$

Hence, g'' does change signs at $p = -1$. We have $g(-1) = -e^{-2}$. Therefore, g has a single inflection point $(-1, -e^{-2})$.

Example 4: Determine the inflection points for $f(x) = (x-5)^{2/3}$.

Solution We first take the derivatives:

$$f'(x) = \frac{2}{3}(x-5)^{-1/3}, \quad \text{and}$$

$$f''(x) = \frac{2}{3}\frac{-1}{3}(x-5)^{-4/3} = \frac{-2}{9(x-5)^{4/3}}.$$

Notice that f'' is not defined at $x = 5$. That is, $f''(5)$ does not exist. Notice that $x = 5$ is in the domain of f. Hence, $x = 5$ is a candidate for the first coordinate of an inflection point. Since the numerator of f'' is a negative constant, there is no solution for $f''(x) = 0$. We now check the signs of f'' around $x = 5$. We have

$$f''(4) = \frac{-2}{9(-1)^{4/3}} = \frac{-2}{9} < 0,$$

$$f''(6) = \frac{-2}{9(1)^{4/3}} = \frac{-2}{9} < 0.$$

Hence, f'' does not change signs at $x = 5$. Therefore, f has no inflection point.

The next example summarizes the studies presented in this section.

Example 5: For the function $f(x) = x^3 - 3x + 1$, do the following:

(a) Find all x-values at which the first derivative is zero;

(b) Use the signs of the first derivative to determine the intervals on which f is increasing;

(c) Use the signs of the first derivative to determine the intervals on which f is decreasing;

(d) Use the information from parts (b) and (c) to determine the relative extrema of f.

(e) Find all inflection points;

(f) Use the signs of the second derivative to determine the intervals on which f is concave up;

(g) Use the signs of the second derivative to determine the intervals on which f is concave down.

Solution (a) We first take the derivative:

$$f'(x) = 3x^2 - 3.$$

We then solve the equation $f'(x) = 0$:

$$3x^2 - 3 = 0 \quad \text{or} \quad x^2 - 1 = 0 \quad \text{or} \quad (x-1)(x+1) = 0.$$

Hence, $x = 1$ or $x = -1$. Hence, the roots are -1 and 1.

(b) The roots of the first derivative divide the real line into three intervals: $(-\infty, -1)$, $(-1, 1)$, and $(1, \infty)$. On each interval the derivative keeps the same sign. We determine the signs of f' in each interval by plugging in some numbers:

$$f'(-2) = 3(-2)^2 - 3 = 9 > 0; \quad f'(0) = -3 < 0; \quad f'(2) = 3(2)^2 - 3 = 9 > 0.$$

Therefore, f' is positive on the intervals $(-\infty, -1)$ and $(1, \infty)$. By Theorem 3, f is increasing on the intervals $(-\infty, -1)$ and $(1, \infty)$.

(c) Since f' is negative on the interval $(-1, 1)$, by Theorem 3, f is decreasing on the interval $(-1, 1)$.

(d) Since f changes from increasing to decreasing at $c_1 = -1$, and $f(-1) = (-1)^3 - 3(-1) + 1 = -1 + 3 + 1 = 3$, f has a relative maximum at $(-1, 3)$. Since f changes from decreasing to increasing at $c_2 = 1$, and $f(1) = (1)^3 - 3(1) + 1 = 1 - 3 + 1 = -1$, f has a relative minimum at $(1, -1)$.

(e) We take the second derivative:

$$f''(x) = 6x.$$

Since f'' exists everywhere, the first coordinate p of all inflection points satisfies $f''(p) = 0$. So we have $p = 0$. We now check the signs of f'' around 0. We have

$$f''(-1) = -6 < 0, \quad \text{and} \quad f''(1) = 6 > 0,$$

Hence, f'' does change signs at 0. We have

$$f(0) = (0)^3 - 3(0) + 1 = 1.$$

Therefore, f has a single inflection point $(0, 1)$.

(f) Since f'' is positive on $(0, \infty)$, then by Theorem 5, f is concave up on the interval $(0, \infty)$.

(g) Since f'' is negative on $(-\infty, 0)$, then by Theorem 5, f is concave down on the interval $(-\infty, 0)$.

Notice that the algebraic computations in this example confirm the features shown in the graph of $f(x) = x^3 - 3x + 2$ in Figure 3.3.5.

Exercise Set 3.3

For each function graphed in Exercises #1 to #4, determine the interval(s) where the function is increasing, the interval(s) where it is decreasing, and the interval(s) where it is a constant.

1.

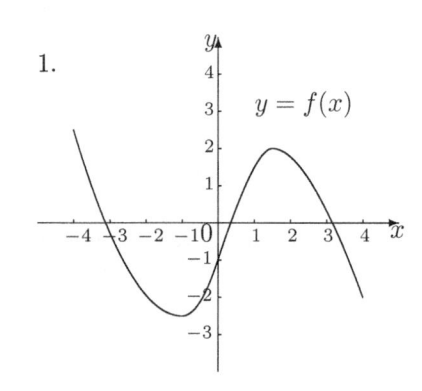

3.3. HOW DERIVATIVES AFFECT THE SHAPE OF A GRAPH

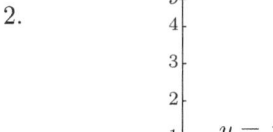

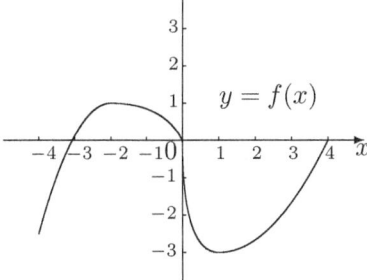

2. $y = f(x)$

3. $y = g(x)$

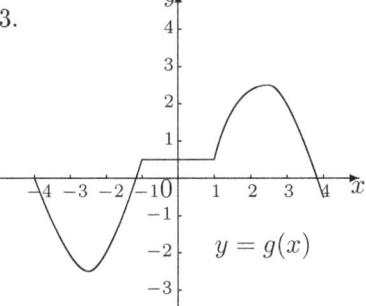

4. $y = g(x)$

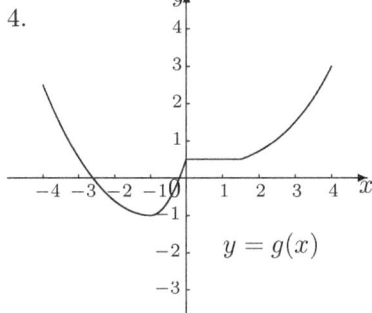

For each function graphed in Exercises #5 to #8, determine the interval(s) where the **derivative** is (a) positive; (b) negative; and the x-value(s) at which the **derivative** is (c) equals zero; (d) does not exist.

5. $y = f(x)$

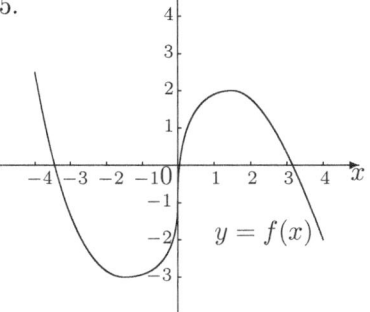

6. $y = f(x)$

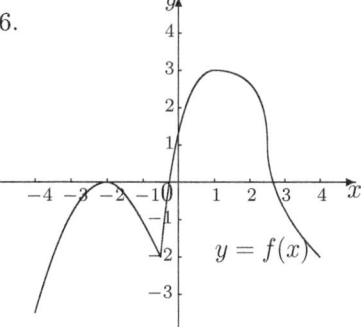

7. $y = h(x)$

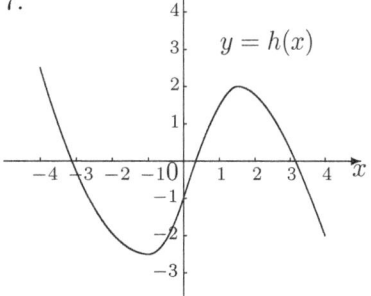

8. $y = h(x)$

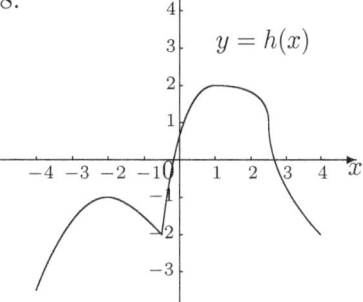

9. For the function graphed in Exercise #7, determine

    (a) The point(s) at which h has a relative maximum;

    (b) The point(s) at which h has a relative minimum;

    (c) The absolute maximum of h;

    (d) The absolute minimum of h.

10. For the function graphed in Exercise #8, determine

    (a) The point(s) at which h has a relative maximum;

    (b) The point(s) at which h has a relative minimum;

    (c) The absolute maximum of h;

    (d) The absolute minimum of h.

For each function graphed in Exercises #11 and #12, determine the interval(s) where the function is concave up, the interval(s) where it is concave down, and the point(s) at which it has a inflection point.

11.

12.

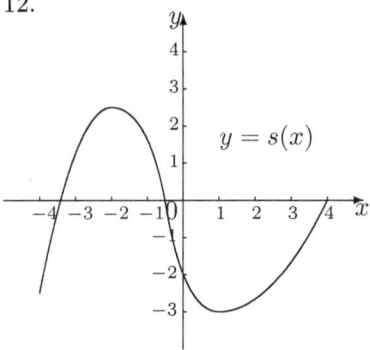

For each function graphed in Exercises #13 and #14, determine the interval(s) where the **derivative** *is increasing, and the interval(s) where the* **derivative** *is decreasing.*

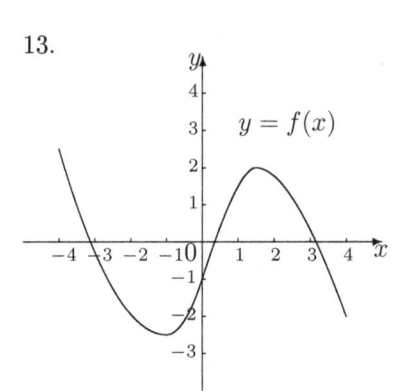

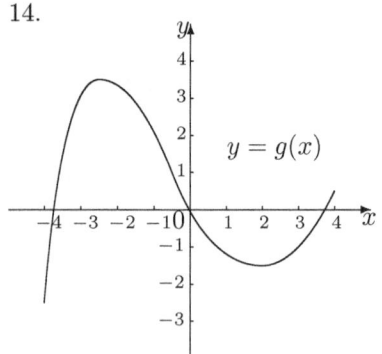

13.

14.

For each function graphed in Exercises #15 and #16, determine the interval(s) where the **second derivative** *is (a) positive; (b) negative.*

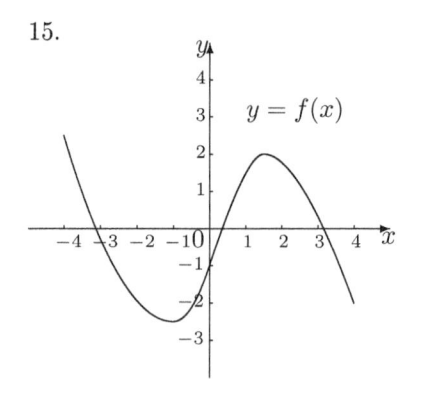

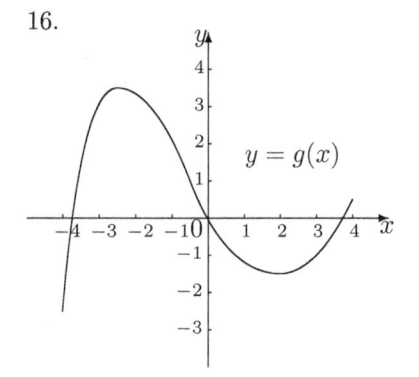

15.

16.

Determine the inflection point(s) of each function in Exercises #17 to #24.

17. $f(x) = x^2 - 5x + 6$

18. $f(x) = 2x^2 + 3x - 5$

19. $f(x) = x^4 - 2x^2$

20. $f(x) = 3x^4 - 4x^3$

21. $h(x) = \frac{1}{3}x^3 - 3x^2 + 5x + 6$

22. $h(x) = -x^3 - 3x^2 + 15x + 3$

23. $y = x^2 e^{-x}$

24. $y = xe^{3x}$

For the functions graphed in Exercises #25 and #26, determine the following: (Round answers to nearest one-half.)

(a) The interval(s) on which f is increasing;

(b) The interval(s) on which f is decreasing;

(c) The point(s) at which f' either equals zero or does not exist;

(d) The point(s) at which f has a relative maximum;

(e) The point(s) at which f has a relative minimum;

(f) The absolute maximum of f;

(g) The absolute minimum of f;

(h) The inflection point(s) of f;

(i) The interval(s) on which f is concave up;

(j) The interval(s) on which f is concave down.

25.

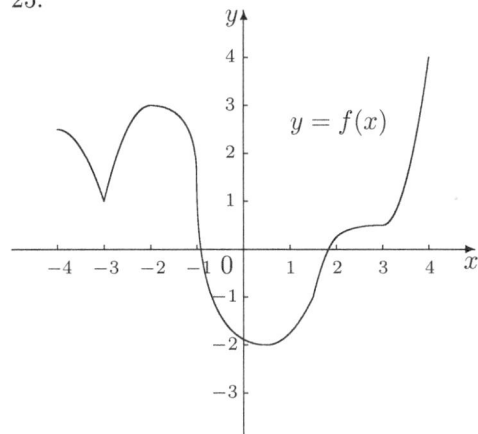

26.

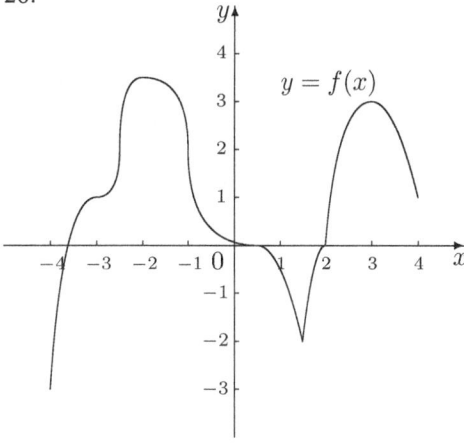

For each function in Exercises #27 to #30, do the following:

(a) Find all x-values at which the first derivative is zero;

(b) Use the signs of the first derivative to determine the intervals on which f is increasing;

(c) Use the signs of the first derivative to determine the intervals on which f is decreasing;

(d) Use the information from parts (b) and (c) to determine the relative extrema of f.

(e) Determine the inflection points;

(f) Use the signs of the second derivative to determine the intervals on which f is concave up;

(g) Use the signs of the second derivative to determine the intervals on which f is concave down.

27. $f(x) = x^3 - 12x - 4$

28. $f(x) = x^3 + 3x^2 - 45x + 1$

29. $f(x) = x^4 - 2x^3$

30. $f(x) = x^4 - 2x^2$

3.4 Derivative Tests and Relative Extrema

From Section 3.3, we learned how the sign of the first derivative on an interval indicates whether the function is increasing or decreasing, and how the sign of the second derivative on an interval indicates whether the function is concave up or concave down. In this section, we study where possible relative extrema may occur, as well as how to use the derivatives to determine whether a function has a relative maximum or a relative minimum at a particular candidate location.

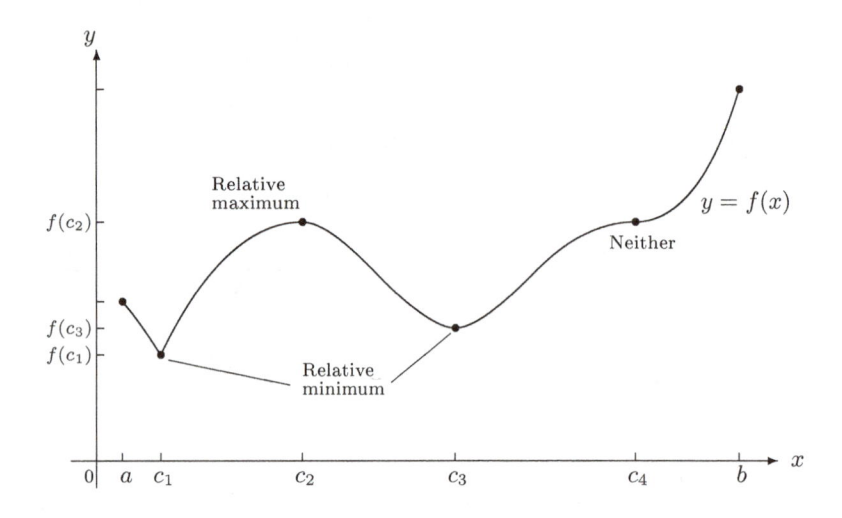

Figure 3.4.1: Relative extrema vs. critical points

Relative Extrema and Critical Numbers

Let's observe the graph of the function in Figure 3.4.1. Notice that, if at a point $(c, f(c))$ the function has a relative extremum, then either the graph has a "corner," hence, the function is not differentiable, or the tangent line of the function has *zero* slope, that is, $f'(c) = 0$. Since such number c indicates the possible location of a relative extremum, it is called a **critical number**. We formally define a critical number.

> **Definition**   *critical numbers and points*
>
> Assume that an open interval (a, b) is contained in the domain of a function f, and c is contained in the interval (a, b). Then c is called a **critical number** if either
>
> $f'(c)$ does not exist, or $f'(c) = 0$.
>
> In addition, the point $(c, f(c))$ is called a **critical point**.

The function f in Figure 3.4.1 has four critical numbers c_1, c_2, c_3, and c_4. The graphs in Figures 3.4.2 to 3.4.4 give some additional examples of critical points.

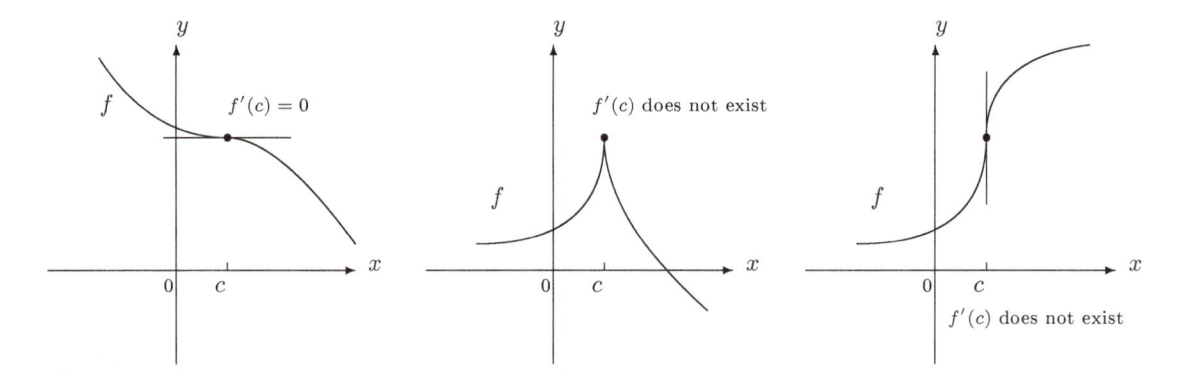

Figure 3.4.2: A critical point   **Figure 3.4.3:** A critical point   **Figure 3.4.4:** A critical point

3.4. DERIVATIVE TESTS AND RELATIVE EXTREMA

Notice that f may or may not have a relative extremum at a critical number. However, if f has a relative extremum at a point $(c, f(c))$, then that point must be a critical point, or equivalently, c must be a critical number. We state the conclusion in the following theorem.

> **Theorem 1**   *Relative Extrema and Critical Numbers*
>
> Suppose that the function f is continuous on an open interval (a, b). If f has a relative extremum at $x = c$, then c is a critical number. That is, either $f'(c) = 0$ or $f'(c)$ does not exist.

Example 1: Determine the critical numbers for each function:

(a) $f(x) = x^4 - 8x^2$ (b) $g(x) = xe^{2x}$

Solution (a) We first take the derivative. We have $f'(x) = 4x^3 - 16x$. Since f' exists everywhere, all its critical numbers satisfy $f'(x) = 0$:

$$4x^3 - 16x = 0$$
$$x^3 - 4x = 0$$
$$x(x - 2)(x + 2) = 0$$
$$x = 0 \quad \text{or } x - 2 = 0 \quad \text{or } x + 2 = 0$$
$$x = 0 \quad \text{or } x = 2 \quad \text{or } x = -2.$$

Hence, the critical numbers are -2, 0, and 2.

(b) We first take the derivative. Applying the Product Rule and the Chain Rule, we obtain

$$g'(x) = e^{2x} + 2xe^{2x} = (1 + 2x)e^{2x}.$$

Since g' exists everywhere, all the critical numbers satisfy $g'(x) = 0$. Since an exponential function has all positive values, the only solution is from $1 + 2x = 0$, or $x = -\frac{1}{2}$. Hence, g has a single critical number $c = -\frac{1}{2}$.

Example 2:   Determine the critical numbers for each function:

(a) $f(x) = (x - 5)^{2/3}$ (b) $g(x) = \dfrac{2x}{x^2 + 1}$

Solution (a) We first take the derivative:

$$f'(x) = \frac{2}{3}(x - 5)^{-1/3} = \frac{2}{3(x - 5)^{1/3}}.$$

Notice that f' is not defined at $x = 5$. That is, $f'(5)$ does not exist. Notice that $x = 5$ is in the domain of f. Hence, $c = 5$ is a critical number for f. Since the numerator of f' is a positive constant, there is no solution for $f'(x) = 0$. Therefore, f has a single critical number $c = 5$.

(b) We first take the derivative. Applying the Quotient Rule we obtain:

$$g'(x) = \frac{2 \cdot (x^2 + 1) - 2x \cdot 2x}{(x^2 + 1)^2} = \frac{2(1 - x^2)}{(x^2 + 1)^2}.$$

Notice that g' is defined everywhere. To look for the critical number c such that $g'(c) = 0$, we only need to solve for the root of the numerator.

$$1 - x^2 = 0$$

$$(1 - x)(1 + x) = 0$$

$$1 - x = 0 \quad \text{or} \quad 1 + x = 0$$

$$x = 1 \quad \text{or} \quad x = -1.$$

Hence, g has critical numbers -1 and 1.

Example 3: Determine the critical numbers for $f(x) = x + \dfrac{9}{x}$ on the interval $(1, 6)$.

Solution We first take the derivative:

$$f'(x) = 1 - \frac{9}{x^2}.$$

Notice that f' is not defined at $x = 0$. However, 0 is not in the domain of f. We solve for $f'(x) = 0$:

$$1 - \frac{9}{x^2} = 0$$

$$\frac{9}{x^2} = 1$$

$$9 = x^2$$

$$x = 3 \quad \text{or} \quad x = -3.$$

We drop -3 since it is not in the interval $(1, 6)$. Hence, f has a critical number $c = 3$ on the interval $(1, 6)$.

The First-Derivative Test

So far we have learned that if a function has a relative extremum $f(c)$, then c must be a critical number. We also learned how to use the first derivative to determine the critical numbers. Notice that a critical number c is only a candidate for the location of a relative extremum. $f(c)$ may be a relative maximum, a relative minimum, or neither, as illustrated in Figure 3.4.1.

Let us observe the graph at each critical point. Notice that at $x = c_1$ and $x = c_3$, the function changes from decreasing to increasing; therefore, f has relative minimums. At $x = c_2$, the function changes from increasing to decreasing; therefore, f has a relative maximum. At $x = c_4$, the function keeps increasing; therefore, f does not have a relative extremum.

Recall that by Theorem 3.1 in Section 3.3, the positive sign of the first derivative, f', indicates that the function is increasing, and the negative sign of f' indicates that the function is decreasing. Hence, we can use the signs of f' to determine whether f has a relative maximum, a relative minimum, or neither, at a critical number $x = c$. If f' changes from positive to negative when x moves from the left side of c to the right side, then f has a relative maximum at $x = c$. If f' changes from negative to positive when x moves from the left side of c to the right side, then f has a relative minimum at $x = c$. If f' does not change signs (i.e., either keeps positive or keeps negative) when x moves from the left side of c to the right side, then f has no relative extremum at $x = c$. The following theorem describes the First-Derivative Test, as illustrated in the following diagrams (Figures 3.4.5-3.4.8):

3.4. DERIVATIVE TESTS AND RELATIVE EXTREMA

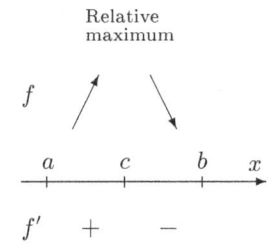

Figure 3.4.5: An f' Test (Relative maximum: f' + on (a,c), − on (c,b))

Figure 3.4.6: An f' Test (Relative minimum: f' − on (a,c), + on (c,b))

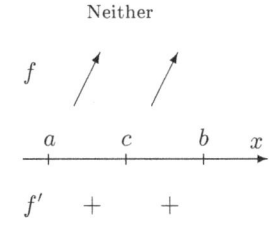

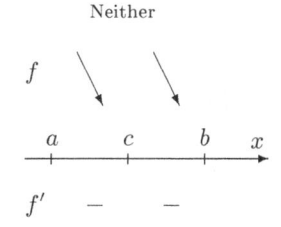

Figure 3.4.7: An f' Test (Neither: f' + on both sides)

Figure 3.4.8: An f' Test (Neither: f' − on both sides)

Theorem 2   The First-Derivative Test for Relative Extrema

Suppose that a function f is continuous on an open interval (a,b) that contains exactly one critical number c:

1. If $f'(x) > 0$ on (a,c) and $f'(x) < 0$ on (c,b), then $f(c)$ is a relative maximum.
2. If $f'(x) < 0$ on (a,c) and $f'(x) > 0$ on (c,b), then $f(c)$ is a relative minimum.
3. If $f'(x)$ has the same sign on both (a,c) and (c,b), then f has no relative extremum on (a,b).

Reality Check   *How to perform the first-derivative test*

1. Mark all critical numbers on the real line. Record the derivative (either zero or none) under each critical number.
2. These numbers divide the real line into intervals. In each interval, determine the sign of the derivative by substituting a sample value into the derivative. Mark the sign in each interval.
3. In each interval, according to the sign of the derivative, determine whether the function is increasing or decreasing, and use an arrow to mark the behavior.
4. At each critical number, determine whether the function has a relative maximum (if f' is positive on the left side and negative on the right side), a relative minimum (if f' is negative on the left side and positive on the right side), or neither (if f' has the same sign on both sides). Mark the shape of the function around the critical number according to the information.

Example 4: For the function $f(x) = x^3 - 3x + 1$, use the First-Derivative Test to determine the relative extrema.

Solution   We first take the derivative:

$$f'(x) = 3x^2 - 3.$$

Since f' exists everywhere, all its critical numbers satisfy $f'(x) = 0$:

$3x^2 - 3 = 0$

$x^2 - 1 = 0$

$(x-1)(x+1) = 0$

$x = 1$ or $x = -1$.

Hence, the critical numbers are -1 and 1.

The critical numbers divide the real line into three intervals: $(-\infty, -1)$, $(-1, 1)$, and $(1, \infty)$. On each interval the derivative keeps the same sign. We determine the signs of f' in each interval by plugging in some numbers:

$f'(-2) = 3(-2)^2 - 3 = 9 > 0; \quad f'(0) = -3 < 0; \quad f'(2) = 3(2)^2 - 3 = 9 > 0.$

Therefore, f' is positive on the intervals $(-\infty, -1)$ and $(1, \infty)$, and negative on the interval $(-1, 1)$. We can make a First-Derivative Test diagram (Figure 3.4.9):

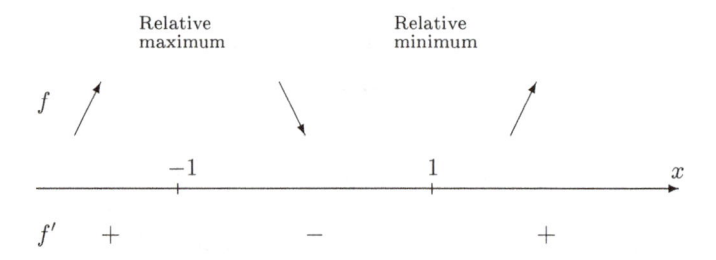

Figure 3.4.9: f' Test for $f(x) = x^3 - 3x + 1$

We have

$f(-1) = (-1)^3 - 3(-1) + 1 = -1 + 3 + 1 = 3,$ and

$f(1) = (1)^3 - 3(1) + 2 = 1 - 3 + 1 = -1.$

Hence, $f(x)$ has a relative maximum at $(-1, 3)$, and a relative minimum at $(1, -1)$. The graph of this function is given in Figure 3.3.5.

━ The Second-Derivative Test ━

We now illustrate how to use the second derivative at a critical number $x = c$ to judge whether the function has a relative maximum or minimum. In this case, the second derivative must exist at the critical number. That means the first derivative must exist and $f'(c) = 0$. Observe the graphs in Figures 3.4.1 and 3.3.5. Notice that the function graphed in Figure 3.4.1 is concave down around the critical number $x = c_2$ where $f'(c_2) = 0$, and f has a relative maximum at $x = c_2$. It is concave up around the critical number $x = c_3$ where $f'(c_3) = 0$, and it has a relative minimum at $x = c_3$. Similarly, in Figure 3.3.5, the function $f(x) = x^3 - 3x + 1$ is concave down around $x = -1$ and concave up around $x = 1$, and it has a relative maximum at $x = -1$ and a relative minimum at $x = 1$. This observation leads us to conclude that, around a critical number $x = c$ where $f'(c) = 0$, if the function is concave down, then it has a relative maximum at $x = c$; and if the function is concave up, then it has a relative minimum at $x = c$.

Recall that by Theorem 3.2 in Section 3.3, the positive sign of the second derivative, f'', indicates

3.4. DERIVATIVE TESTS AND RELATIVE EXTREMA

that the function is concave up, and the negative sign of f'' indicates that the function is concave down. Hence, we can use the signs of f'' to determine whether f has a relative maximum or a relative minimum at a critical number $x = c$ where $f'(c) = 0$: If the $f''(c)$ is positive then f has a relative maximum at $x = c$. If the $f''(c)$ is negative, then f has a relative minimum at $x = c$. The following theorem describes the Second-Derivative Test:

Theorem 3 *The Second-Derivative Test for Relative Extrema*

Suppose that a function f is differentiable on an open interval (a, b) that contains a critical number c for which $f'(c) = 0$:

1. If $f''(c) > 0$, then $f(c)$ is a relative minimum.
2. If $f''(c) < 0$, then $f(c)$ is a relative maximum.
3. If $f''(c) = 0$, then the First-Derivative Test must be used to determine whether $f(c)$ is a relative extremum.

Reality Check *How to perform the second-derivative test*

1. First, notice that the test can only be performed for the critical numbers at which the first derivative is zero.
2. Substitute each such critical number c into the second derivative to obtain $f''(c)$.
3. If $f''(c) > 0$, then $f(c)$ is a relative minimum of the function. The shape of the function around c is similar to that in Figure 3.4.11.
4. If $f''(c) < 0$, then $f(c)$ is a relative maximum of the function. The shape of the function around c is similar to that in Figure 3.4.10.
5. If $f''(c) = 0$, then no conclusion can be made. The First-Derivative Test must be used to determine whether $f(c)$ is a relative maximum, relative minimum, or neither.

Reality Check *Why the second-derivative test may fail*

Unlike the First-Derivative Test, the Second-Derivative Test cannot always determine whether the function has a relative extremum at a critical number. First, if $f'(c)$ does not exist, then the Second-Derivative Test cannot be used, since $f''(c)$ does not exist either. Second, if $f''(c) = f'(c) = 0$, then the function may or may not have a relative extremum at $x = c$. The graphs of three functions are illustrated in Figures 3.4.10 to 3.4.12: For $f(x) = x^4 + 1$, $f'(0) = f''(0) = 0$ and f has a relative minimum at $x = 0$. For $f(x) = -x^4 + 1$, $f'(0) = f''(0) = 0$ and f has a relative minimum at $x = 0$. For $f(x) = x^3 + 1$, $f'(0) = f''(0) = 0$ and f has no relative extremum at $x = 0$.

Reality Check *The advantage of the second-derivative test*

When the second derivative is easy to obtain, such as in the cases of polynomials, the Second-Derivative Test is more convenient, as illustrated by Example 5. In addition, the test reveals the concavity of the function on the intervals determined by the inflection points.

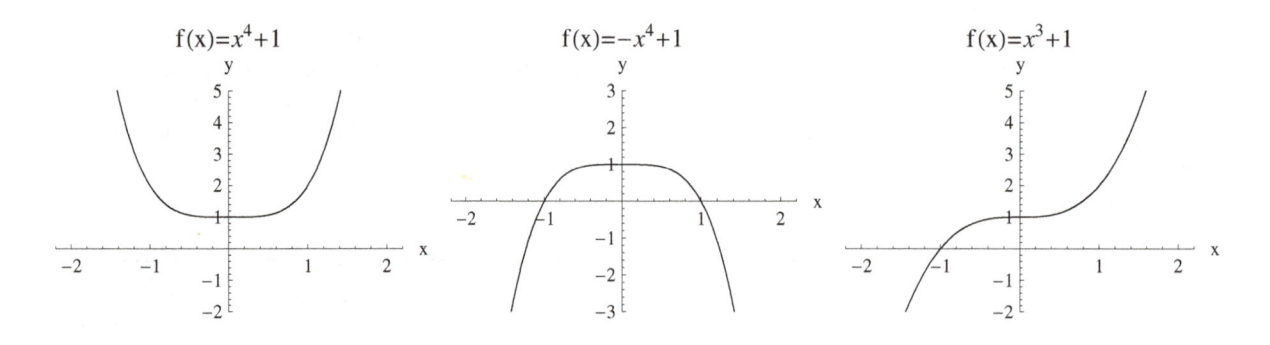

Figure 3.4.10: $f''(0) = 0$ **Figure 3.4.11:** $f''(0) = 0$ **Figure 3.4.12:** $f''(0) = 0$

Example 5: For the function $f(x) = x^3 - 3x + 1$, use the Second-Derivative Test to determine the relative extrema of the function.

Solution We first take the derivatives:

$$f'(x) = 3x^2 - 3 \quad \text{and} \quad f''(x) = 6x.$$

From Example 1, the critical numbers are -1 and 1. We have

$$f''(-1) = 6(-1) = -6 < 0 \quad \text{and} \quad f''(1) = 6(1) = 6 > 0.$$

Therefore, f has a relative maximum at $x = -1$ and f has a relative minimum at $x = 1$.

Determine Relative Extrema: Using the Derivatives

We now summarize the strategy for determine the relative extrema of a given function.

> **Strategy for Determine Relative Extrema Using Derivatives**
>
> **(a)** *Derivatives and Domain.* Find $f'(x)$ and $f''(x)$. Note the domain of f.
>
> **(b)** *Critical numbers of f.* Determine the critical numbers by solving $f'(x) = 0$ and finding where $f'(x)$ does not exist. These numbers form the candidates at which f has a relative maximum or minimum. Calculate the function values at these numbers and plot these critical numbers on the real line.
>
> **(c)** *Increasing and/or decreasing; relative extrema.* Perform either the First-Derivative Test or the Second-Derivative Test for each critical number (see the details for each succeeding test). At a critical number c, if $f(c)$ is a relative maximum, then f is increasing to the left of c and decreasing to the right; if $f(c)$ is a relative minimum, then f is decreasing to the left of c and increasing to the right.

Example 6: Find the relative extrema for the function given by $f(x) = x^4 - 8x^2 + 6$.

Solution (a) *Derivatives and domain.* We take the derivatives:

$$f'(x) = 4x^3 - 16x; \quad f''(x) = 12x^2 - 16.$$

The domain of the function is the set of all real numbers, that is, the whole real line.

(b) *Critical numbers.* Since f' exists everywhere, all its critical numbers satisfy $f'(x) = 0$. We solve the

3.4. DERIVATIVE TESTS AND RELATIVE EXTREMA

equation $f'(x) = 4x^3 - 16x = 0$:

$$4x^3 - 16x = 0 \quad \text{so} \quad 4x(x-2)(x+2) = 0$$

$$x = 0 \quad \text{or} \quad x = 2 \quad \text{or} \quad x = -2.$$

Hence, the critical numbers are 0, 2, and -2. We have $f(0) = 6$, $f(-2) = (-2)^4 - 8(-2)^2 + 2 = 16 - 32 + 6 = -10$, and $f(2) = (2)^4 - 16(2)^2 + 10 = 16 - 32 + 6 = -10$. Hence, the critical points are $(0, 6)$, $(-2, -10)$, and $(2, -10)$.

(c) *Increasing, decreasing, and the relative extrema.* To illustrate the processes, we perform both derivative tests. We first perform the First-Derivative Test:

The critical numbers divide the real line into four intervals: $(-\infty, -2)$, $(-2, 0)$, $(0, 2)$, and $(2, \infty)$. We determine the signs of f' in each interval by plugging in some numbers:

$$f'(-3) = 4(-3)^3 - 16(-3) = -60 < 0; \quad f'(-1) = 4(-1)^3 - 16(-1) = 12 > 0;$$

$$f'(1) = 4(1)^3 - 16(1) = -12 < 0; \quad f'(3) = 4(3)^3 - 16(3) = 60 > 0.$$

Therefore, f' is negative on the intervals $(-\infty, -2)$ and $(0, 2)$, and positive on the intervals $(-2, 0)$ and $(2, \infty)$. By Theorem 3.1 in Section 3.3, f is decreasing on the intervals $(-\infty, -2)$ and $(0, 2)$ and increasing on the intervals $(-2, 0)$ and $(2, \infty)$. We form the First-Derivative Test diagram (Figure 3.4.13):

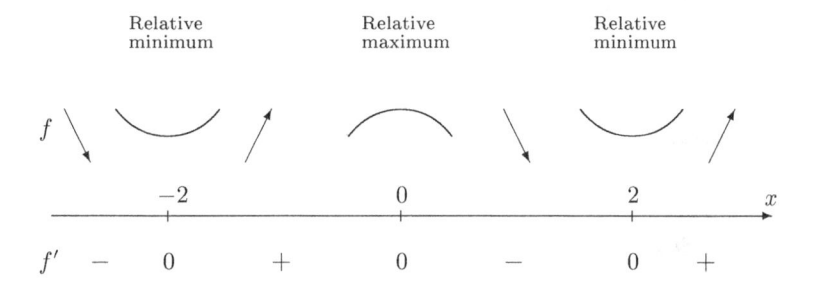

Figure 3.4.13: f' Test for $f(x) = x^4 - 8x^3 + 6$

To perform the Second-Derivative Test, we substitute the three critical numbers $(-2, 0, 2)$ into $f''(x)$:

$f''(-2) = 12(-2)^2 - 16 = 32 > 0$, hence, $f(-2) = -10$ is a relative minimum;

$f''(0) = -16 < 0$, hence, $f(0) = 6$ is a relative maximum;

$f''(2) = 12(2)^2 - 16 = 32 > 0$, hence, $f(2) = -10$ is a relative minimum.

Example 7: Find the relative extrema for the function given by

$$f(x) = -4 + 8x^3 - 3x^4.$$

Solution (a) *Derivatives and domain.* We take the derivatives:

$$f'(x) = 24x^2 - 12x^3; \quad f''(x) = 48x - 36x^2.$$

The domain of the function is the set of all real numbers, that is, the whole real line.

(b) *Critical numbers.* Since f' exists everywhere, all its critical numbers satisfy $f'(x) = 0$. We solve the equation $f'(x) = 24x^2 - 12x^3 = 0$:

$$24x^2 - 12x^3 = 0 \quad \text{or} \quad 12x^2(2-x) = 0, \quad \text{hence} \quad x = 0 \quad \text{or} \quad x = 2.$$

Hence, the critical numbers are 0 and 2. We have $f(0) = -4$ and $f(2) = -4 + 8(2)^3 - 3(2)^4 = -4 + 64 - 48 = 12$. Hence, the critical points are $(0, -4)$ and $(2, 12)$.

(c) *Increasing, decreasing, and the relative extrema.* We perform the Second-Derivative Test. We substitute the two critical numbers 0 and 2 into $f''(x)$:

$$f''(0) = 0,$$

therefore, no conclusion can be made.

$$f''(2) = 48(2) - 36(2)^2 = 96 - 144 = -58 < 0,$$

therefore, $f(2) = 12$ is a relative maximum. We perform the First-Derivative Test around 0. Since

$$f'(-1) = 24(-1)^2 - 12(-1)^3 = 24 + 12 = 36 > 0 \quad \text{and}$$

$$f'(1) = 24(1)^2 - 12(1)^3 = 24 - 12 = 12 > 0,$$

the derivative does not change sign at 0. Therefore, $f(0)$ is not a relative extremum. Combined with the result of the Second-Derivative Test, we conclude that f is increasing on the interval $(-\infty, 2)$ and decreasing on the interval $(2, \infty)$. We can form the diagram (Figure 3.4.14):

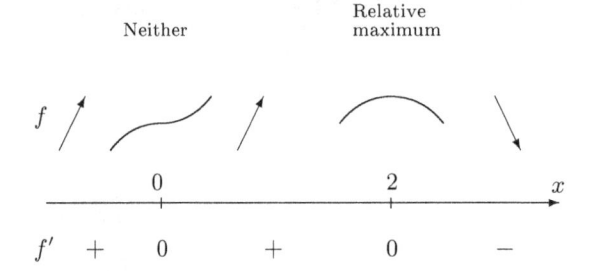

Figure 3.4.14: f' Test for $f(x) = -4 + 8x^3 - 3x^4$

Example 8: Find the relative extrema for the function given by

$$f(x) = 3(x-1)^{\frac{2}{3}} - 2.$$

Solution (a) *Derivatives and domain.* We take the derivatives:

$$f'(x) = 2(x-1)^{-\frac{1}{3}} = \frac{2}{(x-1)^{\frac{1}{3}}}; \quad f''(x) = \frac{-2}{3}(x-1)^{-\frac{4}{3}} = \frac{-2}{3(x-1)^{\frac{4}{3}}}.$$

The domain of the function is the set of all real numbers, that is, the whole real line.

(b) *Critical numbers.* We look for x such that $f'(x) = 0$ or $f'(x)$ does not exist. Since the numerator of f' is a positive constant, there is no solution for $f'(x) = 0$. Notice that f' is not defined at $x = 1$. That is, $f'(1)$ does not exist. Notice that $x = 1$ is in the domain of f. Hence, $c = 1$ is a critical number for f. Therefore, f has a single critical number $c = 1$. We have $f(1) = 3(1-1)^{\frac{2}{3}} - 2 = 3(0)^{\frac{2}{3}} - 2 = -2$. Hence the function has a single critical point $(1, -2)$.

(c) *Increasing, decreasing, and the relative extrema.* Notice that $f'(1)$ does not exist, hence, the Second-Derivative Test cannot be used. We perform the First-Derivative Test around $x = 1$. Since

$$f'(0) = \frac{2}{(0-1)^{\frac{1}{3}}} = \frac{2}{(-1)^{\frac{1}{3}}} = -2 < 0, \quad \text{and} \quad f'(2) = \frac{2}{(2-1)^{\frac{1}{3}}} \frac{2}{(1)^{\frac{1}{3}}} = 2 > 0,$$

3.4. DERIVATIVE TESTS AND RELATIVE EXTREMA

the derivative changes from negative to positive at 1. Therefore, $f(1) = -2$ is a relative minimum. We conclude that f is decreasing on the interval $(-\infty, 1)$ and increasing on the interval $(1, \infty)$ (Figure 3.4.15).

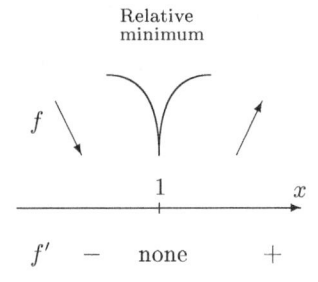

Figure 3.4.15: f' Test for $f(x) = 3(x-1)^{2/3} - 2$

Example 9: Find the relative extrema for the function given by

$$f(x) = 3(2x+3)^{\frac{1}{3}} + 1 \ .$$

Solution (a) *Derivatives and domain.* We take the derivatives:

$$f'(x) = (2x+3)^{-\frac{2}{3}} \cdot 2 = \frac{2}{(2x+3)^{\frac{2}{3}}} \ ; \quad f''(x) = \frac{-4}{3}(2x+3)^{-\frac{5}{3}} \cdot 2 = \frac{-8}{3(2x+3)^{\frac{5}{3}}} \ .$$

The domain of the function is the set of all real numbers, that is, the whole real line.

(b) *Critical numbers.* Similar to Example 8, f has a single critical number $c = -3/2$, where f' does not exist. We have

$$f(-3/2) = 3\big(2(-3/2) + 3\big)^{1/3} + 1 = 3(-3+3)^{1/3} + 1 = 3(0)^{1/3} + 1 = 1 \ .$$

Hence, the function has a single critical point $(-3/2, 1)$.

(c) *Increasing, decreasing, and the relative extrema.* Since $f'(-3/2)$ does not exist, the Second-Derivative Test cannot be used. We perform the First-Derivative Test around $x = -3/2$: Since

$$f'(-2) = \frac{2}{\big(2(-2)+3\big)^{\frac{2}{3}}} = \frac{2}{(-1)^{\frac{2}{3}}} = 2 > 0, \quad \text{and} \quad f'(-1) = \frac{2}{\big(2(-1)+3\big)^{\frac{2}{3}}} = \frac{2}{(1)^{\frac{2}{3}}} = 2 > 0,$$

the derivative does not change signs at $-3/2$. Therefore, $f(-3/2)$ is not a relative extremum. We conclude that f is increasing on the whole interval $(-\infty, \infty)$ (Figure 3.4.16).

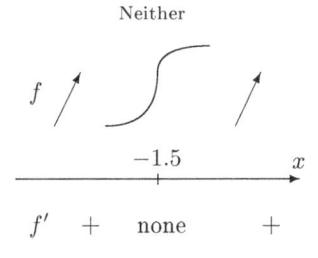

Figure 3.4.16: f' Test for $f(x) = 3(2x+3)^{1/3} + 1$

Example 10: Find the relative extrema for the function given by

$$f(x) = xe^{2x} .$$

Solution (a) *Derivatives and domain.* We take the derivatives: applying the Product Rule and the Chain Rule, we obtain

$$f'(x) = e^{2x} + 2xe^{2x} = (1 + 2x)e^{2x}.$$

Applying the Product and Chain Rules again, we obtain

$$f''(x) = 2e^{2x} + 2(1+2x)e^{2x} = 4(1+x)e^{2x}.$$

The domain of the function is the set of all real numbers, that is, the whole real line.

(b) *Critical numbers.* Since f' exists everywhere, all its critical numbers satisfy $f'(x) = 0$. We solve the equation

$$f'(x) = (1 + 2x)e^{2x} = 0.$$

Since an exponential function has all positive values, the only solution is from $1 + 2x = 0$, or $x = -1/2$. Hence, the function has a single critical number $-1/2$. We have $f(-1/2) = (-1/2)e^{2(-1/2)} = -1/(2e)$. Hence, the critical point is $(-1/2, -1/(2e))$.

(c) *Increasing, decreasing, and the relative extrema.* We perform the Second-Derivative Test. Substitute the critical number $x = -1/2$ into $f''(x)$:

$$f''\left(-\frac{1}{2}\right) = 4(1 + (-\frac{1}{2}))e^{2(-\frac{1}{2})} = 2e^{-1} > 0,$$

therefore, $f(-1/2) = -1/(2e)$ is a relative minimum. f is decreasing on the interval $(-\infty, -1/2)$ and increasing on the interval $(-1/2, \infty)$ (Figure 3.4.17).

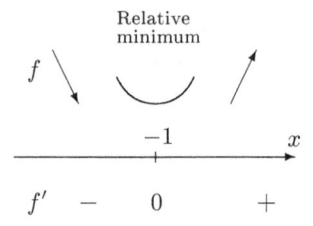

Figure 3.4.17: f' Test for $f(x) = xe^{2x}$

Exercise Set 3.4

Determine the critical numbers of each function in Exercises #1 to #16.

1. $f(x) = x^2 - 5x + 6$
2. $f(x) = 2x^2 + 3x - 5$
3. $g(x) = -2x + 1$
4. $g(x) = 5x - 6$
5. $h(x) = \frac{1}{3}x^3 - 3x^2 + 5x + 6$
6. $h(x) = -x^3 - 3x^2 + 15x + 3$

3.4. DERIVATIVE TESTS AND RELATIVE EXTREMA

7. $y = x^2 e^{-x}$

8. $y = xe^{3x}$

9. $f(x) = \dfrac{2x}{x^2 + 4}$

10. $f(x) = \dfrac{1}{x^2 + 1}$

11. $g(x) = 3(x+3)^{2/3}$

12. $g(x) = -2(x-2)^{2/5}$

13. For $f(x) = x + \dfrac{4}{x}$, find the critical numbers on the interval $(0, \infty)$.

14. For $f(x) = 9x + \dfrac{1}{x}$, find the critical numbers on the interval $(0, \infty)$.

15. For $f(x) = x^4 - 18x^2 + 2$, find the critical numbers on the interval $(-4, 1)$.

16. For $f(x) = 3x^4 - 8x^3 + 4$, find the critical numbers on the interval $(1, 5)$.

For each function in Exercises #17 to #32, find the critical numbers, if they exist. Use the First-Derivative Test to classify each as a relative maximum or minimum, or neither. Draw a diagram as in the examples to illustrate the results.

17. $f(x) = x^2 - 4x + 5$

18. $f(x) = 2x^2 + 3x - 5$

19. $g(x) = -x^2 + 5x - 6$

20. $g(x) = 2 - 3x - 6x^2$

21. $h(x) = \frac{1}{3}x^3 - 3x^2 + 5x + 6$

22. $h(x) = -x^3 - 3x^2 + 15x + 3$

23. $K(x) = x^2 e^{-x}$

24. $K(x) = xe^{3x}$

25. $s(x) = x^4 - x^3$

26. $s(x) = 2x^3 - x^4$

27. $f(x) = x^4 - 8x^2 + 3$

28. $f(x) = x^4 - 18x^2 - 1$

29. $g(x) = 2(x-2)^{2/3}$

30. $g(x) = 4 - 3(x+1)^{2/3}$

31. $h(x) = 1 + 3(x-2)^{1/3}$

32. $h(x) = 2 - (x+1)^{1/3}$

For each function in Exercises #33 to #45, find the critical numbers, if they exist. Use the Second-Derivative Test, if possible, to classify each as a relative maximum or minimum.

33. $f(x) = x^2 - 4x + 5$

34. $f(x) = 2x^2 + 3x - 5$

35. $g(x) = -x^2 + 5x - 6$

36. $g(x) = 2 - 3x - 6x^2$

37. $h(x) = \frac{1}{3}x^3 - 3x^2 + 5x + 6$

38. $h(x) = -x^3 - 3x^2 + 15x + 3$

39. $K(x) = x^2 e^{-x}$

40. $K(x) = xe^{3x}$

41. $s(x) = x^4 - x^3$

42. $s(x) = 2x^3 - x^4$

43. $f(x) = x^4 - 8x^2 + 3$

44. $f(x) = x^4 - 18x^2 - 1$

For each function in Exercises #45 to #70, determine the relative extrema following the process illustrated in the examples, round up to three decimal places if needed:

(i) List the coordinates of where the relative extrema occur;

(ii) Determine the interval(s) where the function is increasing or decreasing;

(iii) Draw a diagram as in the examples to illustrate the results.

45. $f(x) = x^2 - 5x + 6$

46. $f(x) = 2x^2 + 3x - 5$

47. $g(x) = x^3 - 3x^2 + 4$

48. $g(x) = x^3 - 3x + 2$

49. $h(x) = \frac{1}{3}x^3 - 3x^2 + 5x + 6$

50. $h(x) = -x^3 - 3x^2 + 15x + 3$

51. $f(x) = x^3 - 12x$

52. $f(x) = x^3 - 27x$

53. $s(x) = -x^3 + 3x^2 + 1$

54. $s(x) = -x^3 + 6x^2 + 3$

55. $f(x) = x^4 - 6x^2 - 4$

56. $f(x) = x^4 - 2x^2 + 3$

57. $g(x) = -x^4 + 8x^2 + 2$

58. $g(x) = -x^4 + 18x^2 + 3$

59. $h(x) = x^4 - 2x^3 - 2$

60. $h(x) = x^4 - 4x^3 - 5$

61. $f(x) = x^4 - 4x + 2$

62. $f(x) = x^4 - 32x - 1$

63. $s(x) = 5x^3 - 3x^5$

64. $s(x) = 20x^3 - 3x^5$

65. $f(x) = x^2 e^{-x}$

66. $f(x) = xe^{3x}$

67. $g(x) = 3(x+3)^{2/3} + 2$

68. $g(x) = -2(x-2)^{2/3} + 3$

69. $h(x) = -3(x+3)^{1/3} + 2$

70. $h(x) = 2(x-2)^{1/3} + 3$

3.5 Graph Sketching: Using the Derivatives

From Sections 3.3 and 3.4, we learned how the sign of the first derivative on an interval indicates whether the function is increasing or decreasing, and how the sign of the second derivative on an interval indicates whether the function is concave up or concave down. Further, we learned that all relative extrema of a function are located at critical numbers. We also learned how to use the derivatives to determine whether a function has a relative maximum or a relative minimum at a particular critical number. We now study how to use the information obtained though the derivatives to sketch the graph of a given function.

In Section 1.1, several simple graphs are sketched by the following process: The function is evaluated at some spots to create a few ordered pairs $(x_1, f(x_1))$, $(x_2, f(x_2))$, ..., which are represented as the points in the coordinate plane. When enough points are plotted to show some pattern, the points are then linked to form a line or a curve. This process works for sketching lines and some simple curves. However, for complicate curves, the sampled points may miss the important features like "peaks" and "valleys." Even with the help of a graphing device, if the ranges of the two coordinates are not set appropriately, the important feature may still go missing. For illustration purposes, let us consider the function $f(x) = (x-20)^4 - 2(x-20)^2 + 10$. The following three graphs (Figures 3.5.1-3.5.3) all represent this function:

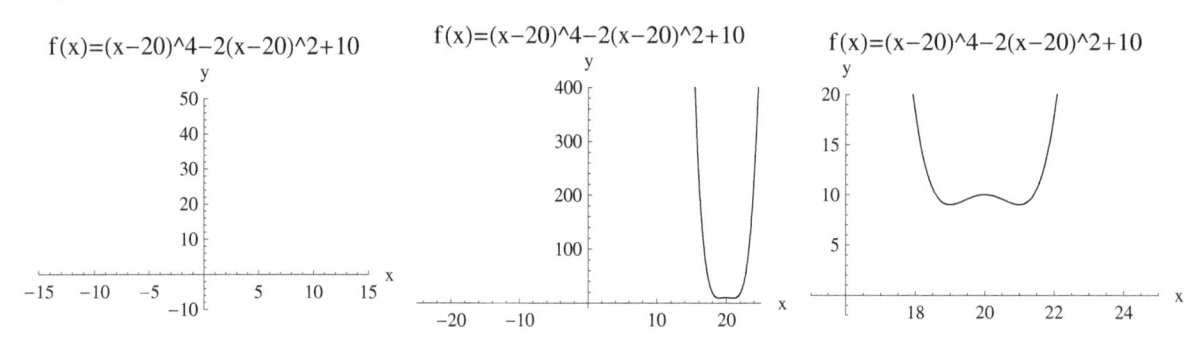

Figure 3.5.1: Improper graph

Figure 3.5.2: Improper graph

Figure 3.5.3: Proper graph

The "graph" in Figure 3.5.1 misses the curve, because the x-coordinate range is too narrow. The graph in Figure 3.5.2 resembles a parabola, missing the details of turns, because the y-coordinate range is too wide. The graph in Figure 3.5.3 properly shows the shape of this curve.

3.5. GRAPH SKETCHING: USING THE DERIVATIVES

A proper graph of the function should include all the points where relative extrema occur. Recall that all relative extrema occur at critical points. Hence, the first step of sketching the graph of a function is to find all these critical numbers using the first derivative. The shapes of a curve around these critical points are illustrated in Section 3.4, Figures 3.4.13 to 3.4.17.

In addition to the relative extrema, the concavity significantly affects the shapes of the curves. The second derivative helps us to determine whether the function is concave up or concave down, and to locate the inflection points. So far we have greatly improved our capability of sketching curves. Additional helpful strategies for graphing rational functions will be introduced in the next section.

The strategies for sketching graphs differ significantly with and without involving a graphing device.

Sketching Graphs Using a Graphing Device

Here, we summarize the strategy for graph sketching sufficient for many functions, such as polynomials, using a graphing device:

Strategy for Sketching Graphs Using a Graphing Device

(a) *Derivatives and Domain.* Find $f'(x)$ and $f''(x)$. Note the domain of f.

(b) *Critical numbers of f.* Determine the critical numbers by solving $f'(x) = 0$ and finding where $f'(x)$ does not exist. These numbers form the candidates at which f has a relative maximum or minimum. Calculate the function values at these numbers.

(c) *Possible Inflection points.* Solve $f''(x) = 0$ and find where $f''(x)$ does not exist. Calculate the function values at these numbers. These points are the candidates for inflection points.

(d) *Set the plotting windows* The horizontal interval should include all critical numbers and all x-values of possible inflection points. The vertical interval should include all all y-values of critical points and possible inflection points.

(e) *Plot the graph.*

Example 1: Use a graphing device to sketch the graph of the function given by $f(x) = x^4 - 8x^2 + 6$.

Solution (a) - (b): This function is the same as in Section 3.4, Example 4.6. The derivatives are:
$$f'(x) = 4x^3 - 16x; \quad f''(x) = 12x^2 - 16.$$
The critical points are $(0, 6)$, $(-2, -10)$, and $(2, -10)$.

(c) *Possible inflection points:* We now determine the candidates of inflection points. Since f'' exists everywhere, we solve the equation $f''(x) = 12x^2 - 16 = 0$:
$$12x^2 - 16 = 0 \quad \text{or} \quad x^2 = \frac{16}{12} = \frac{4}{3} \quad \text{or} \quad x = \pm\frac{2}{\sqrt{3}} \approx \pm 1.15.$$

We substitute $x = \pm 2/\sqrt{3}$ into the function $f(x)$:
$$f\left(\pm\frac{2}{\sqrt{3}}\right) = \left(\pm\frac{2}{\sqrt{3}}\right)^4 - 8\left(\pm\frac{2}{\sqrt{3}}\right)^2 + 6 = \frac{16}{9} - \frac{32}{3} + 6 = -\frac{26}{9} \approx -2.9.$$
Hence, the possible inflection points are $(-1.15, -2.9)$ and $(1.15, -2.9)$.

(d) *Setting the plot windows.* We may set the horizontal plotting interval to be $[-3, 3]$ and the vertical plotting interval to be $[-15, 15]$, to include all the critical points and possible inflection points.

(e) *Plot the graph.* The graph is shown in Figure 3.5.4.

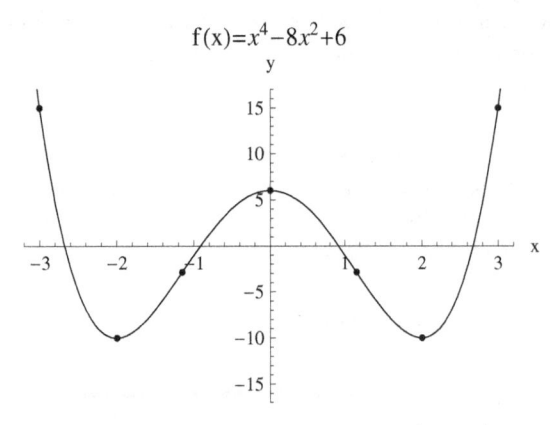

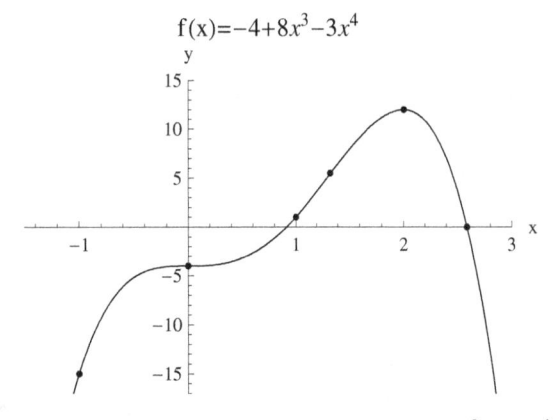

Figure 3.5.4: Graph of $f(x) = x^4 - 8x^2 + 6$

Example 2: Use a graphing device to sketch the graph of the function given by $f(x) = -4 + 8x^3 - 3x^4$.

Solution (a) - (b) This function is the same as in Section 3.4, Example 4.7. The derivatives are:

$$f'(x) = 24x^2 - 12x^3; \quad f''(x) = 48x - 36x^2.$$

The critical points are $(0, -4)$ and $(2, 12)$.

(c) *Possible inflection points:* We now determine the candidates of inflection points. Since f'' exists everywhere, we solve the equation $f''(x) = 48x - 36x^2 = 0$:

$$48x - 36x^2 = 0$$
$$4x - 3x^2 = 0$$
$$x(4 - 3x) = 0$$
$$x = 0 \quad \text{or} \quad x = 4/3.$$

We substitute $x = 0$ and $x = 4/3$ into the function $f(x)$:

$$f(0) = -4 + 8(0)^3 - 3(0)^4 = 0, \quad f(4/3) = -4 + 8(4/3)^3 - 3(4/3)^4 \approx 5.48.$$

Hence, the possible inflection points are $(0, -4)$ and $(1.33, 5.48)$.

(d) *Setting the plot windows.* We may set the horizontal plotting interval to be $[-1, 3]$ and the vertical plotting interval to be $[-15, 15]$, to include all the critical points and possible inflection points.

(e) *Plot the graph.* The graph is shown in Figure 3.5.5.

Figure 3.5.5: Graph of $f(x) = -4 + 8x^3 - 3x^4$

3.5. GRAPH SKETCHING: USING THE DERIVATIVES

Example 3: Use a graphing device to sketch the graph of the function given by $f(x) = 3(x-1)^{\frac{2}{3}} - 2$.

Solution (a) - (b) This function is the same as in Section 3.4, Example 4.8. The derivatives are:

$$f'(x) = 2(x-1)^{-\frac{1}{3}} = \frac{2}{(x-1)^{\frac{1}{3}}}; \quad f''(x) = \frac{-2}{3}(x-1)^{-\frac{4}{3}} = \frac{-2}{3(x-1)^{\frac{4}{3}}}.$$

The function has a single critical point $(1, -2)$.

(c) *Possible inflection points:* We now determine the candidates of inflection points. Since the numerator of f'' is a negative constant, there is no solution for $f''(x) = 0$. Notice that f'' is not defined at $x = 1$, which is in the domain. Therefore, there is a single candidate $x = 1$. Hence, a possible inflection point is $(1, -2)$.

(d) *Setting the plot windows.* We may set the horizontal plotting interval to be $[-2, 3]$ and the vertical plotting interval to be $[-4, 6]$, to include the critical and possible inflection point.

(e) *Plot the graph.* The graph is shown in Figure 3.5.6.

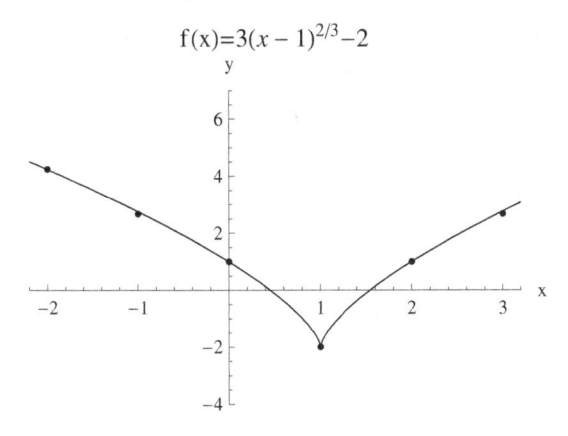

Figure 3.5.6: Graph of $f(x) = 3(x-1)^{\frac{2}{3}} - 2$

Example 4: Use a graphing device to sketch the graph of the function given by $f(x) = 3(2x+3)^{\frac{1}{3}} + 1$.

Solution (a) - (b) This function is the same as in Section 3.4, Example 4.9. The derivatives are:

$$f'(x) = (2x+3)^{-\frac{2}{3}} \cdot 2 = \frac{2}{(2x+3)^{\frac{2}{3}}}; \quad f''(x) = \frac{-4}{3}(2x+3)^{-\frac{5}{3}} \cdot 2 = \frac{-8}{3(2x+3)^{\frac{5}{3}}}.$$

The function has a single critical point $(-1.5, 1)$.

(c) *Possible inflection points:* We now determine the candidates of inflection points. Since the numerator of f'' is a negative constant, there is no solution for $f''(x) = 0$. Notice that f'' is not defined at $x = -1.5$, which is in the domain. Therefore, there is a single candidate $x = -1.5$. Hence, a possible inflection point is $(-1.5, 1)$.

(d) *Setting the plot windows.* We may set the horizontal plotting interval to be $[-3, 1]$ and the vertical plotting interval to be $[-4, 6]$, to include the critical and possible inflection point.

(e) *Plot the graph.* The graph is shown in Figure 3.5.7.

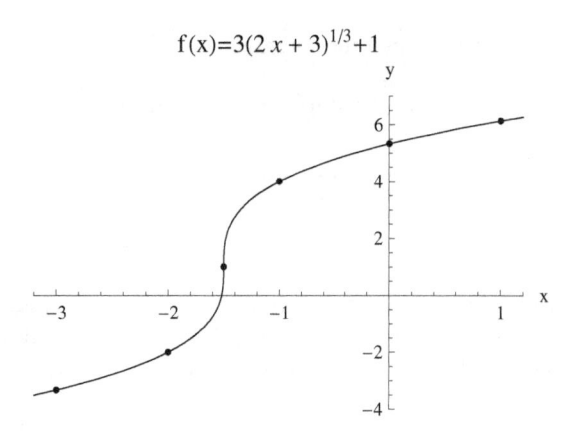

Figure 3.5.7: Graph of $f(x) = 3(2x+3)^{\frac{1}{3}} + 1$

Example 5: Use a graphing device to sketch the graph of the function given by $f(x) = xe^{2x}$.

Solution (a) - (b) This function is the same as in Section 3.4, Example 4.10. The derivatives are: $f'(x) = e^{2x} + 2xe^{2x} = (1+2x)e^{2x}$, $f''(x) = 4(1+x)e^{2x}$. The critical point is $(-1/2, -1/(2e))$.

(c) *Possible inflection points:* We now determine the candidates of inflection points. Since f'' exists everywhere, we solve the equation $f''(x) = 4(1+x)e^{2x} = 0$: Since an exponential function has all positive values, the only solution is from $1 + x = 0$, or $x = -1$. We substitute $x = -1$ into the function $f(x)$: $f(-1) = -e^{-2} \approx -0.14$. Hence, a possible inflection point is $(-1, -0.14)$.

(d) *Setting the plot windows.* We may set the horizontal plotting interval to be $[-2, 0.5]$ and the vertical plotting interval to be $[-0.5, 1.5]$, i to include both the critical point and the possible inflection point.

(e) *Plot the graph.* The graph is shown in Figure 3.5.8.

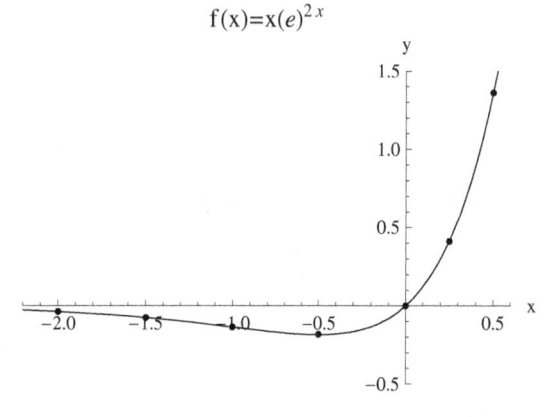

Figure 3.5.8: Graph of $f(x) = xe^{2x}$

Sketching Graphs Without Using a Graphing Device

We now summarize the strategy for graph sketching sufficient for many functions, such as polynomials:

3.5. GRAPH SKETCHING: USING THE DERIVATIVES

Strategy for Sketching Graphs Using Derivatives

(a) *Derivatives and Domain.* Find $f'(x)$ and $f''(x)$. Note the domain of f.

(b) *Critical numbers of f.* Determine the critical numbers by solving $f'(x) = 0$ and finding where $f'(x)$ does not exist. These numbers form the candidates at which f has a relative maximum or minimum. Calculate the function values at these numbers and plot these critical points on the coordinate plane.

(c) *Increasing and/or decreasing; relative extrema.* Perform either the First-Derivative Test or the Second-Derivative Test for each critical number. At a critical number c, if $f(c)$ is a relative maximum, then f is increasing to the left of c and decreasing to the right; if $f(c)$ is a relative minimum, then f is decreasing to the left of c and increasing to the right. Sketch a short segment around each critical point according to the test conclusion.

(d) *Inflection points.* Solve $f''(x) = 0$ and find where $f''(x)$ does not exist. Calculate the function values at these numbers and mark these points on the coordinate plane. These points are the candidates for inflection points.

(e) *Concavity.* Use the first coordinate of the candidates for inflection points from step (d) to define intervals. In each interval, substitute a sample value into $f''(x)$ to determine whether f is concave up (when $f''(x) > 0$) or concave down (when $f''(x) < 0$).

(f) *Sketch the graph.* Use the information obtained in steps (a) – (e) to sketch the graph. Calculate and plot some additional points if needed.

Example 6: Sketch the graph of the function given by $f(x) = -4 + 8x^3 - 3x^4$.

Solution (a) - (c): This function is the same as in Section 3.4, Example 4.7. The derivatives are:

$$f'(x) = 24x^2 - 12x^3; \quad f''(x) = 48x - 36x^2.$$

The critical points are $(0, 4)$ and $(2, 12)$. f is increasing on the interval $(-\infty, 2)$ and decreasing on the interval $(2, \infty)$. f has a relative maximum at $(2, 12)$.

(d) and (e) *Inflection points and concavity.* By Example 2, two possible inflection points are $(0, -4)$ and $(1.33, 5.48)$. It can be verified that f'' is negative when $x < 0$ and $x > 1.33$, and f'' is positive when $0 < x < 1.33$. Hence, f is concave down on the intervals $(-\infty, 0)$ and $(1.33, \infty)$, and f is concave up on the interval $(0, 1.33)$. Hence, both points are inflection points. We plot the known information on the coordinate plane (Figure 3.5.9).

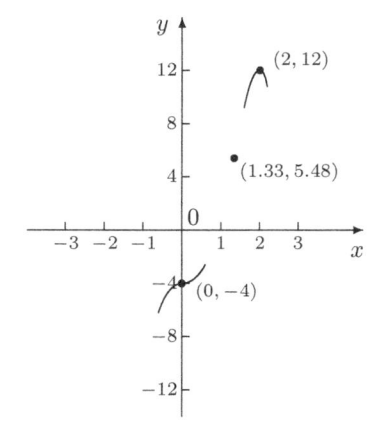

Figure 3.5.9: Partial graph of $f(x) = -4 + 8x^3 - 3x^4$

(f) *Sketch the graph.* The information obtained in steps (a) to (e), as plotted in Figure 3.5.9, has revealed the major features of the graph. We can calculate a few additional function values to help us to complete the sketch. The graph is shown in Figure 3.5.10.

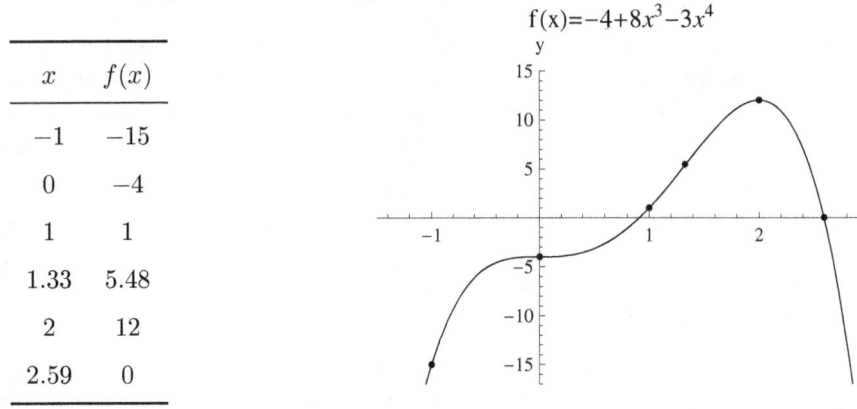

x	$f(x)$
-1	-15
0	-4
1	1
1.33	5.48
2	12
2.59	0

Figure 3.5.10: Graph of $f(x) = -4 + 8x^3 - 3x^4$

Example 7: Sketch the graph of the function given by $f(x) = xe^{2x}$.

Solution (a) - (c) This function is the same as in Section 3.4, Example 4.10. The derivatives are:

$$f'(x) = e^{2x} + 2xe^{2x} = (1+2x)e^{2x}, \quad f''(x) = 2e^{2x} + 2(1+2x)e^{2x} = 4(1+x)e^{2x}.$$

The function has a single critical point $(-0.5, -0.2)$. f is decreasing on the interval $(-\infty, -0.5)$ and increasing on the interval $(-0.5, \infty)$. f has a relative minimum at $(-0.5, -0.2)$.

(d) and (e) *Inflection points and concavity.* By Example 5, one possible inflection points is $(-1, -0.1)$. It can be verified that f'' is negative for all $x < -1$, and positive for all $x > -1$. Hence, f is concave down on $(-\infty, -1)$ and concave up on $(-1, \infty)$. f has one inflection point $(-1, -0.1)$. We plot the known information on the coordinate plane (Figure 3.5.11).

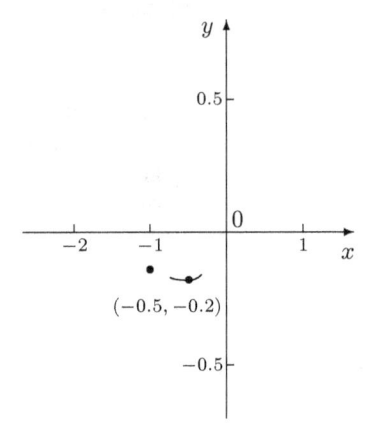

Figure 3.5.11: Partial graph of $f(x) = xe^{2x}$

(f) *Sketch the graph.* We calculate a few additional function values to help us to complete the sketch. The graph is shown in Figure 3.5.12.

x	$f(x)$
-2	-0.037
-1.5	-0.075
-1	-0.135
-0.5	-0.184
0	0
0.25	0.41
0.5	1.36

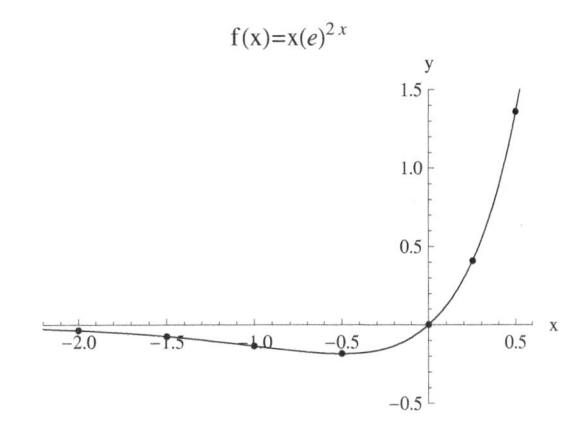

Figure 3.5.12: Graph of $f(x) = xe^{2x}$

Notice that the graph of $f(x) = xe^{2x}$ becomes "flatter" going left, and the curve is approaching the x-axis. This feature is common for rational functions. We study how to determine such features and sketch such graphs in the next section.

Exercise Set 3.5

For each function in Exercises #1 to #28, sketch the graph using a graphing device. Follow the process illustrated in the examples, round up to three decimal places if needed:

(i) List the critical points.

(ii) List the possible inflection points;

(iii) Sketch the graph with appropriate coordinate ranges.

1. $f(x) = x^2 - 5x + 6$
2. $f(x) = 2x^2 + 3x - 5$
3. $g(x) = x^3 - 3x^2 + 4$
4. $g(x) = x^3 - 3x + 2$
5. $h(x) = \frac{1}{3}x^3 - 3x^2 + 5x + 6$
6. $h(x) = -x^3 - 3x^2 + 15x + 3$
7. $f(x) = x^3 - 12x$
8. $f(x) = x^3 - 27x$
9. $s(x) = -x^3 + 3x^2 + 1$
10. $s(x) = -x^3 + 6x^2 + 3$
11. $f(x) = x^4 - 6x^2 - 4$
12. $f(x) = x^4 - 2x^2 + 3$
13. $g(x) = -x^4 + 8x^2 + 2$
14. $g(x) = -x^4 + 18x^2 + 3$
15. $h(x) = x^4 - 2x^3 - 2$
16. $h(x) = x^4 - 4x^3 - 5$
17. $f(x) = x^4 - 4x + 2$
18. $f(x) = x^4 - 32x - 1$
19. $s(x) = 5x^3 - 3x^5$
20. $s(x) = 20x^3 - 3x^5$
21. $f(x) = x^2 e^{-x}$
22. $f(x) = x^2 e^{-2x}$
23. $f(x) = xe^{2x}$
24. $f(x) = xe^{3x}$
25. $g(x) = 3(x+3)^{2/3} + 2$
26. $g(x) = -2(x-2)^{2/3} + 3$
27. $h(x) = -3(x+3)^{1/3} + 2$
28. $h(x) = 2(x-2)^{1/3} + 3$

For each function in Exercises #29 to #54, sketch the graph following the process illustrated in the examples, round up to three decimal places if needed:

(i) List the coordinates of where the relative extrema occur;

(ii) Determine the interval(s) where the function is increasing or decreasing;

(iii) List the coordinates of where the inflection points occur;

(iv) Determine the interval(s) where the function is concave up or concave down;

(v) Sketch the graph with appropriate coordinate ranges.

29. $f(x) = x^2 - 5x + 6$
30. $f(x) = 2x^2 + 3x - 5$
31. $g(x) = x^3 - 3x^2 + 4$
32. $g(x) = x^3 - 3x + 2$
33. $h(x) = \frac{1}{3}x^3 - 3x^2 + 5x + 6$
34. $h(x) = -x^3 - 3x^2 + 15x + 3$
35. $f(x) = x^3 - 12x$
36. $f(x) = x^3 - 27x$
37. $s(x) = -x^3 + 3x^2 + 1$
38. $s(x) = -x^3 + 6x^2 + 3$
39. $f(x) = x^4 - 6x^2 - 4$
40. $f(x) = x^4 - 2x^2 + 3$
41. $g(x) = -x^4 + 8x^2 + 2$
42. $g(x) = -x^4 + 18x^2 + 3$
43. $h(x) = x^4 - 2x^3 - 2$
44. $h(x) = x^4 - 4x^3 - 5$
45. $f(x) = x^4 - 4x + 2$
46. $f(x) = x^4 - 32x - 1$
47. $s(x) = 5x^3 - 3x^5$
48. $s(x) = 20x^3 - 3x^5$
49. $f(x) = x^2 e^{-x}$
50. $f(x) = xe^{3x}$
51. $g(x) = 3(x+3)^{2/3} + 2$
52. $g(x) = -2(x-2)^{2/3} + 3$
53. $h(x) = -3(x+3)^{1/3} + 2$
54. $h(x) = 2(x-2)^{1/3} + 3$

For Exercises #55 to #80, sketch a graph of a function that satisfies the conditions given. Answers will vary. All functions are continuous and have the domain of all real numbers.

55. $f(0) = 1$, $f(x)$ is increasing on the interval $(-\infty, 3)$ and decreasing on the interval $(3, \infty)$.

56. $f(1) = 2$, $f(x)$ is increasing on the interval $(-\infty, -1)$ and decreasing on the interval $(-1, \infty)$.

57. $g(0) = 1$, $g(x)$ is decreasing on the interval $(-\infty, -2)$ and increasing on the interval $(-2, \infty)$.

58. $g(1) = 2$, $g(x)$ is decreasing on the interval $(-\infty, 1)$ and increasing on the interval $(1, \infty)$.

59. $s(0) = 1$, $s(x)$ is increasing on the intervals $(-\infty, -3)$ and $(4, \infty)$, and decreasing on the interval $(-3, 4)$.

60. $s(0) = -1$, $s(x)$ is decreasing on the intervals $(-\infty, 1)$ and $(5, \infty)$, and increasing on the interval $(1, 5)$.

61. $f(0) = 1$, $f(x)$ is decreasing on the interval $(-\infty, -2)$ and increasing on the interval $(-2, \infty)$, concave up on the interval $(-\infty, 0)$ and concave down on the interval $(0, \infty)$.

62. $f(1) = 2$, $f(x)$ is increasing on the interval $(-\infty, 1)$ and increasing on the interval $(1, \infty)$, concave down on the interval $(-\infty, -1)$ and concave up on the interval $(-1, \infty)$.

63. $K(0) = 1$, $K(x)$ is increasing and concave up on the interval $(-\infty, 1)$ and decreasing and concave up on the interval $(1, \infty)$.

64. $K(1) = 2$, $K(x)$ is increasing and concave down on the interval $(-\infty, -1)$ and decreasing and concave up on the interval $(-1, \infty)$.

65. $f(0) = 1$, $f'(x) > 0$ on the interval $(-\infty, 3)$, $f'(3) = 0$, and $f'(x) < 0$ on the interval $(3, \infty)$.

66. $f(1) = 2$, $f'(x) > 0$ on the interval $(-\infty, -1)$, $f'(-1) = 0$, and $f'(x) < 0$ on the interval $(-1, \infty)$.

67. $g(0) = 1$, $g'(x) < 0$ on the interval $(-\infty, -2)$, $g'(-2)$ is undefined, and $g'(x) > 0$ on the interval $(-2, \infty)$.

68. $g(1) = 2$, $g'(x) > 0$ on the interval $(-\infty, 1)$, $g'(1)$ is undefined, and $g'(x) < 0$ on the interval $(1, \infty)$.

69. $s(0) = 1$, $s'(x) > 0$ on the intervals $(-\infty, -3)$ and $(4, \infty)$, and $s'(x) < 0$ on the interval $(-3, 4)$.

70. $s(0) = -1$, $s'(x) < 0$ on the intervals $(-\infty, 1)$ and $(5, \infty)$, and $s'(x) > 0$ on the interval $(1, 5)$.

71. $f(0) = 0$, $f'(x) < 0$ on the interval $(-\infty, -2)$ and $f'(x) > 0$ on the interval $(-2, \infty)$; $f''(x) > 0$ on the interval $(-\infty, 0)$ and $f''(x) < 0$ on the interval $(0, \infty)$.

72. $f(1) = 3$, $f'(x) < 0$ on the interval $(-\infty, 1)$ and $f'(x) > 0$ on the interval $(1, \infty)$, $f''(x) < 0$ on the interval $(-\infty, -1)$ and $f''(x) > 0$ on the interval $(-1, \infty)$.

73. $K(0) = 2$, $K'(x) > 0$ and $K''(x) > 0$ on the interval $(-\infty, 1)$, and $K'(x) < 0$ and $K''(x) > 0$ on the interval $(1, \infty)$.

74. $K(1) = 0$, $K'(x) > 0$ and $K''(x) < 0$ on the interval $(-\infty, -1)$ and $K'(x) > 0$ and $K''(x) > 0$ on the interval $(-1, \infty)$.

75. $f(0) = 1$, $f'(x) > 0$ on the interval $(-\infty, -1)$, $f'(x) = 0$ on the interval $(-1, 1)$, and $f'(x) < 0$ on the interval $(1, \infty)$.

76. $f(1) = 0$, $f'(x) < 0$ on the interval $(-\infty, 0)$, $f'(x) = 0$ on the interval $(0, 2)$, and $f'(x) > 0$ on the interval $(2, \infty)$.

77. $g(0) = 0$, $g'(0) = 0$, $g''(0) > 0$; $g(2) = 4$, $g'(2)$ is not defined, and $g'(x) < 0$ on the interval $(2, \infty)$.

78. $g(-1) = 1$, $g'(-1) = 0$, $g''(-1) < 0$; $g(1) = 0$, $g'(1)$ is not defined, and $g'(x) > 0$ on the interval $(1, \infty)$.

79. $s(0) = 0$, $s'(x) > 0$ on the intervals $(-\infty, -3)$ and $(4, \infty)$, and $s'(x) < 0$ on the interval $(-3, 4)$; $s'(-3) = s'(4) = 0$, and $s''(0) = 0$.

80. $s(0) = -2$, $s'(x) < 0$ on the intervals $(-\infty, 1)$ and $(5, \infty)$, and $s'(x) > 0$ on the interval $(1, 5)$, $s'(1) = s'(5) = 0$, and $s''(2) = 0$.

APPLICATIONS

81. A shoe manufacturer has the cost function

$$C(x) = 1{,}500 + 20x + 0.2x^2$$

(in dollars), $0 \leq x \leq 400$, where x is the number of pairs of shoes produced each week. The revenue function for selling x pairs of shoes is given by

$$R(x) = 90x \text{ (in dollars)}.$$

(a) Find the total profit when x pairs of shoes are produced and sold each week, and sketch the graph of the total profit function.

(b) From the graph of the total profit function, find (approximately) the number of pairs of shoes that should be produced and sold each week for the company to make the maximum profit. Also find the maximum profit.

(c) Find the marginal profit when x pairs of shoes are produced and sold each week, and sketch the graph of the marginal profit function.

(d) From the graph of the marginal profit function, find (approximately) the number of pairs of shoes produced and sold each week where the marginal profit is zero. Does this number match the number obtained in part (b)?

(e) Sketch the total cost function and the total revenue function on the same coordinate plane.

(f) Find the marginal cost and revenue functions and sketch the graphs of the two functions on the same coordinate plane.

(g) From the graphs obtained in part (f), find (approximately) the intersection of the two graphs. Does the first coordinate of the intersection match the numbers obtained in parts (b) and (d)?

82. A clothing manufacturer has the cost function

$$C(x) = 1{,}200 + 30x + 0.5x^2$$

(in dollars), $0 \leq x \leq 250$, where x is the number of suits produced each week. The revenue function for selling x suits is given by

$$R(x) = 120x \text{ (in dollars)}.$$

Follow the steps listed in Exercise #81.

83. An appliance manufacturer has the cost function

$$C(x) = 4{,}000 + 36x + 0.5x^2$$

(in dollars), $0 \leq x \leq 100$, where x is the number of a certain brand of microwave ovens produced each week. The revenue function for selling x microwave ovens is given by

$$R(x) = 500x - 4x^2 \text{ (in dollars)}.$$

Follow the steps listed in Exercise #81.

84. A refrigerator manufacturer determines that its cost, in dollars, of producing x refrigerators is given by

$$C(x) = 2x^2 + 15x + 1{,}500,$$

$0 \leq x \leq 80$. The revenue function for selling x refrigerators is given by

$$R(x) = 1{,}600x - 20x^2 \text{ (in dollars)}.$$

Follow the steps listed in Exercise #81.

Economics *For the demand and supply functions in Exercises #85 to #88, sketch the graphs of the two functions on the same coordinate plane, and find (approximately) the equilibrium price (p_E) and equilibrium quantity (x_E) from the graphs.*

85. Demand: $x = D(p) = 300e^{-0.1p}$;

    Supply: $x = S(p) = 5p - 20$.

86. Demand: $x = D(p) = 200e^{-0.05p}$;

    Supply: $x = S(p) = 2p - 50$.

87. Demand: $x = D(p) = 20 - 2\sqrt{p + 36}$;

    Supply: $x = S(p) = p - 2$.

88. Demand: $x = D(p) = 10 - \sqrt{p + 1}$;

    Supply: $x = S(p) = 2p - 3$.

3.6 Graph Sketching: Asymptotes and Rational Functions

In Section 3.5, we learned how to use the information about the derivatives to sketch the graph of a continuous function. In this section, we study how to sketch some discontinuous functions, mainly the rational functions, using additional tools known as the **asymptotes**.

Asymptotes: Vertical, Horizontal, and Slant

Recall that in Section 1.4 we briefly studied the rational functions. A rational function is defined as the ratio of two polynomials:

$$R(x) = \frac{P(x)}{Q(x)} = \frac{a_n x^n + a_{n-1} x^{n-1} + \ldots + a_1 x + a_0}{b_m x^m + b_{m-1} x^{m-1} + \ldots + b_1 x + b_0}, \quad b_m \neq 0.$$

Unlike polynomials, which all have the domain of all real numbers, a rational function has the domain of the set of all real numbers except those values of x that would make the denominator zero. Consider the rational function $f(x) = \dfrac{2x + 1}{x - 1}$. The domain of f is all real numbers except 1. The graph is shown in Figure 3.6.1. Notice that $x = 1$ is not in the domain, since the denominator becomes zero when $x = 1$. When x gets closer to 1 from the left, the graph is approaching the dotted vertical line $x = 1$ going downward. When x gets closer to 1 from the right, the graph is approaching the dotted vertical line $x = 1$ going upward. In Section 2.1, we have defined such behavior as

$$\lim_{x \to 1^-} f(x) = -\infty \quad \text{and} \quad \lim_{x \to 1^+} f(x) = \infty.$$

Thus the line $x = 1$ acts as a "limiting line." It is called a **vertical asymptote**.

3.6. GRAPH SKETCHING: ASYMPTOTES AND RATIONAL FUNCTIONS

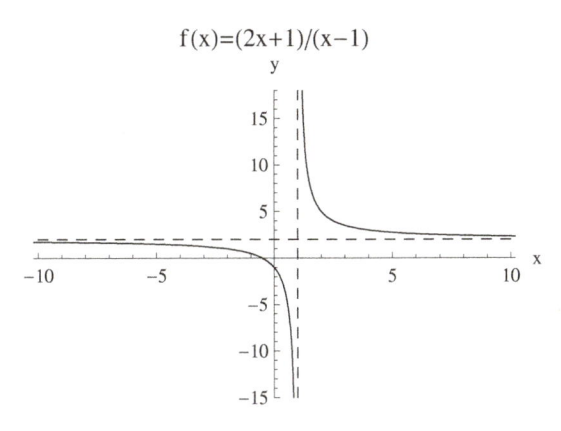

Figure 3.6.1: Graph of $f(x) = (2x+1)/(x-1)$

Definition   *vertical asymptote*

The line $x = a$ is a **vertical asymptote** if any of the following limits holds:

$$\lim_{x \to a^-} f(x) = -\infty, \quad \lim_{x \to a^-} f(x) = \infty, \quad \lim_{x \to a^+} f(x) = -\infty, \quad \text{or} \quad \lim_{x \to a^+} f(x) = \infty.$$

Reality Check   *Vertical asymptotes and rational functions*

For a rational function to have $x = a$ as a vertical asymptote, the denominator must be zero at $x = a$. However, if the numerator is also zero at $x = a$, then the graph may not have $x = a$ as a vertical asymptote. In such a case, the function must be simplified (that is, the common factor $x - a$ being taken away) to determine whether $x = a$ is a vertical asymptote, as illustrated in the following example.

Reality Check   *Vertical asymptotes and non-rational functions*

In addition to the rational functions, other examples of functions that have vertical asymptotes include $f(x) = \ln x$ and $f(x) = 1/\sqrt{x}$ (where the line $x = 0$ is a vertical asymptote for both functions), etc.

Example 1: Determine whether the line $x = 1$ is a vertical asymptote for each function:

(a) $f(x) = \dfrac{x^2 + 2}{x^2 - 1}$ (b) $g(x) = \dfrac{x^2 - 1}{x^2 - 3x + 2}$ (c) $h(x) = \dfrac{x^2 - 1}{x^2 - 2x + 1}$

Solution We substitute $x = 1$ into the denominator and the numerator for each function:

(a) For $f(x) = \dfrac{x^2 + 2}{x^2 - 1}$: $(1)^2 - 1 = 0$, and $(1)^2 + 2 = 3$. Since the denominator is zero at $x = 1$ but the numerator is not zero at $x = 1$, the line $x = 1$ is a vertical asymptote for f.

(b) For $g(x) = \dfrac{x^2 - 1}{x^2 - 3x + 2}$: $(1)^2 - 3(1) + 2 = 0$, and $(1)^2 - 1 = 0$. Since both the denominator and the numerator are zeros at $x = 1$, we factor both and simplify the function:

$$g(x) = \frac{x^2 - 1}{x^2 - 3x + 2} = \frac{(x+1)(x-1)}{(x-2)(x-1)} = \frac{x+1}{x-2}, \quad x \neq 1.$$

Therefore, the line $x = 1$ is not a vertical asymptote for g.

(c) For $h(x) = \frac{x^2 - 1}{x^2 - 2x + 1}$: $(1)^2 - 2(1) + 1 = 0$, and $(1)^2 - 1 = 0$. Since both the denominator and the numerator are zeros at $x = 1$, we factor both and simplify the function:

$$h(x) = \frac{x^2 - 1}{x^2 - 2x + 1} = \frac{(x+1)(x-1)}{(x-1)(x-1)} = \frac{x+1}{x-1}.$$

Therefore, the line $x = 1$ is a vertical asymptote for h.

Example 2: For the function $f(x) = \frac{x^2 - x - 2}{x^3 - 2x^2 - 3x}$, determine the vertical asymptotes.

Solution We factor the numerator and denominator and take away the common factors:

$$f(x) = \frac{(x-2)(x+1)}{x(x+1)(x-3)} = \frac{x-2}{x(x-3)}, \quad x \neq -1.$$

Since the denominator equals zero at $x = 0$ and $x = 3$, and the numerator is nonzero at these numbers, we conclude that $x = 0$ and $x = 3$ are the vertical asymptotes.

There are four ways in which a vertical asymptote can occur, as shown in Figure 3.6.2.

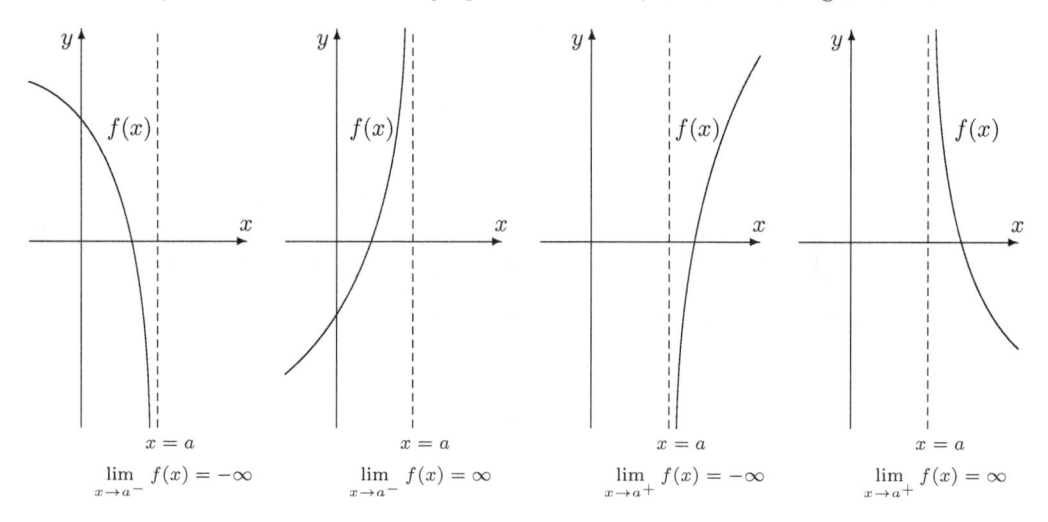

Figure 3.6.2: Four pattens of vertical asymptotes

We now take another look at Figure 3.6.1 and focus on the horizontal dotted line $y = 2$. Note that the graph becomes flatter when x moves either to the far left or to the far right. In Section 2.1, we have defined such behavior as

$$\lim_{x \to -\infty} f(x) = 2 \quad \text{and} \quad \lim_{x \to \infty} f(x) = 2.$$

The horizontal line $y = 2$ is called a **horizontal asymptote**.

3.6. GRAPH SKETCHING: ASYMPTOTES AND RATIONAL FUNCTIONS

Definition *horizontal asymptote*

The line $y = b$ is a **horizontal asymptote** if either or both of the following limits holds:

$$\lim_{x \to -\infty} f(x) = b, \quad \text{or} \quad \lim_{x \to \infty} f(x) = b.$$

Reality Check *Asymptotes and the graph of a function*

The graph of a *continuous* function never crosses a vertical asymptote. The graph of a function may cross a horizontal asymptote $y = b$, as long as when x approaches either positive or negative infinity (towards far left or far right), the function value approaches b. Asymptotes (vertical, as well as horizontal and slant defined later) are not part of the graph of the function. The dotted lines are helpful in illustrating the "approaching" behavior of the function.

Reality Check *Horizontal asymptotes and rational functions*

For a rational function to have $y = b$ as a horizontal asymptote, the degree of the numerator must be less than or equal to the degree of the denominator. If the degree of the numerator is less than the degree of the denominator, then the x-axis, or the line $y = 0$, is a horizontal asymptote. If the degree of the numerator equals the degree of the denominator, then the line $y = a/d$ is a horizontal asymptote, where a is the leading coefficient (i.e., the coefficient of the highest power term) of the numerator and d is the leading coefficient of the denominator.

Reality Check *Horizontal asymptotes and non-rational functions*

In addition to the rational functions, other examples of functions that have horizontal asymptotes include $f(x) = e^x$ and $f(x) = 1/\sqrt{x}$ (where the line $y = 0$ is a horizontal asymptote for both functions), etc.

There are four ways in which a horizontal asymptote can occur, as shown in Figure 3.6.3.

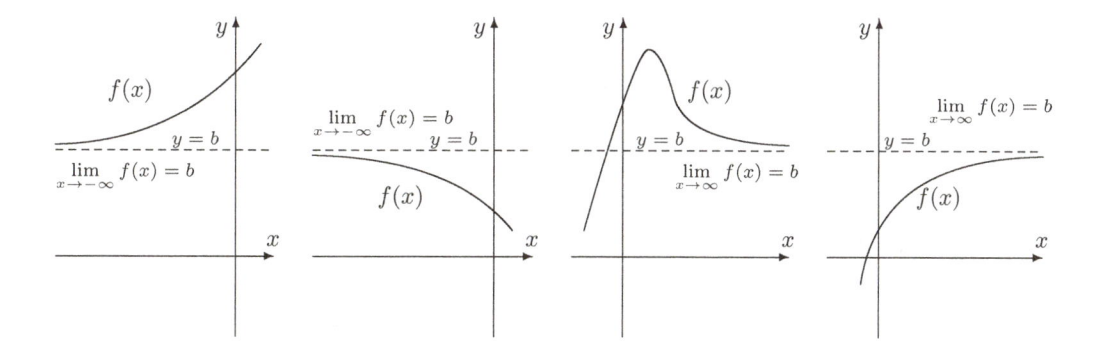

Figure 3.6.3: Four pattens of horizontal asymptotes

Example 3: Determine the horizontal asymptotes of each function:

(a) $f(x) = \dfrac{2x^2 + 5x - 1}{2 - 3x^2}$ (b) $g(x) = \dfrac{3x^3 - 1}{2x^2 + 7}$ (c) $h(x) = \dfrac{2}{x^2 - 4}$

Solution (a) Since the degrees of the numerator and the denominator are equal, we calculate the ratio of the leading coefficients, which is $2/(-3) = -\frac{2}{3}$. Hence, the horizontal asymptote is $y = -\frac{2}{3}$.

(b) Since the degree of the numerator is higher than the degree of the denominator, the function has no horizontal asymptote.

(c) Since the degree of the numerator is lower than the degree of the denominator, the function has horizontal asymptote $y = 0$.

To illustrate how this "judge by comparison of degrees" approach works, we explore the algebraic approach for finding the limit of a rational function $f(x)$ when $x \to \infty$ or $x \to -\infty$. This approach is based on the fact that

$$\lim_{x \to \pm\infty} \frac{1}{x^p} = 0$$

where the power p is a positive number. For example, when $p = 1$, it is straightforward to check numerically that

$$\lim_{x \to \infty} \frac{1}{x} = 0.$$

To find the limit of a rational function $f(x)$ when $x \to \infty$ or $x \to -\infty$, we divide both the numerator and the denominator by x to the *highest power* of the denominator, so that the reformulated function has a denominator that approaches a nonzero number when x approaches infinity.

Example 4: Find the following limits:

(a) $\lim\limits_{x \to \infty} \dfrac{2x+1}{x-3}$ (b) $\lim\limits_{x \to \infty} \dfrac{x^2-4}{2x^3+5x}$

Solution (a) Since the highest power of the denominator is 1, we divide the numerator and denominator of the function by x:

$$\frac{2x+1}{x-3} = \frac{(2x+1)/x}{(x-3)/x} = \frac{2+\frac{1}{x}}{1-\frac{3}{x}}.$$

Now we can apply the limit properties listed in Section 2.2 to find the limit. When $|x|$ gets very large, $1/x$ gets close to zero. Hence,

$$\lim_{x \to \infty} \frac{2x+1}{x-3} = \lim_{x \to \infty} \frac{2+\frac{1}{x}}{1-\frac{3}{x}} = \frac{2+0}{1-0} = 2.$$

(b) Since the highest power of the denominator is 3, we divide the numerator and denominator of the function by x^3:

$$\frac{x^2-4}{2x^3+5x} = \frac{(x^2-4)/x^3}{(2x^3+5x)/x^3} = \frac{\frac{1}{x} - \frac{4}{x^3}}{2 + \frac{5}{x^2}}.$$

When $|x|$ gets very large, all terms with $1/x^p$ get close to zero. Hence

$$\lim_{x \to \infty} \frac{x^2-4}{2x^3+5x} = \lim_{x \to \infty} \frac{\frac{1}{x} - \frac{4}{x^3}}{2 + \frac{5}{x^2}} = \frac{0-0}{2+0} = 0.$$

Some functions have asymptotes that are neither vertical nor horizontal. For example, consider the function

$$f(x) = \frac{x^2-1}{x}.$$

The graph of this function is shown in Figure 3.5.10. Note that the graph gets closer and closer to the line $y = x$ when x moves either to the far left or to the far right. The line $y = x$ is called a **slant** (or

3.6. GRAPH SKETCHING: ASYMPTOTES AND RATIONAL FUNCTIONS

oblique) **asymptote**. Observe that the function can be expressed as

$$f(x) = x - \frac{1}{x} \text{ and we know that } \lim_{x \to \infty} \frac{1}{x} = 0.$$

Therefore, the function behaves in a similar manner to the linear function $y = x$ as $|x|$ gets very large (Figure 3.6.4).

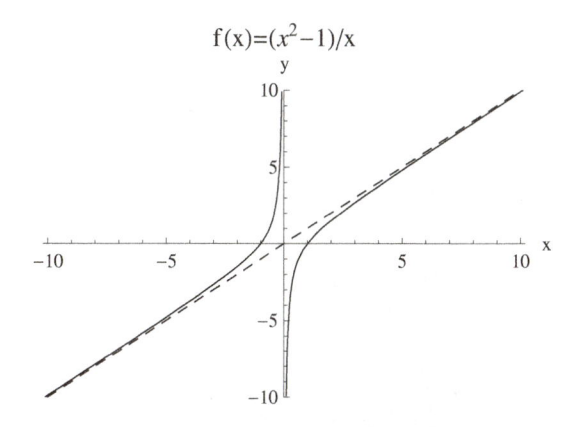

Figure 3.6.4: A function having a slant asymptote

Definition   *slant asymptote*

The line $y = mx + b$ ($m \neq 0$) is a **slant asymptote** if the function $f(x)$ can be expressed as

$$f(x) = mx + b + q(x) \quad \text{where} \quad \lim_{x \to -\infty} q(x) = 0, \quad \text{or} \quad \lim_{x \to \infty} q(x) = 0.$$

Reality Check   *Slant asymptotes and rational functions*

For a rational function to have a slant asymptote, the degree of its numerator must be exactly 1 higher than the degree of its denominator.

Given a rational function with the degree of its numerator exactly 1 higher than the degree of its denominator, how do we find its slant asymptote? One way is by division. Another way is by rewriting the function as shown in the following example.

Example 5: Find the slant asymptote for the function

$$f(x) = \frac{3x^2 - 2}{x + 1}.$$

Solution Since the ratio of the leading coefficients is 3, the slant asymptote has slope $m = 3$. We first re-express the numerator to include the term $3x(x + 1) = 3x^2 + 3x$, that is, the denominator multiplied by $3x$:

$$3x^2 - 2 = 3x^2 + 3x - 3x - 2 = 3x(x + 1) - 3x - 2,$$

hence

$$f(x) = \frac{3x^2 - 2}{x + 1} = \frac{3x(x + 1) - 3x - 2}{x + 1}$$

$$= \frac{3x(x+1)}{x+1} + \frac{-3x-2}{x+1} = 3x + \frac{-3x-2}{x+1}.$$

Notice that the second term of the function has a horizontal asymptote $y = -3$. Therefore, the slant asymptote is $y = 3x - 3$.

Graph Sketching Using a Graphing Device: Refined Strategy

We now summarize the refined strategy for graph sketching with a graphing device:

> **Refined Strategy for Sketching Graphs Using a Graphing Device**
>
> (a) *Derivatives, Asymptotes, and Domain.* Find $f'(x)$ and $f''(x)$. Find all vertical and horizontal asymptotes. Note the domain of f.
>
> (b) *Critical numbers of f.* Determine the critical numbers by solving $f'(x) = 0$ and finding where $f'(x)$ does not exist. These numbers form the candidates at which f has a relative maximum or minimum. Calculate the function values at these numbers.
>
> (c) *Possible Inflection points.* Solve $f''(x) = 0$ and find where $f''(x)$ does not exist. Calculate the function values at these numbers. These points are the candidates for inflection points.

> (d) *Set the plotting windows* The horizontal interval should include all critical numbers, vertical asymptotes, and all x-values of possible inflection points. The vertical interval should include all all horizontal asymptotes and y-values of critical points and possible inflection points.
>
> (e) *Plot the graph.*

> **Warning**
>
> A vertical asymptote is not part of the graph of the function. However, on a graphing device screen, it may show up as a solid vertical line. A graphing device sketches the graph by connecting the dots, hence, may connect two dots crossing the vertical asymptote.

Example 6: Use a graphing device to sketch the graph of $f(x) = \dfrac{2x^2 + 9}{x^2 - 9}$ according to the graphing strategy.

Solution (a) *Domain* The domain of the function is the set of all real numbers except for the x values at which the denominator equals zero. Solving

$$x^2 - 9 = (x-3)(x+3) = 0,$$

we obtain $x = 3$ and $x = -3$. Hence, the domain of the function is $(-\infty, -3) \cup (-3, 3) \cup (3, \infty)$.

Vertical and Horizontal Asymptotes.

Since the denominator equals zero at $x = -3$ and $x = 3$, the graph of the function has the lines $x = -3$ and $x = 3$ as vertical asymptotes.

The degrees of the numerator and the denominator are equal, therefore, the function has a horizontal asymptote $y = 2$.

3.6. GRAPH SKETCHING: ASYMPTOTES AND RATIONAL FUNCTIONS

Derivatives. We take the derivatives. Using the Quotient Rule, we obtain:

$$f'(x) = \frac{4x(x^2-9) - (2x^2+9)(2x)}{(x^2-9)^2} = \frac{4x^3 - 36x - 4x^3 - 18x}{(x^2-9)^2} = \frac{-54x}{(x^2-9)^2};$$

$$f''(x) = \frac{-54(x^2-9)^2 - (-54x)2(x^2-9)(2x)}{(x^2-9)^4} = \frac{54(x^2-9)(-x^2+9+4x^2)}{(x^2-9)^4} = \frac{162(x^2+3)}{(x^2-9)^3}.$$

(b) *Critical numbers.* We first look for values of x such that $f'(x) = 0$. Since the numerator of $f'(x)$ is $-54x$ and is zero at $x = 0$, and the denominator of $f'(x)$ is not zero at $x = 0$, we conclude that 0 is a critical number. Next, we look for values of x at which the derivative does not exist. The derivative does not exist at $x = -3$ and $x = 3$. However, these two values are not in the domain of the function. Therefore, the only critical number is 0. We have $f(0) = -1$. Hence, $(0, -1)$ is a critical point.

(c) *Possible inflection points.* We now determine the possible inflection points. We first look for values of x such that $f''(x) = 0$. Since the numerator of $f''(x)$ is $162(x^2+3)$ and is positive for all values of x, we conclude that there is no value of x such that $f''(x) = 0$. Next, we look for values of x at which the f'' does not exist. The second derivative does not exist at $x = -3$ and $x = 3$. However, these two values are not in the domain of the function. Therefore, the function has no inflection point.

(d) *Setting the plot windows.* We may set the horizontal plotting interval to be $[-5, 5]$ and the vertical plotting interval to be $[-4, 6]$, to include the asymptotes and the critical points.

(e) *Plot the graph.* The graph is shown in Figure 3.6.5.

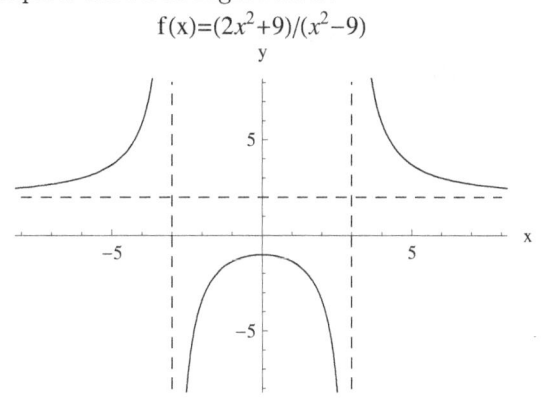

Figure 3.6.5: Graph of $f(x) = (2x^2+9)/(x^2-9)$

Example 7: Use a graphing device to sketch the graph of $f(x) = \dfrac{2x^2 - 4}{x^2 + 1}$ according to the graphing strategy.

Solution (a) *Domain.* The domain of the function is the set of all real numbers since the denominator is positive for all values of x.

Asymptotes. Since the denominator is positive for all values of x, the graph has no vertical asymptote.

The degrees of the numerator and the denominator are equal, therefore, the function has a horizontal asymptote $y = 2$.

Derivatives. We take the derivatives. Using the Quotient Rule, we obtain:

$$f'(x) = \frac{4x(x^2+1) - (2x^2-4)(2x)}{(x^2+1)^2} = \frac{4x^3 + 4x - 4x^3 + 8x}{(x^2+1)^2} = \frac{12x}{(x^2+1)^2};$$

$$f''(x) = \frac{12(x^2+1)^2 - (12x)2(x^2+1)(2x)}{(x^2+1)^4} = \frac{12(x^2+1)(x^2+1-4x^2)}{(x^2+1)^4} = \frac{12(1-3x^2)}{(x^2+1)^3}.$$

(b) *Critical numbers.* Since f' exists everywhere, all its critical numbers satisfy $f'(x) = 0$. Since the denominator is positive for all values of x, $x = 0$ is the only value at which the numerator of $f'(x)$ is zero. We conclude that 0 is the only critical number. We have $f(0) = -4$. Hence, $(0, -4)$ is the only critical point.

(c) *Possible inflection points.* We now determine the possible inflection points. Since f'' exists everywhere, the first coordinate p of all inflection points satisfies $f''(p) = 0$, or equivalently, the numerator of f'' is zero at $x = p$. We solve the equation $f''(x) = 12(1 - 3x^2) = 0$:

$$12(1-3x^2) = 0 \quad \text{or} \quad 1 - 3x^2 = 0 \quad \text{or} \quad x^2 = 1/3 \text{ or } \quad x = -1/\sqrt{3} \quad \text{or} \quad x = 1/\sqrt{3}.$$

We substitute $x = \pm 1/\sqrt{3}$ into the function $f(x)$:

$$f\left(\pm\frac{1}{\sqrt{3}}\right) = \frac{2\left(\frac{1}{3}\right) - 4}{\frac{1}{3} + 1} = \frac{-10}{4} = -\frac{5}{2}.$$

Hence, the possible inflection points are approximately $(-0.58, -2.5)$ and $(0.58, -2.5)$.

(d) *Setting the plot windows.* We may set the horizontal plotting interval to be $[-6, 6]$ and the vertical plotting interval to be $[-6, 6]$, to include the asymptotes and the critical point, and the possible inflection points.

(e) *Plot the graph.* The graph is shown in Figure 3.6.6.

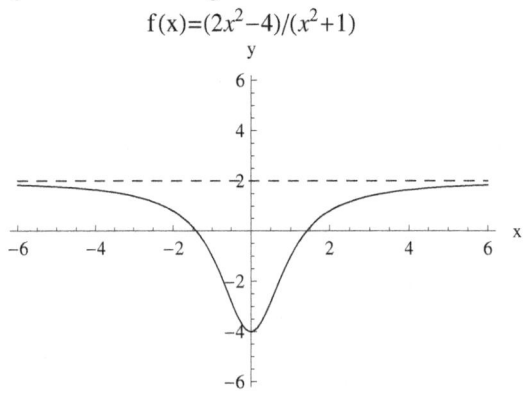

Figure 3.6.6: Graph of $f(x) = (2x^2 - 4)/(x^2 + 1)$

Example 8: Use a graphing device to sketch the graph of $f(x) = \dfrac{4x^2 + 1}{2x}$ according to the graphing strategy.

Solution (a) *Domain.* The domain of the function is the set of all real numbers except for $x = 0$ at which the denominator equals zero. Hence, the domain of the function is $(-\infty, 0) \cup (0, \infty)$.

(b) *Asymptotes.* Since the denominator equals zero at $x = 0$, the graph has the line $x = 0$, the y-axis, as a vertical asymptote.

Since the degrees of the numerator is higher than the degree of the denominator, the function has no horizontal asymptote.

Derivatives. We take the derivatives. Using the new expression $f(x) = 2x + 1/2x$ and the Power Rule, we obtain:

$$f'(x) = 2 - \frac{1}{2x^2} \quad \text{and} \quad f''(x) = \frac{1}{x^3}.$$

(b) *Critical numbers.* We first look for values of x such that $f'(x) = 0$. We express f' as a fraction:

$$f'(x) = 2 - \frac{1}{2x^2} = \frac{4x^2 - 1}{2x^2}.$$

Since the numerator of $f'(x)$ is $4x^2 - 1$ and is zero at $x = -1/2$ and $x = 1/2$, and the denominator of $f'(x)$ is not zero at these values, we conclude that $-1/2$ and $1/2$ are critical numbers. Next, we look for values of x at which the derivative does not exist. The derivative does not exist at $x = 0$. However, 0 is not in the domain of the function. Therefore, the function has two critical numbers $-1/2$ and $1/2$. We have $f(-1/2) = -2$ and $f(1/2) = 2$. Hence, $(-1/2, -2)$ and $(1/2, 2)$ are two critical points.

(d) *Setting the plot windows.* We may set the horizontal plotting interval to be $[-3, 3]$ and the vertical plotting interval to be $[-6, 6]$, to include the asymptote and the critical points.

(e) *Plot the graph.* The graph is shown in Figure 3.6.7.

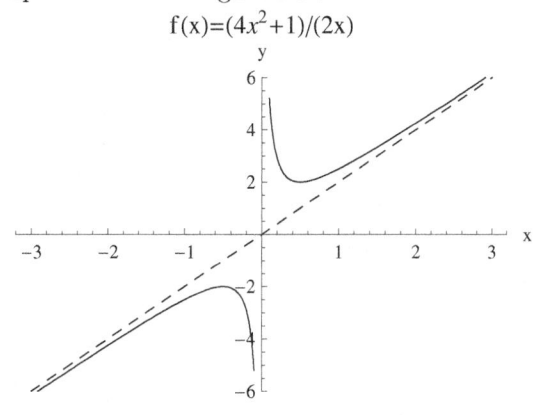

Figure 3.6.7: Graph of $f(x) = (4x^2 + 1)/(2x)$

Graph Sketching: Refined Strategy

We now summarize the strategy for graph sketching without a graphing device:

> **Refined Strategy for Sketching Graphs**
>
> **(a)** *Domain and intercepts.* Find the domain of the function. Find the intercepts if they are easy to calculate.
>
> **(b)** *Asymptotes.* Find all asymptotes—vertical, horizontal, and slant. Plot the asymptotes (if they exist) by dotted lines on the coordinate plane.
>
> **(c)** *Derivatives.* Find $f'(x)$ and $f''(x)$.
>
> **(d)** *Critical numbers of f.* Determine the critical numbers by solving $f'(x) = 0$ and finding where $f'(x)$ does not exist. These numbers form the candidates at which f has a relative maximum or minimum. Calculate the function values at these numbers and plot these critical points on the coordinate plane.
>
> **(e)** *Increasing and/or decreasing; relative extrema.* If a vertical asymptote exists, then use the signs of the first derivative to determine whether f is increasing or decreasing on each side of the vertical asymptote. If critical number(s) exist, then perform either the First-Derivative Test or the Second-Derivative Test for each critical number. At a critical number c, if $f(c)$ is a relative maximum, then f is increasing to the left of c and decreasing to the right; if $f(c)$ is a relative minimum, then f is decreasing to the left of c and increasing to the right. Sketch a short segment around each critical point according to the test conclusion.
>
> **(f)** *Inflection points.* Solve $f''(x) = 0$ and find where $f''(x)$ does not exist. Calculate the function values at these numbers and mark these points on the coordinate plane. These points are the candidates for inflection points.
>
> **(g)** *Concavity.* Use the first coordinate of the candidates for inflection points from step (f), and the vertical asymptotes (if they exist), to define intervals. In each interval, substitute a sample value into $f''(x)$ to determine whether f is concave up (when $f''(x) > 0$) or concave down (when $f''(x) < 0$).
>
> **(h)** *Sketch the graph.* Use the information obtained in steps (a) – (g) to sketch the graph. Calculate and plot some additional points if needed.

Example 9: Sketch the graph of $f(x) = \dfrac{2x^2 + 9}{x^2 - 9}$ according to the graphing strategy.

Solution (a)-(d) The function is given in Example 6. By Example 6, the domain of the function is $(-\infty, -3) \cup (-3, 3) \cup (3, \infty)$. The graph of the function has the lines $x = -3$ and $x = 3$ as vertical asymptotes, and has a horizontal asymptote $y = 2$. The function has no slant asymptote. The y-intercept is easy to obtain:

$$f(0) = \frac{2(0)^2 + 9}{0^2 - 9} = \frac{9}{-9} = -1.$$

Hence the point $(0, -1)$ is on the graph.

By Example 6, the function has a critical point $(0, -1)$. The derivatives are:

$$f'(x) = \frac{-54x}{(x^2 - 9)^2}; \quad f''(x) = \frac{162(x^2 + 3)}{(x^2 - 9)^3}.$$

3.6. GRAPH SKETCHING: ASYMPTOTES AND RATIONAL FUNCTIONS

(e) *Increasing, decreasing, and the relative extrema.* We perform the First-Derivative Test:

The critical number $x = 0$, as well as the two vertical asymptotes $x = -3$ and $x = 3$, divide the real line into four intervals: $(-\infty, -3)$, $(-3, 0)$, $(0, 3)$, and $(3, \infty)$. We determine the signs of f' in each interval: Since the denominator of f' is positive for all values in each interval, the signs of f' are determined by the signs of the numerator $-54x$ only. Therefore, f' is positive on the intervals $(-\infty, -3)$ and $(-3, 0)$, and negative on the intervals $(0, 3)$ and $(3, \infty)$. By Theorem 3 in Section 3.3, f is increasing on the intervals $(-\infty, -3)$ and $(-3, 0)$ and decreasing on the intervals $(0, 3)$ and $(3, \infty)$. We form the First-Derivative Test diagram (Figure 3.6.8):

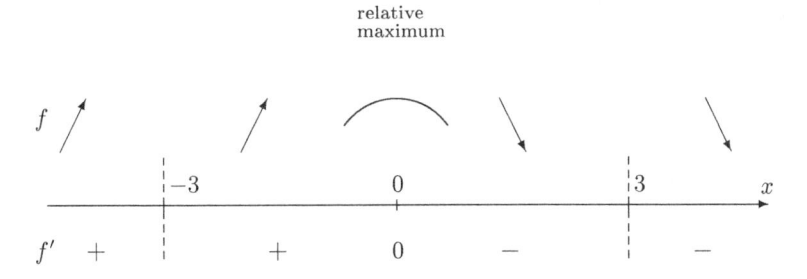

Figure 3.6.8: f' Test with marks of vertical asymptotes

According to the diagram, we conclude that $f(0) = -1$ is a relative maximum.

To help in sketching the graph, we evaluate the function at some values of x in other intervals. We have

$$f(-4) = f(4) = \frac{2(16) + 9}{(16) - 9} \approx 6,$$

hence the two points $(-4, 6)$ and $(4, 6)$ are on the graph, and f is increasing at $(-4, 6)$ and decreasing at $(4, 6)$.

We sketch the corresponding curve segments around the three points on the coordinate plane (Figure 3.6.9).

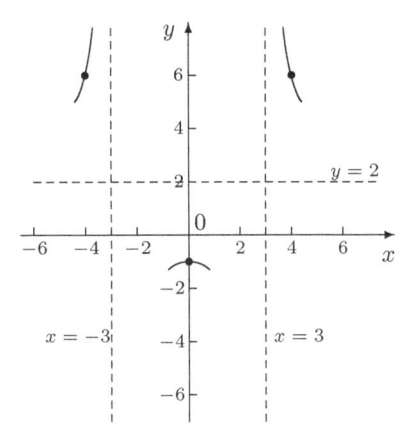

Figure 3.6.9: Partial graph of $f(x) = \left(2x^2 + 9\right) / \left(x^2 - 9\right)$

(f) and (g) *Inflection points and concavity.* By Example 6, the function has no inflection point.

Observe that the denominator of f'' is positive when $x < -3$ and $x > 3$, and negative when $-3 < x < 3$. Hence f'' is positive when $x < -3$ and $x > 3$, and negative when $-3 < x < 3$. Therefore f is concave up on the intervals $(-\infty, -3)$ and $(3, \infty)$, and f is concave down on the interval $(-3, 3)$.

(h) *Sketch the graph.* The information obtained in steps (a) to (g), as plotted in Figure 3.6.15, has revealed the major features of the graph. We can calculate a few additional function values to help us to complete the sketch. The graph is shown in Figure 3.6.10.

x	$f(x)$
-5	3.75
-4	6
...	...
-2	-3.4
0	-1
2	-3.4
...	...
4	6
5	3.75

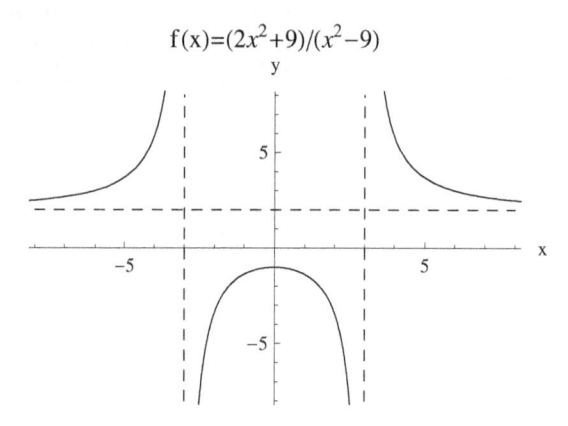

Figure 3.6.10: Graph of $f(x) = (2x^2 + 9) / (x^2 - 9)$

Exercise Set 3.6

For each function in Exercises #1 to #12, find the vertical asymptotes, if they exist. State the fact if no vertical asymptote exists.

1. $f(x) = \dfrac{2x + 5}{x - 2}$

2. $f(x) = \dfrac{x - 3}{x + 4}$

3. $g(x) = \dfrac{x + 1}{x^2 - 16}$

4. $g(x) = \dfrac{2x - 1}{x^2 - 9}$

5. $h(x) = \dfrac{2x}{x^2 - 3x + 2}$

6. $h(x) = \dfrac{3x}{x^2 + 2x - 3}$

7. $K(x) = \dfrac{x + 3}{x^2 + 2x - 3}$

8. $K(x) = \dfrac{x - 1}{x^2 - 3x + 2}$

9. $s(x) = \dfrac{x^2 - 4}{x + 2}$

10. $s(x) = \dfrac{x^2 - 4x + 3}{x - 3}$

11. $f(x) = \dfrac{-3}{x^2 + 4}$

12. $f(x) = \dfrac{2}{x^2 + 5}$

For each function in Exercises #13 to #24, find the horizontal asymptotes, if they exist. State the fact if no horizontal asymptote exists.

13. $f(x) = \dfrac{2x + 5}{x - 2}$

14. $f(x) = \dfrac{x - 3}{x + 4}$

15. $g(x) = \dfrac{x + 1}{x^2 - 16}$

16. $g(x) = \dfrac{2x - 1}{x^2 - 9}$

17. $h(x) = \dfrac{2x + 5}{x}$

18. $h(x) = \dfrac{3x - 2}{2x}$

19. $K(x) = 3 + \dfrac{2}{x^2}$

20. $K(x) = 4 - \dfrac{5}{x}$

21. $s(x) = \dfrac{x^2 - 4}{x + 2}$

22. $s(x) = \dfrac{x^2 - 4x + 3}{x - 3}$

23. $f(x) = \dfrac{1 - 3x - x^4}{x^4 + 7}$

24. $f(x) = \dfrac{2x^2 + x^3}{x^3 + 3}$

For each function in Exercises #25 to #54, use a graphing device to sketch the graph following the process illustrated in the examples; round up to three decimal places if needed:

(a) State the domain of the function and list the asymptotes;

(b) List the critical points and possible inflection points;

(c) Sketch the graph with appropriate coordinate ranges.

25. $f(x) = \dfrac{2}{x}$

26. $f(x) = \dfrac{-3}{x}$

27. $g(x) = \dfrac{-1}{x - 2}$

28. $g(x) = \dfrac{2}{x - 2}$

29. $h(x) = \dfrac{2}{x + 1}$

30. $h(x) = \dfrac{-3}{x + 2}$

31. $K(x) = \dfrac{1}{x^2}$

32. $K(x) = \dfrac{-2}{x^2}$

33. $s(x) = \dfrac{x^2 + 1}{x^2}$

34. $s(x) = \dfrac{x^2 - 1}{x^2}$

35. $f(x) = x + \dfrac{2}{x}$

36. $f(x) = x + \dfrac{3}{x}$

37. $g(x) = \dfrac{2x + 5}{x - 2}$

38. $g(x) = \dfrac{x - 3}{x + 4}$

39. $u(x) = \dfrac{2}{x^2 - 4}$

40. $u(x) = \dfrac{1}{x^2 - 9}$

41. $h(x) = \dfrac{2x + 5}{x}$

42. $h(x) = \dfrac{1 - 4x}{2x}$

43. $K(x) = \dfrac{x^2 + 2}{x}$

44. $K(x) = \dfrac{x^2 - 1}{x}$

45. $u(x) = \dfrac{x + 2}{x^2 - 4}$

46. $u(x) = \dfrac{x + 3}{x^2 - 9}$

47. $f(x) = \dfrac{2}{x^2 + 1}$

48. $f(x) = \dfrac{-1}{x^2 + 2}$

49. $g(x) = \dfrac{-x^2}{x^2 - 1}$

50. $g(x) = \dfrac{x^2}{x^2 - 4}$

51. $h(x) = \dfrac{x^2 + 1}{2x - 4}$

52. $h(x) = \dfrac{x^2 + 2}{x + 1}$

53. $s(x) = \dfrac{x^2 - 4}{x + 2}$ (Hint: Simplify)

54. $s(x) = \dfrac{x^2 - 4x + 3}{x - 3}$

For each function in Exercises #55 to #84, sketch the graph following the process illustrated in the examples; round up to three decimal places if needed:

(a) State the domain of the function and where the intercepts occur;

(b) List the asymptotes;

(c) List the coordinates of where the relative extrema occur;

(d) Determine the interval(s) where the function is increasing or decreasing;

(e) List the coordinates of where the inflection points occur;

(f) Determine the interval(s) where the function is concave up or concave down;

(g) Sketch the graph with appropriate coordinate ranges.

55. $f(x) = \dfrac{2}{x}$

56. $f(x) = \dfrac{-3}{x}$

57. $g(x) = \dfrac{-1}{x - 2}$

58. $g(x) = \dfrac{2}{x - 2}$

59. $h(x) = \dfrac{2}{x + 1}$

60. $h(x) = \dfrac{-3}{x + 2}$

61. $K(x) = \dfrac{1}{x^2}$

62. $K(x) = \dfrac{-2}{x^2}$

63. $s(x) = \dfrac{x^2 + 1}{x^2}$

64. $s(x) = \dfrac{x^2 - 1}{x^2}$

65. $f(x) = x + \dfrac{2}{x}$ 66. $f(x) = x + \dfrac{3}{x}$

67. $g(x) = \dfrac{2x+5}{x-2}$ 68. $g(x) = \dfrac{x-3}{x+4}$

69. $u(x) = \dfrac{2}{x^2-4}$ 70. $u(x) = \dfrac{1}{x^2-9}$

71. $h(x) = \dfrac{2x+5}{x}$ 72. $h(x) = \dfrac{1-4x}{2x}$

73. $K(x) = \dfrac{x^2+2}{x}$ 74. $K(x) = \dfrac{x^2-1}{x}$

75. $s(x) = \dfrac{x^2-4}{x+2}$ (Hint: Simplify)

76. $s(x) = \dfrac{x^2-4x+3}{x-3}$

77. $u(x) = \dfrac{x+2}{x^2-4}$ 78. $u(x) = \dfrac{x+3}{x^2-9}$

79. $f(x) = \dfrac{2}{x^2+1}$ 80. $f(x) = \dfrac{-1}{x^2+2}$

81. $g(x) = \dfrac{-x^2}{x^2-1}$ 82. $g(x) = \dfrac{x^2}{x^2-4}$

83. $h(x) = \dfrac{x^2+1}{2x-4}$ 84. $h(x) = \dfrac{x^2+2}{x+1}$

APPLICATIONS

85. A shoe manufacturer has the cost function

$$C(x) = 1{,}500 + 20x + 0.2x^2$$

(in dollars), $0 \le x \le 400$, where x is the number of pairs of shoes produced each week. The revenue function for selling x pairs of shoes is given by

$$R(x) = 90x \text{ (in dollars).}$$

(a) Find the total profit when x pairs of shoes are produced and sold each week.

(b) The *average profit* is given by
$$A_P(x) = P(x)/x. \text{ Find } A_P(x).$$

(c) Find the slant asymptote for the average profit function.

(d) Sketch the graph of the average profit function.

86. A clothing manufacturer has a cost function

$$C(x) = 1{,}200 + 30x + 0.5x^2$$

(in dollars), $0 \le x \le 250$, where x is the number of suits produced each week. The revenue function for selling x suits is given by

$$R(x) = 120x \text{ (in dollars).}$$

Follow the steps listed in Exercise #85.

Economics *For the demand and supply functions in Exercises #57 to #60, sketch the graphs of the two functions on the same coordinate plane, and find (approximately) the equilibrium point from the graphs.*

87. Demand: $x = D(p) = \dfrac{21}{p+3}$;

   Supply: $x = S(p) = 3p^2 - 2$.

88. Demand: $x = D(p) = \dfrac{30}{p+2}$;

   Supply: $x = S(p) = 2p^2 - 5$.

89. Demand: $x = D(p) = \dfrac{2p+50}{p+10}$;

   Supply: $x = S(p) = 0.01p^2 - 1$.

90. Demand: $x = D(p) = \dfrac{p+100}{p+2}$;

   Supply: $x = S(p) = 0.2p - 3$.

3.7 Using Derivatives to Find Absolute Maximum and Minimum Values

In this section, we continue the study of the **absolute maximum and minimum values** of a function. The concepts were introduced in Section 3.3. The study in this section will lead to the strategy of finding these values using derivatives.

3.7. ABSOLUTE MAXIMUM AND MINIMUM

Absolute Maximum and Minimum Values

The following graph was introduced in Section 3.3. Let us observe the graph again; this time we focus on the highest point and the lowest point on the graph.

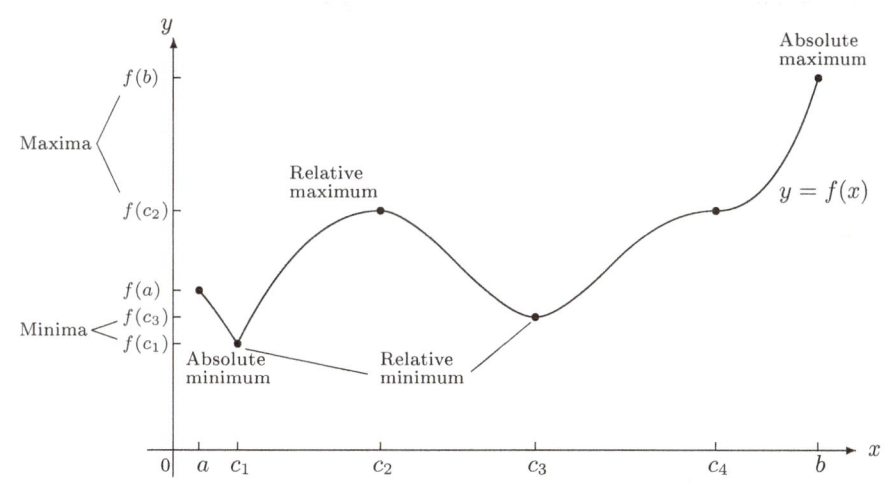

Figure 3.7.1: Maxima and minima of f on interval $[a, b]$

The highest point on this graph occurs at $(b, f(b))$, which is a "boundary point," and the lowest point occurs at $(c_1, f(c_1))$, at which f happens to have a relative minimum. The greatest function value over the whole interval is called the **absolute** (or **global**) **maximum**, and the least function value over the whole interval is called the **absolute** (or **global**) **minimum**. We now describe the absolute maximum and absolute minimum of a function using mathematical formulas.

Definition   *absolute maximum and absolute minimum*

Given a function f:

$f(c)$ is an **absolute maximum** if $f(c) \geq f(x)$ for all x in the domain.

$f(c)$ is an **absolute minimum** if $f(c) \leq f(x)$ for all x in the domain.

Reality Check   *Absolute extrema and constant functions*

If a function has a constant value on the whole domain, then by the definitions, $f(c)$ is both an absolute maximum and an absolute minimum for every number c in the domain.

Reality Check   *Absolute and relative extrema*

An absolute maximum (or minimum) may occur at a "boundary point," as in Figure 3.7.1, where $f(b)$ is the absolute maximum.

If an absolute maximum (or minimum) occurs in the interior of an interval, then it must also be a relative maximum (or minimum).

A function may have several different relative maxima or minima. However, if the absolute maximum (or minimum) exists, it is unique, although it may occur at different locations.

> **Reality Check** *Existence of absolute extrema*
> A function may have both an absolute maximum and an absolute minimum, as the function graphed in Figure 3.7.1, or it may have only an absolute maximum (or minimum), or it may have neither, as indicated in the following graphs (Figures 3.7.2-3.7.4).

$f(c)$ is an absolute maximum
f has no absolute minimum

$f(c)$ is an absolute minimum
f has no absolute maximum

f has no absolute minimum
f has no absolute maximum

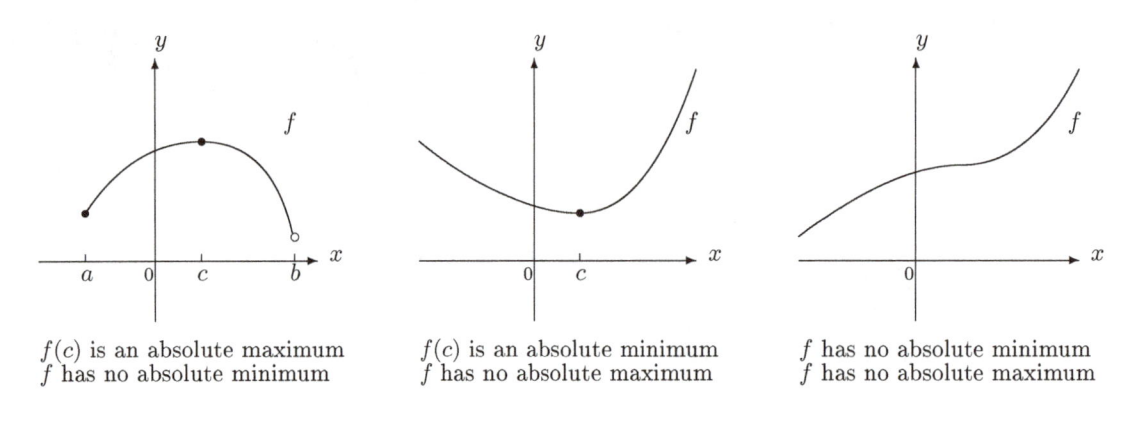

Figure 3.7.2: Case I

Figure 3.7.3: Case II

Figure 3.7.4: Case III

Notice that in Figure 3.7.2, f is not defined at $x = b$ and there is no point on the curve to be the lowest point. In general, determining whether a given function has absolute extrema is not a simple task. However, if we consider a *continuous* function on a *closed* interval $[a, b]$, as the function whose graph is shown in Figure 3.7.1, then both an absolute maximum and an absolute minimum do exist, and they are relatively easy to find.

▬ Absolute Extrema On a Closed Interval

If f is a continuous function defined on a closed (finite) interval $[a, b]$, then f must have both an absolute maximum and an absolute minimum. We state the conclusion in the following theorem.

> **Theorem 1**   *The Extreme Value Theorem*
> Suppose that function f is continuous on an closed interval $[a, b]$. Then f must have an absolute maximum and an absolute minimum.

$f(c)$ is an absolute maximum
$f(b)$ is an absolute minimum

$f(c_1)$ is an absolute maximum
$f(c_2)$ is an absolute minimum

$f(a)$ is an absolute minimum
$f(b)$ is an absolute maximum

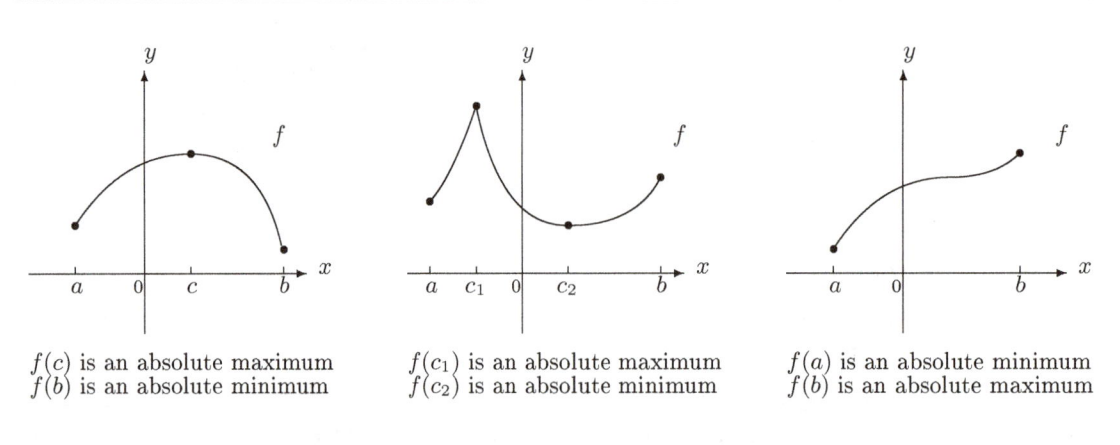

Figure 3.7.5: Case I

Figure 3.7.6: Case II

Figure 3.7.7: Case III

3.7. ABSOLUTE MAXIMUM AND MINIMUM

We now explore the strategy for determining the absolute extrema of a continuous function on a closed interval $[a, b]$. We have noticed that an absolute extremum is either also a relative extremum (if it occurs in the interior of the interval), or a boundary value. Furthermore, recall that Theorem 4.1 in Section 3.4 states that a relative extremum must occur at a critical number. This leads to the conclusion that, to find the absolute extrema, we only need to look among the function values at both boundaries (endpoints) and at all the critical numbers within the interval. We summarize the procedure here:

> **Strategy I: Determining Absolute Extrema on a Closed Interval**
> 1. Verify that function f is continuous on an closed interval $[a, b]$.
> 2. Determine all critical numbers of f on the open interval (a, b).
> 3. Evaluate the function at a, b, and all critical numbers obtained in Step 2.
> 4. The greatest (largest) value obtained in Step 3 is the absolute maximum of f on $[a, b]$, and the least (smallest) value obtained in Step 3 is the absolute minimum of f on $[a, b]$.

> **Warning**
> The function values in Step 3 should include both boundary values, and those at the critical numbers that locate **inside** the open interval (a, b).

Example 1: Determine the absolute maximum and absolute minimum of
$$f(x) = 2x^3 - 3x^2 - 12x + 5$$
on the interval $[-2, 3]$.

Solution **Step 1:** Since f is a polynomial, it is continuous on the closed interval $[-2, 3]$. Hence, by the Extreme Value Theorem, it must have an absolute maximum and an absolute minimum.

Step 2: To determine the critical numbers, we first take the derivative. We have $f'(x) = 6x^2 - 6x - 12$. Since f' exists everywhere, all its critical numbers satisfy $f'(x) = 0$. We solve the equation $f'(x) = 6x^2 - 6x - 12 = 0$:

$$6x^2 - 6x - 12 = 0$$
$$x^2 - x - 2 = 0$$
$$(x - 2)(x + 1) = 0$$
$$x = 2 \quad \text{or} \quad x = -1.$$

Hence, the critical numbers are -1 and 2. Notice that both numbers are in the open interval $(-2, 3)$, hence, we keep both.

Step 3: We evaluate the function at $x = -2, -1, 2,$ and 3:

$$f(-2) = 2(-2)^3 - 3(-2)^2 - 12(-2) + 5 = 1,$$
$$f(-1) = 2(-1)^3 - 3(-1)^2 - 12(-1) + 5 = 12,$$
$$f(2) = 2(2)^3 - 3(2)^2 - 12(2) + 5 = -15,$$
$$f(3) = 2(3)^3 - 3(3)^2 - 12(3) + 5 = -4.$$

Step 4: Since 12 is the largest number and -15 is the smallest number from Step 3, we conclude that on the interval $[-2, 3]$, the absolute maximum is 12 at $x = -1$ and the absolute minimum is -15 at $x = 2$.

Example 2: Determine the absolute maximum and absolute minimum of
$$f(x) = 2x^3 - 3x^2 - 12x + 5$$
on the following intervals: (a) $[0, 4]$ (b) $[-1, 1]$.

Solution (a) **Steps 1 and 2:** Since the function is the same as that in Example 1, we have already determined that the critical numbers of f, on the whole real line, are $x = -1$ and $x = 2$. However, since $x = -1$ is not in the open interval $(0, 4)$, there is only one critical number, $x = 2$, in $(0, 4)$.

Step 3: We evaluate the function at $x = 0, 2$, and 4:

$$f(0) = 5, \quad f(2) = -15,$$
$$f(4) = 2(4)^3 - 3(4)^2 - 12(4) + 5 = 37.$$

Step 4: Since 37 is the largest number and -15 is the smallest number from Step 3, we conclude that on the interval $[0, 4]$, the absolute maximum is 37 at $x = 4$ and the absolute minimum is -15 at $x = 2$.

(b) **Steps 1 and 2:** Similar to part (a), the critical numbers of f, on the whole real line, are $x = -1$ and $x = 2$. However, since neither number is in the open interval $(-1, 1)$, there is no critical number in $(-1, 1)$.

Step 3: We evaluate the function at the two endpoints $x = -1$, and 1:

$$f(-1) = 12, \quad f(1) = -8$$

Step 4: Since 12 is the largest number and -8 is the smallest number from Step 3, we conclude that on the interval $[-1, 1]$, the absolute maximum is 12 at $x = -1$ and the absolute minimum is -8 at $x = 1$.

The following three figures (Figures 3.7.8-3.7.10) illustrate the absolute extrema of the function in the two examples on different closed intervals:

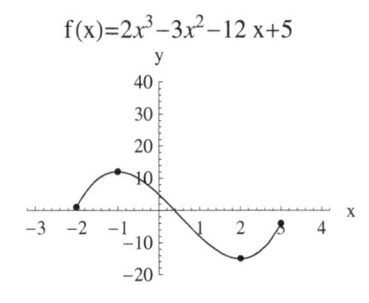

Figure 3.7.8: On $[-2, 3]$

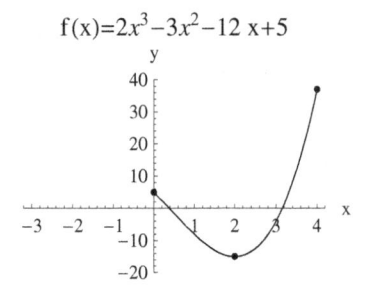

Figure 3.7.9: On $[0, 4]$

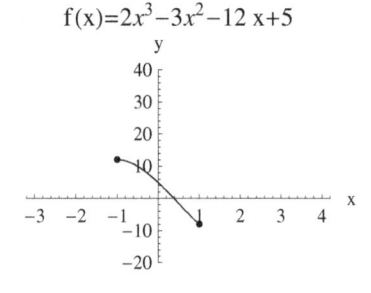

Figure 3.7.10: On $[-1, 1]$

▬ Absolute Extrema On Other Intervals

We now explore the procedure for finding absolute extrema of a continuous function on a general interval I. If the interval I takes the form of $[a, b]$, then we can apply Strategy I. If the interval takes other forms, such as (a, b), $(a, b]$, $(-\infty, b]$, (a, ∞), etc., then an absolute maximum or absolute minimum may not exist, as illustrated in Figures 3.7.2 – 3.7.4. However, if a function f has **exactly one** critical number, c, in the given interval, then we can further explore the behavior of the function. Since c is the only critical point in I, it divides I into two subintervals, and the first derivative of the function must keep the same sign in each subinterval. That means the function keeps increasing or decreasing in each subinterval. Therefore, if f has a relative maximum (or minimum) at c, $f(c)$ must be

3.7. ABSOLUTE MAXIMUM AND MINIMUM

an absolute maximum (or minimum). The following figures (Figures 3.7.11-3.7.13) illustrate the three cases:

$f(c)$ is a relative maximum
$f(c)$ is an absolute maximum

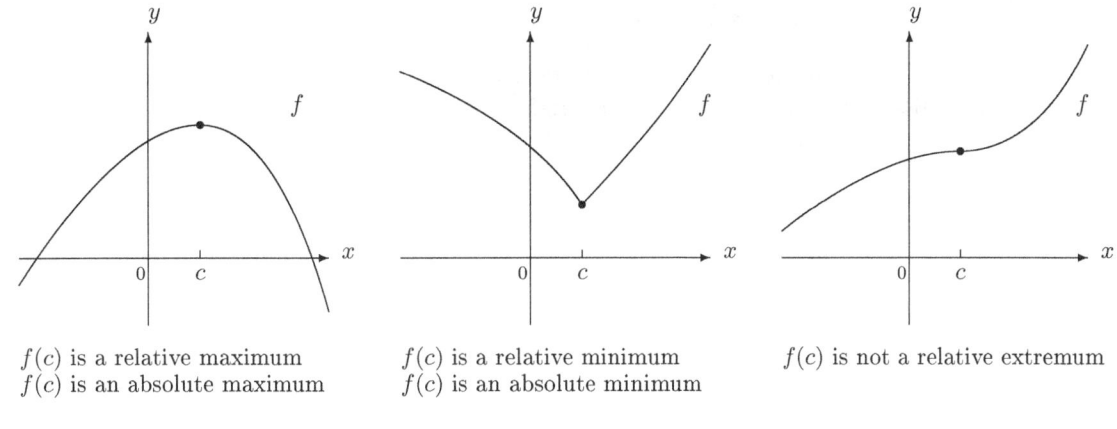

Figure 3.7.11: Case I

$f(c)$ is a relative minimum
$f(c)$ is an absolute minimum

Figure 3.7.12: Case II

$f(c)$ is not a relative extremum

Figure 3.7.13: Case III

We state the conclusions in the following theorem.

Theorem 2   *Unique Critical Number and Absolute Extrema*

Suppose that function f is continuous on an interval I, and that there is *exactly one* critical number c in I. Then:

If $f(c)$ is a relative maximum, then it must be an absolute maximum.

If $f(c)$ is a relative minimum, then it must be an absolute minimum.

To determine whether $f(c)$ is a relative extremum, we can apply either the First-Derivative Test or the Second-Derivative Test. We summarize the procedure here:

Strategy II: Determining Absolute Extremum in Case of Unique Critical Number

1. Verify that function f is continuous on interval I and has exactly one critical number, c, on I.

2. Apply either the First-Derivative Test or the Second-Derivative Test. If f has a relative maximum at c, then $f(c)$ is an absolute maximum. If f has a relative minimum at c, then $f(c)$ is an absolute minimum.

Applying Strategy II

Strategy II can be applied to any interval I, including the closed interval $[a, b]$. Unlike Procedure I, this procedure only determines either an absolute maximum or an absolute minimum, and does not reveal whether the other absolute extremum exists. However, Procedure II is very convenient to apply if we are only looking for either an absolute maximum or an absolute minimum, as in solving the optimization problems in Section 3.8.

Example 3: Determine the absolute extrema, if exist, of $g(x) = xe^{2x}$.

Solution When no interval is specified, we need to consider the whole domain of the function. The domain of this function is $(-\infty, \infty)$. Hence, only Strategy II may be applied.

Step 1: The function is continuous. To determine the critical numbers, we first take the derivative. Applying the Product Rule and the Chain Rule, we obtain

$$g'(x) = e^{2x} + 2xe^{2x} = (1 + 2x)e^{2x}.$$

Since g' exists everywhere, all its critical numbers satisfy $g'(x) = 0$. Since an exponential function has all positive values, the only solution is from $1 + 2x = 0$, or $x = -1/2$. Hence, g has a single critical number $c = -1/2$.

Step 2: We perform the Second-Derivative Test. Applying the Product Rule and the Chain Rule to g', we obtain

$$g''(x) = 2e^{2x} + 2(1 + 2x)e^{2x} = 4(1 + x)e^{2x}.$$

Substitute the the critical number $x = -1/2$ into $g''(x)$:

$$g''\left(-\frac{1}{2}\right) = 4\left(1 + \left(-\frac{1}{2}\right)\right)e^{2(-\frac{1}{2})} = 2e^{-1} > 0,$$

therefore, g has a relative minimum at $-1/2$. Hence $g(-1/2) = -1/(2e)$ is an absolute minimum. g has no absolute maximum, as the function approaches infinity when x approaches infinity, as is also illustrated by the graph (Figure 3.7.14).

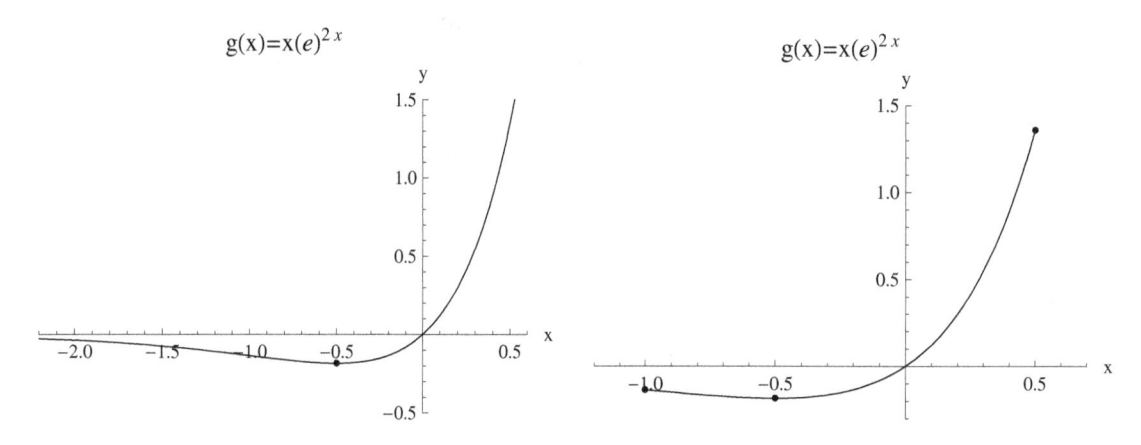

Figure 3.7.14: On $(-\infty, \infty)$

Figure 3.7.15: On $[-1, 0.5]$

Example 4: Determine the absolute extrema of $g(x) = xe^{2x}$ on the interval $[-1, 1/2]$.

Solution By the discussions in Example 3, we know that the absolute minimum of g is $-1/(2e)$ and occurs at $x = -1/2$. Since $-1/2$ is in the interval $[-2, 1]$, the absolute minimum is $g(-1/2) = -1/(2e)$. Since $[-2, 1]$ is a closed interval, the absolute maximum also exists. We evaluate the function at the two endpoints $x = -1$, and $1/2$:

$$f(-1) = -e^{-2} \approx -0.135, \quad f(1/2) = e/2 \approx 1.36.$$

Since $f(1/2)$ is the largest, we conclude that on the interval $[-1, 1/2]$, the absolute maximum is $e/2$ at $x = 1/2$ and the absolute minimum is $-1/(2e)$ at $x = -1/2$ (Figure 3.7.15).

3.7. ABSOLUTE MAXIMUM AND MINIMUM

Example 5: Determine the absolute extrema, if exist, of $f(x) = 4 - x^5$.

Solution The domain of this function is $(-\infty, \infty)$. The function is continuous. To determine the critical numbers, we first take the derivative. We obtain $f'(x) = -5x^4$. The only solution for $f'(x) = 0$ is $x = 0$. Hence, f has a single critical number $c = 0$.

We apply Strategy II. We need to decide whether f has a relative extremum at $c = 0$. The Second-Derivative Test does not work in this case since $f''(0) = 0$. We apply the First-Derivative Test. We have $f'(-1) = f'(1) = -5 < 0$, hence f' does not change sign around $c = 0$. Therefore, the function does not have a relative extremum at 0. Notice that f' is negative for all $x \neq 0$. That means the function is decreasing everywhere. Therefore, f has no absolute maximum or minimum, as illustrated by the graph (Figure 3.7.16).

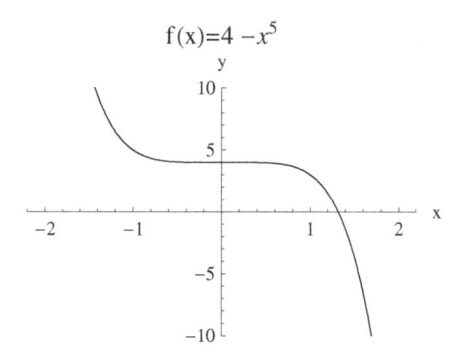

Figure 3.7.16: Function has no absolute extremum

Example 6: Determine absolute extrema, if exist, of of $f(x) = 9x + \dfrac{1}{x} - 4$ on the interval $(0, \infty)$.

Solution The function is continuous on the interval $(0, \infty)$. To determine the critical numbers, we first take the derivative. We obtain

$$f'(x) = 9 - \frac{1}{x^2}.$$

Since $f'(x)$ exists for all values of x in $(0, \infty)$, we solve the equation $f'(x) = 0$:

$$9 - \frac{1}{x^2} = 0, \quad \text{hence } 9 = \frac{1}{x^2}$$

$$9x^2 = 1, \quad \text{hence } x^2 = \frac{1}{9}, \quad \text{or } x = \pm\frac{1}{3}.$$

Only $x = 1/3$ is in $(0, \infty)$. Hence, f has a single critical number $c = 1/3$.

We apply Strategy II. We need to decide whether f has a relative extremum at $c = 1/3$. We perform the Second-Derivative Test. We have

$$f''(x) = \frac{2}{x^3}.$$

Substitute the the critical number $x = 1/3$ into $f''(x)$:

$$f''\left(\frac{1}{3}\right) = \frac{2}{(1/3)^3} > 0,$$

therefore, f has a relative minimum at $1/3$. Hence

$$f\left(\frac{1}{3}\right) = 9\left(\frac{1}{3}\right) + \frac{1}{(1/3)} - 4 = 3 + 3 - 4 = 2$$

is an absolute minimum. f has no absolute maximum, as illustrated by the graph (Figure 3.7.17).

$$f(x) = 9x + 1/x - 4$$

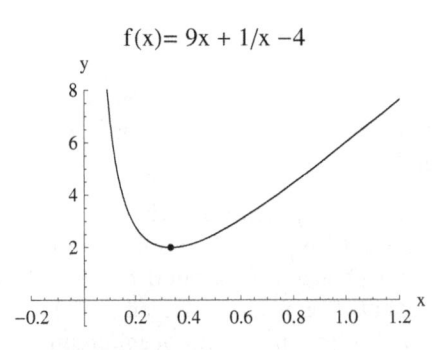

Figure 3.7.17: Function has absolute minimum only

Reality Check *Using a graphing device to determine absolute extrema*

The strategy introduced in sections 3.5 is very helpful for determining the extrema in a given interval. The horizontal plotting window should not exceed the given interval.

Exercise Set 3.7

1. For the function graphed here, determine its absolute maximum and absolute minimum on the following intervals. Indicate the x-values at which they occur.

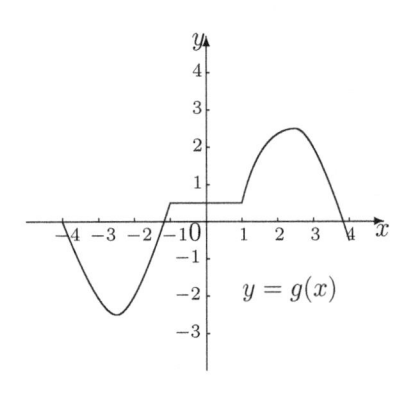

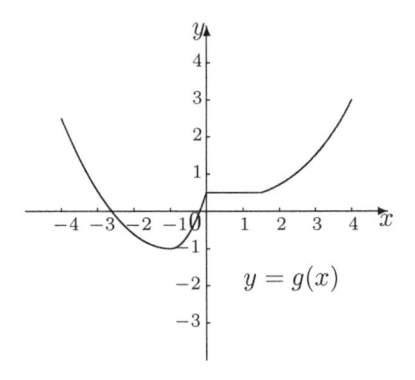

   (a) $[-4, 4]$ (b) $[-2, 2]$

   (c) $[-1, 1]$ (d) $[-2, -1]$

   For the function graphed here ($y = g(x)$):

   (a) $[-4, 4]$ (b) $[0, 1]$

   (c) $[-2, 0]$ (d) $[2, 4]$

2. For the function graphed here, determine its absolute maximum and absolute minimum on the following intervals. Indicate the x-values at which they occur.

3. For the function graphed here, determine its absolute maximum and absolute minimum on the following intervals. Indicate the x-values at which they occur. State the fact if either does not exist.

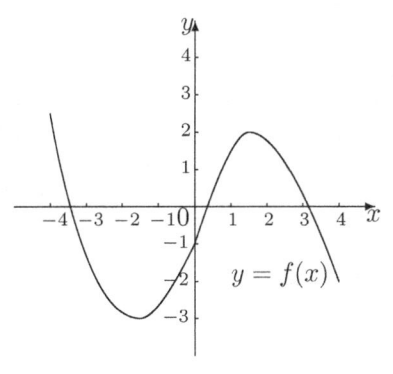

3.7. ABSOLUTE MAXIMUM AND MINIMUM

(a) $(-3,-1)$ (b) $[0,2]$

(c) $[-3,-1.5)$ (d) $(-2,0]$

4. For the function graphed here, determine its absolute maximum and absolute minimum on the following intervals. Indicate the x-values at which they occur. State the fact if either does not exist.

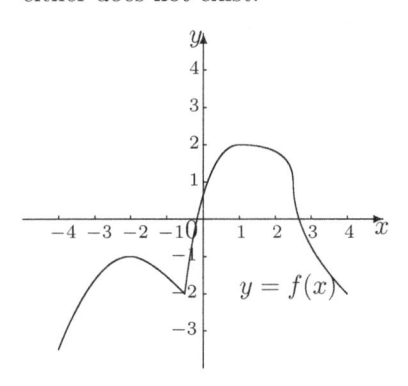

(a) $(-3,-0.5)$ (b) $[0,2]$

(c) $(0,3)$ (d) $(-2,0]$

Determine the absolute maximum and minimum values of each function over the indicated interval. Indicate the x-values at which they occur.

5. $h(x) = \frac{1}{3}x^3 - 3x^2 + 5x + 6$; $[0,6]$
6. $h(x) = -x^3 - 3x^2 + 15x + 3$; $[-6,3]$
7. $y = x^2 e^{-x}$; $[0,4]$
8. $y = xe^{3x}$; $[-1,1]$
9. $f(x) = \frac{2x}{x^2+4}$; $[-2,2]$
10. $f(x) = \frac{1}{x^2+1}$; $[-2,2]$
11. $f(x) = x^2 - 5x + 6$; $[0,3]$
12. $f(x) = 2x^2 + 3x - 5$; $[0,4]$
13. $g(x) = -2x + 1$; $[-1,2]$
14. $g(x) = 5x - 6$; $[-2,3]$
15. $h(x) = \frac{1}{3}x^3 - 3x^2 + 5x + 6$; $[2,6]$
16. $h(x) = -x^3 - 3x^2 + 15x + 3$; $[0,2]$
17. $y = x^2 e^{-x}$; $[-2,2]$
18. $y = xe^{3x}$; $[0,2]$
19. $f(x) = \frac{2x}{x^2+4}$; $[0,4]$
20. $f(x) = \frac{1}{x^2+1}$; $[1,4]$
21. $g(x) = 3(x+3)^{2/3}$; $[-4,4]$
22. $g(x) = -2(x-2)^{2/5}$; $[-4,4]$
23. $f(x) = x + \frac{4}{x}$; $[1,4]$
24. $f(x) = 9x + \frac{1}{x}$; $[0.1, 1]$
25. $f(x) = x^4 - 18x^2 + 2$; $[-4,4]$
26. $f(x) = 3x^4 - 8x^3 + 4$; $[-4,4]$
27. $y = x^2 e^{-x}$; $[1,3]$
28. $y = xe^{3x}$; $[-1,0]$
29. $f(x) = \frac{2x}{x^2+4}$; $[-2,0]$
30. $f(x) = \frac{1}{x^2+1}$; $[-2,0]$
31. $g(x) = 3(x+3)^{2/3}$; $[-2,3]$
32. $g(x) = -2(x-2)^{2/5}$; $[1,4]$
33. $f(x) = x + \frac{4}{x}$; $[-3,-1]$
34. $f(x) = 9x + \frac{1}{x}$; $[-2,-1]$

Determine the absolute maximum and minimum values of each function, if they exist, over the indicated interval. Indicate the x-values at which they occur. When no interval is specified, use the real line $(-\infty, \infty)$.

35. $h(x) = \frac{1}{3}x^3 - 3x^2 + 5x + 6$; $(0,4)$
36. $h(x) = -x^3 - 3x^2 + 15x + 3$; $(0,4)$
37. $y = x^2 e^{-x}$; $(0, \infty)$

38. $y = xe^{3x}$; $(-\infty, 0)$

39. $f(x) = \dfrac{2x}{x^2+4}$; $(0, 4)$

40. $f(x) = \dfrac{1}{x^2+1}$; $(-1, 1)$

41. $f(x) = x^2 - 5x + 6$

42. $f(x) = 2x^2 + 3x - 5$

43. $g(x) = -2x + 1$; $(-1, 4]$

44. $g(x) = 5x - 6$; $(0, 3]$

45. $h(x) = \frac{1}{3}x^3 - 3x^2 + 5x + 6$; $(-\infty, 0)$

46. $h(x) = -x^3 - 3x^2 + 15x + 3$; $(-\infty, 0)$

47. $y = x^2 e^{-x}$; $(-1, 1)$

48. $y = xe^{3x}$

49. $f(x) = \dfrac{2x}{x^2+4}$; $(-4, -1]$

50. $f(x) = \dfrac{1}{x^2+1}$; $[-1, 2)$

51. $g(x) = 3(x+3)^{2/3}$

52. $g(x) = -2(x-2)^{2/5}$

53. $f(x) = x + \dfrac{4}{x}$; $(-\infty, 0)$

54. $f(x) = 9x + \dfrac{1}{x}$; $(-\infty, 0)$

55. $f(x) = x + \dfrac{4}{x}$; $(0, \infty)$

56. $f(x) = 9x + \dfrac{1}{x}$; $(0, \infty)$

APPLICATIONS

57. An appliance store estimates that each month, the number of a certain brand of television the store sells depends on the amount it spends on advertising, with the following equation:

$N(x) = -0.05x^2 + 20x + 12, \; 0 \le x \le 250,$

where x is in hundreds of dollars and N is the television sets the store sells. Find the maximum number of the television sets that can be sold each month, and the amount that must be spent on advertising to achieve that amount.

58. An appliance store estimates that each month, the number of a certain brand of microwave oven it sells depends on the amount it spends on advertising, with the following equation:

$N(x) = 50xe^{-0.05x} + 200, \; 0 \le x \le 25,$

where x is in hundreds of dollars and N is the number of microwave ovens the store sells. Find the maximum number of the microwave ovens that can be sold each month, and the amount that must be spent on advertising to achieve that amount.

59. A shoe manufacturer has the cost function

$C(x) = 1{,}500 + 20x + 0.2x^2$

(in dollars), $0 \le x \le 400$, where x is the number of pairs of shoes produced each week. The revenue function for selling x pairs of shoes is given by

$R(x) = 90x$ (in dollars).

(a) Find the total profit when x pairs of shoes are produced and sold each week.

(b) Find the number of pairs of shoes that should be produced and sold each week for the company to make the maximum profit. Also find the maximum profit.

60. A clothing manufacturer has the cost function

$C(x) = 1{,}200 + 30x + 0.5x^2$

(in dollars), $0 \le x \le 250$, where x is the number of suits produced each week. The revenue function for selling x suits is given by

$R(x) = 120x$ (in dollars).

(a) Find the total profit when x suits are produced and sold each week.

(b) Find the number of suits that should be produced and sold each week for the company to make the maximum profit. Also find the maximum profit.

61. An appliance manufacturer has the cost function
$$C(x) = 4{,}000 + 36x + 0.5x^2$$
(in dollars), $0 \leq x \leq 100$, where x is the number of a certain brand of microwave ovens produced each week. The revenue function for selling x microwave ovens is given by
$$R(x) = 500x - 4x^2 \text{ (in dollars)}.$$

   (a) Find the number of microwave ovens that should be sold each week for the company to make the maximum revenue. Also find the maximum revenue.

   (b) Find the total profit when x microwave ovens are produced and sold each week.

   (c) Find the number of microwave ovens that should be produced and sold each week for the company to make the maximum profit. Also find the maximum profit.

62. A refrigerator manufacturer determines that its weekly cost, in dollars, of producing x refrigerators is given by
$$C(x) = 2x^2 + 15x + 1{,}500,\ 0 \leq x \leq 80.$$
The weekly revenue function for selling x refrigerators is given by
$$R(x) = 1{,}600x - 20x^2 \text{ (in dollars)}.$$

   (a) Find the number of refrigerators that should be sold each week for the company to make the maximum revenue. Also find the maximum revenue.

   (b) Find the total profit when x refrigerators are produced and sold each week.

   (c) Find the number of refrigerators that should be produced and sold each week for the company to make the maximum profit. Also find the maximum profit.

3.8 Optimization Problems

A very important application of calculus is to solve optimization problems, also called maximum-minimum problems. To solve an optimization problem is to find the absolute maximum or minimum value of some targeted varying quantity, and the "optimal solution" at which that maximum or minimum occurs. In Section 3.2, we used the marginal functions, which are the derivatives of cost, revenue, and profit functions, to determine the level of productivity at which the profit reaches its maximum. Hence, we have already studied a special type of optimization problem, where the targeted varying quantity is the profit, and the optimal solution is the number of units produced and sold to maximize the profit. In this section, we extend the study to more general settings.

Example 1: A farmer has 3,600 ft of fencing and wants to fence off a rectangular field along a straight river. The side along the river needs no fence. What is the largest area that can be enclosed? What are the dimensions of the field that has the largest area?

Solution We first test some specific dimensions to calculate the yield areas. Figures 3.8.1 to 3.8.3 (not to scale) illustrate three possible ways of laying out the 3,600 ft of fence. We can observe that either shallow, wide fields or deep, narrow fields yield relatively small areas. It seems plausible that there are some intermediate dimensions that yields the largest area.

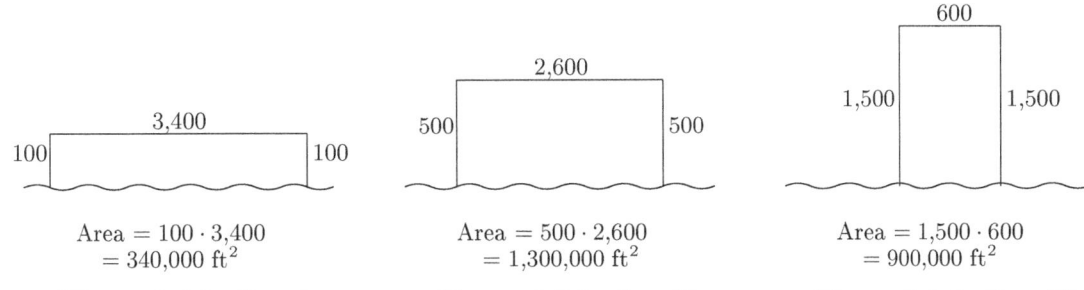

Area = 100 · 3,400
    = 340,000 ft²

Figure 3.8.1: Case I

Area = 500 · 2,600
    = 1,300,000 ft²

Figure 3.8.2: Case II

Area = 1,500 · 600
    = 900,000 ft²

Figure 3.8.3: Case III

To find the "optimal" dimensions, we need to study the general case. Figure 3.8.4 illustrates the general case. We are seeking the dimensions of a rectangle which yields the largest area. Since the area depends the depth and width, we denote them to be x and y. Then we express the area A in terms of x and y: $A = xy$.

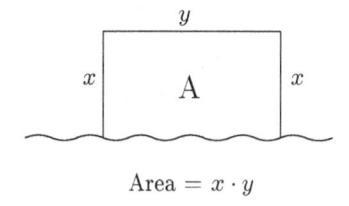

Area $= x \cdot y$

Figure 3.8.4: General case

Notice that the quantity A depends on two variables. The next step is to express A as a function of just one variable. We can replace y by an expression of x using the given information. Since the total length of the fencing is 3,600 ft, we have $2x + y = 3,600$. From this equation we have $y = 3,600 - 2x$. Substituting into the expression of A we have

$$A = x(3,600 - 2x) = 3,600x - 2x^2.$$

Next, we need to find the domain of this function. Note that both x and y must take nonnegative values, hence from the total length equation we conclude that the largest value x can take is 1,800. Therefore, the domain of the function $A(x)$ is $0 \leq x \leq 1,800$.

Now the optimization problem is turned into a maximizing problem which we studied in Section 3.7. We are looking for the absolute maximum of $A(x) = 3,600x - 2x^2$ over the closed interval $[0, 1,800]$. First, we find the derivative of $A(x)$:

$$A'(x) = 3,600 - 4x.$$

To find the critical numbers we solve the equation

$$3,600 - 4x = 0.$$

The only solution is $x = 900$, which is in the open interval $(0, 1,800)$. Hence there is exactly one critical number in $(0, 1,800)$. We apply the Strategy II: We check the second derivative of $A(x)$:

$$A''(x) = -4 < 0.$$

By the Second-Derivative Test, A has a relative maximum at $x = 900$, and we conclude that $A(900)$ is an absolute maximum.

From the equation $y = 3,600 - 2x$, substituting $x = 900$ we obtain $y = 3,600 - 2(900) = 1,800$. The pair $x = 900$ and $y = 1,800$ is called the optimal dimensions, since the area reaches the maximum at these dimensions. The maximum area is $900 \times 1,800 = 1,620,000$ ft^2.

Thus, the largest area that can be enclosed is 1,620,000 ft^2, and the rectangular field should be 900 ft deep and 1,800 ft wide, with the longer fence parallel to the river.

Example 1 illustrates the steps involved in solving optimization problems. We summarize these steps:

Steps for Solving Optimization Problems

1. **Understand the Problem.** Read the problem carefully until you understand it clearly. Ask the question: What is going to be maximized or minimized?

2. **Draw a Diagram and Introduce Variables.** In most problems it is useful to draw a diagram and denote the related quantities on the diagram. Assign symbols to the quantity that is to be maximized or minimized and all other unknowns.

3. **Write an Equation.** Write an equation for the quantity to be maximized or minimized.

4. **Eliminate Extra Variables.** If in the equation obtained in Step 3, the quantity to be optimized depends on more than one variable, then use the given conditions to eliminate all but one of the variables. Now the quantity to be optimized is a function of one variable. Determine the domain of the function.

5. **Determine the Absolute Extremum.** Apply Strategy I or II introduced in Section 3.7 to determine the absolute extremum.

6. **Interpret the Solution.** Write a sentence that answers the question posed in the problem.

Warning

An optimization problem often includes more than one equations. As in Example 1, there are two equations $A = xy$ and $2x + y = 3,600$. The first equation is for the quantity (namely the area A) needs to be maximized. The second equation is for replacing the second variable y by x so that the area equation depends x only. These two types of equations may be confusing.

Example 2: A firm receives an order for a rectangular storage container with an open top. The container has a volume of 20 m^3, and the length of the base is twice the width. Material for the base costs \$15 per square meter. Material for the sides costs \$12 per square meter. What is the lowest cost of materials for making such a container? What are dimensions of the container that require the lowest cost for materials?

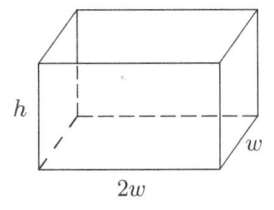

Figure 3.8.5: Container with variables

Solution This problem seeks the lowest cost for the material to make a rectangular container. We draw a diagram and introduce variables as illustrated in Figure 3.8.5. Since the length of the base is twice the width, we can use $2w$ for the length.

Since the cost of the material is the quantity that needs to be minimized, we express the cost (in dollars) in terms of h and w. Note that the materials needed include the material for the base and the material for four sides. The base has area $2w \times w$, hence the cost for the base is $15 \times 2w \times w$. Two sides have area $2w \times h$ and other two have area $w \times h$. Hence, the cost for the sides is $12 \times 2 \times (2wh + wh)$. We now have the equation for the quantity to be minimized:

Cost $C = 15 \times 2w \times w + 12 \times 2 \times (2wh + wh) = 30w^2 + 72wh$.

The next step is to express the cost C as a function of just one variable. We can replace h by an expression of w using the given information. Since the volume of the container is 20 m^3, we have

$$2w \cdot w \cdot h = 20.$$

From this equation, we have $h = 10/w^2$. Substituting this into the expression of C we have

$$C(w) = 30w^2 + 72w \left(\frac{10}{w^2}\right) = 30w^2 + \frac{720}{w}.$$

Next we find the domain of the function $C(w)$. Note that w must take nonnegative values, but there is no restriction on how large a value w can take (large values for w mean a very shallow container). However, since w occurs in the denominator, it cannot take 0. Therefore, the domain is $0 < w < \infty$.

We are looking for the absolute minimum of

$$C(w) = 30w^2 + \frac{720}{w}$$

over the interval $(0, \infty)$. First we find the derivative of $C(w)$:

$$C'(w) = 60w - \frac{720}{w^2}.$$

To find the critical numbers we solve the equation

$$60w - \frac{720}{w^2} = 0 \text{ or } 60w = \frac{720}{w^2} \text{ or } w^3 = 12 \text{ or } w = \sqrt[3]{12}.$$

Hence, the only solution is $w = \sqrt[3]{12} \approx 2.29$, which is in the interval $(0, \infty)$. Therefore, there is exactly one critical number in $(0, \infty)$. We apply the Strategy II. We check the second derivative of $C(w)$. Since for all $w > 0$ we have

$$C''(w) = 60 + \frac{1,440}{w^3} > 0,$$

by the Second-Derivative Test, C has a relative minimum at $\sqrt[3]{12}$, and we conclude that $C(\sqrt[3]{12})$ is an absolute minimum.

From the equation $h = 10/w^2$, substituting $w = \sqrt[3]{12}$ we obtain $h = 10/(\sqrt[3]{12})^2 \approx 1.91$. The minimum cost is $30(\sqrt[3]{12})^2 + 720/\sqrt[3]{12} \approx \$471.73$.

Thus, the minimum cost of materials for making such a container is \$471.73; and the rectangular container should be about 1.91 m deep with a base that is 4.58 m long and 2.29 m wide.

Example 3: An appliance store has been selling 200 microwave ovens at \$120 each every month. A market survey indicates that for each \$10 rebate offered to buyers, the number of microwave ovens sold will increase by 25 per month.

(a) Find the demand function, expressing p, the price charged for each microwave oven, as a function of x, the number of microwave ovens sold each month.

(b) Find the revenue function $R(x)$.

(c) Find the rebate amount the store should offer to buyers to maximize its revenue.

(d) It costs the store \$56 to purchase each microwave oven from the manufacturer. Find the rebate

3.8. OPTIMIZATION PROBLEMS

amount the store should offer to buyers to maximize its profit.

Solution (a) From the given conditions, the demand function is linear. We write

$$p = mx + b.$$

For each $10 rebate, the actual amount a buyer pays is reduced by $10. Hence, when p decreased by $10, x is increased by 25. Therefore, the slope is

$$m = \frac{-10}{25} = -0.4.$$

Next we plug $x = 200$ and $p = 120$ into the equation:

$$120 = -0.4(200) + b.$$

Hence $b = 120 + 0.4(200) = 200$. We obtain the demand function:

$$p(x) = -0.4x + 200.$$

(b) The revenue is the price times the quantity sold, or $R = p \cdot x$. From part (a) we have $p = -0.4x + 200$, hence the revenue function is

$$R(x) = p \cdot x = (-0.4x + 200) \cdot x = -0.4x^2 + 200x.$$

(c) The quantity to be maximized is the revenue. In part (b) we expressed the revenue as a function of x. We now find the domain of the revenue function. Since both x and p have to be nonnegative, from part (a) we obtain $x \leq 500$. Hence, the domain is $0 \leq x \leq 500$.

We are looking for the absolute maximum of

$$R(x) = -0.4x^2 + 200x$$

over the interval $[0, 500]$. First we find the derivative of $R(x)$:

$$R'(x) = -0.8x + 200.$$

To find the critical numbers we solve the equation

$$R'(x) = -0.8x + 200 = 0.$$

The only solution is $x = 250$, which is in the interval $[0, 500]$. Since there is exactly one critical number in $[0, 500]$, we apply the Strategy II. We check the second derivative of $R(x)$:

$$R''(x) = -0.8 < 0.$$

By the Second-Derivative Test, R has a relative maximum at 250, and we conclude that $R(250)$ is an absolute maximum. Therefore, 250 microwave ovens should be sold in order the maximize the revenue.

From the equation $p = -0.4x + 200$, substituting $x = 250$ we obtain $p = -0.4(250) + 200 = 100$. Therefore, the rebate is $120 - 100 = 20$. That is, to maximize its revenue, the store should offer a rebate of $20.

(d) The quantity to be maximized is the profit. Since it costs the store $56 for each microwave oven, the cost function is

$$C(x) = 56x.$$

Therefore, the profit function is

$$P(x) = R(x) - C(x) = -0.4x^2 + 200x - 56x = -0.4x^2 + 144x.$$

We now find the domain of the profit function. Note that the restrictions are applied only to x and $R(x)$, as in part (c). Hence $0 \le x \le 500$.

Therefore, we are looking for the absolute maximum of

$$P(x) = -0.4x^2 + 144x$$

over the interval $[0, 500]$.

By similar reasoning as in part (c), there is exactly one critical number $x = 180$ in $[0, 500]$. By the Second-Derivative Test, P has a relative maximum at 180, and we conclude that $P(180)$ is an absolute maximum.

From the equation $p = -0.4x + 200$, substituting $x = 180$ we obtain $p = -0.4(180) + 200 = 128$. Since it is higher than the selling price, no rebate should be offered. The store actually should consider raising the selling price to \$128 to maximize its profit.

Example 4: A motel owner observes that when a room is priced at \$50 per day, all 40 rooms of the motel are occupied. For every \$2 rise in the charge per room per day, one more room is vacant. Each occupied room costs an additional \$12 per day to maintain.

(a) Find the demand function that expresses p, the price charged for each room per day, as a function of x, the number of rooms occupied.

(b) Find the revenue function $R(x)$.

(c) Find the price per room per day the motel should charge to maximize its revenue.

(d) Find the profit function and the price per room per day the motel should charge to maximize its profit.

Solution (a) Similar to Example 3, part (a), we assume

$$p = mx + b.$$

The slope is $m = -2/1 = -2$. Using $x = 40$ and $p = 50$ we have

$$50 = -2(40) + b,$$

and we obtain $b = 50 + 2(40) = 130$. Hence the demand function is

$$p = -2x + 130.$$

(b) The revenue is $R = p \cdot x$. From part (a) we have $p = -2x + 130$, hence the revenue function is

$$R(x) = p \cdot x = (-2x + 130) \cdot x = -2x^2 + 130x.$$

(c) The quantity to be maximized is the revenue. In part (b), we expressed the revenue as a function of x. We now find the domain of the revenue function. Since both x and p have to be nonnegative, we have $0 \le x \le 65$. We also have $0 \le x \le 40$ since the motel has total of 40 rooms. Combining these conditions, the domain is $[0, 40]$.

We are looking for the absolute maximum of

$$R(x) = -2x^2 + 130x$$

over the interval $[0, 40]$. By reasoning similar to Example 3, part (b), there is exactly one critical number $c = 32.5$ in $[0, 40]$. By the Second-Derivative Test, R has a relative maximum at 32.5, and we conclude that $R(32.5)$ is an absolute maximum. However, x, the number of rooms occupied, must be a positive integer. We examine both $R(32)$ and $R(33)$. We have

$$R(32) = -2(32)^2 + 130(32) = 2{,}112 \quad \text{and} \quad R(33) = -2(33)^2 + 130(33) = 2{,}112.$$

Hence, either 32 or 33 rooms should be occupied to the maximize the motel's revenue.

From the equation $p(x) = -2x + 130$, substitute $x = 32$ and $x = 33$ we obtain $p(32) = 66$ and $p(33) = 64$. Therefore, to maximize its revenue, the motel should charge either $64 or $66 per room per day.

(d) The quantity to be maximized is the profit. Since it costs the motel additional $12 to maintain each occupied room, the cost function is

$$C(x) = 12x.$$

Therefore, the profit function is

$$P(x) = R(x) - C(x) = -2x^2 + 130x - 12x = -2x^2 + 118x.$$

We now find the domain of the profit function. Note that the restrictions are the same as in part (c). Hence, $0 \leq x \leq 40$.

Therefore, we are looking for the absolute maximum of

$$P(x) = -2x^2 + 118x$$

over the interval $[0, 40]$. By reasoning similar to Example 3, part (b), there is exactly one critical number $c = 29.5$ in $[0, 40]$. By the Second-Derivative Test, P has a relative maximum at 29.5, and we conclude that $P(29.5)$ is an absolute maximum. Since x must be a positive integer, we examine both $P(29)$ and $P(30)$. We have $P(29) = P(30) = 1{,}740$. Therefore, either 29 or 30 rooms should be occupied in order the maximize the motel's profit. However, the cost is lower for maintaining 29 rooms, hence, 29 rooms should be occupied.

From the equation $p(x) = -2x + 130$, substituting $x = 29$ we obtain $p(29) = 72$. Therefore, to maximize its profit, the motel should charge $72 per room per day.

Example 5: A mail-order firm begins a television advertising campaign to market a new product. The "target market" is approximately 10,000,000 people. A market survey from a previous campaign concludes that the percentage, as a decimal, of the market that buys a product is a function of the number of days of the campaign, and the function can be expressed as

$$f(t) = 1 - e^{-0.05t}.$$

The product sells for $25 each and costs $10 each to purchase from the manufacturer. The campaign costs $10,000 per day.

(a) Find the profit function $P(t)$, where t is the number of days of the campaign.

(b) Find the number of the days the campaign should last to maximize the firm's profit.

Solution (a) Since the profit function is the difference of the revenue function and the cost function, we find the revenue function first. The revenue is the price times the quantity sold, or $R = p \cdot x$. The price is $25 per unit. The quantity sold, x, depends on t, the number of days of the campaign. After t days of the advertising campaign, $f(t) \times 100\%$ of the target market of 10,000,000 buys the product. Hence

$$x = 10{,}000{,}000 \times \left(1 - e^{-0.05t}\right).$$

Therefore, the revenue, as a function of t, is

$$R(t) = 25 \times 10{,}000{,}000 \times \left(1 - e^{-0.05t}\right) = 250{,}000{,}000 \left(1 - e^{-0.05t}\right).$$

Next we find the cost function. The total cost consists of two parts: the cost of purchasing the product from the manufacturer and the cost of the campaign. Each unit of the product costs $10, hence the cost of purchasing x units is $10x$. The campaign costs $10,000 per day, hence the cost of t days of advertising is $10{,}000t$. Therefore, the total cost is

$$\text{Cost} = 10x + 10{,}000t.$$

Since $x = 10{,}000{,}000 \times \left(1 - e^{0.05t}\right)$, the total cost as a function of t is

$$C(t) = 10 \times 10{,}000{,}000 \times \left(1 - e^{-0.05t}\right) + 10{,}000t = 100{,}000{,}000 \left(1 - e^{-0.05t}\right) + 10{,}000t.$$

Therefore, the profit function is

$$P(t) = R(t) - C(t) = 250{,}000{,}000 \left(1 - e^{-0.05t}\right) - [100{,}000{,}000 \left(1 - e^{-0.05t}\right) + 10{,}000t]$$

$$= 250{,}000{,}000 \left(1 - e^{-0.05t}\right) - 100{,}000{,}000 \left(1 - e^{-0.05t}\right) - 10{,}000t$$

$$= 150{,}000{,}000 \left(1 - e^{-0.05t}\right) - 10{,}000t = 150{,}000{,}000 - 150{,}000{,}000 e^{-0.05t} - 10{,}000t.$$

(b) The quantity to be maximized is the profit. In part (a), we expressed the profit as a function of time t. We now find the domain of the revenue function. Since t must be nonnegative, we have $t \geq 0$.

Therefore, we are looking for the absolute maximum of

$$P(t) = 150{,}000{,}000 - 150{,}000{,}000 e^{-0.05t} - 10{,}000t$$

over the interval $[0, \infty)$. First we find the derivative of $P(t)$:

$$P'(t) = -150{,}000{,}000(-0.05)e^{-0.05t} - 10{,}000 = 7{,}500{,}000 e^{-0.05t} - 10{,}000.$$

To find the critical numbers we solve the equation

$$P'(t) = 7{,}500{,}000 e^{-0.05t} - 10{,}000 = 0.$$

This is an exponential equation. We follow the procedure introduced in Section 1.6:

$$7{,}500{,}000 e^{-0.05t} = 10{,}000$$

$$e^{-0.05t} = \frac{10{,}000}{7{,}500{,}000} = \frac{1}{750}$$

$$-0.05t = \ln \frac{1}{750} = -\ln 750 \quad \text{hence} \quad t = \frac{-\ln 750}{-0.05} \approx 132.$$

Therefore, there is exactly one critical number $t \approx 132$ in $(0, \infty)$. We apply the Strategy II. We check the second derivative of $P(t)$:

$$P''(t) = 7{,}500{,}000(-0.05)e^{-0.05t} = -375{,}000 e^{-0.05t} < 0.$$

By the Second-Derivative Test, P has a relative maximum at 132, and we conclude that $P(132)$ is an absolute maximum. Therefore, to maximize the firm's profit, the advertising campaign should run for 132 days.

Example 6: A furniture store sells 600 couches each year. It is impractical to buy all 600 couches

3.8. OPTIMIZATION PROBLEMS

at once as they would need a warehouse that stored all 600, and it costs $30 per year to rent the space to store one couch. On the other hand, every time the furniture store reorders, they have to pay a procesing fee of $80. No matter when they order the couches, each couch ordered requires an additional $20 shipping fee.

(a) Assume that the store will always reorder the same number of couches each time and that this number equals the total lot size that the store maintains.. Let x denote the number of couches the store reorders each time (and, therefore, x is also the lot size the store must maintain to store the loads). Find the cost function, expressed as a function of x.

(b) To minimize the inventory costs, how large is the lot size the store should maintain, and how many orders should be placed each year?

Solution (a) The total annual inventory costs consist of three parts: the processing cost of reordering, the cost of shipping, and the cost of storage. Let n be the number of reorders, then the processing cost of reordering is $80n$. The cost of shipping is $20 \times 600 = 12{,}000$, hence it does not depend on either n or x. The cost of storage is $30x$.

Therefore, the total inventory cost is

$$\text{Cost} = 80n + 12{,}000 + 30x.$$

Since the total number of couches the store sells is 600, we have $nx = 600$, hence $n = 600/x$. Therefore, the total inventory cost as a function of x is

$$C(x) = 80 \times \frac{600}{x} + 12{,}000 + 30x = \frac{48{,}000}{x} + 12{,}000 + 30x.$$

(b) The quantity to be minimized is the total inventory cost. In part (a), we have expressed the cost as a function of x. We now find the domain of the cost function. Note that x must be positive, but not exceed 600. Consider also that we are interested in integer solutions only, the domain is $1 \leq x \leq 600$.

Therefore, we are looking for the absolute minimum of

$$C(x) = \frac{48{,}000}{x} + 12{,}000 + 30x$$

over the interval $[1, 600]$. First we find the derivative of $C(x)$:

$$C'(x) = \frac{-48{,}000}{x^2} + 30.$$

To find the critical numbers we solve the equation

$$C'(x) = \frac{-48{,}000}{x^2} + 30 = 0 \quad \text{or}$$

$$\frac{48{,}000}{x^2} = 30 \quad \text{or} \quad 30x^2 = 48{,}000 \quad \text{or}$$

$$x^2 = 1{,}600; \quad \text{hence} \quad x = \pm 40.$$

The only positive solution is $x = 40$, which is in the interval $[1, 600]$. Therefore, there is exactly one critical number in $[1, 600]$. We apply the Strategy II. We check the second derivative of $C(x)$:

$$C''(x) = \frac{96{,}000}{x^3} > 0 \quad \text{for all} \quad x > 0.$$

By the Second-Derivative Test, C has a relative minimum at 40, and we conclude that $C(40)$ is an absolute minimum.

From the equation $n = 600/x$, substituting $x = 40$ we obtain $n = 15$. Therefore, to minimize its inventory costs, the store should maintain a lot size of 40 couches and 15 orders should be placed annually.

Example 7: The furniture store in Example 6 has been informed by the storage firm that the cost of storage has changed to $25 for one couch for a year. To minimize its inventory costs, how large is the lot size the store should maintain, and how many orders should be placed each year?

Solution Using the same reasoning as in Example 6, the inventory cost function is now

$$C(x) = 80 \times \frac{600}{x} + 12{,}000 + 25x = \frac{48{,}000}{x} + 12{,}000 + 25x$$

with the same domain as in Example 6. Therefore, we are looking for the absolute minimum of

$$C(x) = \frac{48{,}000}{x} + 12{,}000 + 25x$$

over the interval $[1, 600]$. First we find the derivative of $C(x)$:

$$C'(x) = \frac{-48{,}000}{x^2} + 25.$$

To find the critical numbers we solve the equation

$$C'(x) = \frac{-48{,}000}{x^2} + 25 = 0 \text{ or}$$

$$\frac{48{,}000}{x^2} = 25 \text{ or } 25x^2 = 48{,}000 \text{ or}$$

$$x^2 = 1{,}920; \text{ hence, } x = \sqrt{1{,}920} \approx 43.81.$$

Similar to Example 6, we conclude that $C(43.81)$ is an absolute minimum. However, it is impossible to order 43.81 couches. Hence, we check both $C(43)$ and $C(44)$:

$$C(43) = 14{,}191.3, \quad C(44) = 14{,}190.9.$$

The difference is rather insignificant, with $C(44)$ slightly lower. Next we check the reordering number n, which also takes positive integer values only.

From the equation $n = 600/x$, substituting $x = 43$ and $x = 44$ we obtain $n \approx 13.95$ and $n \approx 13.63$. To satisfy the total annual need of 600 couches, 14 orders have to be placed in both cases. With 14 orders, a lot size of 43 sets is sufficient. Note that $43 \times 14 = 602$, with 2 couches exceeding the need. Therefore, to minimize the inventory costs, the store should maintain a lot size of 43 couches and place 14 orders annually, with 12 orders of 43 couches and 2 orders of 42 couches.

Reality Check   *Discrete data vs. continuous model*

In Examples 4 and 7, the solutions from the mathematical models did not immediately meet the required condition. Strategy II is applied under the assumption that the function to be optimized is continuous on a domain consisting of an interval. However, in both examples, only the integer values of the involved variables (number of rooms, number of couches, number of orders) make sense. When seeking integer solutions only, the procedure applied in the examples, that is, finding a (non-integer) solution by applying Strategy II and checking the nearby integer solutions, does not work for all functions. However, for the types of the functions considered in this course, the procedure will work.

Exercise Set 3.8

1. Find two numbers whose sum is 30 and whose product is a maximum; that is, maximize $P = xy$ where $x + y = 30$.

2. Find two numbers whose sum is 40 and whose product is a maximum; that is, maximize $P = xy$ where $x + y = 40$.

3. Find two integer numbers whose sum is 23 and whose product is a maximum; that is, maximize $P = xy$ where x and y are integers and $x + y = 23$.

4. Find two integer numbers whose sum is 17 and whose product is a maximum; that is, maximize $P = xy$ where x and y are integers and $x + y = 17$.

5. Find two numbers whose difference is five and whose product is a minimum.

6. Find two numbers whose difference is eight and whose product is a minimum.

7. Find two positive numbers whose product is 100 and whose sum is a minimum.

8. Find two positive integer numbers whose product is 50 and whose sum is a minimum.

9. A rectangle has a perimeter of 20 in. Find the length and width of the rectangle under which the area is the largest. Follow the steps listed here:

   (a) Draw a diagram to show the rectangle. Label with variables.

   (b) Determine a function (depends on ONE variable) for the area to be maximized.

   (c) Determine the domain of the function.

   (d) Maximize the function.

   (e) Determine the dimensions under which the area is the largest.

10. Repeat the steps listed in Exercise 9, but assume that the perimeter of the rectangle is 32 in.

11. A rectangle has an area of 64 in.² Find the length and width of the rectangle under which the perimeter is the minimum. Follow the steps listed here:

    (a) Draw a diagram to show the rectangle. Label with variables.

    (b) Determine a function (depends on one variable) for the perimeter to be minimized.

    (c) Determine the domain of the function.

    (d) Minimize the function.

    (e) Determine the dimensions under which the perimeter is the minimum.

12. Repeat the steps listed in Exercise 11, but assume that the area of the rectangle is 16 in.²

13. A homeowner plans to enclose a 200 ft² rectangular playground in his garden, with one side along the boundary of his property. His neighbor will pay for half of the cost of materials on that side.

    (a) Draw a diagram to show the playground. Label with variables.

    (b) Determine a function (depends on one variable) for the cost to be minimized.

    (c) Determine the domain of the function.

    (d) Minimize the function.

    (e) Find the dimensions of the playground that will minimize the homeowner's total cost for materials.

14. Repeat Exercise 13 using the same data, but assume that the neighbor will pay for one-third of the cost on the boundary side.

15. A landscape architect plans to enclose a 2,000 square feet rectangular region in a botanical garden. She will use shrubs costing $20 per foot along three sides and fencing costing $30 per foot along the fourth side.

    (a) Draw a diagram to show the garden. Label with variables.

    (b) Determine a function (depends on one variable) for the cost to be minimized.

    (c) Determine the domain of the function.

    (d) Minimize the function.

(e) Find the dimensions of the garden that will minimize the total cost.

16. Repeat Exercise 15 using the same data, but assume that the garden is 2,500 square feet, the shrubs cost $18 per foot, and the fencing costs $32 per foot.

17. A farmer wants to fence an area of 1,000 square yards in a rectangular field and then divide it in half with a fence parallel to one of the sides of the rectangle. How can he do this so that the cost of the fence is minimized?

18. A rancher wants to fence an area of 1,500 square yards in a rectangular field and then divide it in half with a fence parallel to one of the sides of the rectangle. The dividing fence costs two-thirds as much as the surrounding fence. How can he do this so that the cost of the fence is minimized?

19. A farmer wants to fence an area of 1,000 square yards in a rectangular field that borders a straight river, and then divide it in half with a fence perpendicular to the river. He needs no fence along the river. How can he do this so that the cost of the fence is minimized?

20. A rancher wants to fence an area of 1,500 square yards in a rectangular field that borders a straight river, and then divide it in half with a fence perpendicular to the river. He needs no fence along the river. The dividing fence costs half as much as the surrounding fence. How can he do this so that the cost of the fence is minimized?

21. A firm receives an order for a square-base rectangular storage container with an open top. The container has a volume of 10 m³. Material for the base costs $10 per square meter. Material for the sides costs $6 per square meter. What is the lowest cost of materials for making such a container? What are the dimensions of the container that require the lowest cost for materials?

22. A firm receives an order for a square-base rectangular storage container with a lid. The container has a volume of 20 m³. Material for the base costs $20 per square meter. Material for the sides and the lid costs $10 per square meter. What is the lowest cost of materials for making such a container? What are the dimensions of the container that require the lowest cost for materials?

23. A firm receives an order for a rectangular poster. The top and bottom margins of the poster are 2 in. each and the side margins are each 1.5 in. The area of printed material on the poster is 60 in.² Find the dimensions of the poster with the smallest area.

24. A firm receives an order for a rectangular poster with an area of 180 in.² The top margin of the poster is 2 in. and the bottom and side margins are each 1 in. Find the dimensions of the poster with the largest printing area.

25. Mary has a piece of cardboard 12 in. by 12 in. She wants to make an open top box from it by cutting a square piece from each corner and folding up. How large are the square pieces she should cut to maximize the volume of the box? What is the maximum volume?

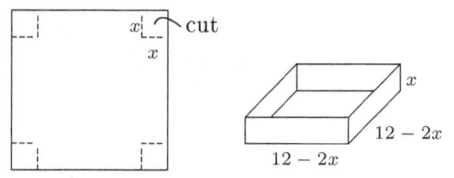

26. Karen has a piece of cardboard 16 in. by 10 in. She wants to make an open top box from it by cutting a square piece from each corner and folding up. How large are the square pieces she should cut to maximize the volume of the box? What is the maximum volume?

27. An appliance store has been selling 240 microwave ovens at $100 each every month. A market survey indicates that for each $6 increase in the price, the number of microwave ovens sold will decrease by 30 per month. It costs the store $46 to purchase each microwave oven from the manufacturer.

(a) Find the demand function, expressing p, the price charged for each microwave oven, as a function of x, the number of microwave ovens sold each month.

(b) Find the revenue function $R(x)$.

(c) Find the price the store should charge for each microwave oven to maximize its revenue.

(d) Find the price the store should charge for each microwave oven to maximize its profit.

3.8. OPTIMIZATION PROBLEMS

28. A furniture store has been selling 120 barstools at $60 each every month. A market survey indicates that for each $3 increase in the price, the number of barstools sold will decrease by 10 per month. It costs the store $32 to purchase each barstool from the manufacturer.

(a) Find the demand function, expressing p, the price charged for a barstool, as a function of x, the number of barstools sold each month.

(b) Find the revenue function $R(x)$.

(c) Find the price the store should charge for each barstool to maximize its revenue.

(d) Find the price the store should charge for each barstool to maximize its profit.

29. A motel owner observes that when a room is priced at $60 per day, all 80 rooms of the motel are occupied. For every $3 rise in the charge per room per day, one more room is vacant. Each occupied room costs an additional $15 per day to maintain.

(a) Find the demand function, expressing p, the price charged for each room per day, as a function of x, the number of rooms occupied.

(b) Find the profit function $P(x)$.

(c) Find the price of per room per day the motel should charge to maximize its profit.

30. Repeat Exercise 29 if the additional maintenance cost is $16 per room per day.

31. A theater owner observes that when he charges $25 for each ticket, there is an average attendance of 200 people. For every $2 increase in the admission price, there is a loss of five customers from the average attendance. The theater has 240 seats.

(a) Find the demand function, expressing p, the price of each ticket, as a function of x, the average attendance.

(b) Find the revenue function $R(x)$.

(c) Find the price the theater should charge for each ticket to maximize its revenue.

32. A concert organizer learned from a market survey that when the admission price is $30, there is an average attendance of 800 people. For every $1 drop in price, there is a gain of 20 customers. Each customer spends an average of $5 on concessions. The concert hall has 1,000 seats.

(a) Find the demand function, expressing p, the price of each ticket, as a function of x, the average attendance.

(b) Find the revenue function $R(x)$.

(c) Find the price the concert organizer should charge for each ticket to maximize his revenue.

33. An orange grower determines that when 25 orange trees are planted on an acre of ground, the yield is 40 bushels of oranges per tree. Each additional tree planted decreases the yield by 1 bushel per tree due to overcrowding. How many orange trees should be planted per acre to produce the highest yield for the grower?

34. Repeat Exercise 33 if each additional tree planted decreases the yield by 1.2 bushels per tree.

35. Repeat Example 5 if the campaign costs $12,000 per day.

36. Repeat Example 5 if the campaign costs $9,000 per day.

37. Repeat Example 5 if the percentage, as a decimal, of the market that buys a product is
$$f(t) = 1 - e^{-0.04t}.$$

38. Repeat Example 5 if the percentage, as a decimal, of the market that buys a product is
$$f(t) = 1 - e^{-0.06t}.$$

39. A retail appliance store sells 800 refrigerators each year. It costs $25 to store one refrigerator for a year. To reorder, there is a fixed processing cost of $50, plus a $20 shipping fee for each refrigerator.

(a) Find the cost function, expressed as a function of x, the number of refrigerators the store reorders each time (x is also the lot size the store must maintain).

(b) To minimize the inventory costs, how large is the lot size the store should maintain, and how many orders should be placed?

40. A sporting goods store sells 180 surfboards each year. It costs $9 to store one surfboard for a year. To reorder, there is a fixed processing cost of $20, plus a $6 shipping fee for each surfboard.

    (a) Find the cost function, expressed as a function of x, the number of surfboards the store reorders each time (x is also the lot size the store must maintain).

    (b) To minimize the inventory costs, how large is the lot size the store should maintain, and how many orders should be placed?

41. Repeat Exercise 39 using the same data, but assume that the processing fee for each reorder is $60.

42. Repeat Exercise 40 using the same data, but assume that the number of surfboards sold each year is 200.

43. A furniture store sells 200 dining sets each year. It costs $20 to store one dining set for a year. To reorder, there is a fixed processing cost of $40, plus a $10 shipping fee for each dining set. To minimize the inventory costs, how large is the lot size the store should maintain, and how many orders should be placed?

44. A retail appliance store sells 4000 television sets each year. It costs $10 to store one television set for a year. To reorder, there is a fixed processing cost of $25, plus a $8 shipping fee for each television set. To minimize the inventory costs, how large is the lot size the store should maintain, and how many orders should be placed?

45. Repeat Exercise 43 using the same data, but assume that the storage cost for each dining set is $25 per year.

46. Repeat Exercise 44 using the same data, but assume that the storage cost for each television set is $8 per year.

3.9 Elasticity of Demand

One example of the type of optimization problem studied in Section 3.8 is to find the optimal price of a piece of merchandise which will maximize revenue (as in Examples 8.3 and 8.4). We obtain the revenue function through the demand function, with the price, p, expressed as a function of x, the quantity of the merchandise. We then express the revenue function as a function of x. After we find the optimal x, we plug it back into the demand function to obtain the optimal price. The main reason for going through the indirect route is that, in the two examples, in addition to the maximum revenue, the maximum profit is also of interest. Since the cost function only depends on the quantity x, it is much more convenient to express the profit as a function of x. However, in many cases, retailers or manufacturers are interested in maximizing their revenue only. In this section, we focus on this particular type of the optimization problem. Instead of determining the revenue function through the demand function, we use a very helpful function of the price p, called the **elasticity of demand**, to help us solve such problems directly. We start by revisiting the optimization problem introduced in Example 8.3 of Section 3.8.

Example 1: An appliance store has been selling 200 microwave ovens at $120 each every month. A market survey indicates that for each $10 reduction in the price, the number of microwave ovens sold will increase by 25 per month.

(a) Find the demand function, expressing x, the number of microwave ovens sold each month, as a function of p, the price charged for each microwave oven.

(b) Find the revenue function $R(p)$.

(c) Find the price of the microwave oven the store should set to maximize its revenue.

3.9. ELASTICITY OF DEMAND

Solution (a) From the given condition, the demand function is linear. We write

$$x = mp + b.$$

When p is decreased by \$10, x is increased by 25. Therefore, the slope is

$$m = \frac{25}{-10} = -2.5.$$

Next we plug $p = 120$ and $x = 200$ into the equation:

$$200 = -0.2.5(120) + b.$$

Hence $b = 200 + 2.5(120) = 500$. We obtain the demand function:

$$x = D(p) = -2.5p + 500.$$

(b) The revenue is the price times the quantity sold, or $R = p \cdot x$. From part (a) we have $x = -2.5p + 500$, hence the revenue function is

$$R(p) = p \cdot x = p \cdot (-2.5p + 500) = -2.5p^2 + 500p.$$

(c) The quantity to be maximized is the revenue. In part (b), we have expressed the revenue as a function of p. We now find the domain of the revenue function. Since both x and p have to be nonnegative, from part (a) we obtain $p \leq 200$. Hence, the domain is $0 \leq p \leq 200$.

We are looking for the absolute maximum of

$$R(p) = -2.5p^2 + 500p$$

over the interval $[0, 200]$. First we find the derivative of $R(p)$:

$$R'(p) = -5p + 500.$$

To find the critical numbers we solve the equation

$$R'(p) = -5p + 500 = 0.$$

The only solution is $p = 100$, which is in the interval $[0, 200]$. By the Second-Derivative Test, R has a relative maximum at 100, and we conclude that $R(100)$ is an absolute maximum. Therefore, to maximize the revenue, the store should set the price at \$100.

Now let us observe what happens to the revenue when the current price is near \$100 and there is a small rise in the price.

Case I: Suppose that the current price is \$96. Then the current demand is $x = -2.5(96) + 500 = 260$, and the revenue is $96 \times 260 = \$24{,}960$. If the price is raised by \$2, to \$98, then the demand is $x = -2.5(98) + 500 = 255$, and the store loses five customers. However, the total revenue becomes $98 \times 255 = \$24{,}990$, it is actually increased by \$30.

Case II: Suppose that the current price is \$104. Then the current demand is $x = -2.5(104) + 500 = 240$, and the revenue is $104 \times 240 = \$24{,}960$. If the price is raised by \$2, to \$106, then the demand is $x = -2.5(106) + 500 = 235$, and the store again loses five customers. The total revenue becomes $104 \times 235 = \$24{,}440$, it is decreased by \$520.

What is happening here? Since the demand function is linear, the number of customers the store loses is the same for every dollar of increase in the price. However, the *percentage* of the changes are different:

When the price changes from $96 to $98, it increases by $2/96 \approx 0.0208 = 2.08\%$. At the same time, the demand changes from 260 to 255, decreased by $5/260 \approx 0.0192 = 1.92\%$. Hence, the demand is *less sensitive* to the change in the price, and in such a situation the revenue increases. When the price changes from $104 to $106, it increases by $2/104 \approx 0.0192 = 1.92\%$. At the same time, the demand changes from 240 to 235, decreased by $5/240 \approx 0.0208 = 2.08\%$. Hence the demand is *more sensitive* to the change in the price, and in such a situation the revenue decreases.

From the aforementioned exploration, it seems that the sensitivity, measured by a percentage, of the demand with respect to the percent change in the price, is a useful indicator. It helps retailers and manufacturers to judge whether the price should be raised (or reduced) from the current setting, to increase revenue. When the percent decrease in the demand for a product is larger than the percent increase in its price, the total revenue is going down. In such a case we say that the demand for the product is **elastic** at that price. Conversely, when the percent decease in the demand for a product is smaller than the percent increase in its price, the total revenue is going up. In such a case we say that the demand for the product is **inelastic** at that price.

We now explore how such sensitivity can be measured. We are comparing the percent change in demand, x, with the percent change in price, p. We can compare the ratio of the two percent changes.

If the price changes from p to $p + \Delta p$, then the percent change in p is

$$\frac{\Delta p}{p} = \frac{\Delta p}{p} \cdot \frac{100}{100} = \frac{\Delta p \cdot 100}{p}\%.$$

A change in the price produces a change Δx in the demand, hence the demand changes from x to $x + \Delta x$. Therefore, the percent change in x is

$$\frac{\Delta x}{x} = \frac{\Delta x}{x} \cdot \frac{100}{100} = \frac{\Delta x \cdot 100}{x}\%.$$

The ratio of the percent change in demand to the percent change in price is

$$\frac{\frac{\Delta x \cdot 100}{x}\%}{\frac{\Delta p \cdot 100}{p}\%} = \frac{\Delta x / x}{\Delta p / p} = \frac{p}{x} \cdot \frac{\Delta x}{\Delta p}.$$

Since the demand and the price are linked by the demand function

$$x = D(p),$$

we have

$$\Delta x = D(p + \Delta p) - D(p),$$

therefore, if we let Δp approach 0, by the definition of the derivatives we have

$$\lim_{\Delta p \to 0} \frac{\Delta x}{\Delta p} = \lim_{\Delta p \to 0} \frac{D(p + \Delta p) - D(p)}{\Delta p} = D'(p).$$

Therefore, when Δp approaches 0, the ratio of the percent change in demand to the percent change in price becomes

$$\lim_{\Delta p \to 0} \frac{p}{x} \cdot \frac{\Delta x}{\Delta p} = \frac{p}{x} \cdot D'(p) = \frac{p}{D(p)} \cdot D'(p).$$

Notice that this quantity is always negative, since p and $D(p)$ are both positive, but $D(p)$ is a decreasing function, therefore, $D'(p)$ is negative. We call the negative of this quantity the **elasticity of demand**.

3.9. ELASTICITY OF DEMAND

> **Definition** *elasticity*
>
> Given the demand function $x = D(p)$, the **elasticity of demand**, denoted by E, of a product at a price p, is defined by
>
> $$E(p) = \frac{-p \cdot D'(p)}{D(p)}.$$

According to the previous discussions, now that the measurement $E(p)$ has been introduced, we say that the demand for the product is **elastic** at the price p if $E(p) > 1$. We say that the demand for the product is **inelastic** at the price p if $E(p) < 1$. We say that the demand for the product is **unitary** at the price p if $E(p) = 1$.

Example 2: Consider the appliance store in Example 1. The demand x for microwave ovens every month is given by

$$x = D(p) = -2.5p + 500,$$

where p is the price, in dollars, per microwave oven.

(a) Find the elasticity function, $E(p)$.

(b) Find the elasticity at $p = 96$ and $p = 102$. Interpret the meaning of these values for the elasticity.

(c) Find the price of the microwave oven at which $E(p) = 1$. Interpret the meaning of this value for the elasticity.

Solution (a) The elasticity $E(p)$ is

$$E(p) = \frac{-p \cdot D'(p)}{D(p)}.$$

Since $D(p) = -2.5p + 500$, we have $D'(p) = -2.5$. Hence

$$E(p) = \frac{-p \cdot (-2.5)}{-2.5p + 500} = \frac{2.5p}{500 - 2.5p} = \frac{p}{200 - p}.$$

(b) We have

$$E(96) = \frac{96}{200 - 96} \approx 0.923.$$

The elasticity is less than 1. That means if there is a small percent increase in the price from $96 per microwave oven, the percent decrease in the demand is even smaller. Hence, the demand is inelastic at the price of $96. We have

$$E(102) = \frac{102}{200 - 102} \approx 1.04.$$

The elasticity is greater than 1. That means if there is a small percent increase in the price from $102 per microwave oven, the percent decrease in the demand is slightly larger. Hence, the demand is elastic at the price of $102.

(c) We solve the equation $E(p) = 1$:

$$\frac{p}{200 - p} = 1$$

$p = (200 - p)$, hence $2p = 200$, or $p = 100$.

When the price is $100 per microwave oven, the percent decrease in the rate of the demand is the same as the percent increase in the rate of the price.

Recall that in Example 1, we concluded that to maximize its revenue, the store should set the price to be $100. It seems that solving for $E(p) = 1$ corresponds to maximizing $R(p)$. We now prove that it is indeed the case. Actually, the increasing and decreasing behavior of the revenue is closely related to whether the product is elastic or inelastic. We prove the relationship mathematically. Note that

$$R(p) = p \cdot D(p).$$

Taking the derivative using the Product Rule we have:

$$R'(p) = 1 \cdot D(p) + p \cdot D'(p).$$

Hence by the definition of $E(p)$,

$$R'(p) = D(p)\left[1 + \frac{p \cdot D'(p)}{D(p)}\right] = D(p)[1 - E(p)].$$

Therefore $R'(p) < 0$ if $E(p) > 1$, $R'(p) > 0$ if $E(p) < 1$, and $R'(p) = 0$ if $E(p) = 1$. Recalling that a function is increasing (decreasing) if its derivative is positive (negative), we conclude that the revenue is increasing if the product is inelastic at the price, hence slightly raising the price will increase the revenue. If the product is elastic at the price, slightly raising the price will decrease the revenue. When the elasticity is unitary, then the revenue is at a maximum.

We summarize the relationship between the elasticity and the revenue of a product (Figure 3.9.1):

Theorem 1   *Elasticity-Revenue Relationship*

If the demand is elastic, then $E(p) > 1$, and $R'(p) < 0$, the total revenue is decreasing.

If the demand is inelastic, then $E(p) < 1$, and $R'(p) > 0$, the total revenue is increasing.

If the elasticity of demand is unitary, then $E(p) = 1$, and $R'(p) = 0$, the total revenue is at a maximum.

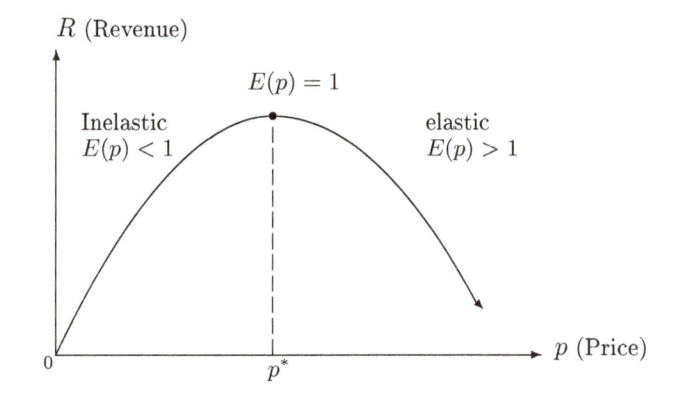

Figure 3.9.1: Maximum revenue vs. unitary elasticity

Example 3: A clothing manufacturer is selling its polo shirts at a price of $16. The demand function for the polo shirts is

$$x = D(p) = 180{,}000\sqrt{22 - p},$$

3.9. ELASTICITY OF DEMAND

where p is the price of a polo shirt in dollars and x is in items.

(a) Find the elasticity function, $E(p)$.

(b) Find the elasticity at $p = 16$. Should the manufacturer lower or raise the price to increase its revenue?

(c) Find the price of the polo shirts at which the revenue is greatest.

Solution (a) The elasticity $E(p)$ is

$$E(p) = \frac{-p \cdot D'(p)}{D(p)}.$$

Since $D(p) = 180{,}000\sqrt{22-p}$, we have

$$D'(p) = \frac{180{,}000(-1)}{2\sqrt{22-p}} = \frac{-90{,}000}{\sqrt{22-p}}.$$

Hence

$$E(p) = \frac{-p \cdot \frac{-90{,}000}{\sqrt{22-p}}}{180{,}000\sqrt{22-p}} = \frac{90{,}000\,p}{\sqrt{22-p}} \cdot \frac{1}{180{,}000\sqrt{22-p}} = \frac{p}{2(22-p)}.$$

(b) We have

$$E(16) = \frac{16}{2(22-16)} = \frac{16}{12} = \frac{4}{3}.$$

Since $E(16) > 1$, the demand is elastic and the prices should be lowered to increase the revenue.

(c) We solve the equation $E(p) = 1$:

$$\frac{p}{2(22-p)} = 1$$

$p = 2(22-p) = 44 - 2p$, hence $3p = 44$, or $p \approx 14.67$.

Thus, to maximize the revenue, the price of the polo shirts should be $14.67.

Exercise Set 3.9

1. The elasticity of a good is $E = 2$. What is the effect on the demand of:

   (a) A 3% price increase?

   (b) A 3% price decrease?

2. The elasticity of a good is $E = 0.5$. What is the effect on the demand of:

   (a) A 3% price increase?

   (b) A 3% price decrease?

For the demand functions (the prices are in dollars) in each of Exercises 3–18, do the following:

(a) Find the elasticity $E(p)$.

(b) Find the elasticity at the given price, and classify the demand as elastic, inelastic, or unitary.

(c) Find the price at which the total revenue is at a maximum.

3. $x = D(p) = 220 - 5p; \quad p = 10.$

4. $x = D(p) = 50 - 4p$; $p = 5$.

5. $x = D(p) = 250 - p^2$; $p = 10$.

6. $x = D(p) = 200 - p^2$; $p = 8$.

7. $x = D(p) = \sqrt{150 - 3p}$; $p = 30$.

8. $x = D(p) = \sqrt{100 - 2p}$; $p = 25$.

9. $x = D(p) = \dfrac{100}{p}$; $p = 30$.

10. $x = D(p) = \dfrac{500}{p}$; $p = 20$.

11. $x = D(p) = 4{,}500e^{-0.02p}$; $p = 200$.

12. $x = D(p) = 6{,}500e^{-0.04p}$; $p = 100$.

13. $x = D(p) = 100e^{-0.05p}$; $p = 40$.

14. $x = D(p) = 3{,}000e^{-0.03p}$; $p = 100$.

15. $x = D(p) = 30 - 2p - p^2$; $p = 2$.

16. $x = D(p) = 50 - p - p^2$; $p = 4$.

17. $x = D(p) = \dfrac{200}{(p+2)^2}$; $p = 1$.

18. $x = D(p) = \dfrac{300}{(3p+10)^2}$; $p = 10$.

APPLICATIONS

19. The manager of the city metro train system finds that the number of people, x, willing to take public transportation depends on the price, p, in dollars, of the ticket, by the demand function

    $x = D(p) = 600(5 - \sqrt{p})$.

    Currently there are about 1,800 people riding the trains per day with a price of $4 for each ticket.

    (a) Find the elasticity function, $E(p)$.

    (b) Find the elasticity at $p = 4$. Is the demand elastic or inelastic at the current $4 ticket price?

    (c) Should the city lower or raise the price of a ticket to increase its revenue?

20. The manager of the city bus system finds that the number of people, x, willing to take public transportation depends on the price, p, in cents, of the ticket, by the demand function

    $x = D(p) = 2{,}000\sqrt{900 - p}$.

    Currently there are about 54,722 people riding the buses per day with a price of $1.50 for each ticket.

    (a) Find the elasticity function, $E(p)$.

    (b) Find the elasticity at $p = 150$. Is the demand elastic or inelastic at the current $1.50 ticket price?

    (c) Should the city lower or raise the price of a ticket to increase its revenue?

21. A clothing manufacturer is selling its blouses at a price of $18. The demand function for the blouses is

    $x = D(p) = 120 - 4p$,

    where p is the price of a blouse in dollars and x is in hundred items.

    (a) Find the elasticity function, $E(p)$.

    (b) Find the elasticity at $p = 18$. Should the manufacturer lower or raise the price in order to increase its revenue?

    (c) Find the price of the blouses at which the revenue is greatest by solving the equation $E(p) = 1$.

22. A furniture manufacturer is selling its barstools at a price of $60. The demand function for the barstools is

    $x = D(p) = 270 - 2.5p$,

    where p is the price of a barstool in dollars and x is in hundred items.

    (a) Find the elasticity function, $E(p)$.

    (b) Find the elasticity at $p = 60$. Should the manufacturer lower or raise the price to increase its revenue?

    (c) Find the price of the barstools at which the revenue is greatest by solving the equation $E(p) = 1$.

23. A furniture store is selling 200 barstools at $65 each every month. A market survey indicates that for each $2 increase in the price, the number of barstools sold will decrease by 10 per month.

3.9. ELASTICITY OF DEMAND

(a) Find the demand function, expressing x, the number of barstools sold each month, as a function of p, the price charged for a barstool.

(b) Find the revenue function $R(p)$.

(c) Find the price the store should charge per barstool to maximize its revenue.

(d) Find the elasticity function, $E(p)$.

(e) Find the elasticity at $p = 60$. Should the store lower or raise the price to increase its revenue?

(f) Find the price of the barstools at which the revenue is greatest by solving the equation $E(p) = 1$. Compare the result with that in part (c).

24. An appliance store has been selling 240 microwave ovens at $100 each every month. A market survey indicates that for each $6 increase in the price, the number of microwave ovens sold will decrease by 30 per month.

(a) Find the demand function, expressing x, the number of microwave ovens sold each month, as a function of p, the price charged per microwave oven.

(b) Find the revenue function $R(p)$.

(c) Find the price the store should charge per microwave oven to maximize its revenue.

(d) Find the elasticity function, $E(p)$.

(e) Find the elasticity at $p = 60$. Should the store lower or raise the price to increase its revenue?

(f) Find the price of the microwave ovens at which the revenue is greatest by solving the equation $E(p) = 1$. Compare the result with that in part (c).

25. A potato chip company determines that the demand function for one brand of its potato chips is given by

$$x = D(p) = 30 - p - p^2$$

where p is the price, in dollars, of a bag of potato chips, and x is in hundreds of bags of potato chips sold each month.

(a) Find the elasticity function, $E(p)$.

(b) Find the elasticity at a price of $2.50 per bag, stating whether the demand is elastic or inelastic at that price.

(c) Find the elasticity at a price of $3.60 per bag, stating whether the demand is elastic or inelastic at that price.

(d) Find the price of a bag of potato chips at which the revenue is greatest.

(e) What quantity of bags of potato chips will the company sell each month at the price that maximizes revenue?

(f) At a price of $3.60 per bag, will a small increase in price cause the total revenue to increase or decrease?

26. A night-light manufacturer determines that the demand function for one brand of its night-lights is given by

$$x = D(p) = 20 - 2p^2$$

where p is the price, in dollars, of a night-light, and x is in hundreds of night-lights sold each month.

(a) Find the elasticity function, $E(p)$.

(b) Find the elasticity at a price of $1.50 per night-light, stating whether the demand is elastic or inelastic at that price.

(c) Find the elasticity at a price of $2.20 per night-light, stating whether the demand is elastic or inelastic at that price.

(d) Find the price of a night-light at which the revenue is greatest.

(e) What quantity of night-lights will the manufacturer sell each month at the price that maximizes revenue?

(f) At a price of $1.50 per night-light, will a small increase in price cause the total revenue to increase or decrease?

27. A radio manufacturing firm determines that the demand function for one brand of its radios is given by

$$x = D(p) = 100 \ln(150 - p)$$

where p is the price, in dollars, of a radio, and x is in radios sold each month.

(a) Find the elasticity function, $E(p)$.

(b) Find the elasticity at a price of $100 per radio, stating whether the demand is elastic or inelastic at that price. Will a small increase in price cause the total revenue to increase or decrease?

28. A sports shoes manufacturing firm determines that the demand function for one brand of its sports shoes is given by

$$x = D(p) = 150\ln(120 - p)$$

where p is the price, in dollars, of a pair of shoes, and x is in pairs of shoes sold each month.

(a) Find the elasticity function, $E(p)$.

(b) Find the elasticity at a price of $100 per pair of shoes, stating whether the demand is elastic or inelastic at that price. Will a small increase in price cause the total revenue to increase or decrease?

29. A day-pack manufacturing firm determines that the demand function for a certain brand of day-pack is

$$x = D(p) = 2{,}000e^{-0.04p}$$

where p is the price, in dollars, of a day-pack, and x is in number of day-packs sold each month.

(a) Find the elasticity function, $E(p)$.

(b) Find the elasticity at $p = 20$. Should the manufacturer lower or raise the price to increase its revenue?

(c) Find the price of a day-pack at which the revenue is greatest by solving the equation $E(p) = 1$.

30. A sporting goods manufacturer determines that the demand function for a certain brand of skateboards is

$$x = D(p) = 1{,}000e^{-0.01p}$$

where p is the price, in dollars, of a skateboard and x is in number of skateboards sold each month.

(a) Find the elasticity function, $E(p)$.

(b) Find the elasticity at $p = 90$. Should the manufacturer lower or raise the price to increase its revenue?

(c) Find the price of a skateboard at which the revenue is greatest by solving the equation $E(p) = 1$.

31. Assume that the demand function for a certain brand of cigarettes is

$$x = D(p) = 5{,}500p^{-0.8}$$

where p is the price, in dollars, of a pack of cigarettes, and x is in number of packs of cigarettes sold each month.

(a) Find the elasticity function, $E(p)$.

(b) Is the value of the elasticity dependent on the price?

(c) Does the total revenue have a maximum? When?

(d) Will raising cigarette prices by imposing a tax decrease the *total* revenue?

(e) Will raising cigarette prices by imposing a tax decrease the *cigarettes manufacturer's* revenue?

32. Assume that there is only one company offering cell phone service in a city. Would you expect elasticity of demand for cell phone service to be high or low? Explain.

Chapter 3 Summary

This chapter illustrates how derivatives can be used in real-world situations. The chapter starts by showing how to use the derivative to estimate the function values around a given point. This approximation is called **linear approximation**, since on the graph of the function, the approximating point is on the tangent line.

CHAPTER 3 SUMMARY

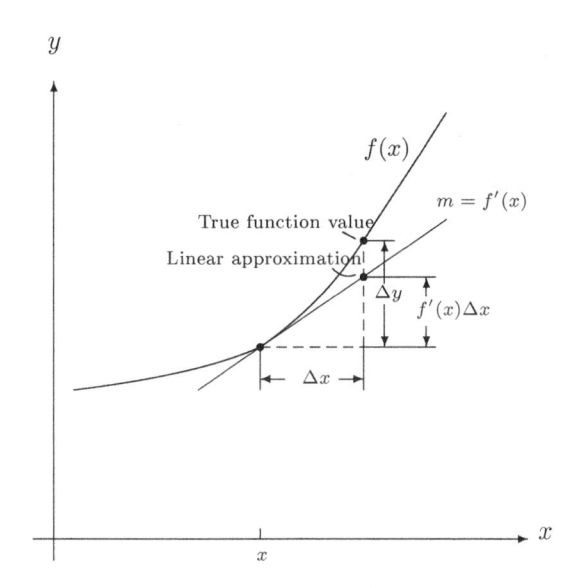

The formula for linear approximation can be expressed by

$$f(x + \Delta x) \approx f(x) + f'(x)\Delta x,$$

where Δx is the (small) change in variable x. Denoting the change in variable y as Δy, that is,

$$\Delta y = f(x + \Delta x) \approx f(x),$$

the linear approximation can be expressed as

$$\frac{dy}{dx} \approx \frac{\Delta y}{\Delta x}.$$

This expression leads to the concept of the **differentials**, where dy and dx are treated as two separate symbols: We define the **differential of** x, denoted by dx, as

$$dx = \Delta x.$$

We define the **differential of** y, denoted by dy, as

$$dy = f'(x)dx.$$

Therefore, the linear approximation can also be expressed as

$$\Delta y \approx dy, \quad \text{or} \quad f(x + dx) \approx f(x) + dy.$$

Example 1: For $y = f(x) = \ln(x^2 - 3)$:

(a) Find dy.

(b) Find dy when $x = 2$ and $dx = 0.01$.

(c) Find the linear approximation of $f(2.01)$.

Solution (a) We first find the derivative $\dfrac{dy}{dx}$. Applying the Extended Logarithmic Rule, we obtain

$$\frac{dy}{dx} = \frac{2x}{x^2 - 3}.$$

Hence,
$$dy = \frac{dy}{dx} \cdot dx = \frac{2x}{x^2 - 3} \cdot dx.$$

(b) When $x = 2$, and $dx = 0.01$, we have
$$dy = \left(\frac{2(2)}{(2)^2 - 3}\right)(0.01) = (4)(0.01) = 0.04.$$

(c) We use the expression of linear approximation
$$f(x + dx) \approx f(x) + dy.$$

We have $x = 2$, $dx = 0.01$, $f(2) = \ln(2^2 - 3) = \ln 1 = 0$, and $dy = 0.04$. Hence,
$$f(2.01) = f(2 + 0.01) \approx f(2) + dy = 0 + 0.04 = 0.04.$$

The true value of $f(2.01)$ is $0.039317\ldots$. Therefore, linear approximation works well for small changes in x.

Applying the concept of linear approximation to business functions, **marginal analysis** provides a tool for analyzing productivity behaviors. The marginal cost, revenue, and profit are simply the derivatives of the cost, revenue, and profit functions. They approximate the cost, revenue, and profit for producing and selling one more item of a product. Using the marginal functions, we can determine the number of items that must be produced and sold to result in the maximum profit. This is just one application of the general **optimization** problems studied in Section 3.8. Many examples that illustrate how to solve theses problems are presented in Sections 3.2 and 3.8. However, to be able to use derivatives to determine where the **maximum** or **minimum** occur for a general function, we first need to study how to use the derivative to determine where a function is **increasing** or **decreasing**. Suppose that a function f is continuous and differentiable on an open interval (a, b), then

If $f'(x) > 0$ for all x in (a, b), then f is increasing on (a, b).

If $f'(x) < 0$ for all x in (a, b), then f is decreasing on (a, b).

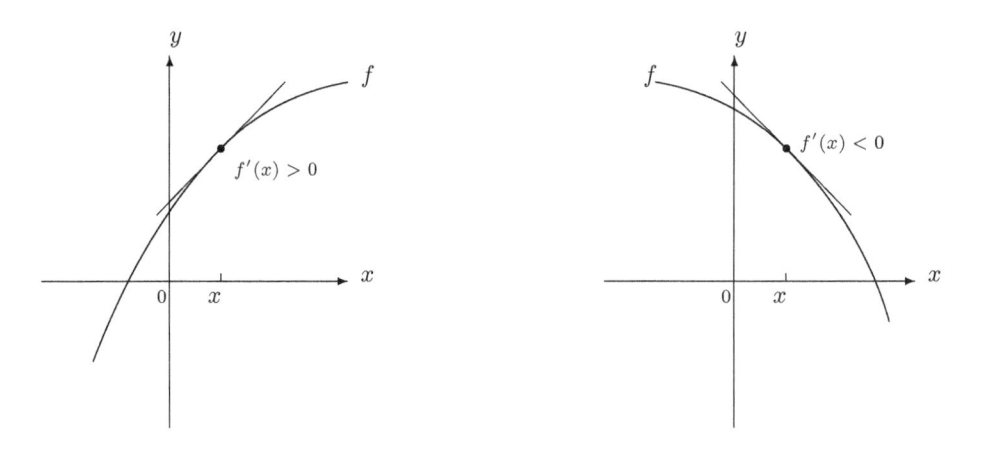

To determine where the maximum or minimum values occur, we look for the x-values around which the sign of $f'(x)$ changes. For a continuous function, such an x-value $x = c$ is called a **critical number**. It must satisfy either

$f'(c)$ does not exist, or $f'(c) = 0$.

For a continuous function, any **relative** (or **local**) **maximum** or **minimum** must occur at a critical number. However, an absolute extremum may occur either at a critical number or at a boundary number.

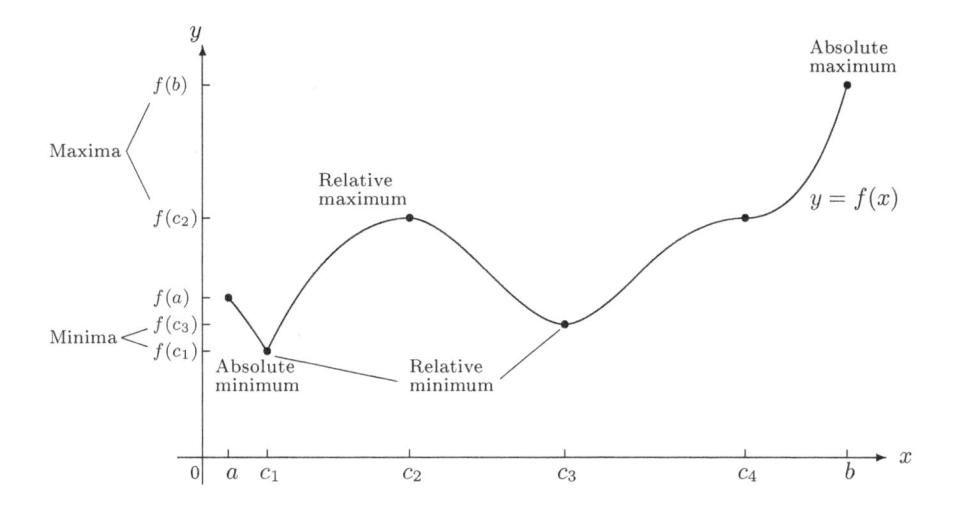

For the aforementioned figure, on the interval $[a, b]$, $f(c_2)$ is a relative maximum, and $f(c_1)$ and $f(c_3)$ are relative maxima. $f(b)$ is an absolute maximum, and $f(c_1)$ is an absolute minimum. The critical numbers are c_1, c_2, c_3, and c_4. Note that $f(c_4)$ is neither a relative maximum nor a relative minimum. To judge whether the function has a relative extremum at a critical value, the **First-Derivative Test** or the **Second-Derivative Test** can be performed.

Example 2: For the function $f(x) = x^3 - 3x + 1$, use both the First-Derivative Test and the Second-Derivative Test to determine the relative extrema of the function.

Solution We first take the derivative:

$$f'(x) = 3x^2 - 3.$$

Since f' exists everywhere, all its critical numbers satisfy $f'(x) = 0$:

$$3x^2 - 3 = 0$$

$$x^2 - 1 = 0$$

$$(x-1)(x+1) = 0$$

$$x = 1 \quad \text{or} \quad x = -1.$$

Hence, the critical numbers are -1 and 1.

The critical numbers divide the real line into three intervals: $(-\infty, -1)$, $(-1, 1)$, and $(1, \infty)$. On each interval the derivative keeps the same sign. We determine the signs of f' in each interval by plugging in some numbers:

$$f'(-2) = 3(-2)^2 - 3 = 9 > 0; \quad f'(0) = -3 < 0; \quad f'(2) = 3(2)^2 - 3 = 9 > 0.$$

Therefore, f' is positive on the intervals $(-\infty, -1)$ and $(1, \infty)$, and negative on the interval $(-1, 1)$. We can make a First-Derivative Test diagram:

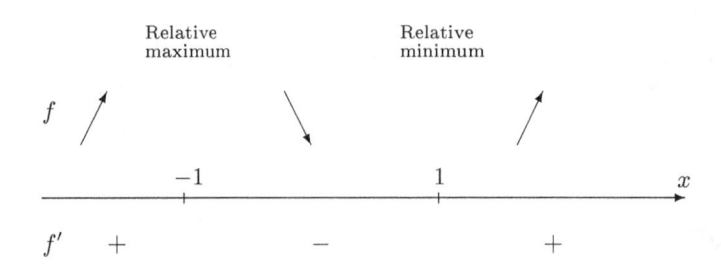

We have

$$f(-1) = (-1)^3 - 3(-1) + 1 = -1 + 3 + 1 = 3, \text{ and}$$

$$f(1) = (1)^3 - 3(1) + 2 = 1 - 3 + 1 = -1.$$

Hence, $f(x)$ has a relative maximum at $(-1, 3)$, and a relative minimum at $(1, -1)$.

To perform the Second-Derivative Test, we check the sign of the second derivative at each of the critical numbers -1 and 1: We have $f''(-1) = 6(-1) = -6 < 0$ and $f''(1) = 6(1) = 6 > 0$. Therefore, f has a relative maximum at $x = -1$ and f has a relative minimum at $x = 1$.

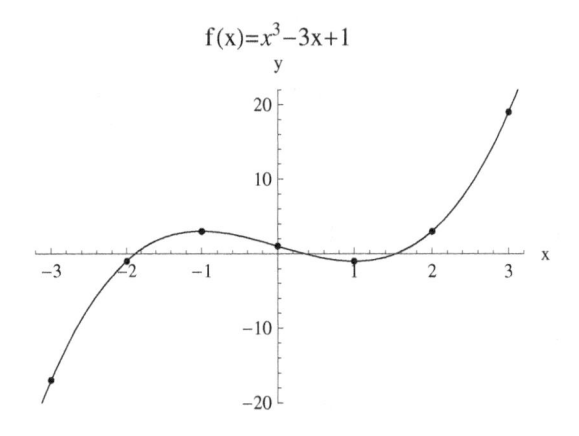

The sign of the second derivative indicates the **concavity** of the function. For the aforementioned example, it is obvious that the function is concave up around the relative minimum point $(1, -1)$, and it is concave down around the relative maximum point $(-1, 3)$. Points at which a function changes its concavity are called **inflection points**. Such points have the property that either $f''(p) = 0$ or $f''(p)$ does not exist.

The information about a function revealed by its derivatives allows us to develop procedures for sketching its graph. The graph of a function often supplies very important information about the behavior of the function. It is still important to learn to use calculus to sketch graphs, even in the age of graphing devices such as graphing calculators, since calculus helps us to grasp the major features of a function its graph should include.

To sketch the graph of a continuous function, with the help of a graphing device, the following procedure is helpful:

(a) Find $f'(x)$ and $f''(x)$. Note the domain of f.

(b) Determine the critical numbers by solving $f'(x) = 0$ and finding where $f'(x)$ does not exist.

(c) Solve $f''(x) = 0$ and find where $f''(x)$ does not exist. Calculate the function values at these numbers.

These points are the candidates for inflection points.

(d) Set the plotting windows. The horizontal interval should include all critical numbers and all x-values of possible inflection points. The vertical interval should include all all y-values of critical points and possible inflection points.

(e) Plot the graph.

To sketch the graph of a continuous function, without the help of a graphing device, the following procedure is helpful:

(a) Find $f'(x)$ and $f''(x)$. Note the domain of f.

(b) Determine the critical numbers by solving $f'(x) = 0$ and finding where $f'(x)$ does not exist.

(c) Perform either the First- or Second-Derivative Test for each critical number to determine whether the function has a relative extremum there. The information also identifies the intervals on which the function is increasing or decreasing.

(d) Solve $f''(x) = 0$ and find where $f''(x)$ does not exist. Calculate the function values at these numbers. These points are the candidates for inflection points.

(e) Use the sign of the second derivative to determine the intervals on which the function is concave up or concave down.

(f) Use the information obtained in steps (a)-(e) to sketch the graph. Calculate and plot some additional points if needed.

Many detailed examples of sketching the graph of a continuous function are presented in Section 3.5. To sketch the graph of rational functions, additional information such as the **asymptotes** is of great help. The types of asymptotes are **vertical**, **horizontal**, and **slant**. Given a simplified rational function $f(x) = P(x)/Q(x)$, the vertical asymptotes can be located by solving the equation $Q(x) = 0$. The function has $y = 0$ as the horizontal asymptote if the degree of P is lower than the degree of Q. It has $y = a/b$ as the horizontal asymptote if the degree of P is equal to the degree of Q, where a is the leading coefficient (i.e., the coefficient of the highest power term) of the numerator and d is the leading coefficient of the denominator. If the degree of P is exactly one higher than the degree of Q, the graph of f has a slant asymptote. A detailed strategy and many examples of sketching the graph of a rational function are presented in Section 3.6.

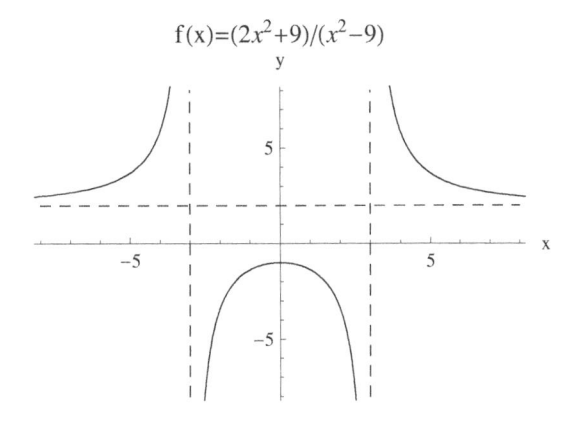

If a function is continuous on a closed interval $[a, b]$, then it has both an **absolute maximum** and an **absolute minimum** on $[a, b]$. To find the absolute extrema, it is sufficient to locate all the critical numbers on (a, b), then compare the function values at these critical numbers as well as at the boundary numbers a and b. The largest of these values is the absolute maximum and the least is the absolute minimum.

If a function is continuous on an interval such as $[a, b)$, (a, b), $(-\infty, b)$, (a, ∞), or $(-\infty, \infty)$, etc., then it may or may not have absolute extrema. However, if the function has exactly one critical number

Chapter 3 Review Exercises

1. The quantity, q, of a certain skateboard sold depends on the selling price, p, in dollars, so we write $q = D(p)$, where p is in dollars per skateboard and q is in number of skateboard sold.

   (a) Using units, explain the meaning of $D(120) = 8,000$.

   (b) Using units, explain the meaning of $D'(120) = -80$.

   (c) Using the information in parts (a) and (b), estimate the number of skateboard sold at the price of $122 dollars per skateboard.

2. For $y = f(x) = 3x^3 + 2x + 1$:

   (a) Find Δy and dy when $x = 1$ and $\Delta x = dx = 0.01$.

   (b) Use linear approximation to approximate $f(1.01)$.

3. A clothing manufacturer has the cost function
   $$C(x) = 1,200 + 18x + 0.1x^2 \text{ (in dollars)},$$
   where x is the number of suits produced each week. The revenue function for selling x suits is given by
   $$R(x) = 120x \text{ (in dollars)}.$$

   (a) Find the total profit when x suits are produced and sold each week.

   (b) Find the total cost, revenue, and profit when 70 suits are produced and sold each week.

   (c) Find the marginal cost, revenue, and profit when 70 suits are produced and sold each week.

   (d) Use the information in parts (b) and (c) to estimate the total cost, revenue, and profit when 74 suits are produced and sold each week.

   (e) Find the number of suits that should be produced and sold each week for the company to make the maximum profit.

4. For the function graphed here:

   (a) Determine where the function is increasing and where it is decreasing.

   (b) Determine where the **derivative** is (i) positive; (ii) negative; (iii) equals zero; (iv) does not exist.

   (c) Determine the critical number(s) of f.

   (d) Determine the point(s) at which f has a relative maximum;

   (e) Determine the point(s) at which f has a relative minimum;

   (f) Determine the absolute maximum of f;

   (g) Determine the absolute minimum of f.

   (h) Determine where the function is concave up, where it is concave down, and where it has an inflection point.

CHAPTER 3 REVIEW EXERCISES

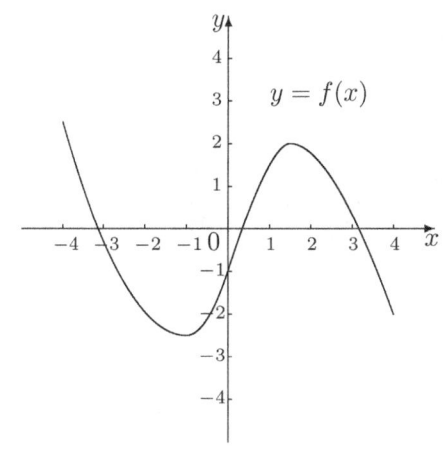

5. For $f(x) = \frac{1}{3}x^3 - 3x^2 + 5x + 6$, do the following:

    (a) Determine the critical points;

    (b) Use the signs of the first derivative to determine the intervals on which f is increasing;

    (c) Use the signs of the first derivative to determine the intervals on which f is decreasing;

    (d) Use the information from parts (a) to (c) to determine the relative extrema of f.

    (e) Determine the inflection points;

    (f) Use the signs of the second derivative to determine the intervals on which f is concave up;

    (g) Use the signs of the second derivative to determine the intervals on which f is concave down.

    (h) Sketch a graph of the function.

6. Repeat the steps (a) – (h) listed in Exercise #5 for $f(x) = x^4 - 2x^3 - 2$.

7. For $f(x) = 3(x+3)^{2/3} + 2$, find all relative maxima or minima if they exist, and sketch a graph of this function.

For each function in Exercises #8 to #10, sketch the graph following the process listed here:

   (a) State the domain of the function and where the intercepts occur;

   (b) List the asymptotes if they exist;

   (c) List the coordinates of where the relative extrema occur;

   (d) State where the function is increasing or decreasing;

   (e) List the coordinates of where the inflection points occur;

   (f) State where the function is concave up or concave down;

   (g) Sketch the graph with appropriate coordinate ranges.

8. $f(x) = \dfrac{2}{x^2 - 4}$

9. $f(x) = \dfrac{4x^2}{x^2 + 1}$

10. $g(x) = \dfrac{x+1}{x^2 - x - 2}$

For each function in Exercises #11 to #16, determine the absolute maximum and minimum values of each function, if they exist, over the indicated interval. Indicate the x-values at which they occur.

11. $h(x) = \frac{1}{3}x^3 - 3x^2 + 5x + 6$; $[2, 6]$

12. $y = x^2 e^{-x}$; $[0, 4]$

13. $f(x) = \dfrac{2x}{x^2 + 4}$; $[-2, 2]$

14. $y = x^2 e^{-x}$; $(0, \infty)$

15. $f(x) = \dfrac{2x}{x^2 + 4}$; $(0, 4)$

16. $f(x) = x + \dfrac{4}{x}$; $(0, \infty)$

17. Find two numbers whose difference is 8 and whose product is a minimum.

18. A homeowner plans to enclose a 243 square foot rectangular playground in his garden, with one side along the boundary of his property. His neighbor will pay for half of the cost on that side.

    (a) Draw a diagram to show the playground. Label with variables.

    (b) Determine a function (depends on one variable) for the cost to be minimized.

    (c) Determine the domain of the function.

    (d) Minimize the function.

    (e) Find the dimensions of the playground that will minimize the total cost.

19. A rancher wants to fence an area of 1,500 square yards in a rectangular field and then divide it in half with a fence parallel to one of the sides of the rectangle. The dividing fence costs two-thirds as much as the surrounding fence. How can he do this so as to minimize the cost of the fence?

20. A retail appliance store sells 800 refrigerators each year. It costs $25 to store one refrigerator for a year. To reorder, there is a fixed processing cost of $50, plus a $20 shipping fee for each refrigerator.

(a) Find the cost function, expressed as a function of x, for the number of refrigerators the store reorders each time (x is also the lot size the store must maintain).

(b) To minimize the inventory costs, how large is the lot size the store should maintain, and how many orders should be placed?

21. A clothing manufacturer is selling its blouses at a price of $18. The demand function for the blouses is $x = D(p) = 120 - 4p$, where p is the price of a blouse in dollars and x is in hundred items.

(a) Find the elasticity function, $E(p)$.

(b) Find the elasticity at $p = 18$. Should the manufacturer lower or raise the price to increase its revenue?

(c) Find the price of the blouses at which the revenue is greatest by solving the equation $E(p) = 1$.

22. A concert organizer learned from a market survey that when the admission price is $30, there is an average attendance of 800 people. For every $1 drop in price, there is a gain of 20 customers. Each customer spends an average of $5 on concessions. The concert hall has 1,000 seats.

(a) Find the demand function; expressing p, the price of each ticket, as a function of x, the average attendance.

(b) Find the revenue function $R(x)$.

(c) Find the price of each ticket the concert organizer should charge to maximize his revenue.

Chapter 3 Test (150 points)

SCORE _____

NAME _____

ANSWERS

1. (10pts) For $y = f(x) = 2x^3 + x^2 - x$:

(a) Find Δy and dy when $x = 1$ and $\Delta x = dx = 0.02$.

1. (a) $\Delta y =$ _____

$dy =$ _____

(b) Use linear approximation to approximate $f(1.02)$.

(b) $f(1.02) \approx$ _____

2. (16pts) A clothing manufacturer has the cost function
$$C(x) = 900 + 80x - \frac{5}{9}x^2 + \frac{1}{135}x^3 \text{ (in dollars)},$$
where x is the number of suits produced each week.
The revenue function for selling x suits is given by
$R(x) = 160x$ (in dollars).

(a) Find the total profit when 80 suits are produced and sold each week.

2. (a) _____

(b) Find the marginal profit when 80 suits are produced and sold each week.

(b) _____

(c) Use the information in parts (a) and (b) to estimate the total profit when 82 suits are produced and sold each week.

(c) _____

(d) Find the number of suits that should be produced and sold each week for the company to make the maximum profit.

(d) _____

3. (20pts) Given $f(x) = 2x^3 - x^4 + 1$:

(a) Find all the critical points
(if they exist) of the function.
List your answers in terms of ordered pairs.   3. (a) _____

(b) Apply the First-Derivative Test to
all the critical points. Illustrate the test with a diagram.

(c) Apply the Second-Derivative Test to
all the critical points.

(d) Find the relative minimum points
(if they exist) of the function.
List your answers in terms of ordered pairs.   (d) _____

(e) Find the relative maximum points
(if they exist) of the function.
List your answers in terms of ordered pairs.   (e) _____

(f) Find the inflection points
(if they exist) of the function.
List your answers in terms of ordered pairs.   (f) _____

(g) Sketch the graph of the function $f(x) = 2x^3 - x^4 + 1$.

CHAPTER 3 TEST   269

4. (10pts) Given the function $f(x) = \dfrac{1-3x}{x-2}$:

(a) Find the horizontal asymptote

   (if it exists) of the function.   4. (a) _____

(b) Find the vertical asymptote

   (if it exists) of the function.   (b) _____

(c) Sketch the graph of the function $f(x) = \dfrac{1-3x}{x-2}$.

For problems 5–8: find the absolute maximum and minimum values of the function, if they exist, over the indicated interval.

5. (8pts) $f(x) = 2x^3 - 3x^2 - 12x + 20$; $[-2, 2.1]$ 5. abs. max. = ____ at $x =$ ____
   Show your work.

                                                              abs. min. = ____ at $x =$ ____

6. (6pts) $f(x) = 2x^3 - 3x^2 - 12x + 20$; (same as #5) $[0, 1]$ 6. abs. max. = ____ at $x =$ ____
   Show your work.

                                                              abs. min. = ____ at $x =$ ____

7. (8pts) $f(x) = 4x^2 - x - 3$; $(-\infty, \infty)$
   Show your work.

7. abs. max. = ____ at $x =$ ____

   abs. min. = ____ at $x =$ ____

8. (8pts) $f(x) = 32x^2 + \frac{64}{x}$; $(0, \infty)$
   Show your work.

8. abs. max. = ____ at $x =$ ____

   abs. min. = ____ at $x =$ ____

9. (10pts) Minimize $Q = x^2 + 2y^2$, where $x - y = 4$.
   Show your work.

9. _____

CHAPTER 3 TEST

10. (12pts) A carpenter is building a rectangular room with a fixed perimeter of 54 ft.

    (a) What are the dimensions of the largest room that can be built?   10. (a) _____

    (b) What is the largest area?   (b) _____

    Show your work.

11. (15pts) A sporting goods store sells 1,000 basketballs per year. It cost $2 to store one basketball for 1 year. To reorder basketballs there is a fixed processing fee of $20, plus a $3 shipping fee for each basketball.
    To minimize inventory costs,
    (a) how large is lot size the store should maintain?
    (b) how many times per year should the store reorder?

    11. (a) _____

    (b) _____

    Show your work.

12. (15pts) A university is trying to determine what price to charge for football tickets. At a price of $6 per ticket, it averages 70,000 people per game. For every increase of $1, it loses 10,000 people from the average number. Every person at the game spends an average of $1.50 on concessions. (a) What price per ticket should be charged to maximize revenue? (b) How many people will attend at that price?
Show your work.

12. (a) _____

(b) _____

13. (12pts) A clothing manufacturer is selling its blouses at a price of $20. The demand function for the blouses is

$$x = D(p) = 400 - 4p,$$

where p is the price of a blouse in dollars and x is in items.

(a) Find the elasticity function, $E(p)$.

13. (a) _____

(b) Find the elasticity at $p = 20$. Should the manufacturer lower or raise the price to increase its revenue?

(b) _____

(c) Find the price of the blouses at which the revenue is greatest by solving the equation $E(p) = 1$.

(c) _____

Chapter 3 Project

Suppose that you are in your last semester at college and have entered the job market. Today you interviewed with the company Take-a-Hike Outdoor Supplies (THOS), and you are extremely interested in their opening in their product development section. As part of their job application process, they have invited you to participate in the preparation of a feasibility report on a new line of camping tents. They are interested in producing high-quality tents, but at reasonable costs. They are asking for your impressions based on a cost analysis of a dome tent. What follows is the tent worksheet that THOS provided. Your task is to write a report in which you respond to the issues raised on the worksheet and then to write a short summary of your findings and recommendations.

The Spherical Cap Dome Tents

Imaging making a tent in the shape of a spherical cap (a sphere with a lower portion sliced away by a plane). Assume we want the volume to be $6.8 M^3$, to sleep three or four people.

Make a sketch of this tent, identifying all appropriate variables.

Hint: If the tent has height h and bottom radius r and is from a sphere with radius R, then $0 \leq h \leq 2R$. Draw a diagram to illustrate the relationship among R, h, and r. This relationship differs according to whether you slice away less or more than half of the sphere (i.e., $0 \leq h \leq R$ or $R \leq h \leq 2R$).

The First Tent: The floor of the tent is cheaper material than the rest. Assume that the material making up the dome of the tent is two times as expensive per square meter as the material touching the ground.

(a) Determine and graph a function which gives the cost of the tent material as a function of an appropriate variable.

Hint: If the tent has height h and bottom radius r and is from a sphere with radius R, then the volume is

$$V = \pi R h^2 - \frac{\pi}{3} h^3 \text{ for } 0 \leq h \leq 2R,$$

and the surface area of the top is

$$S = 2\pi R h \text{ for } 0 \leq h \leq 2R.$$

Use the condition on the volume and the relationship among R, h, and r (refer to the hint given in part (a)) to reduce the number of variables in the cost function to one, that is, Cost $= C(h)$. Plot the function within its domain.

(b) Use the methods you have learned in class involving the critical number and derivative test to find the value of h at which the cost function reaches its absolute minimum.

(c) What should the dimensions (i.e., R, r, and h) of the tent be so that the cost of the material is a minimum? Draw a picture to show the dimensions.

(d) What are the areas of the dome and bottom materials used for this minimum cost tent?

(e) How practical would this tent be? If it is practical, then search for suitable prices and find the total cost.

The Second Tent: This time consider that the floor of the tent uses more expensive material than the rest. Assume that the material touching the ground is 1.5 times as expensive per square meter as the material making up the dome of the tent.

(a)–(e) Repeat the same process you used for the first tent.

The completed report should have the following parts: a cover page with appropriate title; a table of content page; an introduction including a description of the tent and its graph; two chapters; and a conclusion. Each chapter has five sections corresponding to the five parts (a)–(e).

Chapter 4

Integration

Introduction

In Chapters 2 and 3, our focus was on differentiation, the first of the two main building blocks of calculus, and how we can use derivatives to determine maximum and minimum values of functions on different intervals.

We now turn our focus to **integration**, the second main branch of calculus. The concept of integration has two seemingly very different forms: A **definite integral** finds the **accumulation** of certain quantity, or the **area** under the graph of a function; while an **indefinite integral** undoes the process of differentiation, and hence is also called an **antiderivative**. Throughout this chapter we will study how these two integrals are linked together by the **Fundamental Theorem of Calculus**, which is the glue that holds the two main branches of calculus, differential and integral calculus, together.

We start the chapter with the topic of area under a graph of a function, which leads to the linkage of the area and the function through differentiation. From there the Fundamental Theorem of Calculus is introduced. The second half of the chapter is devoted to the properties and techniques of integrating.

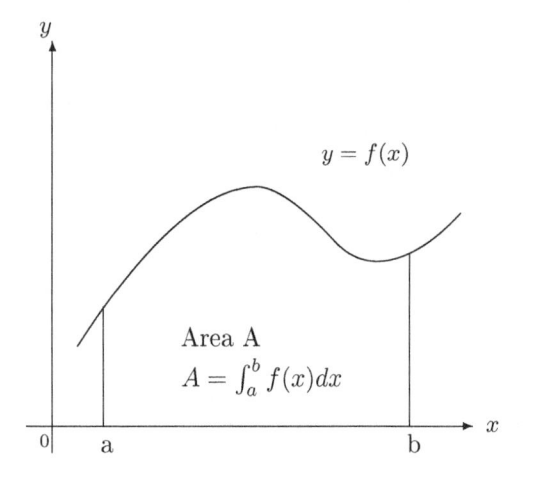

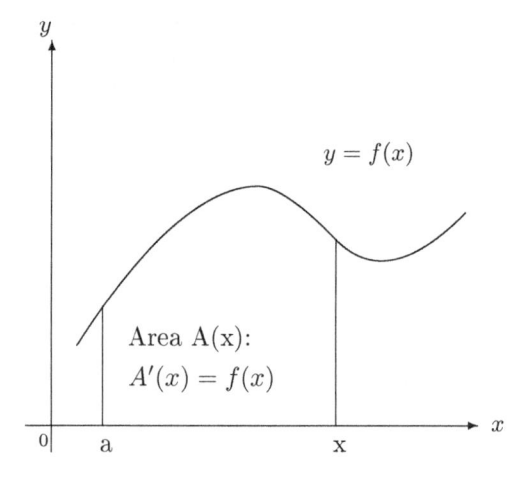

Study Objectives

1. Use anti-derivatives to determine areas.

2. Use numerical method(s) to estimate areas.

3. Use areas to determine the cumulative change of a function.

4. Apply the anti-derivative techniques, such as substitution or tables, to determine the anti-derivative of a function.

5. Understand the relationship among the anti-derivative of a function, the definite integral, and the areas formed by the function.

4.1 The Area under a Graph

Let us explore the area under a graph of a function, as illustrated in Figure 4.1.1 and Figure 4.1.2.

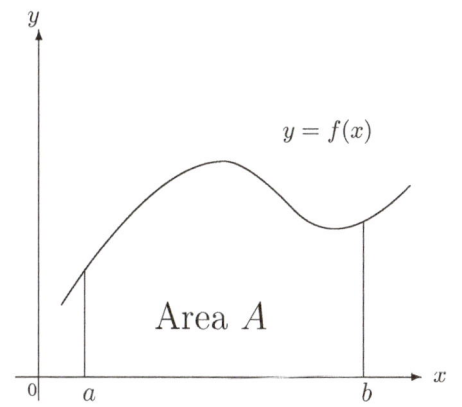

Figure 4.1.1: Area as a constant

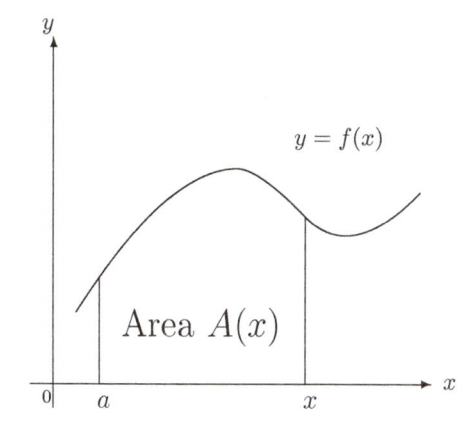

Figure 4.1.2: Area as a function of x

Observe that in Figure 4.1.1, the area A is a constant. The region is enclosed by the graph of the function and the lines of the x-axis, $x = a$, and $x = b$. It is generally described as "the area under the graph of $f(x)$ on the interval $[a, b]$," and it is assumed that the function is nonnegative, that is, $f(x) \geq 0$. In Figure 4.1.2, the right end of the interval is not specified, but changes with the value of x. Therefore the area is a function of x, and is denoted by $A(x)$. If x takes value b, then $A(b) = A$, exactly the area specified in Figure 4.1.1.

Questions about A, $A(x)$, and $f(x)$

1. What does the area represent?

2. What is the relationship between the area $A(x)$ and the function $f(x)$?

3. How can the exact area under the graph of $f(x)$ on the interval $[a, b]$ be calculated?

4. How do we estimate the area A?

Recall that in differential calculus, the tangent lines illustrate the graphic feature of the derivatives. Similarly, the area under a graph plays a very important role in integral calculus. We explore the answers to the four related questions presented here. In this section we focus on the first question, and introduce a primitive approach to the fourth question. The answers to remaining questions are revealed in other sections in this chapter.

What Does the Area Under a Graph Represent?

Suppose that you have been driving on an interstate highway for two and half hours at a constant speed of 65 mph. That is, the velocity v is

$$v(t) = 65, \quad 0 \leq t \leq 2.5.$$

A graph of the velocity function is in Figure 4.1.3.

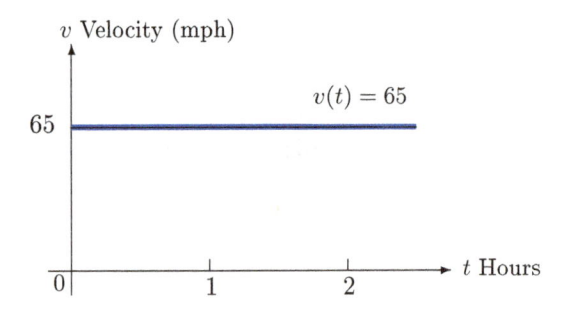

Figure 4.1.3: Graph of velocity

Suppose that you are interested in knowing the distance traveled in the following three periods: From 0.5 to 1 hour and from 1 to 2 hours, and for the whole 2.5 hours of the trip. Since the speed is a constant, the distance traveled from time t_1 to time t_2 is simply

Distance = Constant speed × Time passed = $65 \times (t_2 - t_1)$.

Hence from 0.5 to 1 hour, the distance traveled is $65(1 - 0.5) = 32.5$ miles (Figure 4.1.4); from 1 to 2 hours, the distance traveled is $65(2 - 1) = 65$ miles (Figure 4.1.5); and for the whole $2.5 \ (= 2.5 - 0)$ hours of the trip, the distance traveled is $65(2.5 - 0) = 162.5$ miles (Figure 4.1.6). We now compare these computed results with the areas under the graph of the velocity function on the corresponding time intervals:

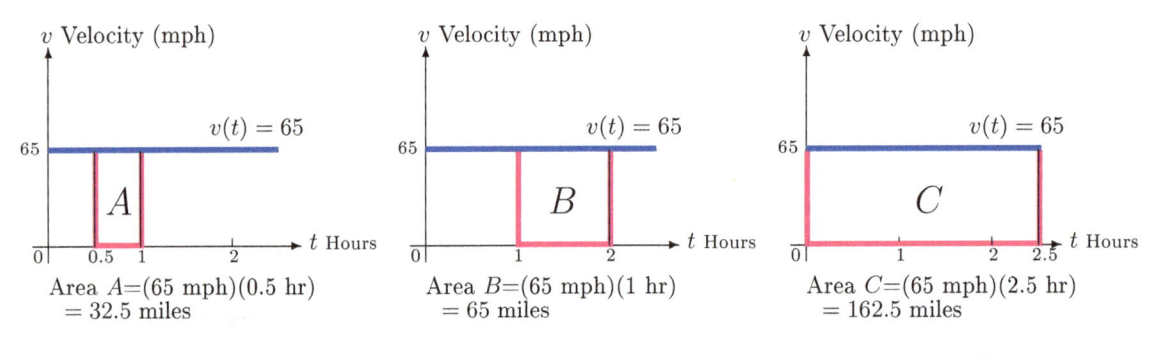

Area A=(65 mph)(0.5 hr)
   = 32.5 miles

Area B=(65 mph)(1 hr)
   = 65 miles

Area C=(65 mph)(2.5 hr)
   = 162.5 miles

Figure 4.1.4: Case I **Figure 4.1.5:** Case II **Figure 4.1.6:** Case III

It confirms that the three marked areas match the distances traveled in the three time intervals. Hence, we reach the conclusion that, if the graph represents the (positive) velocity function, then the area

4.1. THE AREA UNDER A GRAPH

under the graph on the interval $[a, b]$ represents the distance traveled in the time interval $a \leq t \leq b$. This conclusion is also true when the velocity is not a constant (Figure 4.1.7).

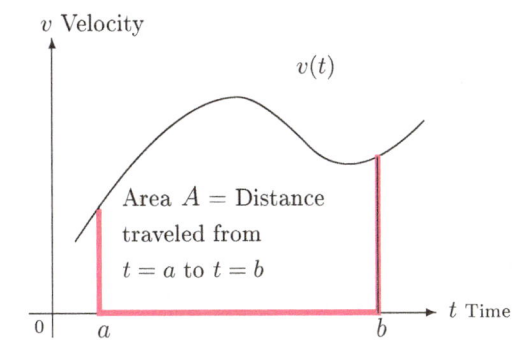

Figure 4.1.7: Distance vs. area

Example 1: A truck driver has been driving on an interstate highway for 2 hours at the following speeds: For the first 15 minutes he is speeding up linearly from 0 to 60 mph; for the next 1.5 hours his speed is 60 mph; and for the last 15 minutes he is slowing down linearly from 60 to 0 mph.

(a) Draw a graph that represents the velocity of the trip;

(b) Determine the distance traveled in the whole trip.

Solution (a) We draw the graph using the given information (Figure 4.1.8):

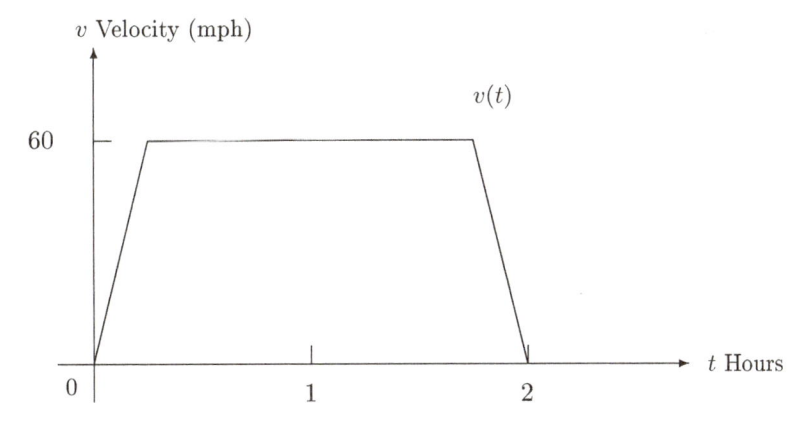

Figure 4.1.8: Graph of velocity

(b) Since the distance traveled is the area under the graph on the interval $[0, 2]$, we calculate the area (Figure 4.1.9). We can cut the involved region into three pieces: two triangles and a rectangle. Now we can determine the area of each region. The two triangles both have base $= 0.25$ and height $= 60$, hence each area is

$$\frac{1}{2}(60 \text{ mph})(0.25 \text{ hour}) = 7.5 \text{ miles}.$$

The rectangle has base $= 1.5$ and height 60, hence the area is

$$(60 \text{ mph})(1.5 \text{ hour}) = 90 \text{ miles}.$$

Therefore, the distance traveled is $7.5 + 7.5 + 90 = 105$ miles.

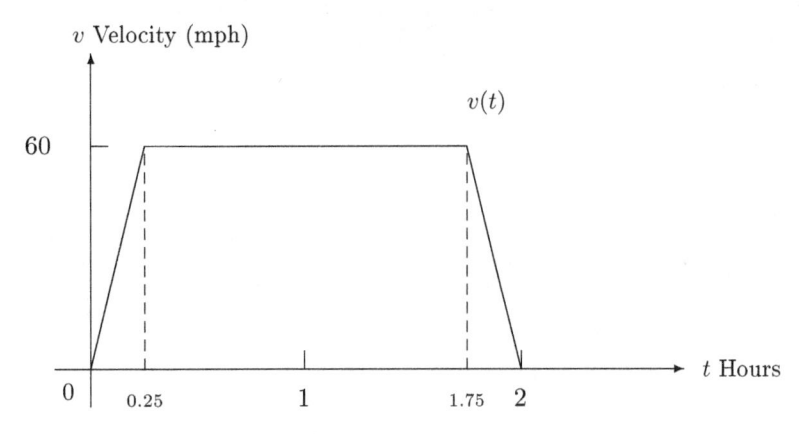

Figure 4.1.9: Area for distance

Before we generalize the meaning of the area under a graph, let us look at another example.

Example 2: A coffeehouse determines that it costs $0.8 to make each cup of coffee, hence the marginal cost function, in dollars, is

$$MC(x) = 0.8,$$

where x represents the number of cups of coffee the coffeehouse makes.

(a) Draw a graph that represents the marginal cost;

(b) Determine the total **variable cost** of making x cups of coffee, and compare the result with the area under the graph.

Solution (a) We draw the graph of the marginal cost according to the given function (Figure 4.1.10):

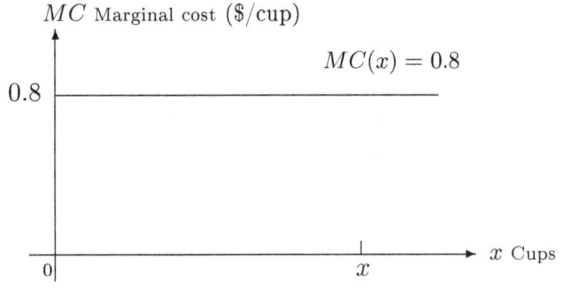

Figure 4.1.10: Graph of MC

(b) Since it costs $0.8 to make each cup of coffee, the total variable cost of making x cups of coffee is $0.8x$ dollars. It is the same as the area under the graph of the marginal cost function on the interval $[0, x]$ (Figure 4.1.11).

4.1. THE AREA UNDER A GRAPH

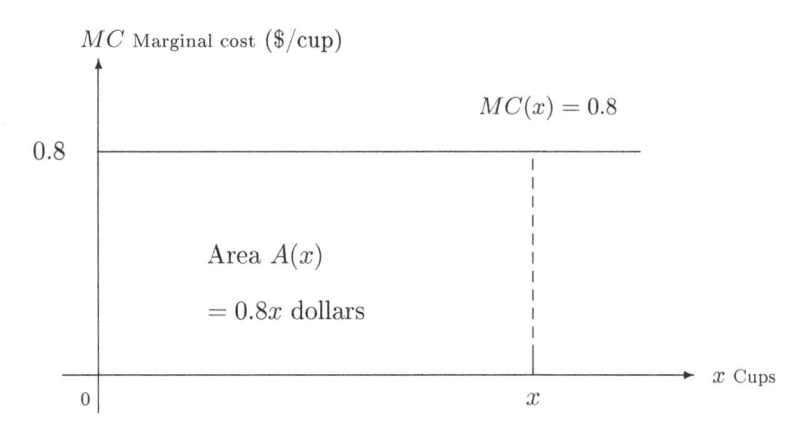

Figure 4.1.11: Area vs. variable cost

> **Reality Check**   *Incomplete information*
> When only the marginal cost function is given, we do not have information about the fixed cost.

Observing the two examples, we notice that the velocity represents the rate of change of the position function, and the marginal cost represents the rate of change of the total cost. Meanwhile, the area under the graph of the velocity represents the distance traveled, which is the *total change* of the position function. Similarly, the area under the graph of the marginal cost represents the total variable cost, which is the *total change* of the total cost function. In general we have the following conclusion:

> ### Area and Cumulative Change
> If a positive function f represents the *rate of change* of a quantity on an interval $[a, b]$, then the area under the graph of f on the interval $[a, b]$ represents the *total (or cumulative) change* of the quantity from the variable values a to b.

■ Estimate the Area Under a Curve

The discussions in the first part of this section establish the importance of calculating the area under a graph. In the examples, the regions take shapes whose areas are easy to calculate. When the graph is a general curve, the task of calculating the area becomes quite challenging. With the help of integral calculus introduced in the next few sections, we will be able to calculate some of these areas. However, in many applications, the area cannot be exactly calculated with integral formulas. Therefore being able to *estimate* the area is quite essential in practice. More advanced methods for estimating the areas can be found under the topic of numerical methods beyond this course. Here we introduce a rather intuitive and primitive method for the estimation, called the **left-rectangle approximation**. Although this is not an efficient approach, it will lead to the definition of the **definite integral** in Section 4.3.

Let us look at the area illustrated in Figure 4.1.12. The area is under the graph of $f(x) = x^2 + 1$ on the interval $[0, 2]$. The area cannot be calculated using elementary geometric formulas. We now illustrate how the area can be estimated. First, we divide the interval $[0, 2]$ into four equal subintervals, each subinterval with length $\Delta x = (2 - 0)/4 = 0.5$, as illustrated in Figure 4.1.13.

Figure 4.1.12: Area under curve

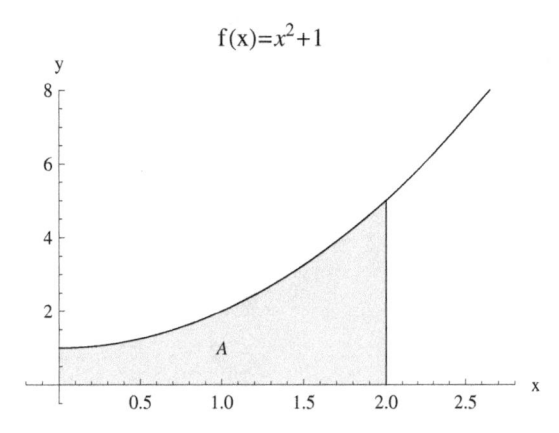

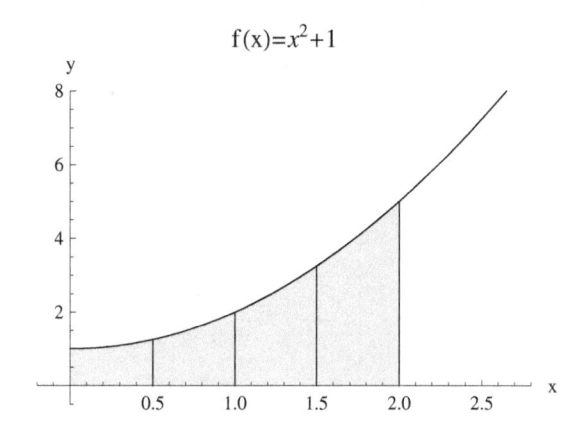

Figure 4.1.13: Divided region

For each subregion, we approximate the area by a rectangle, with the height of the rectangle equaling the function value at the *left end* of the subregion (Figure 4.1.14):

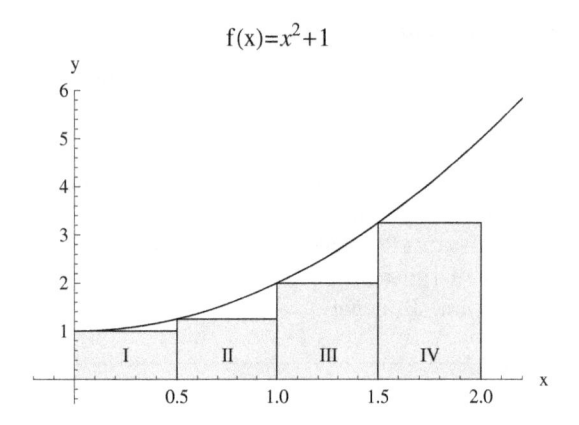

Figure 4.1.14: Approximated subregions

We now approximate the original area by the sum of the areas of the four rectangles. Since each rectangle has base $\Delta x = 0.5$, it becomes a common factor in the sum. The first rectangle has height $f(0) = 1$. The second rectangle has height $f(0.5) = 1.25$. Similarly, the third and fourth rectangles have heights $f(1) = 2$ and $f(1.5) = 3.25$. Therefore the sum of the areas of the four rectangles is

$$A \approx L_4 = \text{Area I} + \text{Area II} + \text{Area III} + \text{Area IV}$$

$$= f(0) \cdot \Delta x + f(0.5) \cdot \Delta x + f(1) \cdot \Delta x + f(1.5) \cdot \Delta x$$

$$= (f(0) + f(0.5) + f(1) + f(1.5))\Delta x$$

$$= (1 + 1.25 + 2 + 3.25)0.5 = 3.75.$$

It is obvious that the approximated area, 3.75, underestimates the exact area A. We can improve the approximation by using eight rectangles with equal bases. This time, we have each rectangle with base $\Delta x = (2 - 0)/8 = 0.25$, and the heights are listed in the calculations:

$$A \approx L_8 = (f(0) + f(0.25) + f(0.5) + f(0.75) + f(1) + f(1.25) + f(1.5) + f(1.75))\Delta x$$

$$= (1 + 1.0625 + 1.25 + 1.5625 + 2 + 2.5625 + 3.25 + 4.0625)0.25 = 4.1875.$$

The approximation is illustrated in Figure 4.1.15.

4.1. THE AREA UNDER A GRAPH

$$f(x) = x^2 + 1$$

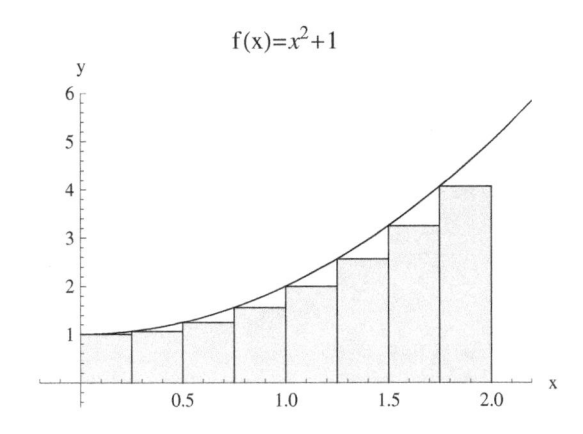

Figure 4.1.15: Approximated subregions

Reality Check   *Estimating areas*

Although it is true that we can improve our approximation by dividing the region into more and more equal parts and approximate each subregion by a rectangle, the process is rather tedious and labor-consuming without the help of a computing device. Therefore, in this section, we only apply the left-rectangle approximation using a small number of rectangles. Other more efficient methods do exist and can be found under the topic of numerical methods beyond this course.

Example 3: A truck driver has been driving on an interstate highway for 1 hour at a velocity represented by the function

$$v(t) = 240t - 240t^2, \quad 0 \leq t \leq 1,$$

where t is measured in hours and the velocity is measured in miles per hour.

(a) Draw a graph that represents the velocity of the trip;

(b) Estimate the distance traveled on the trip using four left-rectangles.

Solution (a) We draw the graph of the velocity function (Figure 4.1.16):

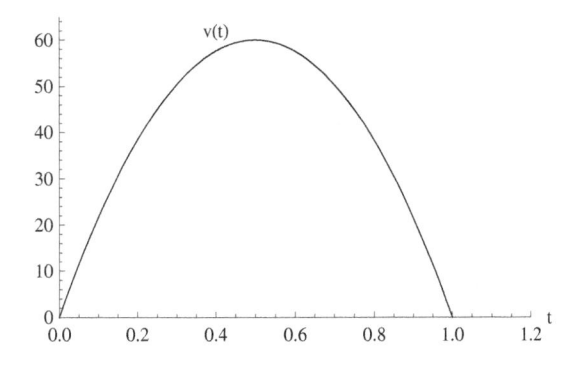

Figure 4.1.16: Graph of velocity

(b) Since the distance traveled is the area under the graph on the interval $[0, 1]$, we estimate the area by the left-rectangle approximation with four rectangles (Figure 4.1.17):

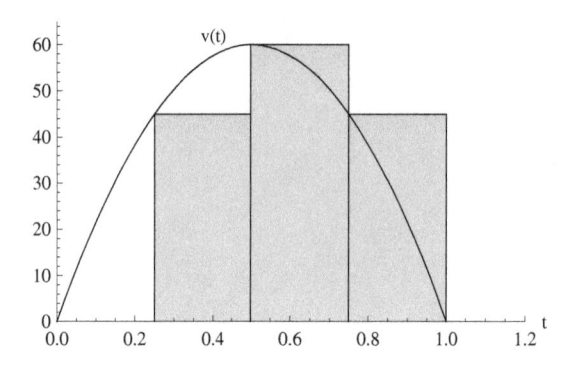

Figure 4.1.17: Approximated subregions

We have each rectangle with base $\Delta t = (1 - 0)/4 = 0.25$ hour, and the heights are listed in the calculations:

$$A \approx L_4 = (v(0) + v(0.25) + v(0.5) + v(0.75))\Delta t$$

$$= (0 + 45 + 60 + 45)0.25 = 32.5 \text{ miles.}$$

Therefore, the distance traveled in the trip is approximately 32.5 miles.

Example 4: A furniture manufacturer has determined that its monthly fixed cost is $500, and the marginal cost, in dollars, of producing the xth barstool is

$$MC(x) = 0.0000625x^2 - 0.05x + 25, 0 \leq x \leq 550.$$

(a) Draw a graph that represents the marginal cost function;

(b) Estimate the total monthly cost of producing 500 barstools using five left-rectangles.

Solution (a) We draw the graph of the marginal cost function (Figure 4.1.18):

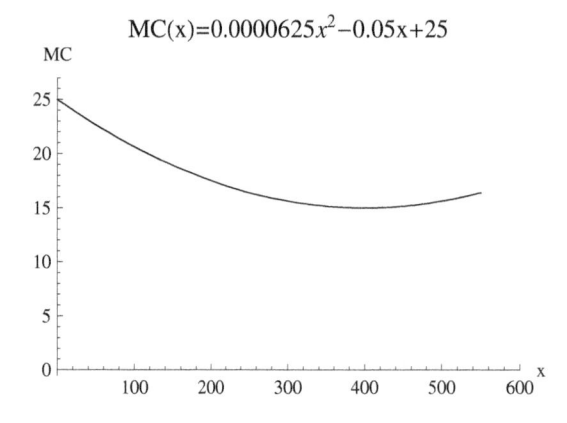

Figure 4.1.18: Graph of MC

(b) Since the total variable cost of producing 500 barstools is the area under the graph on the interval

4.1. THE AREA UNDER A GRAPH

[0, 500], we estimate the area by the left-rectangle approximation with five rectangles (Figure 4.1.19):

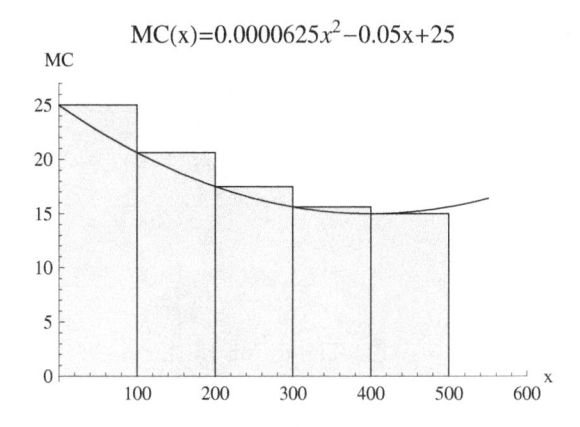

Figure 4.1.19: Approximated subregions

We have each rectangle with base $\Delta x = (500 - 0)/5 = 100$, and the heights are listed in the calculations:

$$A \approx L_5 = (MC(0) + MC(100) + MC(200) + MC(300) + MC(400))\Delta x$$

$$= (25 + 20.625 + 17.5 + 15.625 + 15)100 = 9375.$$

Therefore the total variable cost of producing 500 barstools is approximately $9,375. Since the total cost is the sum of the fixed cost and the total variable cost, the total monthly cost of producing 500 barstools is approximately $9,875.

Exercise Set 4.1

For Exercises #1 to #4, find the area under the graph on the following intervals:
(a) [0, 4]; (b) [−2, 2]; (c) [−2, 0]; (d) [−4, 4].

1. $y = f(x)$

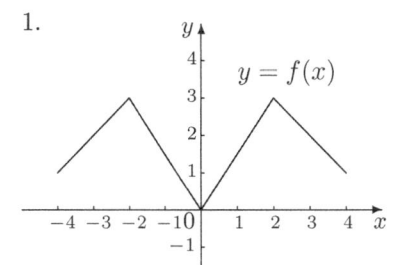

2. $y = f(x)$

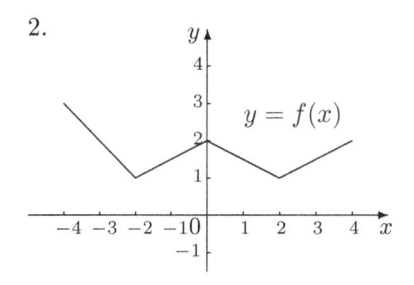

3. $y = g(x)$

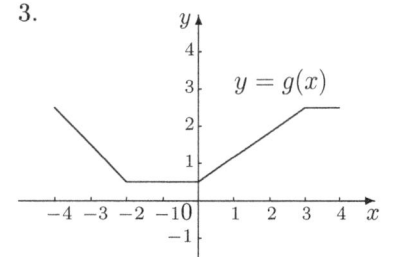

4. $y = g(x)$

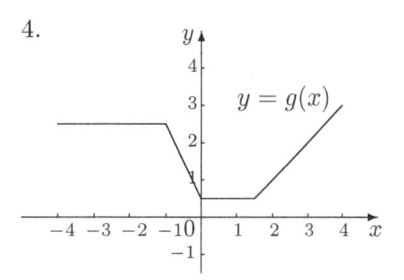

For Exercises #5 to #8: estimate the area under the graph on the interval $[-4, 4]$ using left-rectangle approximation with four rectangles.

5.

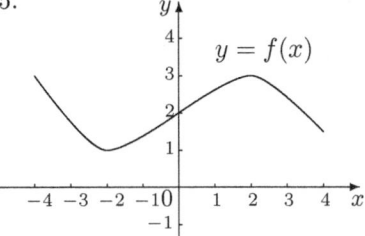

6.

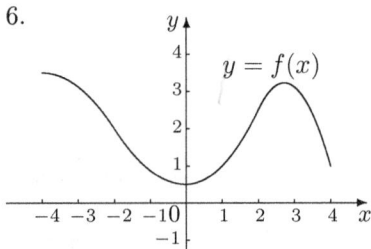

7.

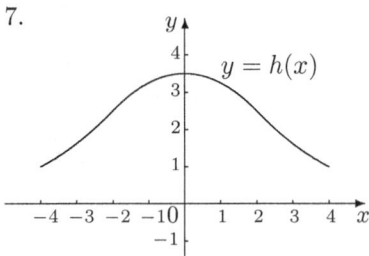

8.

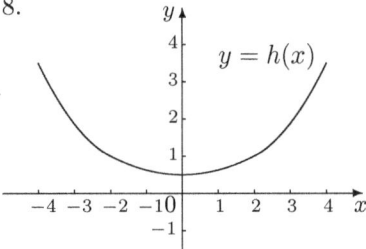

9. The graph of the velocity function $v = v(t)$ is given below. Determine the distance traveled during the following time intervals:

   (a) $[0, 4]$; (b) $[0, 2]$; (c) $[1, 3]$; (d) $[2, 4]$.

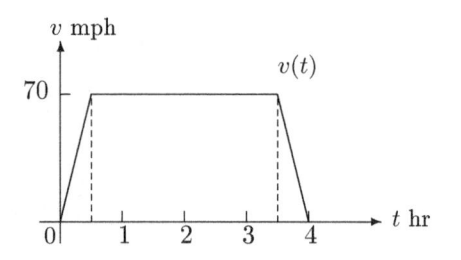

10. The graph of the velocity function $v = v(t)$ is given below. Determine the distance traveled during the following time intervals:

    (a) $[0, 8]$; (b) $[0, 6]$; (c) $[2, 4]$; (d) $[4, 8]$.

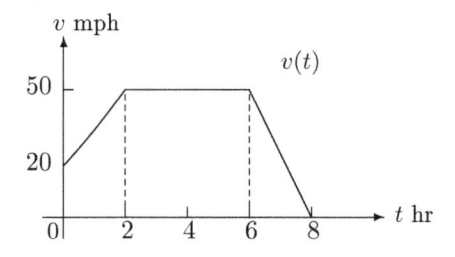

11. The graph of the marginal cost function $MC(x)$ of a product is given below. Assuming that the fixed cost is $600, determine the total cost of producing x items, where x is

    (a) 20; (b) 40; (c) 60; (d) 80.

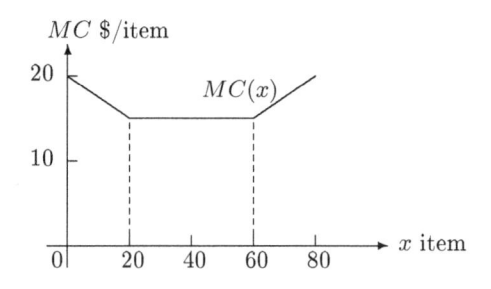

12. The graph of the marginal cost function $MC(x)$ of a product is given below. Assuming that the fixed cost is $400, determine the total cost of producing x items, where x is

    (a) 10; (b) 20; (c) 30; (d) 40.

4.1. THE AREA UNDER A GRAPH

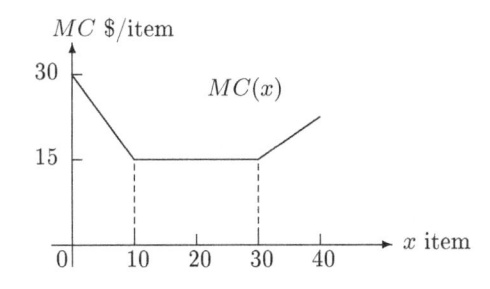

For Exercises #13 to #16, refer to the graph below and explain what the area A represents. Specify the units.

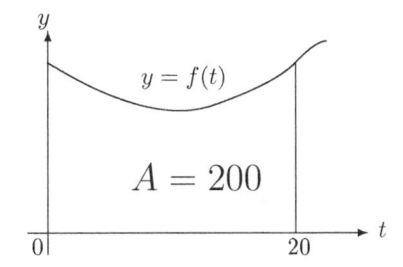

13. Where t represents the time in weeks, and y represents the rate of sales per week in thousands of dollars.

14. Where t represents the time in days, and y represents the cost of heating per day in dollars.

15. Where t represents the time in minutes, and y represents the velocity in meters per minute.

16. Where t represents the time in hours, and y represents the velocity in kilometers per hour.

For Exercises #17 to #20, refer to the graph below, explain what the area A represents. Specify the units.

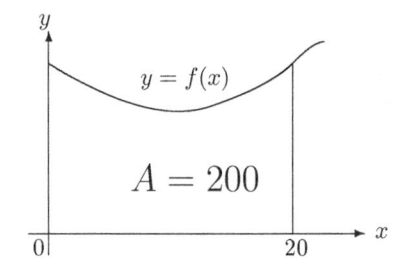

17. Where x represents the number of TV sets produced, and y represents the marginal cost in dollars per TV set.

18. Where x represents the number of chairs produced and sold, and y represents the marginal profit in dollars per chair. Assume $P(0) = 0$.

19. Where x represents shoes produced and sold in hundreds of pairs, and y represents the marginal profit in thousands of dollars per hundred pair. Assume $P(0) = 0$.

20. Where x represents screwdrivers produced in thousands, and y represents the marginal cost in hundreds of dollars per thousand screwdrivers.

━━ Applications ━━

21. A car has been traveling on an interstate highway for 10 minutes at a velocity represented by the function

    $$v(t) = 0.4 + 0.05t, \quad 0 \leq t \leq 10,$$

    where t is measured in minutes and the velocity is measured in miles per minute.

    (a) Draw a graph that represents the velocity of the trip;

    (b) Determine the distance traveled during the 10 minutes.

22. A car has been traveling on an interstate highway for 15 minutes at a velocity represented by the function

    $$v(t) = 1.2 - 0.02t, \quad 0 \leq t \leq 15,$$

    where t is measured in minutes and the velocity is measured in miles per minute.

(a) Draw a graph that represents the velocity of the trip;

(b) Determine the distance traveled during the 15 minutes.

23. A driver hits the brakes when the car is going at a speed of 30 miles per hour(i.e., 44 feet per second). The car stops in 10 seconds, and the velocity function is

$$v(t) = 44 - 4.4t, \quad 0 \leq t \leq 10,$$

where t is measured in seconds and the velocity is measured in feet per second.

(a) Draw a graph that represents the velocity of the stopping period;

(b) Determine the distance traveled during the 10 seconds.

24. A driver accelerates when the car is going at a speed of 30 miles per hour (i.e., 44 feet per second). The car reaches a speed of 60 miles per hour (i.e., 88 feet per second) in 20 seconds, and the velocity function is

$$v(t) = 44 + 2.2t, \quad 0 \leq t \leq 20,$$

where t is measured in seconds and the velocity is measured in feet per second.

(a) Draw a graph that represents the velocity of the acceleration period;

(b) Determine the distance traveled during the 20 seconds.

25. A shoe manufacturer has the marginal cost function $MC(x) = 20 - 0.02x$, $0 \leq x \leq 300$, (in dollars per pair), where x is the number of pairs of shoes produced each week. The fixed cost is $900.

(a) Draw a graph that represents the marginal cost;

(b) Determine the total cost of producing 250 pairs of shoes.

26. A clothing manufacturer has the marginal cost function

$MC(x) = 18 - 0.05x$, $0 \leq x \leq 100$, (in dollars per suit), where x is the number of suits produced each week. The fixed cost is $500.

(a) Draw a graph that represents the marginal cost;

(b) Determine the total cost of producing 60 suits.

27. The marginal cost function, in dollars per item, of producing the xth item of a certain brand of skateboard is given by $MC(x) = 54 - 0.05x$, $0 \leq x \leq 200$. The fixed cost is $1,200.

(a) Draw a graph that represents the marginal cost;

(b) Determine the total cost of producing 100 skateboards.

28. The marginal cost function, in dollars per item, of producing the xth item of a certain brand of barstool is given by $MC(x) = 15 - 0.06x$, $0 \leq x \leq 100$. The fixed cost is $200.

(a) Draw a graph that represents the marginal cost;

(b) Determine the total cost of producing 80 barstools.

29. A shoe manufacturer has the marginal revenue function for selling the xth pair of shoes $MR(x) = 90 - 0.5x$, $0 \leq x \leq 120$ (in dollars per pair).

(a) Draw a graph that represents the marginal revenue;

(b) Determine the total revenue from selling 100 pairs of shoes.

30. A clothing manufacturer has the marginal revenue function

$MR(x) = 30 - 0.2x$, $0 \leq x \leq 100$, (in dollars per suits), where x is the number of suits sold each week.

(a) Draw a graph that represents the marginal revenue;

(b) Determine the total revenue from selling 60 suits.

31. A car has been traveling on an interstate highway for 10 minutes at a velocity represented by the function

$$v(t) = 0.4 + 0.5t - 0.025t^2, \quad 0 \leq t \leq 10,$$

where t is measured in minutes and the velocity is measured in miles per minute.

(a) Draw a graph that represents the velocity of the trip;

4.1. THE AREA UNDER A GRAPH

(b) Estimate the distance traveled during the 10 minutes using the left-rectangle approximation with five rectangles.

(c) Does the approximation in part (b) overestimate or underestimate the exact distance?

32. A car has been traveling on an interstate highway for 15 minutes at a velocity represented by the function

$$v(t) = 1.2 - 0.05t + 0.001t^2, \quad 0 \le t \le 15,$$

where t is measured in minutes and the velocity is measured in miles per minute.

(a) Draw a graph that represents the velocity of the trip;

(b) Estimate the distance traveled during the 15 minutes using the left-rectangle approximation with five rectangles.

(c) Does the approximation in part (b) overestimate or underestimate the exact distance?

33. A driver hits the brakes when the car is going at a speed of 30 miles per hour (i.e., 44 feet per second). The car stops in 10 seconds, and the velocity function is

$$v(t) = 44 - 0.44t^2, \quad 0 \le t \le 10,$$

where t is measured in seconds and the velocity is measured in feet per second.

(a) Draw a graph that represents the velocity of the stopping period;

(b) Estimate the distance traveled during the 10 seconds using the left-rectangle approximation with four rectangles.

34. A driver accelerates when the car is going at a speed of 30 miles per hour (i.e., 44 feet per second). The car reaches a speed of 60 miles per hour (i.e., 88 feet per second) in 20 seconds, and the velocity function is

$$v(t) = 44 + 0.11t^2, \quad 0 \le t \le 20,$$

where t is measured in seconds and the velocity is measured in feet per second.

(a) Draw a graph that represents the velocity of the acceleration period;

(b) Estimate the distance traveled during the 20 seconds using the left-rectangle approximation with four rectangles.

35. The marginal cost function, in dollars per item, for producing the xth item of a certain brand of skateboards is given by $MC(x) = 84 - 0.9x + 0.008x^2$, $0 \le x \le 100$. The fixed cost is $800.

(a) Draw a graph that represents the marginal cost.

(b) Estimate the total cost of producing 80 skateboards using the left-rectangle approximation with four rectangles.

36. The marginal cost function, in dollars per item, for producing the xth item of a certain brand of barstool is given by $MC(x) = 20 - 0.5\sqrt{x}$, $0 \le x \le 100$. The fixed cost is $200.

(a) Draw a graph that represents the marginal cost;

(b) Estimate the total cost of producing 100 barstools using the left-rectangle approximation with five rectangles.

37. A mutual fund has a growth rate recorded each year, expressed by the function $r(t) = 200e^{0.04t}$, $0 \le t \le 10$ (in thousands of dollars per year), where t is the number of years passed from year 1997.

(a) Draw a graph that represents the growth rate;

(b) Estimate the total change in the fund over 10 years using the left-rectangle approximation with five rectangles.

(c) Does the approximation in part (b) overestimate or underestimate the total change?

38. A mutual fund has a growth rate recorded each year, expressed by the function $r(t) = 200e^{-0.04t}$, $0 \le t \le 6$ (in thousands of dollars per year), where t is the number of years passed from year 2000.

(a) Draw a graph that represents the growth rate;

(b) Estimate the total change in the fund over 6 years using the left-rectangle approximation with three rectangles.

(c) Does the approximation in part (b) overestimate or underestimate the total change?

39. A shoe manufacturer has the marginal profit function for producing and selling the xth pair of shoes $MP(x) = 90 - 10\ln(x+1)$, $0 \le x \le 120$ (in dollars per pair).

(a) Draw a graph that represents the marginal profit;

(b) Estimate the total profit from producing and selling 100 pairs of shoes using the left-rectangle approximation with four rectangles.

40. A clothing manufacturer has the marginal profit function

$MP(x) = 30 - 2x^{1/3}$, $0 \le x \le 100$, (in dollars per suit), where x is the number of suits produced and sold each week.

(a) Draw a graph that represents the marginal profit;

(b) Estimate the total profit from producing and selling 60 suits using the left-rectangle approximation with three rectangles. Assume $P(0) = 0$.

4.2 Areas and Antiderivatives

In this section, we focus on the area $A(x)$, as illustrated in Figure 4.2.1, and seek the answer to the question: How are the two functions, $A(x)$ and $f(x)$, related? This relationship will lead to the topic of **antiderivatives**, with notations and basic formulas that are used frequently in the remainder of this chapter and in Chapter 5.

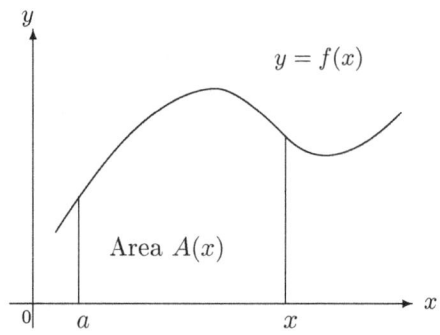

Figure 4.2.1: $A(x)$ vs. $f(x)$

In the previous section two types of application functions occurred in the examples. Let us reexamine these examples.

Reality Check *Distance vs. positive velocity*

In Example 1.1 of Section 4.1, the graphed function is the positive velocity, and the area under the graph represents the distance traveled during the time interval. Notice that if an object is moving in one direction, then the distance traveled from time a to time b can be expressed by the difference of the position function values, $s(b) - s(a)$. Observe that the velocity and the position have the relationship $v(t) = s'(t)$.

Reality Check *Total variable cost vs. marginal cost*

In Example 1.2 of Section 4.1, the graphed function is the marginal cost, and the area under the graph represents the total variable cost. Notice that the total variable cost of producing x items can be obtained by $C(x) - C(0)$, where $C(x)$ is the total cost function. Observe again that the marginal and total costs have the relationship $MC(x) = C'(x)$.

Before we claim that this relationship holds for every case, let us observe a few more examples.

4.2. AREAS AND ANTIDERIVATIVES

Example 1: Let $A_1(x)$ and $A_2(x)$ be the areas under the graph of the constant function $f(x) = 3$ on the intervals $[1, x]$ and $[2, x]$ respectively.

(a) Draw graphs to illustrate the two areas and determine the formulas for $A_1(x)$ and $A_2(x)$.

(b) Find the derivatives $A_1'(x)$ and $A_2'(x)$ and compare them with $f(x)$.

Solution (a) The areas are illustrated in Figures 4.2.2 and 4.2.3.

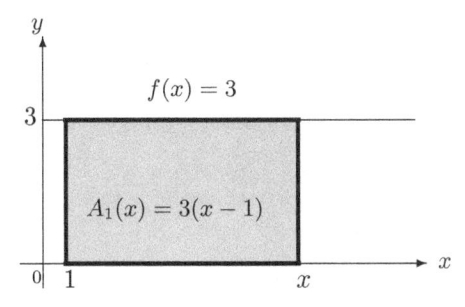

Figure 4.2.2: $A_1(x)$ vs. $f(x)$

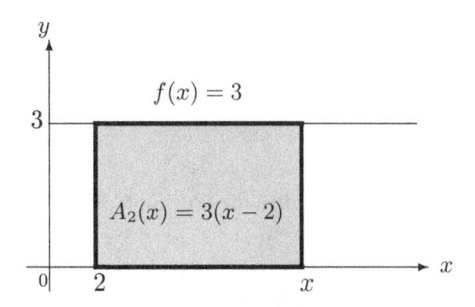

Figure 4.2.3: $A_2(x)$ vs. $f(x)$

Since both regions are rectangles, their areas are easy to calculate. The first region has base $(x-1)$ and height 3, hence the area is

$$A_1(x) = 3(x-1) = 3x - 3.$$

The second region has base $(x-2)$ and height 3, hence the area is

$$A_2(x) = 3(x-2) = 3x - 6.$$

(b) Although $A_1(x)$ and $A_2(x)$ are different, they only differ by a constant; hence they have the same derivative:

$$A_1'(x) = A_2'(x) = 3 = f(x).$$

Example 2: Let $A_3(x)$ and $A_4(x)$ be the areas under the graph of the linear function $f(x) = 2x$ on the intervals $[1, x]$ and $[2, x]$ respectively. Repeat the process listed in Example 1.

Solution (a) The areas are illustrated in Figures 4.2.4 and 4.2.5.

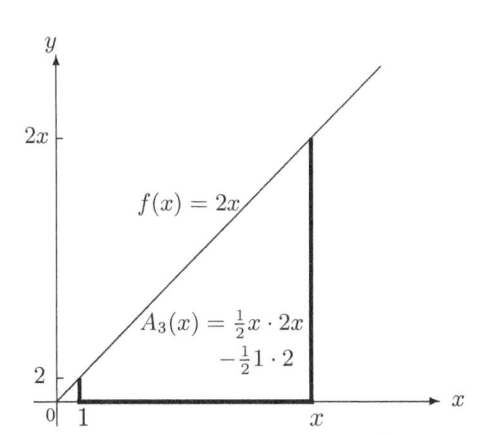

Figure 4.2.4: $A_3(x)$ vs. $f(x)$

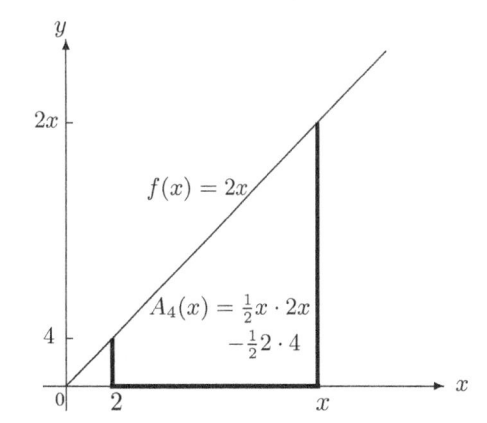

Figure 4.2.5: $A_4(x)$ vs. $f(x)$

Both regions can be viewed as cutting a small triangle from the large triangle with base x and height

$2x$. In the first region, the small triangle has base 1 and height 2, hence the area is

$$A_3(x) = \frac{1}{2}x \cdot 2x - \frac{1}{2}1 \cdot 2 = x^2 - 1.$$

In the second region, the small triangle has base 2 and height 4, hence the area is

$$A_4(x) = \frac{1}{2}x \cdot 2x - \frac{1}{2}2 \cdot 4 = x^2 - 4.$$

(b) As in Example 1, although A_3 and A_4 are different, they only differ by a constant and therefore have the same derivative:

$$A_3'(x) = A_4'(x) = 2x = f(x).$$

We now consider a general case. Let $A(x)$ represent the area under a nonnegative continuous function $f(x)$ over the interval $[a, x]$, as illustrated in Figure 4.2.1. We want to determine $A'(x)$. By definition,

$$A'(x) = \lim_{h \to 0} \frac{A(x+h) - A(x)}{h}.$$

Assume that $h > 0$, then the areas $A(x+h)$ and $A(x+h) - A(x)$ are illustrated in Figures 4.2.6 and 4.2.7:

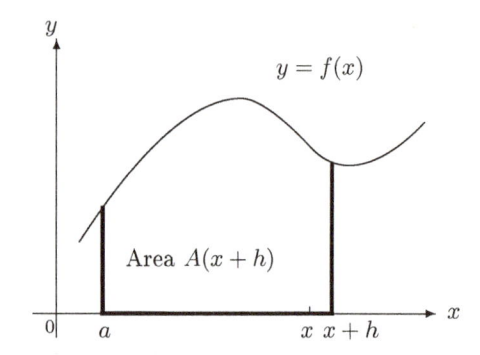

Figure 4.2.6: $A(x+h)$

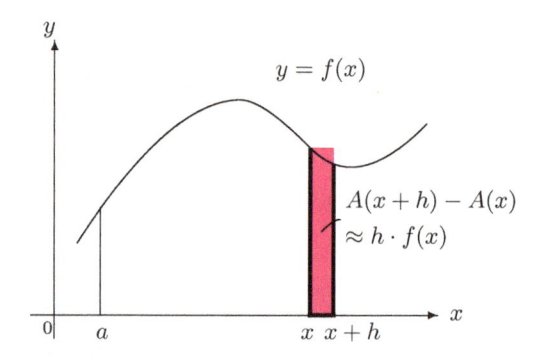

Figure 4.2.7: $A(x+h) - A(x)$

When h is near zero, the area $A(x+h) - A(x)$ is approximately the area of the rectangle with base h and height $f(x)$. That is,

$$A(x+h) - A(x) \approx h \cdot f(x), \quad \text{hence} \quad \frac{A(x+h) - A(x)}{h} \approx f(x).$$

Therefore

$$A'(x) = \lim_{h \to 0} \frac{A(x+h) - A(x)}{h} = f(x).$$

We state this remarkable conclusion in the following theorem:

Theorem 1   *The Derivative of $A(x)$*

Assume that f is a positive continuous function on an interval $[a, b]$, and $A(x)$ is the area under the graph of f on the interval $[a, x]$, where $a < x < b$. Then $A(x)$ is differentiable and $A'(x) = f(x)$.

Theorem 1 suggests a promising approach for determining the exact area under a curved graph. If we can find a function $F(x)$ such that $F'(x) = f(x)$, we may have obtained the area function $A(x)$. However, as the aforementioned examples illustrate, different functions can have the same derivative.

4.2. AREAS AND ANTIDERIVATIVES

This is not surprising, since in differentiation, a constant function has the derivative zero, therefore it is impossible to "recover" the constant without knowing other conditions. It turns out that a constant is the only difference permitted for two functions to have the same derivative. We state this property in the following theorem:

Theorem 2   *Functions Having the Same Derivatives*

Assume that F and G are two differentiable functions on an interval and that $F(x) = G(x)$ for all x in the interval. Then

$$F(x) = G(x) + C,$$

where C is a constant.

Let us return to the task introduced in Section 4.1: Given a function $f(x)$, we want to determine $A(x)$, the area under the graph of $f(x)$. By Theorem 1 we know that $A'(x) = f(x)$. If we can find a function $F(x)$ such that $F(x) = f(x)$ (e.g., if $f(x) = 2x$, then $F(x) = x^2$ is such a candidate), then by Theorem 2 we know that $A(x) = F(x) + C$ (C can be zero). How can we find the constant C? We find the answer in Section 4.3. Our focus is now turned to the process of finding such a $F(x)$. We need to introduce some new terminologies to efficiently describe the process of finding a function whose derivative is given.

▬ Antiderivatives, Integrals, and Integration

Given a function $f(x)$, we call the function $F(x)$ an **antiderivative** of $f(x)$ if $F(x)$ satisfies $F(x) = f(x)$. Since there are many antiderivatives to the same function $f(x)$, all differing by a constant, we denote the general antiderivative

$$F(x) + C,$$

where C is an arbitrary constant.

Example 3: Find the (general) antiderivative of $f(x) = 3x^2$.

Solution It is easy to check that $y = x^3$ satisfies $y' = 3x^2 = f(x)$. Therefore the general antiderivative of $f(x) = 3x^2$ is

$$F(x) = x^3 + C.$$

Given a function $f(x)$, the process of finding its antiderivative(s) is called **antidifferentiating**. However, it is also often called **integrating**, and the notation,

$$\int f(x)\, dx,$$

called an **indefinite integral**, is used to denote the antiderivative of $f(x)$. If $F(x)$ is an antiderivative of $f(x)$, then we denote

$$\int f(x)\, dx = F(x) + C.$$

For example, in Example 3, we can write

$$\int 3x^2\, dx = x^3 + C.$$

It is read "the indefinite integral of $3x^2$ is $x^3 + C$."

Let us observe the new notation in the equation $\int f(x)\,dx = F(x)+C$. The symbol "$\int$" is called an **integral sign**. The function $f(x)$ is called the **integrand**. The meaning of dx in the integral is clear in Section 4.3. Currently we treat it as an indicator that the integral is a function of x. The constant C is called the **constant of integration**. Since C can be any fixed value, the use of the word "indefinite" is appropriate. If we can determine the indefinite integral of a function $f(x)$, we say that $f(x)$ is **integrable**.

Example 4: Determine the indefinite integral $\int x^5\,dx$.

Solution It is easy to check that $y = x^6/6$ satisfies $y' = x^5$. Therefore, we have

$$\int x^5\,dx = \frac{1}{6}x^6 + C.$$

Example 5: Determine the indefinite integral $\int e^{2x}\,dx$.

Solution Applying the Extended Exponential Rule we can check that $y = e^{2x}/2$ satisfies $y' = e^{2x}$. Therefore, we have

$$\int e^{2x}\,dx = \frac{1}{2}e^{2x} + C.$$

Since we have already established the basic formulas of differentiation for all the functions that occur in this course, the task of establishing the basic formulas of integration becomes straightforward. Let us take a look at the integration of a power function $f(x) = x^r$ where r is any real number except -1. Since the Power Rule in differentiation reduces the power by 1, the integration should increase the power by 1. Hence, we can try $y = x^{r+1}$. The derivative is

$$\frac{d}{dx}\left(x^{r+1}\right) = (r+1)x^r.$$

Now we realize that the coefficient $(r+1)$ is not in the formula of $f(x)$. We divide this coefficient from $y = x^{r+1}$ (remember that $r \neq -1$) and try again:

$$\frac{d}{dx}\left(\frac{x^{r+1}}{r+1}\right) = \frac{1}{r+1}\frac{d}{dx}\left(x^{r+1}\right) = \frac{1}{r+1}(r+1)x^r = x^r = f(x).$$

Therefore, we obtain the **Integration Formula for Powers**:

$$\int x^r\,dx = \frac{1}{r+1}x^{r+1} + C, \quad \text{provide } r \neq -1$$

> **Reality Check** *The integration formula for powers*
>
> The Integration Formula for Powers is important for you to remember, since it is the most often used formula in integration. Sometimes a formula needs to be rewritten before the rule is applied.

Example 6: Determine the following indefinite integrals:

(a) $\int x^7\,dx$ (b) $\int \sqrt{x}\,dx$ (c) $\int \frac{1}{x^2}\,dx$

Solution (a) Applying the Integration Formula for Powers with $r = 7$, we get

$$\int x^7\,dx = \frac{1}{8}x^8 + C.$$

(b) Since $\sqrt{x} = x^{1/2}$, we apply the Integration Formula for Powers with $r = 1/2$. Then the power of the antiderivative is $r + 1 = \frac{1}{2} + 1 = \frac{3}{2}$, and the coefficient is $\frac{1}{r+1} = \frac{2}{3}$:

$$\int \sqrt{x}\, dx = \int x^{\frac{1}{2}}\, dx = \frac{1}{\frac{1}{2}+1} x^{\frac{1}{2}+1} + C = \frac{2}{3} x^{\frac{3}{2}} + C.$$

(c) Since $\frac{1}{x^2} = x^{-2}$, we apply the Integration Formula for Powers with $r = -2$. Then the power of the antiderivative is $r + 1 = (-2) + 1 = -1$, and the coefficient is $\frac{1}{r+1} = \frac{1}{-1} = -1$:

$$\int \frac{1}{x^2}\, dx = \int x^{-2}\, dx = \frac{1}{(-2)+1} x^{(-2)+1} + C$$

$$= \frac{1}{-1} x^{-1} + C = -\frac{1}{x} + C.$$

We list some of the basic integration formulas. Each formula can be verified by differentiating the right side of the equation to obtain the same function as the integrand. Furthermore, recall that, in addition to the basic rules for differentiation, in Section 2.5 we also have the basic properties of differentiation. These properties are carried over to integration.

Theorem 3 *Basic Integration Formulas*

1. (**Integration Formula for Constants**) $\quad \int k\, dx = kx + C \quad$ (k is a constant)

2. (**Integration Formula for Powers**) $\quad \int x^r\, dx = \frac{1}{r+1} x^{r+1} + C$, provided that $r \ne -1$

3. (**Integration Formula for $1/x$**) $\quad \int \frac{1}{x}\, dx = \ln |x| + C$, or $\int \frac{1}{x}\, dx = \ln x + C$, if $x > 0$

   (We generally assume that $x > 0$)

4. (**Integration Formula for ae^{kx}**) $\quad \int ae^{kx}\, dx = \frac{a}{k} e^{kx} + C$

Theorem 4 *Basic Integration Properties*

1. (**Constant Multiple Property**) $\quad \int kf(x)\, dx = k \int f(x)\, dx \quad (k \ne 0)$

   (The integral of a nonzero constant times a function is the constant times the integral of the function.)

2. (**Sum and Difference Properties**) $\quad \int [f(x) \pm g(x)]\, dx = \int f(x)\, dx \pm \int g(x)\, dx$

   (The integral of a sum or a difference of two functions is the sum or the difference of the integrals.)

Example 7: Determine the indefinite integral $\int (3x^4 + 6x^2 - 2x + 8)\, dx$.

Solution Applying the basic formulas and properties of integration, we obtain

$$\int (3x^4 + 6x^2 - 2x + 8)\, dx = 3\int x^4\, dx + 6\int x^2\, dx - 2\int x\, dx + \int 8\, dx$$

$$= 3\left(\frac{1}{5}x^5\right) + 6\left(\frac{1}{3}x^3\right) - 2\left(\frac{1}{2}x^2\right) + 8x + C = \frac{3}{5}x^5 + 2x^3 - x^2 + 8x + C.$$

Example 8: Determine the indefinite integral $\int \left(3e^{-2x} - \dfrac{5}{\sqrt[3]{x^2}}\right) dx$.

Solution We first apply the basic properties of integration, then we rewrite the second term to a power formula and apply the basic integration formulas. We obtain

$$\int \left(3e^{-2x} - \frac{5}{\sqrt[3]{x^2}}\right) dx = 3\int e^{-2x}\, dx - 5\int \frac{1}{\sqrt[3]{x^2}}\, dx$$

$$= 3\int e^{-2x}\, dx - 5\int x^{-\frac{2}{3}}\, dx$$

$$= \frac{3}{-2}e^{-2x} - 5 \cdot \frac{1}{\left(-\frac{2}{3}\right)+1} x^{-\frac{2}{3}+1} + C$$

$$= -\frac{3}{2}e^{-2x} - 5 \cdot \frac{1}{1/3} x^{\frac{1}{3}} + C = -\frac{3}{2}e^{-2x} - 15\sqrt[3]{x} + C.$$

Example 9: Determine the indefinite integral $\int \left(\dfrac{2}{x} + \dfrac{4}{x^3} - \dfrac{3}{x^5}\right) dx$.

Solution We first apply the basic properties of integration, then we rewrite the second and third terms to power formulas and apply the basic integration formulas. We obtain

$$\int \left(\frac{2}{x} + \frac{4}{x^3} - \frac{3}{x^5}\right) dx = 2\int \frac{1}{x}\, dx + 4\int \frac{1}{x^3}\, dx - 3\int \frac{1}{x^5}\, dx$$

$$= 2\int \frac{1}{x}\, dx + 4\int x^{-3}\, dx - 3\int x^{-5}\, dx$$

$$= 2\ln x + 4 \cdot \frac{1}{(-3)+1} x^{-3+1} - 3 \cdot \frac{1}{(-5)+1} x^{-5+1} + C$$

$$= 2\ln x - 2x^{-2} + \frac{3}{4}x^{-4} + C = 2\ln x - \frac{2}{x^2} + \frac{3}{4x^4} + C.$$

Specific Antiderivatives

We have seen that any integrable function has infinitely many antiderivatives. For example, $f(x) = 2x$ has a general antiderivative form of $F(x) = x^2 + C$. That means $x^2 - 3$, $x^2 - 1$, x^2, $x^2 + 2$, $x^2 + 3$, ... are all antiderivatives of $2x$ (Figure 4.2.8).

Suppose that we look for an antiderivative of $2x$ with a specified value at a certain point, then we will find that there is only one such antiderivative. For example, if we look for a function $F(x)$ such that $F'(x) = 2x$ and $F(1) = 3$, then from

$$F(x) = x^2 + C$$

we have

$$(1)^2 + C = F(1) = 3,$$

therefore, $C = 2$, and the specific antiderivative is

$$F(x) = x^2 + 2.$$

4.2. AREAS AND ANTIDERIVATIVES

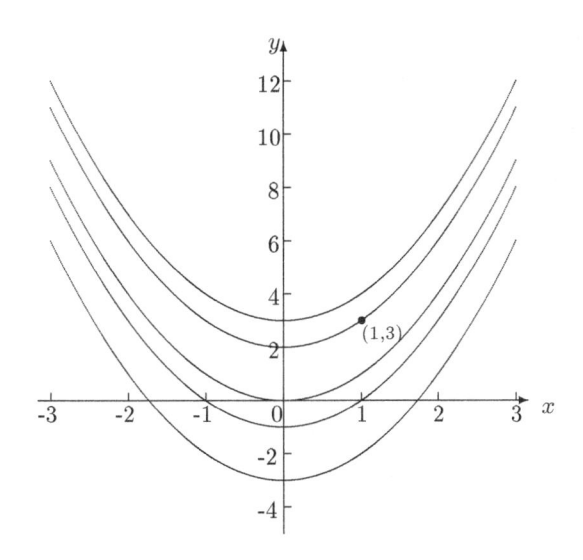

Figure 4.2.8: Infinitely many antiderivatives of $f(x) = 2x$

Example 10: Determine function F such that

$$F'(x) = x^3 \quad \text{and} \quad F(2) = 1.$$

Solution We first find $F(x)$ by integration:

$$F(x) = \int x^3 \, dx = \frac{1}{4}x^4 + C.$$

Then we use the condition $F(2) = 1$ to find C:

$$\frac{1}{4}(2)^4 + C = F(2) = 1$$

$$4 + C = 1, \quad \text{hence} \quad C = -3.$$

Therefore

$$F(x) = \frac{1}{4}x^4 - 3.$$

Applications

Example 11: Suppose that the acceleration function is given by $a(t) = 6t$, the initial velocity is $v(0) = 5$ and the initial position is $s(0) = 12$. Determine the position function $s(t)$.

Solution Since $s(t)$, $v(t)$, and $a(t)$ have the relationships

$$s'(t) = v(t);$$

$$s''(t) = v'(t) = a(t),$$

we find $v(t)$ by integrating $a(t)$:

$$v(t) = \int a(t)\, dt = \int 6t\, dt = 3t^2 + C_1.$$

Then we use the condition $v(0) = 5$ to find C_1:

$$3(0)^2 + C_1 = v(0) = 5, \quad \text{hence} \quad C_1 = 5.$$

Therefore

$$v(t) = 3t^2 + 5.$$

Next we find $s(t)$ by integrating $v(t)$:

$$s(t) = \int v(t)\, dt = \int (3t^2 + 5)\, dt = t^3 + 5t + C_2.$$

Then we use the condition $s(0) = 12$ to find C_2:

$$(0)^3 + 5(0) + C_2 = s(0) = 12, \quad \text{hence} \quad C_2 = 12.$$

Therefore

$$s(t) = t^3 + 5t + 12.$$

Example 12: A container firm receives an order for 100 small containers to be used for disaster relief. The containers are going to be dropped from a helicopter 500 feet above the ground. The containers in inventory can withstand an impact speed of 200 feet per second. Should the firm use these containers, or should it make stronger ones?

Solution We need to find the impact velocity when the containers hit the ground. We begin with the equation of acceleration for a free-falling object: $a(t) = -32$ ft/sec^2. We know that $v(0) = 0$ feet per second and $s(0) = 500$ feet.

We find $v(t)$ by integrating $a(t)$:

$$v(t) = \int a(t)\, dt = \int -32\, dt = -32t + C_1.$$

Then we use the condition $v(0) = 0$ to find that $C_1 = 0$. Therefore

$$v(t) = -32t.$$

Next we find $s(t)$ by integrating $v(t)$:

$$s(t) = \int v(t)\, dt = \int (-32t)\, dt = -16t^2 + C_2.$$

Then we use the condition $s(0) = 500$ to find C_2:

$$-16(0)^2 + C_2 = s(0) = 500, \quad \text{hence} \quad C_2 = 500.$$

Therefore

$$s(t) = -16t^2 + 500.$$

When the containers hit the ground, we have $s(t) = 0$. Solving for t from the equation

$$-16t^2 + 500 = 0$$

4.2. AREAS AND ANTIDERIVATIVES

we have

$$t = \sqrt{\frac{500}{16}} = 5.59 \text{ seconds.}$$

Therefore the containers hit the ground 5.59 seconds after being dropped. At that time, the velocity is

$$v(5.59) = -32(5.59) = -178.9 \text{ feet per second.}$$

Since the containers in inventory can withstand an impact speed of 200 feet per second, the firm should use these containers.

Example 13: A furniture manufacturer has determined that the marginal cost, in dollars per item, of producing the xth barstool is

$$MC(x) = 0.00006x^2 - 0.04x + 25, \quad 0 \le x \le 550.$$

The fixed cost is $500. Determine the total cost of producing 300 barstools.

Solution Since $MC(x) = C'(x)$, we find $C(x)$ by integrating $MC(x)$:

$$C(x) = \int MC(x) \, dx = \int (0.00006x^2 - 0.04x + 25) \, dx$$

$$= 0.00002x^3 - 0.02x^2 + 25x + C.$$

Then we use the condition $C(0) = 500$ to find that $C = 500$. Therefore

$$C(x) = 0.00002x^3 - 0.02x^2 + 25x + 500.$$

Plugging in $x = 300$ we obtain

$$C(300) = 0.00002(300)^3 - 0.02(300)^2 + 25(300) + 500 = \$6,740.$$

Hence the total cost of producing 300 barstools is $6,740.

Exercise Set 4.2

1. Let $A_1(x)$ and $A_2(x)$ be the areas under the graph of the linear function $f(x) = 2 + 3x$ on the intervals $[1, x]$ and $[2, x]$ respectively.

(a) Draw graphs to illustrate the two areas and determine the formulas of $A_1(x)$ and $A_2(x)$ (hint: both regions can be viewed either as a rectangle plus a triangle, or as a trapezoid);

(b) Find the derivatives $A_1'(x)$ and $A_2'(x)$ and compare them with $f(x)$. Do the results conform with Theorem 1?

2. Let $A_3(x)$ and $A_4(x)$ be the areas under the graph of the linear function $f(x) = 3 + 0.5x$ on the intervals $[1, x]$ and $[2, x]$ respectively. Repeat the process listed in Exercise #1.

For Exercises #3 to #50: Determine the indefinite integrals.

3. $\int x^3 \, dx$

4. $\int x^4 \, dx$

5. $\int 3 \, dx$

6. $\int 6 \, dx$

7. $\int x^{1.4} \, dx$

8. $\int x^{1.2} \, dx$

9. $\int x^{-\pi} \, dx$

10. $\int x^{-e} \, dx$

11. $\int (2x^3 - 4) \, dx$

12. $\int (3x^4 - 5x^2 + 2) \, dx$

13. $\int 3\sqrt{x} \, dx$

14. $\int 4\sqrt[3]{x} \, dx$

15. $\int 12\sqrt[5]{x^7} \, dx$

16. $\int 7\sqrt{x^5} \, dx$

17. $\int \left(-3\sqrt[5]{x^3}\right) dx$

18. $\int 2\sqrt[3]{x^2} \, dx$

19. $\int \frac{2}{x^2} \, dx$

20. $\int \frac{3}{x^4} \, dx$

21. $\int \frac{2}{\sqrt{x}} \, dx$

22. $\int \frac{3}{\sqrt[3]{x^2}} \, dx$

23. $\int \frac{1}{x} \, dx$

24. $\int \frac{3}{x} \, dx$

25. $\int \frac{4}{x^{0.6}} \, dx$

26. $\int \frac{3}{x^{0.4}} \, dx$

27. $\int \frac{-3}{\sqrt[5]{x^3}} \, dx$

28. $\int \frac{-4}{\sqrt[3]{x^2}} \, dx$

29. $\int \left(\frac{1}{x^3} + \frac{4}{x}\right) dx$

30. $\int \left(\frac{1}{x^2} - \frac{2}{x}\right) dx$

31. $\int \left(\frac{1}{\sqrt{x}} - \frac{3}{\sqrt[3]{x}}\right) dx$

32. $\int \left(\frac{2}{\sqrt[4]{x}} + \frac{1}{\sqrt[3]{x}}\right) dx$

33. $\int \frac{1}{x^\pi} \, dx$

34. $\int \frac{3}{x^e} \, dx$

35. $\int \frac{-4}{x^{2.5}} \, dx$

36. $\int \frac{-3}{x^{1.2}} \, dx$

37. $\int e^x \, dx$

38. $\int 3e^x \, dx$

39. $\int 2e^{3x} \, dx$

40. $\int 3e^{2x} \, dx$

41. $\int (2e^{5x} + 5) \, dx$

42. $\int (3x^4 - 5e^{2x} + 2) \, dx$

43. $\int 6e^{-0.02x} \, dx$

44. $\int 3e^{-0.04x} \, dx$

45. $\int \frac{2}{3} e^{x/2} \, dx$

46. $\int \frac{1}{4} e^{x/3} \, dx$

47. $\int \left(\frac{-3}{x} + 5e^{-10x} - 2\sqrt[3]{x^2}\right) dx$

48. $\int \left(\frac{-2}{x} + 4e^{-8x} + 3\sqrt{x^3}\right) dx$

49. $\int \left(\frac{3}{\sqrt{x}} + \frac{4}{x} - \frac{2}{e^{3x}}\right) dx$

50. $\int \left(\dfrac{5}{\sqrt[4]{x}} - \dfrac{2}{x} + \dfrac{3}{e^{2x}} \right) dx$

For Exercises #51 to #64, find the function $f(x)$ satisfying the given conditions.

51. $f'(x) = 2x + 4$, $f(1) = 3$
52. $f'(x) = 3 - 4x$, $f(2) = 1$
53. $f'(x) = x^2 + 4$, $f(0) = 3$
54. $f'(x) = x^2 - 3$, $f(1) = -1$
55. $f'(x) = 2x^2 - x + 5$, $f(0) = 2$
56. $f'(x) = 3x^2 + 2x - 4$, $f(0) = 1$
57. $f'(x) = x^2 - 4x + 3$, $f(1) = 2$
58. $f'(x) = 3x^2 - 2x - 5$, $f(1) = 3$
59. $f'(x) = 2e^{3x}$, $f(0) = 4$
60. $f'(x) = 3e^{2x}$, $f(0) = 2$
61. $f'(x) = 3/\sqrt{x}$, $f(1) = 3$
62. $f'(x) = 2/\sqrt[4]{x}$, $f(1) = 4$
63. $f'(x) = 3/x$, $f(1) = 2$
64. $f'(x) = -2/x$, $f(1) = 1$

For Exercises #65 to #70, find the distance function $s(t)$ satisfying the given conditions.

65. $v(t) = 4t^2$, $s(0) = 4$
66. $v(t) = 3t$, $s(0) = 5$
67. $a(t) = 3$, $v(0) = 2$, $s(0) = 5$
68. $a(t) = 5$, $v(0) = 1$, $s(0) = 3$
69. $a(t) = 4t$, $v(0) = 3$, $s(0) = 10$
70. $a(t) = 2t$, $v(0) = 2$, $s(0) = 6$

APPLICATIONS

71. A car has been traveling on an interstate highway for 10 minutes at a velocity represented by the function
    $v(t) = 0.4 + 0.2t - 0.02t^2$, $0 \le t \le 10$,
    where t is measured in minutes and the velocity is measured in miles per minute. At $t = 0$, the car passed a sign saying "10 miles to Peggy's Diner."

    (a) Determine the distance function $s(t)$, set $s(0) = 0$.

    (b) How far away is Peggy's Diner at the end of the 10-minute drive?

72. A car has been traveling on an straight interstate highway for 15 minutes at a velocity (in miles per minute) represented by the function
    $v(t) = 1.2 - 0.05t + 0.001t^2$, $0 \le t \le 15$,
    At $t = 0$, the car crossed the state borderline.

    (a) Determine the distance function $s(t)$, set $s(0) = 0$.

    (b) How far is the car from the state borderline at the end of the 15-minute drive?

73. A driver hits the brakes when the car is going at a speed of 30 miles per hour (i.e., 44 feet per second). The velocity (in feet per second) function is $v(t) = 44 - 4.4t$.

    (a) Determine the distance function $s(t)$, set $s(0) = 0$.

    (b) How far has the car traveled between the time the driver hits the brakes and the time the car stops? Compare your result to that of Exercise 23 in Section 4.1.

74. A driver accelerates when the car is traveling at a speed of 30 miles per hour (i.e., 44 feet per second). The velocity (in feet per second) function is $v(t) = 44 + 2.2t$.

    (a) Determine the distance function $s(t)$, set $s(0) = 0$.

    (b) How far does the car travel between the time the driver hits the gas pedal and the time the car reaches the speed of 60 miles per hour (i.e., 88 feet per second)? Compare your result to that of Exercise 24 in Section 4.1.

75. A shoe manufacturer has the marginal cost function
    $MC(x) = 20 - 0.02x$, $0 \le x \le 300$,
    (in dollars per pair), where x is the number of pairs of shoes produced each week. The fixed cost is $900.

    (a) Determine the total cost function $C(x)$;

(b) Determine the total cost of producing 250 pairs of shoes. Compare your result to that of Exercise 25 in Section 4.1.

76. A clothing manufacturer has the marginal cost function

$MC(x) = 18 - 0.05x$, $0 \leq x \leq 100$,

(in dollars per suit), where x is the number of suits produced each week. The fixed cost is $500.

(a) Determine the total cost function $C(x)$;

(b) Determine the total cost of producing 60 suits. Compare your result to that of Exercise 28 in Section 4.1.

77. The marginal cost function, in dollars per item, for producing the xth item of a certain brand of skateboards is given by

$MC(x) = 84 - 0.9x + 0.008x^2$,

$0 \leq x \leq 200$. The fixed cost is $1,200. Determine the total cost function $C(x)$ and the total cost of producing 100 skateboards.

78. The marginal cost function, in dollars per item, for producing the xth item of a certain brand of barstool is given by

$MC(x) = 20 - 0.5\sqrt{x}$, $0 \leq x \leq 100$. The fixed cost is $200. Determine the total cost function $C(x)$ and the total cost of producing 80 barstools.

79. A shoe manufacturer has the marginal revenue function for selling the xth pair of shoes

$MR(x) = 90 - 0.5x$, $0 \leq x \leq 120$

(in dollars per pair).

(a) Determine the total revenue function $R(x)$, assuming $R(0) = 0$;

(b) Is $R(0) = 0$ a reasonable assumption? Why?

(c) Determine the total revenue from selling 100 pairs of shoes. Compare your result to that of Exercise 29 in Section 4.1.

80. A clothing manufacturer has the marginal revenue function

$MR(x) = 30 - 0.2x$, $0 \leq x \leq 100$

(in dollars per suit), where x is the number of suits sold each week.

(a) Determine the total revenue function $R(x)$, assuming $R(0) = 0$;

(b) Is $R(0) = 0$ a reasonable assumption? Why?

(c) Determine the total revenue from selling 60 suits. Compare your result to that of Exercise 30 in Section 4.1.

81. A mutual fund has a growth rate recorded each year, expressed by the function

$V'(t) = r(t) = 200e^{0.04t}$, $0 \leq t \leq 10$

(in thousands of dollars per year), where t is the number of years passed from year 1997. The value of the fund in year 1997 is $1,000,000.

(a) Determine the function $V(t)$ which represents the value of the fund t years after year 1997.

(b) Determine the value of the fund in year 2007. Compare your result to that of Exercise 37 in Section 4.1.

82. A mutual fund has a growth rate recorded each year, expressed by the function

$V(t) = r(t) = 200e^{-0.04t}$, $0 \leq t \leq 7$

(in thousands of dollars per year), where t is the number of years passed from year 2000. The value of the fund in year 2000 is $1,500,000.

(a) Determine the function $V(t)$ which represents the value of the fund t years after year 2000.

(b) Determine the value of the fund in year 2007. Compare your result to that of Exercise 38 in Section 4.1.

83. A clothing manufacturer has the marginal profit function

$MP(x) = 30 - 0.02x^{1/3}$, $0 \leq x \leq 100$

(in dollars per suit), where x is the number of suits produced and sold each week. The initial profit is $P(0) = -300$ due to the fixed costs. Determine the total profit function $P(x)$ and the total profit from producing and selling 60 suits.

84. A clothing manufacturer has the marginal profit function

$MP(x) = 30 - 0.02\sqrt{x}$, $0 \leq x \leq 100$

(in dollars per suits), where x is the number of suits produced and sold each week. The initial profit is $P(0) = -200$ due to the fixed costs. Determine the total profit from producing and selling 70 suits.

85. A container firm receives an order for 80 small containers to be used for disaster relief. The containers are going to be dropped from a helicopter 600 feet above the ground. The containers in inventory can withstand an impact speed of 200 feet per second. Should the firm use these containers, or should it make stronger ones?

86. A container firm receives an order for 120 small containers to be used for disaster relief. The containers are going to be dropped from a helicopter 400 feet above the ground. The containers in inventory can withstand an impact speed of 150 feet per second. Should the firm use these containers, or should it make stronger ones?

87. A clothing store finds that when the price of a brand of suits is $70 per suit, the number of suits the manufacturer is willing to supply is 200 suits per week. The rate at which the number of suits that manufacturer supply changes with respect to price is given by

$$S'? = 0.00018p^2 + 0.04p + 0.6,$$

where p is the price of each suit in dollars. Determine the supply function $x = S(p)$.

4.3 Areas and Integrals: The Fundamental Theorem of Calculus

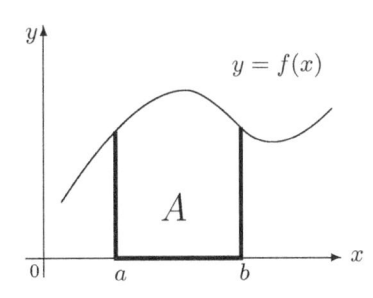

Figure 4.3.1: Case I

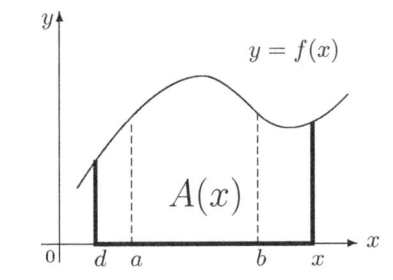

Figure 4.3.2: Case II

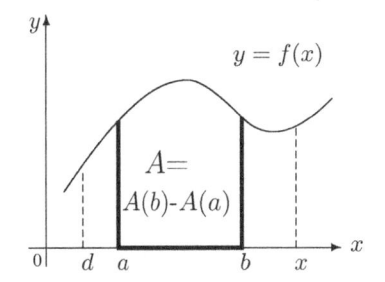

Figure 4.3.3: Case III

We are now ready to use integration to find the area under a graph on a certain interval $[a, b]$. In most of the applications, we seek the cumulative change of a quantity on a certain interval. From Section 4.1 we concluded that the cumulative change on the interval is represented by the area under the graph of the function that represents the rate of change of the quantity. Therefore being able to find the particular area is very important. Let us observe the three graphs shown in Figures 4.3.1 to 4.3.3. The area A in Figure 4.3.1 is the region under the graph of $f(x)$ on the interval $[a, b]$. The area $A(x)$ in Figure 4.3.2 is the region under the graph of $f(x)$ on the interval $[d, x]$. Figure 4.3.3 illustrates that the area A can be obtained by $A(b) - A(a)$, that is, the area on $[d, b]$ minus the area on $[d, a]$.

In Section 4.2, we concluded that if we can find a function $F(x)$ that is an antiderivative to $f(x)$, that is, $F(x) = f(x)$, then $A(x) = F(x) + C$, where C is a constant. Since we are looking for $A = A(b) - A(a)$, we have

$$A = A(b) - A(a) = (F(b) + C) - (F(a) + C)$$

$$= F(a) + C - F(b) - C = F(b) - F(a).$$

It means that the constant C does not play a role in this particular case! We have obtained the answer to the question raised in Section 4.1: "How can the exact area A, that is, the area under the graph of $f(x)$ on the interval $[a, b]$, be calculated?" The solution is to integrate $f(x)$ and then calculate the difference of the antiderivative values at $x = b$ and $x = a$.

Example 1: Determine the area under the graph of
$$f(x) = x^2 + 1$$
(a) on the interval $[0, 2]$ and (b) on the interval $[-1, 3]$ (Figures 4.3.4-4.3.5).

Solution We first integrate $f(x) = x^2 + 1$:
$$\int (x^2 + 1) \, dx = \frac{1}{3}x^3 + x + C$$

Hence, $F(x) = x^3/3 + x$ is an antiderivative of $x^2 + 1$.

(a) On $[0, 2]$, the area is
$$F(2) - F(0) = \left[\frac{1}{3}(2)^3 + (2)\right] - \left[\frac{1}{3}(0)^3 + (0)\right] = 4\frac{2}{3}.$$

(b) On $[-1, 3]$, the area is
$$F(3) - F(-1) = \left[\frac{1}{3}(3)^3 + (3)\right] - \left[\frac{1}{3}(-1)^3 + (-1)\right] = 13\frac{1}{3}.$$

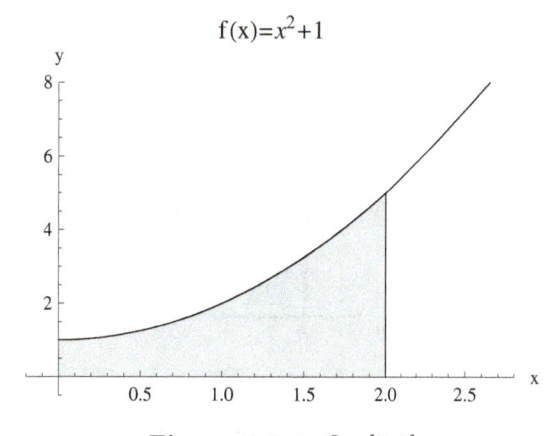

Figure 4.3.4: On $[0, 2]$

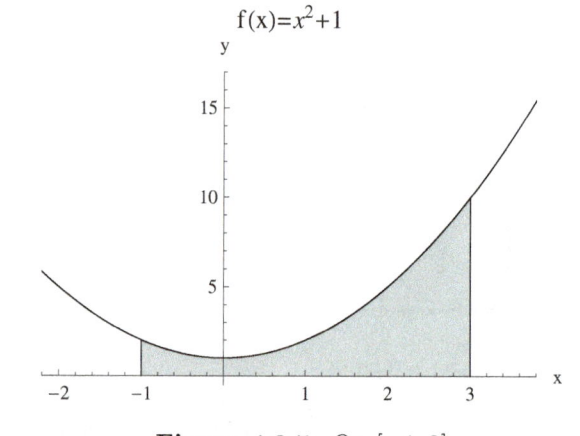

Figure 4.3.5: On $[-1, 3]$

Reality Check   *Process for calculating area*

We will use a more convenient process for calculating the area after we have learned how to evaluate the definite integrals.

Definite Integrals

It is true that only after the definite integral is introduced, do the word "integrate" (means "bring together") and the symbols "$\int$" and "dx" start to make sense. The concept of the definite integral arises when we try to calculate a cumulative quantity. In particular, it arises when we try to calculate the area of a curved region. In order to introduce this concept, we need to reexamine the problem of finding the area under a curve, and approach it from another angle.

We bring back the problem raised in Section 4.1: Estimating the area under $f(x) = x^2 + 1$ on the interval $[0, 2]$. The area was estimated by the left-rectangle approximation, first using four, then using eight rectangles (Figures 4.3.6-4.3.7):

4.3. THE FUNDAMENTAL THEOREM OF CALCULUS

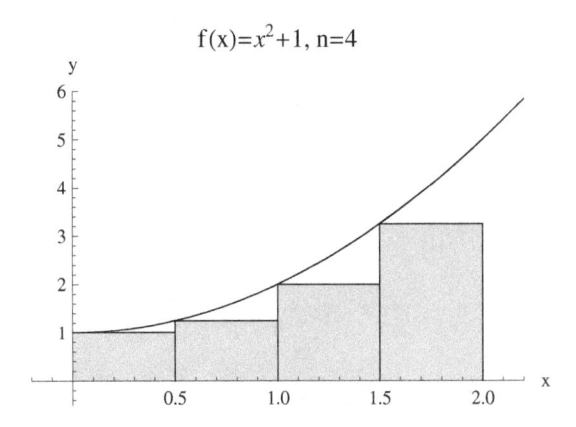

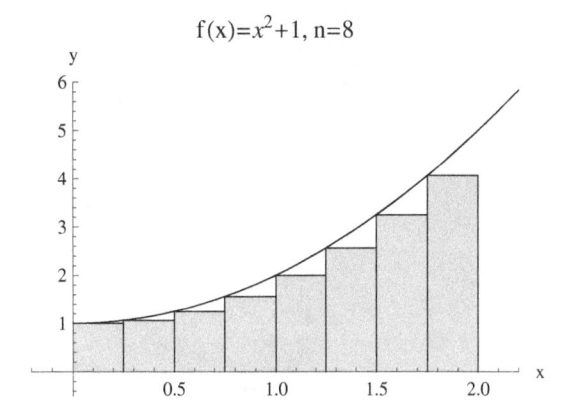

Figure 4.3.6: Estimating using four rectangles

Figure 4.3.7: Estimating using eight rectangles

We can observe that the eight-rectangle approximation approximates the area better than the four-rectangle approximation. We can continue to refine this approach by using 16 rectangles (Figure 4.3.8):

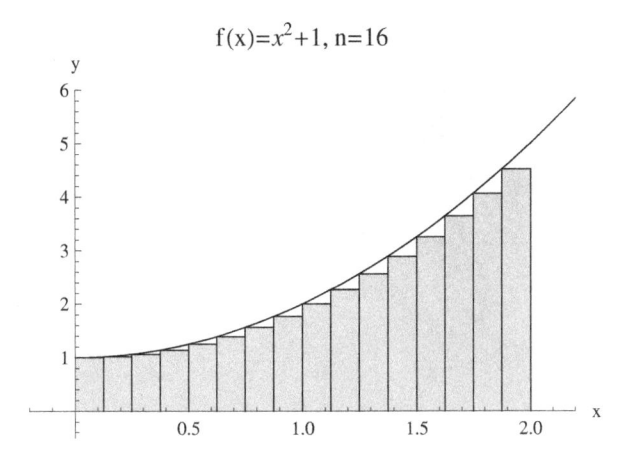

Figure 4.3.8: Estimating using 16 rectangles

We can conclude that, as more and more rectangles are used in the approximation, the estimated area gets closer to the exact area. This idea can be summarized into the limit:

$$A = \lim_{n \to \infty} \sum_{k=1}^{n} f(x_k) \Delta x.$$

In this equation, n represents the number of rectangles used; Δx represents the length of each rectangle; $f(x_k)$ represents the height of the kth rectangle; and the summation symbol is used to make writing the formula more efficient.

> **Reality Check** *The limit and sum expression of a definite integral*
>
> This limit is used to define a definite integral, and it relates to the symbols used in an integral. However, we will not use the limit to calculate an area or a definite integral.

We introduce the **definite integral** now:

> **Definition** *definite integral*
>
> For a function $y = f(x)$ defined on an interval $[a, b]$, the **definite integral of f from a to b**, denoted as
> $$\int_a^b f(x)\, dx,$$
> is defined by
> $$\int_a^b f(x)\, dx = \lim_{n \to \infty} \sum_{k=1}^n f(x_k)\Delta x,$$
> if such a limit exists. If $\int_a^b f(x)\, dx$ exists, then we say that f is **integrable** on $[a, b]$.

We have already encountered the symbols "$\int$," the integral sign; $f(x)$, the integrand; and dx when an indefinite integral is introduced. In the definite integral, a is called the **lower limit**, and b is called the **upper limit**. The integral sign $\int$ resembles S for summation, and dx is from Δx, since when n gets larger and larger, Δx approaches zero.

> **Reality Check** *Definite and indefinite integrals*
>
> - The definite integral $\int_a^b f(x)\, dx$ is a *number*, while the indefinite integral $\int f(x)\, dx$ is a *function*.
> - The definite integral $\int_a^b f(x)\, dx$ does not depend on x. We can use any letter in place of x without changing the value of the integral:
> $$\int_a^b f(x)\, dx = \int_a^b f(t)\, dt = \int_a^b f(r)\, dr.$$
> - If $f(x)$ is nonnegative on $[a, b]$, then the definite integral $\int_a^b f(x)\, dx$ is the same as the area under the graph of $f(x)$ on $[a, b]$.
> - All of the functions that occur in this course are integrable.

Observing how the definite integral and the indefinite integral are defined, it seems that they have very different features (one is a number obtained via a limit of a sum, the other is an antiderivative), yet they share the same name and symbols. We will explore the answers to the following four related questions:

> **Questions about $\int_a^b f(x)\, dx$ and areas**
>
> 1. Why the definite integral and the indefinite integral share the same name and symbols?
> 2. What is the relationship between the areas formed by the graph of $f(x)$ and the definite integral $\int_a^b f(x)\, dx$, when the function $f(x)$ is negative, or partially negative?
> 3. How can the definite integral $\int_a^b f(x)\, dx$ be calculated?
> 4. How do we estimate a definite integral $\int_a^b f(x)\, dx$?

The answers to the first and third questions lie in the Fundamental Theorem of Calculus. The answer to the second question is illustrated through graphs. The answer to the fourth question is related to the

4.3. THE FUNDAMENTAL THEOREM OF CALCULUS

answer to the second question, and is briefly explored in Section 4.1 (see comments in "Reality Check: estimating areas").

The Fundamental Theorem of Calculus

The Fundamental Theorem of Calculus is appropriately named because it establishes a connection between the two branches of calculus: differential calculus and integral calculus. The Fundamental Theorem of Calculus is unquestionably the most important theorem in calculus. It ranks as one of the great accomplishments of the human mind. Before it was discovered by Isaac Newton and Gottfried Wilhelm Leibniz, finding areas and other geometric quantities was so difficult that only a genius could meet the challenge. Now we can use this great theorem to solve these problems systematically.

Theorem 1 *The Fundamental Theorem of Calculus*
Assume that f is continuous on $[a, b]$, then

$$\int_a^b f(x)\, dx = F(b) - F(a)$$

where F is any antiderivative of f, that is, $F'(x) = f(x)$.

Returning to Example 1, we can rewrite the process of calculating the area under the graph of $f(x) = x^2 + 1$ on the interval $[0, 2]$ as follows:

$$A = \int_0^2 (x^2 + 1)\, dx = \left[\frac{1}{3}x^3 + x\right]_0^2$$

$$= \left[\frac{1}{3}(2)^3 + (2)\right] - \left[\frac{1}{3}(0)^3 + (0)\right] = 4\frac{2}{3}.$$

We have introduced an intermediate notation

$$\int_a^b f(x)\, dx = [F(x)]_a^b = F(b) - F(a),$$

where $F(x)$ is an antiderivative of $f(x)$.

Example 2: Evaluate the following definite integrals:

(a) $\int_{-2}^1 x^3\, dx$ (b) $\int_0^4 \sqrt{x}\, dx$ (c) $\int_0^2 2e^x\, dx$

Solution

(a) $\int_{-2}^1 x^3\, dx = \left[\frac{1}{4}x^4\right]_{-2}^1$

$$= \frac{1}{4}(1)^4 - \frac{1}{4}(-2)^4 = \frac{1}{4} - \frac{1}{4}(16) = -3\frac{3}{4}.$$

(b) $\int_0^4 \sqrt{x}\, dx = \left[\frac{2}{3}x^{\frac{3}{2}}\right]_0^4$

$$= \frac{2}{3}(4)^{\frac{3}{2}} - \frac{2}{3}(0)^{\frac{3}{2}} = \frac{2}{3}(8) = 5\frac{1}{3}.$$

(c) $\displaystyle\int_0^2 2e^x\,dx = [2e^x]_0^2$

$= 2e^2 - 2e^0 = 2(e^2 - 1).$

Areas and Definite Integrals

When we evaluate a definite integral $\int_a^b f(x)\,dx$ and the integrand $f(x)$ is nonnegative, we obtain the area under the graph of $f(x)$ on the interval $[a,b]$. For example, in Example 2 (b), we have

$$\int_0^4 \sqrt{x}\,dx = \frac{16}{3}.$$

The area is illustrated in Figure 4.3.9:

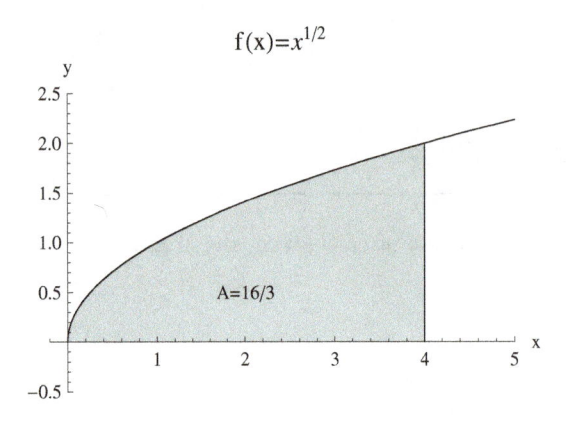

Figure 4.3.9: Integral vs. area

When the integrand $f(x)$ is not all nonnegative on the interval, as in Example 2 (a), we have

$$\int_{-2}^1 x^3\,dx = -\frac{15}{4}.$$

What does that number represent? Observe that the region bounded by the graph of $f(x)$, the x-axis, and the lines $x = -2$ and $x = 1$ consists of two parts (Figure 4.3.10). The darker shaded region is above the x-axis, and the lighter shaded region is below the x-axis. If we evaluate $\int_{-2}^0 x^3\,dx$ and $\int_0^1 x^3\,dx$, we obtain

$$\int_{-2}^0 x^3\,dx = \left[\frac{1}{4}x^4\right]_{-2}^0$$

$$= \frac{1}{4}(0)^4 - \frac{1}{4}(-2)^4 = 0 - \frac{1}{4}(16) = -4, \text{ and}$$

$$\int_0^1 x^3\,dx = \left[\frac{1}{4}x^4\right]_0^1$$

$$= \frac{1}{4}(1)^4 - \frac{1}{4}(0)^4 = \frac{1}{4} - 0 = \frac{1}{4}.$$

That means the lighter shaded region has area $B = 4$ and the darker shaded region has area $A = 1/4$. The integral $\int_{-2}^1 x^3\,dx$ is the same as Area B subtracted from Area A.

4.3. THE FUNDAMENTAL THEOREM OF CALCULUS

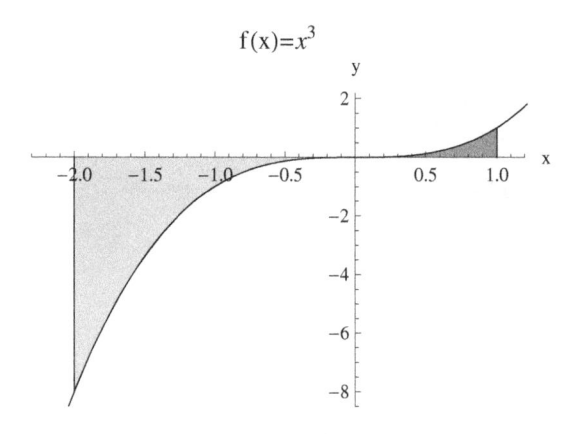

Figure 4.3.10: Integral vs. areas

In general, we have the following conclusions:

Areas and Definite Integral

- If the integrand $f(x)$ is nonnegative on $[a, b]$, then $\int_a^b f(x)\, dx$ equals the area under the graph of $f(x)$ on $[a, b]$;

- If the integrand $f(x)$ is negative on $[a, b]$, then $\int_a^b f(x)\, dx$ equals the *negative* of the area of the region between the x-axis and the graph of $f(x)$ on $[a, b]$;

- If the integrand $f(x)$ is partially positive and partially negative on $[a, b]$, then $\int_a^b f(x)\, dx$ equals the the area of the regions above the x-axis minus the area of the regions below the x-axis on $[a, b]$.

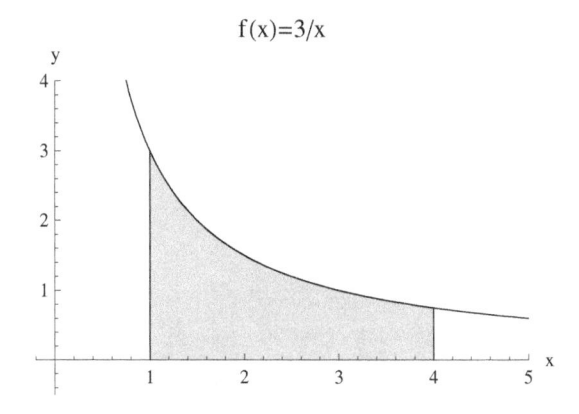

Figure 4.3.11: Integral vs. area

Example 3: The graph of $f(x) = 3/x$ with some shaded region is presented in Figure 4.3.11.

(a) Evaluate the definite integral $\displaystyle\int_1^4 \frac{3}{x}\, dx$.

(b) Interpret the result in part (a) with the area involved.

Solution (a) We have
$$\int_1^4 \frac{3}{x}\, dx = [3\ln x]_1^4$$

$$= 3\ln 4 - 3\ln 1 = 3\ln 4 = 6\ln 2.$$

(b) The number $6\ln 2$ obtained in part (a) is equal to the area of the shaded region, since the function is positive on $[1, 4]$, or the region is above the x-axis.

Example 4: The graph of $f(x) = 1 - x^2$ with some shaded region is presented in Figure 4.3.12.

(a) Evaluate the definite integral $\int_1^3 (1 - x^2)\, dx$.

(b) Interpret the result in part (a) with the area involved.

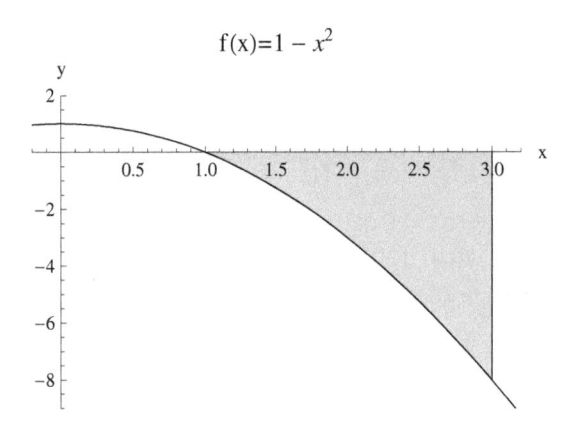

Figure 4.3.12: Integral vs. area

Solution (a) We have

$$\int_1^3 (1 - x^2)\, dx = \left[x - \frac{1}{3}x^3 \right]_1^3$$

$$= \left[(3) - \frac{1}{3}(3)^3 \right] - \left[1 - \frac{1}{3}(1)^3 \right] = -6 - \frac{2}{3} = -6\frac{2}{3}.$$

(b) The number $-6\frac{2}{3}$ obtained in part (a) is equal to the negative of the area of the shaded region, since the function is negative on $[1, 3]$, or the region is below the x-axis.

Example 5: The graph of $f(x) = x^2 - 4$ with more than one shaded region is presented in Figure 4.3.13.

(a) Evaluate the definite integral $\int_0^3 (x^2 - 4)\, dx$.

(b) Interpret the result in part (a) with the areas involved.

(c) Determine the *total* area of the shaded regions.

4.3. THE FUNDAMENTAL THEOREM OF CALCULUS

$$f(x) = x^2 - 4$$

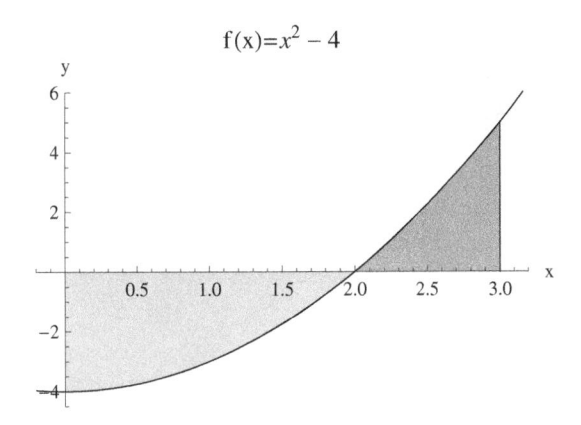

Figure 4.3.13: Find area using integrals

Solution (a) We have

$$\int_0^3 (x^2 - 4) \, dx = \left[\frac{1}{3}x^3 - 4x\right]_0^3$$

$$= \left[\frac{1}{3}(3)^3 - 4(3)\right] - \left[\frac{1}{3}(0)^3 - 4(0)\right] = -3 - 0 = -3.$$

(b) The number -3 obtained in part (a) is equal to the area of the darker shaded region minus the area of the lighter shaded region.

(c) To obtain the total shaded area, we integrate $f(x) = x^2 - 4$ on $[0, 2]$ and $[2, 3]$ (Since $x = 2$ is where $f(x)$ changes from negative to positive).

$$\int_0^2 (x^2 - 4) \, dx = \left[\frac{1}{3}x^3 - 4x\right]_0^2$$

$$= \left[\frac{1}{3}(2)^3 - 4(2)\right] - \left[\frac{1}{3}(0)^3 - 4(0)\right] = -5\frac{1}{3} - 0 = -5\frac{1}{3},$$

$$\int_2^3 (x^2 - 4) \, dx = \left[\frac{1}{3}x^3 - 4x\right]_2^3$$

$$= \left[\frac{1}{3}(3)^3 - 4(3)\right] - \left[\frac{1}{3}(2)^3 - 4(2)\right] = -3 - \left(-5\frac{1}{3}\right) = 2\frac{1}{3}.$$

Therefore, the total shaded area is

$$5\frac{1}{3} + 2\frac{1}{3} = 7\frac{2}{3}.$$

Using the relationship between a definite integral and the areas, we can extend the conclusion obtained in Section 4.1:

Definite Integral and Cumulative Change

If a function f represents the *rate of change* of a quantity on an interval $[a, b]$, then the definite integral, $\int_a^b f(x) \, dx$, represents the *total (or cumulative) change* of the quantity from the variable values a to b.

Notice that the rate function $f(x)$ is no longer restricted to be positive. For example, if $f(t)$ represents the record in the daily change in a checking account (deposits and withdrawals) for a month, then $\int_0^{30} f(t)\, dt$ represents the change in the checking account between the beginning and the end of the month. If the total deposits exceed the total withdrawals, by the end of the month the balance in the account will be higher (i.e., the total change is positive). If the total withdrawals exceed the total deposits, by the end of the month the balance in the account will be lower (that is, the total change is negative).

Example 6: A driver driving on a straight east-west highway records the velocity of the car in the hours after he leaves home at 8:00 a.m.:

$$v(t) = 40t - 30t^2,$$

where t, in hours, measures the time passed after 8:00 a.m., and v is in miles per hour. Positive velocity means going east, and negative velocity means going west. Determine the location of the driver, in relation to his home, at 11:00 a.m..

Solution Since the velocity is the rate of change in the position function s, the change in the position can be obtained by the definite integral:

$$s(3) - s(0) = \int_0^3 v(t)\, dt = \int_0^3 \left(40t - 30t^2\right)\, dt$$

$$= \left[20t^2 - 10t^3\right]_0^3 = 20(3)^2 - 10(3)^3 - 0 = -90.$$

That is, the change of the position is -90 miles. Since the driver starts from home, at 11:00 a.m. the driver is 90 miles west of home.

Integrals and Areas: A Summary

In the first three sections of this chapter we have been dealing with the areas of regions bounded by the graph of a function and the x-axis on intervals like $[a, b]$ and $[a, x]$. Through the discussions and the Fundamental Theorem of Calculus, we have obtained the following:

- The area $A(x)$ of a region, bounded by the x-axis and the graph of a positive function $f(x)$, on the interval $[a, x]$, satisfies

$$A'(x) = f(x).$$

That is, the area is an antiderivative of $f(x)$.

- The antiderivatives of $f(x)$ in general form can be obtained by the basic formulas and properties of an indefinite integral:

$$\int f(x) = F(x) + C.$$

- The definite integral, $\int_a^b f(x)\, dx$, is a number which is independent of x. It can be evaluated by

$$\int_a^b f(x)\, dx = [F(x)]_a^b = F(b) - F(a)$$

where $F(x)$ is any antiderivative of $f(x)$.

- The area A of a region, bounded by the x-axis and the graph of a *positive* function $f(x)$, on the interval $[a, b]$, can be obtained by a definite integral:

$$A = \int_a^b f(x)\, dx.$$

CHAPTER 4 SUMMARY

- The definite integral, $\int_a^b f(x)\, dx$, represents the "signed sum" of the areas of the regions, bounded by the x-axis and the graph of a function $f(x)$, on the interval $[a, b]$. That is, the value of $\int_a^b f(x)\, dx$ equals the area of the regions above the x-axis minus the area of the regions below the x-axis.

- If the function $f(x)$ represents the *rate of change* of a quantity on an interval $[a, b]$, then the definite integral, $\int_a^b f(x)\, dx$, represents the *total (or cumulative) change* of the quantity from the variable values a to b.

Exercise Set 4.3

For Exercises #1 to #4, use definite integrals to find the area under the graph on the following intervals:
(a) $[0, 2]$; (b) $[1, 3]$; (c) $[-2, 0]$; (d) $[-3, 1]$.
In each case, mark the region on the graph.

1. $f(x) = 3x^2 + 1$

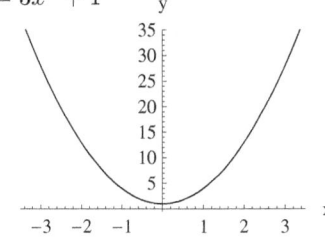

2. $f(x) = 0.2x^2 + 2$

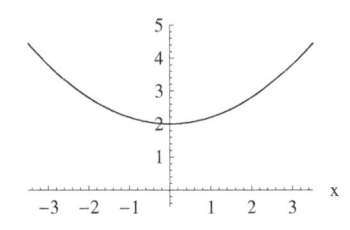

3. $f(x) = 2e^{0.2x}$

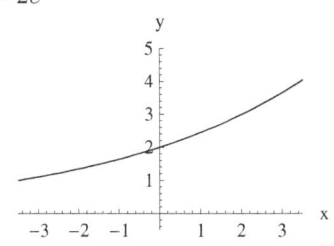

4. $f(x) = 2e^{-0.3x}$

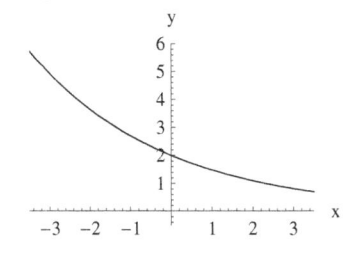

For Exercises #5 to #8, use definite integrals to find the area under the graph on the following intervals: (a) $[1, 2]$; (b) $[1, 4]$. In each case, mark the region on the graph.

5. $f(x) = 1 + 3\sqrt{x}$

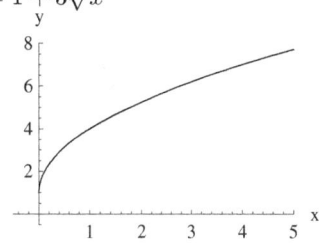

6. $f(x) = 0.5x + \dfrac{3}{\sqrt{x}}$

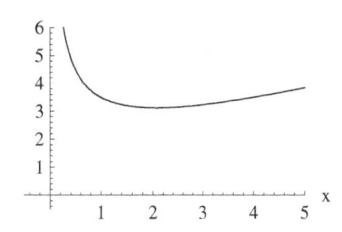

7. $f(x) = x + \dfrac{2}{x}$

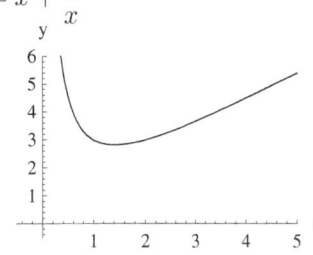

8. $f(x) = 2 + \dfrac{3}{x^2}$

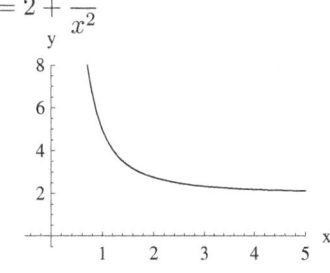

For Exercises #9 to #24: Evaluate the definite integrals.

9. $\int_0^3 (3x^2 + 2)\, dx$

10. $\int_0^2 (2x^3 - 4)\, dx$

11. $\int_{-1}^1 (4x^2 + 5x)\, dx$

12. $\int_{-1}^1 (3x^4 - 5x^2 + 2)\, dx$

13. $\int_1^9 \left(\sqrt{t} - 2\right)\, dt$

14. $\int_1^8 \left(\sqrt[3]{t} + 1\right)\, dt$

15. $\int_{-2}^1 2e^{3r}\, dr$

16. $\int_{-3}^1 3e^{-0.05r}\, dr$

17. $\int_1^5 \frac{1}{x^2}\, dx$

18. $\int_1^2 \frac{2}{x^3}\, dx$

19. $\int \frac{2}{x^2}\, dx$

20. $\int \frac{3}{x^4}\, dx$

21. $\int_1^4 \frac{2}{\sqrt{x}}\, dx$

22. $\int_1^{27} \frac{3}{\sqrt[3]{x^2}}\, dx$

23. $\int_1^e \left(3x + \frac{2}{x}\right)\, dx$

24. $\int_1^e \left(4x - \frac{3}{x}\right)\, dx$

25. (a) Evaluate the definite integral $\int_1^3 (2/x)\, dx$.

(b) Interpret the result in part (a) with the area(s) involved. Illustrate with a graph.

26. (a) Evaluate the definite integral $\int_1^2 (4/x)\, dx$.

(b) Interpret the result in part (a) with the area(s) involved. Illustrate with a graph.

27. (a) Evaluate the definite integral $\int_1^2 (1 - 5/x^2)\, dx$.

(b) Interpret the result in part (a) with the area(s) involved. Illustrate with a graph.

28. (a) Evaluate the definite integral $\int_1^2 (1 - 8/x^3)\, dx$.

(b) Interpret the result in part (a) with the area(s) involved. Illustrate with a graph.

29. (a) Evaluate the definite integral $\int_0^3 (4 - x^2)\, dx$.

(b) Interpret the result in part (a) with the area(s) involved. Illustrate with a graph.

30. (a) Evaluate the definite integral $\int_0^2 (1 - 4x^2)\, dx$.

(b) Interpret the result in part (a) with the area(s) involved. Illustrate with a graph.

APPLICATIONS

31. The marginal cost function, in dollars per item, for producing the xth item of a certain brand of skateboard is given by
$$MC(x) = 84 - 0.9x + 0.008x^2,$$
$0 \le x \le 200$. Use a definite integral to find the total variable cost of producing 100 skateboards. Compare your result to that of Exercise 77 in Section 4.2.

32. The marginal cost function, in dollars per item, for producing the xth item of a certain brand of barstool is given by
$MC(x) = 20 - 0.5\sqrt{x}$, $0 \le x \le 100$. Use a definite integral to find the total variable cost of producing 80 barstools. Compare your result to that of Exercise 78 in Section 4.2.

33. A car has been traveling on an interstate highway for 10 minutes at a velocity represented by the function
$$v(t) = 0.4 + 0.2t - 0.02t^2,\ 0 \le t \le 10,$$
where t is measured in minutes and the velocity is measured in miles per minute. At $t = 0$, the car passed a sign saying "10 miles to Peggy's Diner." How far away is Peggy's Diner at the end of the 10-minute drive? Use a definite integral to find the answer. Compare your result to that of Exercise 71 in Section 4.2.

34. A car has been traveling on a straight interstate highway for 12 minutes at a velocity represented by the function
$$v(t) = 1.2 - 0.06t + 0.005t^2,\ 0 \le t \le 12,$$
where t is measured in minutes and the velocity is measured in miles per minute. At $t = 0$, the car crossed the state borderline. How far is the car from the state borderline at the end of the 12-minute drive? Use a definite integral to find the answer. Compare your result to that of Exercise 72 in Section 4.2.

35. A driver hits the brakes when the car is going at a speed of 30 miles per hour (i.e., 44 feet per second). The velocity function is

$$v(t) = 44 - 4.4t,$$

where t is measured in seconds and the velocity is measured in feet per second. How far does the car travel between the time the driver hits the brakes and the time the car stops? Use a definite integral to find the answer. Compare your result to that of Exercise 23 in Section 4.1 and Exercise 73 in Section 4.2.

36. A driver accelerates when the car is going at a speed of 30 miles per hour (i.e., 44 feet per second). The velocity function is

$$v(t) = 44 + 2.2t,$$

where t is measured in seconds and the velocity is measured in feet per second. How far does the car travel between the time the driver hits the gas pedal and the time the car reaches the speed of 60 miles per hour (i.e., 88 feet per second)? Use a definite integral to find the answer. Compare your result to that of Exercise 24 in Section 4.1 and Exercise 74 in Section 4.2.

37. A shoe manufacturer has the marginal cost function

$MC(x) = 20 - 0.02x$, $0 \leq x \leq 300$

(in dollars per pair), where x is the number of pairs of shoes produced each week. The fixed cost is $900. Determine the total cost of producing 250 pairs of shoes using a definite integral. Compare your result to that of Exercise 25 in Section 4.1 and Exercise 75 in Section 4.2.

38. A clothing manufacturer has the marginal cost function

$MC(x) = 18 - 0.05x$, $0 \leq x \leq 100$

(in dollars per suit), where x is the number of suits produced each week. The fixed cost is $500. Determine the total cost of producing 60 suits using a definite integral. Compare your result to that of Exercise 28 in Section 4.1 and Exercise 76 in Section 4.2.

39. A shoe manufacturer has the marginal revenue function for selling the xth pair of shoes

$MR(x) = 90 - 0.5x$, $0 \leq x \leq 120$

(in dollars per pair). Determine the total revenue from selling 100 pairs of shoes using a definite integral. Compare your result to that of Exercise 29 in Section 4.1 and Exercise 79 in Section 4.2.

40. A clothing manufacturer has the marginal revenue function

$MR(x) = 30 - 0.2x$, $0 \leq x \leq 100$

(in dollars per suit), where x is the number of suits sold each week. Determine the total revenue from selling 60 suits using a definite integral. Compare your result to that of Exercise 30 in Section 4.1 and Exercise 80 in Section 4.2.

41. A mutual fund has a growth rate recorded each year, expressed by the function

$V'(t) = r(t) = 200e^{0.04t}$, $0 \leq t \leq 10$

(in thousands of dollars per year), where t is the number of years passed from year 1997. The value of the fund in year 1997 is $1,000,000. Determine the value of the fund in year 2007 using a definite integral. Compare your result to that of Exercise 81 in Section 4.2.

42. A mutual fund has a growth rate recorded each year, expressed by the function

$V'(t) = r(t) = 200e^{-0.04t}$, $0 \leq t \leq 7$

(in thousands of dollars per year), where t is the number of years passed from year 2000. The value of the fund in year 2000 is $1,500,000. Determine the value of the fund in year 2007 using a definite integral. Compare your result to that of Exercise 82 in Section 4.2.

43. A clothing manufacturer has the marginal profit function

$MP(x) = 30 - 0.02x^{1/3}$, $0 \leq x \leq 100$

(in dollars per suit), where x is the number of suits produced and sold each week. The initial profit is $P(0) = -300$ due to the fixed costs. Determine the total profit from producing and selling 60 suits using a definite integral. Compare your result to that of Exercise 83 in Section 4.2.

44. A clothing manufacturer has the marginal profit function

$MP(x) = 40 - 0.02\sqrt{x}$, $0 \leq x \leq 100$

(in dollars per suit), where x is the number of suits produced and sold each week. The initial profit is $P(0) = -400$ due to the fixed costs. Determine the total profit from producing and selling 80 suits using a definite integral. Compare your result to that of Exercise 84 in Section 4.2.

45. A driver driving on a straight east-west highway records the velocity of the car in the hours after he leaves home at 7:00 a.m.:

$$v(t) = 60t - 15t^2,$$

where t, in hours, measures the time passed after 7:00 a.m., and v is in miles per hour. Positive velocity means going east, and negative velocity means going west. Determine the location of the driver, in relation to his home, at 10:00 a.m..

46. A driver driving on a straight south-north highway records the velocity of the car in the hours after he leaves home at 11:00 a.m.:

$$v(t) = 54t - 24t^2,$$

where t, in hours, measures the time passed after 11:00 a.m., and v is in miles per hour. Positive velocity means going north, and negative velocity means going south. Determine the location of the driver, in relation to his home, at 1:00 p.m..

4.4 Areas and Definite Integrals

In the previous sections we have been dealing with the areas of regions bounded by the graph of a function and the x-axis on intervals $[a, b]$, and have discovered the relationship between the areas and the definite integrals. In this section, we use the properties of areas to illustrate an important property of definite integrals and the formula for calculating the average value of a function on an interval $[a, b]$, and explore how definite integrals can be used to find the areas of regions bounded by graphs of two functions.

We first introduce an important and unique property of a definite integral.

Theorem 1   **The Splitting Property of $\int_a^b f(x)\,dx$**

For any number $a < c < b$,

$$\int_a^b f(x)\,dx = \int_a^c f(x)\,dx + \int_c^b f(x)\,dx.$$

That is, for any number c between a and b, the integral of $f(x)$ from a to b is the sum of the integral of $f(x)$ from a to c and the integral of $f(x)$ from c to b.

The splitting property can be illustrated by Figure 4.4.1. Since the integrals $\int_a^c f(x)\,dx$, $\int_c^b f(x)\,dx$, and $\int_a^b f(x)\,dx$ can be regarded as the areas under the graph of $f(x)$ on the intervals $[a, c]$, $[c, b]$, and $[a, b]$ respectively, the conclusion becomes obvious.

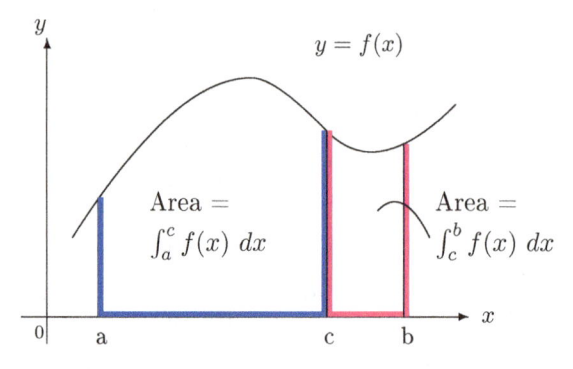

Figure 4.4.1: The splitting property of a definite integral

4.4. AREAS AND DEFINITE INTEGRALS

Theorem 1 is very useful when we want to find the area under the graph of a piecewise defined function.

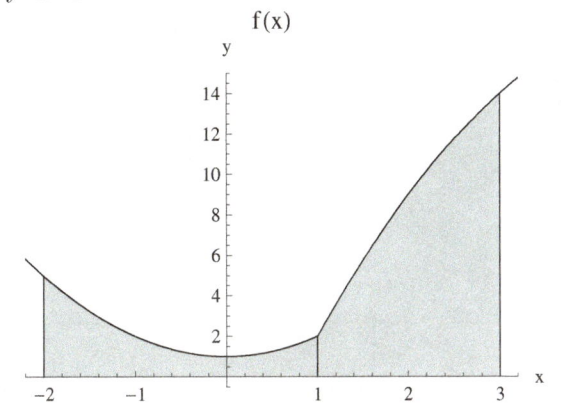

Figure 4.4.2: Applying the splitting property

Example 1: Determine the area under the graph of $f(x)$ on the interval $[-2, 3]$ (Figure 4.4.2), where

$$f(x) = \begin{cases} x^2 + 1 & \text{if } x \leq 1, \\ -x^2 + 10x - 7 & \text{if } x > 1. \end{cases}$$

Solution We apply Theorem 1. Since the function $f(x)$ changes formulas at $x = 1$, we have $a = -1$, $c = 1$, and $b = 3$:

$$\int_{-2}^{3} f(x)\, dx = \int_{-2}^{1} f(x)\, dx + \int_{1}^{3} f(x)\, dx\ .$$

Since

$$\int_{-2}^{1} (x^2 + 1)\, dx = \left[\frac{1}{3}x^3 + x\right]_{-2}^{1} = \left[\frac{1}{3}(1)^3 + (1)\right] - \left[\frac{1}{3}(-2)^3 + (-2)\right] = \frac{4}{3} - (-\frac{14}{3}) = 6,$$

and $\displaystyle\int_{1}^{3} (-x^2 + 10x - 7)\, dx = \left[-\frac{1}{3}x^3 + 5x^2 - 7x\right]_{1}^{3}$

$$= \left[-\frac{1}{3}(3)^3 + 5(3)^2 - 7(3)\right] - \left[-\frac{1}{3}(1)^3 + 5(1)^2 - 7(1)\right] = (15) - \left(-\frac{7}{3}\right) = 17\frac{1}{3},$$

We have the total area as $6 + 17\frac{1}{3} = 23\frac{1}{3}$.

Area Between Curves

We now turn to exploring how to use definite integrals to find the area of a region bounded by the graphs of two functions on a certain interval $[a, b]$, as illustrated in Figures 4.4.3 to 4.4.5.

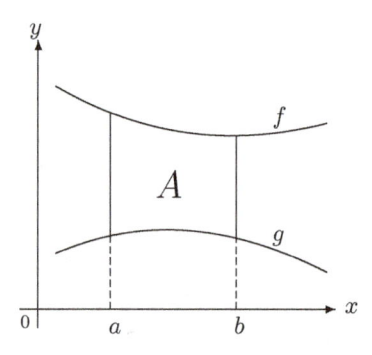

Figure 4.4.3: Area A 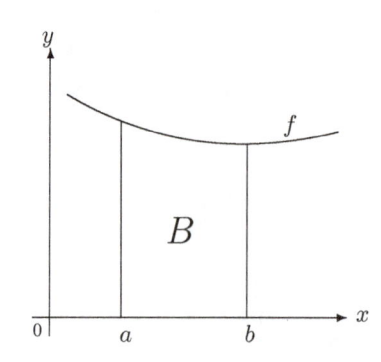   **Figure 4.4.4:** Area B 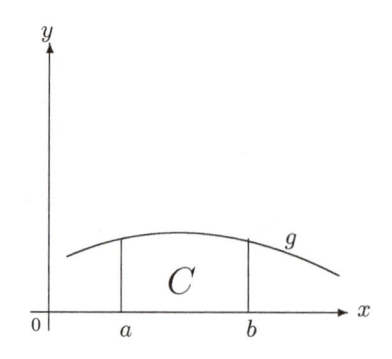   **Figure 4.4.5:** Area C

Notice that the relationship among the areas A, B, and C is $A = B - C$. Since B is the area under the graph of $f(x)$ on $[a,b]$, and C is the area under the graph of $g(x)$ on $[a,b]$, we have

$$B = \int_a^b f(x)\,dx, \quad \text{and} \quad C = \int_a^b g(x)\,dx.$$

Therefore

$$A = \int_a^b f(x)\,dx - \int_a^b g(x)\,dx, \quad \text{or} \quad A = \int_a^b [f(x) - g(x)]\,dx.$$

We state the conclusion in the following theorem:

> **Theorem 2** **Area Between Two Curves**
>
> Assume that f and g are two continuous functions on the interval $[a,b]$. If $f(x) \geq g(x)$ for all x in $[a,b]$, then the area of the region bounded by the two curves on $[a,b]$ is
>
> $$A = \int_a^b [f(x) - g(x)]\,dx.$$
>
> Conversely, if $g(x) \geq f(x)$ for all x in $[a,b]$, then the area of the region bounded by the two curves on $[a,b]$ is
>
> $$A = \int_a^b [g(x) - f(x)]\,dx.$$

> **Warning**
>
> The formulas only work when the two functions do not exchange positions on the whole interval. Therefore a graph of the two functions is very helpful.

Example 2: Determine the area between the graphs of $g(x) = -x^2 + 3$ and $f(x) = 2x + 8$ on the interval $[-2, 3]$.

4.4. AREAS AND DEFINITE INTEGRALS

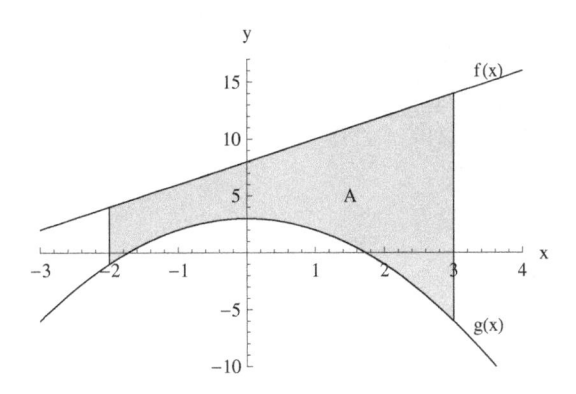

Figure 4.4.6: Area between two curves

Solution We begin with a graph of f and g and shade the area. See Figure 4.4.6. Since $f(x) \geq g(x)$ for all x in $[-2, 3]$, the area is

$$A = \int_{-2}^{3} [f(x) - g(x)]\, dx = \int_{-2}^{3} \left[(2x+8) - (-x^2 + 3)\right]\, dx$$

$$= \int_{-2}^{3} (x^2 + 2x + 5)\, dx = \left[\frac{1}{3}x^3 + x^2 + 5x\right]_{-2}^{3}$$

$$= \frac{1}{3}\left((3)^3 - (-2)^3\right) + \left((3)^2 - (-2)^2\right) + 5((3) - (-2))$$

$$= \frac{35}{3} + (5) + (25) = 41\frac{2}{3}.$$

Example 3: Determine the area of the region bounded by the graphs of $g(x) = x^2 + 1$ and $f(x) = 2x + 4$.

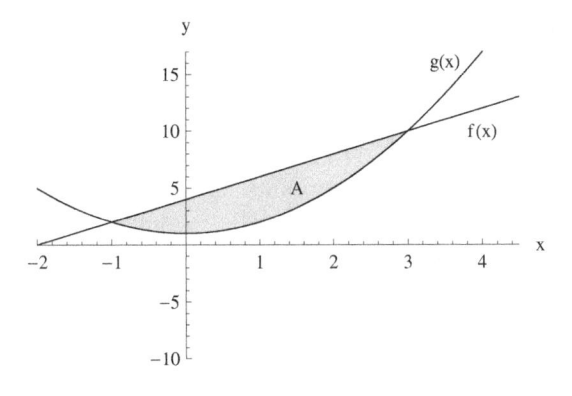

Figure 4.4.7: Area between two curves

Solution We begin with a graph of f and g and shade the area. See Figure 4.4.7. To determine the intersection points, we solve the equation $f(x) = g(x)$:

$x^2 + 1 = 2x + 4$

$x^2 - 2x - 3 = 0$

$(x + 1)(x - 3) = 0$

$x = -1 \quad \text{or} \quad x = 3.$

Hence, the limits of the integral are -1 and 3. Since $f(x) \geq g(x)$ for all x in $[-1, 3]$, the area is

$$A = \int_{-1}^{3} [f(x) - g(x)]\, dx = \int_{-1}^{3} \left[(2x + 4) - (x^2 + 1)\right]\, dx$$

$$= \int_{-1}^{3} (-x^2 + 2x + 3) \, dx = \left[-\frac{1}{3}x^3 + x^2 + 3x \right]_{-1}^{3}$$

$$= -\frac{1}{3}\left((3)^3 - (-1)^3\right) + \left((3)^2 - (-1)^2\right) + 3((3) - (-1))$$

$$= -\frac{28}{3} + (8) + (12) = 10\frac{2}{3}.$$

Example 4: Determine the area of the region bounded by the graphs of $f(x) = \sqrt{x} + 1$, $g(x) = x - 1$, $x = 0$, and $x = 9$.

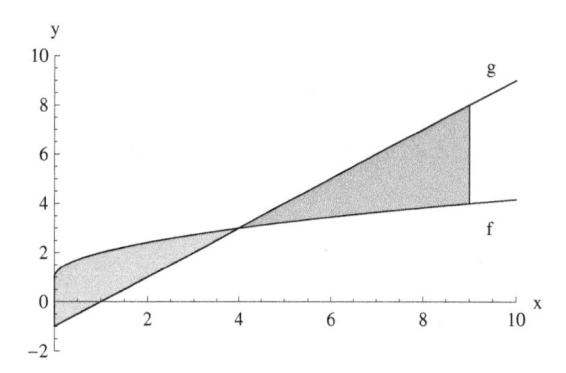

Figure 4.4.8: Area between two curves

Solution We begin with a graph of f, g, $x = 0$, and $x = 9$, and shade the area. See Figure 4.4.8. It indicates that the graphs of f and g intersect and switch positions. To determine the intersection point, we solve the equation $f(x) = g(x)$:

$$\sqrt{x} + 1 = x - 1$$

$$\sqrt{x} = x - 2, \quad \text{so } x = (x - 2)^2 = x^2 - 4x + 4$$

$$x^2 - 5x + 4 = 0$$

$$(x - 1)(x - 4) = 0$$

$$x = 1 \quad \text{or} \quad x = 4.$$

Since the original equation involves a radical, we need to check both numbers. It turns out that $x = 1$ is not a solution, so we drop it. Since f and g change positions at $x = 4$, we determine the area in separate intervals: On $[0, 4]$, f is above g, and on $[4, 9]$, g is above f. Hence, the area is

$$A = \int_{0}^{4} [f(x) - g(x)] \, dx + \int_{4}^{9} [g(x) - f(x)] \, dx$$

We have

$$\int_{0}^{4} \left[(\sqrt{x} + 1) - (x - 1) \right] dx = \int_{0}^{4} (\sqrt{x} - x + 2) \, dx$$

$$= \left[\frac{2}{3}x^{\frac{3}{2}} - \frac{1}{2}x^2 + 2x \right]_{0}^{4} = \frac{2}{3}\left(4^{\frac{3}{2}} - 0\right) - \frac{1}{2}(4^2 - 0) + 2(4 - 0) = \frac{16}{3},$$

and

$$\int_{4}^{9} \left[(x - 1) - (\sqrt{x} + 1) \right] dx = \int_{4}^{9} (x - \sqrt{x} - 2) \, dx$$

$$= \left[\frac{1}{2}x^2 - \frac{2}{3}x^{\frac{3}{2}} - 2x \right]_{4}^{9} = \frac{1}{2}(9^2 - 4^2) - \frac{2}{3}\left(9^{\frac{3}{2}} - 4^{\frac{3}{2}}\right) - 2(9 - 4)$$

4.4. AREAS AND DEFINITE INTEGRALS

$$= \frac{65}{2} - \frac{38}{3} - 10 = 9\frac{5}{6}.$$

Hence, the total area is $5\frac{1}{3} + 9\frac{5}{6} = 15\frac{1}{6}.$

An Application

Example 5: A market analyst for a potato chip company estimates that with no promotion the weekly sales of one brand of its potato chips can be modeled by

$$s_1(t) = 0.8t + 5.2$$

in thousands of dollars per week, t weeks from now. This analyst estimates that, with a small-scale promotional campaign, the weekly sales can be modeled as

$$s_2(t) = 8.6 e^{0.06t}.$$

In the period of the first 20 weeks, how much can the promotional campaign increase the total sales?

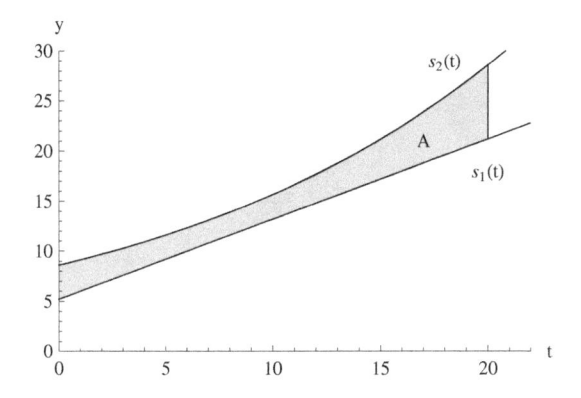

Figure 4.4.9: Area between two curves

Solution The increase in the total sales in the first 20 weeks is the difference between the total sales with and without the promotional campaign. Since the total sales are represented by the areas under the graphs of the sales rate functions, we are looking for the area between the two curves (Figure 4.4.9). Since $s_2(t) \geq s_1(t)$ on $[0, 20]$, the area is

$$A = \int_0^{20} [s_2(t) - s_1(t)] \, dt = \int_0^{20} \left[8.6 e^{0.06t} - (0.8t + 5.2)\right] \, dx$$

$$= \int_0^{20} \left(8.6 e^{0.06t} - 0.8t - 5.2\right) \, dx$$

$$= \left[\frac{8.6}{0.06} e^{0.06t} - 0.4 t^2 - 5.2 t\right]_0^{20}$$

$$= \frac{8.6}{0.06} \left(e^{0.06(20)} - e^{0.06(0)}\right) - 0.4 \left(20^2 - 0\right) - 5.2(20 - 0)$$

$$= \frac{8.6}{0.06} \left(e^{1.2} - 1\right) - (160) - (104) \approx 68.55.$$

Hence, in the period of the first 20 weeks, the promotional campaign can increase the total sales by about \$68,550.

Average Value of a Function

In the remaining part of this section we examine another important use of the area under a curve: The average value of a continuous function on an interval $[a, b]$. In order to better understand this concept, let us start with a discrete case:

Suppose that a car dealer sells 22 cars in the first week, 40 cars in the second week, and 28 cars in the third week. How many cars on average does the dealer sell each week? The answer is

$$\text{Average sales per week} = \frac{22 + 40 + 28}{3} = 30 \text{ cars per week.}$$

As is illustrated by the graph in Figure 4.4.10, the average sales, 30 cars per week, can be viewed as if the dealer sells 30 cars each week for the 3 weeks. The total sales during the 3 weeks equal 90 cars and is the same as the original.

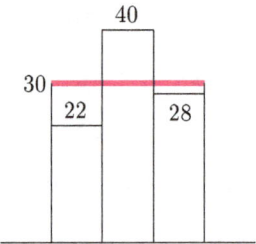

Figure 4.4.10: Average sales

Next, we consider a daily-life example. Mary has been driving on an interstate highway for 2 hours. The speed (the velocity stays positive over the 2 hours of travel) of the car is recorded as the following curve (Figure 4.4.11):

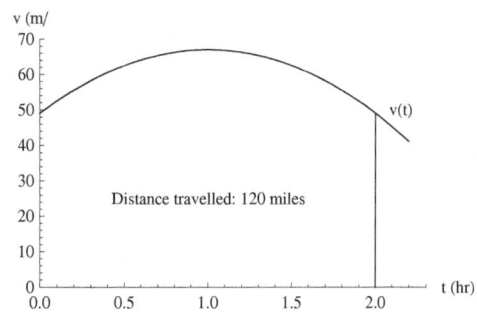

Figure 4.4.11: Total distance I

By the end of the 2-hour drive, her odometer indicates that she has traveled for 120 miles. So her *average speed* in the 2 hours of travel is

$$\text{Average speed} = \frac{120}{2} = 60 \text{ miles per hour.}$$

As is illustrated by the graph in Figure 4.4.12, the average speed, 60 miles per hour, can be viewed as if Mary drives 60 miles per hour constantly for the 2 hours. The total distance traveled during the 2 hours equal 120 miles, the same as the original.

4.4. AREAS AND DEFINITE INTEGRALS

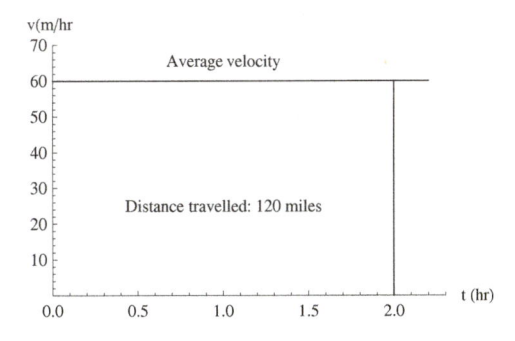

Figure 4.4.12: Total distance II

Since the total distance traveled in the 2 hours equals the area under the graph of the velocity on interval $[0, 2]$, the average velocity represents the height of the rectangle with base 2 and area $\int_0^2 v(t)\, dt$ (Figure 4.4.13).

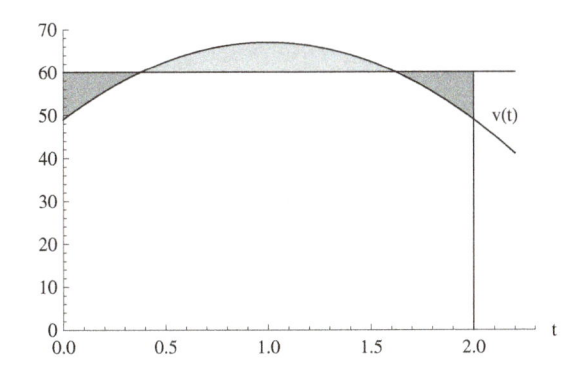

Figure 4.4.13: Total distance I and II

We now extend this concept to a general positive function. Given a positive and continuous function $f(x)$ on an interval $[a, b]$, the **average value** of $f(x)$ on $[a, b]$, denoted by f_{av}, is the y value that equals the height of the rectangle with base $b - a$ and area $\int_a^b f(x)\, dx$. Hence, it is calculated by

$$f_{\text{av}} = \frac{1}{b-a} \int_a^b f(x)\, dx.$$

This formula can be extended to cases when $f(x)$ is not always positive. We formally define the average value of a continuous function on an interval $[a, b]$:

Definition *the average value*

The **average value** f_{av} average of a continuous function f, on the interval $[a, b]$, is given by

$$f_{\text{av}} = \frac{1}{b-a} \int_a^b f(x)\, dx.$$

Example 6: Find the average value of $f(x) = 3/x$ on the interval $[1, 4]$.

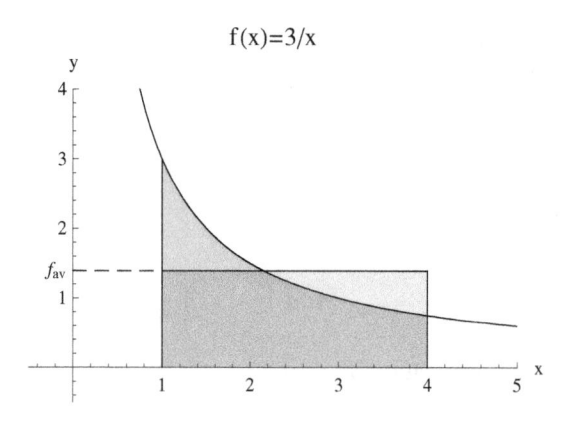

Figure 4.4.14: Average value of $f(x) = 3/x$ on $[1, 4]$

Solution We have

$$f_{\text{av}} = \frac{1}{4-1}\int_1^4 \frac{3}{x}\, dx = \frac{1}{3}\left[3\ln x\right]_1^4 = \ln 4 - \ln 1 = \ln 4 \approx 1.3863.$$

It is illustrated in Figure 4.4.14.

Applications

Example 7: Sarah has been driving on an interstate highway for 20 minutes. The speed of her car is recorded as the following function:

$$v(t) = -\frac{1}{180}t^3 + \frac{1}{8}t^2 + 50,\ 0 \le t \le 20,$$

where t is in minutes and v is in miles per hour. What is Sarah's average speed so far? How far has she traveled by the end of 20 minutes?

Solution We find Sarah's average speed by the formula

$$v_{\text{av}} = \frac{1}{20-0}\int_0^{20} v(t)\, dt = \frac{1}{20-0}\int_0^{20}\left(-\frac{1}{180}t^3 + \frac{1}{8}t^2 + 50\right)\, dt$$

$$= \frac{1}{20}\left[-\frac{1}{720}t^4 + \frac{1}{24}t^3 + 50t\right]_0^{20}$$

$$= \frac{1}{20-0}\left[-\frac{1}{720}(20)^4 + \frac{1}{24}(20)^3 + 50(20)\right] - 0 \approx 55.56\text{mph}.$$

Since Sarah's average speed is 55.56 mph and she has driven for 20 minutes $= 1/3$ hour, the distance traveled in the 20 minutes is $55.56/3 \approx 18.52$ miles.

Example 8: Betty deposits $2,000 into a money market account earning 5.5% interest compounded continuously.

(a) Determine Betty's average balance over 1 year.

(b) Suppose that at end of the year the fund pays a bonus that is equal to 1.5% of the average balance. Determine how much bonus Betty will receive.

4.4. AREAS AND DEFINITE INTEGRALS

Solution (a) By the balance formula provided in Section 1.5, the balance of the continuously compounded account is given by

$$A(t) = 2{,}000 e^{0.055t}.$$

Hence, the average value of $A(t)$ during one year is

$$A_{\text{av}} = \frac{1}{1-0} \int_0^1 2{,}000 e^{0.055t}\, dt$$

$$= \left[\frac{2{,}000}{0.055} e^{0.055t}\right]_0^1 = \frac{2{,}000}{0.055}\left(e^{0.055(1)} - e^{0.055(0)}\right) = \frac{2{,}000}{0.055}\left(e^{0.055} - 1\right) \approx \$2{,}056.02.$$

(b) Since Betty's average balance over 1 year is $2,056.02, her bonus is $0.015(2{,}056.02) \approx \$30.84$.

Exercise Set 4.4

For Exercises #1 to #4, use definite integral(s) to find the area under the graph on the following intervals:
(a) $[1,4]$; (b) $[-4,-1]$; (c) $[-4,4]$.
In each case, mark the region on the graph.

1. $f(x) = \begin{cases} 1 - x & \text{if } x \le 0, \\ 1 + 2x & \text{if } x > 0. \end{cases}$

2. $f(x) = \begin{cases} 3 & \text{if } x \le 0, \\ 3 + x^2 & \text{if } x > 0. \end{cases}$

3. $f(x) = \begin{cases} 2 - 0.5x & \text{if } x \le 0, \\ 0.2x^2 + 2 & \text{if } x > 0. \end{cases}$

4. $f(x) = \begin{cases} 1 + x^2 & \text{if } x \le 0, \\ 1 + x & \text{if } x > 0. \end{cases}$

For Exercises #5 to #8, use definite integrals to find the area under the graph on the interval $[0,6]$. Mark the region on the graph.

5. $f(x) = \begin{cases} 2 - x & \text{if } x \le 1, \\ 0.2x^2 + 0.8 & \text{if } x > 1. \end{cases}$

6. $f(x) = \begin{cases} 1 + x^2 & \text{if } x \le 3, \\ 10 & \text{if } x > 3. \end{cases}$

7. $f(x) = \begin{cases} 4 & \text{if } x \le 2, \\ x^2 & \text{if } x > 2. \end{cases}$

8. $f(x) = \begin{cases} x^2 + 3 & \text{if } x \le 2, \\ 1 + 3x & \text{if } x > 2. \end{cases}$

For Exercises #9 to #16: Find the x-value(s) at which the graphs of f and g cross. State the fact if no such x-value exists.

9. $f(x) = 4,\ g(x) = x^2$

10. $f(x) = 10,\ g(x) = x^2 + 1$

11. $f(x) = x - 1,\ g(x) = x^2 - 3$

12. $f(x) = 4 - 2x,\ g(x) = x^2 - 3x - 2$

13. $f(x) = 3,\ g(x) = \sqrt{x} + 1$

14. $f(x) = 4,\ g(x) = \sqrt{x} - 5$

15. $f(x) = -2,\ g(x) = x^2 + 1$

16. $f(x) = x - 1,\ g(x) = x^2$

For Exercises #17 to #20: Find the area of the shaded region: (a) Determine the integral(s) representing the area; (b) Evaluate.

17. $f(x) = x + 2x^2 - x^3,\ g(x) = 2$

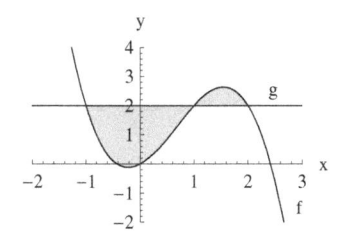

18. $f(x) = 2 + \sqrt{x}$, $g(x) = 4 - x$

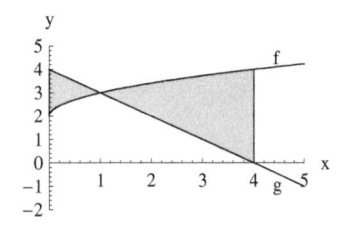

19. $g(x) = x^2 - 6x + 12$, $f(x) = -x^2 + 6x + 2$

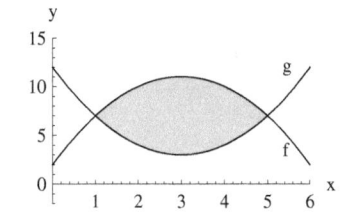

20. $g(x) = 0$, $f(x) = x^2 - 1$

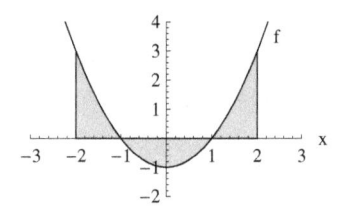

For Exercises #21 to #30: Find the area between the graphs of f and g on the given interval: (a) Draw a graph to illustrate the region; (b) Determine the integral(s) representing the area; (c) Evaluate.

21. $f(x) = x + 1$, $g(x) = x^2 + 3$, on $[-2, 2]$

22. $f(x) = x^2 - 2x - 2$, $g(x) = 5$, on $[-1, 2]$

23. $f(x) = 2x - 1$, $g(x) = 1 - x$, on $[-2, 3]$

24. $f(x) = x - 1$, $g(x) = -2$, on $[-3, 1]$

25. $f(x) = \sqrt{x} + 2$, $g(x) = x + 3$, on $[1, 4]$

26. $f(x) = e^x + 1$, $g(x) = 1 - x$, on $[0, 2]$

27. $f(x) = \sqrt{x} + 2$, $g(x) = x - 3$, on $[1, 4]$

28. $f(x) = 4$, $g(x) = 1 - e^{-x}$, on $[-1, 2]$

29. $f(x) = \dfrac{2}{x}$, $g(x) = \frac{1}{4}x^2$, on $[1, 4]$

30. $f(x) = \dfrac{4}{x^2}$, $g(x) = \frac{1}{2}x$, on $[1, 5]$

For Exercises #31 to #42: Find the area of the region bounded by the graphs of the given equations: (a) Draw a graph to illustrate the region; (b) Determine the integral(s) representing the area; (c) Evaluate.

31. $y = x$ and $y = x^2$

32. $y = 8\sqrt{x}$ and $y = x^2$

33. $y = -2x$ and $y = 3 - x^2$

34. $y = x^3$ and $y = x^2$

35. $y = x^3$ and $y = 4x$

36. $y = x^3$ and $y = x$

37. $y = \sqrt{x}$, $y = 3$, and $x = 0$

38. $y = \sqrt{x}$, $y = 2$, and $x = 0$

39. $y = 4 - x^2$ and $y = x^2 - 4$

40. $y = 4 - x^2$ and $y = 4 - 4x$

41. $y = x^2$, $y = x^2 + 1$, $x = -1$, and $x = 2$

42. $y = x^2$, $y = x^2 - 2$, $x = 0$, and $x = 2$

For Exercises #43 to #52: Find the average value of f on the given interval. Draw a graph to illustrate the result.

43. $f(x) = e^x$; $[0, 2]$

44. $f(x) = e^{-x}$; $[0, 1]$

45. $f(x) = 4 - x^2$; $[-2, 2]$

46. $f(x) = x^2 - 2x + 1$; $[0, 2]$

47. $f(x) = 2x + 1$; $[0, 4]$

48. $f(x) = 2 - 0.5x$; $[0, 4]$

49. $f(x) = \dfrac{3}{x^2}$; $[1, 3]$

50. $f(x) = \dfrac{2}{x^2}$; $[1, 4]$

51. $f(x) = \sqrt{x} + 1$; $[0, 4]$

52. $f(x) = 4 - \sqrt{x}$; $[0, 4]$

APPLICATIONS

53. Two cars enter a freeway at the same time. The velocities, measured by miles per minute, of Car A and Car B t minutes after entering the freeway are given by

    $v_A(t) = 0.4 + 0.2t - 0.02t^2$, $0 \leq t \leq 10$,
    $v_B(t) = 0.2 + 0.25t - 0.02t^2$, $0 \leq t \leq 10$.

    At the end of 10 minutes of driving, which car is ahead and by how far?

54. Two cars enter a freeway at the same time. The velocities, measured by miles per minute, of Car A and Car B t minutes after entering the freeway are given by

$v_A(t) = 0.4 + 0.2t - 0.02t^2$, $0 \le t \le 5$,
$v_B(t) = 0.1 + 0.3t - 0.02t^2$, $0 \le t \le 5$.

At the end of 5 minutes of driving, which car is ahead and by how far?

55. A market analyst for a potato chip company estimates that with no promotion the weekly sales of one brand of its potato chips can be modeled by

$s_1(t) = 0.6t + 4.5$ thousand dollars per week, t weeks from now. This analyst estimates that, with a small-scale promotional campaign, the weekly sales can be modeled as $s_2(t) = 7.2e^{0.06t}$. In the period of the first 20 weeks, how much can the promotional campaign increase the total sales?

56. A market analyst for a chocolate company estimates that with no promotion the weekly sales of one brand of its chocolate bars can be modeled by

$s_1(t) = 0.6t + 6.2$ thousand dollars per week, t weeks from now. This analyst estimates that, with a small-scale promotional campaign, the weekly sales can be modeled as $s_2(t) = 8.8e^{0.05t}$. In the period of the first 16 weeks, how much can the promotional campaign increase the total sales?

57. A service station orders 80 cases of motor oil every three months. The number of cases of oil remaining t months after the order arrives is modeled by

$f(t) = 80e^{-0.9t}$.

(a) How many cases are there at the start of the 3-month period? How many cases are left at the end of the 3-month period?

(b) Determine the average number of cases in inventory over the 3-month period.

58. A service station orders 150 cases of motor oil every six months. The number of cases of oil remaining t months after the order arrives is modeled by

$f(t) = 150e^{-0.6t}$.

(a) How many cases are there at the start of the 6-month period? How many cases are left at the end of the 6-month period?

(b) Determine the average number of cases in inventory over the 6-month period.

59. David has been driving on an interstate highway for 15 minutes. The speed of the car is recorded as the following function:

$v(t) = -\frac{1}{180}t^3 + \frac{1}{8}t^2 + 50$, $0 \le t \le 15$, where t is in minutes and v is in miles per hour. What is David's average speed so far? How far has he traveled by the end of 15 minutes?

60. Sara has been driving on an interstate highway for 10 minutes. The speed of the car is recorded as the following function:

$v(t) = -\frac{1}{180}t^3 + \frac{1}{8}t^2 + 50$, $0 \le t \le 10$, where t is in minutes and v is in miles per hour. What is Sara's average speed so far? How far has she traveled by the end of 10 minutes?

61. Karen deposits $5,000 into a money market account earning interest at an annual rate of 6% compounded continuously.

(a) Determine Karen's average balance over one year.

(b) Suppose that at end of the year the fund pays a bonus that is equal to 1.2% of the average balance. Determine how much bonus Karen will receive.

62. Jim deposits $6,000 into a money market account interest at an annual rate of 5.5% compounded continuously.

(a) Determine Jim's average balance over one year.

(b) Suppose that at end of the year the fund pays a bonus that is equal to 1.2% of the average balance. Determine how much bonus Jim will receive.

63. Jane has a checking account in a bank that requires an average daily balance of $200 to avoid an $8 monthly fee. If the average daily balance is above $200, then a monthly interest payment equal to 1.2% of the average balance will be added to the account. Jane's daily balance, in dollars, over the month can be modeled as

$f(t) = -\frac{1}{200}t^3 + \frac{3}{20}t^2 - \frac{3}{8}t + 210$, $0 \le t \le 30$.

(a) Determine Jane's average daily balance over the month.

(b) Determine whether Jane will pay the $8 fee, or she will receive an interest payment. If it is the latter, then find how much interest she will receive.

64. Kathy has a checking account in a bank that requires an average daily balance of $300 to avoid a $10 monthly fee. If the average daily balance is above $300, then a monthly interest payment equal to 1.4% of the average balance will be added to the account. Kathy's daily balance, in dollars, over the month can be modeled as
$$f(t) = \frac{1}{160}t^3 - \frac{3}{20}t^2 + \frac{1}{4}t + 285,\ 0 \le t \le 30.$$

(a) Determine Kathy's average daily balance over the month.

(b) Determine whether Kathy will pay the $10 fee, or she will receive an interest payment. If it is the latter, then find how much interest she will receive.

65. An ice cream store owner determines that its sales, $S(t)$, in dollars, t weeks after the beginning of the summer, can be estimated by
$$S(t) = 800e^{0.1t},\ 0 \le t \le 12.$$

Find the average weekly sales for the first 6 weeks of the summer.

66. A coffee shop owner determines that its revenue, $R(t)$, in dollars, t weeks after the beginning of the year, can be estimated by
$$S(t) = 500e^{0.15t},\ 0 \le t \le 10.$$

Find the average weekly revenue for the first 5 weeks of the year.

67. Refer to Exercise #65. Find the average weekly sales of the ice cream store from week 3 through week 7 of the summer ($t = 2$ to $t = 7$).

68. Refer to Exercise #66. Find the average weekly revenue of the coffee shop from week 2 through week 6 of the year ($t = 1$ to $t = 6$).

69. A mutual fund has a growth rate recorded each year, expressed by the function
$$r(t) = 200e^{0.04t},\ 0 \le t \le 10$$
(in thousands of dollars per year), where t is the number of years passed from year 1997. Determine the average growth rate of the fund from year 1997 to year 2007.

70. A mutual fund has a growth rate recorded each year, expressed by the function
$$r(t) = 200e^{-0.04t},\ 0 \le t \le 7$$
(in thousands of dollars per year), where t is the number of years passed from year 2000. Determine the average growth rate of the fund from year 2000 to year 2007.

4.5 Integration by Substitution or Algebraic Manipulation

The first four sections of this chapter are devoted to concepts related to areas and integrals. Thus far we have learned how to apply the basic integration formulas and properties to integrate a group of basic functions. However, we need additional integration techniques to integrate more complicated functions, such as $f(x) = x\sqrt{x^2 + 1}$ or $f(x) = \ln x$. The remaining sections are devoted to enhancing our capacity for integration. In this section, we first illustrate how to use algebraic manipulation to rewrite an integral so that basic integration formulas can be applied. We then focus on the integration of a *composite* function.

▬ Integration by Algebraic Manipulation

Often an integral with a product or a quotient can be rewritten as an expression that allows basic integration formulas to be applied. We illustrate such cases with a few examples.

Example 1: Determine the following integrals:

(a) $\int x\left(3x^2 + 2\right)\,dx$ (b) $\int \sqrt{x}(2x + 1)\,dx$ (c) $\int \left(x^3 + 1\right)^2\,dx$

4.5. INTEGRATION BY SUBSTITUTION OR ALGEBRAIC MANIPULATION

Solution (a) We rewrite $x\left(3x^2 + 2\right)$ as $3x^3 + 2x$, then apply the power formulas:

$$\int x\left(3x^2 + 2\right)\, dx = \int \left(3x^3 + 2x\right)\, dx = \frac{3}{4}x^4 + x^2 + C.$$

(b) We rewrite $\sqrt{x}(2x + 1)$ as $2x^{3/2} + x^{1/2}$, then apply the power formulas:

$$\int \sqrt{x}(2x + 1)\, dx = \int \left(2x^{3/2} + x^{1/2}\right)\, dx = \frac{4}{5}x^{\frac{5}{2}} + \frac{2}{3}x^{\frac{3}{2}} + C.$$

(c) We rewrite $\left(x^3 + 1\right)^2$ as $x^6 + 2x^3 + 1$, then apply the power formulas:

$$\int \left(x^3 + 1\right)^2 dx = \int \left(x^6 + 2x^3 + 1\right)\, dx = \frac{1}{7}x^7 + \frac{1}{2}x^4 + x + C.$$

Example 2: Determine the following integrals:

(a) $\displaystyle\int \frac{3x^2 + 2}{x}\, dx$ (b) $\displaystyle\int \frac{x + 1}{\sqrt{x}}\, dx$ (c) $\displaystyle\int \frac{4}{e^{2x}}\, dx$

Solution (a) We rewrite

$$\frac{3x^2 + 2}{x} \text{ as } \frac{3x^2}{x} + \frac{2}{x} = 3x + \frac{2}{x},$$

then apply the power and logarithmic formulas:

$$\int \frac{3x^2 + 2}{x}\, dx = \int \left(3x + \frac{2}{x}\right)\, dx = \frac{3}{2}x^2 + 2\ln|x| + C.$$

(b) We rewrite

$$\frac{x + 1}{\sqrt{x}} \text{ as } \frac{x}{\sqrt{x}} + \frac{1}{\sqrt{x}} = \sqrt{x} + \frac{1}{\sqrt{x}},$$

then apply the power formulas:

$$\int \frac{x + 1}{\sqrt{x}}\, dx = \int \left(\sqrt{x} + \frac{1}{\sqrt{x}}\right)\, dx = \frac{2}{3}\sqrt{x^3} + 2\sqrt{x} + C.$$

(c) We rewrite $\dfrac{4}{e^{2x}}$ as $4e^{-2x}$, then apply the exponential formula:

$$\int \frac{4}{e^{2x}}\, dx = \int 4e^{-2x}\, dx = -2e^{-2x} + C.$$

Integration by Substitution

We now turn our attention to the integration of composite functions. For example, how do we approach

$$\int 6x\sqrt{x^2+1}\, dx\ ?$$

This time we cannot rewrite $6x\sqrt{x^2+1}$ in power terms, since we cannot separate terms within a radical. Recall that, in differentiation, the Chain Rule must be applied to a composite function. If we let

$$F(x) = 2\sqrt{(x^2+1)^3} = 2\left(x^2+1\right)^{3/2}, \quad \text{then}$$

$$F'(x) = 2\,\frac{3}{2}\left(x^2+1\right)^{\frac{1}{2}} \cdot 2x = 6x\left(x^2+1\right)^{\frac{1}{2}} = 6x\sqrt{x^2+1}.$$

Hence, we have

$$\int 6x\sqrt{x^2+1}\, dx = 2\sqrt{(x^2+1)^3} + C.$$

The aforementioned example illustrates that integrating a composite function involves reversing the use of the Chain Rule. However, finding $F(x)$ directly is not straightforward. In order to carry out the process, we need to recall some basic integration formulas, the definition of a differential, and the Leibniz notation of the Chain Rule.

We first recall the three basic integration formulas we are going to apply:

Basic Integration Formulas

A. (Power Formula for Integration) $\int u^r\, du = \dfrac{1}{r+1} u^{r+1} + C$, provide $r \neq -1$

B. Integration Formula for $\dfrac{1}{u}$ $\quad \int \dfrac{1}{u}\, du = \ln|u| + C$, or $\int \dfrac{1}{u}\, du = \ln u + C$, if $u > 0$
(We generally assume that $u > 0$)

C. (Exponential Formula for Integration) $\int e^u\, du = e^u + C$

Reality Check   *The middle variable u*

Why does the variable u replace the variable x in the list? Because u will play the role as the "middle function."

Recall that in Section 3.1 the differential is defined as

$$du = \frac{du}{dx} \cdot dx = g'(x)dx$$

if $u = g(x)$. Also recall that in Leibniz notation, if $y = f(u)$ and $u = g(x)$, then the Chain Rule is stated as

$$\frac{dy}{dx} = \frac{dy}{du} \cdot \frac{du}{dx} = f'(u)\frac{du}{dx}.$$

Hence, if

$$y = F(x) = f(g(x)),$$

then

$$F'(x) = f'(u)g'(x), \quad \text{or}\quad F'(x)dx = f'(u)g'(x)dx = f'(u)du.$$

4.5. INTEGRATION BY SUBSTITUTION OR ALGEBRAIC MANIPULATION

If we can use the basic integration formulas to find $f(u)$, then we can find $F(x)$ by replacing $u = g(x)$ back into the formula. This process is called **integration by substitution**. We illustrate this process with the following examples.

Example 3: Determine the indefinite integral $\int 6x\sqrt{x^2+1}\, dx$.

Solution Let $u = x^2 + 1$, then $\dfrac{du}{dx} = 2x$, hence $du = 2x\, dx$. We substitute $x^2 + 1$ by u and $2x\, dx$ by du in the integral to obtain

$$\int 6x\sqrt{x^2+1}\, dx = \int 3\sqrt{x^2+1}\, 2x\, dx = \int 3\sqrt{u}\, du.$$

We then apply the power formula to obtain

$$\int 3\sqrt{u}\, du = 3\int u^{\frac{1}{2}}\, du = 3 \cdot \frac{2}{3} u^{\frac{3}{2}} + C = 2\sqrt{u^3} + C.$$

Finally we substitute $u = x^2 + 1$ back into the formula:

$$2\sqrt{u^3} + C = 2\sqrt{(x^2+1)^3} + C.$$

Therefore $\int 6x\sqrt{x^2+1}\, dx = 2\sqrt{(x^2+1)^3} + C.$

How to determine u and du

How do we decide which part of the integrand should be u? Note that in the three basic formulas, u is either "inside" of a power, or "above" base e. Hence, if the integrand has an exponential expression, then u should be the whole exponent part. Otherwise, u should be the expression inside a power. After u is picked, we take its derivative with respect to x then multiply it with dx to obtain du. Now return to the original integral to do the "substitution" step. The resulting integral should fit in one of the three basic formulas and there should have no variable x left.

Example 4: Determine the indefinite integral $\int 3x^2 \left(x^3 + 5\right)^3\, dx$.

Solution We let $u = x^3 + 5$, then $du = 3x^2\, dx$, hence,

$$\int 3x^2 \left(x^3+5\right)^3\, dx = \int \left(x^3+5\right)^3 3x^2\, dx = \int u^3\, du$$

$$= \frac{1}{4}u^4 + C = \frac{1}{4}(x^3+5)^4 + C.$$

Example 5: Determine the indefinite integral $\int \dfrac{3e^x}{e^x+2}\, dx$.

Solution We let $u = e^x + 2$, then $du = e^x\, dx$, hence,

$$\int \frac{3e^x}{e^x+2}\, dx = 3\int \frac{1}{e^x+2} e^x\, dx = 3\int \frac{1}{u}\, du$$

$$= 3\ln u + C = 3\ln(e^x+2) + C.$$

Example 6: Determine the indefinite integral $\int \frac{x^3}{(x^4+1)^2}\, dx$.

Solution We let $u = x^4 + 1$, then $du = 4x^3\, dx$. We have $x^3\, dx$ in the integral, hence, we need the coefficient 4. We can rewrite the equation

$$du = 4x^3\, dx \quad \text{to} \quad \frac{1}{4}\, du = x^3\, dx.$$

Hence

$$\int \frac{x^3}{(x^4+1)^2}\, dx = \int \frac{1}{(x^4+1)^2}\, x^3\, dx$$

$$= \int \frac{1}{u^2}\, \frac{1}{4}\, du = \frac{1}{4} \int \frac{1}{u^2}\, du$$

$$= \frac{1}{4}\left(-\frac{1}{u}\right) + C = \frac{-1}{4(x^4+1)} + C.$$

Example 7: Determine the indefinite integral $\int x e^{-0.03x^2}\, dx$.

Solution We let $u = -0.03x^2$, then $du = -0.06x\, dx$. We have $x\, dx$ in the integral, hence, we need the coefficient -0.06. We can rewrite the equation

$$du = -0.06x\, dx \quad \text{to} \quad \frac{1}{-0.06}\, du = x\, dx, \quad \text{or} \quad \frac{-1}{0.06}\, du = x\, dx.$$

Hence,

$$\int x e^{-0.03x^2}\, dx = \int e^{-0.03x^2}\, x\, dx$$

$$= \int e^u\, \frac{-1}{0.06}\, du = \frac{-1}{0.06} \int e^u\, du$$

$$= \frac{-1}{0.06} e^u + C = \frac{-1}{0.06} e^{-0.03x^2} + C.$$

Example 8: Determine the indefinite integral $\int \frac{\ln x}{x}\, dx$.

Solution We let $u = \ln x$, then $du = \frac{1}{x}\, dx$, hence

$$\int \frac{\ln x}{x}\, dx = \int \ln x\, \frac{1}{x}\, dx = \int u\, du$$

$$= \frac{1}{2}u^2 + C = \frac{1}{2}(\ln x)^2 + C.$$

We now explore the technique of substitution applied to evaluating definite integrals. There are two approaches that accomplish this task. They are illustrated in the following examples.

Example 9: Evaluate the definite integral $\int_0^4 \frac{x}{\sqrt{x^2+9}}\, dx$.

Solution **Approach 1:** We first determine the indefinite integral $\int \frac{x}{\sqrt{x^2+9}}\, dx$: We let $u = x^2 + 9$, then $du = 2x\, dx$, or $\dfrac{1}{2} du = x\, dx$. Hence,

$$\int \frac{x}{\sqrt{x^2+9}}\, dx = \int \frac{1}{\sqrt{x^2+9}}\, x\, dx$$

$$= \int \frac{1}{\sqrt{u}}\, \frac{1}{2}\, du = \frac{1}{2} \int \frac{1}{\sqrt{u}}\, du$$

$$= \frac{1}{2}\left(2\sqrt{u}\right) + C = \sqrt{u} + C = \sqrt{x^2+9} + C.$$

Hence,

$$\int_0^4 \frac{x}{\sqrt{x^2+9}}\, dx = \left[\sqrt{x^2+9}\right]_0^4$$

$$= \sqrt{4^2+9} - \sqrt{0^2+9} = \sqrt{25} - \sqrt{9} = 5 - 3 = 2.$$

Approach 2: We let $u = x^2 + 9$, then $du = 2x\, dx$, or $\dfrac{1}{2} du = x\, dx$. When $x = 0$, we have $u = 9$. When $x = 4$, we have $u = 25$. We now substitute all formulas **and the limits** of x in the definite integral by that of u:

$$\int_0^4 \frac{x}{\sqrt{x^2+9}}\, dx = \int_0^4 \frac{1}{\sqrt{x^2+9}}\, x\, dx$$

$$= \int_9^{25} \frac{1}{\sqrt{u}}\, \frac{1}{2}\, du = \frac{1}{2} \int_9^{25} \frac{1}{\sqrt{u}}\, du$$

$$= \left[\frac{1}{2}\left(2\sqrt{u}\right)\right]_9^{25} = \sqrt{25} - \sqrt{9} = 5 - 3 = 2.$$

Notice that in this approach, there is no back-substitution.

Example 10: Evaluate the definite integral $\displaystyle\int_{-1}^0 3x^2 \left(x^3+2\right)^3\, dx$.

Solution Approach 1: From Example 4,

$$\int 3x^2 \left(x^3+2\right)^3\, dx = \frac{1}{4}\left(x^3+2\right)^4 + C. \text{ Hence,}$$

$$\int_{-1}^0 3x^2 \left(x^3+2\right)^3\, dx = \left[\frac{1}{4}\left(x^3+2\right)^4\right]_{-1}^0$$

$$= \frac{1}{4}\left[(0+2)^4 - ((-1)^3+2)^4\right] = \frac{1}{4}[16-1] = \frac{15}{4}.$$

Approach 2: Refer to Example 4, We let $u = x^3 + 2$, then $du = 3x^2\, dx$. When $x = -1$, we have $u = 1$. When $x = 0$, we have $u = 2$. We now substitute all formulas **and the limits** of x in the definite integral by that of u:

$$\int_{-1}^0 3x^2 \left(x^3+2\right)^3\, dx = \int_{-1}^0 \left(x^3+2\right)^3 3x^2\, dx$$

$$= \int_1^2 u^3\, du = \left[\frac{1}{4}u^4\right]_1^2 = \frac{1}{4}\left[2^4 - 1^4\right] = \frac{1}{4}[16-1] = \frac{15}{4}.$$

Notice that in this approach, there is no back-substitution.

Example 11: Evaluate the definite integral $\int_0^1 \frac{x^3}{(x^4+1)^2}\,dx$.

Solution **Approach 1:** From Example 6,

$$\int \frac{x^3}{(x^4+1)^2}\,dx = \frac{-1}{4(x^4+1)} + C.$$

Hence,

$$\int_0^1 \frac{x^3}{(x^4+1)^2}\,dx = \left[\frac{-1}{4(x^4+1)}\right]_0^1$$

$$= \frac{-1}{4(1+1)} - \frac{-1}{4(0+1)} = -\frac{1}{8} + \frac{1}{4} = \frac{1}{8}.$$

Approach 2: Refer to Example 6, We let $u = x^4 + 1$, then $du = 4x^3\,dx$, or $\frac{1}{4}du = x^3\,dx$. When $x = 0$, we have $u = 1$. When $x = 1$, we have $u = 2$. We now substitute all formulas **and the limits** of x in the definite integral by that of u:

$$\int_0^1 \frac{x^3}{(x^4+1)^2}\,dx = \int_0^1 \frac{1}{(x^4+1)^2}\,x^3\,dx$$

$$= \int_1^2 \frac{1}{u^2}\frac{1}{4}\,du = \left[\frac{1}{4}\left(-\frac{1}{u}\right)\right]_1^2 = \frac{1}{4}\left[-\frac{1}{2} - (-1)\right] = \frac{1}{4}\cdot\frac{1}{2} = \frac{1}{8}.$$

Notice that in this approach, there is no back-substitution.

Before we leave this section, we should discuss other integration techniques. Recall that in differentiation, besides the basic formulas/rules and the Chain Rule, we also have the Product Rule and the Quotient Rule. Since the integration technique of substitution is the reverse use of the Chain Rule, what integration techniques correspond to the other two rules? Since the Quotient Rule can be replaced by the combination of the Product Rule and the Chain Rule (a quotient is the product of the numerator and the negative power of the denominator), there is only the Product Rule which has a corresponding technique in integration. It is called **integration by parts**. However, we are not going to be applying this technique, since the Integration Table introduced in Section 4.6 includes all the formulas used in this course which may be derived by that technique.

Exercise Set 4.5

For Exercises #1 to #8: Determine the indefinite integrals by algebraic manipulation.

1. $\int x^2\left(4x^3 + 3x - 5\right)\,dx$

2. $\int \sqrt{x}(2x+1)\,dx$

3. $\int \left(\sqrt{x} + 2\right)(3x - 1)\,dx$

4. $\int \left(2x^4 - 1\right)^2\,dx$

5. $\int \frac{2x^3 - 4}{x^2}\,dx$

6. $\int \frac{1 + 2x^3}{x}\,dx$

7. $\int \frac{2}{e^{3x}}\,dx$

8. $\int \frac{e^{2x} + 2}{e^x}\,dx$

4.5. INTEGRATION BY SUBSTITUTION OR ALGEBRAIC MANIPULATION

For Exercises #9 to #50: Determine the indefinite integrals by substitution. (It is recommended that you check your results by differentiation.) Assume $u > 0$ when $\ln u$ appears.

9. $\displaystyle\int 2x\left(x^2+1\right)^6 dx$

10. $\displaystyle\int 4x\left(2x^2+3\right)^4 dx$

11. $\displaystyle\int 6x^2\left(2x^3-5\right)^5 dx$

12. $\displaystyle\int 3x^2\left(x^3-2\right)^8 dx$

13. $\displaystyle\int \sqrt{x}\left(\sqrt{x^3}+2\right)^3 dx$

14. $\displaystyle\int \sqrt[3]{x}\left(\sqrt[3]{x^4}-1\right)^4 dx$

15. $\displaystyle\int xe^{3x^2} dx$ 16. $\displaystyle\int 2x^2 e^{4x^3} dx$

17. $\displaystyle\int \frac{4x^3}{x^4+1} dx$ 18. $\displaystyle\int \frac{4x^3}{(x^4+1)^2} dx$

19. $\displaystyle\int \frac{1}{3x+4} dx$ 20. $\displaystyle\int \frac{1}{2x-3} dx$

21. $\displaystyle\int \frac{(2\sqrt{x}+3)^3}{\sqrt{x}} dx$ 22. $\displaystyle\int \frac{(3\sqrt{x}-2)^5}{\sqrt{x}} dx$

23. $\displaystyle\int \frac{\ln 3x}{x} dx$ 24. $\displaystyle\int \frac{3(\ln x)^2}{x} dx$

25. $\displaystyle\int \frac{4}{\sqrt{4-x}} dx$ 26. $\displaystyle\int \frac{3}{\sqrt{2x+1}} dx$

27. $\displaystyle\int \frac{-3e^x}{(2e^x+1)^3} dx$ 28. $\displaystyle\int \frac{-4e^{-x}}{3e^{-x}-2} dx$

29. $\displaystyle\int x\sqrt{x^2+4}\, dx$ 30. $\displaystyle\int x^2\sqrt{x^3+2}\, dx$

31. $\displaystyle\int \frac{e^{2x}}{\sqrt{e^{2x}+1}} dx$ 32. $\displaystyle\int \frac{2}{x\sqrt{\ln x+4}} dx$

33. $\displaystyle\int \frac{1}{(3x+5)^2} dx$ 34. $\displaystyle\int \frac{3}{(2x-3)^2} dx$

35. $\displaystyle\int \frac{\ln x^4}{x} dx$ 36. $\displaystyle\int \frac{\ln \sqrt{x}}{x} dx$

37. $\displaystyle\int e^{-2x+1} dx$ 38. $\displaystyle\int 3e^{-3x-2} dx$

39. $\displaystyle\int 2e^{x/3} dx$ 40. $\displaystyle\int 3e^{x/2} dx$

41. $\displaystyle\int \frac{e^{\sqrt{x}}}{\sqrt{x}} dx$ 42. $\displaystyle\int \frac{1}{\sqrt{x}e^{\sqrt{x}}} dx$

43. $\displaystyle\int \frac{1}{x\ln x} dx$ 44. $\displaystyle\int \frac{1}{x\ln x^3} dx$

45. $\displaystyle\int \frac{2}{x^2}e^{2/x} dx$ 46. $\displaystyle\int \frac{1}{x^3}e^{3/x^2} dx$

47. $\displaystyle\int \left(3x^2+2x+1\right)\left(x^3+x^2+x+1\right)^6 dx$

48. $\displaystyle\int \left(6x^2+8x+5\right)\left(2x^3+4x^2+5x+1\right)^5 dx$

49. $\displaystyle\int \left(\sqrt{x}\left(1+\sqrt{x^3}\right)^3 + 2xe^{3x^2}\right) dx$

50. $\displaystyle\int \left(x^2\sqrt{1+x^3} + 4xe^{-3x^2}\right) dx$

For Exercises #51 to #70: Evaluate the definite integrals.

51. $\int_0^1 \dfrac{x}{(x^2+1)^3}\,dx$

52. $\int_0^1 \dfrac{8x}{(2x^2+1)^2}\,dx$

53. $\int_{-1}^1 x\left(x^2-1\right)^5\,dx$

54. $\int_1^2 x\left(x^2-1\right)^5\,dx$

55. $\int_1^9 \dfrac{\left(\sqrt{t}-2\right)^2}{\sqrt{t}}\,dt$

56. $\int_1^8 \dfrac{\left(\sqrt[3]{t}+1\right)^3}{\sqrt[3]{t^2}}\,dt$

57. $\int_0^1 2re^{r^2}\,dr$

58. $\int_0^1 3r^2 e^{r^3}\,dr$

59. $\int_1^2 \dfrac{1}{3x+4}\,dx$

60. $\int_1^2 \dfrac{2}{2x+1}\,dx$

61. $\int_1^e \dfrac{\ln x^2}{x}\,dx$

62. $\int_1^e \dfrac{\ln\sqrt{x}}{x}\,dx$

63. $\int_1^4 \dfrac{2}{\sqrt{x}\left(\sqrt{x}+2\right)^2}\,dx$

64. $\int_1^{27} \dfrac{3}{\sqrt[3]{x^2}\left(\sqrt[3]{x}+1\right)^2}\,dx$

65. $\int_0^3 \dfrac{x}{\sqrt{x^2+16}}\,dx$

66. $\int_2^{10} \dfrac{1}{\sqrt{5x-1}}\,dx$

67. $\int_0^2 3x^2\sqrt{x^3+1}\,dx$

68. $\int_0^1 x^3\sqrt{x^4+1}\,dx$

69. $\int_1^2 \dfrac{1}{(3x+4)^2}\,dx$

70. $\int_1^2 \dfrac{2}{(2x+1)^2}\,dx$

APPLICATIONS

71. A clothing manufacturer has the marginal cost function
$$MC(x) = \dfrac{40x}{\sqrt{x^2+100}},\ 0 \le x \le 100$$
(in dollars per suit), where x is the number of suits produced each week. The fixed cost is $500.

    (a) Determine the total cost function $C(x)$;

    (b) Determine the total cost of producing 60 suits.

72. A shoe manufacturer has the marginal cost function
$$MC(x) = 40 - 0.02x\sqrt{0.0001x^2+1},$$
$0 \le x \le 200$ (in dollars per pair), where x is the number of pairs of shoes produced each week. The fixed cost is $900.

    (a) Determine the total cost function $C(x)$;

    (b) Determine the total cost of producing 150 pairs of shoes.

73. A mutual fund has a growth rate recorded each year, expressed by the function
$$V'(t) = r(t) = 200te^{-0.04t^2},\ 0 \le t \le 10$$
(in thousands of dollars per year), where t is the number of years passed from year 1997. The value of the fund in year 1997 is $1,000,000.

    (a) Determine the function $V(t)$ which represents the value of the fund t years after year 1997.

    (b) Determine the value of the fund in year 2007.

74. A mutual fund has a growth rate recorded each year, expressed by the function
$$V'(t) = r(t) = 100te^{-0.01t^2},\ 0 \le t \le 7$$
(in thousands of dollars per year), where t is the number of years passed from year 2000. The value of the fund in year 2000 is $1,500,000.

    (a) Determine the function $V(t)$ which represents the value of the fund t years after year 2000.

    (b) Determine the value of the fund in year 2007.

75. A clothing manufacturer has the marginal profit function
$$MP(x) = \dfrac{40x}{\sqrt{2x^2-400}},\ 15 \le x \le 100$$
(in dollars per suit), where x is the number of suits produced and sold each week.

    (a) Evaluate $\int_{20}^{60} MP(x)\,dx$ and interpret.

    (b) If the clothing manufacturer breaks even when 40 suits are produced and sold each week, determine the total profit function $P(x)$.

76. A clothing manufacturer finds that when the price of a brand of suits is $100 per suit, the consumers demand is 200 suits per week. The rate at which the number of suits that consumers demand changes with respect to price is given by

$$D'(p) = \frac{-0.2p}{\sqrt{260{,}000 - p^2}},$$

where p is the price of each suit in dollars. Determine the demand function $x = D(p)$.

4.6 Integration by Tables

In general, integration is more difficult than differentiation. Often tedious and complicated algebraic manipulations are required. For that reason, integral formulas for a wide class of important functions are gathered into tables. These tables can be found in most calculus books, libraries, and online. In this section we provide a brief list of integral formulas. This integral table is sufficient for this course. We illustrate, through examples, how to properly match a given integral with a formula in the table. Many formulas have parameters, and some formulas need to be applied repeatedly. Sometimes algebra or the substitution technique may be required before a formula in the table can be matched.

> **Warning**   *How to pick the correct integration formula*
>
> In order to apply one of the formulas in the table, the expression of the variable must match exactly. That includes the position, power, and coefficients of the variable. Hence, sometimes the integral needs to be re-expressed to fit a formula. If a formula (such as Formula 5 or Formula 7) needs to be applied repeatedly, then do not stop until no integral is left.

Example 1: Determine the indefinite integral $\int \sqrt{4x^2 + 5}\, dx$.

Solution The integral resembles *Formula 17* in Table 1,

$$\int \sqrt{x^2 \pm a^2}\, dx = \frac{x}{2}\sqrt{x^2 \pm a^2} \pm \frac{a^2}{2} \ln\left|x + \sqrt{x^2 \pm a^2}\right| + C.$$

However, to apply this formula, the coefficient of x^2 must be 1. We factor out 4 and then apply *Formula 17*.

$$\int \sqrt{4x^2 + 5}\, dx = \int \sqrt{4\left(x^2 + \frac{5}{4}\right)}\, dx$$

$$= \int 2\sqrt{x^2 + \frac{5}{4}}\, dx \quad (\text{So } a^2 = \tfrac{5}{4})$$

$$= 2\left[\frac{x}{2}\sqrt{x^2 + \frac{5}{4}} + \frac{5/4}{2} \ln\left|x + \sqrt{x^2 + \frac{5}{4}}\right|\right] + C$$

$$= x\sqrt{x^2 + \frac{5}{4}} + \frac{5}{4} \ln\left|x + \sqrt{x^2 + \frac{5}{4}}\right| + C.$$

Notice that in the formula only $a^2 = 5/4$, instead of a, is used.

Example 2: Determine the indefinite integral $\int \frac{3}{2x^2 - 7x}\, dx$.

Solution We first factor out the coefficient 3 and the x in the denominator:

$$\int \frac{3}{2x^2 - 7x}\, dx = 3\int \frac{1}{x(2x - 7)}\, dx.$$

Table 1 Integration Formulas

1. $\displaystyle\int x^n \, dx = \frac{1}{n+1}x^{n+1} + C, \, n \neq -1$

2. $\displaystyle\int \frac{1}{x} \, dx = \ln x + C, \, x > 0$

3. $\displaystyle\int e^{ax} \, dx = \frac{1}{a}e^{ax} + C$

4. $\displaystyle\int xe^{ax} \, dx = \frac{1}{a^2} \cdot e^{ax}(ax - 1) + C$

5. $\displaystyle\int x^n e^{ax} \, dx = \frac{x^n e^{ax}}{a} - \frac{n}{a}\int x^{n-1}e^{ax} \, dx + C$

6. $\displaystyle\int \ln x \, dx = x \ln x - x + C$

7. $\displaystyle\int (\ln x)^n \, dx = x(\ln x)^n - n\int (\ln x)^{n-1} \, dx + C, \, n > 1$

8. $\displaystyle\int x^n \ln x \, dx = x^{n+1}\left[\frac{\ln x}{n+1} - \frac{1}{(n+1)^2}\right] + C, \, n \neq -1$

9. $\displaystyle\int a^x \, dx = \frac{a^x}{\ln a} + C, \, a > 0, \, a \neq 1$

10. $\displaystyle\int \frac{1}{x^2 - a^2} \, dx = \frac{1}{2a}\ln\left|\frac{x-a}{x+a}\right| + C$

11. $\displaystyle\int \frac{1}{\sqrt{x^2 \pm a^2}} \, dx = \ln\left|x + \sqrt{x^2 \pm a^2}\right| + C$

12. $\displaystyle\int \frac{1}{x\sqrt{a^2 \pm x^2}} \, dx = -\frac{1}{a}\ln\left|\frac{a + \sqrt{a^2 \pm x^2}}{x}\right| + C$

13. $\displaystyle\int \frac{x}{a+bx} \, dx = \frac{a}{b^2} - \frac{x}{b} - \frac{a}{b^2}\ln|a+bx| + C$

14. $\displaystyle\int \frac{x}{(a+bx)^2} \, dx = \frac{a}{b^2(a+bx)} + \frac{1}{b^2}\ln|a+bx| + C$

15. $\displaystyle\int \frac{1}{x(a+bx)} \, dx = \frac{1}{a}\ln\left|\frac{x}{a+bx}\right| + C$

16. $\displaystyle\int \frac{1}{x(a+bx)^2} \, dx = \frac{1}{a(a+bx)} + \frac{1}{a^2}\ln\left|\frac{x}{a+bx}\right| + C$

17. $\displaystyle\int \sqrt{x^2 \pm a^2} \, dx = \frac{x}{2}\sqrt{x^2 \pm a^2} \pm \frac{a^2}{2}\ln\left|x + \sqrt{x^2 \pm a^2}\right| + C$

18. $\displaystyle\int x\sqrt{a+bx} \, dx = \frac{2}{15b^2}(3bx - 2a)(a+bx)^{3/2} + C$

19. $\displaystyle\int x^n \sqrt{a+bx} \, dx = \frac{2}{b(2n+3)}\left[x^n(a+bx)^{3/2} - na\int x^{n-1}\sqrt{a+bx} \, dx\right] + C$

20. $\displaystyle\int \frac{x}{\sqrt{a+bx}} \, dx = \frac{2}{3b^2}(bx - 2a)\sqrt{a+bx} + C$

4.6. INTEGRATION BY TABLES

The integral resembles *Formula 15* in Table 1,

$$\int \frac{1}{x(a+bx)} \, dx = \frac{1}{a} \ln \left| \frac{x}{a+bx} \right| + C.$$

We have $a = -7$ and $b = 2$. Hence

$$\int \frac{3}{2x^2 - 7x} \, dx = 3 \int \frac{1}{x(2x-7)} \, dx = 3 \int \frac{1}{x(-7+2x)} \, dx$$

$$= 3 \left(\frac{1}{-7} \right) \ln \left| \frac{x}{-7+2x} \right| + C = -\frac{3}{7} \ln \left| \frac{x}{2x-7} \right| + C.$$

Example 3: Determine the indefinite integral $\int \frac{1}{x^2 - 5} \, dx$.

Solution The integral resembles *Formula 10* in Table 1,

$$\int \frac{1}{x^2 - a^2} \, dx = \frac{1}{2a} \ln \left| \frac{x-a}{x+a} \right| + C.$$

We have $a^2 = 5$, hence $a = \sqrt{5}$. Therefore by the formula,

$$\int \frac{1}{x^2 - 5} \, dx = \frac{1}{2\sqrt{5}} \ln \left| \frac{x - \sqrt{5}}{x + \sqrt{5}} \right| + C.$$

Example 4: Determine the indefinite integral $\int \ln 4x \, dx$.

Solution The integral resembles *Formula 6* in Table 1,

$$\int \ln x \, dx = x \ln x - x + C.$$

However, the coefficient of x must be 1. We use the substitution $u = 4x$ and $du = 4 \, dx$ (so $\frac{1}{4} du = dx$) to rewrite the integral:

$$\int \ln 4x \, dx = \int \ln u \, \frac{1}{4} \, du = \frac{1}{4} \int \ln u \, du.$$

We then can apply the formula:

$$\frac{1}{4} \int \ln u \, du = \frac{1}{4} [u \ln u - u] + C = \frac{u}{4} \ln u - \frac{u}{4} + C.$$

Replace u by $4x$ we obtain

$$\int \ln 4x \, dx = \frac{1}{4} \int \ln u \, du$$

$$= \frac{u}{4} \ln u - \frac{u}{4} + C = x \ln 4x - x + C.$$

Example 5: Determine the indefinite integral $\int (\ln x)^2 \, dx$.

Solution The integral fits *Formula 7* in Table 1,

$$\int (\ln x)^n \, dx = x(\ln x)^n - n \int (\ln x)^{n-1} \, dx + C$$

with $n = 2$. Hence

$$\int (\ln x)^2 \, dx = x(\ln x)^2 - 2 \int (\ln x)^1 \, dx + C.$$

We can then apply *Formula 6* in Table 1 for the remainder of the integral:

$$\int (\ln x)^2 \, dx$$

$$= x(\ln x)^2 - 2 \int (\ln x)^1 \, dx + C$$

$$= x(\ln x)^2 - 2[x \ln x - x] + C$$

$$= x[(\ln x)^2 - 2 \ln x + 2] + C.$$

Example 6: Determine the indefinite integral $\int x^2 e^{-4x} \, dx$.

Solution The integral fits *Formula 5* in Table 1,

$$\int x^n e^{ax} \, dx = \frac{x^n e^{ax}}{a} - \frac{n}{a} \int x^{n-1} e^{ax} \, dx + C$$

with $n = 2$ and $a = -4$. Hence

$$\int x^2 e^{-4x} \, dx = \frac{x^2 e^{-4x}}{-4} - \frac{2}{-4} \int x^1 e^{-4x} \, dx + C$$

$$= -\frac{1}{4} x^2 e^{-4x} + \frac{1}{2} \int x^1 e^{-4x} \, dx + C$$

We can apply *Formula 4* in Table 1,

$$\int x e^{ax} \, dx = \frac{1}{a^2} \cdot e^{ax} (ax - 1) + C$$

with $a = -4$ for the remaining integral:

$$\int x^2 e^{-4x} \, dx$$

$$= -\frac{1}{4} x^2 e^{-4x} + \frac{1}{2} \int x^1 e^{-4x} \, dx + C$$

$$= -\frac{1}{4} x^2 e^{-4x} + \frac{1}{2} \left[\frac{1}{16} \cdot e^{-4x} (-4x - 1) \right] + C$$

$$= -\frac{1}{4} x^2 e^{-4x} - \frac{1}{32} (4x + 1) e^{-4x} + C.$$

Exercise Set 4.6

Determine the indefinite integrals using Table 1.

1. $\int 2xe^{-3x}\, dx$

2. $\int x^2 e^{0.5x}\, dx$

3. $\int \ln(5x)\, dx$

4. $\int \ln(7x)\, dx$

5. $\int x\ln(2x)\, dx$

6. $\int x\ln(3x)\, dx$

7. $\int 2x\ln\left(x^2+1\right)\, dx$

8. $\int 3x^2 \ln\left(x^3+2\right)\, dx$

9. $\int \dfrac{x}{3x+4}\, dx$

10. $\int \dfrac{5x}{2x-3}\, dx$

11. $\int \dfrac{1}{x\sqrt{x^2+4}}\, dx$

12. $\int \dfrac{3}{x\sqrt{25+x^2}}\, dx$

13. $\int \sqrt{4x^2-9}\, dx$

14. $\int \sqrt{2x^2+5}\, dx$

15. $\int \dfrac{1}{\sqrt{x^2+4}}\, dx$

16. $\int \dfrac{2}{\sqrt{4x^2+1}}\, dx$

17. $\int \dfrac{2x}{x^4-9}\, dx$

18. $\int \dfrac{3}{16-x^2}\, dx$

19. $\int \dfrac{x}{(3x+4)^2}\, dx$

20. $\int \dfrac{x}{(5-3x)^2}\, dx$

21. $\int \dfrac{\ln x}{x^3}\, dx$

22. $\int \dfrac{\ln x}{x^2}\, dx$

23. $\int \dfrac{2}{5x-3x^2}\, dx$

24. $\int \dfrac{3}{2x-5x^2}\, dx$

25. $\int \dfrac{4}{x\sqrt{4-x^2}}\, dx$

26. $\int \dfrac{3}{x\sqrt{1-4x^2}}\, dx$

27. $\int 3x\sqrt{2x+1}\, dx$

28. $\int 2x\sqrt{5x-4}\, dx$

29. $\int x^2\sqrt{x+4}\, dx$

30. $\int x^2\sqrt{x+1}\, dx$

31. $\int \dfrac{x}{\sqrt{2x-1}}\, dx$

32. $\int \dfrac{x}{\sqrt{x+4}}\, dx$

Chapter 4 Summary

This chapter explores the second main branch of calculus, **integration**. The concept of integration has two forms: A **definite integral**, denoted by $\int_a^b f(a)\,dx$, is a number. An **indefinite integral**, denoted by $\int f(x)\,dx$, also called an **antiderivative**, is a function with an undetermined constant, $F(x) + C$, that satisfies $F'(x) = f(x)$. These two integrals are linked together by the **Fundamental Theorem of Calculus**: If f is continuous on the interval $[a, b]$, and F is an antiderivative of f, then

$$\int_a^b f(x)\,dx = [F(x)]_a^b = F(b) - F(a).$$

The Fundamental Theorem of Calculus is the glue that holds the two main branches of calculus, differential and integral calculus, together.

── Relationships Among Areas, Definite Integrals, and Antiderivatives ──

Integrals can be visually illustrated as the areas of regions bounded by the graph of the function $f(x)$ and the x-axis on intervals like $[a, b]$ and $[a, x]$.

Through the discussions in the first three sections, especially through the Fundamental Theorem of Calculus, the relationships among the areas of regions bounded by the graph of the function $f(x)$ and the x-axis on intervals like $[a, b]$ and $[a, x]$, the definite integral $\int_a^b f(a)\,dx$, and the antiderivative, can be summarized by the following:

- The area $A(x)$ of a region, bounded by the x-axis and the graph of a positive function $f(x)$, on the interval $[a, x]$, satisfies

$$A'(x) = f(x).$$

That is, the area is an antiderivative of $f(x)$.

- The antiderivatives of $f(x)$ in general form can be obtained by the basic formulas and properties of an indefinite integral:

$$\int f(x) = F(x) + C.$$

- The definite integral, $\int_a^b f(x)\,dx$, is a number which is independent of x. It can be evaluated by

$$\int_a^b f(x)\,dx = [F(x)]_a^b = F(b) - F(a)$$

where $F(x)$ is any antiderivative of $f(x)$.

- The area A of a region, bounded by the x-axis and the graph of a *positive* function $f(x)$, on the interval $[a, b]$, can be obtained by a definite integral:

$$A = \int_a^b f(x)\,dx.$$

- The definite integral, $\int_a^b f(x)\,dx$, represents the "signed sum" of the areas of the regions, bounded by the x-axis and the graph of a function $f(x)$, on the interval $[a, b]$. That is, the value of $\int_a^b f(x)\,dx$ equals the area of the regions above the x-axis minus the area of the regions below the x-axis.

- If the function $f(x)$ represents the rate of change of a quantity on an interval $[a, b]$, then the definite integral, $\int_a^b f(x)\,dx$, represents the total (or cumulative) change of the quantity from the variable values a to b.

4.6. INTEGRATION BY TABLES

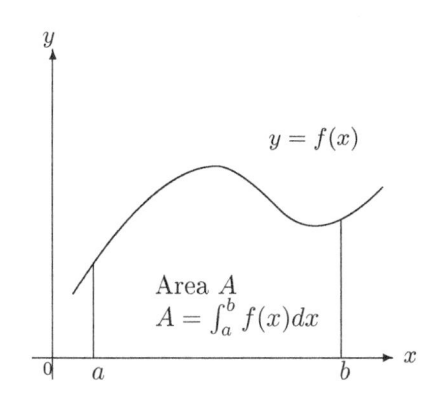

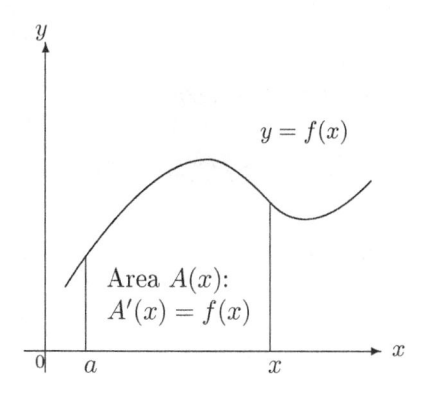

Area A
$A = \int_a^b f(x)dx$

Area $A(x)$:
$A'(x) = f(x)$

Using the relationship between the areas and definite integrals, many important properties of definite integrals can be visualized.

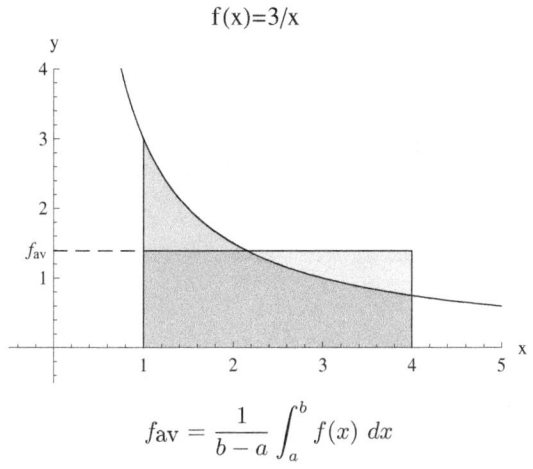

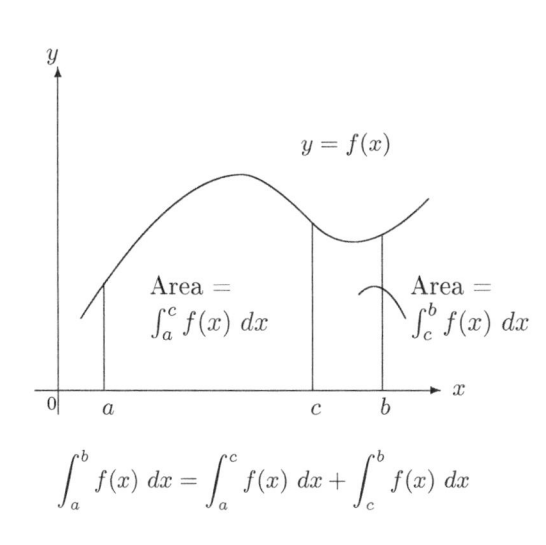

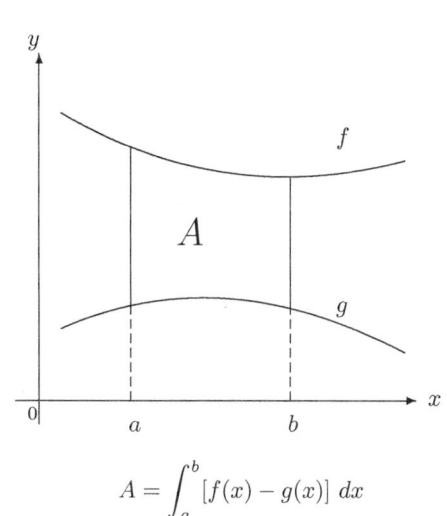

f(x)=3/x

$$f_{av} = \frac{1}{b-a} \int_a^b f(x)\,dx$$

Area $= \int_a^c f(x)\,dx$

Area $= \int_c^b f(x)\,dx$

$$\int_a^b f(x)\,dx = \int_a^c f(x)\,dx + \int_c^b f(x)\,dx$$

$$A = \int_a^b [f(x) - g(x)]\,dx$$

■ Calculating Areas, Definite Integrals, and Antiderivatives ■

The indefinite integrals are the basic tools for evaluating the areas and other accumulations presented as definite integrals. Some important formulas and properties for integration are summarized here:

1. $\int k\, dx = kx + C$ (k is a constant)

2. $\int x^r\, dx = \dfrac{1}{r+1} x^{r+1} + C$, provide $r \neq -1$

3. $\int \dfrac{1}{x}\, dx = \ln|x| + C$, or $\int \dfrac{1}{x}\, dx = \ln x + C$, if $x > 0$

4. $\int ae^{kx}\, dx = \dfrac{a}{k} e^{kx} + C$

5. $\int kf(x)\, dx = k \int f(x)\, dx$

6. $\int [f(x) \pm g(x)]\, dx = \int f(x)\, dx \pm \int g(x)\, dx$

Example 1: Determine the indefinite integral

$$\int \left(2e^{-2x} + \dfrac{1}{\sqrt[3]{x^2}} - \dfrac{4}{x} - 6x^2 + 3 \right) dx, \quad \text{for } x > 0.$$

Solution Applying the basic rules and properties of integration, we obtain

$$\int \left(2e^{-2x} + \dfrac{1}{\sqrt[3]{x^2}} - \dfrac{4}{x} - 6x^2 + 3 \right) dx$$

$$= 2 \int e^{-2x}\, dx + \int x^{-\frac{2}{3}}\, dx - 4 \int \dfrac{1}{x}\, dx - 6 \int x^2\, dx + \int 3\, dx$$

$$= \dfrac{2}{-2} e^{-2x} + \dfrac{1}{(-\frac{2}{3})+1} x^{-\frac{2}{3}+1} - 4 \ln x - 6 \dfrac{1}{3} x^3 + 3x + C$$

$$= -e^{-2x} + 3x^{\frac{1}{3}} - 4 \ln x - 2x^3 + 3x + C.$$

If a particular value of the antiderivative is given, then the constant C can be determined. In the aforementioned example, if we are looking for a function $F(x)$ such that

$$F'(x) = f(x) = 2e^{-2x} + \dfrac{1}{\sqrt[3]{x^2}} - \dfrac{4}{x} - 6x^2 + 3 \text{ and } F(1) = 4,$$

then we have

$$-e^{-2(1)} + 3(1)^{\frac{1}{3}} - 4\ln(1) - 2(1)^3 + 3(1) + C = 4.$$

Solving for C we obtain $C = e^{-2}$. Hence, the particular antiderivative of f is

$$F(x) = -e^{-2x} + 3x^{\frac{1}{3}} - 4\ln x - 2x^3 + 3x + e^{-2}.$$

Applying the Fundamental Theorem of Calculus, the definite integrals can be evaluated through the indefinite integrals using the basic formulas and properties.

Example 2: Evaluate $\displaystyle\int_0^4 (\sqrt{x} + 1)\, dx$.

4.6. INTEGRATION BY TABLES

Solution

$$\int_0^4 (\sqrt{x}+1)\ dx = \left[\frac{2}{3}x^{\frac{3}{2}} + x\right]_0^4$$

$$= \left[\frac{2}{3}(4)^{\frac{3}{2}} + 4\right] - \left[\frac{2}{3}(0)^{\frac{3}{2}} + 0\right] = \frac{2}{3}(8) + 4 = 9\frac{1}{3}.$$

The basic formulas and properties can be applied, combined with the relationship between integrals and areas, to find the area of regions bounded by the graph of the function $f(x)$ and the x-axis on an interval $[a, b]$.

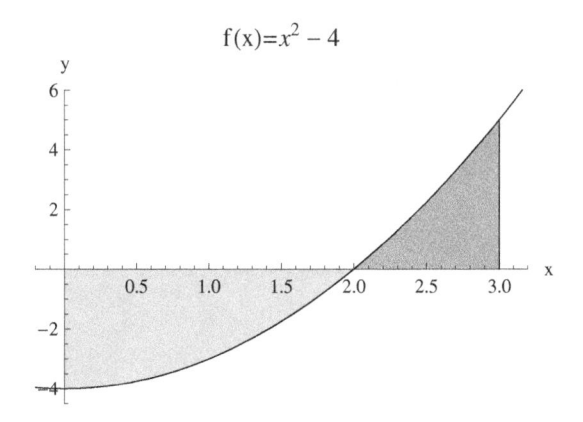

Example 3: The graph of $f(x) = x^2 - 4$ with some shaded region is given above.

(a) Evaluate the definite integral $\int_0^3 (x^2 - 4)\ dx$.

(b) Interpret the result in part (a) with the areas involved.

(c) Determine the *total* shaded area.

Solution (a) We have

$$\int_0^3 (x^2 - 4)\ dx = \left[\frac{1}{3}x^3 - 4x\right]_0^3$$

$$= \left[\frac{1}{3}(3)^3 - 4(3)\right] - \left[\frac{1}{3}(0)^3 - 4(0)\right] = -3 - 0 = -3.$$

(b) The number -3 obtained in part (a) is equal to the area of the darker shaded region minus the area of the lighter shaded region.

(c) To obtain the total shaded areas, we integrate $f(x) = x^2 - 4$ on $[0, 2]$ and $[2, 3]$ (since $x = 2$ is where $f(x)$ changes from negative to positive).

$$\int_0^3 (x^2 - 4)\ dx = \left[\frac{1}{3}x^3 - 4x\right]_0^2$$

$$= \left[\frac{1}{3}(2)^3 - 4(2)\right] - \left[\frac{1}{3}(0)^3 - 4(0)\right] = -5\frac{1}{3} - 0 = -5\frac{1}{3},$$

$$\int_1^3 (x^2 - 4)\ dx = \left[\frac{1}{3}x^3 - 4x\right]_2^3$$

$$= \left[\frac{1}{3}(3)^3 - 4(3)\right] - \left[\frac{1}{3}(2)^3 - 4(2)\right] = -3 - \left(-5\frac{1}{3}\right) = 2\frac{1}{3}.$$

Therefore the total shaded area is

$$5\frac{1}{3} + 2\frac{1}{3} = 7\frac{2}{3}.$$

In addition to the basic formulas, there are other techniques that enable us to evaluate a broader group of integrals. Sometimes **algebraic manipulations** can turn an integral into an expression to which basic integration formulas can be applied. Another approach is to use an **integral table**. Integral tables include formulas for a wide class of important functions. These tables can be found in most calculus books, libraries, and online. Finally, some integrals need the technique of **substitution** before an existing formula can be applied.

Example 4: Evaluate $\int \frac{5e^x}{3e^x + 2} \, dx$ using substitution.

Solution We let $u = 3e^x + 2$, then $du = 3e^x \, dx$, or $\frac{1}{3} du = e^x \, dx$. Hence

$$\int \frac{5e^x}{3e^x + 2} \, dx = 5 \int \frac{1}{e^x + 2} \, e^x \, dx$$

$$= 5 \int \frac{1}{u} \cdot \frac{1}{3} \, du = \frac{5}{3} \int \frac{1}{u} \, du$$

$$= \frac{5}{3} \ln u + C = \frac{5}{3} \ln (e^x + 2) + C.$$

Estimating Areas and Definite Integrals

Even with the help of all the integration techniques and tables, we will encounter integrals that cannot be anti-differentiated by any of these methods or formulas. The first such example is $\int e^{-x^2} \, dx$, which has no elementary antiderivative. In practice, a function (e.g., data obtained through a survey) often cannot be easily modeled by a formula. When a function does have an antiderivative, the antiderivative can be much more complicated than the integrand.

At the other hand, with more powerful computers, performing large amounts of basic calculations becomes much easier. This makes numerical integration more attractive. **Numerical integration** means to approximate a definite integral $\int_a^b f(x) \, dx$ by interpreting it as an area under (or above) a curve. In Section 4.1, left-rectangles are used to approximate a definite integral. Other more efficient numerical methods exist but are not included in this book.

Applications

Definite integrals have many important applications in the real world. One such application is in motion problems. Others include accumulated production or sales, and average values.

Example 5: A driver driving on an straight east-west highway records the velocity of the car in the hours after he leaves home at 7:00 a.m.:

$$v(t) = 80t - 3t^2,$$

where t, in hours, measures the time passed after 7:00 a.m., and v is in miles per hour. Positive velocity means going east, and negative velocity means going west. Determine the location of the driver, in relation to home, at 8:00 a.m..

Solution Since the velocity is the rate of change in the position function s, the change in the position can be obtained by the definite integral:

$$s(1) - s(0) = \int_0^1 v(t)\, dt = \int_0^1 \left(80t - 3t^2\right)\, dt$$

$$= \left[40t^2 - t^3\right]_0^1 = 40(1)^2 - (1)^3 - 0 = 39.$$

That is, the change in the position is 39 miles. Since the driver starts from home, at 8:00 a.m. the driver is 39 miles east of home.

Example 6: Betty deposits \$2,000 into a money market account earning 5.5% interest compounded continuously.

(a) Determine Betty's average balance over one year.

(b) Suppose that at end of the year the fund pays a bonus that is equal to 1.5% of the average balance. Determine how much bonus Betty will receive.

Solution (a) By the balance formula supplied in Section 1.5, the balance in the continuously compounded account is given by

$$A(t) = 2{,}000 e^{0.055t}.$$

Hence, the average value of $A(t)$ during 1 year is

$$A_{\text{av}} = \frac{1}{1-0} \int_0^1 2{,}000 e^{0.055t}\, dx$$

$$= \left[\frac{2{,}000}{0.055} e^{0.055t}\right]_0^1 = \frac{2{,}000}{0.055}\left(e^{0.055(1)} - e^{0.055(0)}\right) = \frac{2{,}000}{0.055}\left(e^{0.055} - 1\right) \approx \$2{,}056.02.$$

(b) Since Betty's average balance over 1 year is \$2,056.02, her bonus is $0.015(2056.02) \approx \$30.84$.

Chapter 4 Review Exercises

1. The graph of the marginal cost function $MC(x)$ of a product is given below. Assuming that the fixed cost is \$600, determine the total cost of producing x items, where x is

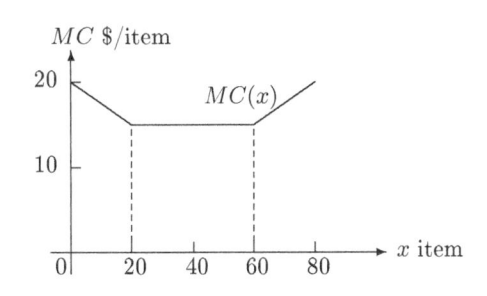

    (a) 20; (b) 40; (c) 60; (d) 80.

2. Refer to the graph below; explain what the area A represents. Specify the units:

    (a) Where x represents the number of chairs produced and sold, and y represents the

marginal profit in dollars per chair.

(b) Where x represents screwdrivers produced in thousands, and y represents the marginal cost in hundreds of dollars per thousand screwdrivers.

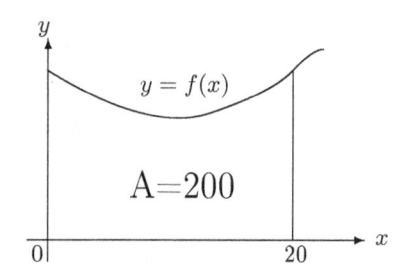

[Graph: $y = f(x)$ with area $A = 200$ between $x = 0$ and $x = 20$]

3. The marginal cost function, in dollars, for producing the xth item of a certain brand of barstools is given by $MC(x) = 20 - 0.5\sqrt{x}$, $0 \leq x \leq 100$. The fixed cost is $200.

(a) Draw a graph that represents the marginal cost;

(b) Estimate the total cost of producing 100 barstools using the left-rectangle approximation with five rectangles.

(c) Use a definite integral to find the exact total cost of producing 100 barstools.

For Exercises 4–7, evaluate the integrals.

4. $\int \left(6x^5 + 8e^{3x} - 4\right) dx$

5. $\int \left(\dfrac{-3}{\sqrt[5]{x^3}} + \dfrac{1}{x^3} + \dfrac{4}{x}\right) dx \ (x > 0)$

6. $\int_1^9 \left(\sqrt{t} - 2\right) dt$

7. $\int_1^e \left(4x - \dfrac{3}{x}\right) dx$

8. Find the function $f(x)$ satisfying $f'(x) = 3e^{2x}$, $f(0) = 2$.

9. Find the area under the graph of $f(x) = 4 - x^2$ on the following intervals:

  (a) $[0, 2]$; (b) $[1, 2]$; (c) $[-2, 2]$.

10. (a) Evaluate the definite integral

$$\int_0^3 (4 - x^2) \, dx.$$

(b) Interpret the result in part (a) with the area(s) involved. Illustrate with a graph.

11. A car has been traveling on an interstate highway for 10 minutes at a velocity represented by the function

$$v(t) = 0.4 + 0.2t - 0.02t^2, \ 0 \leq t \leq 10,$$

where t is measured in minutes and the velocity is measured in miles per minute. At $t = 0$, the car passed a sign saying "10 miles to Peggy's Diner." How far away is Peggy's Diner at the end of the 10-minute drive? Use a definite integral to determine the answer.

12. Find the area bounded by $y = 8\sqrt{x}$ and $y = x^2$.

13. A market analyst for a potato chip company estimates that with no promotion the weekly sales of one brand of its potato chips can be modeled by

$s_1(t) = 0.6t + 4.5$ thousands dollars per week, t weeks from now. This analyst estimates that, with a small-scale promotional campaign, the weekly sales can be modeled as $s_2(t) = 7.2e^{0.06t}$. In the period of the first 20 weeks, how much can the promotional campaign increase the total sales?

For Exercises 14–15, evaluate using algebraic manipulation.

14. $\int \sqrt{x}(2x + 1) \, dx$

15. $\int \dfrac{2x^3 - 4}{x^2} \, dx$

For Exercises 16–21, evaluate using substitution.

16. $\int 6x^2 \left(2x^3 - 5\right)^5 dx$

17. $\int xe^{3x^2} \, dx$

18. $\int \dfrac{4x^3}{x^4 + 1} \, dx$

19. $\int \dfrac{-3e^x}{(2e^x + 1)^3} \, dx$

20. $\int_0^1 x^3 \sqrt{x^4 + 1} \, dx$

21. $\int_0^1 \dfrac{8x}{(2x^2 + 1)^2} \, dx$

For Exercises 22–24, evaluate using Table 1.

CHAPTER 4 REVIEW EXERCISES

22. $\displaystyle\int 2xe^{-3x}\,dx$

23. $\displaystyle\int \ln(5x)\,dx$

24. $\displaystyle\int 2x\sqrt{5x-4}\,dx$

For Exercises 25–32, evaluate using any method.

25. $\displaystyle\int x^2\left(4x^3+3x-5\right)\,dx$

26. $\displaystyle\int x^2\left(2x^3+7\right)^6\,dx$

27. $\displaystyle\int \frac{x}{3x+4}\,dx$

28. $\displaystyle\int \frac{\ln x^4}{x}\,dx$

29. $\displaystyle\int \frac{\ln x}{x^3}\,dx$

30. $\displaystyle\int xe^{-4x}\,dx$

31. $\displaystyle\int 2xe^{-4x^2}\,dx$

32. $\displaystyle\int \frac{1}{3x+4}\,dx$

33. Karen deposits $5,000 into a money market account earning interest at an annual rate of 6%, compounded continuously.

   (a) Determine Karen's average balance over 1 year.

   (b) Suppose that at end of the year the fund pays a bonus that is equal to 1.2% of the average balance. Determine how much bonus Karen will receive.

Chapter 4 Test (100 points)

SCORE _____

NAME _____

ANSWERS

1.(5pts) Refer to the graph below, explain what the area A represents. Specify the units:

(a) Where x represents the number of shirts produced and sold in hundreds, and y represents the marginal cost in hundreds of dollars per hundred shirts.

1. (a) _____

(b) Where x represents screwdrivers produced in thousands, and y represent the marginal profit in hundreds of dollars per thousand screwdrivers. Assume $P(0) = 0$.

(b) _____

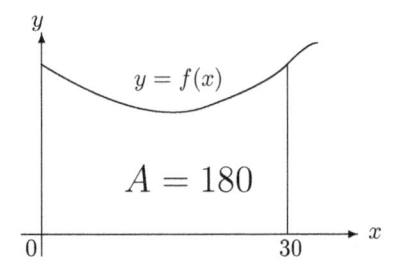

$y = f(x)$

$A = 180$

0, 30

For problems 2–5: Evaluate the integrals.

2. (5pts) $\int \left(3x^4 - 2e^{3x} + 2\right) dx$

2. _____

3. (5pts) $\int \left(\dfrac{-3}{\sqrt[5]{x^3}} + \dfrac{1}{x^3} + \dfrac{4}{x}\right) dx \ (x > 0)$

3. _____

4. (5pts) $\int_{1}^{9} \left(\sqrt{t} - 2\right) dt$

4. _____

5. (5pts) $\int_{1}^{e} \left(4x - \dfrac{3}{x}\right) dx$

5. _____

6. (6pts) Find the area under the graph of $f(x) = 2 - 0.3x^2$ on the following intervals: (a) $[-1, 2]$; (b) $[0, 1]$.

6. (a) _____

(b) _____

7. (8pts) (a) Evaluate the definite integral $\int_0^3 (4 - x^2)\ dx$.

7. (a) _____

(b) Interpret the result in part (a) with the area(s) involved. Illustrate with a graph.

(b) _____

8. (8pts) Find the area bounded by $y = x^2 - 4$ and $y = x + 2$.

(a) Draw a graph to illustrate the region.

(b) Determine the integral(s) representing the area.

8. (b) _____

(c) Evaluate. Show your work below.

(c) _____

For problems 9–10, evaluate using substitution. Do not use Table 1.

9. (5pts) $\displaystyle\int x\left(x^2-2\right)^4 dx$

9. _____

10. (5pts) $\displaystyle\int \frac{6x-15}{x^2-5x}dx$

10. _____

For problems 11–12, evaluate using Table 1.

11. (5pts) $\displaystyle\int \sqrt{x^2+4}\, dx$

11. _____

12. (5pts) $\displaystyle\int \frac{x}{(2-x)^2}dx$

12. _____

For problems 13–15, evaluate using any method.

13. (5pts) $\displaystyle\int (\ln x)^2 dx$

13. _____

14. (5pts) $\displaystyle\int t^2 e^{-t^3} dt$

14. _____

15. (5pts) $\displaystyle\int \frac{(\ln x)^2}{x} dx$

15. _____

16. (8pts) Karen deposits $3,000 into a money market account earning interest at an annual rate of 5%, compounded continuously.

(a) Determine Karen's average balance over 1 year.

16. (a) _____

(b) Suppose that at end of the year the fund pays a bonus that is equal to 1.5% of the average balance. Determine how much bonus Karen will receive.

(b) _____

17. (10pts) The marginal cost function, in dollars, for producing the xth item of a certain brand of barstool is given by

$MC(x) = 30 - 0.2\sqrt{x}$, $0 \leq x \leq 100$.

The fixed cost is $200.

(a) Draw a graph that represents the marginal cost.

(b) Estimate the total cost of producing 100 barstools using the left-rectangle approximation with four rectangles. Illustrate the rectangles on the graph.

17. (b) _____

(c) Use a definite integral to find the exact total cost of producing 100 bar stools.

(c) _____

Chapter 4 Project

The following graph represents your velocity, v (in miles per hour) with respect to time t (in hours), on a bicycle trip along a straight road. Suppose that you started out 25 miles west from home, and positive velocities take you towards home (i.e., eastbound) and negative velocities take you away from home. Answer each of the questions, explaining clearly how you reached each answer.

(a) At what time are you closest to home?

(b) At what time are you farthest from home?

(c) Write a definite integral that expresses the total distance you have covered for the LAST hour. Illustrate your answer using the graph.

(d) Write a definite integral that expresses the total distance you have covered for the LAST two and half hours. Illustrate your answer using the graph.

(e) At $t = 4.5$ hour, how far away from home are you? Are you at the east side or at the west side of home? Estimate using an appropriate method.

(f) Estimate the average speed of your pedaling for the LAST two hours.

(g) About how many miles in total have you pedaled for the four and half hours?

(h) What does the integral $\int_0^{4.5} v(t)\, dt$ represent with respect to your bike ride?

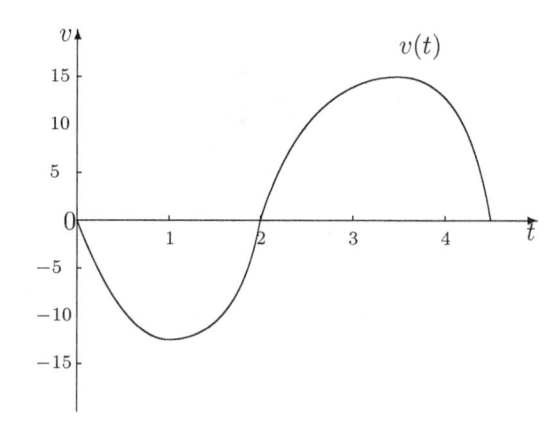

Chapter 5

Applications of Integration

Introduction

In Chapter 4, we saw how definite integrals can be used to calculate cumulative changes. We also learned a variety of integration techniques. Now we are ready to explore the applications of integration in economics, finance, and other related fields. The topics in this chapter include consumers' and producers' surpluses, financial models with continuous money flow, probability and important distributions, and differential equations.

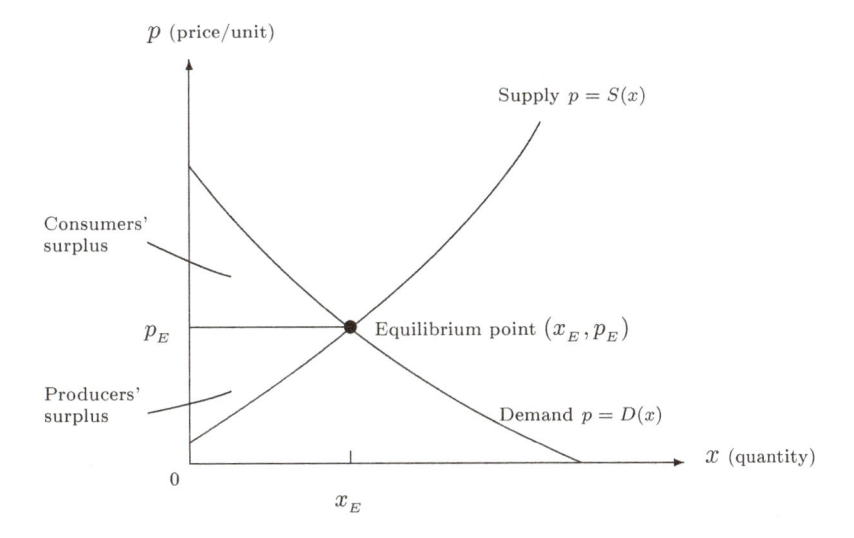

Study Objectives

1. Use areas and integrals to illustrate and compute consumers' and producers' surpluses.

2. Use integrals to set up financial models and solve financial problems.

3. Use areas and integrals to set up probability models and solve probability problems.

4. Solve differential equations.

5.1 Consumers' and Producers' Surpluses

In this section, we explore the economic applications of area between two curves: consumers' and producers' surpluses. We start with a review of the supply and demand functions, which were first introduced in Section 1.3. There the functions are expressed as $x = D(p)$ and $x = S(p)$. The expression $x = D(p)$ also occurs in Section 3.9 for establishing the elasticity function $E(p)$.

However, in Sections 3.2 and 3.7, the expression $p = D(x)$ is used to maximize the revenue or the profit. In this section, we again use the expressions $p = D(x)$ and $p = S(x)$ to study consumers' and producers' surpluses. That is, we treat the price as a function of the quantity. Such an interpretation is common in economics.

In a free market for a certain product, at the price p_E, manufacturers are willing to produce q_E quantity of the good, while the consumers are willing to purchase q_E quantity of the good. It is assumed that the market naturally settles to this equilibrium point (Figure 5.1.1).

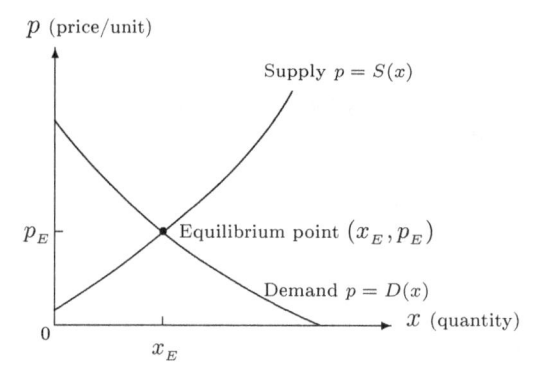

Figure 5.1.1: Demand and supply curves in a free market

Consumers' Utility, Expenditure, and Surplus

We now focus on the demand curve $p = D(x)$ and explore the meaning of the area under this curve. We start with an example.

x	subgroup	p
0 – 10	10	$5.00
10 – 30	20	$4.50
30 – 60	30	$4.00
60 – 90	30	$3.50
90 – 100	10	$3.00

Table 5.1.1: Result of demand survey

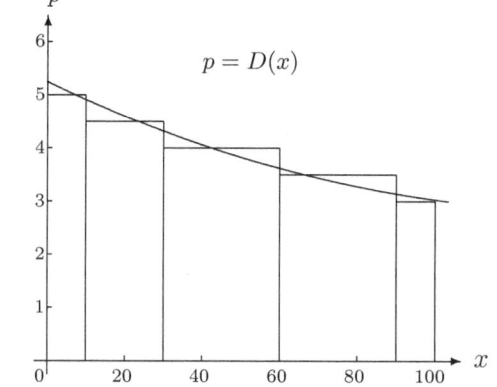

Figure 5.1.2: Result of demand survey

Jane starts a small company, *Sweet Dreams Chocolate*, making gourmet chocolates using her secret recipes. To test the demand, she conducts a survey, asking 100 potential customers to taste the chocolate bars and determine the highest price they are willing to pay. The result of the survey is presented in

5.1. CONSUMERS' AND PRODUCERS' SURPLUSES

Table 5.1.1 and Figure 5.1.2. If Jane sells her chocolate bars at $4.00 each, 60 out of 100 customers will buy it. Notice that among the 60 customers, 10 actually are willing to purchase the chocolate bar at $5.00 each, and 20 customers are willing to purchase the chocolate bar at $4.50 each. Hence, the total amount the 60 customers are *willing and able to spend* is

$$5(10) + 4.5(20) + 4(30) = \$260.$$

This amount is the same as the area under the three bars. Meanwhile, the amount the 60 customers *actually spend*, at the price of $4.00, is $4(60) = \$240$. Hence, the 30 customers who are willing to spend more on the chocolate bar are "richer" by $20.

In economics, the total amount consumers are willing and able to spend for a certain quantity x_0 of goods or service is called **utility**, and is represented by the area under the demand curve on the interval $[0, x_0]$. The amount that consumers spend at a certain market price p_0 is called **consumers' expenditure**, and is represented by the rectangular area $p_0 \times x_0$. Furthermore, the amount that consumers are willing to spend, but do not spend, for x_0 items at a market price p_0 is called **consumers' surplus**, and is represented by the area of the region between the demand curve and $p = p_0$ on the interval $[0, x_0]$ (Figure 5.1.3).

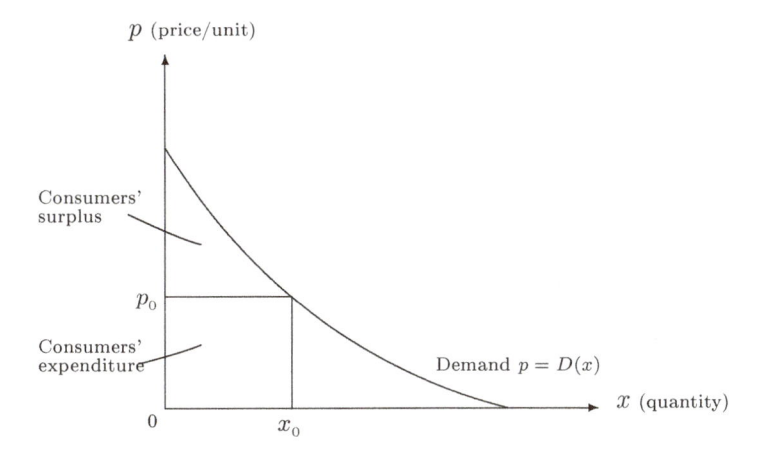

Figure 5.1.3: Consumers' surplus at price p_0

Since the area under a curve on an interval $[a, b]$ is obtained by the definite integral $\int_a^b f(x)\, dx$, we state the definitions mathematically:

Definition *consumers' surplus*

Suppose that $p = D(x)$ describes the demand function for a commodity. Then at the point (x_0, p_0), the **utility**, or the amount that consumers are willing and able to spend, is defined by

$$\int_0^{x_0} D(x)\, dx.$$

The **consumers' expenditure** is defined by $p_0 \times x_0$. And the **consumers' surplus** is defined by

$$\text{Utility} - \text{Consumers' expenditure} = \int_0^{x_0} D(x)\, dx - (p_0 \times x_0).$$

Example 1: For the demand function $p = D(x) = (x - 5)^2$, determine the consumers' surplus at the demand level $x = 3$.

Solution We substitute $x = 3$ into $D(x)$ to obtain $p = 4$. By the definition of the consumers' surplus, given that $x_0 = 3$ and $p_0 = 4$, we have

$$\text{C.S.} = \int_0^3 (x-5)^2 \, dx - 4 \times 3 = \int_0^3 \left(x^2 - 10x + 25\right) dx - 12$$

$$= \left[\frac{x^3}{3} - 5x^2 + 25x\right]_0^3 - 12 = (9 - 45 + 75) - 0 - 12 = 27.$$

Example 2: Assume that the demand for a certain model of car in the United States is described by

$$p = D(x) = -14.4 \ln(0.07x + 0.018)$$

where x is measured in million cars and p is measured in thousand dollars per car.

(a) Determine the price at which consumers will purchase 2 million cars.

(b) Determine the consumers' expenditure when purchasing 2 million cars.

(c) What is the total amount that consumers are willing to spend on the 2 million cars?

(d) Determine the consumers' surplus when purchasing 2 million cars.

Solution (a) Setting $x_0 = 2$ we obtain

$$p = D(2) = -14.4 \ln(0.07(2) + 0.018) \approx 26.57.$$

That is, at a market price p_0 of approximately \$26,570 per car, consumers will purchase 2 million cars.

(b) The consumers' expenditure when purchasing 2 million cars is

$$p_0 \times x_0 \approx 26.57(\text{ thousand dollars per car}) \times 2(\text{ million cars}) = \$53.14 \text{ (billion)}.$$

(c) We evaluate the integral

$$\int_0^{x_0} D(x) \, dx = \int_0^2 -14.4 \ln(0.07x + 0.018) \, dx.$$

We need to find the antiderivative first. The corresponding indefinite integral resembles *Formula 6* in Table 1,

$$\int \ln x \, dx = x \ln x - x + C.$$

We use the substitution $u = 0.07x + 0.018$ and $du = 0.07 \, dx$ (so $\frac{1}{0.07} \, du = dx$) to rewrite the integral:

$$\int -14.4 \ln(0.07x + 0.018) \, dx = \int -14.4 \ln u \, \frac{1}{0.07} \, du = \frac{-14.4}{0.07} \int \ln u \, du.$$

We then can apply the formula:

$$\frac{-14.4}{0.07} \int \ln u \, du = \frac{-14.4}{0.07} [u \ln u - u] + C = \frac{-14.4}{0.07} u(\ln u - 1) + C.$$

Replacing u by $0.07x + 0.018$ we obtain

$$\int -14.4 \ln(0.07x + 0.018) \, dx = \frac{-14.4}{0.07} \int \ln u \, du$$

$$= \frac{-14.4}{0.07} u(\ln u - 1) + C$$

5.1. CONSUMERS' AND PRODUCERS' SURPLUSES

$$= \frac{-14.4}{0.07}(0.07x + 0.018)[\ln(0.07x + 0.018) - 1] + C.$$

Hence,

$$\int_0^2 -14.4\ln(0.07x + 0.018)\,dx = \left[\frac{-14.4}{0.07}(0.07x + 0.018)(\ln(0.07x + 0.018) - 1)\right]_0^2$$

$$= -\frac{1440}{7}[(0.158)(\ln 0.158 - 1) - (0.018)(\ln 0.018 - 1)] \approx \$73.90 \text{ (billion)}.$$

(d) By the definition and parts (b) and (c), the consumers' surplus when purchasing 2 million cars is approximately $73.9 - 53.14 = \$20.76$ billion (Figure 5.1.4).

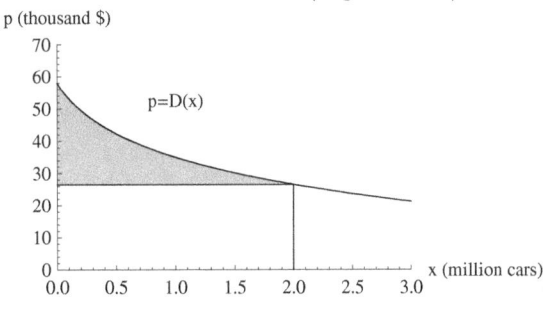

Figure 5.1.4: Consumers' surplus at demand of 2 million cars

Producers' Willingness to Receive, Revenue, and Surplus

We now focus on the supply curve $p = S(x)$ and explore the meaning of the area under this curve. We start with the example of Jane's company, *Sweet Dreams Chocolate*.

x	p
0	$2.00
200	$3.00
300	$3.50
400	$4.00

Table 5.1.2: Jane's supply curve

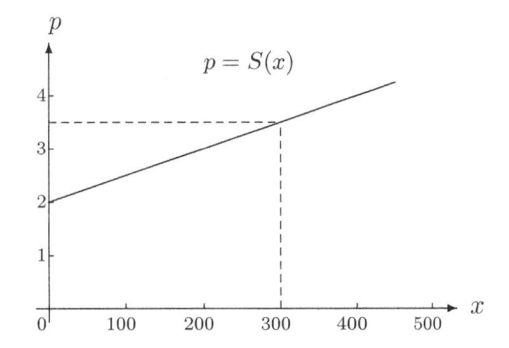

Figure 5.1.5: Jane's supply curve

Jane has to make a profit from producing and selling her chocolate bars. If the price is at or below $2.00 for each chocolate bar, she is not willing or able to supply any. The quantity of chocolate bars she is willing to supply per week can be described by $x = 200p - 400$. Table 5.1.4 and Figure 5.1.5 illustrate the supply curve.

Assume that the market price is $3.50 for each chocolate bar. Then Jane will supply 300 chocolate bars weekly, and receive the total revenue of

$$3.50(300) = \$1,150.$$

Notice that Jane can make a profit at prices lower than $3.50 until the price drops to $2.00. Hence, her

willingness and ability to receive is the area below the supply curve, which is
$$\frac{2+3.5}{2} \cdot 300 = \$825.$$
Therefore, by receiving \$1,150 as the actual revenue, she makes an "extra profit" of \$325.

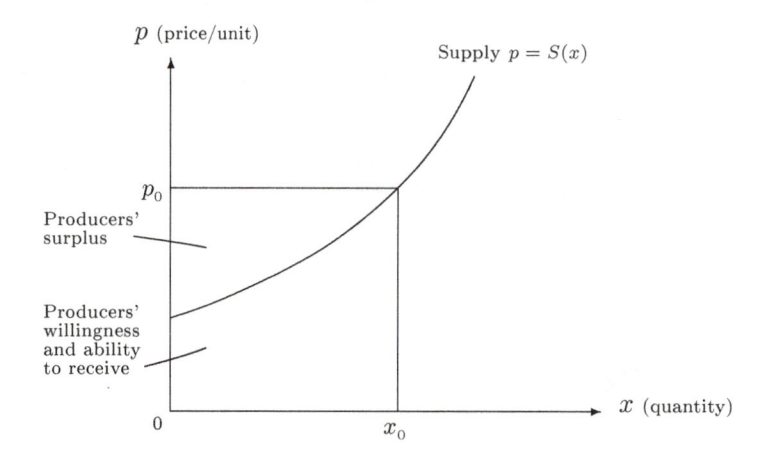

Figure 5.1.6: Producers' surplus at price p_0

In economics, the total amount that producers are willing or able to receive for a certain quantity x_0 of goods or services is represented by the area under the supply curve on the interval $[0, x_0]$. The amount that producers receive at a certain market price p_0 is the revenue, represented by the rectangular area $p_0 \times x_0$. Furthermore, the amount that producers receive above the amount that they are willing or able to receive for x_0 items at a market price p_0 is called **producers' surplus**, and is represented by the area of the region between $p = p_0$ and the demand curve on the interval $[0, x_0]$ (Figure 5.1.6). Similar to the consumers' surplus, we state the definition of producers' surplus mathematically:

Definition *producers' surplus*

Suppose that $p = S(x)$ describes the supply function for a commodity. Then at the point (x_0, p_0), the amount that **producers are willing and able to receive** is defined by
$$\int_0^{x_0} S(x)\, dx.$$
The **producers' revenue** is defined by $p_0 \times x_0$. And the **producers' surplus** is defined by
$$(p_0 \times x_0) - \int_0^{x_0} S(x)\, dx.$$

Example 3: For the supply function $p = S(x) = x^2 + 2x + 1$, determine the producer' surplus at the supply level $x = 3$.

Solution We substitute $x = 3$ into $S(x)$ to obtain $p = 16$. By the definition of the producers' surplus, given that $x_0 = 3$ and $p_0 = 16$, we have

$$\text{P.S.} = 16 \times 3 - \int_0^3 (x^2 + 2x + 1)\, dx = 48 - \left[\frac{x^3}{3} + x^2 + x\right]_0^3$$
$$= 48 - (9 + 9 + 3 - 0) = 27.$$

5.1. CONSUMERS' AND PRODUCERS' SURPLUSES

Example 4: Assume that the function for the average monthly supply of a certain brand of cellular phone can be described by

$$p = S(x) = 1{,}500 e^{0.04x}$$

where x is measured in hundreds of cellular phones and p is measured in dollars per hundred cellular phones.

(a) Determine the price at which producers will supply 2,500 cellular phones each month.

(b) Determine the producers' monthly revenue from selling 2,500 cellular phones at the corresponding market price.

(c) Determine the monthly amount producers are willing and able to receive for supplying 2,500 cellular phones.

(d) Determine the producers' monthly surplus for supplying 2,500 cellular phones at the corresponding market price.

Solution (a) Setting $x_0 = 25$ we obtain

$$p = S(2) = 1{,}500 e^{0.04(25)} \approx \$4{,}077 \text{ per hundred cellular phones.}$$

That is, at a market price p_0 of approximately \$4,077 per hundred cellular phones, or \$40.77 per cellular phone, producers will supply 2,500 cellular phones each month.

(b) The producers' monthly revenue from selling 2,500 cellular phones at the corresponding market price

$$p_0 \times x_0 \approx \$40.77 \text{ (per cellular phone)} \times 2{,}500 \text{ (cellular phones)} = \$101{,}925.$$

(c) We evaluate the integral

$$\int_0^{x_0} S(x)\,dx = \int_0^{25} 1{,}500 e^{0.04x}\,dx$$

$$= \left[\frac{1{,}500}{0.04} e^{0.04x}\right]_0^{25} = \frac{1{,}500}{0.04}\left[e^{0.04(25)} - e^0\right] \approx \$64{,}435.6.$$

(d) By the definition and parts (b) and (c), the producers' monthly surplus from supplying 2,500 cellular phones at the corresponding market price is approximately

$$\$101{,}925 - \$64{,}435.6 = \$37{,}489.4 \text{ (Figure 5.1.7)}.$$

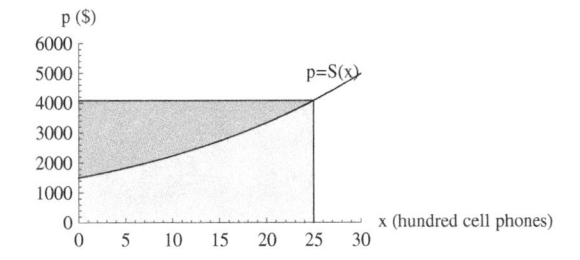

Figure 5.1.7: Producers' surplus at demand of 2,500 phones

Consumers' and Producers' Surpluses at the Equilibrium Point

As we discussed in Section 1.3, in a free market for a certain product, the market naturally settles at the equilibrium point (x_E, p_E). In such a case, the consumers' expenditure and the producers' revenue are equal, both being $p_E \times x_E$, and the sum of the consumers' and producers' surpluses equals the area of the region between the demand curve and the supply curve, on interval $[0, x_E]$ (Figure 5.1.8).

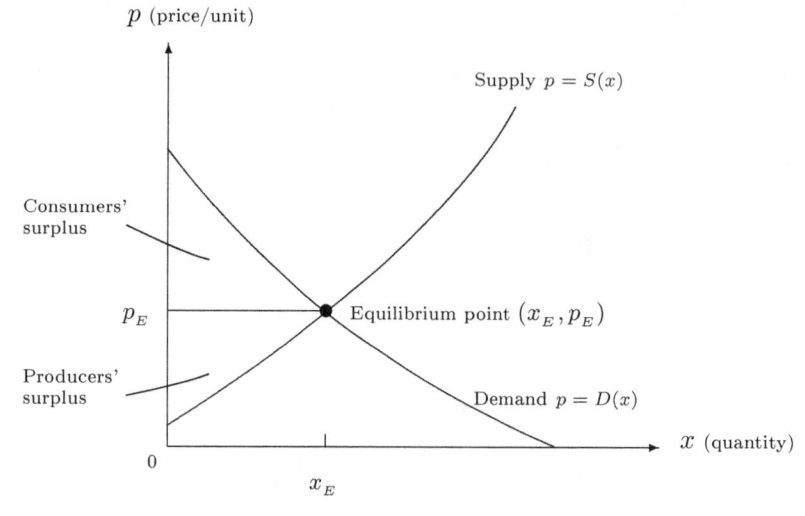

Figure 5.1.8: Consumers' and producers' surpluses at equilibrium point

Example 5: For the demand and supply functions
$$p = D(x) = (x-5)^2; \quad p = S(x) = x^2 + 6x + 4,$$
do the following:

(a) Determine the equilibrium point (x_E, p_E);

(b) Determine the consumers' surplus at the equilibrium point.

(c) Determine the producers' surplus at the equilibrium point.

Solution (a. We solve the equation $D(x) = S(x)$:
$$(x-5)^2 = x^2 + 2x + 1, \quad \text{or} \quad x^2 - 10x + 25 = x^2 + 2x + 1,$$
so $24 = 12x$, hence $x_E = 2$.

We substitute $x = 2$ into $S(x)$ to obtain
$$p_E = S(x_E) = 2^2 + 2(2) + 1 = 9.$$

Therefore, the equilibrium point is $(2, 9)$.

(b) By the definition of the consumers' surplus, given that $x = 2$ and $p = 9$, we have

$$\text{C.S.} = \int_0^2 (x-5)^2 \, dx - (9 \times 2) = \int_0^2 (x^2 - 10x + 25) \, dx - 18$$

$$= \left[\frac{x^3}{3} - 5x^2 + 25x\right]_0^2 - 18 = \left(\frac{8}{3} - 20 + 50\right) - 0 - 18 = 14\frac{2}{3}.$$

5.1. CONSUMERS' AND PRODUCERS' SURPLUSES

(c) By the definition of the producers' surplus, given that $x = 2$ and $p = 9$, we have

$$\text{P.S.} = 9 \times 2 - \int_0^2 (x^2 + 2x + 1)\, dx = 18 - \left[\frac{x^3}{3} + x^2 + x\right]_0^2$$

$$= 18 - \left(\frac{8}{3} + 4 + 2 - 0\right) = 9\frac{1}{3}.$$

When the demand and supply functions are linear, it is often easier to obtain the surpluses at the equilibrium point by evaluating the areas of the involved triangles directly, as illustrated by the following example.

Example 6: Jane, the owner of *Sweet Dreams Chocolate*, hired a market survey company that determined that the weekly demand for her chocolate bars can be described by

$$p = D(x) = 6.5 - 0.004x$$

where x is the number of chocolate bars and p is in dollars. The weekly supply function can described by

$$p = S(x) = 0.005x + 2.$$

(a) Determine the equilibrium point (x_E, p_E).

(b) Determine the producers' weekly surplus at the equilibrium point.

(c) Determine the consumers' weekly surplus at the equilibrium point.

(d) Determine the total (consumers' and producers') weekly surplus at the equilibrium point.

Solution (a) We solve the equation $S(x) = D(x)$:

$$0.005x + 2 = 6.5 - 0.004x, \quad \text{so} \quad 0.009x = 4.5, \quad \text{hence} \quad x_E = 500.$$

We substitute $x = 500$ into $S(x)$ to obtain

$$p_E = S(x_E) = 0.005(500) + 2 = \$4.5.$$

Therefore, the equilibrium point is $(500, \$4.5)$.

(b) From the graph shown in Figure 5.1.9, the producers' weekly surplus is the triangular area

$$\frac{1}{2}(500)(4.5 - 2) = \$625.$$

(c) From the graph shown in Figure 5.1.9, the consumers' weekly surplus is the triangular area

$$\frac{1}{2}(500)(6.5 - 4.5) = \$500.$$

(d) The total surplus is $625 + 500 = \$1{,}125$.

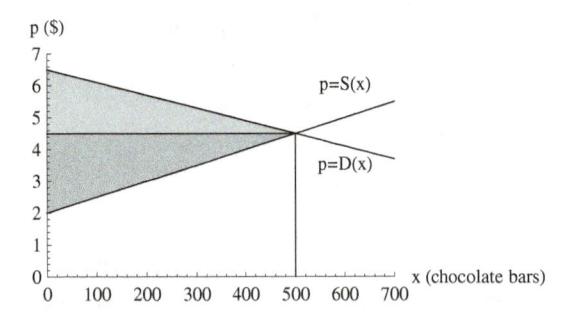

Figure 5.1.9: Consumers' and producers' surpluses at equilibrium point

Effects of Taxes and Price Control on Surpluses

In the next few examples, we illustrate how imposing taxes or price controls can affect surpluses.

Example 7: Jane, the *Sweet Dreams Chocolate* owner, was notified that a "sugary tax" of $0.25 for each chocolate bar will be imposed on her chocolate bars.

(a) Determine the new supply function and the new equilibrium point (x_E, p_E).

(b) Determine the producers' weekly surplus at the new equilibrium point.

(c) Determine the consumers' weekly surplus at the new equilibrium point.

(d) Determine the total (consumers' and producers') weekly surplus at the new equilibrium point.

Solution (a) For each chocolate bar sold at a market price p, $0.25 will be deducted for the tax. Therefore, Jane will only receive $p - 0.25$. Hence, her new supply curve is

$$p - 0.25 = 0.005x + 2.$$

Solving the equation for p we obtain

$$p = S(x) = 0.005x + 2.25.$$

Next, we find the new equilibrium point. We solve the equation $S(x) = D(x)$:

$$0.005x + 2.25 = 6.5 - 0.004x, \quad \text{so} \quad 0.009x = 4.25, \quad \text{hence,} \quad x_E \approx 472.$$

We substitute $x = 472$ into $S(x)$ to obtain

$$p_E = S(x_E) = 0.005(472) + 2.25 = \$4.61.$$

Therefore, the equilibrium point is $(472, \$4.61)$. Fewer chocolate bars will be sold.

(b) From the graph shown in Figure 5.1.10, the producers' weekly surplus is the triangular area

$$\frac{1}{2}(472)(4.61 - 2.25) = \$556.96.$$

Jane's surplus will be reduced.

(c) From the graph shown in Figure 5.1.10, the consumers' weekly surplus is the triangular area

$$\frac{1}{2}(472)(6.5 - 4.61) = \$446.04.$$

5.1. CONSUMERS' AND PRODUCERS' SURPLUSES

The consumers' surplus will be reduced, too.

(d) The total surplus is $556.96 + 446.04 = \$1{,}003$.

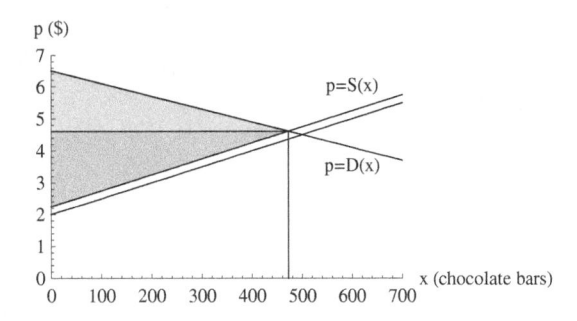

Figure 5.1.10: Consumers' producers' surpluses with tax on producer

Example 8: Suppose that the "sugary tax" will not be imposed. Instead, there will be a 9% sales tax for the Sweet Dreams Chocolate bars.

(a) Determine the new demand function and the new equilibrium point (x_E, p_E).

(b) Determine the producers' weekly surplus at the new equilibrium point.

(c) Determine the consumers' weekly surplus at the new equilibrium point.

(d) Determine the total (consumers' and producers') weekly surplus at the new equilibrium point.

Solution (a) For each chocolate bar purchased at a market price p, consumers will pay an additional $9\%p$ for the chocolate bars, hence the actual payment will be $(1.09)p$ for each chocolate bar. Hence, the new demand curve is

$$(1.09)p = 6.5 - 0.004x.$$

Solving the equation for p we obtain

$$p = D(x) \approx 5.9633 - 0.00367x.$$

Next, we find the new equilibrium point. We solve the equation $S(x) = D(x)$:

$$0.005x + 2 = 5.9633 - 0.00367x, \quad \text{so} \quad 0.00867x = 3.9633, \quad \text{hence} \quad x_E \approx 457.$$

We substitute $x = 457$ into $S(x)$ to obtain

$$p_E = S(x_E) = 0.005(457) + 2 = \$4.29.$$

Therefore, the equilibrium point is $(457, \$4.29)$. Fewer chocolate bars will be sold.

(b) From the graph shown in Figure 5.1.11, the producers' weekly surplus is the triangular area

$$\frac{1}{2}(457)(4.29 - 2) = \$523.27.$$

Jane's surplus will be reduced.

(c) From the graph shown in Figure 5.1.11, the consumers' weekly surplus is the triangular area

$$\frac{1}{2}(457)(5.9633 - 4.29) = \$382.35.$$

The consumers' surplus will be reduced, too.

(d) The total surplus is $523.27 + 382.35 = \$905.61$.

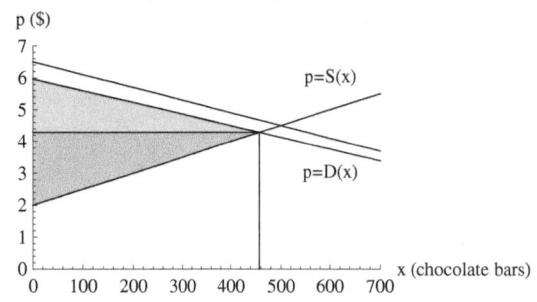

Figure 5.1.11: Consumers' and producers' surpluses with tax on consumers

Example 9: Suppose that no taxes will be imposed on the Sweet Dreams Chocolate bars, but a price control will be imposed that caps the price of the chocolate bars at no higher than $4.

(a) How will the price control affect the market?

(b) Determine the producers' weekly surplus at the market price of $4.

(c) Determine the consumers' weekly surplus at the market price of $4.

(d) Determine the total (consumers' and producers') weekly surplus at the market price of $4.

Solution (a) From the graph shown in Figure 5.1.12, there will be a shortage of supply, since Jane will only supply 400 chocolate bars, but the consumers will want 625 chocolate bars. Therefore, only 400 chocolate bars will be sold each week.

(b) From the graph shown in Figure 5.1.12, the producers' weekly surplus is the triangular area

$$\frac{1}{2}(400)(4-2) = \$400.$$

Jane's surplus will be reduced.

(c) From the graph shown in Figure 5.1.12, the consumers' weekly surplus is the sum of the triangular area and the rectangular area

$$\frac{1}{2}(400)(6.5-4.9) + (400)(4.9-4) = 320 + 360 = \$680.$$

The consumers' surplus will be increased.

(d) The total surplus is $400 + 680 = \$1,080$. Hence, the total surplus will be reduced.

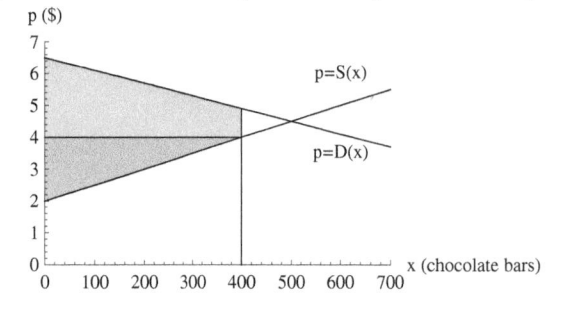

Figure 5.1.12: Consumers' and producers' surpluses with price control

Exercise Set 5.1

1. Refer to the graph here.

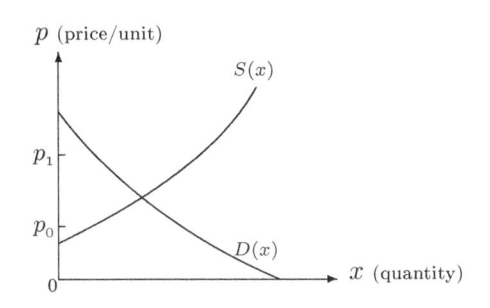

    (a) Shade the region that corresponds to the consumers' surplus at the market price p_0. Write an integral that can determine the consumers' surplus.

    (b) Shade the region that corresponds to the consumers' surplus at the market price p_1.

    (c) Shade the region that corresponds to the producers' surplus at the market price p_0. Write an integral that can determine the producers' surplus.

    (d) Shade the region that corresponds to the producers' surplus at the market price p_1.

    (e) Shade the region that corresponds to the consumers' expenditure at the market price p_1.

    (f) Shade the region that corresponds to the consumers' surplus at the equilibrium price.

    (g) Shade the region that corresponds to the producers' surplus at the equilibrium price.

    (h) At which market price is the total surplus (i.e., the sum of the consumers' and the producers' surpluses) the largest?

2. Refer to the graph here; repeat Exercise #1, (a) – (h).

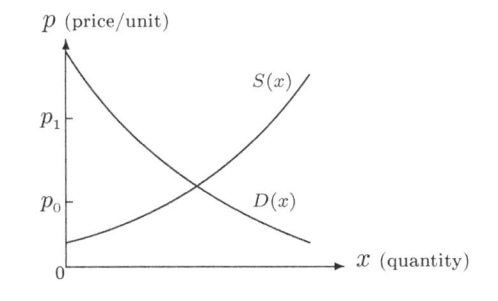

3. Refer to the graph here.

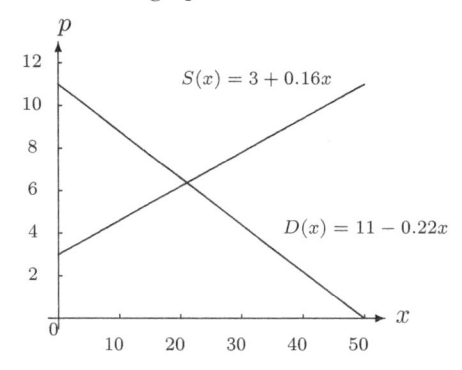

    (a) Shade the region that corresponds to the consumers' surplus at the market price $p = 8$. Write an integral to evaluate the consumers' surplus.

    (b) Shade the region that corresponds to the producers' surplus at the market price $p = 8$. Write an integral to evaluate the producers' surplus.

    (c) Evaluate the consumers' surplus at the equilibrium price.

    (d) Evaluate the producers' surplus at the equilibrium price.

4. Refer to the graph here; repeat Exercise #3, (a) – (d).

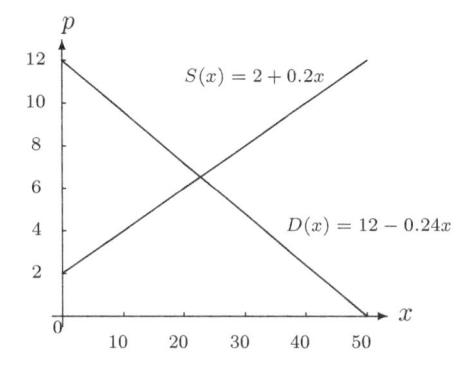

For Exercises #5 to #18, for each demand function and the demand level x given, find the consumers' surplus.

5. $D(x) = 360 - 3x$, $x = 40$

6. $D(x) = 360 - 3x$, $x = 80$

7. $D(x) = 800 - 0.8x$, $x = 200$

8. $D(x) = 800 - 0.8x$, $x = 400$

9. $D(x) = 50 - \sqrt{x+4}$, $x = 21$

10. $D(x) = 40 - \sqrt{x+5}$, $x = 11$

11. $D(x) = (x-6)^2$, $x = 2$

12. $D(x) = 1600 - 0.08x^2$, $x = 120$

13. $D(x) = 1200 - x^2$, $x = 30$

14. $D(x) = (x-90)^2$, $x = 40$

15. $D(x) = 300e^{-0.02x}$, $x = 100$

16. $D(x) = 200e^{-0.04x}$, $x = 80$

17. $D(x) = 360e^{-0.01x}$, $x = 200$

18. $D(x) = 540e^{-0.02x}$, $x = 150$

For Exercises #19 to #32, for each supply function and the demand level x given, find the producers' surplus.

19. $S(x) = 700 + 1.5x$, $x = 200$

20. $S(x) = 160 + 0.4x$, $x = 60$

21. $S(x) = 120 + 0.1x$, $x = 100$

22. $S(x) = 40 + 0.08x$, $x = 200$

23. $S(x) = 4\sqrt{x+5}$, $x = 4$

24. $S(x) = 6\sqrt{x+16}$, $x = 9$

25. $S(x) = 20 + 0.01x^2$, $x = 30$

26. $S(x) = 40 + 0.03x^2$, $x = 20$

27. $S(x) = x^2 + 2x + 8$, $x = 3$

28. $S(x) = x^2 + 4x + 1$, $x = 2$

29. $S(x) = 20e^{0.02x}$, $x = 40$

30. $S(x) = 30e^{0.01x}$, $x = 120$

31. $S(x) = 10e^{0.02x}$, $x = 100$

32. $S(x) = 40e^{0.01x}$, $x = 80$

For Exercises #32 to #44, do the following:

(a) Determine the equilibrium point (x_E, p_E);

(b) Determine the consumers' surplus at the equilibrium point.

(c) Determine the producers' surplus at the equilibrium point.

33. $D(x) = 320 - 3x$; $S(x) = 120 + 5x$

34. $D(x) = 368 - 0.4x$; $S(x) = 17 + 0.5x$

35. $D(x) = 94 - 0.25x$; $S(x) = 24 + 0.75x$

36. $D(x) = 300 - 1.7x$; $S(x) = 20 + 0.3x$

37. $D(x) = 220 - 0.04x^2$; $S(x) = 26.4 + 0.06x^2$

38. $D(x) = 90 - 0.01x^2$; $S(x) = 15 + 0.02x^2$

39. $D(x) = (x-4)^2$; $S(x) = x^2 + 2x + 6$

40. $D(x) = (x-6)^2$; $S(x) = x^2 + 4x + 4$

41. $D(x) = 6 - x$, for $0 \le x \le 6$; $S(x) = \sqrt{x+6}$

42. $D(x) = 8 - x$, for $0 \le x \le 8$; $S(x) = \sqrt{2x+8}$

43. $D(x) = \dfrac{81}{\sqrt{x+1}}$; $S(x) = \sqrt{x+1}$

44. $D(x) = \dfrac{200}{\sqrt{x+3}}$; $S(x) = 2\sqrt{x+3}$

APPLICATIONS

45. A demand curve is given by
$$45p + 50x = 3,000,$$
where p is the price of the product, in dollars, and x is the quantity demanded at that price.

(a) Assuming that the equilibrium price is $8 per item, determine the equilibrium demand.

(b) Determine the consumers' surplus at the equilibrium demand.

46. A demand curve is given by
$$x = D(p) = 1,500 - 20p,$$
where p is the price of the product, in dollars, and x is the quantity demanded at that price.

(a) Assuming that the equilibrium price is $50 per item, determine the equilibrium demand.

(b) Determine the consumers' surplus at the equilibrium demand.

47. A city tour company finds that if the price p, charged for a 2-hour bus tour, is $30, the average number of passengers per week, x, is 200. When the price is reduced to $25, the average number of passengers per week increases to 250.

(a) Assuming that the demand curve $p = D(x)$ is linear, find its formula.

(b) Assuming that the equilibrium price is $27 per tour, determine the equilibrium demand.

(c) Determine the consumers' surplus at the equilibrium demand.

48. A boat tour company finds that if the price p, charged for an 1-hour harbor tour, is $20, the average number of passengers per week, x, is 300. When the price is reduced to $18, the average number of passengers per week increases to 360.

(a) Assuming that the demand curve $p = D(x)$ is linear, find its formula.

(b) Assuming that the equilibrium price is $22 per tour, determine the equilibrium demand.

(c) Determine the consumers' surplus at the equilibrium demand.

49. The demand and supply curves of a certain brand of sunglasses are given by

$$p = D(x) = 200 - 0.1x, \text{ and}$$

$$p = S(x) = 20 + 0.05x$$

where p is the price in dollars and x is the quantity.

(a) Find the equilibrium quantity and price.

(b) Determine the producers' surplus at the equilibrium demand.

(c) Determine the consumers' surplus at the equilibrium demand.

(d) Determine total surplus (i.e., the sum of the consumers' and producers' surpluses) at the equilibrium demand.

50. The demand and supply curves of a certain brand of running shoes are given by

$$p = D(x) = 112 - 0.04x, \text{ and}$$

$$p = S(x) = 0.06x + 42$$

where p is the price in dollars and x is the quantity sold.

(a) Find the equilibrium quantity and price.

(b) Determine the producers' surplus at the equilibrium demand.

(c) Determine the consumers' surplus at the equilibrium demand.

(d) Determine total surplus (i.e., the sum of the consumers' and producers' surpluses) at the equilibrium demand.

51. Refer to Exercise #49: A tax of $2.50 will be imposed on the producer for each pair of the sunglasses sold.

(a) Determine the new supply function and the new equilibrium quantity and price.

(b) Determine the producers' surplus at the new equilibrium point.

(c) Determine the consumers' surplus at the new equilibrium point.

(d) Determine the total (consumers' and producers') surplus at the new equilibrium point.

(e) Compare the results from (b) – (d) to those in Exercise #49.

52. Refer to Exercise #50: A tax of $2.00 will be imposed on the producer for each pair of the running shoes sold.

(a) Determine the new supply function and the new equilibrium quantity and price.

(b) Determine the producers' surplus at the new equilibrium point.

(c) Determine the consumers' surplus at the new equilibrium point.

(d) Determine the total (consumers' and producers') surplus at the new equilibrium point.

(e) Compare the results from (b) – (d) to those in Exercise #50.

53. Refer to Exercise #49: An 8% sales tax will be imposed on each pair of the sunglasses sold.

(a) Determine the new demand function and the new equilibrium quantity and price.

(b) Determine the producers' surplus at the new equilibrium point.

(c) Determine the consumers' surplus at the new equilibrium point.

(d) Determine the total (consumers' and producers') surplus at the new equilibrium point.

(e) Compare the results from (b) – (d) to those in Exercise #49.

54. Refer to Exercise #50: A 7% sales tax will be imposed on each pair of the running shoes sold.

(a) Determine the new demand function and the new equilibrium quantity and price.

(b) Determine the producers' surplus at the new equilibrium point.

(c) Determine the consumers' surplus at the new equilibrium point.

(d) Determine the total (consumers' and producers') surplus at the new equilibrium point.

(e) Compare the results from (b) – (d) to those in Exercise #50.

55. Refer to Exercise #49: A price control will be imposed that caps the price of each pair of the sunglasses at no more than $70.

(a) How will the price control affect the market (i.e., will there be an over-supply or short-supply)? How many pairs of sunglasses will be sold at that price?

(b) Determine the producers' weekly surplus at the market price of $70.

(c) Determine the consumers' weekly surplus at the market price of $70.

(d) Determine the total (consumers' and producers') weekly surplus at the market price of $70.

(e) Compare the results from (b) – (d) to those in Exercise #49.

56. Refer to Exercise #50: A price control will be imposed that caps the price of each pair of the running shoes at no more than $65.

(a) How will the price control affect the market (i.e., will there be an over-supply or short-supply)? How many pairs of running shoes will be sold at that price?

(b) Determine the producers' weekly surplus at the market price of $65.

(c) Determine the consumers' weekly surplus at the market price of $65.

(d) Determine the total (consumers' and producers') weekly surplus at the market price of $65.

(e) Compare the results from (b) – (d) to those in Exercise #50.

57. Refer to Exercise #49: A price control will be imposed that sets the minimum price of each pair of the sunglasses at no less than $90.

(a) How will the price control affect the market (i.e., will there be an over-supply or short-supply)? How many pairs of sunglasses will be sold at that price?

(b) Determine the producers' weekly surplus at the market price of $90.

(c) Determine the consumers' weekly surplus at the market price of $90.

(d) Determine the total (consumers' and producers') weekly surplus at the market price of $90.

(e) Compare the results from (b) – (d) to those in Exercise #49.

58. Refer to Exercise #50: A price control will be imposed that sets the minimum price of each pair of the running shoes at no less than $90.

(a) How will the price control affect the market (i.e., will there be an over-supply or short-supply)? How many pairs of running shoes will be sold at that price?

(b) Determine the producers' weekly surplus at the market price of $90.

(c) Determine the consumers' weekly surplus at the market price of $90.

(d) Determine the total (consumers' and producers') weekly surplus at the market price of $90.

(e) Compare the results from (b) – (d) to those in Exercise #50.

5.2 Definite Integrals in Finance

In section 1.5, some financial phenomena that can be modeled as exponential functions are presented. The models involve finding future or present value of lump-sum investments. In this section, we explore other financial phenomena which involve **continuous money flows**.

We start this section by reviewing some of the financial models raised in Section 1.5.

Suppose that Mary invests $1,000 dollars in a fund paying interest at an annual rate of 6%, compounded continuously, for t years. Then her future balance is

$$B(t) = 1,000 e^{0.06t}.$$

Now suppose that Mary wants to have a balance of $5,000 by the end of 5 years with the fund. Then she needs to invest

$$P(5) = 5,000 e^{-0.06(5)} \approx \$3,704.09.$$

Before we go on, let us clarify how realistic this "continuous" model is. A bank does not continuously place money (the interest earned) into an invested account. Similarly, in this section, we deal with a "continuous money flow" (also called a "continuous income stream" in economics), but no one can deposit or withdraw money continuously.

> **Discrete vs. continuous models**
>
> A continuous model often can describe a discrete circumstance very well, and then calculus can be used to seek related solutions.

We have seen such applications in marginal analysis in Section 3.2 and in optimization in Section 3.7. In those cases, a commodity often involves discrete quantities, such as televisions or number of hotel rooms. Despite the difference, calculus can still help us to analyze such circumstances by treating the discrete functions as continuous. Let us compare the following models:

Example 1: Suppose that Mary invests $1,000 dollars in a fund paying interest at an annual rate of 6% for 3 years. What is the balance of her investment at the end of the 3 years if the interest is compounded (a) daily; (b) continuously?

Solution (a) We apply the formula

$$B(t) = P_0 \left(1 + \frac{r}{n}\right)^{nt}.$$

We have $P_0 = 1,000$, $r = 0.06$, $n = 365$, and $t = 3$. Hence, the amount in the account is

$$B(3) = 1,000 \left(1 + \frac{0.06}{365}\right)^{365 \times 3} \approx \$1,197.20.$$

(b) We apply the formula

$$B(t) = P_0 e^{rt}.$$

We have $P_0 = 1,000$, $r = 0.06$, and $t = 3$. Hence, her future balance is

$$B(3) = 1,000 e^{0.06(3)} \approx \$1,197.22.$$

Hence, the future values computed by the two models are almost the same.

We now consider other financial situations:

A. Sam, a 20-year-old agent for an insurance firm, wants to invest part of his weekly wage in order to have $1,000,000 when he retires at 67. Suppose that he invests the money in a fund paying interest at an annual rate of 6%, compounded continuously. How much should Sam invest each week?

B. Karen has just purchased her first house. She has a $250,000, 30-year mortgage with interest at an annual rate of 5%, compounded continuously. How much is her monthly payment? If she pays an extra $150 each month, how soon can she pay off the mortgage? How much money can she save by paying an extra $150 each month?

These two cases deal with a "flow of money," and fit into two financial models. To answer Sam's question, we need to use the model of **future value of a continuous money flow**.

Future Value of a Continuous Money Flow

Suppose that Jane, the owner of *Sweet Dreams Chocolate*, invests part of her profit, $7,300 a year, into a fund paying interest at an annual rate of 5%, compounded continuously, for 10 years. She deposits the money daily, hence, the daily deposit is $7,300/365 = \$20$. Now let us see how the money grows.

The value of the first day's deposit will grow to

$$20e^{0.05(10)}.$$

The value of the second day's deposit will grow to

$$20e^{0.05(10-1/365)}.$$

The value of the third day's deposit will grow to

$$20e^{0.05(10-2/365)}.$$

And so on, the value of the last day's deposit will grow to

$$20e^{0.05(10-9\frac{364}{365})} = 20e^{0.05(1/365)}.$$

The total value in the account at the end of 10 years is the sum of all the future values:

$$20e^{0.05(10)} + 20e^{0.05(10-1/365)} + 20e^{0.05(10-2/365)} \ldots 20e^{0.05(10-9\frac{364}{365})}.$$

Using the summation symbol, with $\Delta t = 1/365$ (hence $1 = 365\Delta t$, or $20 = 7{,}300\Delta t$) and $n = 10 \cdot 365$, we can write the sum as

$$\sum_{k=1}^{n} 7{,}300 e^{0.05(10-(k-1)/365)} \Delta t = \sum_{k=1}^{n} f(t_k) \Delta t.$$

The function $f(t_k)$ is $7{,}300 e^{0.05(10-t_k)}$, where $t_k = (k-1)/365$. Comparing this summation with the definition of the definite integral in Section 4.3, we note that if n approaches infinity (which means that the deposit is going to be made continuously), the value of Jane's account at the end of 10 years is

$$\int_0^{10} 7{,}300 e^{0.05(10-t)} \, dt.$$

5.2. DEFINITE INTEGRALS IN FINANCE

We can evaluate this integral using the technique introduced in Section 4.5: We let $u = 0.05(10 - t)$, then $du = -0.05\,dt$, or $du/(-0.05) = dt$. When $t = 0$, we have $u = 0.5$. When $t = 10$, we have $u = 0$. We now substitute all formulas **and the limits** of t in the definite integral by that of u:

$$\int_0^{10} 7{,}300 e^{0.05(10-t)}\, dt = \int_{0.5}^{0} 7{,}300 e^u \frac{1}{-0.05}\, du = \frac{7{,}300}{-0.05} \int_{0.5}^{0} e^u\, du$$

$$= -146{,}000 \left[e^u\right]_{0.5}^{0} = -146{,}000 \left[1 - e^{0.5}\right] \approx \$94{,}713.$$

Therefore, Jane's investment at the end of 10 years will grow to \$94,713.

We generalize this result to the following case: Suppose that $R(t)$ per year is invested continuously into a fund paying interest at an annual rate of $100\,r\%$ compounded continuously, for T years, then the balance of the account at the end of T years can be calculated by the integral $\int_0^T R(t)e^{r(T-t)}\, dt$. This future sum is called the **future value of a continuous money flow (or income stream)** in economics.

Future Value of a Continuous Money Flow

Suppose that a yearly amount of $R(t)$, where t is measured in years, is invested continuously into an interest-bearing account at the annual interest rate of $100\,r\%$ compounded continuously. Then the future value of the account at the end of T years is

$$\int_0^T R(t) e^{r(T-t)}\, dt.$$

When the money flow is a constant, there is no need to use the integral every time.

Future Value of a Constant Continuous Money Flow

When the yearly investment $R(t)$ is a constant S, the future value can be calculated by the following formula:

$$\mathbf{FV} = \frac{S}{r}\left(e^{rT} - 1\right).$$

We now derive the formula provided above. If $R(t) = S$, the integral becomes

$$\int_0^T S e^{r(T-t)}\, dt$$

It can be evaluated using the same technique introduced in Section 4.5: We let $u = r(T - t)$, then $du = -r\,dt$, or $\frac{1}{-r} du = dt$. When $t = 0$, we have $u = rT$. When $t = T$, we have $u = 0$. We now substitute all formulas **and the limits** of t in the definite integral by that of u:

$$\int_0^T S e^{r(T-t)}\, dt = \int_{rT}^{0} S e^u \frac{1}{-r}\, du = \frac{S}{-r} \int_{rT}^{0} e^u\, du = \frac{S}{-r} \left[e^u\right]_{rT}^{0}$$

$$= \frac{S}{-r}\left[e^0 - e^{rT}\right] = \frac{S}{-r}\left[1 - e^{rT}\right] = \frac{S}{r}\left(e^{rT} - 1\right).$$

Example 2: The grandparents of a 1-year-old girl plan to invest for her college education. They will invest \$2,000 per year continuously into an account paying interest at an annual rate of 5%, compounded continuously. How much will the account have when their granddaughter is 18 years old?

Solution We have $T = 18-1 = 17$, $S = 2{,}000$, and $r = 0.05$. Therefore, the future value of the account is

$$\frac{2{,}000}{0.05}\left(e^{0.05(17)} - 1\right) \approx \$53{,}586.$$

Example 3: Sam, a 20-year-old agent for an insurance firm, wants to invest part of his weekly wage in order to have \$1,000,000 when he retires at 67. Suppose that he invests the money in a fund paying interest at a rate of 6% a year, compounded continuously. Assuming a continuous money flow, how much should Sam invest each week?

Solution Assume that Sam invests S dollars per year continuously. He has $67-20 = 47$ years to invest up to his retirement. We have $r = 0.06$ and $T = 47$, and the future value is \$1,000,000. Substituting the known numbers into the future value formula we have

$$1{,}000{,}000 = \frac{S}{0.06}\left(e^{0.06(47)} - 1\right) = S(262.95).$$

Solving for S we obtain

$$S = 1{,}000{,}000/262.95 \approx 3803.$$

Hence, Sam needs to save \$3,803 per year, or $3{,}803/52 = \$73.14$ per week.

Note: When the yearly investment $R(t)$ is not a constant, some integration technique is required, as illustrated in the following example.

Example 4: Refer to Example 3. Suppose that Sam estimates that, taking into account future promotions and raises, he can invest $R(t) = 3{,}600 + 180t$ at the tth year from now on, into the same account. Assuming a continuous money flow, how much will Sam have for retirement at 67?

Solution We need to use the integral formula

$$\int_0^T R(t)e^{r(T-t)}\,dt.$$

We have $r = 0.06$, $T = 47$, and $R(t) = 3{,}600 + 180t$. Hence the future value is

$$\int_0^{47} (3{,}600 + 180t)e^{0.06(47-t)}\,dt\,.$$

We let $u = 0.06(47 - t)$, then $du = -0.06\,dt$, or $\dfrac{1}{-0.06}du = dt$. When $t = 0$, we have $u = 2.82$. When $t = 47$, we have $u = 0$. We also need to change $3{,}600 + 180t$ into a formula of u. Since $u = 0.06(47 - t)$, solving for t we have $t = 47 - u/0.06$. Hence,

$$3{,}600 + 180t = 3{,}600 + 180(47 - u/0.06) = 12{,}060 - 3{,}000u.$$

We now substitute all formulas **and the limits** of t in the definite integral by that of u:

$$\int_0^{47} (3{,}600 + 180t)e^{0.06(47-t)}\,dt = \int_{2.82}^0 (12{,}060 - 3{,}000u)e^u \frac{1}{-0.06}\,du$$

$$= \int_{2.82}^0 (-201{,}000 + 50{,}000u)e^u\,du = -201{,}000\int_{2.82}^0 e^u\,du + 50{,}000\int_{2.82}^0 ue^u\,du.$$

The first integral is a basic formula:

$$-201{,}000\int_{2.82}^0 e^u\,du = -201{,}000\left(1 - e^{2.82}\right) \approx \$3{,}171{,}150.$$

5.2. DEFINITE INTEGRALS IN FINANCE

To evaluate the second integral, we can use the *Formula 4* in Table 1,

$$\int xe^{ax}\,dx = \frac{1}{a^2}\cdot e^{ax}(ax-1) + C$$

with $a = 1$ and $x = u$:

$$\int ue^u\,du = e^u(u-1) + C.$$

Hence,

$$50{,}000\int_{2.82}^{0} ue^u\,du = 50{,}000\left[e^u(u-1)\right]_{2.82}^{0}$$

$$= 50{,}000\left[e^0(-1) - e^{2.82}(1.82)\right] \approx -\$1{,}576{,}690.$$

Therefore, when Sam is 67, his account will have

$$\$3{,}171{,}150 - \$1{,}576{,}690 = \$1{,}594{,}450,$$

or about 1.6 million dollars.

We now explore the second model: the **(accumulated) present value of a continuous money flow**.

▬ Accumulated Present Value of a Continuous Money Flow ▬

Suppose that Jane, the owner of *Sweet Dreams Chocolate*, wants to expand her business and open a second production line. She plans to apply for a 10-year loan that charges interest at an annual rate of 6%, compounded continuously. Jane estimates that she can pay the amount of \$3,650 per year from the income of the expanded business. She deposits the money daily, hence the daily payment is $3{,}650/365 = \$10$. Now let us calculate how much she can borrow. Notice that each of her \$10 payments in the future includes the principal and the interest accumulated up to that time. Hence, the principal part is the present value of the payment.

The first day's payment will pay back the loan for

$$10e^{-0.06(1/365)}.$$

The second day's payment will pay back the loan for

$$10e^{-0.06(2/365)}.$$

The third day's payment will pay back the loan for

$$10e^{-0.06(3/365)}.$$

And so on. The last day's payment will pay back the loan for

$$10e^{-0.06(10)}.$$

The total amount of the principals paid over the 10 years is the sum of all the present values:

$$10e^{-0.06(1/365)} + 10e^{-0.06(2/365)} + 10e^{-0.06(3/365)} \ldots 10e^{-0.06(10)}.$$

Using the summation symbol, with $\Delta t = 1/365$ and $n = 10\cdot 365$, we can write the sum as

$$\sum_{k=1}^{n} 3{,}650 e^{-0.06(k/365)}\Delta t = \sum_{k=1}^{n} f(t_k)\Delta t.$$

The function $f(t_k)$ is $3{,}650e^{-0.06t_k}$. If n approaches infinity (which means that the payment is going to be made continuously), the amount of Jane's loan is

$$\int_0^{10} 3{,}650 e^{-0.06t}\, dt = 3{,}650 \left[\frac{1}{-0.06} e^{-0.06t}\right]_0^{10}$$

$$= \frac{3{,}650}{-0.06}\left[e^{-0.06(10)} - e^0\right] = \frac{3{,}650}{0.06}\left[1 - e^{-0.6}\right] \approx \$27{,}447.$$

We generalize this result to the following case: Suppose that $R(t)$ per year is paid continuously into a fund with annual interest rate of $100\,r\%$ compounded continuously, for T years, then the present value can be calculated by the integral $\int_0^T R(t)e^{-rt}\, dt$. This is called the **present value of a continuous money flow (or income stream)**, or the **accumulated present value** in economics.

Accumulated Present Value of a Continuous Money Flow

Suppose that a yearly amount of $R(t)$, where t is measured in years, is deposited continuously into an interest-bearing account at the annual interest rate $100\,r\%$ compounded continuously, for T years. Then the accumulated present value of the account is

$$\int_0^T R(t)e^{-rt}\, dt.$$

When the money flow is a constant, there is no need to use the integral every time.

Accumulated Present Value of a Constant Continuous Money Flow

When the yearly payment $R(t)$ is a constant D, the accumulated present value can be calculated by the following formula:

$$\mathbf{APV} = \frac{D}{r}\left(1 - e^{-rT}\right).$$

We now derive the formula provided above. If $R(t) = D$, then the integral can be computed directly:

$$\int_0^T D e^{-rt}\, dt = \left[\frac{D}{-r} e^{-rt}\right]_0^T$$

$$= \frac{D}{-r}\left[e^{-rT} - e^0\right] = \frac{D}{r}\left[1 - e^{-rT}\right].$$

Example 5: Refer to Example 2. Find the accumulated present value of the grandparents' investment, and explain to the grandparents what this value means.

Solution We apply the accumulated present value formula with $T = 17$, $D = 2{,}000$ and $r = 0.05$:

$$\frac{2{,}000}{0.05}\left(1 - e^{-0.05(17)}\right) \approx \$22{,}903.$$

This value is equal to a lump sum investment of $P_0 = 22{,}903$. That is, if the grandparents invests $\$22{,}903$ into the same account when the granddaughter is 1-year old, then when she is 18, the balance in that account should be the same as the future value of their continuous money flow investment, $\$53{,}586$. We

5.2. DEFINITE INTEGRALS IN FINANCE

can verify this:

$$22{,}903 e^{0.05(17)} = \$53{,}586.$$

Example 6: Sam wants to purchase a car. The car dealer offers a 5-year loan that charges interest at an annual rate of 12%, compounded continuously. Sam can pay $300 each month. Assuming a continuous money flow, how large a loan can Sam apply for?

Solution We have $D = 300(12) = 3{,}600$, $r = 0.12$, and $T = 5$. Substituting into the accumulated present value formula we obtain

$$\frac{3{,}600}{0.12}\left(1 - e^{-0.12(5)}\right) = \$13{,}535.7.$$

Hence, Sam can apply for a car loan of $13,535.70 from the dealer.

Example 7: Karen has just purchased her first house. She has a $250,000, 30-year mortgage with interest at an annual rate of 5%, compounded continuously. Assuming Karen pays her mortgage continuously with a fixed amount.

(a) How much is her monthly payment?

(b) If she pays an extra $150 each month, how soon can she pay off the mortgage?

(c) How much money can Karen save by paying an extra $150 each month?

Solution (a) Assume that Karen pays D dollars per year continuously. We have $r = 0.05$ and $T = 30$, and the accumulated present value is $250,000. Substituting the known numbers into the accumulated present value formula we have

$$250{,}000 = \frac{D}{0.05}\left(1 - e^{-0.05(30)}\right) = D(15.5374).$$

Solving for D we obtain

$$D = 250{,}000/15.5374 \approx 16{,}090.$$

Hence, Karen needs to pay $16,090 per year, or $16{,}090/12 = \$1{,}340.85$ per month.

(b) If Karen pays $1{,}340.85 + 150 = 1{,}490.85$ per month, or $1{,}490.85(12) = 17{,}890.2$ per year, then the the accumulated present value formula is

$$250{,}000 = \frac{17{,}890.2}{0.05}\left(1 - e^{-0.05T}\right).$$

This is an exponential equation. We solve for T:

$$250{,}000 = 357{,}804\left(1 - e^{-0.05T}\right)$$

$$0.69871 = \frac{250{,}000}{357{,}804} = 1 - e^{-0.05T}$$

$$e^{-0.05T} = 1 - 0.69871 = 0.301297$$

$$T = \frac{\ln 0.301297}{-0.05} \approx 24\text{yr}.$$

Therefore, Karen will pay off the mortgage in 24 years.

(c) The total payment of $16,090 per year for 30 years is $16{,}090(30) = \$482{,}700$, while the total payment of $17,890.20 per year for 24 years is $17{,}890.20(24) = \$429{,}364.80$. Hence, Karen saves $53,335.20 by paying an extra $150 each month.

Example 8: Joan is 65 and is considering retiring soon. She has $840,000 in a fund paying interest at an annual rate of 4% compounded continuously. She would like to withdraw a fixed amount continuously after she retires, and have a balance of $100,000 when she is 90 years old. Assuming a continuous money flow, how much can she spend each month?

Solution We first calculate the present value of the $100,000 Joan wishes to have when she is 90. We have $T = 90 - 65 = 25$ and $r = 0.04$. Hence,

$$P = 100{,}000 e^{-0.04(25)} \approx \$36{,}788.$$

Hence, Joan has to set $36,788 aside from the fund. She can spend the remaining $840,000 - \$36,788 = \$803,212$ for the 25 years. We have the accumulated present value $= 803{,}212$, $r = 0.04$, and $T = 25$. Substituting into the accumulated present value formula we obtain

$$803{,}212 = \frac{D}{0.04}\left(1 - e^{-0.04(25)}\right) = D(15.8).$$

Solving for D we have

$$D = 803{,}212/15.8 = 50{,}836.$$

Therefore, Joan can withdraw $50,836 per year, or $4,236 per month.

Example 9: Paul is 25 years old and plans to retire at 65. He wants to have a fund at 65 that will let him spend $5,000 a month after retirement, and last for 50 years. Assume a continuous money flow:

(a) Suppose that after his retirement Paul puts the money in a fund paying interest at an annual rate of 4%, compounded continuously. How much will Paul need for his retirement?

(b) Suppose that Paul starts to invest a fixed amount each month from now until he retires, in a fund that pays interest at an annual rate of 6%, compounded continuously. How much should he invest each month?

Solution (a) We have $D = 5{,}000(12) = 60{,}000$, $r = 0.04$, and $T = 50$. Substituting into the accumulated present value formula we obtain

$$\frac{60{,}000}{0.04}\left(1 - e^{-0.04(50)}\right) = 1{,}296{,}997.$$

Therefore, Paul will need $1,296,997 when he is 65.

(b) We need to use the formula for future value of a continuous money flow. Assume that Paul invests S dollars per year continuously. We have the future value $= 1{,}296{,}997$, $r = 0.06$, and $T = 65 - 25 = 40$. Substituting the known numbers into the future value formula we have

$$1{,}296{,}997 = \frac{S}{0.06}\left(e^{0.06(40)} - 1\right) = S(167.053).$$

Solving for S we obtain

$$S = 1{,}296{,}997/167.053 \approx 7{,}764.$$

Hence, Paul needs to save $7,764 per year, or $7{,}764/12 = \$645$ per month.

Exercise Set 5.2

1. Suppose that Sam deposits $2,000 per year into an account that has a 5% annual interest rate compounded continuously. Assuming a continuous money flow, how much will the account be worth at the end of 20 years?

2. Suppose that Mike deposits $8,000 per year into an account that has a 6% annual interest rate compounded continuously. Assuming a continuous money flow, how much will the account be worth at the end of 16 years?

3. Suppose that Joe deposits $200 per month into an account that has a 4% annual interest rate compounded continuously. Assuming a continuous money flow, how much will the account be worth at the end of 6 years?

4. Suppose that Jane deposits $80 per week into an account that has a 4.5% annual interest rate compounded continuously. Assuming a continuous money flow, how much will the account be worth at the end of 8 years?

5. Suppose that Sara deposits $2,500 per year into an account that has a 5% annual interest rate compounded continuously. Assuming a continuous money flow, how long will it take for the account to be worth $80,000?

6. Suppose that Mark deposits $4,000 per year into an account that has a 5.5% annual interest rate compounded continuously. Assuming a continuous money flow, how long will it take for the account to be worth $200,000?

7. Suppose that Paul deposits $400 per month into an account that has a 4.5% annual interest rate compounded continuously. Assuming a continuous money flow, how long will it take for the account to be worth $100,000?

8. Suppose that Mary deposits $80 per week into an account that has a 4.2% annual interest rate compounded continuously. Assuming a continuous money flow, how long will it take for the account to be worth $160,000?

9. Jenny is 24. She wants to invest part of her salary in order to have $580,000 when she retires at 67. Suppose that she invests the money to a fund paying interest at an annual rate of 6%, compounded continuously. Assuming a continuous money flow, how much should Jenny invest each year?

10. David is 23. He wants to invest part of his salary in order to have $680,000 when he retires at 66. Suppose that he invests the money in a fund paying interest at an annual rate of 5.6%, compounded continuously. Assuming a continuous money flow, how much should David invest each year?

11. Jack is 20. He wants to invest part of his weekly wage in order to have $600,000 when he retires at 65. Suppose that he invests the money in a fund paying interest at an annual rate of 6%, compounded continuously. Assuming a continuous money flow, how much should Jack invest each week?

12. Debbie is 22. She wants to invest part of her monthly salary in order to have $760,000 when she retires at 66. Suppose that she invests the money in a fund paying interest at an annual rate of 5.4%, compounded continuously. Assuming a continuous money flow, how much should Debbie invest each month?

13. Mark is 26. He plans to retire at 65. He estimates that with future promotions and raises, he can invest $R(t) = 3,600 + 160t$ at the tth year from now on. Suppose that he invests the money in an account paying interest at an annual rate of 5%, compounded continuously. Assuming a continuous money flow, how much will Mark have for retirement at 65?

14. Matthew is 22. He plans to retire at 66. He estimates that with future promotions and raises, he can invest $R(t) = 3,200 + 220t$ at the tth year from now on. Suppose that he invests the money into an account paying interest at an annual rate of 5.4%, compounded continuously. Assuming a continuous money flow, how much will Matthew have for retirement at 66?

15. Lucy plans to invest for her retirement in 40 years. She plans to invest 12% of her salary each month into a fund paying interest at an annual rate of 5.6%, compounded continuously. Lucy currently makes $2,500

per month, and expects her income to increase by 3% per year. Assuming a continuous money flow, how much will Lucy have for retirement?

16. Elizabeth plans to invest for her retirement in 36 years. She plans to invest 14% of her salary each month into a fund paying interest at an annual rate of 6%, compounded continuously. Elizabeth currently makes $3,800 per month, and expects her income to increase by 2.5% per year. Assuming a continuous money flow, how much will Elizabeth have for retirement?

17. Refer to Exercise #1. How much would Sam have to invest now, in one lump sum instead of in a continuous money flow, into the same account, in order to have the same future (20 year) value?

18. Refer to Exercise #2. How much would Mike have to invest now, in one lump sum instead of in a continuous money flow, into the same account, in order to have the same future (16 year) value?

19. Refer to Exercise #13. How much would Mark have to invest now, in one lump sum instead of in a continuous money flow, into the same account, in order to have the same amount of money when he retires?

20. Refer to Exercise #14. How much would Matthew have to invest now, in one lump sum instead of in a continuous money flow, into the same account, in order to have the same amount of money when he retires?

21. Karen wants to purchase a car. The car dealer offers a 5-year loan that charges interest at an annual rate of 11.5%, compounded continuously. Karen can pay $400 each month. Assuming a continuous money flow, how large of a loan can Karen afford?

22. Joe wants to purchase a car. The car dealer offers a 4-year loan that charges interest at an annual rate of 12.5%, compounded continuously. Joe can pay $360 each month. Assuming a continuous money flow, how large of a loan can Joe afford?

23. James wants to purchase a particular new car. The price of the car is $38,000. The car dealer requires at least a 20% initial payment, and offers a 5-year loan that charges interest at an annual rate of 12.5%, compounded continuously. James has $8,000 in savings, and he can pay $550 each month for the loan. Assuming a continuous money flow, can James purchase this car with a loan from the dealer?

24. Mark wants to purchase a particular new car. The price of the car is $26,000. The car dealer requires at least a 15% initial payment, and offers a 5-year loan that charges an interest with an annual rate of 12%, compounded continuously. Mark has $5,000 in savings, and he can pay $450 each month for the loan. Assuming a continuous money flow, can Mark purchase this car with a loan from the dealer?

25. Marian has just purchased a house. She has a $220,000, 15-year mortgage with interest at an annual rate of 4.5% a year, compounded continuously. Assuming Marian pays her mortgage continuously with a fixed amount:

(a) How much is her monthly payment?

(b) If she pays an extra $100 each month, how soon can she pay off the mortgage?

(c) How much money can Marian save by paying an extra $100 each month?

26. Sara has just purchased a house. She has a $360,000, 30-year mortgage with interest at an annual rate of 5.5% a year, compounded continuously. Assuming Sara pays her mortgage continuously with a fixed amount:

(a) How much is her monthly payment?

(b) If she pays an extra $300 each month, how soon can she pay off the mortgage?

(c) How much money can Sara save by paying an extra $300 each month?

27. Richard is 66 and is considering retiring soon. He has $540,000 in a fund paying interest at an annual rate of 4.5% compounded continuously. He would like to withdraw a fixed amount continuously after he retires, and have a balance of $50,000 when he is 92 years old. Assuming a continuous money flow, how much can he spend each month?

28. Jan is 62 and is considering retiring soon. She has $680,000 in a fund paying interest at

an annual rate of 4.2% compounded continuously. She would like to withdraw a fixed amount continuously after she retires, and have a balance of $80,000 when she is 90 years old. Assuming a continuous money flow, how much can she spend each month?

29. Rick is 26 years old and plans to retire at 64. He wants to have a fund at 64 that will let him spend $3,000 a month after retirement, and last for 40 years. Assume a continuous money flow.

(a) Suppose that after his retirement Rick puts the money in a fund paying interest at an annual rate of 4.5%, compounded continuously. How much will Rick need for his retirement?

(b) Suppose that Rick starts to invest a fixed amount each month from now until he retires, in a fund that pays interest at an annual rate of 6.8%, compounded continuously. How much should he invest each month?

30. John is 28 years old and plans to retire at 67. He wants to have a fund at 67 that will let him spend $4,500 a month after retirement, and last for 45 years. Assume a continuous money flow.

(a) Suppose that after his retirement John puts the money in a fund paying interest at an annual rate of 4.2%, compounded continuously. How much will John need for his retirement?

(b) Suppose that John starts to invest a fixed amount each month from now until he retires, in a fund that pays interest at an annual rate of 6.2%, compounded continuously. How much should he invest each month?

31. Refer to Exercise #27. If Richard decides to postpone his retirement for 2 years, how much can he spend each month after retirement?

32. Refer to Exercise #28. If Jan decides to postpone her retirement for 4 years, how much can she spend each month after retirement?

33. Refer to Exercise #29. If Rick waits until he is 36 to start investing for his retirement, how much should he invest each month in order to have the same amount when he retires at 64?

34. Refer to Exercise #30. If John waits until he is 35 to start investing for his retirement, how much should he invest each month in order to have the same amount when he retires at 67?

35. The parents of a 1-year old boy plan to invest for his college education. Their target is that when he is 18 years old, the fund should have the amount of $80,000. Assume that they invest the money in a fund paying interest at an annual rate of 5%, compounded continuously.

(a) If they invest now in one lump sum, how much should they invest?

(b) If they invest continuously with a fixed amount, from now on until the boy is 18, how much should they invest each month?

(c) If they invest continuously with a fixed amount, starting when the boy is 10, until he is 18, how much should they invest each month?

36. The parents of a 2-year old boy plan to invest for his college education. Their target is that when he is 18 years old, the fund should have the amount of $100,000. Assume that they invest the money in a fund paying interest at an annual rate of 5.5%, compounded continuously.

(a) If they invest now in one lump sum, how much should they invest?

(b) If they invest continuously with a fixed amount, from now on until the boy is 18, how much should they invest each month?

(c) If they invest continuously with a fixed amount, starting when the boy is 12, until he is 18, how much should they invest each month?

5.3 Improper Integrals

So far all the integrals in our applications are on a finite closed interval $[a, b]$. However, in many applications we need to deal with integrals on an unbounded interval, such as $(-\infty, b]$, $[a, \infty)$, or $(-\infty, \infty)$. Integrals of the forms $\int_a^\infty f(x)\,dx$, $\int_{-\infty}^b f(x)\,dx$, and $\int_{-\infty}^\infty f(x)\,dx$ are called **improper integrals**. In this section, we explore how to define and evaluate such integrals.

We start with an example of perpetual accumulated present value of a continuous money flow.

Suppose that Amanda inherits $1,000,000. She wants to set up a fund to help inner-city children get after-school tutoring. The money is invested in an account paying interest at an annual rate of 5%, compounded continuously. Amanda wants to withdraw a fixed amount continuously, in such a manner that the account will not run out of money. How much at most can be withdrawn from the account each year?

Suppose that D is withdrawn each year. According to the accumulated present value formula, if the money lasts for T years, then we have

$$1{,}000{,}000 = \int_0^T De^{-0.05}dt = \frac{D}{0.05}\left(1 - e^{-0.05T}\right).$$

Since Amanda does not want to exhaust the fund, we let T approach infinity. Then we have

$$\lim_{T \to \infty} \frac{D}{0.05}\left(1 - e^{-0.05T}\right) = \frac{D}{0.05} \lim_{T \to \infty} \left(1 - e^{-0.05T}\right).$$

Recall that

$$\lim_{T \to \infty} e^{-0.05T} = 0,$$

we have

$$\lim_{T \to \infty} \frac{D}{0.05}\left(1 - e^{-0.05T}\right) = \frac{D}{0.05}.$$

Therefore from the equation

$$1{,}000{,}000 = \frac{D}{0.05}$$

we solve for D to obtain

$$D = 1{,}000{,}000(0.05) = 50{,}000.$$

That is, the fund can use $50,000 each year.

This example illustrates how we can evaluate an integral on an unbounded interval: We can start by evaluating it on a finite interval, and then let the limit approach infinity. Does this process always produce a finite number? Let us explore other cases.

Suppose that we are looking for the area under each of the two curves:

$$f(x) = \frac{1}{x^{1.1}} \quad \text{and} \quad g(x) = \frac{1}{x^{0.9}},$$

both on the interval $[1, \infty)$, as shown in Figures 5.3.1 and 5.3.2. They look almost identical. We follow the process of first evaluating the two integrals

$$\int_1^b f(x)\,dx \quad \text{and} \quad \int_1^b g(x)\,dx.$$

5.3. IMPROPER INTEGRALS

We have

$$\int_1^b f(x)\,dx = \int_1^b \frac{1}{x^{1.1}}\,dx = \left[\frac{1}{-0.1}x^{-0.1}\right]_1^b = -10\left[b^{-0.1} - 1\right] = 10 - \frac{10}{b^{0.1}}, \text{ and}$$

$$\int_1^b g(x)\,dx = \int_1^b \frac{1}{x^{0.9}}\,dx = \left[\frac{1}{0.1}x^{0.1}\right]_1^b = 10\left[b^{0.1} - 1\right].$$

Next, we let $b \to \infty$:

$$\lim_{b \to \infty}\left[10 - \frac{10}{b^{0.1}}\right] = 10 \quad \text{and} \quad \lim_{b \to \infty} 10\left[b^{0.1} - 1\right] = \infty.$$

That means that the area under the graph of $f(x)$ on the interval $[1, \infty)$ can be defined to be 10, but the region under the graph of $g(x)$ on the interval $[1, \infty)$ does not have a finite area.

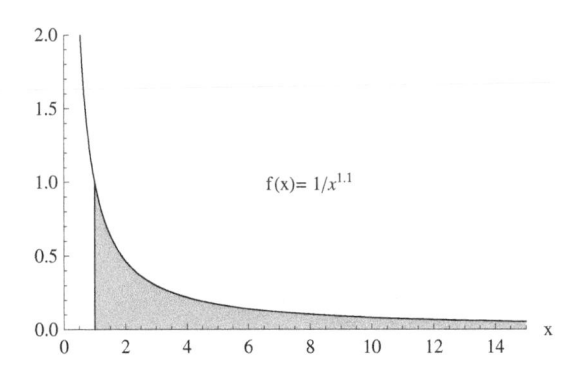

Figure 5.3.1: Finite area of shaded region

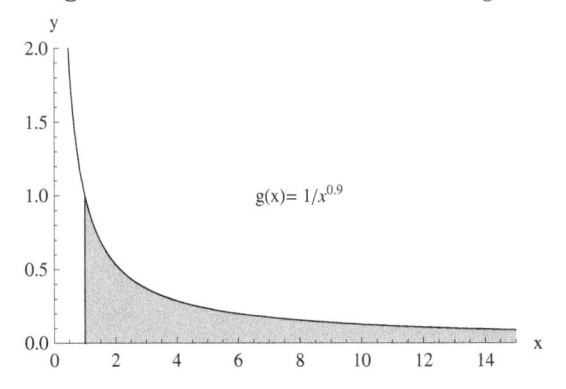

Figure 5.3.2: Infinite area of shaded region

We define the improper integrals according to the limit process.

> **Definition** *improper integrals*
>
> 1. $\displaystyle\int_a^\infty f(x)\,dx = \lim_{b\to\infty} \int_a^b f(x)\,dx$
>
> If the limit (a finite number) exists, we say that the integral is **convergent**, or the integral **converges**. If the limit does not exist, we say that the integral is **divergent**, or the integral **diverges**.
>
> 2. Similarly, $\displaystyle\int_{-\infty}^b f(x)\,dx = \lim_{a\to -\infty} \int_a^b f(x)\,dx$
>
> 3. $\displaystyle\int_{-\infty}^\infty f(x)\,dx = \int_{-\infty}^c f(x)\,dx + \int_c^\infty f(x)\,dx$
>
> where c can be any real number. For this integral to be convergent, both $\displaystyle\int_{-\infty}^c f(x)\,dx$ and $\displaystyle\int_c^\infty f(x)\,dx$ have to be convergent.

According to the definition,

$$\int_1^\infty \frac{1}{x^{1.1}}\,dx = 10 \text{ is convergent and } \int_1^\infty \frac{1}{x^{0.9}}\,dx \text{ is divergent}.$$

> **Reality Check** *Convergent or divergent?*
>
> For a nonnegative function $f(x)$, for the limit $\displaystyle\lim_{b\to\infty}\int_a^b f(x)\,dx$ to exist, the region under the graph of $f(x)$ must have a "thin tail." Even in such a case, it is not necessary for the region to have a finite area, as in the case of $\displaystyle\int_1^\infty \frac{1}{x^{0.9}}\,dx$. Hence, if the integrand $f(x)$ does not approach zero as x approaches infinity, then the improper integral $\displaystyle\int_a^\infty f(x)\,dx$ must be divergent.

Example 1: Determine whether the improper integral $\displaystyle\int_5^\infty x^{1.2}\,dx$ is convergent or divergent. If it is convergent, determine its value.

Solution Notice that when $x \to \infty$, we have $x^{1.2} \to \infty$. Therefore $f(x) = x^{1.2} \not\to 0$ as $x \to \infty$. Therefore, the improper integral $\displaystyle\int_5^\infty x^{1.2}\,dx$ is divergent.

Example 2: For each of the following improper integrals, determine whether it is convergent or divergent. If it is convergent, determine its value.

(a) $\displaystyle\int_0^\infty e^{-0.2x}\,dx$ (b) $\displaystyle\int_1^\infty \ln x\,dx$ (c) $\displaystyle\int_1^\infty \frac{4}{x^3}\,dx$ (d) $\displaystyle\int_2^\infty \frac{1}{x}\,dx$

Solution (a) We first evaluate the integral $\displaystyle\int_0^b e^{-0.2x}\,dx$:

$$\int_0^b e^{-0.2x}\,dx = \left[\frac{1}{-0.2}e^{-0.2x}\right]_0^b = -5\left[e^{-0.2b} - e^0\right] = 5 - 5e^{-0.2b}.$$

We then take the limit. Notice that when b becomes larger and larger, so does $e^{0.2b}$, hence $e^{-0.2b} = 1/e^{0.2b}$

approaches 0:

$$\lim_{b \to \infty} \int_0^b e^{-0.2x}\, dx = \lim_{b \to \infty} \left(5 - 5e^{-0.2b}\right) = 5.$$

Therefore, the improper integral $\int_0^\infty e^{-0.2x}\, dx$ is convergent. Its value is 5.

(b) Notice that when $x \to \infty$, we have $\ln x \to \infty$. Therefore $f(x) = \ln x \not\to 0$ as $x \to \infty$. Therefore, the improper integral $\int_1^\infty \ln x\, dx$ is divergent.

(c) We first evaluate the integral $\int_1^b \dfrac{4}{x^3}\, dx$:

$$\int_1^b \frac{4}{x^3}\, dx = \left[\frac{4}{-2}\frac{1}{x^2}\right]_1^b = -2\left[\frac{1}{b^2} - \frac{1}{1^2}\right] = 2 - \frac{2}{b^2}.$$

We then take the limit. Notice that as $b \to \infty$, we have $2/b^2 \to 0$:

$$\lim_{b \to \infty} \int_1^b \frac{4}{x^3}\, dx = \lim_{b \to \infty}\left(2 - \frac{2}{b^2}\right) = 2.$$

Therefore, the improper integral $\int_1^\infty \dfrac{4}{x^3}\, dx$ is convergent. Its value is 2.

(d) We first evaluate the integral $\int_2^b \dfrac{1}{x}\, dx$:

$$\int_2^b \frac{1}{x}\, dx = [\ln x]_2^b = \ln b - \ln 2 = \ln b - \ln 2.$$

We then take the limit. Notice that when b becomes larger and larger, so does $\ln b$. Hence,

$$\lim_{b \to \infty} \int_2^b \frac{1}{x}\, dx = \lim_{b \to \infty}(\ln b - \ln 2) = \infty.$$

Therefore, the improper integral $\int_2^\infty \dfrac{1}{x}\, dx$ is divergent.

An Application

> **Perpetual Present Value of a Continuous Money Flow**
>
> Suppose that a yearly amount of $R(t)$, where t is measured in years, is **perpetually** deposited continuously into an interest-bearing account at the annual interest rate $100r\%$ compounded continuously. Then the perpetual accumulated present value of the account is
>
> $$\int_0^\infty R(t)e^{-rt}\,dt.$$
>
> If $R(t)$ is a constant $R(t) = D$, then
>
> $$\int_0^\infty De^{-rt}\,dt = \frac{D}{r}.$$

We can now use the definition of improper integrals to describe the perpetual accumulated present value. Suppose that a yearly amount of D is **perpetually** deposited continuously into an interest-bearing account at the annual interest rate $100r\%$, compounded continuously. Then the perpetual accumulated present value of the account is

$$\int_0^\infty De^{-rt}\,dt = \lim_{T\to\infty} \int_0^T De^{-rt}\,dt = \lim_{T\to\infty} \frac{D}{r}\left[1 - e^{-rT}\right] = \frac{D}{r}.$$

Example 3: Paul is 25 years old and plans to retire at 65. He wants to have a fund at 65 that will let him spend \$5,000 a month perpetually after retirement. Assume a continuous money flow:

(a) Suppose that after his retirement Paul puts the money in a fund paying interest at an annual rate of 4%, compounded continuously. How much will Paul need for his retirement?

(b) Suppose that Paul starts to invest a fixed amount each month from now until he retires, to a fund that pays interest at an annual rate of 6%, compounded continuously. How much should he invest each month?

Solution (a) We have $D = 5{,}000(12) = 60{,}000$, $r = 0.04$. Substituting into the perpetual accumulated present value formula we obtain

$$\frac{60{,}000}{0.04} = 1{,}500{,}000.$$

Therefore, Paul will need \$1,500,000 when he is 65.

(b) Assume that Paul invests S dollars per year continuously, from 25 to 65. We have the future value $= 1{,}500{,}000$, $r = 0.06$, and $T = 40$. Substituting the known numbers into the future value formula we have

$$1{,}500{,}000 = \frac{S}{0.06}\left(e^{0.06(40)} - 1\right) = S(167.053).$$

Solving for S we obtain

$$S = 1{,}500{,}000/167.053 \approx 8{,}979.19.$$

Hence, Paul needs to save \$8,979 per year, or $8{,}979/12 = \$740.27$ per month.

5.3. IMPROPER INTEGRALS

Exercise Set 5.3

For Exercises #1 to #28: Determine whether the improper integral is convergent or divergent. If it is convergent, determine its value.

1. $\int_1^\infty \frac{2}{x^2}\, dx$
2. $\int_2^\infty \frac{5}{x^2}\, dx$
3. $\int_1^\infty \frac{2}{x}\, dx$
4. $\int_2^\infty \frac{5}{x}\, dx$
5. $\int_1^\infty \frac{1}{x^{1.4}}\, dx$
6. $\int_1^\infty \frac{1}{x^{0.8}}\, dx$
7. $\int_4^\infty \frac{1}{\sqrt{x}}\, dx$
8. $\int_9^\infty \frac{3}{\sqrt{x^3}}\, dx$
9. $\int_0^\infty 2e^{-2x}\, dx$
10. $\int_0^\infty 3e^{-0.5x}\, dx$
11. $\int_0^\infty \frac{2x}{x^2+1}\, dx$
12. $\int_0^\infty \frac{3x^2}{x^3+2}\, dx$
13. $\int_0^\infty \frac{1}{x+1}\, dx$
14. $\int_0^\infty \frac{5}{x+2}\, dx$
15. $\int_1^\infty x^{1.4}\, dx$
16. $\int_1^\infty x^{0.8}\, dx$
17. $\int_0^\infty xe^x\, dx$
18. $\int_0^\infty x^2 e^x\, dx$
19. $\int_0^\infty 2e^{2x}\, dx$
20. $\int_0^\infty 3e^{0.05x}\, dx$
21. $\int_{-\infty}^{-1} \frac{4}{x^2}\, dx$
22. $\int_{-\infty}^{-2} \frac{1}{x^4}\, dx$
23. $\int_0^\infty \frac{2x}{(x^2+1)^2}\, dx$
24. $\int_0^\infty \frac{4x}{(x^2+4)^3}\, dx$
25. $\int_{-\infty}^3 x^5\, dx$
26. $\int_{-\infty}^1 x^4\, dx$
27. $\int_{-\infty}^\infty x\, dx$
28. $\int_{-\infty}^\infty x^3\, dx$

29. Find the area, if it is finite, of the region under the graph of $y = 3/x^4$ on the interval $[2, \infty)$.

30. Find the area, if it is finite, of the region under the graph of $y = 1/x^{1.5}$ on the interval $[1, \infty)$.

31. Find the area, if it is finite, of the region bounded by $y = 2xe^{-0.2x^2}$, $x = 0$, and $y = 0$.

32. Find the area, if it is finite, of the region bounded by $y = 2x/\sqrt{(x^2+16)^3}$, $x = 3$, and $y = 0$.

APPLICATIONS

33. A mutual fund has a growth rate recorded each year, expressed by the function
$$V'(t) = r(t) = 200e^{-0.04t}$$
(in thousands of dollars per year), where t is the number of years passed from year 1980. The value of the fund in year 1980 is $1,000,000.

   (a) Determine the function $V(t)$ which represents the value of the fund t years after year 1980.

   (b) Determine the highest possible value of the fund in the future.

34. A mutual fund has a growth rate recorded each year, expressed by the function
$$V'(t) = r(t) = 100e^{-0.01t}$$
(in thousands of dollars per year), where t is the number of years passed from year 1990. The value of the fund in year 1990 is $1,500,000.

   (a) Determine the function $V(t)$ which represents the value of the fund t years after year 1990.

   (b) Determine the highest possible value of the fund in the future.

35. Lisa has inherited $800,000. She wants to set up a fund to help needy families. The money is invested in an account paying interest at a 6% annual rate, compounded continuously. Lisa wants to be able to perpetually withdraw a fixed amount continuously. How much at most can be withdrawn from the account each year?

36. Jeff won a lottery of $500,000. The money is invested in an account paying interest at a 5.5% annual rate, compounded continuously. Jeff wants to be able to perpetually withdraw a fixed amount continuously. How much at most can be withdrawn from the account each month?

37. Rick is 26 years old and plans to retire at 64. He wants to have a fund at 64 that will let him perpetually spend $3,000 a month after retirement. Assume a continuous money flow.

    (a) Suppose that after his retirement Rick puts the money in a fund paying interest at an annual rate of 4.5%, compounded continuously. How much will Rick need for his retirement?

    (b) Suppose that Rick starts to invest a fixed amount each month from now until he retires, in a fund that pays interest at an annual rate of 6.8%, compounded continuously. How much should he invest each month?

38. John is 28 years old and plans to retire at 67. He wants to have a fund at 67 that will let him perpetually spend $4,500 a month after retirement. Assume a continuous money flow.

    (a) Suppose that after his retirement John puts the money in a fund paying interest at an annual rate of 4.2%, compounded continuously. How much will John need for his retirement?

    (b) Suppose that John starts to invest a fixed amount each month from now until he retires, in a fund that pays interest at an annual rate of 6.2%, compounded continuously. How much should he invest each month?

5.4 Probability Distributions and Density Functions (I)

In this section, we explore another important field involving applications of definite integrals: theoretical probability. We will use the area under a curve to measure the likelihood that a certain event will occur, in situations that involve some degree of uncertainty. We will need to use terminology from the field of probability.

Probability and Probability Density Functions

A **probability model** deals with situations that are random in character and attempts to predict the outcomes of events with a certain degree of accuracy. For example, if we toss a coin, it is impossible to predict in advance whether the outcome will be heads or tails. To create a probability model for this situation is to assign a number p, $0 \leq p \leq 1$, called the "probability of an event" to each of the possible outcomes, that is, heads or tails, which will predict the relative frequency in a *sufficiently large* number of repetitions. For example, suppose that we assign $P(\text{heads}) = 0.5$, and $P(\text{tails}) = 0.5$. If we only toss the coin 10 times, we may have three heads and seven tails, or even all heads. However, when the number of repetitions gets larger and larger, then the numbers of heads and tails will be closer and closer to being equal. That means, if we toss a coin 10,000 times, then *roughly* 5,000 heads and 5,000 tails may occur.

Probabilities can be obtained in two ways. The first way is through experiment. For example, if a coin is tossed 1,000 times and 487 are heads, then we can assign the probability of heads to be $P(\text{heads}) = 487/1{,}000 = 0.487$. Such a probability is called an **empirical probability**. The second way is through logical reasoning according to stated conditions. For example, $P(\text{heads}) = P(\text{tails}) = 0.5$, because there are two possible outcomes, each with an equal chance of occurring, so heads should appear 50% of the time. Such a probability is called a **theoretical probability**.

5.4. PROBABILITY DISTRIBUTIONS AND DENSITY FUNCTIONS (I)

Our focus in this section is on theoretical probabilities. In addition, we focus on the outcomes that can be described by numbers, and the numbers comprise an interval. Such a quantity is called a **continuous random variable**. We illustrate such a variable in the following example.

Suppose that Emily rides the school bus every weekday morning. She always arrives at 8:00 at the conner of her street to wait for the bus to arrive. The bus has an equal chance of arriving at the corner of her street between 8:00 and 8:12. So Emily's waiting time, t, is a continuous random variable, since it can randomly take any value between 0 and 12. We know that $P(0 \leq t \leq 12) = 1$, since the bus is certain to arrive no later than 8:12. Suppose that one rainy day Emily wishes that her waiting time is no more than 4 minutes. What is the probability of the bus arriving between 8:00 and 8:04? Since the bus has an equal chance of arriving between 8:00 and 8:12, it must have 1/3 of the chance of arriving between 8:00 and 8:04. Hence $P(0 \leq t \leq 4) = \dfrac{1}{3}$. We can describe the probabilities by the following graph (Figure 5.4.1):

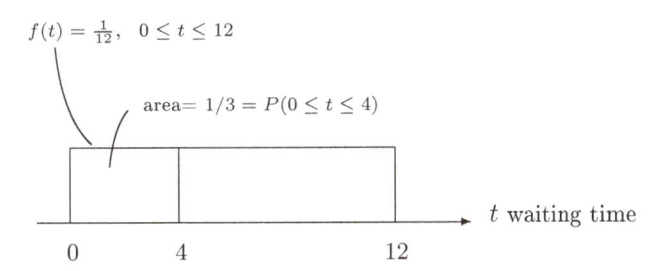

Figure 5.4.1: Probability vs. area

Notice that the function $f(t) = 1/12$, $0 \leq t \leq 12$ can be used to describe the probability of Emily's waiting time between any two values. Such a function is called a **probability density function**. It is also called a **probability distribution**, since it describes how probabilities are distributed over an interval associated with the random variable t. We formally define probability density functions of a random variable x:

Definition   *probability density function*

Suppose that x is a continuous random variable. A function $f(x)$ is called a **probability density function** if it satisfies the following:
1. $f(x) \geq 0$ for all x in the domain.

2. The area under the graph on the domain of f is 1.

The probability that x will be in an interval $[c, d]$ in the domain is then obtained by

$$P(c \leq x \leq d) = \int_c^d f(x)\, dx.$$

Example 1: Suppose that Emily rides the school bus every weekday morning. The bus has an equal chance of arriving at the corner of her street between 8:00 and 8:12. Suppose that Emily always arrives at the corner of her street at 8:00. Let t be her waiting time in minutes. Then, the probability density function of t is given by $f(t) = \dfrac{1}{12}$, $0 \leq t \leq 12$.

(a) Verify that f is a probability density function.

(b) Find the probability that Emily's waiting time is longer than 9 minutes.

Solution (a) It is obvious that $f(t) \geq 0$ for all t on the domain $[0, 12]$. We check second condition:

$$\int_0^{12} f(t)\, dt = \int_0^{12} \frac{1}{12}\, dt = \left[\frac{1}{12} t\right]_0^{12} = \frac{1}{12}(12) - \frac{1}{12}(0) = 1.$$

Hence, f is a probability density function.

(b) We have

$$P(9 \leq t \leq 12) = \int_9^{12} f(t)\, dt = \int_9^{12} \frac{1}{12}\, dt = \left[\frac{1}{12} t\right]_9^{12} = \frac{1}{12}(12) - \frac{1}{12}(9) = \frac{3}{12} = 0.25.$$

Hence, the probability that Emily's waiting time is longer than 9 minutes is 25%.

Reality Check   *Constructing a probability density function*

Given a nonnegative function $g(x)$ such that the area under its graph, A, is finite, and the shape of the graph describes the probability distribution of a continuous random variable, we can construct a probability density function $f(x)$ from $g(x)$ by setting $f(x) = kg(x)$, where $k = 1/A$.

Example 2: Refer to Example 1. Suppose that the the probability density function of t is given by $f(t) = kt$, $0 \leq t \leq 12$.

(a) Determine the value of k.

(b) Find the probability that Emily's waiting time is longer than 9 minutes.

Solution (a) Since $f(t)$ is a probability density function on $[0, 12]$, it must satisfy the condition $\int_0^{12} f(t)\, dt = 1$. We have

$$1 = \int_0^{12} f(t)\, dt = \int_0^{12} kt\, dt = \left[\frac{k}{2} t^2\right]_0^{12} = \frac{k}{2}[144 - 0] = k \cdot 72.$$

Hence, $k = 1/72$, and

$$f(t) = \frac{1}{72} t, \quad 0 \leq t \leq 12 .$$

(b) We have

$$P(9 \leq t \leq 12) = \int_9^{12} f(t)\, dt = \int_9^{12} \frac{1}{72} t\, dt = \left[\frac{1}{144} t^2\right]_9^{12}$$

$$= \frac{1}{144}(144) - \frac{1}{144}(81) = \frac{63}{144} \approx 0.44.$$

Hence, the probability that Emily's waiting time is longer than 9 minutes is 44%.

Example 3: A large school district purchased 1,000 projectors for classroom use. The manufacturer told the district manager that the life span of the projector light bulbs, t, can be described by the following probability density function: $f(t) = k\sqrt{t}$, $1 \leq t \leq 4$, where t is measured in thousands of hours.

(a) Determine the value of k.

(b) Determine the probability that a projector light bulb lasts at least 3,000 hours.

5.4. PROBABILITY DISTRIBUTIONS AND DENSITY FUNCTIONS (I)

(c) Determine the probability that a projector light bulb lasts at most 2,000 hours.

(d) The projectors are on for about 800 hours per year. In the third year after the purchase, how many projector light bulbs should the manager plan on buying to replace the burned-out ones?

Solution (a) Since $f(t)$ is a probability density function on $[1,4]$, it must satisfy the condition $\int_1^4 f(t)\,dt = 1$. We have

$$1 = \int_1^4 f(t)\,dt = \int_1^4 k\sqrt{t}\,dt = k\left[\frac{2}{3}\sqrt{t^3}\right]_1^4 = k\frac{2}{3}\left[\sqrt{4^3} - \sqrt{1^3}\right] = k \cdot \frac{14}{3}.$$

Hence, $k = 3/14$ and

$$f(t) = \frac{3}{14}\sqrt{t}, \quad 1 \le t \le 4.$$

(b) A projector light bulb lasts at least 3,000 hours means that $3 \le t \le 4$. Hence, the probability is

$$P(3 \le t \le 4) = \int_3^4 \frac{3}{14}\sqrt{t}\,dt = \frac{3}{14}\left[\frac{2}{3}\sqrt{t^3}\right]_3^4$$

$$= \frac{1}{7}\left[\sqrt{4^3} - \sqrt{3^3}\right] \approx 0.40.$$

Hence, the probability that a projector light bulb lasts at least 3,000 hours is about 40%.

(c) A projector light bulb lasts at most 2,000 hours means that $1 \le t \le 2$. Hence, the probability is

$$P(1 \le t \le 2) = \int_1^2 \frac{3}{14}\sqrt{t}\,dt = \frac{3}{14}\left[\frac{2}{3}\sqrt{t^3}\right]_1^2$$

$$= \frac{1}{7}\left[\sqrt{2^3} - \sqrt{1^3}\right] \approx 0.2612.$$

Hence, the probability that a projector light bulb lasts at most 2,000 hours is about 26.12%.

(d) From the beginning to the end of the third year after the purchase, the projector light bulbs have been used for 1,600 to 2,400 hours. We have

$$P(1.6 \le t \le 2.4) = \int_{1.6}^{2.4} \frac{3}{14}\sqrt{t}\,dt = \frac{3}{14}\left[\frac{2}{3}\sqrt{t^3}\right]_{1.6}^{2.4}$$

$$= \frac{1}{7}\left[\sqrt{(2.4)^3} - \sqrt{(1.6)^3}\right] \approx 0.242.$$

Hence, about 24.2% of the 1,000 projector light bulbs are likely to be burned out. Therefore, the manager needs to purchase about 242 replacement projector light bulbs.

Even though probability density functions can take a variety of shapes, there are a few probability distributions that are used most often. We explore some of those probability distributions.

The Uniform Probability Distribution

The simplest density function is one that assumes a constant value on a finite interval $[a,b]$ and zero elsewhere, such as the density functions in Example 1. This probability distribution is called the **uniform density function** and has the equation $f(x) = \dfrac{1}{b-a}$, $a \le x \le b$ (Figure 5.4.2). In this case, we call the probability distribution a **uniform distribution**.

$$f(x) = \frac{1}{b-a}, \quad a \le x \le b.$$

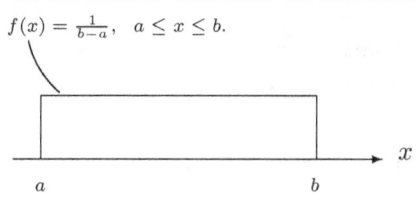

Figure 5.4.2: Uniform distribution

The uniform density function provides a good model for random variables that are evenly distributed over an interval. For example, consider the experiment of selecting a number at random from among the numbers between 0 and 6. If x denotes the number being selected, then x has a uniform distribution on $[0, 6]$.

> **Reality Check** *Mean equals median in the uniform model*
>
> Recall that the mean of set of numbers is the average, while the median is the number in the middle if the numbers are placed in increasing order (or the average of the two middle numbers if there are an even number of numbers). For a uniform distribution, the mean and median are the same. Thus, if the uniform density function is given by $f(x) = \dfrac{1}{b-a}$, $a \le x \le b$, then the median, $\dfrac{b-a}{2}$, is also the mean (or average).

Example 4: The average amount of time Star-Lord waits for Rocket Raccoon to formulate an escape plan is four minutes.

(a) Assume that the waiting time has a uniform distribution. Determine the uniform density function.

(b) Determine the probability that Star-Lord waits no more than two minutes for the escape plan assuming a uniform distribution.

(c) Determine the probability that Star-Lord waits between two minutes and three minutes for the escape plan assuming a uniform distribution.

Solution

(a) In principle, Rocket could formulate an escape plan any time between now and forever. However, the knowledge that the average wait time is four minutes and that the waiting time has a uniform distribution implies that four minutes must be the median (or middle) of the interval on which the waiting time uniform density function is nonzero. The interval starts at 0, so for 4 to be in the middle, the interval must end at 8. That is, if $f(t)$ is the uniform density function, then we must have $f(t) = \dfrac{1}{8-0} = \dfrac{1}{8}, 0 \le t \le 8$.

(b) As $0 \le t \le 2$ makes up one-fourth of the entire interval, the probability that Star-Lord waits no more than two minutes for the escape plan is 0.25. One can rephrase this solution with an integral by noting that

$$P(0 \le t \le 2) = \int_0^2 \frac{1}{8}\, dt = \left[\frac{1}{8}t\right]_0^2 = 0.25.$$

(c) As $2 \le t \le 3$ make us one-eighth of the entire interval, the probability that Star-Lord waits between two and three minutes for the escape plan is 0.125. Again, we could instead use an integral:

$$P(2 \le t \le 3) = \int_2^3 \frac{1}{8}\, dt = \left[\frac{1}{8}t\right]_2^3 = 0.125.$$

The Exponential Probability Distribution

The exponential density functions are used to describe random variables that are more likely to take small values than large, like the waiting time in line at the post office, the duration of a phone call, the distance between successive cars on a freeway, etc. An **exponential density function** has the equation $f(x) = ke^{-kx}$, $x \geq 0$ (Figure 5.4.3). We can check that f is a probability density function:

$$\int_0^\infty ke^{-kx}\, dx = \lim_{b \to \infty} \int_0^b ke^{-kx}\, dx$$

$$= \lim_{b \to \infty} \left[\frac{k}{-k} e^{-kx} \right]_0^b$$

$$= \lim_{b \to \infty} \left[(-e^{-kb}) - (-1) \right] = 1.$$

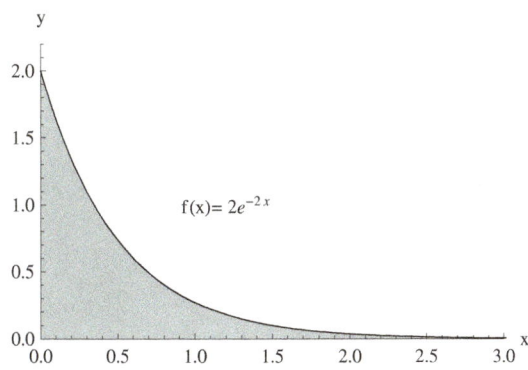

Figure 5.4.3: An exponential distribution

> **Reality Check**   *The value of k in an exponential model*
>
> The value of k represents the average occurrence frequency of occurrence of an event per unit. It is also the reciprocal of the average value of the random variable.

Example 5: Determine the exponential density functions for the following distributions:

(a) x is the random variable measuring the distance between two concessive cars on a certain section of a freeway and the average distance between the cars is 150 feet.

(b) t is the random variable measuring the time between successive arrivals at the drive-through window of a fast-food restaurant during the lunch hours, and on average five cars arrive at the drive-through window every 2 minutes.

Solution (a) The value of k is $k = 1/150$. Hence the exponential density function is

$$f(x) = \frac{1}{150} e^{-x/150}, \quad x \geq 0.$$

(b) The value of k is $k = 5/2 = 2.5$. Hence the exponential density function is

$$f(t) = 2.5 e^{-2.5t}, \quad t \geq 0.$$

Example 6: The distribution of the time between successive arrivals at a post office in a large city around noon on a weekday can be approximated by an exponential density function. Suppose that three customers arrive at the post office every 2 minutes.

(a) Determine the exponential density function.

(b) Determine the probability that the time between successive arrivals will be no more than 1 minute.

(c) Determine the probability that there is no arrival for at least 2 minutes.

Solution (a) We have on average three customers arrive at the post office every 2 minutes. Therefore, $k = 3/2 = 1.5$. Hence, the exponential density function is

$$f(t) = 1.5e^{-1.5t}, \quad t \geq 0.$$

(b) The probability that the time between successive arrivals will be no more than 1 minute can be calculated by

$$P(0 \leq t \leq 1) = \int_0^1 f(t)\, dt = \int_0^1 1.5e^{-1.5t}\, dt$$

$$= \left[\frac{1.5}{-1.5} e^{-1.5t}\right]_0^1 = -\left[e^{-1.5(1)} - e^{1.5(0)}\right] = 1 - e^{-1.5} \approx 0.777.$$

Hence, the probability that the time between successive arrivals will be no more than 1 minute is about 77.7%.

(c) We are looking for the probability that the time between successive arrivals will be more than 2 minutes. We can first find the probability that the time between successive arrivals will be no more than 2 minutes by

$$P(0 \leq t \leq 2) = \int_0^2 f(t)\, dt = \int_0^2 1.5e^{-1.5t}\, dt$$

$$= \left[\frac{1.5}{-1.5} e^{-1.5t}\right]_0^2 = -\left[e^{-1.5(2)} - e^{1.5(0)}\right] = 1 - e^{-3} \approx 0.950.$$

Hence, the probability that the time between successive arrivals will be no more than 2 minutes is about 95%. Therefore, the probability that the time between successive arrivals will be more than 2 minutes is $1 - 0.95 = 0.05 = 5\%$.

Exercise Set 5.4

For Exercises #1 to #16, for each function on the given interval $[a, b]$, verify that the integral $\int_a^b f(x)\, dx$ is 1 (so the function is a probability density function).

1. $f(x) = \frac{1}{3}$, $[1, 4]$

2. $f(x) = \frac{1}{5}$, $[0, 5]$

3. $f(x) = \frac{1}{2}x$, $[0, 2]$

4. $f(x) = \frac{1}{4}x$, $[1, 3]$

5. $f(x) = \frac{1}{9}x^2$, $[0, 3]$

6. $f(x) = \frac{1}{21}x^2$, $[1, 4]$

7. $f(x) = \dfrac{3}{2x^2}$, $[1, 3]$

8. $f(x) = \dfrac{4}{3x^2}$, $[1, 4]$

9. $f(x) = \dfrac{1}{x}$, $[1, e]$

10. $f(x) = \dfrac{1}{x}$, $[e, e^2]$

11. $f(x) = \dfrac{3}{16}x^2$, $[-2, 2]$

12. $f(x) = \dfrac{3}{2}x^2$, $[-1, 1]$

13. $f(x) = 0.2e^{-0.2x}$, $[0, \infty)$

14. $f(x) = 3e^{-3x}$, $[0, \infty)$

15. $f(x) = \dfrac{3}{x^4}$, $[1, \infty)$

16. $f(x) = \dfrac{24}{x^4}$, $[2, \infty)$

For Exercises #17 to #24, find k such that the function is a probability density function on the given interval.

17. $f(x) = \dfrac{k}{x^2}$, $[1, 2]$

18. $f(x) = \dfrac{k}{x^2}$, $[2, 4]$

19. $f(x) = kx$, $[1, 5]$

20. $f(x) = kx$, $[0, 3]$

21. $f(x) = kx^2$, $[-1, 2]$

22. $f(x) = kx^2$, $[-2, 1]$

23. $f(x) = ke^x$, $[0, 2]$

24. $f(x) = ke^{2x}$, $[0, 1]$

APPLICATIONS

25. Suppose that Kathy rides the school bus every weekday morning. The bus has an equal chance of arriving at the corner of her street at any time between 7:20 and 7:30. Suppose that Kathy always arrives at the corner of her street at 7:20. Let t be her waiting time in minutes. Then the probability density function of t is given by
$$f(t) = \frac{1}{10}, \quad 0 \le t \le 10.$$

(a) Determine the probability that Kathy needs to wait for at least 6 minutes.

(b) Determine the probability that Kathy needs to wait for at most 8 minutes.

26. Suppose that the traffic light at a certain location remains red for 30 seconds at a time. Suppose that Sam arrives at that spot and finds the light is red. Let t be Sam's waiting time in seconds before the light turns green.

(a) Determine the probability density function of t.

(b) Determine the probability that Sam needs to wait for at least 20 seconds.

(c) Determine the probability that Sam needs to wait for at most 12 seconds.

27. Refer to Exercise #25. Suppose that Kathy's waiting time t has the following probability density function:
$$f(t) = \frac{1}{50}t, \quad 0 \le t \le 10.$$

Repeat parts (a) and (b).

28. Refer to Exercise #26. Suppose that Sam's waiting time t has the following probability density function:
$$f(t) = k(30 - t), \quad 0 \le t \le 30.$$

(a) Determine the value of k.

Repeat parts (b) and (c).

29. A large school district purchased 800 projectors for classroom use. The manufacturer told the district manager that the life span of the projector light bulbs, t, can be described by the following probability density function:
$$f(t) = kt^2, \quad 0.5 \le t \le 2,$$
where t is measured in thousands of hours.

(a) Determine the value of k.

(b) Determine the probability that a bulb lasts at least 1,200 hours.

(c) Determine the probability that a bulb lasts at most 1,000 hours.

(d) The projectors are on for about 700 hours per year. In the second year after the purchase, how many projector light bulbs should the manager budget to purchase for replacing the burned-out ones?

30. The distribution of the time between successive arrivals at the drive-through window of a fast food restaurant around noon on a weekday can be approximated by an exponential density function. Suppose that seven vehicles arrive at the drive-through window every 2 minutes.

(a) Determine the exponential density function.

(b) Determine the probability that the time between successive arrivals will be no more than half a minute.

(c) Determine the probability that there is no arrival for at least 1 minute.

31. The distance x, in feet, between successive vehicles on a certain stretch of freeway has an exponential probability distribution. The average distance between successive vehicles is 120 feet.

(a) Determine the exponential density function.

(b) Determine the probability that the distance between successive vehicles is 50 feet or less.

32. The average waiting time at a checkout line in a supermarket is 2 minutes.

(a) Assume that the waiting time has a uniform distribution. Determine the uniform density function.

(b) Assume that the waiting time has an exponential distribution. Determine the exponential density function.

(c) Determine the probability that a customer waits no more than 1 minute to check out assuming a uniform distribution.

(d) Determine the probability that a customer waits no more than 1 minute to check out assuming an exponential distribution.

(e) Determine the probability that a customer waits between 2 minutes and 3 minutes to check out assuming a uniform distribution.

(f) Determine the probability that a customer waits between 2 minutes and 3 minutes to check out assuming an exponential distribution.

(g) Which model does a customer prefer?

5.5 Probability Distributions and Density Functions (II)

In this section, we continue to explore the field of theoretical probability. We will introduce two important measures of a continuous random variable, and study the most important distribution in statistics – the normal distribution.

Measures of Center and Variability

Two important characteristics are often described with a continuous random variable: the probability distribution's "central" value and how widely the variable values are scattered about the central value. Take the example of Emily's school bus introduced in the previous section . Suppose that Emily always arrives at the corner of her street at 8:00. Let t be her waiting time in minutes. Then she has an equal chance of waiting between 0 minutes and 12 minutes. What is her average waiting time? Common sense tells that it is 6 minutes. In theoretical probability, the "central" value is called the **mean**, or the **expected value**, and is denoted by the Greek letter μ. When the graph of a probability distribution is symmetric, the mean μ is always the center. However, if a random variable x has a higher chance of taking values in one end of the distribution, the mean will be located towards that end (Figure 5.5.1):

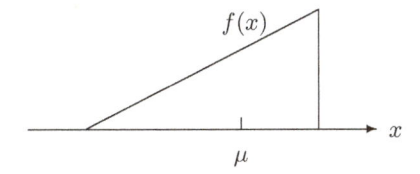

Figure 5.5.1: Location of the mean

Probability distributions can have the same mean but completely different spreads, or variabilities, about the center, as illustrated in the graphs shown in Figures 5.5.2 and 5.5.3. One measurement of how closely the values of the distribution cluster about its mean is called the **standard deviation**, and is denoted by the Greek letter σ. If a random variable x has most of its values close to the mean, as in Figure 5.5.3, then the standard deviation is small. On the other hand, if it is likely that the values of a random variable x are widely scattered about the mean, then the standard deviation is large. The mean and standard deviation of a known density function are determined by the following definitions:

Definition *mean (expected value) and standard deviation*

Let $f(x)$ be the probability density function of a continuous random variable x defined on an interval $[a, b]$. Then:

- The **mean**, μ, of x is $\mu = \int_a^b x f(x)\, dx$.

- The **standard deviation**, σ, of x is $\sigma = \sqrt{\int_a^b (x-\mu)^2 f(x)\, dx} = \sqrt{\int_a^b x^2 f(x)\, dx - \mu^2}$.

If the domain of f is an unbounded interval, then one or both limits of the integral can be infinity.

Reality Check *Formulas for computing σ*

The first formula in the definition of the standard deviation σ reveals the property of this quantity, that is, it measures how far away from the mean μ the random variable x takes its value. However, to compute the standard deviation σ, it is more convenient to apply the second formula.

Figure 5.5.2: Large variability

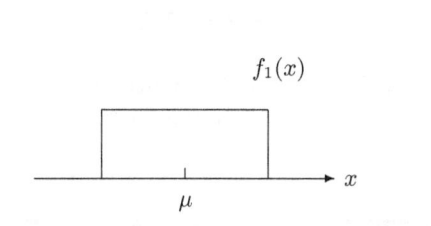

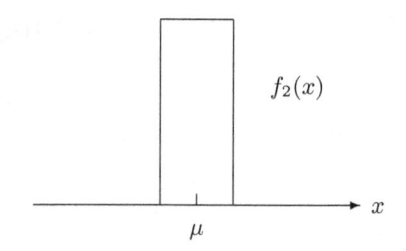

Figure 5.5.3: Small variability

The following is a brief proof that the two formulas are equal, using the definition of μ and the property that $f(x)$ is a probability density function:

$$\sigma^2 = \int_a^b (x-\mu)^2 f(x)\,dx = \int_a^b \left(x^2 - 2\mu x + \mu^2\right) f(x)\,dx$$

$$= \int_a^b x^2 f(x)\,dx - 2\mu \int_a^b x f(x)\,dx + \mu^2 \int_a^b f(x)\,dx$$

$$= \int_a^b x^2 f(x)\,dx - 2\mu \cdot \mu + \mu^2 \cdot 1 = \int_a^b x^2 f(x)\,dx - \mu^2.$$

Example 1: Suppose that Emily rides the school bus every weekday morning. The bus has an equal chance of arriving at the corner of her street between 8:00 and 8:12. Suppose that Emily always arrives at the corner of her street at 8:00. Let t be her waiting time in minutes. Then the probability density function of t is given by $f(t) = \dfrac{1}{12}$, $0 \leq t \leq 12$.

(a) Determine the mean waiting time.

(b) Determine the standard deviation of the waiting times.

Solution (a) Since $f(t) > 0$ only on $[0, 12]$, we can treat the domain of f as $[0, 12]$. By the definition of the mean, we have

$$\mu = \int_0^{12} t f(t)\,dt = \int_0^{12} t \frac{1}{12}\,dx$$

$$= \left[\frac{1}{12} \cdot \frac{t^2}{2}\right]_0^{12} = \frac{1}{24}\left[12^2 - 0\right] = 6.$$

Hence, the mean waiting time is 6 minutes.

(b) To find the standard deviation of the waiting times, we first evaluate

$$\int_0^{12} t^2 f(t)\,dt = \int_0^{12} t^2 \frac{1}{12}\,dx$$

$$= \left[\frac{1}{12} \cdot \frac{t^3}{3}\right]_0^{12} = \frac{1}{36}\left[12^3 - 0\right] = 48.$$

We then use the definition to find σ:

$$\sigma = \sqrt{\int_0^{12} t^2 f(t)\,dt - \mu^2} = \sqrt{48 - 6^2} = \sqrt{12} \approx 3.46.$$

Hence, the standard deviation of the waiting times is about 3.46 minutes.

5.5. PROBABILITY DISTRIBUTIONS AND DENSITY FUNCTIONS (II)

Example 2: A large school district purchased 1,000 projectors for classroom use. The manufacturer told the district manager that the life span of the projector light bulbs, t, can be described by the following probability density function: $f(t) = \dfrac{3}{14}\sqrt{t}$, $1 \leq t \leq 4$, where t is measured in thousands of hours.

(a) Determine the mean life span of the projector light bulbs.

(b) Determine the standard deviation of the life span of the projector light bulbs.

Solution (a) By the definition of the mean, we have

$$\mu = \int_1^4 tf(t)\, dt = \int_1^4 t\frac{3}{14}\sqrt{t}\, dx$$

$$= \frac{3}{14}\int_1^4 t^{3/2}\, dt = \frac{3}{14}\left[\frac{2}{5}t^{5/2}\right]_1^4$$

$$= \frac{3}{35}\left[(4)^{5/2} - (1)^{5/2}\right] = \frac{3}{35}(31) \approx 2.657.$$

Hence, the mean life span of the projector light bulbs is 2,657 hours.

(b) To find the standard deviation, we first evaluate

$$\int_1^4 t^2 f(t)\, dt = \int_1^4 t^2 \frac{3}{14}\sqrt{t}\, dx$$

$$= \frac{3}{14}\int_1^4 t^{5/2}\, dt = \frac{3}{14}\left[\frac{2}{7}t^{7/2}\right]_1^4$$

$$= \frac{3}{49}\left[(4)^{7/2} - (1)^{7/2}\right] = \frac{3}{49}(127) \approx 7.7755.$$

We then use the definition to find σ:

$$\sigma = \sqrt{\int_1^4 t^2 f(t)\, dt - \mu^2} = \sqrt{7.7755 - (2.657)^2} \approx 0.846.$$

Hence, the standard deviation of the life spans of the projector light bulbs is about 846 hours.

Recall that an exponential distribution has the exponential density function $f(t) = ke^{-kt}$, $0 < t$. It can be verified that the mean is $\mu = 1/k$ and the standard deviation is $\sigma = 1/k$. The verifications are left as an exercise problem.

▬ The Normal Distribution

The **normal distribution** is the most important distribution in statistics. Suppose that the average height of adult males in the United States is 5 foot 9 inches, or 69 inches. Then there are about as many U.S. adult males taller than 69 inches as there are about as many U.S. adult males shorter than 69 inches, and the farther away from this average height is, the fewer adult U.S. males having this height. There are probably more U.S. adult males having heights around 6 feet (72 inch) than around 7 feet (84 inch); and more U.S. adult males having heights around 5.5 feet (66 inch) than around 4.5 feet (54 inch). Hence the distribution is a bell-shaped curve presented in Figure 5.5.4.

Figure 5.5.4: A normal distribution

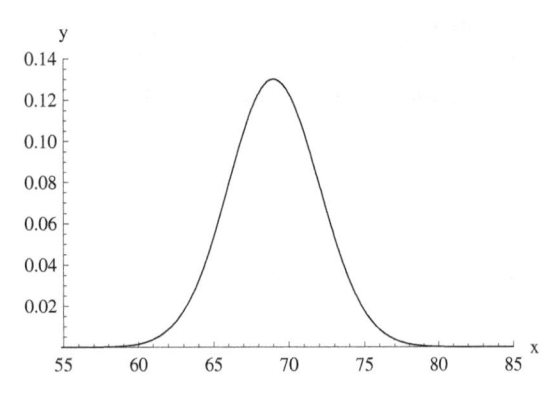

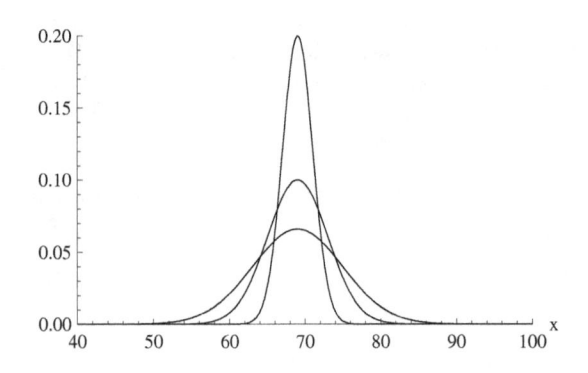

Figure 5.5.5: Normal distributions with same means

We can observe that this distribution is symmetric about the line $x = 69$, and by the previous discussion, this center is the mean: $\mu = 69$. In this curve, the standard deviation is $\sigma = 3$. For comparison, Figure 5.5.5 illustrates three normal distributions, all having the mean $\mu = 69$, but with standard deviations 2, 4, and 6. Recall that the larger the standard deviation is, the wider the curve spreads.

The **normal density function with mean** μ **and standard deviation** σ is given by

$$f(x) = \frac{1}{\sigma\sqrt{2\pi}}\, e^{-\frac{1}{2}\left(\frac{x-\mu}{\sigma}\right)^2}, \quad -\infty < x < \infty.$$

This rather complicated function is not used for calculating probabilities. Instead, any normal distribution can be transformed into the **standard normal distribution**, which has mean $\mu = 0$ and standard deviation $\sigma = 1$ (Figure 5.5.6):

$$f(x) = \frac{1}{\sqrt{2\pi}}\, e^{-\frac{x^2}{2}}, \quad -\infty < x < \infty.$$

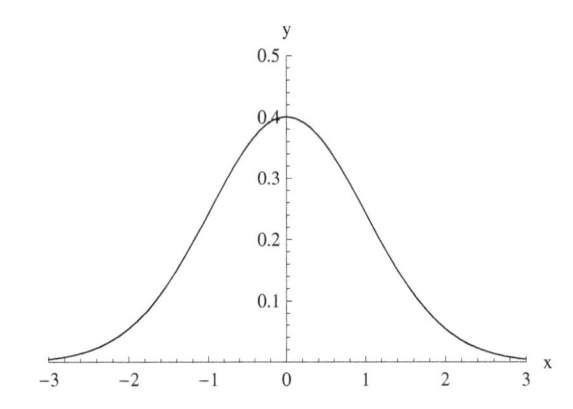

Figure 5.5.6: Standard normal distribution

This function has an antiderivative. However, the antiderivative cannot be expressed by a basic algebraic formula. Since the normal distribution is the most important probability distribution in statistics, numerical methods are applied to approximate the associated definite integrals, and the results are made into tables.

Table 2 lists the numerical approximations of the definite integral

$$P(0 \leq x \leq z) = \int_0^z \frac{1}{\sqrt{2\pi}}\, e^{-\frac{x^2}{2}}\, dx \quad \text{for various values of } z.$$

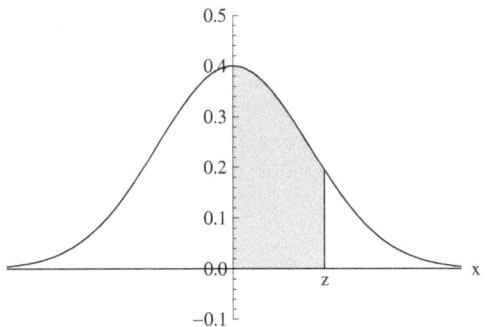

Table 2 Areas of a Standard Normal Distribution

z-score	0.00	0.01	0.02	0.03	0.04	0.05	0.06	0.07	0.08	0.09
0.0	0.0000	0.0040	0.0080	0.0120	0.0160	0.0199	0.0239	0.0279	0.0319	0.0359
0.1	0.0398	0.0438	0.0478	0.0517	0.0557	0.0596	0.0636	0.0675	0.0714	0.0753
0.2	0.0793	0.0832	0.0871	0.0910	0.0948	0.0987	0.1026	0.1064	0.1103	0.1141
0.3	0.1179	0.1217	0.1255	0.1293	0.1331	0.1368	0.1406	0.1443	0.1480	0.1517
0.4	0.1554	0.1591	0.1628	0.1664	0.1700	0.1736	0.1772	0.1808	0.1844	0.1879
0.5	0.1915	0.1950	0.1985	0.2019	0.2054	0.2088	0.2123	0.2157	0.2190	0.2224
0.6	0.2257	0.2291	0.2324	0.2357	0.2389	0.2422	0.2454	0.2486	0.2517	0.2549
0.7	0.2580	0.2611	0.2642	0.2673	0.2704	0.2734	0.2764	0.2794	0.2823	0.2852
0.8	0.2881	0.2910	0.2939	0.2967	0.2995	0.3023	0.3051	0.3078	0.3106	0.3133
0.9	0.3159	0.3186	0.3212	0.3238	0.3264	0.3289	0.3315	0.3340	0.3365	0.3389
1.0	0.3413	0.3438	0.3461	0.3485	0.3508	0.3531	0.3554	0.3577	0.3599	0.3621
1.1	0.3643	0.3665	0.3686	0.3708	0.3729	0.3749	0.3770	0.3790	0.3810	0.3830
1.2	0.3849	0.3869	0.3888	0.3907	0.3925	0.3944	0.3962	0.3980	0.3997	0.4015
1.3	0.4032	0.4049	0.4066	0.4082	0.4099	0.4115	0.4131	0.4147	0.4162	0.4177
1.4	0.4192	0.4207	0.4222	0.4236	0.4251	0.4265	0.4279	0.4292	0.4306	0.4319
1.5	0.4332	0.4345	0.4357	0.4370	0.4382	0.4394	0.4406	0.4418	0.4429	0.4441
1.6	0.4452	0.4463	0.4474	0.4484	0.4495	0.4505	0.4515	0.4525	0.4535	0.4545
1.7	0.4554	0.4564	0.4573	0.4582	0.4591	0.4599	0.4608	0.4616	0.4625	0.4633
1.8	0.4641	0.4649	0.4656	0.4664	0.4671	0.4678	0.4686	0.4693	0.4699	0.4706
1.9	0.4713	0.4719	0.4726	0.4732	0.4738	0.4744	0.4750	0.4756	0.4761	0.4767
2.0	0.4772	0.4778	0.4783	0.4788	0.4793	0.4798	0.4803	0.4808	0.4812	0.4817
2.1	0.4821	0.4826	0.4830	0.4834	0.4838	0.4842	0.4846	0.4850	0.4854	0.4857
2.2	0.4861	0.4864	0.4868	0.4871	0.4875	0.4878	0.4881	0.4884	0.4887	0.4890
2.3	0.4893	0.4896	0.4898	0.4901	0.4904	0.4906	0.4909	0.4911	0.4913	0.4916
2.4	0.4918	0.4920	0.4922	0.4925	0.4927	0.4929	0.4931	0.4932	0.4934	0.4936
2.5	0.4938	0.4940	0.4941	0.4943	0.4945	0.4946	0.4948	0.4949	0.4951	0.4952
2.6	0.4953	0.4955	0.4956	0.4957	0.4959	0.4960	0.4961	0.4962	0.4963	0.4964
2.7	0.4965	0.4966	0.4967	0.4968	0.4969	0.4970	0.4971	0.4972	0.4973	0.4974
2.8	0.4974	0.4975	0.4976	0.4977	0.4977	0.4978	0.4979	0.4979	0.4980	0.4981
2.9	0.4981	0.4982	0.4982	0.4983	0.4984	0.4984	0.4985	0.4985	0.4986	0.4986
3.0	0.4987	0.4987	0.4987	0.4988	0.4988	0.4989	0.4989	0.4989	0.4990	0.4990

Example 3: Let x be a continuous random variable with standard normal distribution. Use Table 2 to find the following probabilities:

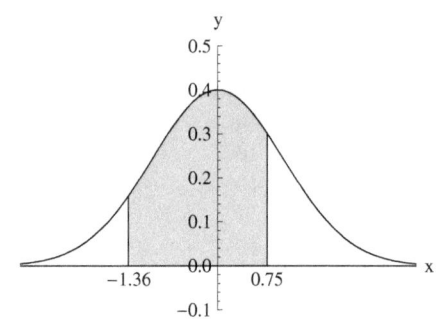

(a) $P(0 \leq x \leq 1.23)$ (b) $P(-0.58 \leq x \leq 0)$

(c) $P(-1.36 \leq x \leq 0.75)$ (d) $P(0.25 \leq x \leq 2.03)$

(e) $P(-0.2 \leq x)$ (f) $P(-1.46 \leq x \leq -0.9)$

(g) $P(0.8 \leq x)$ (h) $P(x \leq -1.3)$

Solution (a) $P(0 \leq x \leq 1.23)$ is the area under the standard normal curve on the interval $[0, 1.23]$. It can be found directly from Table 2: We go down the left column to 1.2, then move to the right to the column headed 0.03. There we read 0.3907. Hence,

$$P(0 \leq x \leq 1.23) = 0.3907.$$

The shaded area represents the probability.

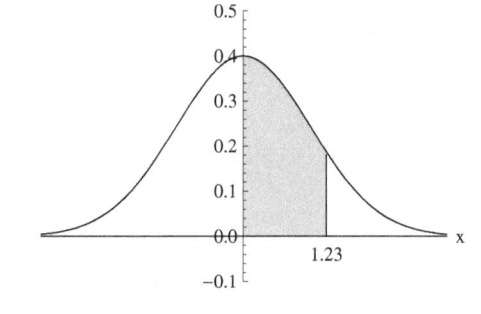

(b) $P(-0.58 \leq x \leq 0)$ is the area under the standard normal curve on the interval $[-0.58, 0]$. Since the normal curve is symmetric around the y-axis, the area is the same as the area under the standard normal curve on the interval $[0, 0.58]$. From Table 2, we read 0.2190. Hence,

$$P(-0.58 \leq x \leq 0) = P(0 \leq x \leq 0.58) = 0.2190.$$

The shaded area represents the probability.

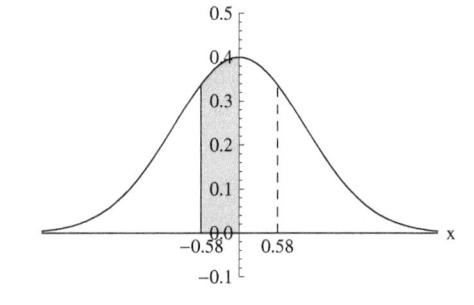

(c) $P(-1.36 \leq x \leq 0.75)$ is the area under the standard normal curve on the interval $[-1.36, 0.75]$. It can be viewed as the sum of the areas under the standard normal curve on the intervals $[-1.36, 0]$ and $[0, 0.75]$. Using the same reasoning as in part (b), the total area is the same as the sum of the areas under the standard normal curve on the intervals $[0, 1.36]$ and $[0, 0.75]$. From Table 2, we read 0.4131 and 0.2734. Hence,

$$\begin{aligned} &P(-1.36 \leq x \leq 0.75) \\ &= P(-1.36 \leq x \leq 0) + P(0 \leq x \leq 0.75) \\ &= P(0 \leq x \leq 1.36) + P(0 \leq x \leq 0.75) \\ &= 0.4131 + 0.2734 = 0.6865. \end{aligned}$$

The shaded area represents the probability.

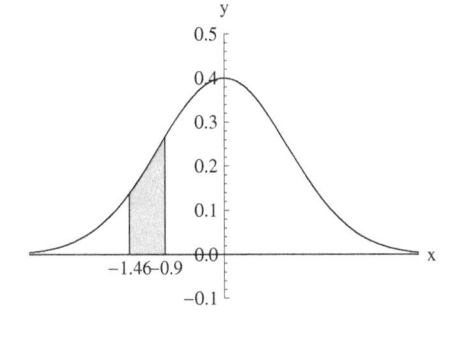

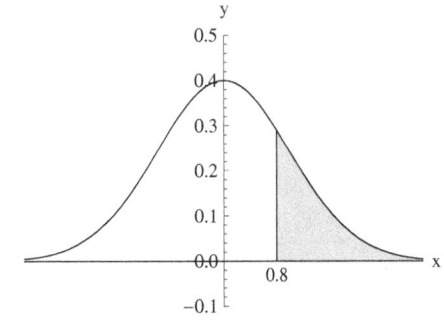

5.5. PROBABILITY DISTRIBUTIONS AND DENSITY FUNCTIONS (II)

(d) $P(0.25 \leq x \leq 2.03)$ is the area under the standard normal curve on the interval $[0.25, 2.03]$. It can be viewed as the difference of the areas under the standard normal curve on the intervals $[0, 2.03]$ and $[0, 0.25]$. From Table 2, we read 0.4788 and 0.0987. Hence,

$$P(0.25 \leq x \leq 2.03)$$
$$= P(0 \leq x \leq 2.03) - P(0 \leq x \leq 0.25)$$
$$= 0.4788 - 0.0987 = 0.3801.$$

The shaded area represents the probability.

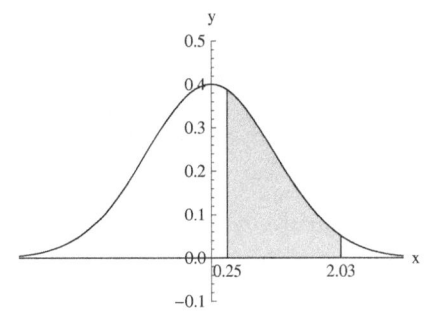

(e) $P(-0.2 \leq x)$ is the area under the standard normal curve on the interval $[-0.2, \infty)$. It can be viewed as the sum of the areas under the standard normal curve on the intervals $[-0.2, 0]$ and $[0, \infty)$. We know that the whole region has area 1. Hence, the positive part has area 0.5. From Table 2, we read $P(0 \leq x \leq 0.2) = 0.0793$. Hence,

$$P(-0.2 \leq x) = P(-0.2 \leq x \leq 0) + P(0 \leq x)$$
$$= P(0 \leq x \leq 0.2) + P(0 \leq x)$$
$$= 0.0793 + 0.5 = 0.5793.$$

The shaded area represents the probability.

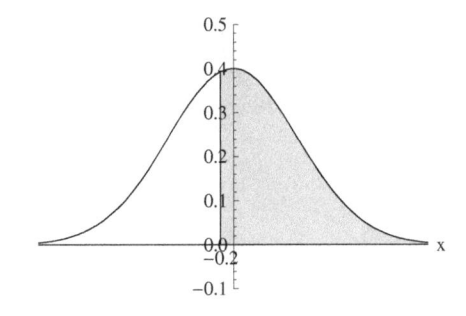

(f) $P(-1.46 \leq x \leq -0.9)$ is the area under the standard normal curve on the interval $[-1.46, -0.9]$. It is the same as the area under the standard normal curve on the interval $[0.9, 1.46]$. Similar to part (d),

$$P(-1.46 \leq x \leq -0.9) = P(0.9 \leq x \leq 1.46)$$
$$= P(0 \leq x \leq 1.46) - P(0 \leq x \leq 0.9)$$
$$= 0.4279 - 0.3159 = 0.1120.$$

The shaded area represents the probability.

(g) $P(0.8 \leq x)$ is the area under the standard normal curve on the interval $[0.8, \infty)$. It can be viewed as the difference of the areas under the standard normal curve on the intervals $[0, \infty)$ and $[0, 0.8]$. Hence,

$$P(0.8 \leq x) = P(0 \leq x) - P(0 \leq x \leq 0.8)$$
$$= 0.5 - 0.2881 = 0.2119.$$

The shaded area represents the probability.

(h) $P(x \leq -1.3)$ is the area under the standard normal curve on the interval $(-\infty, -1.3]$. It is the same as the area under the standard normal curve on the interval $[1.3, \infty)$. Similar to part (g),

$$P(x \leq -1.3) = P(1.3 \leq x)$$
$$= P(0 \leq x) - P(0 \leq x \leq 1.3)$$
$$= 0.5 - 0.4032 = 0.0968.$$

The shaded area represents the probability.

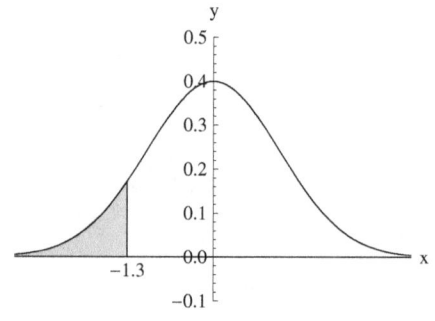

So what is the usefulness of Table 2, since it only lists the areas of a standard normal distribution? In most of normal distributions, the mean is not zero and the standard deviation is not 1. In fact, Table 2 can be used to find probabilities of any normal distribution with mean μ and standard deviation σ. All that is required is a transformation that will turn a normal distribution with mean μ and standard deviation σ into a standard normal distribution.

Reality Check   *Normal vs. standard normal*

Suppose that a continuous random variable X follows a normal distribution with mean μ and standard deviation σ. If we let $x = \dfrac{X - \mu}{\sigma}$, then x is a continuous random variable that follows the standard normal distribution. Therefore,

$$P(A \leq X \leq B) = P(a \leq x \leq b), \quad \text{where} \quad a = \dfrac{A - \mu}{\sigma} \quad \text{and} \quad b = \dfrac{B - \mu}{\sigma}.$$

Example 4: Let X be a continuous random variable that has a normal distribution with mean $\mu = 200$ and standard deviation $\sigma = 40$. Use Table 2 to find the following probabilities:

(a) $P(200 \leq X \leq 250)$ (b) $P(180 \leq X \leq 240)$

(c) $P(220 \leq X \leq 250)$ (d) $P(X \leq 180)$

(e) $P(160 \leq X)$ (f) $P(220 \leq X)$

Solution Let $x = \dfrac{X - 200}{40}$, then x has standard normal distribution.

(a). Let

$$a = \dfrac{200 - 200}{40} = 0, \quad \text{and} \quad b = \dfrac{250 - 200}{40} = 1.25, \text{ then}$$

$$P(200 \leq X \leq 250) = P(0 \leq x \leq 1.25)$$

by Table 2, $= 0.3944.$

Hence, $P(200 \leq X \leq 250) = 0.3944.$

(b). Let

$$a = \dfrac{180 - 200}{40} = -0.5, \quad \text{and} \quad b = \dfrac{240 - 200}{40} = 1, \text{ then}$$

$$P(180 \leq X \leq 240) = P(-0.5 \leq x \leq 1)$$

$$= P(0 \leq x \leq 0.5) + P(0 \leq x \leq 1)$$

by Table 2, $= 0.1915 + 0.3413 = 0.5328.$

Hence, $P(180 \leq X \leq 240) = 0.5328.$

(c). Let

$$a = \frac{220 - 200}{40} = 0.5, \text{ and } b = \frac{250 - 200}{40} = 1.25, \text{ then}$$

$$P(220 \leq X \leq 250) = P(0.5 \leq x \leq 1.25)$$
$$= P(0 \leq x \leq 1.25) - P(0 \leq x \leq 0.5)$$

by Table 2, $= 0.3944 - 0.1915 = 0.2029.$

Hence, $P(220 \leq X \leq 250) = 0.2029.$

(d). Let

$$b = \frac{180 - 200}{40} = -0.5, \text{ then}$$

$$P(X \leq 180) = P(x \leq -0.5)$$
$$= P(0 \leq x) - P(0 \leq x \leq 0.5)$$

by Table 2, $= 0.5 - 0.1915 = 0.3085.$

Hence, $P(X \leq 180) = 0.3085.$

(e). Let

$$a = \frac{160 - 200}{40} = -1, \text{ then}$$

$$P(160 \leq X) = P(-1 \leq x)$$
$$= P(0 \leq x) + P(0 \leq x \leq 1)$$

by Table 2, $= 0.5 + 0.3413 = 0.8413.$

Hence, $P(160 \leq X) = 0.8413.$

(f). Let

$$a = \frac{220 - 200}{40} = 0.5, \text{ then}$$

$$P(220 \leq X) = P(0.5 \leq x)$$
$$= P(0 \leq x) - P(0 \leq x \leq 0.5)$$

by Table 2, $= 0.5 - 0.1915 = 0.3085..$

Hence, $P(220 \leq X) = 0.3085.$

Example 5: A large company has just moved into a new building that has 500 compact fluorescent light bulbs of a particular brand. The manufacturer told the building manager that the life span of the compact fluorescent light bulbs, t, follows a normal distribution with mean $\mu = 10,000$ hours and standard deviation $\sigma = 2,000$ hours.

(a) Determine the probability that a compact fluorescent light bulb lasts at least 12,000 hours.

(b) Determine the probability that a compact fluorescent light bulb lasts at most 9,000 hours.

(c) The building is on automatic energy control so the lights are on for 3,000 hours per year. In the third year after the move, how many compact fluorescent light bulbs should the manager budget to purchase for replacing the burned-out ones?

Solution (a) A compact fluorescent light bulb lasts at least 12,000 hours means that $12{,}000 \leq t$. Hence the probability is $P(12{,}000 \leq t)$. Let $x = \dfrac{t - 10{,}000}{2{,}000}$, then x has standard normal distribution. Let

$$a = \dfrac{12{,}000 - 10{,}000}{2{,}000} = 1, \quad \text{then}$$

$$P(12{,}000 \leq t) = P(1 \leq x) = P(0 \leq x) - P(0 \leq x \leq 1)$$

by Table 2, $= 0.5 - 0.3413 = 0.1587$.

Hence, the probability that a compact fluorescent light bulb lasts at least 12,000 hours is about 15.9%.

(b) A compact fluorescent light bulb lasts at most 9,000 hours means that $t \leq 9{,}000$. Hence the probability is $P(t \leq 9{,}000)$. Let

$$b = \dfrac{9{,}000 - 10{,}000}{2{,}000} = -0.5, \quad \text{then}$$

$$P(t \leq 9{,}000) = P(x \leq -0.5) = P(0.5 \leq x) = P(0 \leq x) - P(0 \leq x \leq 0.5)$$

by Table 2, $= 0.5 - 0.1915 = 0.3085$.

Hence, the probability that a compact fluorescent light bulb lasts at most 9,000 hours is about 30.9%.

(c) From the beginning to the end of the third year after the move, the compact fluorescent light bulbs have been used for between 6,000 to 9,000 hours. Let

$$a = \dfrac{6{,}000 - 10{,}000}{2{,}000} = -2, \quad b = \dfrac{9{,}000 - 10{,}000}{2{,}000} = -0.5, \quad \text{then}$$

$$P(6{,}000 \leq t \leq 9{,}000) = P(-2 \leq x \leq -0.5) = P(0.5 \leq x \leq 2) = P(0 \leq x \leq 2) - P(0 \leq x \leq 0.5)$$

by Table 2, $= 0.4772 - 0.1915 = 0.2857$.

about 28.57% of the 500 compact fluorescent light bulbs are likely to be burned out. Hence, the manager needs to purchase about 143 compact fluorescent light bulbs for replacement.

Example 6: A food company estimates that w, the weight of a package of one type of the cereal, marked "Net Weight 12 oz," is normally distributed with a mean of 12 oz and a standard deviation of 0.2 oz.

(a) Determine the probability that a package weighs less than 11.7 oz.

(b) Determine the weight that will be exceeded by 2.2% of the packages.

Solution (a) A package weigh less than 11.7 oz means that $w \leq 11.7$. Hence, the probability is $P(w \leq 11.7)$. Let $x = \dfrac{w - 12}{0.2}$, then x has standard normal distribution. Let

$$b = \dfrac{11.7 - 12}{0.2} = -1.5, \quad \text{then}$$

$$P(w \leq 11.7) = P(x \leq -1.5) = P(1.5 \leq x) = P(0 \leq x) - P(0 \leq x \leq 1.5)$$

by Table 2, $= 0.5 - 0.4332 = 0.0668$.

Hence, the probability that a package weighs less than 11.7 oz is about 6.7%.

(b) We are looking for the weight A such that $P(A \leq w) = 2.2\% = 0.022$. Let

$$a = \frac{A-12}{0.2}, \text{ then}$$

$$0.022 = P(A \leq w) = P(a \leq x) = P(0 \leq x) - P(0 \leq x \leq a)$$
$$= 0.5 - P(0 \leq x \leq a).$$

Hence,

$$P(0 \leq x \leq a) = 0.5 - 0.022 = 0.478.$$

By Table 2, we have $a \approx 2.01$. Therefore,

$$A = 12 + 0.2a = 12 + 0.2(2.01) = 12.402.$$

That is, about 2.2% packages will exceed 12.4 oz.

Exercise Set 5.5

For Exercises #1 to #16, for each function on the given interval $[a, b]$, do the following:

(a) Find the mean μ.

(b) Find the standard deviation σ.

1. $f(x) = \frac{1}{3}$, $[1, 4]$

2. $f(x) = \frac{1}{5}$, $[0, 5]$

3. $f(x) = \frac{1}{2}x$, $[0, 2]$

4. $f(x) = \frac{1}{4}x$, $[1, 3]$

5. $f(x) = \frac{1}{9}x^2$, $[0, 3]$

6. $f(x) = \frac{1}{21}x^2$, $[1, 4]$

7. $f(x) = \frac{3}{2x^2}$, $[1, 3]$

8. $f(x) = \frac{4}{3x^2}$, $[1, 4]$

9. $f(x) = \frac{1}{x}$, $[1, e]$

10. $f(x) = \frac{1}{x}$, $[e, e^2]$

11. $f(x) = \frac{3}{16}x^2$, $[-2, 2]$

12. $f(x) = \frac{3}{2}x^2$, $[-1, 1]$

13. $f(x) = 0.2e^{-0.2x}$, $[0, \infty)$

14. $f(x) = 3e^{-3x}$, $[0, \infty)$

15. $f(x) = \frac{3}{x^4}$, $[1, \infty)$

16. $f(x) = \frac{24}{x^4}$, $[2, \infty)$

For Exercises #17 to #32, assume that x is a continuous random variable with a standard normal distribution. Use Table 2 to find each probability, and illustrate by an area under the standard normal curve.

17. $P(0 \leq x \leq 2.13)$

18. $P(0 \leq x \leq 1.54)$

19. $P(-1.04 \leq x \leq 0)$

20. $P(-1.58 \leq x \leq 0)$

21. $P(-1.2 \leq x \leq 0.9)$

22. $P(-0.8 \leq x \leq 1.4)$

23. $P(0.36 \leq x \leq 2.95)$

24. $P(0.45 \leq x \leq 2.6)$

25. $P(-2.3 \leq x \leq -0.4)$

26. $P(-1.46 \leq x)$

27. $P(-1.25 \leq x)$

28. $P(5 \leq x)$

29. $P(0 \leq x)$

30. $P(x \leq 0)$

31. $P(1.8 \leq x)$

32. $P(x \leq -0.83)$

For Exercises #33 to #38, assume that x is a continuous random variable that has a normal distribution with mean $\mu = 16$ and standard deviation $\sigma = 4$. Use Table 2 to find each probability.

33. $P(12 \leq x \leq 18)$ 34. $P(15 \leq x \leq 20)$

35. $P(18 \leq x \leq 22)$ 36. $P(17 \leq x \leq 23)$

37. $P(14 \leq x)$ 38. $P(x \leq 14)$

39. Let t be a continuous random variable with exponential density function
$$f(t) = ke^{-kt}.$$

(a) Verify that the mean is $\mu = 1/k$.

(b) Find the standard deviation σ.

40. (a) Verify that for the exponential density function $f(t) = ke^{-kt}$, the mean is $\mu = 1/k$ and the standard deviation is $\sigma = 1/k$.

(b) Let t_1 and t_2 be two continuous random variables with exponential density functions
$$f_1(t) = k_1 e^{-k_1 t}, \; f_2(t) = k_2 e^{-k_2 t},$$
and $k_1 > k_2$.

Which random variable has the larger mean?

Which random variable has the larger standard deviation?

APPLICATIONS

41. Suppose that Kathy rides the school bus every weekday morning. The bus has an equal chance of arriving at the corner of her street at any time between 7:20 and 7:30. Suppose that Kathy always arrives at the corner of her street at 7:20. Let t be her waiting time in minutes. Then the probability density function of t is given by
$$f(t) = \frac{1}{10} \quad 0 \leq t \leq 10.$$

(a) Determine the mean waiting time.

(b) Determine the standard deviation of the waiting times.

42. Suppose that the traffic light at a certain location remains red for 30 seconds at a time. Suppose that Sam arrives at that spot and finds the light is red. Let t be Sam's waiting time in seconds before the light turns green.

(a) Determine the probability density function of t.

(b) Determine the mean waiting time.

43. Refer to Exercise #41. Suppose that Kathy's waiting time t has the following probability density function:
$$f(t) = \frac{1}{50}t, \; 0 \leq t \leq 10.$$

Repeat parts (a) and (b).

44. Refer to Exercise #42. Suppose that Sam's waiting time t has the following probability density function:
$$f(t) = k(30 - t), \; 0 \leq t \leq 30.$$

(a) Determine the value of k.

(b) Determine the mean waiting time.

45. A large school district purchased 800 projectors for classroom use. The manufacturer told the district manager that the life span of the projector light bulbs, t, can be described by the following probability density function:
$$f(t) = kt^2, \; 0.5 \leq t \leq 2,$$

where t is measured in thousands of hours.

(a) Determine the value of k.

(b) Determine the mean life span of the bulbs.

(c) Determine the standard deviation of the life span of the light bulbs.

46. The average waiting time at a checkout line in a supermarket is 2 minutes. Which distribution will a customer prefer: uniform or exponential? Why?

47. A food company estimates that w, the weight of a package of one type of frozen dinner, marked "Net Weight 20 oz," is normally distributed with a mean of 20 oz and a standard deviation of 0.25 oz.

(a) Determine the probability that a package weighs less than 19.5 oz.

(b) Determine the weight that will be exceeded by 2.3% of the packages.

48. A food company estimates that w, the weight of a package of one type of frozen dinner, marked "Net Weight 18 oz," is normally distributed with a mean of 18 oz and a standard deviation of 0.15 oz.

(a) Determine the probability that a package weighs more than 18.5 oz.

(b) Determine the weight that will be exceeded by 2% of the packages.

49. A certain brand of home-delivery dinner comes in packages with weights that are normally distributed, with a standard deviation of 0.3 oz. Suppose that 2.4% of the dinners weigh more than 22.5 oz. Determine the mean weight of the packages.

50. A certain brand of home-delivery dinner comes in packages with weights that are normally distributed, with a standard deviation of 0.25 oz. Suppose that 2.2% of the dinners weigh more than 16.5 oz. Determine the mean weight of the packages.

51. A building changes all its 600 lights to compact fluorescent light bulbs. The life span of the compact fluorescent light bulbs, t, follows a normal distribution with mean $\mu = 11,000$ hours and standard deviation $\sigma = 2,500$ hours.

(a) Determine the probability that a bulb lasts at least 12,000 hours.

(b) Determine the probability that a bulb lasts at most 9,000 hours.

(c) The building is on automatic energy control so the lights are on for 3,000 hours per year. In the third year after the conversion, how many compact fluorescent light bulbs need to be replaced?

52. Refer to Exercise #51: If $\mu = 9,600$ and $\sigma = 1,500$, repeat parts (a) – (c).

5.6 Differential Equations

Many mathematical models involve an equation that has both an unknown function and its derivative. Such an equation is called an **differential equation**. The following is a comparison between an algebraic equation and a differential equation:

	Algebraic Equations	Differential Equations
Example	$x^2 - 4 = 0$	$y' - 2y = 0$
Unknown	number(s)	function(s)
Solution(s)	$x = -2, 2$	$y = Ce^{2x}$
Verify solution(s)	$(-2)^2 - 4 = 0$, and $(2)^2 - 4 = 0$	$(Ce^{2x})' - 2(Ce^{2x})$ $= C(2e^{2x}) - 2Ce^{2x} = 0$ for all x

Strictly speaking, we have been dealing with differential equations all through the second half of this textbook. For example, if the velocity of a moving object is known as

$$v(t) = 2t + 1,$$

and we look for the position function $s(t)$, then we have the differential equation

$$s'(t) = v(t) = 2t + 1.$$

We solve this equation by integrating $v(t)$:

$$s(t) = \int v(t)\, dt = \int (2t + 1)\, dt = t^2 + t + C.$$

Another example is the balance in a savings account with annual interest rate r, compounded continuously. Since the change in the balance is from the interest added, and the interest is calculated by the rate times the current balance, we can model the rate of change of the current balance as

$$\frac{dP}{dt} = rP.$$

This is a differential equation with both the function $P(t)$ and its derivative $P'(t)$ occurring in the equation. We can check that the formula introduced in Section 1.5,

$$P(t) = P_0 e^{rt},$$

satisfies this differential equation:

$$P'(t) = P_0\left(re^{rt}\right) = r\left(P_0 e^{rt}\right) = rP(t).$$

The study of the properties of differential equations, such as whether a solution exists; how to solve a given differential equation; how to interpret a solution; etc., forms an important branch of mathematics. In this section, we explore one of the techniques for solving a certain type of differential equation. It is called **separation of variables**. We illustrate this technique by an example.

Consider the following differential equation:

$$\frac{dy}{dx} = \frac{2x}{3y^2}, \quad y \neq 0.$$

Recall that after the differentials dx and dy are defined in Section 3.1, we can treat dy/dx as a quotient of the two differentials. Hence we can rewrite the equation, with all expressions involving y on the left side, and all expressions involving x on the right side:

$$\frac{dy}{dx} = \frac{2x}{3y^2}$$

$$3y^2 \frac{dy}{dx} = 2x$$

$$3y^2 \, dy = 2x \, dx.$$

We then integrate both sides of this equation:

$$\int 3y^2 \, dy = \int 2x \, dx$$

$$y^3 = x^2 + C, \quad \text{hence} \quad y = \sqrt[3]{x^2 + C}.$$

Note that only one constant is needed. Therefore, we have found that $y = \sqrt[3]{x^2 + C}$ is the solution to the differential equation:

$$\frac{dy}{dx} = \frac{2x}{3y^2}, \quad y \neq 0.$$

Is this function indeed the solution? We can verify it by taking the derivative of y, and substituting both y and y' into the differential equation. We have on the left side of the equation,

$$\frac{dy}{dx} = \frac{d}{dx}\sqrt[3]{x^2 + C} = \frac{d}{dx}\left(x^2 + C\right)^{1/3}$$

$$= \frac{1}{3}\left(x^2 + C\right)^{-2/3}(2x) = \frac{2x}{\left(x^2 + C\right)^{2/3}}.$$

Meanwhile, on the right side of the equation,

$$\frac{2x}{3y^2} = \frac{2x}{3(\sqrt[3]{x^2 + C})^2} = \frac{2x}{\left(x^2 + C\right)^{2/3}}.$$

Hence, the differential equation is balanced.

5.6. DIFFERENTIAL EQUATIONS

> **Reality Check** *Differential equations and solutions*
>
> - A differential equation that can be solved by this method is called a **separable equation**. A separable equation must have the form of $dy/dx = f(x)g(y)$. Some differential equations, such as the simple equation $y' = x + y$, cannot be solved by the method of separation of variables.
>
> - The solution, $y = \sqrt[3]{x^2 + C}$, is called a **general solution**. If a particular pair of x and y values are given, then the value of C can be determined.

Example 1: Determine the general solution(s) for $\dfrac{dy}{dx} = 4xy$, then find the particular solution with condition $y(0) = 3$.

Solution We start by separating variables,

$$\frac{dy}{dx} = 4xy$$

$$\frac{1}{y}\frac{dy}{dx} = 4x$$

$$\frac{1}{y}\,dy = 4x\,dx.$$

Integrating both sides, we have

$$\int \frac{1}{y}\,dy = \int 4x\,dx$$

$\ln|y| = 2x^2 + C_1$,   hence, $|y| = e^{2x^2 + C_1} = e^{2x^2} \cdot e^{C_1}$.

Letting $C_2 = e^{C_1}$, we have

$$|y| = C_2 e^{2x^2}, \quad \text{or} \quad y = \pm C_2 e^{2x^2}.$$

Since $\pm C_2$ represents an arbitrary constant, we can let $C = \pm C_2$ to obtain the general solution

$$y = Ce^{2x^2}.$$

For the particular solution, we substitute $x = 0$ and $y = 3$ into the general solution, and solve for C:

$$3 = Ce^{2(0)^2} = Ce^0 = C.$$

Hence, $C = 3$, and the particular solution is

$$y = 3e^{2x^2}.$$

> **Reality Check** *Simplifying constants of integration*
>
> The details illustrated in Example 1 indicate that we can skip the absolute value bars (i.e., assuming $y > 0$) and later compensate for the possible negative values of y by the general constant C.

Example 2: Determine the general solutions for
$$y' = \frac{3x^2}{y}, \quad y \neq 0,$$
then find the particular solution with condition $y(1) = 2$.

Solution We first replace y' by dy/dx, then separate the variables:
$$\frac{dy}{dx} = \frac{3x^2}{y}$$
$$y\frac{dy}{dx} = 3x^2$$
$$y\, dy = 3x^2\, dx.$$
Integrating both sides, we have
$$\int y\, dy = \int 3x^2\, dx$$
$$\frac{1}{2}y^2 = x^3 + C, \quad \text{hence} \quad y^2 = 2x^3 + 2C_1.$$
Letting $C = 2C_1$, we have the general solutions
$$y = \sqrt{2x^3 + C} \quad \text{and} \quad y = -\sqrt{2x^3 + C}.$$
For the particular solution, we substitute $x = 1$ and $y = 2$ into the general solution. Since $y(1) > 0$, we take the positive branch, and solve for C:
$$2 = \sqrt{2(1)^3 + C}$$
$$4 = (2)^2 = 2 + C,$$
hence $C = 2$, and the particular solution is
$$y = \sqrt{2x^3 + 2}.$$

Example 3: Determine the general solution for $y' - x + 2xy = 0$.

Solution This equation needs to be re-expressed before we can separate the variables:
$$y' - x + 2xy = 0$$
$$y' = x - 2xy$$
$$\frac{dy}{dx} = x(1 - 2y).$$
Now we can separate x and y:
$$\frac{dy}{dx} = x(1 - 2y)$$
$$\frac{dy}{1 - 2y} = x\, dx.$$
Integrating both sides, we have
$$\int \frac{dy}{1 - 2y} = \int x\, dx.$$

5.6. DIFFERENTIAL EQUATIONS

To integrate the left side, we need to apply substitution. Let $u = 1 - 2y$, the $du = -2dy$, or $-\frac{1}{2} du = dy$. Hence,

$$\int \frac{dy}{1-2y} = \int \frac{1}{1-2y} dy$$

$$= \int \frac{1}{u}\left(-\frac{1}{2}\right) du = -\frac{1}{2}\int \frac{1}{u} du$$

$$= \frac{-1}{2}\ln u + C = \frac{-1}{2}\ln(1-2y) + C.$$

Therefore, we have

$$\frac{-1}{2}\ln(1-2y) = \frac{1}{2}x^2 + C.$$

We now solve for y:

$$\frac{-1}{2}\ln(1-2y) = \frac{1}{2}x^2 + C$$

$$\ln(1-2y) = -x^2 - 2C$$

$$1 - 2y = e^{-x^2 - 2C} = e^{-x^2} \cdot e^{-2C}$$

$$y = \frac{1}{2} + \left(-\frac{1}{2}e^{-2C}\right)e^{-x^2}.$$

We follow the same steps as in Example 1 to use C for the general constant and obtain the general solution:

$$y = \frac{1}{2} + Ce^{-x^2}.$$

Applications

Example 4: Continuous Compounding Suppose that $\$P_0$ is invested in a fund that offers interest at an annual rate of 4%, compounded continuously. That is, the balance at time t is given by the differential equation

$$\frac{dP}{dt} = 0.04P.$$

(a) Determine the particular solution for this differential equation.

(b) Suppose that $2,000 is invested. Find the balance in the account at the end of 2 years.

Solution (a) We start by separating the variables, then integrate both sides:

$$\frac{dP}{dt} = 0.04P$$

$$\frac{dP}{P} = 0.04 \, dt$$

$$\int \frac{dP}{P} = \int 0.04 \, dt$$

$$\ln P = 0.04t + C_1$$

$$P = e^{0.04t+C_1} = e^{C_1}e^{0.04t} = Ce^{0.04t}.$$

Hence, the general solution is

$$P = Ce^{0.04t}.$$

When $t = 0$, $P = P_0$. Hence,

$$P_0 = Ce^{0.04(0)} = Ce^0 = C.$$

Therefore, the particular solution is

$$P(t) = P_0 e^{0.04t}.$$

This is exactly the formula for the future balance of a continuous compounding account, introduced in Section 1.5.

(b) When $P_0 = 2{,}000$ and $t = 2$, we have the balance as

$$P(2) = 2{,}000 e^{0.04(2)} \approx \$2{,}166.57.$$

Example 5: Newton's Law of Cooling This law is stated as the following: The temperature, T, of a cooling object, drops at a rate that is proportional to the difference $T - C$, where C is the constant temperature of the surrounding medium. Thus,

$$\frac{dT}{dt} = -k(T - C).$$

Find the particular solution for this differential equation given that $T(0) = T_0$.

Solution We start by separating the variables:

$$\frac{dT}{dt} = -k(T - C)$$

$$\frac{dT}{T - C} = -k\,dt.$$

Integrating both sides, we obtain

$$\int \frac{dT}{T - C} = \int -k\,dt$$

$$\ln(T - C) = -kt + C_1$$

$$T - C = e^{-kt + C_1} = e^{C_1} e^{-kt} = ae^{-kt},$$

where $a = e^{C_1}$. Therefore

$$T = C + ae^{-kt}.$$

If the initial temperature of the cooling object is $T(0) = T_0$, then

$$T_0 = C + ae^{-k(0)} = C + ae^0 = C + a.$$

Hence $a = T_0 - C$ and we obtain the particular solution

$$T(t) = (T_0 - C)e^{-kt} + C.$$

This is exactly the formula introduced in Section 1.6.

5.6. DIFFERENTIAL EQUATIONS

Example 6: Elasticity A certain product has elasticity $E(p) = 2$ and $D(2) = 3$. Determine the demand function $x = D(p)$.

Solution By the definition of $E(p)$ introduced in Section 3.9, the elasticity is defined as

$$E(p) = -\frac{pD'(p)}{D(p)}.$$

Let $x = D(p)$, then $D'(p) = dx/dp$ and the equation $E(p) = 2$ becomes

$$-\frac{p}{x}\frac{dx}{dp} = 2.$$

This differential equation can be solved by separation of variables.

$$-\frac{p}{x}\frac{dx}{dp} = 2$$

$$\frac{dx}{x} = -2\frac{dp}{p}$$

$$\int \frac{dx}{x} = -2\int \frac{dp}{p}$$

$$\ln x = -2\ln p + \ln C = \ln p^{-2} + \ln C = \ln\left(\frac{C}{p^2}\right)$$

(Note: The constant $\ln C$ instead of C is used to make the expression simple.) Hence

$$x = \frac{C}{p^2}.$$

Since $D(2) = 3$, we have

$$3 = \frac{C}{2^2},$$

so $C = 12$. Therefore the demand function is

$$x = D(p) = \frac{12}{p^2}.$$

Exercise Set 5.6

For Exercises 1–6, verify that the given function is a solution for the companion differential equation.

1. $y = x - 1/x;\quad xy' + y = 2x$
2. $y = e^{-x};\quad y' + y = 0$
3. $y = e^{2x};\quad y'' + y' - 6y = 0$
4. $y = e^{-3x};\quad y'' + y' - 6y = 0$
5. $y = \dfrac{1}{x+1};\quad y' + y^2 = 0$
6. $y = xe^{-x};\quad y'' + 2y' + y = 0$

For Exercises 7–24, determine the general solution for the given differential equation.

7. $\dfrac{dy}{dx} = \dfrac{x}{y}$
8. $\dfrac{dy}{dx} = \dfrac{2x}{y}$
9. $\dfrac{dy}{dx} = 3x^2 y$
10. $\dfrac{dy}{dx} = 4x^3 y$
11. $4y^3 \dfrac{dy}{dx} = 4x$
12. $2x^2 y \dfrac{dy}{dx} = 1$
13. $\dfrac{dy}{dx} = \dfrac{y}{x}$
14. $\dfrac{dy}{dx} = \dfrac{2y}{x^2}$

15. $\dfrac{dy}{dx} = \dfrac{e^x}{4y^3}$

16. $\dfrac{dy}{dx} = \dfrac{2e^{2x}}{y^2}$

17. $\dfrac{dy}{dx} = \dfrac{3}{y}$

18. $\dfrac{dy}{dx} = \dfrac{4}{y}$

19. $\dfrac{dP}{dt} = 0.02P$

20. $\dfrac{dP}{dt} = -0.04P$

21. $y' = 4x + xy$

22. $y' = 3x - xy$

23. $y' - xy - 2y = 0$

24. $y' + 3x + xy = 0$

For Exercises 25–34, determine the particular solution for the given differential equation and the given condition.

25. $\dfrac{dy}{dx} = \dfrac{x}{y}$; $y = 2$ when $x = 1$

26. $\dfrac{dy}{dx} = \dfrac{2x}{y}$; $y = 5$ when $x = 2$

27. $\dfrac{dy}{dx} = 3x^2 y$; $y(0) = 3$

28. $\dfrac{dy}{dx} = 4x^3 y$; $y(0) = 2$

29. $\dfrac{dy}{dx} = \dfrac{3}{y^2}$; $y(0) = 1$

30. $\dfrac{dy}{dx} = \dfrac{4}{y^2}$; $y(0) = 2$

31. $\dfrac{dP}{dt} = 0.02P$; $P(0) = 100$

32. $\dfrac{dP}{dt} = -0.04P$; $P(0) = 200$

33. $y' = 4x + xy$; $y(0) = 5$

34. $y' = 3x - xy$; $y(0) = 2$

For Exercises 35–40, determine the demand function $x = D(p)$ that satisfies the given elasticity and the given condition.

35. $E(p) = 1$ for all $p > 0$, and $D(2) = 3$.

36. $E(p) = 0.5$ for all $p > 0$, and $D(4) = 6$.

37. $E(p) = \dfrac{2}{p}$ and $D(2) = 5e$.

38. $E(p) = \dfrac{5}{p}$ and $D(5) = 3e$.

39. $E(p) = \dfrac{p}{100 - p}$ and $D(50) = 500$.

40. $E(p) = \dfrac{2p}{400 - p}$ and $D(300) = 50$.

APPLICATIONS

41. Karen deposits $10,000 in a bank account that earns interest at an annual rate of 4%, compounded continuously. She withdraws from the account continuously at a rate of $800 a year. Therefore the balance, P, in the account satisfies the differential equation $\dfrac{dP}{dt} = 0.04P - 800$.

    (a) Determine the solution of this differential equation.

    (b) How much is in the account after 6 years? After 8 years?

    (c) How long does it take for the balance to reduce to zero?

42. David deposits $30,000 in a bank account that earns interest at an annual rate of 5%, compounded continuously. He withdraws from the account continuously at a rate of $2,000 a year. Therefore the balance, P, in the account satisfies the differential equation $\dfrac{dP}{dt} = 0.05P - 2,000$.

    (a) Determine the solution of this differential equation.

    (b) How much is in the account after 6 years? After 8 years?

    (c) How long does it take for the balance to reduce to zero?

43. Refer to Exercise #41. Suppose that the initial deposit is $24,000, repeat parts (a) – (c).

44. Refer to Exercise #42. Suppose that the initial deposit is $50,000, repeat parts (a) – (c).

45. Refer to Exercise #41. Find the least amount of the initial deposit needed such that the account will never run out of money.

46. Refer to Exercise #42. Find the least amount of the initial deposit needed such that the account will never run out of money.

Chapter 5 Summary

This chapter explores the applications of integration in economics, finance, and other related fields.

Consumers' and Producers' Surpluses

The **consumers' surplus** for a certain product or service in the market measures the difference between the total amount the consumers are willing and able to spend and the amount the consumers actually spend for a certain quantity x_0 of goods or services at a certain market price p_0. It is represented by the area of the region between the demand curve and $p = p_0$ on the interval $[0, x_0]$.

> **Consumers' Surplus** Suppose that $p = D(x)$ describes the demand function for a commodity. Then at the point (x_0, p_0), the **consumers' surplus** is defined by
> $$\int_0^{x_0} D(x)\,dx - (p_0 \times x_0).$$

The **producers' surplus** for a certain product or service in the market measures the amount the producers receive above the amount the producers are willing or able to receive for x_0 items at a market price p_0. It is represented by the area of the region between $p = p_0$ and the demand curve on the interval $[0, x_0]$.

> **Producers' Surplus** Suppose that $p = S(x)$ describes the supply function for a commodity. Then at the point (x_0, p_0), the **producers' surplus** is defined by
> $$(p_0 \times x_0) - \int_0^{x_0} S(x)\,dx.$$

In a free market, for a certain product, the market naturally settles to this equilibrium point (x_E, p_E). In such a case, the consumers' expenditure and the producers' revenue are equal, both being $p_E \times x_E$, and the sum of the consumers' and producers' surpluses equals the area of the region between the demand curve and the supply curve, on interval $[0, x_E]$.

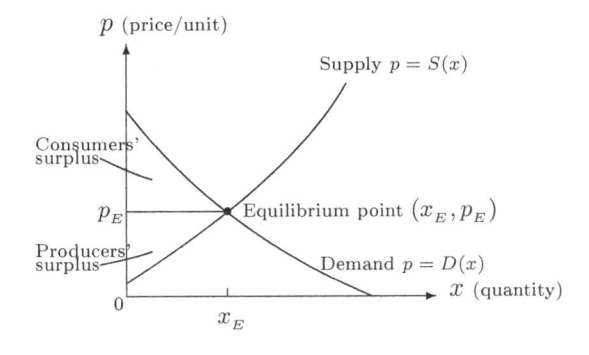

Example 1: Given the demand function

$$p = D(x) = 6.5 - 0.004x$$

where x is the number of chocolate bars sold per week, and p is in dollars; and the weekly supply function

$$p = S(x) = 0.005x + 2.$$

(a) Determine the equilibrium point (x_E, p_E).

(b) Determine the producers' weekly surplus at the equilibrium point.

(c) Determine the consumers' weekly surplus at the equilibrium point.

Solution (a) We solve the equation $S(x) = D(x)$:

$$0.005x + 2 = 6.5 - 0.004x, \quad \text{so} \quad 0.009x = 4.5, \quad \text{hence} \quad x_E = 500.$$

We substitute $x = 500$ to $S(x)$ to obtain

$$p_E = S(x_E) = 0.005(500) + 2 = \$4.5.$$

Therefore, the equilibrium point is $(500, \$4.5)$.

(b) The producers' weekly surplus is

$$4.5(500) - \int_0^{500}(0.005x + 2)\,dx = 2{,}250 - \left[0.0025x^2 + 2x\right]_0^{500}$$

$$= 2{,}250 - [0.0025(500)^2 + 2(500)] = 2{,}250 - 1{,}625 = \$625.$$

(c) The consumers' weekly surplus is

$$\int_0^{500}(6.5 - 0.004x)\,dx - 4.5(500) = \left[6.5x - 0.002x^2\right]_0^{500} - 2{,}250$$

$$= \left[6.5(500) - 0.002(500)^2\right] - 2{,}250 = 2{,}750 - 2{,}250 = \$500.$$

▬ Future and Present Values of a Continuous Money Flow ▬

In addition to the investment model of a lump-sum, an important mathematics model in finance deals with a continuous money flow.

Future Value of a Continuous Money Flow Suppose that a yearly amount of $R(t)$, where t is measured in years, is invested continuously into an interest-bearing account at the annual interest rate of $100\,r\%$ compounded continuously. Then the future value of the account at the end of T years is

$$\int_0^T R(t)e^{r(T-t)}\,dt.$$

When the yearly investment $R(t)$ is a constant S, the integral can be evaluated directly and the future value is

$$\int_0^T Se^{r(T-t)}\,dt = \frac{S}{r}\left(e^{rT} - 1\right).$$

CHAPTER 5 SUMMARY

Example 2: The grandparents of a 1-year old girl plan to invest for her college education. They will invest $2,000 per year continuously into an account paying interest at an annual rate of 5% compounded continuously. How much will the account have when their granddaughter is 18 years old?

Solution We have $T = 18 - 1 = 17$, $S = 2,000$, and $r = 0.05$. Therefore, the future value of the account is

$$\frac{2,000}{0.05}\left(e^{0.05(17)} - 1\right) \approx \$53{,}586.$$

Example 3: Sam, a 20-year old agent for an insurance firm, wants to invest part of his weekly wage in order to have $1,000,000 when he retires at 67. Suppose that he invests the money in a fund paying interest at an annual rate of 6% a year, compounded continuously. Assuming a continuous money flow, how much should Sam invest each week?

Solution Assume that Sam invests S dollars per year continuously. He has $67 - 20 = 47$ years to invest up to his retirement. We have $r = 0.06$ and $T = 47$, and the future value is $1,000,000. Substituting the known numbers into the future value formula we have

$$1{,}000{,}000 = \frac{S}{0.06}\left(e^{0.06(47)} - 1\right) = S(262.95).$$

Solving for S we obtain

$$S = 1{,}000{,}000/262.95 \approx 3{,}803.$$

Hence, Sam needs to save $3,803 per year, or $3,803/52 = \$73.14$ per week.

Often the present value of a continuous money flow is of interest. Such a model is used when a loan is involved.

Present Value of a Continuous Money Flow Suppose that a yearly amount of $R(t)$, where t is measured in years, is deposited continuously into an interest-bearing account at the annual interest rate of $100\,r\%$ compounded continuously, for T years. Then the accumulated present value of the account is

$$\int_0^T R(t)e^{-rt}\, dt.$$

When the yearly payment $R(t)$ is a constant D, then the integral can be computed directly:

$$\int_0^T D e^{-rt}\, dt = \frac{D}{r}\left[1 - e^{-rT}\right].$$

Example 4: Sam wants to purchase a car. The car dealer offers a 5-year loan that charges an interest at an annual rate of 12%, compounded continuously. Sam can pay $300 each month. Assuming a continuous money flow, how large a loan can Sam afford?

Solution We have $D = 300(12) = 3{,}600$, $r = 0.12$, and $T = 5$. Substituting into the accumulated present value formula we obtain

$$\frac{3{,}600}{0.12}\left(1 - e^{-0.12(5)}\right) = \$13{,}535.7.$$

Hence, Sam can apply for a car loan of $13,535.70 from the dealer.

When the time length of a continuous money flow becomes perpetual, the mathematics model involved turns to an **improper integral**.

> **Perpetual Present Value of a Continuous Money Flow** Suppose that a yearly amount of $R(t)$, where t is measured in years, is **perpetually** deposited continuously into an interest-bearing account at the annual interest rate of $100\,r\%$ compounded continuously. Then the perpetual accumulated present value of the account is
>
> $$\int_0^\infty R(t)e^{-rt}\,dt.$$
>
> If $R(t)$ is a constant $R(t) = D$, then
>
> $$\int_0^\infty De^{-rt}\,dt = \frac{D}{r}.$$

In general, any integral with one or both of its limits being infinity is called an improper integral. To evaluate such an integral, we evaluate it with finite limits, then let the involved limit approach infinity.

$$\int_a^\infty f(x)\,dx = \lim_{b\to\infty}\int_a^b f(x)\,dx;$$

$$\int_{-\infty}^b f(x)\,dx = \lim_{a\to-\infty}\int_a^b f(x)\,dx.$$

If the limit (a finite number) exists, we say that the integral is **convergent**, or the integral **converges**. If the limit does not exist, we say that the integral is **divergent**, or the integral **diverges**.

Finally, $\displaystyle\int_{-\infty}^\infty f(x)\,dx = \int_{-\infty}^c f(x)\,dx + \int_c^\infty f(x)\,dx$

where c can be any real number. For this integral to be convergent, both $\displaystyle\int_{-\infty}^c f(x)\,dx$ and $\displaystyle\int_c^\infty f(x)\,dx$ have to be convergent.

Example 5: For each of the following improper integrals, determine whether it is convergent or divergent. If it is convergent, determine its value.

(a) $\displaystyle\int_0^\infty e^{-0.2x}\,dx$ (b) $\displaystyle\int_0^\infty \ln x\,dx$ (c) $\displaystyle\int_2^\infty \frac{1}{x}\,dx$

Solution (a) We first evaluate the integral $\displaystyle\int_0^b e^{-0.2x}\,dx$:

$$\int_0^b e^{-0.2x}\,dx = \left[\frac{1}{-0.2}e^{-0.2x}\right]_0^b = -5\left[e^{-0.2b} - e^0\right] = 5 - 5e^{-0.2b}.$$

We then take the limit. Notice that when b becomes larger and larger, so does $e^{0.2b}$, hence $e^{-0.2b} = 1/e^{0.2b}$ approaches 0:

$$\lim_{b\to\infty}\int_0^b e^{-0.2x}\,dx = \lim_{b\to\infty}\left(5 - 5e^{-0.2b}\right) = 5.$$

Therefore, the improper integral $\displaystyle\int_0^\infty e^{-0.2x}\,dx$ is convergent. Its value is 5.

(b) Notice that when $x \to \infty$, we have $\ln x \to \infty$. Therefore $f(x) = \ln x \not\to 0$ as $x \to \infty$. Therefore, the improper integral $\displaystyle\int_0^\infty \ln x\,dx$ is divergent.

(c) We first evaluate the integral $\int_2^b \frac{1}{x}\,dx$:

$$\int_2^b \frac{1}{x}\,dx = [\ln x]_2^b = \ln b - \ln 2 = \ln b - \ln 2.$$

We then take the limit. Notice that when b becomes larger and larger, so does $\ln b$. Hence

$$\lim_{b \to \infty} \int_2^b \frac{1}{x}\,dx = \lim_{b \to \infty}(\ln b - \ln 2) = \infty.$$

Therefore, the improper integral $\int_2^\infty \frac{1}{x}\,dx$ is divergent.

Probability Distributions and Density Functions

Another important field involving applications of definite integrals is theoretical probability. The area under a curve can be used to measure the likelihood that certain event will occur, in situations that involve some degree of uncertainty. To use an integral for such events, we focus on the outcomes that can be described by numbers, and the numbers comprise an interval. Such a quantity is called a **continuous random variable**.

Suppose that x is a continuous random variable. A function $f(x)$ is called a **probability density function** if it satisfies the following:

1. $f(x) \geq 0$ for all x in the domain.

2. The area under the graph on the domain of f is 1.

The probability that x will be in an interval $[c, d]$ in the domain is then obtained by

$$P(c \leq x \leq d) = \int_c^d f(x)\,dx.$$

If $g(x)$ is a nonnegative function whose shape describes the probability distribution of a continuous random variable x, and the area A of the region formed by the graph and the x-axis on the interval is finite, then it can be used to form a probability density function $f(x) = k \cdot g(x)$, where $k = 1/A$.

Two important characteristics are often described with a continuous random variable. The first characteristic is the "central" value, the **mean** or the **expected value**, denoted by the Greek letter μ. When the graph of a probability distribution is symmetric, the mean μ is always the center. The second characteristic is how widely the variable values are scattered about the central value. One measurement of how closely the values of the distribution cluster about its mean is called the **standard deviation**, and is denoted by the Greek letter σ. If a random variable x has most of its values close to the mean, then the standard deviation is small. On the other hand, if it is likely that the values of a random variable x are widely scattered about the mean, then the standard deviation is large. The mean and standard deviation of a known density function are determined by the following formulas:

Let $f(x)$ be the probability density function of a continuous random variable x defined on an interval $[a, b]$. Then:

- The **mean**, μ, of x is $\mu = \int_a^b x f(x)\,dx$.

- The **standard deviation**, σ, of x is $\sigma = \sqrt{\int_a^b (x - \mu)^2 f(x)\,dx} = \sqrt{\int_a^b x^2 f(x)\,dx - \mu^2}$.

If the domain of f is an unbounded interval, then one or both limits of the integral can be infinity.

Example 6: A large school district purchased 1,000 projectors for classroom use. The manufacturer told the district manager that the life span of the projector light bulbs, t, can be described by the following probability density function: $f(t) = k\sqrt{t}$, $1 \leq t \leq 4$, where t is measured in thousands of hours.

(a) Determine the value of k.

(b) Determine the mean life span of the projector light bulbs.

(c) Determine the standard deviation of the life span of the projector light bulbs.

(d) Determine the probability that a projector light bulb lasts at least 3,000 hours.

(e) Determine the probability that a bulb projector light lasts at most 2,000 hours.

(f) The projectors are on for about 800 hours per year. In the third year after the purchase, how many projector light bulbs should the manager budget to purchase for replacing the burned-out ones?

Solution (a) Since $f(t)$ is a probability density function on $[1, 4]$, it must satisfy the condition $\int_1^4 f(t)\, dt = 1$. We have

$$1 = \int_1^4 f(t)\, dt = \int_1^4 k\sqrt{t}\, dt = k \left[\frac{2}{3}\sqrt{t^3}\right]_1^4 = k\frac{2}{3}\left[\sqrt{4^3} - \sqrt{1^3}\right] = k \cdot \frac{14}{3}.$$

Hence, $k = 3/14$, and

$$f(t) = \frac{3}{14}\sqrt{t}, \quad 1 \leq t \leq 4.$$

(b) By the definition of the mean, we have

$$\mu = \int_1^4 t f(t)\, dt = \int_1^4 t \frac{3}{14}\sqrt{t}\, dx$$

$$= \frac{3}{14}\int_1^4 t^{3/2}\, dt = \frac{3}{14}\left[\frac{2}{5}t^{5/2}\right]_1^4$$

$$= \frac{3}{35}\left[(4)^{5/2} - (1)^{5/2}\right] = \frac{3}{35}(31) \approx 2.657.$$

Hence, the mean life span of the projector light bulbs is 2,657 hours.

(c) To find the standard deviation, we first evaluate

$$\int_1^4 t^2 f(t)\, dt = \int_1^4 t^2 \frac{3}{14}\sqrt{t}\, dx$$

$$= \frac{3}{14}\int_1^4 t^{5/2}\, dt = \frac{3}{14}\left[\frac{2}{7}t^{7/2}\right]_1^4$$

$$= \frac{3}{49}\left[(4)^{7/2} - (1)^{7/2}\right] = \frac{3}{49}(127) \approx 7.7755.$$

We then use the definition to find σ:

$$\sigma = \sqrt{\int_1^4 t^2 f(t)\, dt - \mu^2} = \sqrt{7.7755 - (2.657)^2} \approx 0.846.$$

Hence, the standard deviation of the life spans of the projector light bulbs is about 846 hours.

CHAPTER 5 SUMMARY

(d) A projector light bulb lasts at least 3,000 hours means that $3 \leq t \leq 4$. Hence, the probability is

$$P(3 \leq t \leq 4) = \int_3^4 \frac{3}{14} \sqrt{t}\, dt = \frac{3}{14}\left[\frac{2}{3}\sqrt{t^3}\right]_3^4 = \frac{1}{7}\left[\sqrt{4^3} - \sqrt{3^3}\right] \approx 0.40.$$

Hence, the probability that a projector light bulb lasts at least 3,000 hours is about 40%.

(e) A projector light bulb lasts at most 2,000 hours means that $1 \leq t \leq 2$. Hence, the probability is

$$P(1 \leq t \leq 2) = \int_1^2 \frac{3}{14}\sqrt{t}\, dt = \frac{3}{14}\left[\frac{2}{3}\sqrt{t^3}\right]_1^2 = \frac{1}{7}\left[\sqrt{2^3} - \sqrt{1^3}\right] \approx 0.2612.$$

Hence, the probability that a projector light bulb lasts at most 9,000 hours is about 26.12%.

(f) From the beginning to the end of the third year after the purchase, the projector light bulbs have been used for 1,600 to 2,400 hours. We have

$$P(1.6 \leq t \leq 2.4) = \int_{1.6}^{2.4} \frac{3}{14}\sqrt{t}\, dt = \frac{3}{14}\left[\frac{2}{3}\sqrt{t^3}\right]_{1.6}^{2.4} = \frac{1}{7}[\sqrt{(2.4)^3} - \sqrt{(1.6)^3}] \approx 0.242.$$

Hence, about 24.2% of the 1,000 projector light bulbs are likely to be burned out. Therefore, the manager needs to purchase about 242 projector light bulbs for replacement.

There are a few probability distributions that are used most often. The simplest density function is one that assumes a constant value on a finite interval $[a, b]$ and zero elsewhere. This probability distribution is called the **uniform density function** and has the equation

$$f(x) = \frac{1}{b-a}, \quad a \leq x \leq b.$$

The **exponential density functions** are used to describe random variables that are more likely to take small values than large, like the waiting time in line at the post office, the duration of a phone call, the distance between successive cars on a freeway, etc. An exponential density function has the equation

$$f(x) = ke^{-kx}, \quad 0 \leq x.$$

The mean of an exponential probability distribution is $\mu = 1/k$. If on average an event occurs at a rate of k arrivals per unit, then the average gap between consecutive arrivals is $1/k$ unit.

The **normal distribution** is the most important distribution in statistics. The distribution is a bell-shaped curve. The **normal density function with mean μ and standard deviation σ** is given by

$$f(x) = \frac{1}{\sigma\sqrt{2\pi}}\, e^{-\frac{1}{2}\left(\frac{x-\mu}{\sigma}\right)^2}, \quad -\infty < x < \infty.$$

This rather complicated function is not used for calculating probabilities. Instead, any normal distribution can be transformed into the **standard normal distribution**, which has mean $\mu = 0$ and standard deviation $\sigma = 1$:

$$f(x) = \frac{1}{\sqrt{2\pi}}\, e^{-\frac{x^2}{2}}, \quad -\infty < x < \infty.$$

This function has an antiderivative. However, the antiderivative cannot be expressed by a basic algebraic formula. Numerical methods are applied to approximate the associated definite integrals, and the results are made into tables. Table 2 lists the numerical approximations of the definite integral

$$P(0 \leq x \leq z) = \int_0^z \frac{1}{\sqrt{2\pi}}\, e^{-\frac{x^2}{2}}\, dx \quad \text{for various values of } z.$$

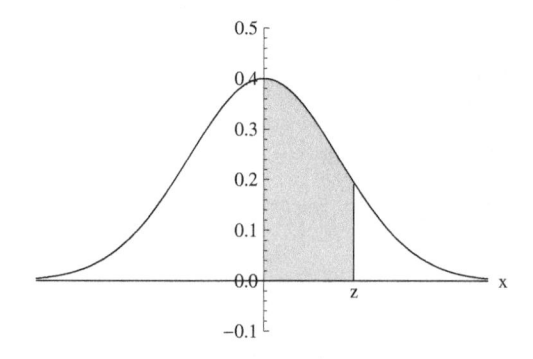

Many detailed examples in Section 5.5 illustrate how, for a continuous random variable x that has a standard normal distribution, Table 2 can be used to find the probabilities of x taking values in various types of intervals. Indeed, Table 2 can be used to find probabilities of any normal distribution with mean μ and standard deviation σ.

Suppose that a continuous random variable X follows a normal distribution with mean μ and standard deviation σ. If we let

$$x = \frac{X - \mu}{\sigma},$$

then x is a continuous random variable that follows the standard normal distribution. Therefore,

$$P(A \leq X \leq B) = P(a \leq x \leq b), \quad \text{where} \quad a = \frac{A - \mu}{\sigma} \quad \text{and} \quad b = \frac{B - \mu}{\sigma}.$$

Example 7: A large company has just moved into a new building that has 500 compact fluorescent light bulbs of a particular brand. The manufacturer told the building manager that the life span of the compact fluorescent light bulbs, t, follows a normal distribution with mean $\mu = 10{,}000$ hours and standard deviation $\sigma = 2{,}000$ hours.

(a) Determine the probability that a compact fluorescent light bulb lasts at least 12,000 hours.

(b) Determine the probability that a compact fluorescent light bulb lasts at most 9,000 hours.

Solution (a) A compact fluorescent light bulb lasts at least 12,000 hours means that $12{,}000 \leq t$. Hence the probability is $P(12{,}000 \leq t)$. Let $x = (t - 10{,}000)/2{,}000$, then x has standard normal distribution. Let

$$a = \frac{12{,}000 - 10{,}000}{2{,}000} = 1, \quad \text{then}$$

$$P(12{,}000 \leq t) = P(1 \leq x) = P(0 \leq x) - P(0 \leq x \leq 1)$$

by Table 2, $= 0.5 - 0.3413 = 0.1587$.

Hence, the probability that a compact fluorescent light bulb lasts at least 12,000 hours is about 15.9%.

(b) A compact fluorescent light bulb lasts at most 9,000 hours means that $t \leq 9{,}000$. Hence the probability is $P(t \leq 9{,}000)$. Let

$$b = \frac{9{,}000 - 10{,}000}{2{,}000} = -0.5, \quad \text{then}$$

$$P(t \leq 9{,}000) = P(x \leq -0.5) = P(0.5 \leq x) = P(0 \leq x) - P(0 \leq x \leq 0.5)$$

by Table 2, $= 0.5 - 0.1915 = 0.3085$.

Hence, the probability that a compact fluorescent light bulb lasts at most 9,000 hours is about 30.9%.

CHAPTER 5 SUMMARY

Differential Equations

The last topic of this chapter is **differential equations**. Many mathematical models involve an equation that has both an unknown function and its derivative. We have been dealing with differential equations all through the second half of the course. For example, if the velocity of a moving object is known as

$$v(t) = 2t + 1,$$

and we look for the position function $s(t)$, then we have the differential equation

$$s'(t) = v(t) = 2t + 1.$$

We solve this equation by integrating $v(t)$:

$$s(t) = \int v(t)\, dt = \int (2t+1)\, dt = t^2 + t + C.$$

This solution involves a constant C, and is called a **general solution**. If more information is given so that a specific value of C can be determined, then we have a **particular solution**. For example, if we also know that $s(0) = 3$, then

$$s(t) = t^2 + t + 3$$

is a particular solution.

One of the techniques for solving a certain type of differential equations is called **separation of variables**.

Example 8: Determine the general solutions for

$$y' = \frac{3x^2}{y},\quad y \neq 0,$$

then find the particular solution with condition $y(1) = 2$.

Solution We first replace y' by dy/dx, then separate the variables:

$$\frac{dy}{dx} = \frac{3x^2}{y}$$

$$y\frac{dy}{dx} = 3x^2$$

$$y\, dy = 3x^2\, dx.$$

Integrating both sides, we have

$$\int y\, dy = \int 3x^2\, dx$$

$$\frac{1}{2}y^2 = x^3 + C, \quad \text{hence}\quad y^2 = 2x^3 + 2C_1.$$

Let $C = 2C_1$, we have the general solutions

$$y = \sqrt{2x^3 + C} \quad \text{and}\quad y = -\sqrt{2x^3 + C}.$$

For the particular solution, we substitute $x = 1$ and $y = 2$ into the general solution. Since $y(1) > 0$, we take the positive branch, and solve for C:

$$2 = \sqrt{2(1)^3 + C}$$

$$4 = (2)^2 = 2 + C,$$

hence $C = 2$, and the particular solution is

$$y = \sqrt{2x^3 + 2}.$$

Chapter 5 Review Exercises

1. Refer to the graph below.

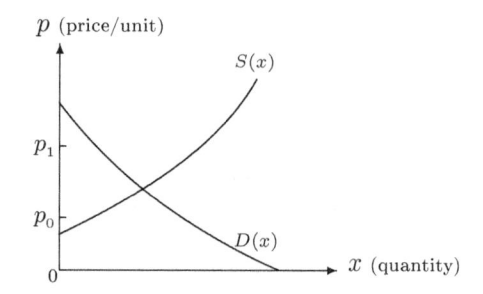

   p (price/unit)

   S(x), p_1, p_0, *D(x)*, *x* (quantity)

   (a) Shade the region that corresponds to the consumers' surplus at the market price p_0. Write an integral that can determine the consumers' surplus.

   (b) Shade the region that corresponds to the consumers' surplus at the market price p_1.

   (c) Shade the region that corresponds to the producers' surplus at the market price p_0. Write an integral that can determine the producers' surplus.

   (d) Shade the region that corresponds to the producers' surplus at the market price p_1.

   (e) Shade the region that corresponds to the consumers' expenditure at the market price p_1.

   (f) Shade the region that corresponds to the consumers' surplus at the equilibrium price.

   (g) Shade the region that corresponds to the producers' surplus at the equilibrium price.

   (h) At which market price is the total surpluses (i.e., the sum of the consumers' and the producers' surpluses) the largest?

2. Given the demand function $D(x) = (x-4)^2$ and the supply function $S(x) = x^2 + 2x + 6$:

   (a) Determine the equilibrium point (x_E, p_E);

   (b) Determine the consumers' surplus at the equilibrium point.

   (c) Determine the producers' surplus at the equilibrium point.

3. Suppose that Sam deposits $2,000 per year into an account that has a 5% annual interest rate compounded continuously. Assuming a continuous money flow, how much will the account be worth at the end of 20 years?

4. Suppose that Sara deposits $2,500 per year into an account that has a 5% annual interest rate compounded continuously. Assuming a continuous money flow, how long will it take for the account to be worth $80,000?

5. Jenny is 24. She wants to invest part of her salary in order to have $580,000 when she retires at 67. Suppose that she invests the money in a fund paying interest at an annual rate of 6%, compounded continuously. Assuming a continuous money flow, how much should Jenny invest each year?

6. Karen wants to purchase a car. The car dealer offers a 5-year loan that charges an annual interest rate of 11.5%, compounded continuously. Karen can pay $400 each month. Assuming a continuous money flow, how large a loan can Karen afford?

7. Marian has just purchased a house. She has a $220,000, 15-year mortgage with an annual interest rate of 4.5%, compounded continuously. Assume that Marian pays her mortgage continuously with a fixed amount. How much is her monthly payment?

8. Rick is 26 years old and plans to retire at 64. He wants to have a fund at 64 that will let him spend $3,000 a month after retirement, and last for 40 years. Assume a continuous money flow:

(a) Suppose that after his retirement Rick puts the money in a fund paying an annual interest rate of 4.5%, compounded continuously. How much will Rick need for his retirement?

(b) Suppose that Rick starts to invest a fixed amount each month from now until he retires, in a fund that pays an annual interest rate of 6.8%, compounded continuously. How much should he invest each month?

For Exercises #9 to #12: Determine whether the improper integral is convergent or divergent. If it is convergent, determine its value.

9. $\int_1^\infty \dfrac{2}{x^2}\, dx$ 10. $\int_2^\infty \dfrac{5}{x}\, dx$

11. $\int_0^\infty 3e^{-0.5x}\, dx$ 12. $\int_1^\infty x^{1.4}\, dx$

13. Find the area, if it is finite, of the region under the graph of $y = 3/x^4$ on the interval $[2, \infty)$.

14. Rick wants to have a fund at retirement that will let him perpetually spend $3,000 a month after retirement (assume a continuous money flow). Suppose that after his retirement Rick puts the money in a fund paying an annual interest rate of 4.5%, compounded continuously. How much will Rick need for his retirement?

15. Find k such that the function $f(x) = \dfrac{k}{x^2}$ is a probability density function on interval $[2, 4]$.

16. Suppose that Kathy rides the school bus every weekday morning. The bus has an equal chance of arriving at the corner of her street between 7:20 and 7:30. Suppose that Kathy always arrives at the corner of her street at 7:20. Let t be her waiting time in minutes. Then the probability density function of t is given by

$$f(t) = \begin{cases} \frac{1}{10} & \text{if } 0 \le t \le 10, \\ 0 & \text{if } t < 0 \text{ or } t > 10. \end{cases}$$

(a) Determine the mean waiting time.

(b) Determine the standard deviation of the waiting times.

(c) Determine the probability that Kathy needs to wait for at least 6 minutes.

(d) Determine the probability that Kathy needs to wait for at most 8 minutes.

17. The distribution of the time between successive arrivals at the drive-through window of a fast-food restaurant around noon on a weekday can be approximated by an exponential density function. Suppose that seven vehicles arrive at the drive-through window every 2 minutes.

(a) Determine the exponential density function.

(b) Determine the probability that the time between successive arrivals will be no more than half a minute.

(c) Determine the probability that there is no arrival for at least 1 minute.

For Exercises #18 to #22: Assume that x is a continuous random variable with a standard normal distribution. Use Table 2 to find each probability.

18. $P(0 \le x \le 2.13)$ 19. $P(-1.58 \le x \le 0)$
20. $P(-1.46 \le x)$ 21. $P(-2.3 \le x \le -0.4)$
22. $P(1.25 \le x)$

23. A food company estimates that w, the weight of a package of one type of frozen dinner, marked "Net Weight 20 oz," is normally distributed with a mean of 20 oz and a standard deviation of 0.25 oz.

(a) Determine the probability that a package weighs less than 19.5 oz.

(b) Determine the weight that will be exceeded by 2.3% of the packages.

24. A building changes all its 600 lights to compact fluorescent light bulbs. The life span of the compact fluorescent light bulbs, t, follows a normal distribution with mean $\mu = 11{,}000$ hours and standard deviation $\sigma = 2{,}500$ hours.

(a) Determine the probability that a bulb lasts at least 12,000 hours.

(b) Determine the probability that a bulb lasts at most 9,000 hours.

(c) The building is on automatic energy control so the lights are on for 3,000 hours per year. In the third year after the conversion, how many compact fluorescent light bulbs need to be replaced?

For Exercises #25 to #30: Solve each differential equation.

25. $\dfrac{dy}{dx} = 3x^2 y$

26. $\dfrac{dy}{dx} = \dfrac{x}{y}$; $y(1) = 2$

27. $2x^2 y \dfrac{dy}{dx} = 1$

28. $\dfrac{dy}{dx} = \dfrac{3}{y^2}$; $y(0) = 1$

29. $y' = 4x + xy$

30. $\dfrac{dP}{dt} = 0.02P$; $P(0) = 100$

31. Determine the demand function $x = D(p)$ that satisfies $E(p) = 1$ for all $p > 0$, and $D(2) = 3$.

32. Karen deposits \$10,000 in a bank account that earns interest at an annual rate of 4% compounded continuously. She withdraws from the account continuously at a rate of \$800 a year. Therefore the balance, P, in the account satisfies the differential equation $\dfrac{dP}{dt} = 0.04P - 800$. Determine the solution of this differential equation.

Chapter 5 Test (150 points)

SCORE _____

NAME _____

ANSWERS

1. (15pts) Given $D(x) = (x-5)^2$ and $S(x) = x^2 + x + 3$, find:

(a) the equilibrium point; 1. (a) _____

(b) the consumers' surplus at the equilibrium point; (b) _____

(c) the producers' surplus at the equilibrium point. (c) _____

2. (10pts) Jim plans to retire 15 years from now and he wants to have $300,000 in his account that pays interest at an annual rate of 6%, compounded continuously. How much per year should he save continuously?

Show your work (write the formula you are using):

2. _____

3. (10pts) John needs $4500 per year for expenses for the next 10 years. He has an account of P_0 that pays interest at an annual rate of 5%, compounded continuously. Assuming continuous money flow, how much is P_0 for the money to last for 10 years?

Show your work (write the formula you are using):

3. _____

4. (10pts) Mark has a loan of $250,000 for 20 years at an annual interest rate of 5.5%, compounded continuously. Assuming continuous money flow, how much is Mark's monthly payment?

   Show your work (write the formula you are using):

   4. _____

For problems 5 and 6: Determine whether each of the improper integrals is convergent or divergent, and calculate its value if it is convergent.

5. (8pts) $\displaystyle\int_1^\infty e^{-3x}\,dx$ 5. _____

6. (8pts) $\displaystyle\int_2^\infty \frac{3}{x}\,dx$ 6. _____

7. (10pts) Find k such that $f(x) = k\dfrac{1}{x^2}$ is a probability density function over the interval $[1,2]$.   7. _____

For problems 8 and 9: Given the probability density function $f(x) = 2x/9$ over $[0,3]$, find:

8. (6pt) The mean μ.   8. _____

9. (6pts) The standard deviation σ.   9. _____

CHAPTER 5 TEST

10. (12pts) The distribution of the time between successive arrivals at the post office around noon on a weekday can be approximated by an exponential density function. Suppose that three people arrive at the office every minute.

(a) Determine the exponential density function.

10. (a) _____

(b) Determine the probability that the time between successive arrivals will be no more than half a minute.

(b) _____

(c) Determine the probability that there is no arrival for at least 1 minute.

(c) _____

For problems 11 and 12: Let x be a continuous random variable with standard normal density. Using Table 2, find:

11. (7pts) $P(0.44 \leq x \leq 1.47)$

11. _____

12. (7pts) $P(-1.69 \leq x)$

12. _____

13. (11pts) The weekly profits of a small business are normally distributed with mean $200 and standard deviation $40. Find the probability that the weekly profit will be $220 or more. Show you work.

13. _____

For problems 14 and 15, solve the differential equations.

14. (10pts) $y' = \dfrac{x}{6y}$

14. _____

15. (10pts) $y' = 4x - xy$, $y = 2$ when $x = 0$.

15. _____

16. (10pts) Karen deposits $6,000 in a bank account that earns interest at an annual rate of 3% compounded continuously. She withdraws from the account continuously at a rate of $500 a year. Therefore the balance, P, in the account satisfies the differential equation

$$\frac{dP}{dt} = 0.03P - 500.$$

Determine the solution of this differential equation.

Show your work:

16. _____

Chapter 5 Projects

Project 1

Three years have passed since Mara graduated from college. She has purchased her dream condominium and received a promotion. Now it is time for her to plan her retirement!

Mara plans to retire when she is 67 years old (she is 25 now). She predicts that the annual interest rate of her account will be at least 4.5%, compounded monthly (she has already considered inflation). Mara has two questions:

(a) How much will she need when she retires at 67 so that she can live on the savings?

(b) How much does she need to save every month from now so that she can reach her target amount by 67?

For question (a), Mara thinks that to live comfortably after retirement she needs $2,600 per month, up to age 94. She also wants to have some balance left by age 94 (in case she continues to be alive!), say $80,000. She also wants to calculate the target amount if she withdraws $3,500 per month.

Instructions:

A. Discrete model (a) Use spreadsheets to help Mara compute the target amount she needs to save by age 67. (Hint: Each month's balance is the balance from the previous month, plus the interest earned in the current month, and minus the withdrawal of the current month.) The initial balance is unknown. Use trial and test (accurate up to $50 at the last cell) to find the initial balance so that the ending balance (when Mara is 94) is $80,000. Print the first 10 and the last 10 months' data only. There are two charts: $2,600 per month withdrawals and $3,500 per month withdrawals.

(b) After finding the target amount Mara needs for her retirement, use Excel to help Mara to find the monthly deposit needed to reach her goal. (Hint: Each month's balance is the balance from the previous month plus the interest earned and the deposit of the current month.) The deposit is unknown. Use trial and test (accurate up to $50 at the last cell) to find the deposit so that the ending balance equals the target amount she needs to save by age 67. Print the first 10 and the last 10 months' data only. There are two charts for the two targets.

B. Continuous model Now repeat the computing in A(a) and A(b) using the formulas in the book, assuming the same interest rate (4.5%) but compounding continuously. The computing should be independent from the discrete model. There are four output results (two targets and two monthly savings).

C. Compare the results in parts A and B and comment.

Project 2

Mara becomes a good friend to her neighbor Sara. After hearing Mara's saving plans, Sara said: "I wish that I were as lucky as you! Instead of a savings account, I have $4,600 debt on my credit card with an annual interest rate of 18%!"

Sara wants to know how long it will take her to pay off her card. The minimum payment each month is $25 or 3% of the balance, whichever is higher, or the total balance if it is below $25. The interest is compounded monthly. Sara wants to know if she pays the minimum payment only, how long will it take her to pay off her card and how much the total interest is. What if she pays at least $50 per month or at least $100 per month? (Note: She still needs to satisfy the company's minimum payment.)

Another credit card company is offering Sara a new card with an annual interest rate of 16%, also compounded monthly, with the same minimum payment policy. It tells Sara that if she transfers all her balance to the new card, she can save a lot of money. Sara wants to know if she switches, how long it will take her to pay off her card and how much the total interest is under each payment strategy (minimum, at least $50, at least $100), and how much she can save by switching, under the three payment strategies.

Instructions: Use spreadsheets to show Sara the balance of her credit card account under each payment strategy (there are six of them!). Then answer Sara's questions (she has EIGHTEEN questions!). Your report must include formulas (with explanation), spreadsheets, and graphs (for comparison of the three payment strategies at the same company, and the same payment strategy at the two companies). The spreadsheets for each payment strategy should only include the first 10 months, the middle 10 months showing the switch, and the last 10 months.

Hint: Each month's balance is the balance from the previous month, plus the interest earned in the current month, and minus the payment of the current month. The monthly payments are either 3% of the previous month's balance or $25, whichever is larger.

Chapter 6

Functions of Two Variables

Introduction

In this chapter, we extend our differential calculus knowledge to functions that involve two independent variables. We continue to see how the concept of derivatives links to instantaneous rate of change, as well as the slope of a tangent line at a specific point. We illustrate how the derivatives help us to solve max-min problems that involve two independent variables.

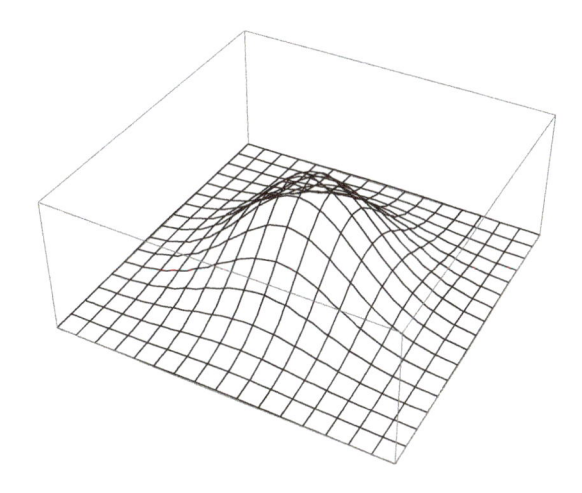

Study Objectives

1. Use partial derivatives to interpret a function behavior near a given point.

2. Use partial derivatives to determine critical point(s) of a function that involve two independent variables.

3. Apply the D-Test to determine relative extrema of a function.

4. Apply the aforementioned skills to solve business related problems, such as maximizing profit, minimizing costs, etc.

6.1 Functions of Two Variables and Partial Derivatives

Sandra rents a car from a car rental company. It costs $30 a day plus 20 cents a mile. So the cost depends on two variables: the number of days, x, and the miles she drives, y. We can write the cost function as

$$C = f(x, y) = 30x + 0.2y.$$

If Sandra rents the car for 6 days and drives a total of 400 miles, then her cost is

$$C = f(6, 400) = 30(6) + 0.2(400) = 180 + 80 = \$260.$$

This example introduces a function that depends on two inputs. In general, a function of two variables, usually denoted as $z = f(x, y)$, can be defined as follows:

Definition *Functions of Two Variables*

A function is a **function of two variables** if its domain is a subset of the xy-plane.

Example 1: For

$$z = f(x, y) = 3x^2 y^5 + 2x^3 y^2 + e^{xy},$$

find $f(0, 1)$, $f(2, 1)$, and $f(1, 1)$.

Solution We have

$$f(0, 1) = 3(0)^2(1)^5 + 2(0)^3(1)^2 + e^{(0)(1)} = 0 + 0 + e^0 = 1;$$

$$f(2, 1) = 3(2)^2(1)^5 + 2(2)^3(1)^2 + e^{(2)(1)} = 12 + 16 + e^2 = 28 + e^2;$$

$$f(1, 1) = 3(1)^2(1)^5 + 2(1)^3(1)^2 + e^{(1)(1)} = 3 + 2 + e^1 = 5 + e.$$

The graph of a function of two variables, $z = f(x, y)$, consists of ordered triples (a, b, c) where $c = f(a, b)$. It is less practical to draw the graph of a function of two variables, even when the function is very simple, like $z = 2x + 3y$. There are many graphics programs for generating graphs of functions of two variables. A few of such graphs generated by *Mathematica* are presented in Figures 6.1.1–6.1.4.

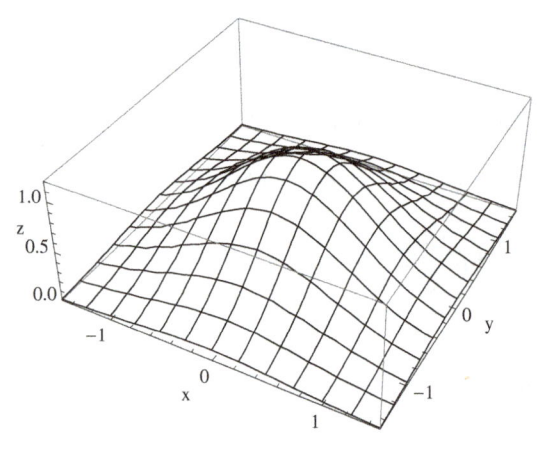

Figure 6.1.1: $z = e^{-(x^2 + y^2)}$

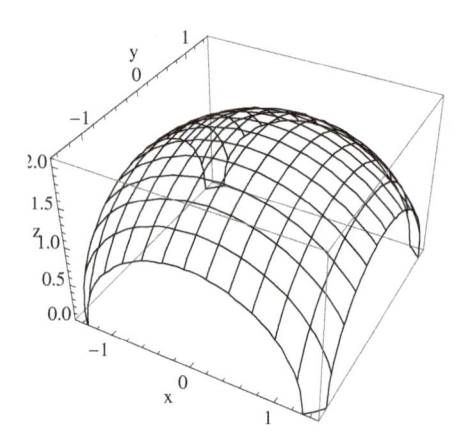

Figure 6.1.2: $z = \sqrt{4 - x^2 - y^2}$

Figure 6.1.3: $z = x^2 + y^2$

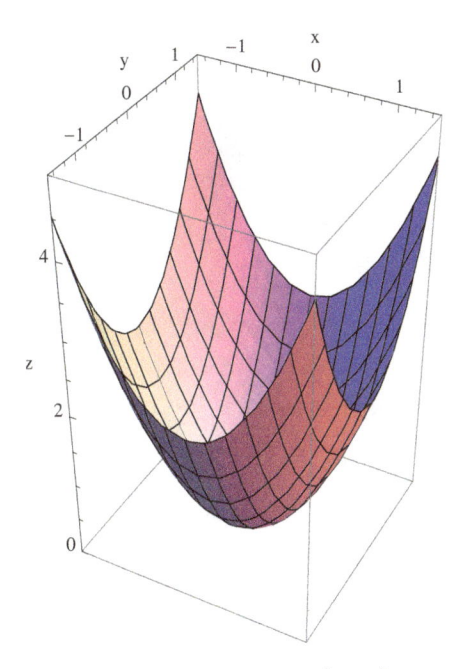

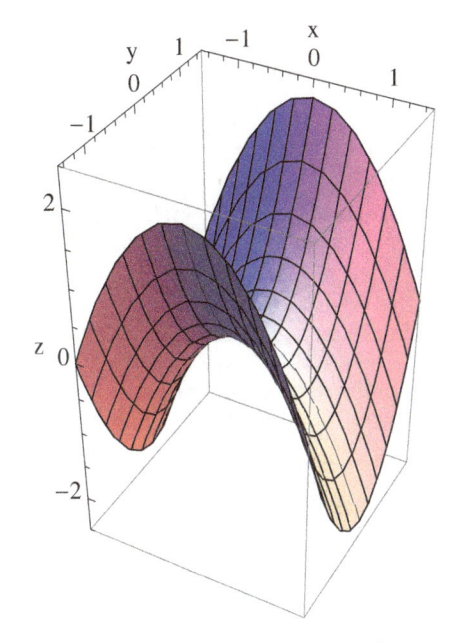

Figure 6.1.4: $z = y^2 - x^2$

Partial Derivatives

We saw in Chapters 2 and 3 how derivatives can reveal important information about a function. Recall that the first derivative of a function measures the rate of change. Now we want to explore the rate of change of a function that depends on two inputs, $z = f(x, y)$. We focus on one of the inputs each time. That is, we study the behavior of z when only x (or y) takes small changes, while the other variable stays the same value. Therefore, there are *two* types of derivatives: the rate of change of z with respect to x, and the rate of change of z with respect to y.

Let us take a look at Sandra's cost function of renting a car,

$$C = f(x, y) = 30x + 0.2y.$$

Suppose that Sandra plans to rent the car for 5 days. Then the cost function becomes

$$C = f(5, y) = 30(5) + 0.2y = 150 + 0.2y = h(y).$$

This is a function of y. The derivative $h'(y) = \$0.2$/mile represents the rate of change of the cost with respect to the change in mileage. At the other hand, suppose that Sandra needs to drive 1,000 miles for the trip. Then the cost function becomes

$$C = f(x, 1{,}000) = 30x + 0.2(1{,}000) = 30x + 200 = g(x).$$

This is a function of x. The derivative $g'(x) = \$30$/day represents the rate of change of the cost with respect to the change in number of days of rental.

The two types of derivatives are called **partial derivatives**, and as in the cases of derivatives, there are several commonly used notations for partial derivatives. We use one of them to define these partial derivatives.

> **Definition — Partial Derivatives**
>
> Let $z = f(x, y)$ be a function of two variables. The **partial derivative of f with respect to x** is
> $$f_x(x, y) = \lim_{h \to 0} \frac{f(x + h, y) - f(x, y)}{h}, \quad \text{if the limit exists.}$$
>
> The **partial derivative of f with respect to y** is
> $$f_y(x, y) = \lim_{k \to 0} \frac{f(x, y + k) - f(x, y)}{k}, \quad \text{if the limit exists.}$$

As with the ordinary derivative, there are several possible alternative notations for the partial derivative. In practice, we use the notation that is more convenient for the given situation.

> **Notation for partial derivatives**
>
> For $z = f(x, y)$, its partial derivative with respect to x can also be denoted as
> $$\frac{\partial z}{\partial x}, \quad \text{or} \quad \frac{\partial f}{\partial x}.$$
>
> This partial derivative is found by treating y as a constant and performing the differentiation techniques introduced in Chapter 2. Similarly, the partial derivative of $z = f(x, y)$ with respect to y can also be denoted as
> $$\frac{\partial z}{\partial y}, \quad \text{or} \quad \frac{\partial f}{\partial y}.$$
>
> This partial derivative is found by treating x as a constant and performing the differentiation techniques introduced in Chapter 2.
>
> The partial derivative at a point (a, b) is denote by appending $\Big|_{(a,b)}$ after the partial derivative.
>
> If $f(x, y) = 3x^2 y$, then we can denote its partial derivative with respect to x by
> $$f_x(x, y) = \frac{\partial f}{\partial x} = \frac{\partial z}{\partial x} = \frac{\partial}{\partial x}(3x^2 y) = 6xy.$$
>
> Similarly, the partial derivative of f with respect to y can be denoted by
> $$f_y(x, y) = \frac{\partial f}{\partial y} = \frac{\partial}{\partial y}(3x^2 y) = 3x^2.$$
>
> The value of the partial derivative with respect to x at $x = 3$ and $y = 2$ can then be denoted as either
> $$f_x(3, 2) \quad \text{or} \quad \frac{\partial f}{\partial x}\bigg|_{(3,2)}.$$
>
> Hence if $f_x(x, y) = \dfrac{\partial f}{\partial x} = 6xy$, then
> $$f_x(3, 2) = \frac{\partial f}{\partial x}\bigg|_{(3,2)} = 6(3)(2) = 36.$$
>
> Similar notations hold for the partial derivative with respect to y.

6.1. FUNCTIONS OF TWO VARIABLES AND PARTIAL DERIVATIVES

> **Warning** *Illegal notation for partial derivatives*
>
> Recall that, for functions of a single variable such as $f(x)$, we used the notation f' to denote the derivative of f. This *prime* notation does not make sense anymore (since the prime does not indicate which variable with respect to which we would take the derivative) and should not be used with partial derivatives. Writing $f'(x,y)$ does not mean anything.

Example 2: For $z = x^2y + 3x - 4y^3$, find $\dfrac{\partial z}{\partial x}$ and $\dfrac{\partial z}{\partial y}$.

Solution We have

$$\frac{\partial z}{\partial x} = \frac{\partial}{\partial x}\left(x^2y + 3x - 4y^3\right) = \frac{\partial}{\partial x}\left(x^2y\right) + \frac{\partial}{\partial x}(3x) - \frac{\partial}{\partial x}\left(4y^3\right) = 2xy + 3 - 0 = 2xy + 3.$$

$$\frac{\partial z}{\partial y} = \frac{\partial}{\partial y}\left(x^2y + 3x - 4y^3\right) = \frac{\partial}{\partial y}\left(x^2y\right) + \frac{\partial}{\partial y}(3x) - \frac{\partial}{\partial y}\left(4y^3\right) = x^2 + 0 - 12y^2 = x^2 - 12y^2.$$

Example 3: For $z = f(x,y) = 3x^2y^5 + 2x^3y^2 + e^{xy}$, find $f_x(0,1)$, and $f_y(1,1)$.

Solution We find the partial derivatives first:

$$f_x(x,y) = 6xy^5 + 6x^2y^2 + ye^{xy},$$

$$f_y(x,y) = 15x^2y^4 + 4x^3y + xe^{xy}.$$

Hence

$$f_x(0,1) = 6(0)(1)^5 + 6(0)^2(1)^2 + (1k)e^{(0)(1)} = 0 + 0 + e^0 = 1,$$

$$f_y(1,1) = 15(1)^2(1)^4 + 4(1)^3(1) + (1)e^{(1)(1)} = 15 + 4 + e^1 = 19 + e.$$

━ Geometric Meaning of Partial Derivatives ━

Recall that if $y = f(x)$, then the derivative at $x = a$, $f'(a)$, represents the slope of the tangent line that passes the point $(a, f(a))$. In the case of $z = f(x,y)$, what do the partial derivatives at a point (a, b), $f_x(a,b)$ and $f_y(a,b)$, represent? We illustrate the geometric meaning of the partial derivatives via an example.
Consider the function $z = f(x,y) = 4 - x^2 - y^2$. We have

$$f_x(x,y) = -2x \quad \text{and} \quad f_y(x,y) = -2y.$$

At the point $(1, -0.5)$, we have

$$f_x(1,-0.5) = -2(1) = -2 \quad \text{and} \quad f_y(1,-0.5) = -2(-0.5) = 1.$$

Now consider the function with its y-value fixed at $y = -0.5$. Then we have

$$z = f(x, -0.5) = 4 - x^2 - (-0.5)^2 = 4 - x^2 - 0.25 = 3.75 - x^2.$$

This is a function of x only. The corresponding graph is the parabolic curve on the vertical plane $y = -0.5$ in the 3D graph. At the point $x = 1$, the curve has the slope of $f_x(1, -0.5) = -2$.
Similarly, consider the function with its x-value fixed at $x = 1$. Then we have

$$z = f(1, y) = 4 - (1)^2 - y^2 = 4 - 1 - y^2 = 3 - y^2.$$

This is a function of y only. The corresponding graph is the parabolic curve on the vertical plane $x = 1$ in the 3D graph. At the point $y = -0.5$, the curve has the slope of $f_y(1, -0.5) = 1$.

In general, for a function $z = f(x, y)$, the partial derivative f_x at a point (a, b) represents the slope of the tangent line that passes the point $(a, b, f(a, b))$ in the 3D surface and stays in the vertical plane $y = b$ (Figure 6.1.5).

Figure 6.1.5: Graph of $z = f(x, b)$

Similarly, the partial derivative f_y at a point (a, b) represents the slope of the tangent line that passes the point $(a, b, f(a, b))$ in the 3D surface and stays in the vertical plane $x = a$ (Figure 6.1.6).

Figure 6.1.6: Graph of $z = f(a, y)$

— **Applications** —

Example 4: The formula for the volume of a cylinder is given by $V(r, h) = \pi r^2 h$, where r is the radius and h is the height, both in inches.

6.1. FUNCTIONS OF TWO VARIABLES AND PARTIAL DERIVATIVES

(a) Determine $V_r(2,5)$ and interpret.

(b) Determine $V_h(2,5)$ and interpret.

Solution

(a) We have $V_r(r,h) = 2\pi rh$. Hence $V_r(2,5) = 2\pi(2)(5) = 20\pi$. This means that when $r = 2$ in and $h = 5$ inch, if the radius increases by 1 inch and the height remains constant at 5 inch, then the volume of the cylinder increases by approximately 20π in^3.

(b) We have $V_h(r,h) = \pi r^2$. Hence $V_h(2,5) = \pi(2)^2 = 4\pi$. This means that when $r = 2$ inch and $h = 5$ inch, if the height increases by 1 inch and the radius remains constant at 2 inch, then the volume of the cylinder increases by 4π in^3.

Example 5: A company sells two models of cellular phones: the basic model and the luxury model. The demand functions for the cellular phones are given by

$$p_b = 160 - 0.15x - 0.05y \quad \text{and} \quad p_l = 220 - 0.15x - 0.15y,$$

where p_b and p_l are the prices of the two models in dollars, x is the units of the basic cellular phones sold per week, and y is the units of the luxury cellular phones sold per week.

(a) Determine the total revenue function $R(x,y)$.

(b) Determine $R_x(200, 400)$ and interpret.

Solution

(a) Since revenue is price times quantity, we have

$$R(x,y) = p_b \cdot x + p_l \cdot y$$
$$= (160 - 0.15x - 0.05y)x + (220 - 0.15x - 0.15y)y$$
$$= 160x - 0.15x^2 - 0.05yx + 220y - 0.15xy - 0.15y^2$$
$$= 160x + 220y - 0.2xy - 0.15x^2 - 0.15y^2.$$

(b) We have $R_x(x,y) = 160 - 0.2y - 0.3x$. Hence

$$R_x(200, 400) = 160 - 0.2(400) - 0.3(200) = 20.$$

This means that when 200 basic models and 400 luxury models of the cellular phones are sold per week, if the sales of the basic model increase by 1 unit per week, and the sales of the luxury model remain constant at 400 units per week, then the total revenue increases by approximately $20 per week.

Example 6: The **Cobb–Douglas production function** states that

$$Q = f(x,y) = Cx^k y^{1-k}, \ 0 < k < 1,$$

where Q is the number of units produced with x units of labor and y units of capital. The partial derivatives

$$f_x(x,y) \quad \text{and} \quad f_y(x,y)$$

are called the **marginal productivity of labor** and the **marginal productivity of capital** respectively. A company that produces compact fluorescent light bulbs has the Cobb–Douglas production function

$$Q = f(x,y) = 15x^{0.6}y^{0.4},$$

where Q is the number of compact fluorescent light bulbs produced per day, x is the number of labor hours per day, and y is the value of the equipment in units of thousands of dollars.

(a) Determine $f(200, 100)$ and interpret.

(b) Determine $f_x(200, 100)$ and interpret.

(c) Determine $f_y(200, 100)$ and interpret.

Solution

(a) We have
$$f(200, 100) = 15(200)^{0.6}(100)^{0.4} = 2{,}273.57.$$

That is, when the company has a labor force of 200 labor hours per day and has \$100,000 worth of equipment, the production is about 2,274 compact fluorescent light bulbs per day.

(b) We have
$$f_x(x, y) = 15(0.6)x^{-0.4}y^{0.4} = 9\left(\frac{y}{x}\right)^{0.4}.$$

Hence $f_x(200, 100) = 9\left(\dfrac{100}{200}\right)^{0.4} \approx 6.82$. This quantity means that when the company has a labor force of 200 labor hours per day and has \$100,000 worth of equipment, if the labor hours increase by 1 hour per day to 201, and the value of equipment remains \$100,000, then the production increases by approximately seven compact fluorescent light bulbs per day.

(c) We have
$$f_y(x, y) = 15(0.4)x^{0.6}y^{-0.6} = 6\left(\frac{x}{y}\right)^{0.6}.$$

Hence $f_y(200, 100) = 6\left(\dfrac{200}{100}\right)^{0.6} \approx 9.09$. This quantity means that when the company has a labor force of 200 labor hours per day and has \$100,000 worth of equipment, if the value of the equipment increases by one thousand dollars to \$101,000, and the labor hours remain 200 per day, then the production increases by approximately nine compact fluorescent light bulbs per day.

Second–Order Partial Derivatives

Consider the function $z = x^2y + 3x - 4y^3$ in Example 2. The partial derivatives are
$$\frac{\partial z}{\partial x} = 2xy + 3 \quad \text{and} \quad \frac{\partial z}{\partial y} = x^2 - 12y^2.$$

Each of the partial derivatives is again a function of x and y. Therefore, we can find their partial derivatives:

$$\frac{\partial}{\partial x}\left(\frac{\partial z}{\partial x}\right) = \frac{\partial}{\partial x}(2xy + 3) = 2y, \quad \frac{\partial}{\partial y}\left(\frac{\partial z}{\partial x}\right) = \frac{\partial}{\partial y}(2xy + 3) = 2x;$$

$$\frac{\partial}{\partial x}\left(\frac{\partial z}{\partial y}\right) = \frac{\partial}{\partial x}(x^2 - 12y^2) = 2x, \quad \frac{\partial}{\partial y}\left(\frac{\partial z}{\partial y}\right) = \frac{\partial}{\partial y}(x^2 - 12y^2) = -24y.$$

They are called **second-order partial derivatives**. There are four second-order partial derivatives of a function of two variables $z = f(x, y)$. The notations are summarized in the following definition.

Definition — Second–Order Partial Derivatives

If $f(x, y)$ is a function of two variables, then

$$f_{xx} = \frac{\partial^2 f}{\partial x \partial x} = \frac{\partial^2 z}{\partial x \partial x} = \frac{\partial^2 f}{\partial x^2} = \frac{\partial^2 z}{\partial x^2} = \frac{\partial}{\partial x}\left(\frac{\partial z}{\partial x}\right)$$

$$f_{xy} = \frac{\partial^2 f}{\partial y \partial x} = \frac{\partial^2 z}{\partial y \partial x} = \frac{\partial}{\partial y}\left(\frac{\partial z}{\partial x}\right)$$

$$f_{yx} = \frac{\partial^2 f}{\partial x \partial y} = \frac{\partial^2 z}{\partial x \partial y} = \frac{\partial}{\partial x}\left(\frac{\partial z}{\partial y}\right)$$

$$f_{yy} = \frac{\partial^2 f}{\partial y \partial y} = \frac{\partial^2 z}{\partial y \partial y} = \frac{\partial^2 f}{\partial y^2} = \frac{\partial^2 z}{\partial y^2} = \frac{\partial}{\partial y}\left(\frac{\partial z}{\partial y}\right)$$

Example 7: For $f(x, y) = 3x^2 y^5 + 2x^3 y^2 + e^{xy}$, find the four second-order partial derivatives.

Solution We include the most often used notations:

(a) $f_{xx} = \dfrac{\partial^2 f}{\partial x^2} = \dfrac{\partial}{\partial x}\left(6xy^5 + 6x^2 y^2 + y e^{xy}\right)$

$= 6y^5 + 12xy^2 + y^2 e^{xy}$,

(b) $f_{xy} = \dfrac{\partial^2 f}{\partial y \partial x} = \dfrac{\partial}{\partial y}\left(6xy^5 + 6x^2 y^2 + y e^{xy}\right)$

$= 30xy^4 + 12x^2 y + e^{xy} + xy e^{xy}$,

(c) $f_{yx} = \dfrac{\partial^2 f}{\partial x \partial y} = \dfrac{\partial}{\partial x}\left(15x^2 y^4 + 4x^3 y + x e^{xy}\right)$

$= 30xy^4 + 12x^2 y + e^{xy} + xy e^{xy}$,

(d) $f_{yy} = \dfrac{\partial^2 f}{\partial y^2} = \dfrac{\partial}{\partial y}\left(15x^2 y^4 + 4x^3 y + x e^{xy}\right)$

$= 60x^2 y^3 + 4x^3 + x^2 e^{xy}$.

Notice that in Example 7, parts (b) and (c) have the same results. That is,

$$\frac{\partial^2 f}{\partial y \partial x} = \frac{\partial^2 f}{\partial x \partial y}, \quad \text{or} \quad f_{xy} = f_{yx}.$$

This is true for most of the functions we encounter, but not all functions. We state the following theorem somewhat imprecisely since, in our situations, the hypotheses will always hold.

Theorem 1 — *Clairaut's Theorem*

Let $z = f(x, y)$ be a function. If the partial derivatives f_{xy} and f_{yx} are continuous, then $f_{xy} = f_{yx}$.

As a consequence of this theorem, we will often refer to the *five* (rather than six) first and second partial derivatives of $f(x, y)$.

Exercise Set 6.1

For Exercises 1–12, determine $f_x(x,y)$ and $f_y(x,y)$.

1. $f(x,y) = 2x^7 y^3$
2. $f(x,y) = -4x^5 y^2$
3. $f(x,y) = 3\sqrt{xy}$
4. $f(x,y) = 5\sqrt[3]{xy}$
5. $f(x,y) = 2xe^{y^2}$
6. $f(x,y) = 2x^2 e^{3y}$
7. $f(x,y) = \ln(x^2 + y^2)$
8. $f(x,y) = \ln(x^2 y^2)$
9. $f(x,y) = \dfrac{2}{yx^2}$
10. $f(x,y) = \dfrac{-3}{x^4 y^2}$
11. $f(x,y) = \dfrac{1}{x+2y}$
12. $f(x,y) = \dfrac{1}{2x-y}$

For Exercises 13–24, determine $\dfrac{\partial z}{\partial x}$ and $\dfrac{\partial z}{\partial y}$.

13. $z = x^2 y + 3x^4 y^2$
14. $z = x^3 y^4 + 2xy^3$
15. $z = e^{2xy^2}$
16. $z = e^{2x^2 y}$
17. $z = e^x \ln(xy)$
18. $z = x^2 \ln(xy)$
19. $z = \dfrac{x}{y}$
20. $z = \dfrac{y}{x}$
21. $z = 2x^3 e^y - 10x^2 y + 6e^x - 3y^2$
22. $z = 5x^4 - 3x \ln y - 7ye^x + 8y^4$
23. $z = 3\sqrt{xy} - 3ye^x + \dfrac{y^2}{x^2}$
24. $z = 2\sqrt[3]{xy} - 5xe^y + \dfrac{x^2}{y^3}$

For Exercises 25–30, determine $f_x(1,2)$ and $f_y(1,2)$.

25. $f(x,y) = 2x^4 + x^3 y^2 - 4y^2 + y$
26. $f(x,y) = 3x^4 - 2y^3 + 5x^2 - 2xy$
27. $f(x,y) = \ln(y^2 + x^2)$
28. $f(x,y) = \ln(y^3 - x^2)$
29. $f(x,y) = e^{x^2 y}$
30. $f(x,y) = e^{x^2 + y}$

For Exercises 31–46, determine $f_{xx}(x,y)$, $f_{xy}(x,y)$ and $f_{yy}(x,y)$.

31. $f(x,y) = 2x^4 + x^3 y^2 - 4y^2 + y$
32. $f(x,y) = 3x^4 - 2y^3 + 5x^2 - 2xy$
33. $f(x,y) = 2x^2 y^3 + e^{x+y}$
34. $f(x,y) = 5x^4 y^2 + e^{xy}$
35. $f(x,y) = y^3 e^x + \ln x$
36. $f(x,y) = x^2 e^y + \ln y$
37. $f(x,y) = \ln(2x + 3y)$
38. $f(x,y) = \ln(3x + 5y)$
39. $f(x,y) = (x^2 - y^2)^3$
40. $f(x,y) = (2x^2 + y^3)^4$
41. $f(x,y) = \dfrac{x^5}{y^3}$
42. $f(x,y) = \dfrac{y^2}{x^7}$
43. $f(x,y) = y^3 e^{xy}$
44. $f(x,y) = x^2 e^{xy}$
45. $f(x,y) = e^{x^2 y}$
46. $f(x,y) = e^{xy^2}$

APPLICATIONS

47. The formula for the volume of a cone is given by $V(r,h) = \dfrac{\pi}{3} r^2 h$, where r is the radius of the base and h is the height, both in inches.

    (a) Determine $V_r(3,5)$ and interpret.

    (b) Determine $V_h(3,5)$ and interpret.

48. The formula for the volume of a box with a square base is given by $V(x,h) = x^2 h$, where x is the length of one side of the base and h is the height, both in feet.

    (a) Determine $V_x(2,3)$ and interpret.

    (b) Determine $V_h(2,3)$ and interpret.

49. Suppose that an amount of P dollars is invested at an annual interest rate of 4% compounded monthly for t years. Then the balance at the end of t years is given by $f(P,t) = P\left(1 + \dfrac{0.04}{12}\right)^{12t}$.

(a) Determine $f_P(2{,}000, 3)$ and interpret.

(b) Determine $f_t(2{,}000, 3)$ and interpret.

50. Suppose that an amount of P dollars is invested at an annual interest rate of 4% compounded continuously for t years. Then the balance at the end of t years is given by $f(P,t) = Pe^{0.04t}$.

(a) Determine $f_P(2{,}000, 3)$ and interpret.

(b) Determine $f_t(2{,}000, 3)$ and interpret.

51. Suppose that an amount of $10{,}000$ is invested at an annual interest rate of $r\%$ compounded monthly for t years. Then, the balance at the end of t years is given by $f(t,r) = 10{,}000(1 + \frac{0.01r}{12})^{12t}$.

(a) Determine $f_t(5, 3)$ and interpret.

(b) Determine $f_r(5, 3)$ and interpret.

52. Suppose that an amount of $10{,}000$ is invested at an annual interest rate of $r\%$ compounded continuously for t years. Then, the balance at the end of t years is given by $f(t,r) = 10{,}000 e^{0.01rt}$.

(a) Determine $f_t(5, 3)$ and interpret.

(b) Determine $f_r(5, 3)$ and interpret.

53. A cereal manufacturer spends x thousand dollars each week on newspaper advertising and y thousand dollars each week on radio advertising. The company has determined that its weekly sales, in thousands of dollars, are given by $R(x,y) = 1.5x^2 + 6y$.

(a) Determine $R_x(5, 2)$ and interpret.

(b) Determine $R_y(5, 2)$ and interpret.

54. A cereal manufacturer spends x thousand dollars each week on television advertising and y thousand dollars each week on radio advertising. The company has determined that its weekly sales, in thousands of dollars, are given by $R(x,y) = 2x^2 + 4y$.

(a) Determine $R_x(4, 1)$ and interpret.

(b) Determine $R_y(4, 1)$ and interpret.

55. A company sells two models of cellular phones: the basic model and the luxury model. The demand functions for the cellular phones are given by

$$p_b = 120 - 0.5x - 0.3y \quad \text{and}$$

$$p_l = 150 - 0.5x - 0.5y,$$

where p_b and p_l are the prices of the two models in dollars, x is the units of the basic cellular phones sold each week, and y is the units of the luxury cellular phones sold each week.

(a) Determine the total revenue function $R(x,y)$.

(b) Determine $R_x(65, 35)$ and interpret.

(c) Determine $R_y(65, 35)$ and interpret.

56. A company sells two sizes of microwave ovens: the small size and the large size. The demand functions for the microwave ovens are given by

$$p_s = 150 - 0.5x - 0.2y \quad \text{and}$$

$$p_l = 250 - 0.1x - 0.5y,$$

where p_s and p_l are the prices of the two sizes in dollars, x is the units of the small microwave ovens sold each week, and y is the units of the large microwave ovens sold each week.

(a) Determine the total revenue function $R(x,y)$.

(b) Determine $R_x(50, 30)$ and interpret.

(c) Determine $R_y(50, 30)$ and interpret.

57. Refer to Exercise #55. The cost function, in dollars, of producing the two types of cellular phones is

$$C(x,y) = 1{,}000 + 10x + 40y.$$

(a) Determine the total profit function $P(x,y)$.

(b) Determine $P_x(65, 35)$ and interpret.

(c) Determine $P_y(65, 35)$ and interpret.

58. Refer to Exercise #56. The cost function, in dollars, of producing the two types of microwave ovens is

$$C(x,y) = 800 + 30x + 40y.$$

(a) Determine the total profit function $P(x,y)$.

(b) Determine $P_x(50, 30)$ and interpret.

(c) Determine $P_y(50, 30)$ and interpret.

59. A company that produces compact fluorescent light bulbs has the Cobb–Douglas production function

$$Q = f(x, y) = 20x^{0.6}y^{0.4},$$

where Q is the number of compact fluorescent light bulbs per day, x is the number of labor hours per day and y is the value of the equipment in units of thousands of dollars.

(a) Determine $f(250, 120)$ and interpret.

(b) Determine $f_x(250, 120)$ and interpret.

(c) Determine $f_y(250, 120)$ and interpret.

60. A company that produces calculators has the Cobb–Douglas production function

$$Q = f(x, y) = 140x^{0.8}y^{0.2},$$

where Q is the number of calculators produced per year, x is the number of thousands of labor hours per year and y is the value of the equipment in units of thousands of dollars.

(a) Determine $f(500, 250)$ and interpret.

(b) Determine $f_x(500, 250)$ and interpret.

(c) Determine $f_y(500, 250)$ and interpret.

61. A company that produces refrigerators has the Cobb–Douglas production function

$$Q = f(x, y) = 160x^{0.65}y^{0.35},$$

where Q is the number of refrigerators produced per year, x is the number of thousand labor hours per year and y is the value of the equipment in units of thousand dollars. The company currently has a labor force of 50,000 labor hours per year and $100,000 of capital.

(a) Determine the current production level.

(b) Determine the current marginal productivity of labor and interpret.

(c) Determine the current marginal productivity of capital and interpret.

62. A company that produces skis has the Cobb–Douglas production function

$$Q = f(x, y) = 180x^{0.7}y^{0.3},$$

where Q is the number of pairs of skis produced per month, x is the number of thousands of labor hours per month and y is the value of the equipment in units of thousands of dollars. The company currently has a labor force of 3,000 labor hours per month and $50,000 of capital.

(a) Determine the current production level.

(b) Determine the current marginal productivity of labor and interpret.

(c) Determine the current marginal productivity of capital and interpret.

6.2 Maximum–Minimum Problems

In Chapter 3, we learned how to maximize or minimize a function of one variable by investigating critical points and derivative tests. In this section, we briefly explore how to find the maximum or minimum values of functions of two variables. We first extend the notions of relative and absolute extrema and critical points.

Relative Extrema and Critical Points

The relative (or local) and absolute (or global) extrema are defined for a function of one variable, $y = f(x)$, in Chapter 3. We say that f has a relative maximum/minimum at $x = x_0$ if all function values $f(x)$ are no greater/lesser than $f(x_0)$ for x near x_0. We say that f has an absolute maximum/minimum at $x = x_0$ if all function values $f(x)$ are no greater/lesser than $f(x_0)$ for x in the whole interval considered. We can similarly define relative and absolute extrema of a function of two variables, $f(x, y)$.

6.2. MAXIMUM–MINIMUM PROBLEMS

> **Definition**   *Relative and Absolute Extrema*
>
> Let $f(x, y)$ be a function of two variables defined on some region $\mathcal{R}$. We say that
> - f has a **relative maximum** at (x_0, y_0) if
>
> $$f(x, y) \leq f(x_0, y_0)$$
>
> for all (x, y) in some circular region containing (x_0, y_0);
> - f has a **relative minimum** at (x_0, y_0) if
>
> $$f(x, y) \geq f(x_0, y_0)$$
>
> for all (x, y) in some circular region containing (x_0, y_0);
> - f has an **absolute maximum** at (x_0, y_0) if
>
> $$f(x, y) \leq f(x_0, y_0)$$
>
> for all (x, y) in the region $\mathcal{R}$;
> - f has an **absolute minimum** at (x_0, y_0) if
>
> $$f(x, y) \geq f(x_0, y_0)$$
>
> for all (x, y) in the region $\mathcal{R}$.

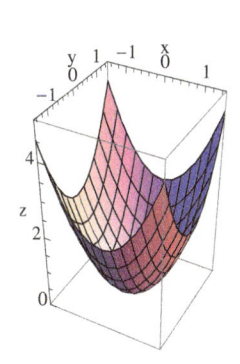

Figure 6.2.1: $z = x^2 + y^2$

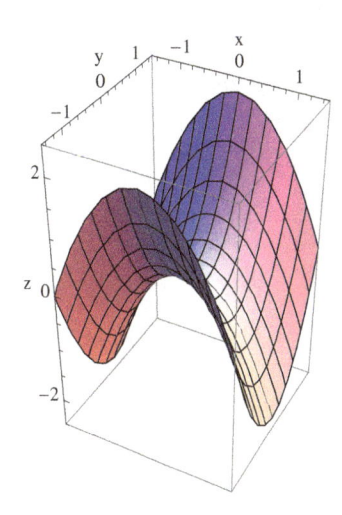

Figure 6.2.2: $z = y^2 - x^2$

One obvious example is $(x_0, y_0) = (0, 0)$ for $z = f(x, y) = x^2 + y^2$, illustrated in Figure 6.2.1. Since $f(0, 0) = 0$, and all other function values are positive, the function has both a relative minimum and an absolute minimum at $(0, 0)$.

Recall that Theorem 4 in Section 3.3 states that, if $y = f(x)$ has a relative extremum at $x = c$, then c must be a critical number. That is, either $f'(c)$ does not exist, or $f'(c) = 0$. In this section, we restrict our attention to the functions that have first and second partial derivatives at all (x, y) in the domain. Therefore, the extended notion of the critical points applies to such functions only.

We will use the word *boundary* in the following definition. The boundary of a region $\mathcal{R}$ intuitively consists of the points on the edge of $\mathcal{R}$. In the following diagrams, the thick black lines denote the boundaries of the regions $\mathcal{R}$ and $\mathcal{S}$.

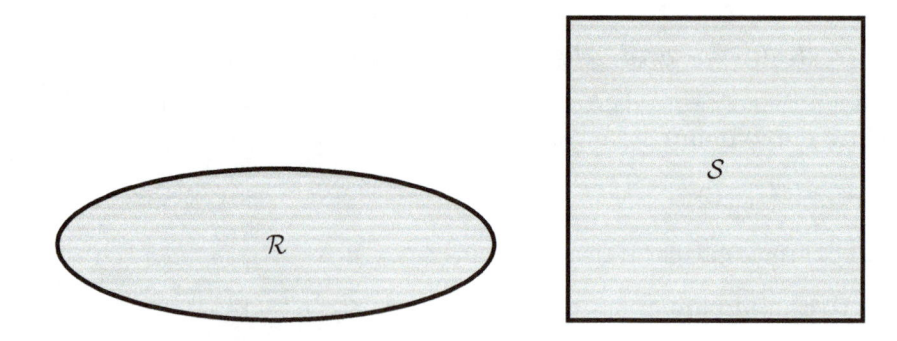

A more precise definition of the boundary of $\mathcal{R}$ is any point (x, y) in $\mathcal{R}$ such that any circle containing (x, y) also contains a point outside of $\mathcal{R}$.

> **Definition** **Critical Point**
>
> Let $f(x, y)$ be a function of two variables on a region $\mathcal{R}$. Assume that the partial derivatives $f_x(x, y)$ and $f_y(x, y)$ exist for all (x, y) in the region $\mathcal{R}$. Assume that the point (x_0, y_0) is not on the boundary of $\mathcal{R}$. We say that (x_0, y_0) is a **critical point** if
>
> $$f_x(x_0, y_0) = 0 \quad \text{and} \quad f_y(x_0, y_0) = 0.$$

Example 1: For

$$z = f(x, y) = x^2 + 2y^2 - 4xy + 2x + 4y + 5,$$

find all critical points.

Solution We have

$$f_x(x, y) = 2x - 4y + 2 \quad \text{and} \quad f_y(x, y) = 4y - 4x + 4.$$

To find the critical point(s), we solve the system

$$\begin{cases} 2x - 4y + 2 = 0 \\ 4y - 4x + 4 = 0. \end{cases}$$

Solving x from the first equation, we obtain

$$x = 2y - 1.$$

Substituting it into the second equation, we have

$$4y - 4(2y - 1) + 4 = 0, \quad \text{or} \quad -4y + 8 = 0.$$

Hence $y = 2$. Substituting back into $x = 2y - 1$, we obtain

$$x = 2(2) - 1 = 3.$$

Therefore, the critical point is $(3, 2)$.

The following theorem resembles Theorem 4 in Section 3.3:

6.2. MAXIMUM–MINIMUM PROBLEMS

> **Theorem 1** *Relative Extrema and Critical Points*
>
> Let $z = f(x,y)$ be a function on a region $\mathcal{R}$. Assume that the partial derivatives f_x and f_y exist for all (x_0, y_0) in the region $\mathcal{R}$. If a point (x_0, y_0) in $\mathcal{R}$ is not on the boundary of $\mathcal{R}$ and if f has a relative extremum at (x_0, y_0), then (x_0, y_0) is a critical point.

This theorem indicates that a critical point (x_0, y_0) is a candidate for the location of a relative extremum of f. However, recall that in the cases of functions of one variable, f may have a relative maximum, relative minimum, or neither, at a critical number $x = c$. Therefore, either the first or the second derivative test is required. For a function of two variables, $z = f(x, y)$, the behavior of f at a critical point (x_0, y_0) is more complicated. The function may or may not have a relative extremum at (x_0, y_0). Other than a relative extremum, the graph of the function may have a particular shape that resembles a saddle at a critical point. In such a case the critical point is called a **saddle point**. Let us observe the graph of the function $z = f(x,y) = y^2 - x^2$, illustrated in Figure 6.2.2, it has a saddle point at $(0,0)$. If we fix the x-value at 0, then $h(y) = f(0,y) = y^2$, and the function h has a relative minimum at $y = 0$. At the other hand, if we fix the y-value at 0, then $g(x) = f(x, 0) = -x^2$, and the function g has a relative maximum at $x = 0$. Notice that the function has neither a maximum nor a minimum at the saddle point.

The D-Test for $z = f(x,y)$

Assume that (x_0, y_0) is a critical point for $z = f(x, y)$. How do we determine whether the function has a relative maximum, a relative minimum, a saddle point, or neither at (x_0, y_0)? In the case of a function of one variable, either the First-Derivative Test or the Second-Derivative Test may be applied to judge whether $y = f(x)$ has a relative maximum or a relative minimum at a critical number $x = c$. Since the First-Derivative Test checks the signs of the first derivative *around* $x = c$, it is difficult to extend the test to the case of a function of two variables. However, the Second-Derivative Test only checks the sign of the second derivative *at the point*. It turns out that similar results exist for a function of two variables. We introduce this test as follows:

> **Theorem 2** *The D-Test*
>
> Suppose that all first and second partial derivatives of a function $z = f(x,y)$ exist on a region that contains a critical point (x_0, y_0) for which $f_x(x_0, y_0) = f_y(x_0, y_0) = 0$. Let
>
> $$D = f_{xx}(x_0, y_0) f_{yy}(x_0, y_0) - [f_{xy}(x_0, y_0)]^2.$$
>
> - If $D > 0$ and $f_{xx}(x_0, y_0) > 0$, then f has a relative minimum at (x_0, y_0).
> - If $D > 0$ and $f_{xx}(x_0, y_0) < 0$, then f has a relative maximum at (x_0, y_0).
> - If $D < 0$, then f has a saddle point at (x_0, y_0).
> - If $D = 0$, then the test is inconclusive.

Example 2: Consider the function

$$z = f(x,y) = x^2 + 2y^2 - 4xy + 2x + 4y + 5$$

in Example 1. It has a critical point $(3, 2)$. Perform the The D-Test to judge whether f has a relative maximum, a relative minimum, or a saddle point at $(3, 2)$.

Solution We have

$$f_{xx}(x,y) = 2, \quad f_{yy}(x,y) = 4, \quad \text{and} \quad f_{xy}(x,y) = -4.$$

Hence

$$D = (2)(4) - 4^2 = -8 < 0.$$

Therefore, the function has a saddle point at $(3, 2)$.

▬ A Strategy for Determining Relative Extrema of $z = f(x, y)$ ▬

We introduce the strategy for finding relative extrema of a function of two variables, $z = f(x, y)$ over a region $\mathcal{R}$. We assume that the function has all first and second partial derivatives on $\mathcal{R}$. Note that a relative extremum may or may not be an absolute extremum. It is significantly more complicated to judge whether a function has an absolute extremum for functions of two variables. However, in most of the application problems, a relative extremum is indeed also an absolute extremum.

A Max-Min Strategy for Determining Relative Extrema of $z = f(x, y)$

(I) Find f_x, f_y, f_{xx}, f_{xy}, and f_{yy}.

(II) Determine the critical points by solving the system

$$f_x(x,y) = 0 \quad \text{and} \quad f_y(x,y) = 0.$$

(III) At each critical point (x_0, y_0), evaluate

$$D = f_{xx}(x_0, y_0) f_{yy}(x_0, y_0) - [f_{xy}(x_0, y_0)]^2.$$

(IV) Perform the D-Test to determine whether f has a relative maximum, a relative minimum, or a saddle point.

Example 3: For $z = f(x, y) = x^3 + y^3 - 6y^2 - 3x + 9$, find all the critical points and determine at each whether f has a relative maximum, a relative minimum, or a saddle point.

Solution We find the six partial derivatives first:

$$f_x(x,y) = 3x^2 - 3,$$

$$f_y(x,y) = 3y^2 - 6y;$$

$$f_{xx}(x,y) = 6x, \quad f_{xy}(x,y) = 0, \quad f_{yy}(x,y) = 6y - 6.$$

Next, we solve the system

$$\begin{cases} 3x^2 - 3 = 0 \\ 3y^2 - 6y = 0. \end{cases}$$

Note that the each equation can be solved separately. Solving for x from the first equation, we obtain

$$3(x+1)(x-1) = 0, \quad \text{hence} \quad x = -1, \quad \text{or} \quad x = 1.$$

Solving for y from the second equation, we obtain

$$3y(y-2) = 0, \quad \text{hence} \quad y = 0, \quad \text{or} \quad y = 2.$$

6.2. MAXIMUM–MINIMUM PROBLEMS

Therefore, there are four critical points: $(-1,0)$, $(-1,2)$, $(1,0)$, and $(1,2)$.

We perform the D-Test at each critical point: at $(-1,0)$, we have

$$D = 6(-1) \cdot (6(0) - 6) - 0^2 = 36 > 0, \quad \text{and} \quad f_{xx}(-1,0) = -6 < 0.$$

Therefore, f has a relative maximum at $(-1,0)$.

At $(-1,2)$, we have

$$D = 6(-1) \cdot (6(2) - 6) - 0^2 = -36 < 0.$$

Therefore, f has a saddle point at $(-1,2)$.

At $(1,0)$, we have

$$D = 6(1) \cdot (6(0) - 6) - 0^2 = -36 < 0.$$

Therefore, f has a saddle point at $(1,0)$.

At $(1,2)$, we have

$$D = 6(1) \cdot (6(1) - 6) - 0^2 = 36 > 0, \quad \text{and} \quad f_{xx}(1,0) = 6 > 0.$$

Therefore, f has a relative minimum at $(1,2)$.

The conclusion is: f has a relative maximum at $(-1,0)$, a relative minimum at $(1,2)$, and two saddle points at $(-1,2)$ and $(1,0)$.

Applications

Example 4: A company sells two models of cellular phones: the basic model and the luxury model. The demand functions for the cellular phones sold are given by

$$p_b = 160 - 0.15x - 0.05y \quad \text{and} \quad p_l = 220 - 0.15x - 0.15y,$$

where p_b and p_l are the prices of the two models in dollars, x is the units of the basic cellular phones sold per week, and y is the units of the luxury cellular phones sold per week. Therefore, the total weekly revenue function $R(x,y)$ is

$$R(x,y) = 160x + 220y - 0.2xy - 0.15x^2 - 0.15y^2.$$

(a) Determine how many units of each model should be sold per week in order to maximize the company's total weekly revenue.

(b) Determine the maximum value of the total weekly revenue.

Solution

(a) We first find the five first and second partial derivatives of $R(x,y)$:

$$R_x(x,y) = 160 - 0.2y - 0.3x,$$
$$R_y(x,y) = 220 - 0.2x - 0.3y;$$
$$R_{xx}(x,y) = -0.3, \quad R_{xy}(x,y) = -0.2, \quad R_{yy}(x,y) = -0.3.$$

Next, we solve the system

$$\begin{cases} 160 - 0.2y - 0.3x = 0 \\ 220 - 0.2x - 0.3y = 0. \end{cases}$$

Solving for y from the first equation, we obtain

$$y = 800 - 1.5x.$$

Substituting this relation into the second equation and solving for x, we obtain

$$220 - 0.2x - 0.3(800 - 1.5x) = 0,$$

$$0.25x - 20 = 0, \quad \text{hence} \quad x = 80.$$

Substituting back into $y = 800 - 1.5x$, we obtain

$$y = 800 - 1.5(80) = 680.$$

Therefore, R has one critical point $(80, 680)$. We perform the D-Test at $(80, 680)$: We have

$$D = (-0.3) \cdot (-0.3) - (-0.2)^2 = 0.09 - 0.04 = 0.05 > 0, \quad \text{and} \quad R_{xx}(80, 680) = -0.3 < 0.$$

Therefore R has a relative maximum at $(80, 680)$. In this case, $R(80, 680)$ is also an absolute maximum. Therefore, to maximize the company's total monthly revenue, 80 basic models and 680 luxury models of the cellular phones should be sold per week.

(b) The maximum total weekly revenue is

$$R(80, 680) = 160(80) + 220(680) - 0.2(80)(680) - 0.15(80)^2 - 0.15(680)^2 = \$81{,}200.$$

Example 5: Refer to Example 4. Assume that the cellular phone company determines that the total cost function of producing x units of basic models and y units of luxury models per week can be expressed by

$$C(x, y) = 5{,}000 + 60x + 100y.$$

(a) Determine the total profit function $P(x, y)$.

(b) Determine how many units of each model should be produced and sold per week in order to maximize the company's total weekly profit.

(c) Determine the maximum value of the total weekly profit.

Solution

(a) Since profit is revenue minus cost, we have

$$P(x, y) = R(x, y) - C(x, y)$$
$$= 160x + 220y - 0.2xy - 0.15x^2 - 0.15y^2 - (5{,}000 + 60x + 100y)$$
$$= 160x + 220y - 0.2xy - 0.15x^2 - 0.15y^2 - 5{,}000 - 60x - 100y$$
$$= 100x + 120y - 0.2xy - 0.15x^2 - 0.15y^2 - 5{,}000.$$

(b) We first find the six partial derivatives of $P(x, y)$:

$$P_x(x, y) = 100 - 0.2y - 0.3x,$$

$$P_y(x, y) = 120 - 0.2x - 0.3y;$$

$$P_{xx}(x, y) = -0.3, \quad P_{xy}(x, y) = -0.2, \quad P_{yy}(x, y) = -0.3.$$

Next, we solve the system

$$\begin{cases} 100 - 0.2y - 0.3x = 0 \\ 120 - 0.2x - 0.3y = 0. \end{cases}$$

6.2. MAXIMUM–MINIMUM PROBLEMS

Solving for y from the first equation, we obtain

$$y = 500 - 1.5x.$$

Substituting this relation into the second equation and solving for x, we obtain

$$120 - 0.2x - 0.3(500 - 1.5x) = 0,$$

$$0.25x - 30 = 0, \text{ hence } x = 120.$$

Substituting back into $y = 500 - 1.5x$, we obtain

$$y = 500 - 1.5(120) = 320.$$

Therefore, P has one critical point $(120, 320)$. We perform the D-Test at $(120, 320)$: We have

$$D = (-0.3) \cdot (-0.3) - (-0.2)^2 = 0.09 - 0.04 = 0.05 > 0, \text{ and } P_{xx}(120, 320) = -0.3 < 0.$$

Therefore P has a relative maximum at $(120, 320)$. In this case, $P(120, 320)$ is also an absolute maximum. Therefore, to maximize the company's total weekly profit, 120 basic models and 320 luxury models of the cellular phones should be produced and sold per week.

(c) The maximum total weekly profit is

$$P(120, 320) = 100(120) + 120(320) - 0.2(120)(320) - 0.15(120)^2 - 0.15(320)^2 - 5{,}000 = \$20{,}200.$$

Exercise Set 6.2

For Exercises 1–10, determine critical points for each given function.

1. $f(x, y) = x^2 + y^2 + 6x - 10y + 8$
2. $f(x, y) = x^2 + y^2 + 10x - 6y + 4$
3. $f(x, y) = x^2 + xy + 10y$
4. $f(x, y) = x^2 + 10x + xy$
5. $f(x, y) = y^3 - 4xy + 8x$
6. $f(x, y) = x^3 - 4xy - 6y$
7. $f(x, y) = x^2 + y^2 + xy - 8y + 1$
8. $f(x, y) = x^2 + y^2 + xy - 6x + 2$
9. $f(x, y) = x^3 + y^3 - 3x^2 - 12y + 2$
10. $f(x, y) = x^3 + y^3 - 6y^2 - 3x + 5$

For Exercises 11–30, determine all the critical points and use the D-Test to determine whether the function has a relative maximum, a relative minimum, or a saddle point at each critical point.

11. $f(x, y) = x^2 + y^2 + xy - y + 1$
12. $f(x, y) = x^2 + y^2 + xy - 5x + 4$
13. $f(x, y) = x^2 + xy + 6y$
14. $f(x, y) = x^2 + 4x + xy$
15. $f(x, y) = 2xy - y^3 - x^2$
16. $f(x, y) = 4xy - x^3 - y^2$
17. $f(x, y) = -x^2 - y^2 + 6x + 4y + 2$
18. $f(x, y) = -x^2 - y^2 + 4x + 6y + 1$
19. $f(x, y) = x^3 + y^3 - 3x^2 - 12y + 2$
20. $f(x, y) = x^3 + y^3 - 6y^2 - 3x + 5$
21. $f(x, y) = x^3 + 3xy - y^3$
22. $f(x, y) = x^3 - 3xy + y^3$
23. $f(x, y) = 4y^3 - 3x^2 - 12y^2 + 6x - 5$
24. $f(x, y) = 4x^3 - 3y^2 - 24x^2 + 6y + 1$
25. $f(x, y) = 2x^3 y - 24x + 16y + 1$
26. $f(x, y) = 2xy^3 + 4y - 2x + 1$
27. $f(x, y) = e^{y^2 + x^2 + 1}$
28. $f(x, y) = e^{3 - x^2 - y^2}$
29. $f(x, y) = e^{x^2 - y^2}$
30. $y = e^{xy}$

APPLICATIONS

31. A manufacturer has the profit function
$$P(x,y) = -x^2 - xy - 1.5y^2 + 80x + 90y - 2{,}000,$$
where x is the amount spent each month on labor and y is the amount spent on equipment upgrade. Both x and y are in thousands of dollars.

(a) Determine how much should be spent on labor and how much should be spent on equipment upgrade to maximize the manufacturer's monthly profit.

(b) Determine the maximum monthly profit.

32. A manufacturer produces two products, Product A and Product B. The weekly profit function, in dollars, is
$$P(x,y) = 560x + 20xy - 20x^2 - 6y^2,$$
where x and y are units of each product in thousands.

(a) Determine how many units of each product should be produced and sold weekly to maximize the manufacturer's total weekly profit.

(b) Determine the maximum value of the total weekly profit.

33. A manufacturer produces two products, Product A and Product B, which are sold in two markets. The weekly demand functions, in dollars, are
$$p_x = 600 - 0.4x \text{ and } p_y = 400 - 0.2y,$$
where x and y are units of each product. The weekly cost function is
$$C(x,y) = 60{,}000 + 30x + 40y.$$

(a) Determine the total weekly revenue function $R(x,y)$.

(b) Determine the total weekly profit function $P(x,y)$.

(c) Determine how many units of each product should be produced and sold weekly to maximize the manufacturer's total weekly profit.

34. A manufacturer produces two products, Product A and Product B, which are sold in two markets. The monthly demand functions, in dollars, are
$$p_x = 500 - 0.2x \text{ and } p_y = 600 - 0.3y,$$
where x and y are hundreds of units of each product. The monthly cost function is
$$C(x,y) = 100{,}000 + 100x + 180y + 0.2xy.$$

(a) Determine the total monthly revenue function $R(x,y)$.

(b) Determine the total monthly profit function $P(x,y)$.

(c) Determine how many units of each product should be produced and sold per month to maximize the manufacturer's monthly total profit.

35. A company sells two models of cellular phones: the basic model and the luxury model. The weekly demand functions for the cellular phones are given by
$$p_b = 200 - 1.5x - 0.3y \text{ and}$$
$$p_l = 300 - 0.1x - 0.8y,$$
where p_b and p_l are the prices of the two models in dollars, x is the units of the basic cellular phones sold per week, and y is the units of the luxury cellular phones sold per week.

(a) Determine the total weekly revenue function $R(x,y)$.

(b) Determine how many units of each model should be sold per week to maximize the company's total weekly revenue.

36. A company sells two sizes of microwave ovens: the small size and the large size. The weekly demand functions for the microwave ovens are given by
$$p_s = 150 - 0.5x - 0.2y \text{ and}$$
$$p_l = 250 - 0.1x - 0.5y,$$
where p_s and p_l are the prices of the two sizes in dollars, x is the units of the small microwave ovens sold per week, and y is the units of the large microwave ovens sold per week. Repeat Exercise #35, parts (a) and (b).

37. Refer to Exercise #35. The weekly cost function, in dollars, of producing the two types of cellular phones is
$$C(x,y) = 1000 + 40x + 80y.$$

(a) Determine the total weekly profit function $P(x,y)$.

(b) Determine how many units of each model should be produced and sold per week to maximize the company's total weekly profit.

38. Refer to Exercise #36. The weekly cost function, in dollars, of producing the two types of microwave ovens is

$$C(x, y) = 8{,}000 + 30x + 40y.$$

Repeat Exercise #37, parts (a) and (b).

39. A company operates two plants, Plant A and Plant B, which produce the same product. The total annual cost function, in thousands of dollars, for each plant is

$$C_A(x) = 9 + 0.04x^2 \text{ and } C_B(y) = 6 + 0.04y^2,$$

where x and y are hundreds of items produced in each plant. The total annual demand for $q = x + y$ can expressed by

$$p = D(q) = 65 - 0.04q.$$

(a) Determine the total annual profit function $P(x, y)$.

(b) Determine how many units of the product each plant should produce and sell in order to maximize the company's total annual profit.

40. A company produces two products, Product A and Product B. The total daily cost function (in dollars) is given by

$$C(x, y) = 15 + 2x^2 + 3y^2,$$

where x and y are items produced for each product. Assume that Product A is sold at a fixed price of $8 and Product B is sold at a fixed price of $12.

(a) Determine the total daily profit function $P(x, y)$.

(b) Determine how many units of the each product should be produced and sold each day in order to maximize the company's daily total profit.

Chapter 6 Summary

This chapter returns to differential calculus and studies the functions that depend on two or more independent variables.

A **function of two variables** is a rule that assigns exactly one element in the range to each pair of elements, (x, y), in the domain.

For example, if $z = f(x, y) = 3x^2y^5 + 2x^3y^2 + e^{xy}$, then

$$f(2, 1) = 3(2)^2(1)^5 + 2(2)^3(1)^2 + e^{(2)(1)} = 12 + 16 + e^2 = 28 + e^2.$$

Partial Derivatives

Similar to the derivative of a function which depends on one variable, which is the rate of change of the function with respect to the variable, the rate of change of a function that depends two inputs, $z = f(x, y)$, is focused on one of the inputs each time. Therefore there are two types of derivatives: The rate of change of z with respect to x, and the rate of change of z with respect to y. They are called **partial derivatives**. When the rate of change of z with respect to x is calculated, the variable y is treated as a constant, and the partial derivative is called the **partial derivative of f with respect to x**. It can be denoted as $f_x(x, y)$, or $\frac{\partial z}{\partial x}$, or $\frac{\partial f}{\partial x}$. Similarly, the **partial derivative of f with respect to y** can be denoted as $f_y(x, y)$, or $\frac{\partial z}{\partial y}$, or $\frac{\partial f}{\partial y}$. This partial derivative is found by treating x as a constant and performing the differentiation techniques introduced in Chapter 2.

If $z = 3yx^2$, then we can denote its partial derivative with respect to x by

$$f_x(x, y) = \frac{\partial f}{\partial x} = \frac{\partial z}{\partial x} = \frac{\partial}{\partial x}\left(3yx^2\right) = 4yx.$$

Similarly, the partial derivative of z with respect to y can be denoted by

$$f_y(x, y) = \frac{\partial f}{\partial y} = \frac{\partial z}{\partial y} = \frac{\partial}{\partial y}\left(3yx^2\right) = 3x^2.$$

The partial derivatives are again functions of x and y. Therefore, we can find their partial derivatives. They are called **second–order partial derivatives**. There are four second-order partial derivatives of a function of two variables $z = f(x, y)$. The notations are summarized in the following.

Second–Order Partial Derivatives

If $z = f(x, y)$ be a function of two variables, then

$$f_{xx} = \frac{\partial^2 f}{\partial x \partial x} = \frac{\partial^2 z}{\partial x \partial x} = \frac{\partial^2 f}{\partial x^2} = \frac{\partial^2 z}{\partial x^2} = \frac{\partial}{\partial x}\left(\frac{\partial z}{\partial x}\right)$$

$$f_{xy} = \frac{\partial^2 f}{\partial y \partial x} = \frac{\partial^2 z}{\partial y \partial x} = \frac{\partial}{\partial y}\left(\frac{\partial z}{\partial x}\right)$$

$$f_{yx} = \frac{\partial^2 f}{\partial x \partial y} = \frac{\partial^2 z}{\partial x \partial y} = \frac{\partial}{\partial x}\left(\frac{\partial z}{\partial y}\right)$$

$$f_{yy} = \frac{\partial^2 f}{\partial y \partial y} = \frac{\partial^2 z}{\partial y \partial y} = \frac{\partial^2 f}{\partial y^2} = \frac{\partial^2 z}{\partial y^2} = \frac{\partial}{\partial y}\left(\frac{\partial z}{\partial y}\right)$$

Provided that f_{xy} and f_{yx} are continuous, it follows that $f_{xy} = f_{yx}$.

Example 1: For $z = f(x, y) = 3x^2 y^5 + 2x^3 y^2 + e^{xy}$, find (a) the two partial derivatives and (b) the four second-order partial derivatives.

Solution We include the most often used notations:

(a) $f_x = \dfrac{\partial f}{\partial x} = \dfrac{\partial}{\partial x}\left(3x^2 y^5 + 2x^3 y^2 + e^{xy}\right)$

$\qquad = 6xy^5 + 6x^2 y^2 + ye^{xy},$

$f_y = \dfrac{\partial f}{\partial y} = \dfrac{\partial}{\partial y}\left(3x^2 y^5 + 2x^3 y^2 + e^{xy}\right)$

$\qquad = 15x^2 y^4 + 4x^3 y + xe^{xy}.$

(b) $f_{xx} = \dfrac{\partial^2 f}{\partial x^2} = \dfrac{\partial}{\partial x}\left(6xy^5 + 6x^2 y^2 + ye^{xy}\right)$

$\qquad = 6y^5 + 12xy^2 + y^2 e^{xy},$

$f_{xy} = \dfrac{\partial^2 f}{\partial y \partial x} = \dfrac{\partial}{\partial y}\left(6xy^5 + 6x^2 y^2 + ye^{xy}\right)$

$\qquad = 30xy^4 + 12x^2 y + e^{xy} + xye^{xy},$

$f_{yx} = \dfrac{\partial^2 f}{\partial x \partial y} = \dfrac{\partial}{\partial x}\left(15x^2 y^4 + 4x^3 y + xe^{xy}\right)$

$\qquad = 30xy^4 + 12x^2 y + e^{xy} + ye^{xy},$

$f_{yy} = \dfrac{\partial^2 f}{\partial y^2} = \dfrac{\partial}{\partial y}\left(15x^2 y^4 + 4x^3 y + xe^{xy}\right)$

$\qquad = 60x^2 y^3 + 4x^3 + x^2 e^{xy}.$

■ **Relative Extrema and Critical Points**

CHAPTER 6 SUMMARY

Just as derivatives can help us to locate and judge the relative extrema for a function of one variable, the partial derivatives can help us to locate and judge the relative extrema of $z = f(x, y)$. We say that f has a relative maximum at (x_0, y_0) if

$$f(x, y) \leq f(x_0, y_0)$$

for all (x, y) in a (small) circular region containing (x_0, y_0);
We say that f has a relative minimum at (x_0, y_0) if

$$f(x, y) \geq f(x_0, y_0)$$

for all (x, y) in a (small) circular region containing (x_0, y_0).
Assume that the partial derivatives $f_x(x, y)$ and $f_y(x, y)$ exist for all (x, y) in the region $\mathcal{R}$. Assume that the point (x_0, y_0) is not on the boundary of $\mathcal{R}$. We say that (x_0, y_0) is a **critical point** if

$$f_x(x_0, y_0) = 0, \quad \text{and} \quad f_y(x_0, y_0) = 0.$$

Theorem 1 indicates that if f has a relative extremum at (x_0, y_0) that is not on the boundary of $\mathcal{R}$, it must be a critical point. Other than a relative extremum, the graph of the function may have a particular shape that resembles a saddle at a critical point. In such a case, the critical point is called a **saddle point**.

Assume that (x_0, y_0) is a critical point for $z = f(x, y)$. The D-Test can help us to determine whether the function has a relative maximum, a relative minimum, a saddle point, or neither, at (x_0, y_0).

The D-Test

Suppose that all first and second partial derivatives of a function $z = f(x, y)$ exist on region that contains a critical point (x_0, y_0) for which $f_x(x_0, y_0) = f_y(x_0, y_0) = 0$. Let

$$D = f_{xx}(x_0, y_0) f_{yy}(x_0, y_0) - [f_{xy}(x_0, y_0)]^2.$$

- If $D > 0$ and $f_{xx}(x_0, y_0) > 0$, then f has a relative minimum at (x_0, y_0).
- If $D > 0$ and $f_{xx}(x_0, y_0) < 0$, then f has a relative maximum at (x_0, y_0).
- If $D < 0$, then f has a saddle point at (x_0, y_0).
- If $D = 0$, then the test is inconclusive.

Combining the aforementioned information, a strategy for finding relative extrema of a function of two variables, $z = f(x, y)$, over a region $\mathcal{R}$ is formed.

A Max–Min Strategy for Determining Relative Extrema of $z = f(x, y)$

(I) Find $f_x(x, y)$, $f_y(x, y)$, $f_{xx}(x, y)$, $f_{xy}(x, y)$, and $f_{yy}(x, y)$.

(II) Determine the critical points by solving the system

$$f_x(x, y) = 0 \quad \text{and} \quad f_y(x, y) = 0.$$

(III) At each critical point (x_0, y_0), evaluate

$$D = f_{xx}(x_0, y_0) f_{yy}(x_0, y_0) - [f_{xy}(x_0, y_0)]^2.$$

(IV) Perform the D-Test to determine whether f has a relative maximum, a relative minimum, or a saddle point.

Example 2: For $z = f(x,y) = x^3 + y^3 - 6y^2 - 3x + 9$, find all the critical points and determine at each whether f has a relative maximum, a relative minimum, or a saddle point.

Solution We find the six partial derivatives first:

$$f_x(x,y) = 3x^2 - 3,$$

$$f_y(x,y) = 3y^2 - 6y;$$

$$f_{xx}(x,y) = 6x, \quad f_{xy}(x,y) = 0, \quad f_{yy}(x,y) = 6y - 6.$$

Next, we solve the system

$$\begin{cases} 3x^2 - 3 = 0 \\ 3y^2 - 6y = 0. \end{cases}$$

Note that the each equation can be solved separately. Solving for x from the first equation, we obtain

$$3(x+1)(x-1) = 0, \quad \text{hence} \quad x = -1, \quad \text{or} \quad x = 1.$$

Solving for y from the second equation, we obtain

$$3y(y-2) = 0, \quad \text{hence} \quad y = 0, \quad \text{or} \quad y = 2.$$

Therefore, there are four critical points: $(-1, 0)$, $(-1, 2)$, $(1, 0)$, and $(1, 2)$.

We perform the D-Test at each critical point: at $(-1, 0)$, we have

$$D = 6(-1) \cdot (6(0) - 6) - 0^2 = 36 > 0, \quad \text{and} \quad f_{xx}(-1, 0) = -6 < 0.$$

Therefore, f has a relative maximum at $(-1, 0)$.

At $(-1, 2)$, we have

$$D = 6(-1) \cdot (6(2) - 6) - 0^2 = -36 < 0.$$

Therefore, f has a saddle point at $(-1, 2)$.

At $(1, 0)$, we have

$$D = 6(1) \cdot (6(0) - 6) - 0^2 = -36 < 0.$$

Therefore, f has a saddle point at $(1, 0)$.

At $(1, 2)$, we have

$$D = 6(1) \cdot (6(1) - 6) - 0^2 = 36 > 0, \quad \text{and} \quad f_{xx}(1, 0) = 6 > 0.$$

Therefore, f has a relative minimum at $(1, 2)$.

We conclude that f has a relative maximum at $(-1, 0)$, a relative minimum at $(1, 2)$, and two saddle points at $(-1, 2)$ and $(1, 0)$.

Chapter 6 Review Exercises

Find the following for the function

$$f(x,y) = 2x^2 e^y + 5y^3 - \ln x:$$

1. $f(1,0)$ 2. f_x 3. f_y
4. f_{xx} 5. f_{xy} 6. f_{yx}
7. f_{yy}

Find the following for the function

$$z = 3e^{xy} + x^2 y^3:$$

8. $\dfrac{\partial z}{\partial x}$ 9. $\dfrac{\partial z}{\partial y}$ 10. $\dfrac{\partial^2 z}{\partial x^2}$

11. $\dfrac{\partial^2 z}{\partial x \partial y}$ 12. $\dfrac{\partial^2 z}{\partial y \partial x}$ 13. $\dfrac{\partial^2 z}{\partial y^2}$

14. A company sells two models of cellular phones: The basic model and the luxury model. The weekly demand functions for the cellular phones are given by

$$p_b = 200 - 0.5x - 0.3y \quad \text{and}$$

$$p_l = 300 - 0.5x - 0.5y,$$

where p_b and p_l are the prices for the two models in dollars, x is the units of the basic cellular phones sold per week, and y is the units of the luxury cellular phones sold per week.

(a) Determine the total weekly revenue function $R(x,y)$.

(b) Determine $R_x(65,35)$ and interpret.

(c) Determine $R_y(65,35)$ and interpret.

For Exercises 15–18, determine all the critical points and use the D-Test to determine whether the function has a relative maximum, a relative minimum, or a saddle point at each critical point.

15. $f(x,y) = 2xy - y^3 - x^2$

16. $f(x,y) = -x^2 - y^2 + 6x + 4y + 2$

17. $f(x,y) = x^3 + 3xy - y^3$

18. $f(x,y) = 2x^3 y - 24x + 16y + 1$

19. A manufacturer has the profit function $P(x,y) = -x^2 - xy - 1.5y^2 + 80x + 90y - 2{,}000$, where x is the amount spent each month on labor and y is the amount spent on equipment upgrade. Both x and y are in thousands of dollars.

(a) Determine how much should be spent on labor and how much should be spent on equipment upgrade in order to maximize the monthly profit.

(b) Determine the maximum monthly profit.

20. A manufacturer produces two products, Product A and Product B, which are sold in two markets. The weekly demand functions, in dollars, are

$$p_x = 600 - 0.4x \quad \text{and} \quad p_y = 400 - 0.2y,$$

where x and y are units of each product. The cost function is

$$C(x,y) = 60{,}000 + 30x + 40y.$$

(a) Determine the total weekly revenue function $R(x,y)$.

(b) Determine the total weekly profit function $P(x,y)$.

(c) Determine how many units of each product should be produced and sold per week so that the total weekly profit is maximized.

(d) Determine the maximum total weekly profit.

Chapter 6 Test (50 points)

SCORE _____

NAME _____

ANSWERS

1. (15pts) For the function $z = f(x,y) = x^3 + 2ye^x - y^2$, find the following.

(a) $f(0,2)$

(b) f_x.

(c) f_y

(d) f_{xx}

(e) f_{xy}

(f) f_{yx}

(g) f_{yy}

1. (a) _____

(b) _____

(c) _____

(d) _____

(e) _____

(f) _____

(g) _____

2. (15pts) For the function $f(x,y) = 4xy - x^3 - y^2$, determine all the critical points and use the D-Test to determine whether the function has a relative maximum, a relative minimum, or a saddle point at each critical point. Show your work.

2. Critical point(s) _____

3. (20pts) A company sells two models of cellular phones: the basic model and the luxury model. The demand functions for the cellular phones sold are given by

$$p_b = 200 - 1.5x - 0.3y \quad \text{and}$$

$$p_l = 300 - 0.1x - 0.8y,$$

where p_b and p_l are the prices of the two models in dollars, x is the units of the basic cellular phones sold, and y is the units of the luxury cellular phones sold.

CHAPTER 6 TEST   459

(a) Determine the total revenue function $R(x,y)$.

                                                3. (a) _____

(b) Determine $R_x(60, 25)$ and interpret.

                                                   (b) _____

(c) Determine $R_y(60, 25)$ and interpret.

                                                   (c) _____

(d) Determine how many units of each model should be sold so that the total revenue is maximized.

                                                   (d) _____

(e) Determine the maximum total revenue.

                                                   (e) _____

Chapter 1 Exercises Answers

Exercises 1.1 [Ref. Example]

1. (a) $(-\infty, \infty)$ (b) 1, 3, 7 [3,4]
   (c) $6a + 3$, $2t^2 + 3$
   (d) $2x + 2h + 3$ (e) 2 $(h \neq 0)$
3. [6]

x	$f(x)$
-1	1
0	3
2	7

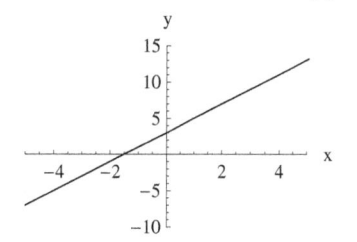

5. (a) $(-\infty, \infty)$ [3,4]
   (b) 6, 0, -2, 0, 6
   (c) $50a^2 - 2$, $2a - 2$
   (d) $2x^2 + 4xh + 2h^2 - 2$
   (e) $4x + 2h$ $(h \neq 0)$
7. [6]

x	$F(x)$
-2	6
-1	0
0	-2
1	0
2	6

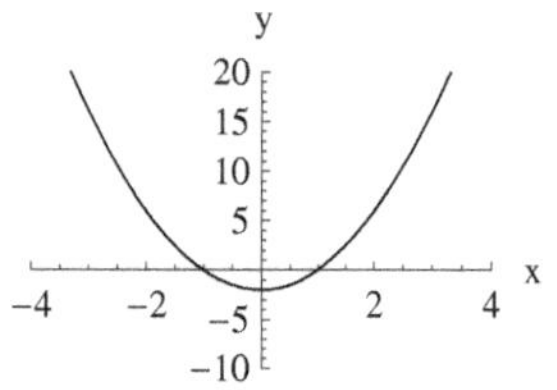

9. (a) $x \neq -1$
   (b) $-2, -4, -20, 20, 2, 1$ [3,4,6]
   (c)

x	$G(x)$
-2	-2
-1.5	-4
-1.1	-20
-0.9	20
0	2
1	1

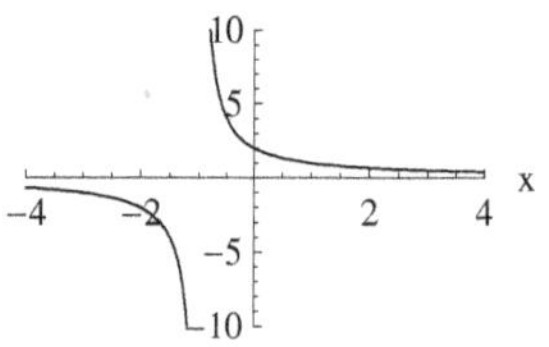

11. (a) 1 (b) 3 (c) -1 [7]
    (d) -1 (e) 1 (f) 0

Exercises 1.2 [Ref. Example]

1. (a) 2 [1,2,3]
   (b)

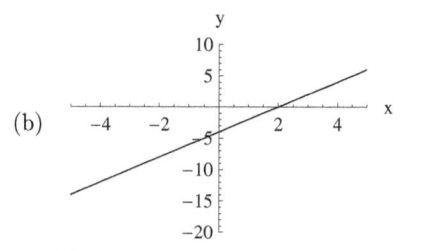

3. (a) $y = 5x + 3$ [1,3]
   (b)

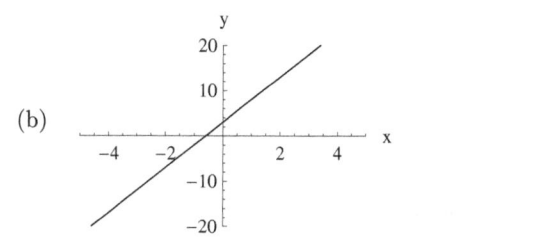

5. (a) -1.5 (b) 3 (c) $y = -1.5x + 3$ [2,3]
7. (a) $y = 2(x - 2)$, $y = 2x - 4$ [3]
9. (a) $y - 1 = -\dfrac{3}{5}(x - 3)$, $y = -\dfrac{3}{5}x + \dfrac{14}{5}$ [3]
11. (a) $y - 1 = -2x$, $y = -2x + 1$ [3]
13. (a) $y - 7 = 0$, $y = 7$ [3]
15. (a) $y - 2 = -3(x - 1)$, $y = -3x + 5$ [3]
17. (a) $y - 2 = 0$, $y = 2$ [3]
19. $10/3$ 21. $3/2$ 23. 0
25. Does not exist.   27. 2
29. 2   31. -4
33. (a) $S = 4x + 680$ (b) $772 per week [4]
    (c) $m = 4$. The value $m = 4$ means that if the number of weeks that the teller has worked for the bank goes up by 1, then his weekly wage goes up by $4. Therefore each additional week raises his weekly wage by 4 dollars.
    (d) The vertical intercept is $b = \$680$. This means that when the teller started working for the bank, his starting weekly wage was $680.
35. (a) $V = -3000t + 31{,}000$ (b) $10,000 [4]
    (c) $m = -3000$. The value $m = -3{,}000$ means that if the age the car goes up by 1, then its resale value goes down by $3{,}000$. Therefore each additional year reduces the resale value by 3,000 dollars.
    (d) The vertical intercept is $b = \$31{,}000$. It represents the value of the car when it was new.
    (e) $T \approx 10.3$ year.
37. (a) $m = \dfrac{9}{4}$ [4]

    The value $m = 9/4$ means for every degree of increase on the Celsius scale, there is a 9/4 degree increase on the Fahrenheit scale.

CHAPTER 6 TEST   461

(b) The vertical intercept is $b = 32$. This means that $0°C = 32°F$.

Exercises 1.3 [Ref. Example]

1. (a) $600, $0.3x$ [1]

(b) [graph]

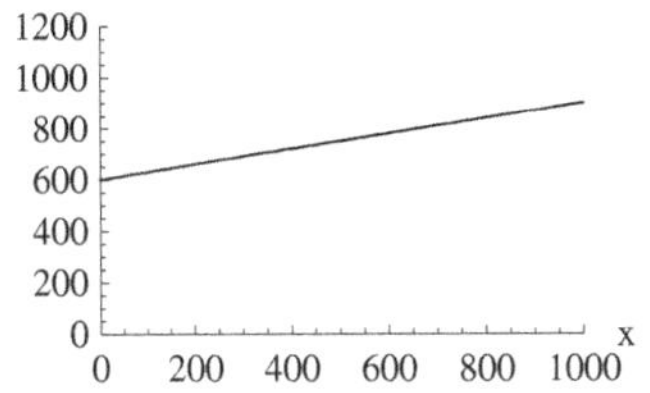

(c) $660, $840

(d) The slope $m = 0.3$ represents the cost of producing an additional gallon of milk.

3. (a) $R(x) = 1.2x$, $0 \leq x \leq 1,000$ [3]

(b) [graph]

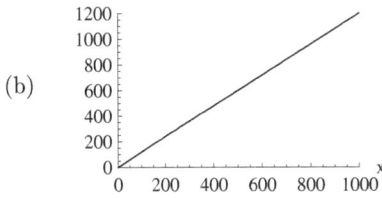

(c) $240, $960

5. (a) $P(x) = 0.9x - 600$ [4]

(b) [graph]

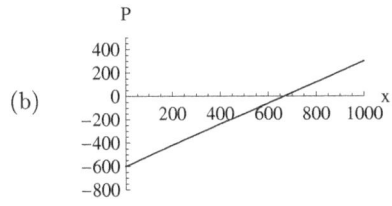

(c) $-$420$, $120   (d) 667 gallons

(e) [graph]

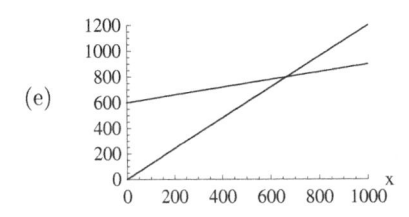

7. (a) $C(x) = 25,000 + 8x$
(b) $R(x) = 16x$ [1-4]
(c) $P(x) = 8x - 25,000$
(d) 3125 radios

(e) [graph]

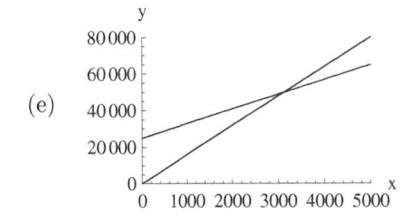

9. (a) $C(x) = 128 + 0.8x$ [2]
(b) The slope $m = \$0.8$ per coffee drink. It represents the cost of producing an additional coffee drink.
(c) $b = \$128$. It represents the fixed daily cost.

11. (a) $R(x) = 2.5x$ (b) $P(x) = 1.7x - 128$ [5]
(c) $x \approx 76$ coffee drinks

13. The p-intercept is $66.67. [6]

It is the maximum price at which the consumers are willing to start purchasing. The x-intercept is 60. It represents the market capacity.

15. $x = D(p) = 500 - 10p$

17. $p_E = \$15.27$, $x_E = 43.65$ [7]

19. (a) $1,500$ (b) 600 (c) higher
(d) $p = \$100$ (e) $p = \$50$
(f) $p_E = \$80$, $x_E = 1,200$

(g) [graph]

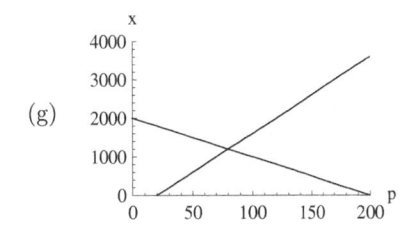

Exercises 1.4 [Ref. Example]

1. vertex $(0, -1)$ [1]

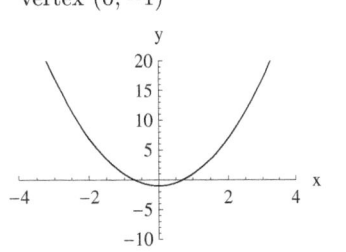

[graph]

3. Vertex $(1, -1)$ [1]

[graph]

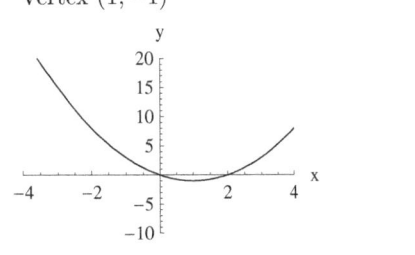

5. Vertex $(-1.5, -5.5)$ [1]

[graph]

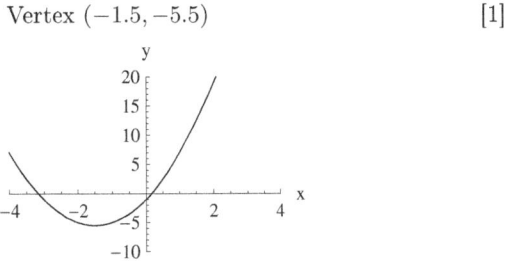

7. Vertex $(1.5, 5.5)$ [1]

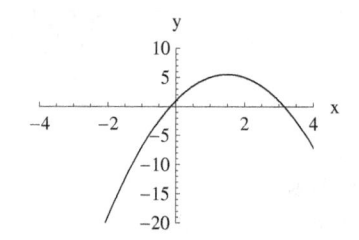

9. $-2, 0, 2$ 11. $-1, 1$ [2]

13. $1, 3$ 15. $2 - \sqrt{6}, 2 + \sqrt{6}$ [2,3]

17. There is no real solution [3]

19. $1/2$ [3]

21. (a) $x \geq 1$ [7,8]

(b)

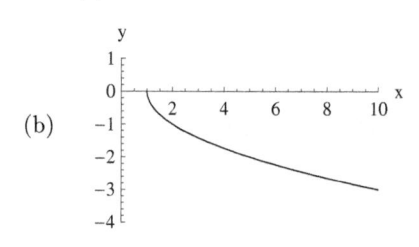

23. (a) $x \neq 3$ [6]

(b)

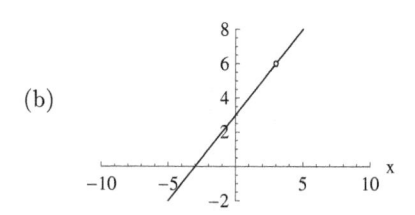

25. (a) $x \neq -1$ [6]

(b)

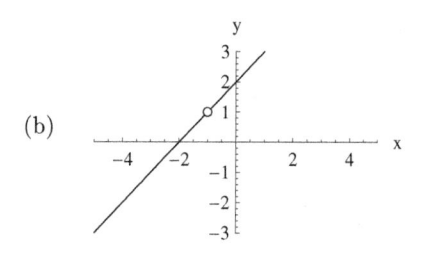

27. (a) $x \leq -1$ or $x \geq 1$ [8]

(b) There is no real solution

29. (a) $x \neq 1, 3$ (b) $x = -3$ [6]

31. (a) $x \neq 2$ (b) $x = 1$ [6]

33. (a) $1, 2, 3, 1, -1$ [10]

(b) (c) (d)

x	$f(x)$
-1	1
0	2
1	3
2	1
3	-1
4	-3

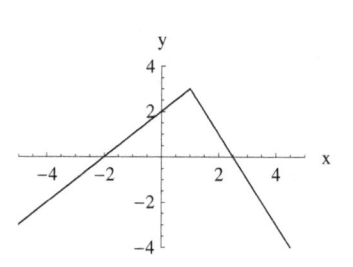

35. [9]

(a) $F(x) = \begin{cases} -x - 2 & \text{if } x < -2, \\ x + 2 & \text{if } x \geq -2. \end{cases}$

(b) $1, 0, 1, 4$

(c) (d) (e)

x	$F(x)$
-4	2
-3	1
-2	0
-1	1
0	2
1	3

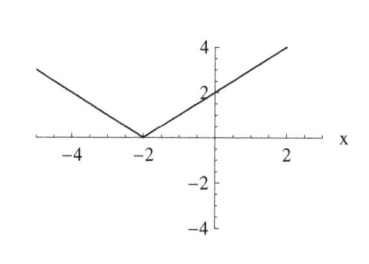

37. (a) $1, 2, 3, 1, 6$ [9]

(b) (c) (d)

x	$f(x)$
-1	1
0	2
1	3
1.1	-1.79
2	1
3	6

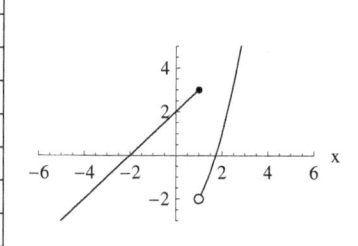

39. (a) $0, 1, 3, 3, 4$ [10]

(b)

x	$f(x)$
-1	0
0	1
2	3
3	4

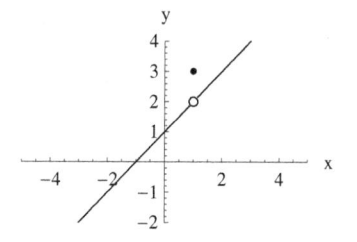

41. (a) $t = 2$ seconds (b) $t \approx 6.9$ seconds [4]

43. $p_E = 2.05, x_E = 4.15$ [5]

45. $p_E = 8.64, x_E = 6.64$

Exercises 1.5 [Ref. Example]

1. (a) (b)

CHAPTER 6 TEST

x	$f(x)$
-4	0.01
-3	0.04
-1	0.33
0	1
1	3
2	9
4	81

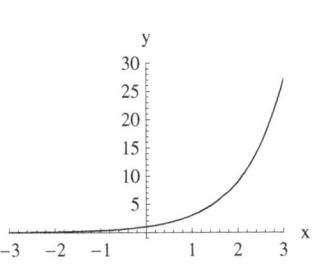

3. (a) (b)

x	$f(x)$
-4	81
-2	9
-1	3
0	1
1	0.33
2	0.11
4	0.01

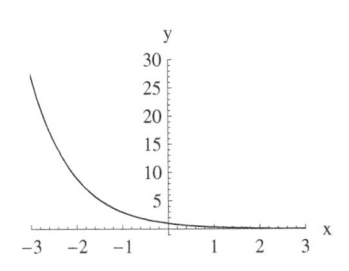

5. (a) (b)

x	$f(x)$
-4	0.45
-3	0.54
-1	0.81
0	1
1	1.22
2	1.49
4	2.22

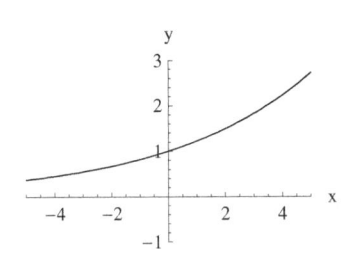

7. (a) (b)

x	$f(x)$
-4	2.22
-3	1.82
-1	1.22
0	1
1	0.81
2	0.67
4	0.45

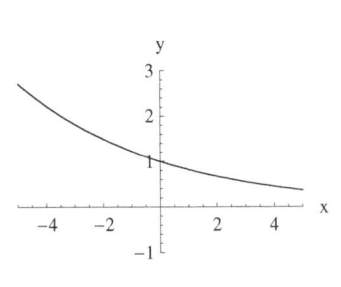

9. $920, $120

11. $928.95, $128.95 [2]

13. $3,382.15, $382.15 [2]

15. $12,778.2, $778.2 [2]

17. (a) $6,400.42 (b) $6,418.46 [2]
(c) $6,419.74 (d) $6,420.07 (e) $6,420.13

19. (a) $7,427.64 (b) $7,445.01 [2]
(c) $7,446.24 (d) $7,446.56 (e) $7,446.61

21. (a) $P(5) = 6,388.89$ [2]
(b) $1,388.89

23. (a) $P(5) = 6,788.45$ (b) $788.45 [2]

25. (a) $P(5) = 10,272.2$ (b) $2,272.2 [2]

27. (a) $P(5) = 3,571.02$ (b) $571.02 [2]

29. (a) $P(5) = 8,549.73$ $1,549.73 [2]

31. $5,470.26 [2]

33. $4,570.17 [4]

35. $22,103.4 [2]

37. $18,096.7 [4]

39. $4,084.43 [4]

41. (a) $37,280.02 [4]
(b) $25,179.4 (c) $16,743.78

43. (a) Take Option A [3]
(Present value of Option A and B: $175,258 and $175,000)
(b) Take Option B (Present value of Option A and B: $174,837 and $175,000)

45. (a) Take Option A [3]
(Present value of Option A and B: $371,307 and $370,000)
(b) Take Option B (Present value of Option A and B: $367,442 and $370,000)

Exercises 1.6 [Ref. Example]

1. $\log 1000 = 3$ 3. $\ln b = k$
5. $\log_a J = h$ 7. $3^2 = 9$
9. $e^k = 10$ 11. $a^G = H$

13. (a)

x	$f(x)$
-2	0.25
-1	0.5
0	1
1	2
2	4
3	8

(b)

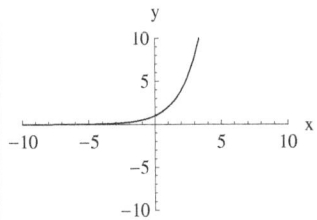

x	$g(x)$
0.25	-2
0.5	-1
1	0
2	1
4	2
8	3

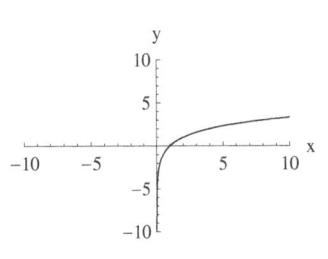

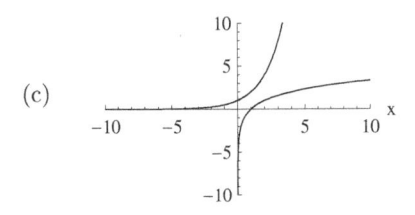

(c)

15. $t = \ln 100$ 17. $t = \dfrac{\ln 8}{\ln 5}$

19. $t = \dfrac{\ln M}{\ln a}$ 21. $t = \dfrac{\ln 100}{3}$

23. $t = \dfrac{\ln 1.4}{4}$ 25. $t = \dfrac{\ln 1.2}{0.02}$

27. $10^b = e^{b \ln 10}$ 29. $3^{-x} = e^{-x \ln 3}$

31. $a^h = e^{h \ln a}$ 33. $\log_3 K = \dfrac{\ln K}{\ln 3}$

35. $\log_2 7 = \dfrac{\ln 7}{\ln 2}$ 37. $\log_a H = \dfrac{\ln H}{\ln a}$

39. 3   41. -3 43. $1/3$

45. $t = \dfrac{\ln 1.5}{0.026} = 15.6$ years

47. $t = \dfrac{\ln 1.5}{0.076} = 5.33$ years

49. $k = \dfrac{\ln 1.25}{4} = 0.056 = 5.6\%$

51. $T = \dfrac{\ln 2}{0.056} = 12.38$ years

53. $k = \dfrac{\ln 2}{5} = 0.139 = 13.9\%$

55. $70/6 = 11.67$, $T = \dfrac{\ln 2}{0.06} = 11.55$ years

57. (a) $A(t) = 5,000 e^{0.045 t}$

(b) $A(5) = \$6,261.61$

(c) $T = \dfrac{\ln 2}{0.045} = 15.4$ years

(d) $T = \dfrac{\ln 1.6}{0.045} = 10.44$ years

59. (a) $\$29,845.3$

(b) $k = \dfrac{\ln 8/3}{17} = 0.0577 = 5.77\%$

61. Take Option (a) if $r < 22\%$, take Option (b) if $r \geq 22\%$

63. $r = 5.1\%$

65. (a) $P(t) = 600 e^{[(\ln 1.5)/1.5] t}$

(b) $P(5) = 2,318$

(c) $T = \dfrac{\ln 2}{(\ln 1.5)/1.5} = 2.564$ hours

(d) $t = \dfrac{\ln(40/3)}{(\ln 1.5)/1.5} = 9.58$ hours

67. (a) $P(t) = 17 e^{(\ln(91/17)/45) t}$
(b) $P(66) = 199$ million tons

69. (a) $T(t) = 55 e^{-(\ln(11/9)/2) t} + 75$

(b) $t = \dfrac{2 \ln(11/8)}{\ln(11/9)} \approx 3.17$ minutes.

71. (a) $T(t) = 110 e^{-(\ln(22/13)/30) t} + 75$

(b) $t = \dfrac{30 \ln(22/5)}{\ln(22/13)} \approx 84.5$ minutes

Chapter 1 Review Exercises

[3]

[3]

[3]

[4]

[4]

[4]

[4]

[4]

[4]

[5]

[5]

[6]

[6]

1. (a) 1   (b) DNE   (c) 0
   (d) -1 (e) 2   (f) 0

2. (a) $(-\infty, \infty)$ (b) 6, 0, -2, 0, 6
   (c) $50a^2 - 2$, $2a - 2$
   (d) $2x^2 + 4xh + 2h^2 - 2$ (e) $4x + 2h$, $h \neq 0$

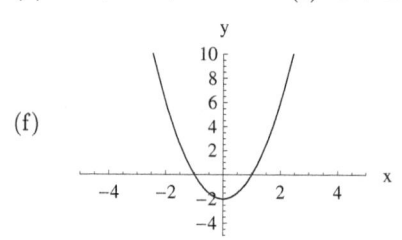

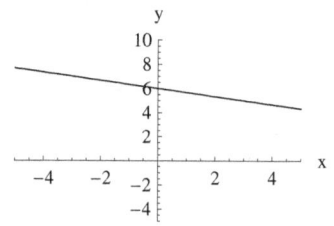

(f) [graph of parabola]

3. (a) $m = -1/3$ (b) [graph of line]

4. $y = 3x + 3$ 5. $y = -3x + 5$

6. (a) $V = -4,000 t + 57,000$.   (b) $\$29,000$

(c) $m = -4,000$. The value $m = -4,000$ means that if the age the car goes up by 1, then its resale value goes down by $\$4,000$. Therefore each additional year reduces the resale value by 4,000 dollars.

(d) The vertical intercept is $b = \$57,000$. It represents the value of the car when it was new.

(e) $T \approx 14.25$ years.

7. (a) $\$6,000$

(b) $15x$, $0 \leq x \leq 1,000$

(c) $R(x) = 25x$, $0 \leq x \leq 1,000$

(d) $P(x) = 10x - 6,000$, $0 \leq x \leq 1,000$

(e) $\$13,500$, $\$12,500$, $-\$1,000$

(f) $\$18,000$, $\$20,000$, $\$2,000$

(g) 600 chairs

(i) $m = \$10$ per chair. It represents the profit from producing and selling an additional chair.

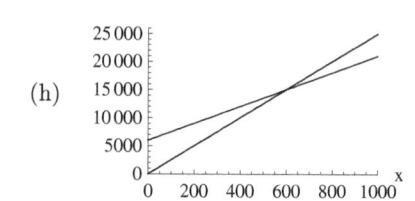

(h) [graph]

8. (a) $C(x) = 128 + 0.8x$

(b) The slope $m = \$0.8$ per coffee drink represents the cost of producing an additional coffee drink.

(c) $b = \$128$ represents the fixed daily cost.

9. (a) vertex $(-1.5, -5.5)$

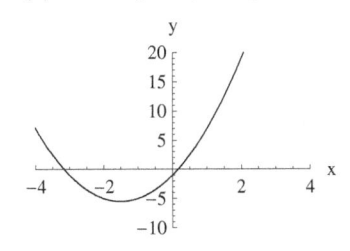

(b) vertex $(1.5, 5.5)$

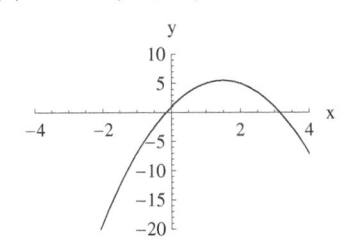

10. (a) $x = -2, 0, 2$ (b) $x = -2, -1, 1, 2$
11. (a) $x = 2 - \sqrt{6},\ 2 + \sqrt{6}$ (b) $x = 3$
12. (a) $x \neq 1, 3$ (b) $x = -3$
13. (a) $x \neq -2, 3$ (b) $x = 2$
14. (a) $x \neq -2$ (b) There is no real solution
15. (a) $x \leq -1$ or $x \geq 1$

(b) There is no real solution

16. (a) 1, 2, 3, 1, 6

(b) (c) (d)

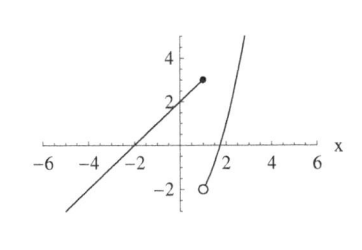

x	$g(x)$
-1	1
0	2
1	3
1.1	-1.79
2	1
3	6

17. (a) $7,764.85 (b) $7,834.96
(c) $7,840.04 (d) $7,841.34 (e) $7,841.56
18. $4,637.31 19. $25,425 20. $16,244.1
21. (a) $37,280.02 (b) $25,179.41 (c) $16,743.78
22. (a) Take Option A (Present value of Option A and B: $175,258 and $175,000)

(b) Take Option B (Present value of Option A and B: $174,837 and $175,000)

23. (a) $D(50) = 1,500$ (b) $S(50) = 600$

(c) higher (d) $p = \$100$

(e) $p = \$50$ (f) $p_E = \$80$, $x_E = 1,200$

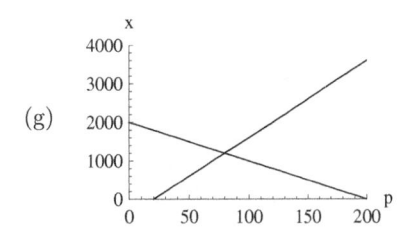

(g)

24. (a) $t = \ln 100$ (b) $t = \dfrac{\ln 120}{\ln 10}$
    (c) $t = \dfrac{\ln 100}{3}$ (d) $t = \dfrac{\ln 1.2}{0.02}$
25. (a) $5^{6x} = e^{(6\ln 5)x}$ (b) $2^{1/3} = e^{(\ln 2)/3}$
    (c) $a^h = e^{h \ln a}$ (d) $10^{-2.5} = e^{-2.5(\ln 10)}$
26. $t = \dfrac{\ln 1.5}{0.036} = 11.3$ years
27. $k = \dfrac{\ln 1.25}{4} = 0.056 = 5.6\%$
28. (a) $A(t) = 8,000 e^{0.051t}$
    (b) $A(5) = \$10,323.7$
    (c) $T = \dfrac{\ln 2}{0.051} = 13.6$ years
    (d) $t = \dfrac{\ln 7/4}{0.051} = 11$ years
29. (a) $29,845.3

(b) $k = \dfrac{\ln 8/3}{17} = 0.0577 = 5.77\%$

30. Take Option (a) if $r < 22\%$, take Option (b) if $r \geq 22\%$
31. (a) $P(t) = 500 e^{(\ln 1.8/1.5)t}$

(b) $P(5) = 3,547$

(c) $T = \dfrac{\ln 2}{\ln 1.8/1.5} = 1.77$ hours

(d) $t = \dfrac{\ln 16}{\ln 1.8/1.5} = 7.08$ hours

32. (a) $T(t) = 58 e^{-(\ln(58/48)/2)t} + 72$

(b) $t = \dfrac{2\ln(58/43)}{\ln(58/48)} \approx 3.16$

Chapter 2 Exercises Answers

Exercises 2.1 [Ref. Example]

1. 1   3. Does not exist
5. -1 7. -1 9. -1
11. 1   13. 1   15. 2
17. -2 19. -2 21. ∞
23. Does not exist   25. ∞ 27. 0
29. 0   31. $-1, 5$ [1]
33. $-1/4$, Does not exist   [5]

35. ∞, 4   [5,7]
37. 6, -5, Does not exist   [2]
39. 0, 0, 0   [2]
41. 0, 8   [1]

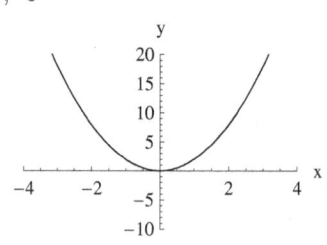

43. Does not exist, -1 [4]

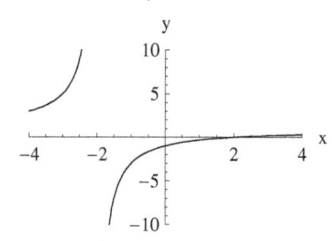

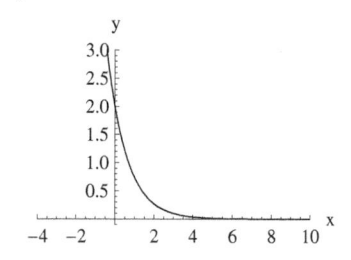

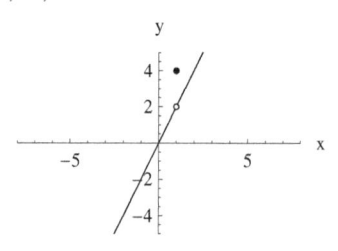

45. 2, 0   [1]

47. 2, 2, 2   [3]

49. -1, 0, Does not exist   [2]

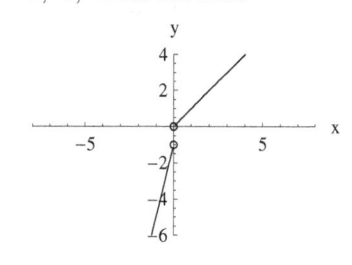

51. 2, 1   53. 1   [4]
55. ∞, -1 [6,7]
57. 4, 4, 4   59. 0, 0, 0   [3]

Exercises 2.2   [Ref. Example]

1. -1 3. 3   5. 3
7. e^3 9. $3x^2$ 11. No   13. No
15. (a) 1 (b) 1, -1, does not exist
    (c) No. Because $\lim_{x \to -1} f(x)$ does not exist.
    (d) 2 (e) 2, 2, 2
    (f) Yes. Because $\lim_{x \to -2} f(x) = 2 = f(-2)$.
17. (a) Does not exist (b) -1, -1, -1
    (c) No. Because $g(-1)$ does not exist.
    (d) 1 (e) 1, 1, 1
    (f) Yes. Because $\lim_{x \to 3} g(x) = 1 = g(3)$.
19. (a) 3 (b) 3, -2, does not exist   [5, 14]
    (c) No. Because $\lim_{x \to 1} f(x)$ does not exist.
21. (a) -3 (b) -3, -3, -3 [5, 14]
    (c) Yes. Because $\lim_{x \to 1} f(x) = -3 = f(1)$.
23. 4   25. 3   27. $1/2$ [9]
29. 4   31. 3   33. $2x$ [11]
35. -1 37. $-1/x^2$ [10, 11]
39. Does not exist   41. ∞ [12,13]
43. Does not exist   45. ∞ [12,13]

Exercises 2.3   [Ref. Example]

1. (a) $6x + 3h, h \neq 0$ [4]
   (b) 21, 18.6, 18.3, 18.01
3. (a) $2x - 1 + h, h \neq 0$ [4]
   (b) 2, 1.2, 1.1, 1.01
5. (a) $5, h \neq 0$ [6]
   (b) 5, 5, 5, 5
7. (a) $\dfrac{-3}{x(x+h)}, h \neq 0$ [7]
   (b) -2, -2.73, -2.97, -2.997
9. (a) $\dfrac{-4x - 2h}{x^2(x+h)^2}, h \neq 0$ [7]
   (b) -0.36, -0.4649, -0.496, -0.4996
11. (a) $0, h \neq 0$ [6]
    (b) 0, 0, 0, 0
13. (a) $-4x - 2h, h \neq 0$ [4]
    (b) -5, -4.2, -4.02, -4.002
15. (a) $-3x^2 - 3xh - h^2, h \neq 0$ [5]
    (b) -19, -13.24, -12.61, -12.06
17. (a) $2x - 2 + h, h \neq 0$ [4]
    (b) 9, 8.2, 8.1, 8.01

CHAPTER 6 TEST 467

19. (a) −128 ft [10]
 The rock drops 128 ft from the 1st second to the 3rd second after it is dropped.
 (b) −64 ft/sec. It represents the average velocity of the rock from the 1st second to the 3rd second after it is dropped.

21. (a) 96 ft [10]
 The rock rises 96 ft in the first 2 seconds after it is thrown.
 (b) 48 ft/sec. It represents the average velocity of the rock in the first 2 seconds after it is thrown.

23. (a) $163.24. It represents [9] the interest earned in the first year.
 (b) $163.24 per year. This average rate of change represents the average annual interest earned by Rich's investment during the first year.
 (c) $333.15. It represents the interest earned in the first two years.
 (d) $166.57 per year. This average rate of change represents the average annual interest earned by Rich's investment during the first 2 years.
 (e) The answer in (d) is larger. In the second year the principal is larger.

25. (a) 112 pairs. It represents the number of sunglasses sold in the first 2 weeks.
 (b) 56 pairs per week. It represents the average weekly sales of the sunglasses in the first 2 weeks.
 (c) 48 pairs. It represents the number of sunglasses sold in the 9th and 10th weeks.
 (d) 24 pairs per week. It represents the average weekly sales of the sunglasses in the 9th and 10th weeks.
 (e) The answer in (b) is larger. Possible market exhaustion.

27. $35.98 per pair. It represents the cost of producing the 101th pair of shoes.

29. $79.9 per pair. It represents the price of selling the 101th pair of sunglasses.

Exercises 2.4 [Ref. Example]

1. (a) $6x + 3h, h \neq 0$ (b) $6x$ [1]
 (c) −6, 0, 6, 12
3. (a) $2x - 1 + h, h \neq 0$ (b) $2x - 1$ [1]
 (c) −5, −1, 0, 1
5. (a) $5, h \neq 0$ (b) 5 [3]
 (c) 5, 5, 5
7. (a) $\frac{-3}{x(x+h)}, h \neq 0$ (b) $\frac{-3}{x^2}$ [4]

 (c) −3, −3, −1/3

9. (a) $\frac{2}{x^2}$ (b) 2, 0.5 [6]
 (c) $y = 2x + 4, y = 0.5x - 2$
 (d)

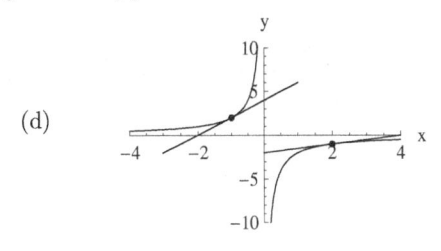

11. (a) $-2x$ (b) 4, 0, −2 [6]
 (c) $y = 4x + 4, y = 0, y = -2x + 1$
 (d)

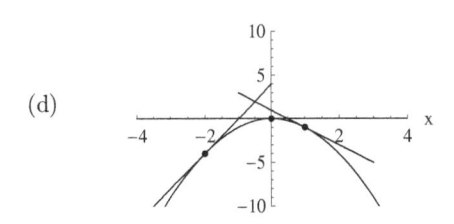

13. (a) $-4x$ (b) 4, 0, −8 [6]
 (c) $y = 4x + 4, y = 2, y = -8x + 10$
 (d)

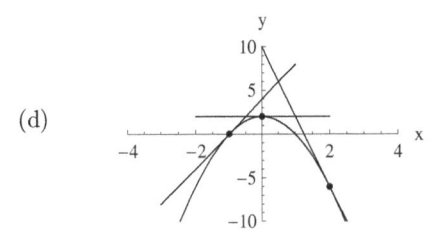

15. (a) $-3x^2$ (b) −3, 0, −3 [2,6]
 (c) $y = -3x - 2, y = 0, y = -3x + 2$
 (d)

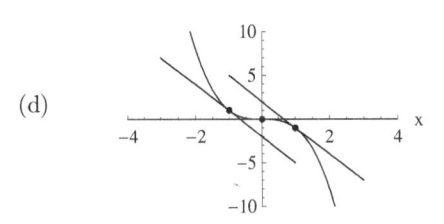

17. (a) $2x - 2$ (b) −6, −2, 0 [6]
 (c) $y = -6x - 1, y = -2x + 3, y = 2$
 (d)

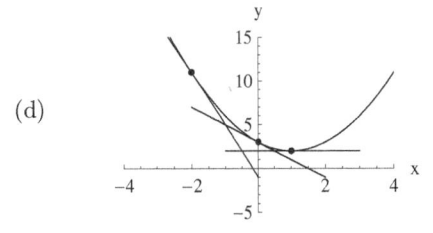

19. x_1: discontinuity; x_2: corner; x_3: corner; x_4: discontinuity; x_6: vertical tangent line; x_7: corner; x_8: corner;

21. (a) $v(t) = -32t$, $0 \le t \le 4$ [5]

    (b) $v(3) = -96$ ft/sec. It represents the velocity of the rock at 3 seconds. That is, at 3 seconds after the rock is dropped, it is falling with a speed of 96 ft/sec.

23. (a) $v(t) = -32t + 80$, $0 \le t \le 5$ [5]

    (b) $v(3) = -16$ ft/sec. It represents the velocity of the rock at 3 seconds. That is, at 3 seconds after the rock is thrown, it is falling with a speed of 16 ft/sec.

    (c) The rock is falling since the velocity is negative.

Exercises 2.5 [Ref. Example]

1. $14x^6$ 3. $\dfrac{3}{2\sqrt{x}}$

5. $\dfrac{-21\sqrt[5]{x^2}}{5}$ 7. $\dfrac{-9}{5\sqrt[5]{x^2}}$

9. $-\dfrac{4}{x^3}$ 11. $\dfrac{-1}{\sqrt{x^3}}$

13. $\dfrac{5}{7}$ 15. $\dfrac{7e^x}{2}$

17. $\dfrac{3}{8\sqrt[4]{x^5}}$ 19. $0.24x^{-1.1}$

21. $6x^2 - 20x + 6e^x$

23. $\dfrac{3}{2\sqrt{x}} - 3e^x - \dfrac{2}{x^2}$

25. $\dfrac{5}{7} + \dfrac{3}{2\sqrt{x}} + \dfrac{10}{3\sqrt[3]{x^5}}$

27. $\dfrac{2x}{3} - \dfrac{6}{x^3}$

29. $5e^x - ex^{e-1}$

31. $16x - 6$ 33. 16

35. $\dfrac{5}{2\sqrt{x}} + \dfrac{3}{2\sqrt{x^3}}$

37. $2520x^2$ 39. $3e^x$

41. $3, 3/8$

43. (a) $y = 13x + 6$ [6]

    (b) $y = x - 1$ (c) $y = x - 2$

45. $(0, -1)$ 47. $(0, -2)$ [7]

49. $(5, 25)$ 51. $(2, 1.8)$ [7]

53. $(-2, 21)$ and $(1, -6)$ [7]

55. $(0, 0)$ and $(4, -10\dfrac{2}{3})$ [7]

57. $(-2, 19)$ and $(2, -13)$ [7]

59. All points on the line: $(x, 7)$

61. No solution

63. (a) $v(t) = -32t$, [9]

    $a(t) = -32$, $0 \le t \le 4$

    (b) $s(3) = 112$ ft, $v(3) = -96$ ft/sec,

    $a(3) = -32$ ft/sec^2. That is, at 3 seconds after the rock is dropped, it is at 112 ft above the ground, is falling with a speed of 96 ft/sec, and the falling speed is increasing by 32 ft/sec^2.

65. (a) $v(t) = -32t + 80$, [9]

    $a(t) = -32$, $0 \le t \le 5$

    (b) About 2.5 seconds after it is thrown.

    (c) 100 ft

    (d) -32 ft/sec^2

Exercises 2.6 [Ref. Example]

1. (a) $f(x) = x^9$, $f'(x) = 9x^8$

    (b) $f'(x) = 7x^6 \cdot x^2 + x^7 \cdot 2x = 7x^8 + 2x^8 = 9x^8$

3. (a) $f(x) = 6x^5 + 3x^4$, $f'(x) = 30x^4 + 12x^3$

    (b) $f'(x) = 9x^2 \cdot (2x^2 + x) + 3x^3 \cdot (4x + 1)$
    $= 18x^4 + 9x^3 + 12x^4 + 3x^3 = 30x^4 + 12x^3$

5. (a) $f(x) = x^{1/2} \cdot x^{1/3} = x^{5/6}$,

    $f'(x) = \dfrac{5}{6}x^{-1/6} = \dfrac{5}{6\sqrt[6]{x}}$

    (b) $f'(x) = \dfrac{1}{2}x^{-1/2} \cdot x^{1/3} + x^{1/2} \cdot \dfrac{1}{3}x^{-2/3}$
    $= \tfrac{1}{2}x^{-1/6} + \tfrac{1}{3}x^{-1/6} = \tfrac{5}{6}x^{-1/6} = \dfrac{5}{6\sqrt[6]{x}}$

7. (a) $f(x) = x^4$, $f'(x) = 4x^3$

    (b) $f'(x) = \dfrac{6x^5 \cdot x^2 - x^6 \cdot 2x}{x^4} = \dfrac{4x^7}{x^4} = 4x^3$

9. (a) $f(x) = \dfrac{1}{x^3} - 2x^3$, $f'(x) = \dfrac{-3}{x^4} - 6x^2$

    (b) $f'(x) = \dfrac{(2x - 16x^7) \cdot x^5 - (x^2 - 2x^8) \cdot 5x^4}{x^{10}}$
    $= \dfrac{-3x^6 - 6x^{12}}{x^{10}} = \dfrac{-3}{x^4} - 6x^2$

11. (a) $f(x) = x - 1$, $x \ne -1$; $f'(x) = 1$, $x \ne -1$

    (b) $f'(x) = \dfrac{2x \cdot (x + 1) - (x^2 - 1) \cdot 1}{(x + 1)^2}$
    $= \dfrac{x^2 + 2x + 1}{x^2 + 2x + 1} = 1$, $x \ne -1$

13. $(14x^6 + 20x^3) \cdot (7x^2 - 6x + 1)$
    $+ (2x^7 + 5x^4 - 3) \cdot (14x - 6)$

15. $\left(\dfrac{3}{2\sqrt{x}} - \dfrac{8}{x^2}\right) \cdot (e^x + x^2 + 4)$
    $+ (3\sqrt{x} + \tfrac{8}{x}) \cdot (e^x + 2x)$

17. $\dfrac{-4x^3 + 3x^2 + 10}{(x^3 + 5)^2}$

19. $\dfrac{2e^x(x - 2)}{x^3}$

21. $\dfrac{2(1 - x)}{e^x}$

23. $\dfrac{dy}{dx} = \dfrac{5}{7}$

25. $\dfrac{7e^x(2x-1)}{(2x+1)^2}$

27. $\dfrac{(2x-3)(2\sqrt[4]{x}+3) - (x^2-3x+2)/(2\sqrt[4]{x^3})}{(2\sqrt[4]{x}+3)^2}$

29. $2 + 3x^2 e^x + x^3 e^x$

31. $(2x^3 - 4x^2 + 12e^x - 20x - 3)e^x$

33. $\left(\dfrac{3}{2\sqrt{x}} + 3\sqrt{x}\right)e^x - \dfrac{2}{x^2}$

35. $\dfrac{-35x^2 + 15}{(7x^2+3)^2}$
    $+ 4.5(x+1)/\sqrt{x} - 2$

37. $\dfrac{3x(2-x)e^x + 4x}{(3e^x+2)^2}$
    $+ \dfrac{3(x-2)e^x - 4}{x^3}$

39. $\dfrac{3e^x(x^3 - 5x^2 + 4x + 4)}{(x^2+4)^2}$

41. $18x - 28$

43. $(x^2 + 4x + 2)e^x$

45. $\dfrac{54}{(3x+1)^3}$

47. $y = 5x + 2$

49. (a) $y = \dfrac{8}{9}x + \dfrac{20}{9}$ (b) $y = 2$
    (c) $y = -\dfrac{8}{9}x + \dfrac{20}{9}$

Exercises 2.7 [Ref. Example]

1. $7(2x^3 + 5x^2 - 3)^6 (6x^2 + 10x)$

3. $\dfrac{3(3x^2 + e^x)}{\sqrt{2x^3 + 2e^x}}$

5. $0.3e^{0.06x}$

7. $2(6x+1)e^{3x^2+x}$

9. $-0.04e^{-0.02x}$

11. $(27x - 8)(x+1)^6 (3x-2)$

13. $4x^4(6x^2 + 5)e^{3x^2}$

15. $\sqrt{x^2+1} + \dfrac{x^2}{\sqrt{x^2+1}}$, or $\dfrac{2x^2+1}{\sqrt{x^2+1}}$

17. (a) $f(x) = (x^3 + 2x)(x^2+1)^{-1}$,
    $f'(x) = (3x^2+2)(x^2+1)^{-1}$
    $- 2x(x^3+2x)(x^2+1)^{-2}$
    (b) $\dfrac{(3x^2+2)(x^2+1) - (x^3+2x)(2x)}{(x^2+1)^2}$

19. (a) $f(x) = (x^2 - 2x)e^{-x}$,
    $f'(x) = (4x - x^2 - 2)e^{-x}$
    (b) $\dfrac{(2x-2)e^x - (x^2 - 2x)e^x}{(e^x)^2}$
    $= \dfrac{4x - x^2 - 2}{e^x}$

21. $\left(\dfrac{5x+3}{7x-2}\right)^2 \cdot \dfrac{-93}{(7x-2)^2}$

23. $\dfrac{1}{2}\sqrt{\dfrac{3x-5}{2x+3}} \cdot \dfrac{-19}{(3x-5)^2}$

25. $-42(2x+1)^{-4}$

27. $5(3x^2 + e^x)^4 (6x + e^x)$

29. $4x + (3x^2 + 4x^3)e^{4x+1}$

31. $2(10x^2 + 6x - 3)(10x^2 + 26x + 3)e^{2x+5}$

33. $\dfrac{3e^x}{\sqrt{2x+1}} + 3\sqrt{2x+1}\,e^x$,
    or $\dfrac{6(x+1)e^x}{\sqrt{2x+1}}$

35. $\tfrac{1}{3}(x^5 - 3x^3 + 2)^{-2/3} \cdot (5x^4 - 9x^2)$

37. $y' = 8x(x^2+1)^3 \cdot (3e^x + 2)^{-1}$
    $- 3(x^2+1)^4 \cdot (3e^x+2)^{-2} \cdot e^x$

39. $6xe^{x^2} + 10(2x+1)^4$

41. $2(1 + 2x^2)e^{x^2}$

43. $24(x^2+1)^2 (7x^2+1)$

45. $y = 2x + 1$

47. (a) $y = 2x + 3$ (b) $y = 4$
    (c) $y = -2x + 3$

Exercises 2.8 [Ref. Example]

1. (a) $(3x+1)^2$ (b) $3x^2 + 1$ [1]

3. (a) $\sqrt{x^2 - 1}$ (b) $x - 1$, $x > 0$ [1]

5. (a) $3e^{2x} + 2e^x + 1$, (b) $e^{3x^2 + 2x + 1}$ [1]

7. $f(u) = u^5$, $g(x) = 2x + 3$

9. $f(u) = \sqrt{u}$, $g(x) = e^x + 2$

11. $f(u) = e^u$, $g(x) = 2x^2 + x$

13. $\dfrac{dy}{du} = 3u^2$, $\dfrac{du}{dx} = 3e^x$ [2]
    $\dfrac{dy}{dx} = 3u^2 \cdot 3e^x = 9(3e^x + 2)^2 e^x$

15. $\dfrac{dy}{du} = e^u$, $\dfrac{du}{dx} = 6x$ [2]
    $\dfrac{dy}{dx} = e^u \cdot 6x = 6xe^{3x^2+4}$

17. $\dfrac{dy}{du} = \dfrac{1}{u}$, $\dfrac{du}{dx} = 4x^3$ [2,4]
    $\dfrac{dy}{dx} = \dfrac{1}{u} \cdot 4x^3 = \dfrac{4x^3}{x^4+1}$

19. $2 + \dfrac{3}{x} + e^x$ [4]

21. $\dfrac{20x}{2x^2+3}$ [5]

23. $\dfrac{5}{5x+3} - \dfrac{4}{4x+1}$ [6]

25. $\dfrac{1}{x}$ [5]

27. $\dfrac{1}{4x}$ [6]

29. $\ln x + 1$ [4]

31. $6(\ln x + 4e^x - 3x^2)^5 \cdot (\dfrac{1}{x} + 4e^x - 6x)$

33. $y' = \dfrac{3(\ln 6x + 2)}{2\sqrt{x}}$

35. $\dfrac{3(\ln x)^2}{x}$

37. $\dfrac{-9}{(3x+1)^2}$

39. $2\ln x + 3$

41. $y = 5x - 5$

43. (a) $y = \dfrac{1}{3}x + (\ln 3 + \dfrac{1}{3})$

    (b) $y = \dfrac{1}{4}x + \ln 4$

    (c) $y = \dfrac{1}{5}x + (\ln 5 - \dfrac{1}{5})$

Chapter 2 Review Exercises

1. (a) 1 (b) 1 (c) 1 (d) −3 (e) 2
   (f) Does not exist (g) −2 (h) −2
   (i) −2 (j) 2

2. (a) 1 (b) ∞, −1 (c) Does not exist, 0

3. (a) 3 (b) $\dfrac{-4}{x^2}$

4. (a) no (b) 1, −1, does not exist, 1
   (c) No. Because $\lim\limits_{x \to -1} f(x)$ does not exist.
   (d) 2, 2, 2, 2
   (e) Yes. Because $\lim\limits_{x \to -2} f(x) = 2 = f(-2)$

5. (a) −1 (b) 1, −1, does not exist
   (c) No. Because $\lim\limits_{x \to 1} f(x)$ does not exist.

6. (a) 3 (b) $-\dfrac{1}{2}$ (c) 12 (d) $10x$ (e) $\dfrac{-2}{x^2}$

7. (a) −128 ft. The rock drops 128 ft from the 1st second to the 3rd second after it is dropped.

   (b) −64 ft/sec. It represents the average velocity of the rock from the 1st second to the 3rd second after it is dropped.

   (c) $v(t) = -32t$, $0 \le t \le 4$

   (d) $v(2) = -64$ ft/sec. It represents the velocity of the rock at 2 seconds. That is, at 2 seconds after the rock is dropped, it is falling down with a speed of 64 ft/sec.

8. (a) $1202, 06$. It represents the increase in Jan's balance in the first 3 years.

   (b) $400.69/yr. It represents the average annual increase in Jan's balance in the first three years.

9. (a) $y = 13x + 6$ (b) $y = x - 1$ (c) $y = x - 2$

10. $(1/3, 98/27)$ and $(2, -1)$

11. $20x^3 - 3 - 7e^x - 2/x$

12. $(\dfrac{3}{2\sqrt{x}} - \dfrac{8}{x^2})(e^x + x^2 + 4) + (3\sqrt{x} + \dfrac{8}{x})(e^x + 2x)$

13. $\dfrac{5}{7} + \dfrac{3}{\sqrt{x}} + \dfrac{10}{3\sqrt[3]{x^5}}$

14. $\dfrac{x^2 - 4x - 6}{(x-2)^2}$

15. $5e^x + ex^{e-1}$

16. $\dfrac{10x}{x^2 + 1} + 0.28e^{-0.04x}$

17. $7(2x^3 + 5x^2 - 3)^6(6x^2 + 10x)$

18. $5xe^{x^2/2}$

19. $14x(x^2+1)^6(3x-2)^2 + 6(x^2+1)^7(3x-2)$

20. $(3x^2 + 3 - x^3 - 3x)e^{-x}$

21. $\dfrac{dy}{dx} = \dfrac{1}{2}\sqrt{\dfrac{7x-3}{5x+1}} \cdot \dfrac{-22}{(7x-3)^2}$

22. $2(10x^2 + 26x + 3)(10x^2 + 6x - 3)e^{2x+5}$

23. $\dfrac{3\ln x}{\sqrt{2x+1}} + \dfrac{3\sqrt{2x+1}}{x}$

24. $840x^3$

25. $3, 3/8$

26. (a) $\dfrac{dy}{du} = 3u^2 + e^u$, $\dfrac{du}{dx} = \dfrac{1}{2\sqrt{x}}$

    $\dfrac{dy}{dx} = \dfrac{3x + e^{\sqrt{x}}}{2\sqrt{x}}$

    (b) $\dfrac{dy}{du} = \dfrac{1}{2\sqrt{u}}$, $\dfrac{du}{dx} = 2x + \dfrac{1}{x}$

    $\dfrac{dy}{dx} = \dfrac{2x + 1/x}{2\sqrt{x^2 + \ln x}}$

27. (a) $v(t) = -32t + 80$, $a(t) = -32$
    (b) $t = 2.5$ second (c) 100 ft (d) -32 ft/sec^2

28. (a) $93.96/yr. At the end of the first year, the balance is increasing at a rate of $93.96 per year.

    (b) $112.45/yr. At the end of the fifth year, the balance is increasing at a rate of $112.45 per year.

    (c) The answer in (b) is greater, since the principal is larger.

    (d) (i) and (iii)

Chapter 3 Exercises Answers

Exercises 3.1 [Ref. Example]

CHAPTER 6 TEST

1. (a) When the outside temperature [1]
is 36°F, it costs $4.50 per day to heat the house.
(b) When the outside temperature is 36°F, the daily heating cost is decreasing at a rate of $0.25/°F.
(c) About $4.00 per day
(d) About $5.25 per day

3. (a) At the end of the 12th month, [1]
the account balance is $1,095.
(b) At the end of the 12th month, the account balance is increasing at a rate of $8 per month.
(c) About $1,099.

5. (a) At the price of $120 per skateboard, [1]
500 skateboards will be sold.
(b) If the price increases from $120 per skateboard, the number of skateboards sold will decrease at the rate of 10 skateboards per dollar.
(c) About 460.

7. (a) 25 minutes after the chicken [1]
is placed in the oven, the temperature of the middle of the chicken is 95°F.
(b) 25 minutes after the chicken is placed in the oven, the temperature of the middle of the chicken is increasing at a rate of 5°F per minute.
(c) About 110°F
(d) About 130°F
(e) The answer in (c) is more accurate. We assume that the temperature changes linearly. But by Newton's law of cooling, it follows an exponential model.

9. (a) $\Delta x = 0.5$, $\Delta y = 3.25$ [2]
(b) $\Delta x = h$, $\Delta y = 6h + h^2$
(c) $\Delta y = 0.61$, $f'(2)\Delta x = 0.6$
(d) 8.6, 8.61

11. $\Delta y = 1.5$, $f'(3)\Delta x = 1.5$ [2]
13. $\Delta y = -0.23077$, $f'(1)\Delta x = -0.3$ [2]
15. $\Delta y = -0.075$, $f'(1)\Delta x = -0.08$ [2]
17. $\Delta y = -0.009975$, $f'(0)\Delta x = -0.01$ [2]
19. $\Delta y = 0.2624$, $f'(0)\Delta x = 0.3$ [2]
21. $\Delta y = 0.12048$, $f'(1)\Delta x = 0.12$ [2]
23. 0.02   25. 2.967 [3]
27. 3.074   29. 2.25 [3]
31. 2.944   33. 0.283 [3]
35. 0.5156 [3]
37. 0.75 [5]
39. 0.1109, 0.11 [5]
41. 3 [5]

Exercises 3.2 [Ref. Example]

1. (a) $MC(x) = C'(x) = 42.4x^{-1/5}$ [1,2]
(b) $MC(77) = \$17.79$/skateboard.
It approximately represents the cost of producing the 78th skateboard.
(c) $17.76
(d) About $2,947.44

3. About $1,424 [2]

5. (a) $P(x) = 275x - 0.05x^2$ [3]
$+300e^{-0.02x} - 3,000$
(b) $15,959.4, $40,000, $24,040.6
(c) $135.8 per set, $400 per set, $264.2 per set
(d) About $16,231, $40,800, about $24,569

7. (a) $P(x) = 70x - 0.2x^2$ [3]
$-100\ln(x+5) - 1,500$
(b) $3,837.44, $5,400, $1,562.56
(c) $45.54 per pair, $90 per pair, $44.46 per pair
(d) About $3,974.06, $5,670, about $1,695.94

9. $-$0.26/skateboard per skateboard [4]
11. $-$0.12/skateboard per skateboard [4]
13. $0.14/skateboard per skateboard [4]
15. $0.29/refrigerator per refrigerator [4]
17. (a) 1,125 sets   (b) $101,563 [5]
19. (a) 175 pairs   (b) $4,625 [5]
21. Since $MP(1,500) = 0.3 > 0$, $P_{\max}$ occurs above 1,500.
23. (a) $p = D(x) = 840 - 2x$ [7]
(b) $R(x) = 840x - 2x^2$
(c) Approximately 103 refrigerators
25. (a) $p = D(x) = 80 - 0.2x$ [7]
(b) $R(x) = 80x - 0.2x^2$
(c) Approximately 179 customers
(d) $44.2

Exercises 3.3 [Ref. Example]

1. f is increasing on $(-1, 1.5)$ [1]
f is decreasing on $(-4, -1)$ and $(1.5, 4)$
3. g is increasing on $(-2.5, -1)$ and $(1, 2.5)$ [1]
g is decreasing on $(-4, -2.5)$ and $(2.5, 4)$
g is constant on $(-1, 1)$
5. $f' > 0$ on $(-1.5, 0)$ and $(0, 1.5)$ [1]
$f' < 0$ on $(-4, -1.5)$ and $(1.5, 4)$

$f' = 0$ at $x = -1.5, 1.5$
At $x = 0$ f' does not exist

7. $h' > 0$ on $(-1, 1.5)$ [1]
$h' < 0$ on $(-4, -1)$ and $(1.5, 4)$
$h' = 0$ at $x = -1, 1.5$

9. (a) $(1.5, 2)$ (b) $(-1.5, -2.5)$
(c) $(-4, 2.5)$ (d) $(-1.5, -2.5)$

11. f is concave up on $(-4, 0)$,
f is concave down on $(0, 4)$,
f has an inflection point at $(0, -1)$

13. f' is increasing on $(-4, 0)$,
f' is decreasing on $(0, 4)$

15. f'' is positive on $(-4, 0)$,
f'' is negative on $(0, 4)$

17. f has no inflection point [3]

19. f has inflection points at
$(-1/\sqrt{3}, -5/9)$ and $(1/\sqrt{3}, -5/9)$ [3]

21. h has an inflection point at $(3, 3)$ [3]

23. The function has inflection points [3]
at $(0.59, 0.19)$ and $(3.41, 0.38)$

25. (a) $(-3, -2)$ and $(0.5, 4)$
(b) $(-4, -3)$ and $(-2, 0.5)$
(c) $(-3, 1), (-2, 3), (-1, 1.5)$,
$(0.5, -2), (3, 0.5)$
(d) $(-2, 3)$
(e) $(-3, 1)$ and $(0.5, -2)$
(f) $(4, 4)$ (g) $(0.5, -2)$
(h) $(-1, 1.5), (1.5, -1), (3, 0.5)$
(i) $(-1, 1.5)$ and $(3, 4)$
(j) $(-4, -3), (-3, -1)$ and $(1.5, 3)$

27. (a) $(-2, 12)$ and $(2, -20)$ [5]
(b) $(-\infty, -2)$ and $(2, \infty)$
(c) $(-2, 2)$
(d) f has a relative maximum at $(-2, 12)$,
and a relative minimum at $(2, -20)$
(e) f has an inflection point at $(0, -4)$
(f) $(0, \infty)$ (g) $(-\infty, 0)$

29. (a) $(0, 0)$ and $(3/2, -27/16)$ [5]
(b) $(3/2, \infty)$
(c) $(-\infty, 0)$ and $(0, 3/2)$
(d) f has a relative minimum
at $(3/2, -27/16)$
(e) f has inflection points
at $(0, 0)$ and $(1, -1)$
(f) $(-\infty, 0)$ and $(1, \infty)$ (g) $(0, 1)$

Exercises 3.4 [Ref. Example]

1. $x = 2.5$ [1]
3. There is no critical number [1]
5. $x = 1, 5$ 7. $x = 0, 2$ [1]
9. $x = -2, 2$ 11. $x = -3$ [2]
13. $x = 2$ 15. $x = -3, 0$ [3]
17. Critical number: $x = 2$ [4]

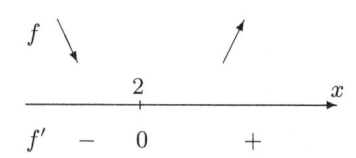

```
                 Relative
                 minimum
     f  ↘           ↗
    ─────────┼──────────→ x
             2
    f'   −   0   +
```

19. Critical number: $x = 2.5$ [4]

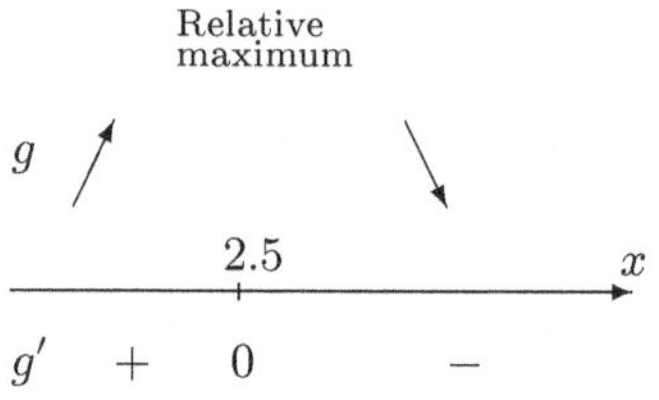

```
                 Relative
                 maximum
     g  ↗           ↘
    ─────────┼──────────→ x
            2.5
    g'   +   0   −
```

21. Critical numbers: $x = 1, 5$ [4]

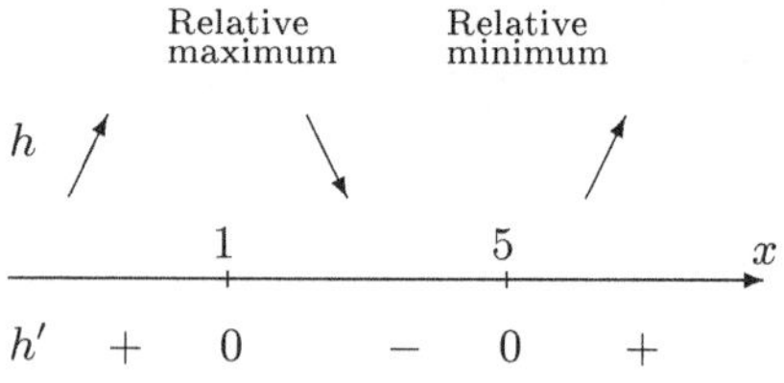

```
              Relative    Relative
              maximum     minimum
     h  ↗        ↘           ↗
    ──────┼──────────┼─────────→ x
          1          5
    h'  +  0    −    0    +
```

23. Critical numbers: $x = 0, 2$ [4]

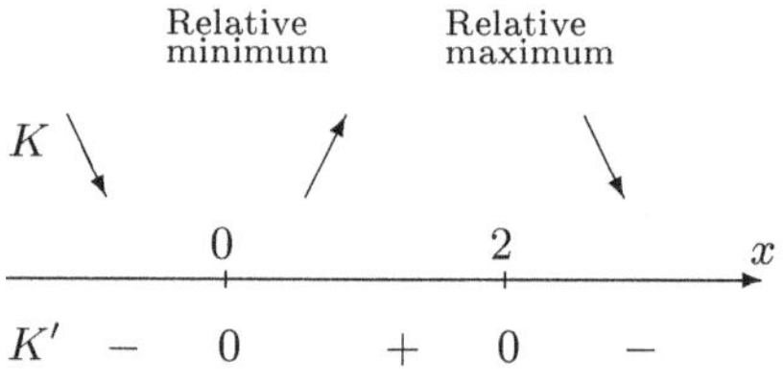

```
              Relative    Relative
              minimum     maximum
     K  ↘        ↗           ↘
    ──────┼──────────┼─────────→ x
          0          2
    K'  −  0    +    0    −
```

25. Critical numbers: $x = 0, 3/4$ [4]

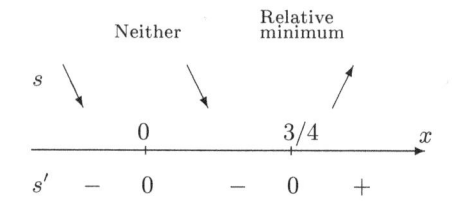

```
                        Relative
              Neither   minimum
     s  ↘        ↘           ↗
    ──────┼──────────┼─────────→ x
          0         3/4
    s'  −  0    −    0    +
```

27. Critical numbers: $x = -2, 0, 2$ [4]

CHAPTER 6 TEST 473

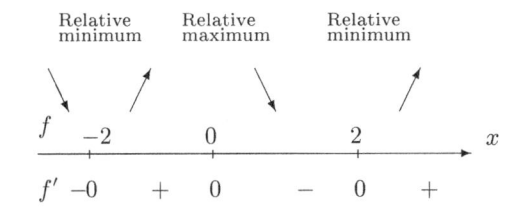

```
              Relative    Relative    Relative
              minimum    maximum     minimum
                ↘          ↗           ↘         ↗
         f     ────┼──────┼──────────┼────────→
                  -2      0          2          x
         f'   -0   +   0   -   0   +
```

29. Critical number: $x = 2$ [4]

```
              Relative
              minimum
         g    ↘         ↗
              ────┼─────────→
                  2          x
         g'    -    none   +
```

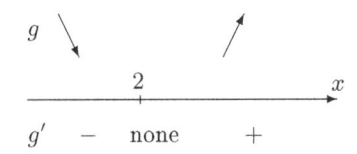

31. Critical number: $x = 2$ [4]

```
              Neither
         h    ↗         ↗
              ────┼─────────→
                  2          x
         h'    +    none   +
```

33. Critical number: $x = 2$ [5]
    $f''(2) = 2 > 0$: f has a relative
    minimum at $x = 2$

35. Critical number: $x = 2.5$ [5]
    $g''(2.5) = -1 < 0$: g has
    a relative maximum at $x = 2.5$

37. Critical numbers: $x = 1, 5$ [5]
    $h''(1) = -4 < 0$: h has
    a relative maximum at $x = 1$
    $h''(5) = 4 > 0$: h has
    a relative minimum at $x = 5$

39. Critical numbers: $x = 0, 2$ [5]
    $K''(0) = 2 > 0$: K has
    a relative minimum at $x = 0$
    $K''(2) = -2e^{-2} < 0$: K has
    a relative maximum at $x = 2$

41. Critical numbers: $x = 0, 3/4$ [5]
    $s''(0) = 0$: No conclusion. $s''(3/4) = 9/4 > 0$:
    s has a relative minimum at $x = 3/4$

43. Critical numbers: $x = -2, 0, 2$ [5]
    $f''(-2) = 32 > 0$: f has a relative minimum at
    $x = -2$. $f''(0) = -16 < 0$: f has a relative
    maximum at $x = 0$. $f''(2) = 32 > 0$: f has a
    relative minimum at $x = 2$.

45. (i) f has a relative minimum at $(2.5, -0.25)$
    f has no relative maximum
    (ii) f is increasing on $(2.5, \infty)$
    f is decreasing on $(-\infty, 2.5)$

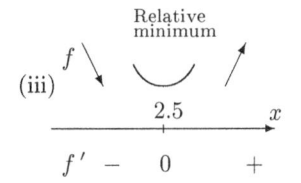

```
              Relative
              minimum
         f   ↘         ↗
    (iii)    ────┼─────────→
                 2.5         x
         f'    -   0    +
```

47. (i) g has a relative maximum at $(0, 4)$
    g has a relative minimum at $(2, 0)$
    (ii) g is increasing on $(-\infty, 0)$ and $(2, \infty)$
    g is decreasing on $(0, 2)$

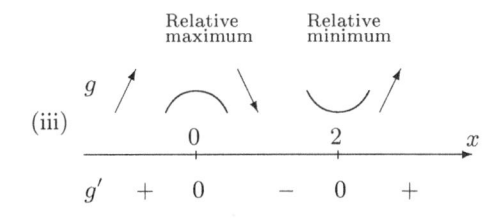

```
               Relative     Relative
               maximum      minimum
         g    ↗       ↘     ↘       ↗
    (iii)    ────┼──────────┼────────→
                 0          2         x
         g'    +   0    -   0    +
```

49. (i) h has a relative maximum at $(1, 8\frac{1}{3})$
    h has a relative minimum at $(5, -2\frac{1}{3})$
    (ii) h is increasing on $(-\infty, 1)$ and $(5, \infty)$
    h is decreasing on $(1, 5)$

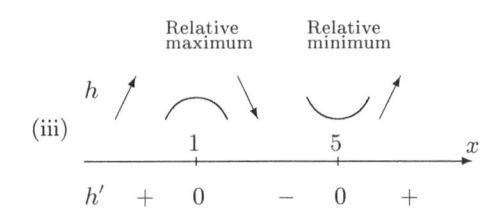

```
               Relative     Relative
               maximum      minimum
         h    ↗       ↘     ↘       ↗
    (iii)    ────┼──────────┼────────→
                 1          5         x
         h'    +   0    -   0    +
```

51. (i) f has a relative maximum at $(-2, 16)$
    f has a relative minimum at $(2, -16)$
    (ii) f is increasing on $(-\infty, -2)$ and $(2, \infty)$
    f is decreasing on $(-2, 2)$

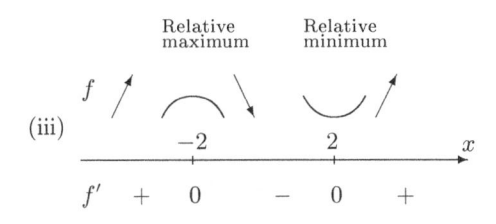

```
               Relative     Relative
               maximum      minimum
         f    ↗       ↘     ↘       ↗
    (iii)    ────┼──────────┼────────→
                 -2         2         x
         f'    +   0    -   0    +
```

53. (i) s has a relative maximum at $(2, 5)$
    s has a relative minimum at $(0, 1)$
    (ii) s is increasing on $(0, 2)$
    s is decreasing on $(-\infty, 0)$ and $(2, \infty)$

(iii) [sign chart for s: Relative minimum at 0, Relative maximum at 2; s': $-$, 0, $+$, 0, $-$]

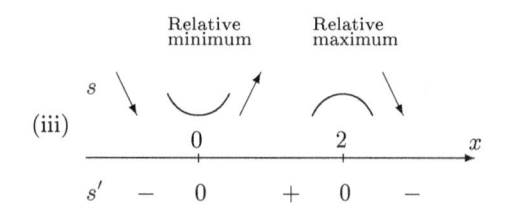

55. (i) f has a relative maximum at $(0, -4)$
    f has relative minima at $(-\sqrt{3}, -13)$ and $(\sqrt{3}, -13)$
    (ii) f is increasing on $(-\sqrt{3}, 0)$ and $(\sqrt{3}, \infty)$
    f is decreasing on $(-\infty, -\sqrt{3})$ and $(0, \sqrt{3})$
    (iii) [sign chart for f: Relative minimum at $-\sqrt{3}$, Relative maximum at 0, Relative minimum at $\sqrt{3}$; f': $-$, 0, $+$, 0, $-$, 0, $+$]

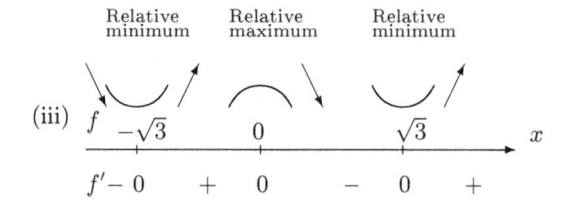

57. (i) g has relative maxima at $(-2, 18)$ and $(2, 18)$, a relative minimum at $(0, 2)$
    (ii) g is increasing on $(-\infty, -2)$ and $(0, 2)$
    g is decreasing on $(-2, 0)$ and $(2, \infty)$
    (iii) [sign chart for g: Relative maximum at -2, Relative minimum at 0, Relative maximum at 2; g': $+$, 0, $-$, 0, $+$, 0, $-$]

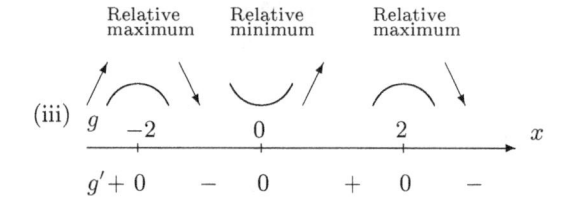

59. (i) h has no maximum
    h has a relative minimum at $(1.5, -3.69)$
    (ii) h is increasing on $(1.5, \infty)$
    h is decreasing on $(-\infty, 1.5)$
    (iii) [sign chart for h: Neither at 0, Relative minimum at 1.5; h': $-$, 0, $-$, 0, $+$]

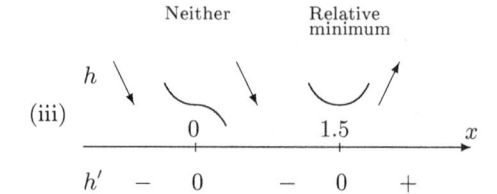

61. (i) f has no maximum
    f has a relative minimum at $(1, -1)$
    (ii) f is increasing on $(1, \infty)$
    f is decreasing on $(-\infty, 1)$
    (iii) [sign chart for f: Relative minimum at 2.5; f': $-$, 1, $+$]

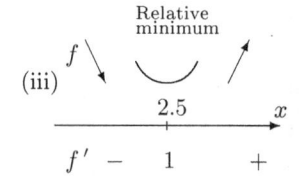

63. (i) s has a relative maximum at $(1, 2)$
    s has a relative minimum at $(-1, -2)$
    (ii) s is increasing on $(-1, 1)$
    s is decreasing on $(-\infty, -1)$ and $(1, \infty)$
    (iii) [sign chart for s: Relative minimum at -1, Neither at 0, Relative maximum at 1; s': $+$, 0, $-$, 0, $+$, 0, $-$]

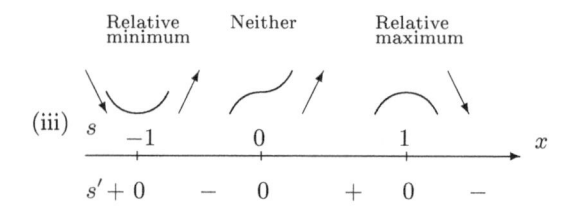

65. (i) f has a relative maximum at $(2, 0.54)$
    f has a relative minimum at $(0, 0)$
    (ii) f is increasing on $(0, 2)$
    f is decreasing on $(-\infty, 0)$ and $(2, \infty)$
    (iii) [sign chart for f: Relative minimum at 0, Relative maximum at 2; f': $-$, 0, $+$, 0, $-$]

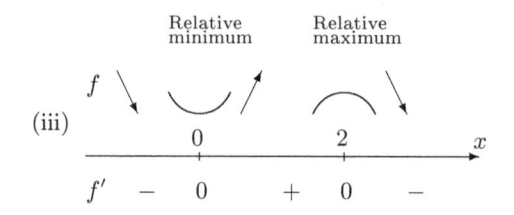

67. (i) g has a relative minimum at $(-3, 2)$
    g has no relative maximum
    (ii) g is increasing on $(-3, \infty)$
    g is decreasing on $(-\infty, -3)$
    (iii) [sign chart for g: Relative minimum at -3; g': $-$, none, $+$]

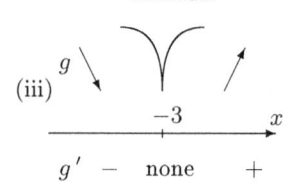

69. (i) h has no relative extremum
    (ii) h is decreasing on $(-\infty, \infty)$
    (iii) [sign chart for h: Neither at -3; h': $-$, none, $-$]

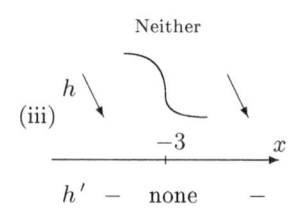

Exercises 3.5 [Ref. Example]

1. (i) f has a critical point $(2.5, -0.25)$.
   (ii) f has no inflection point

CHAPTER 6 TEST

3. (i) g has two critical points $(0, 4)$ and $(2, 0)$.
   (ii) g has a possible inflection point $(1, 2)$.
   (iii)

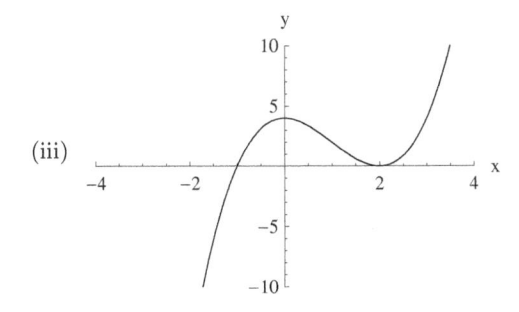

5. (i) h has two critical points $(1, 8\frac{1}{3})$ and $(5, -2\frac{1}{3})$.
   (ii) h has a possible inflection point $(3, 3)$.
   (iii)

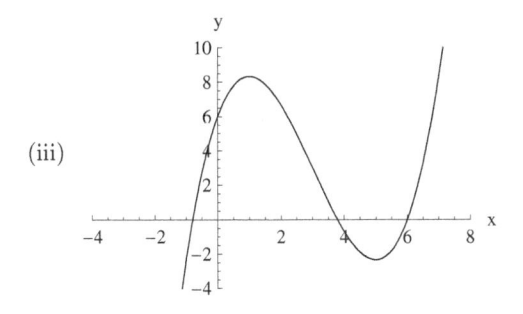

7. (i) f has two critical points $(-2, 16)$ and $(2, -16)$.
   (ii) f has a possible inflection point $(0, 0)$.
   (iii)

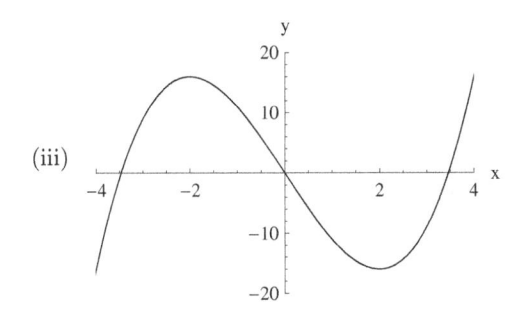

9. (i) s has two critical points $(2, 5)$ and $(0, 1)$.
   (ii) s has a possible inflection point $(1, 3)$.

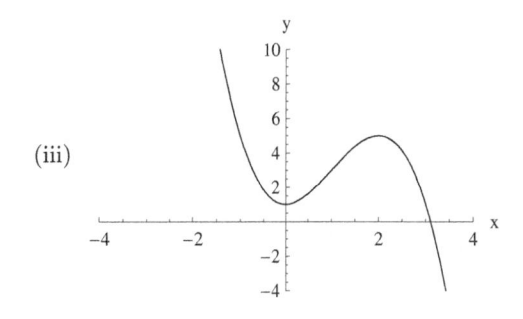

11. (i) f has three critical points $(0, -4)$, $(-\sqrt{3}, -13)$ and $(\sqrt{3}, -13)$.
    (ii) f has possible inflection points $(-1, -9)$ and $(1, -9)$.
    (iii)

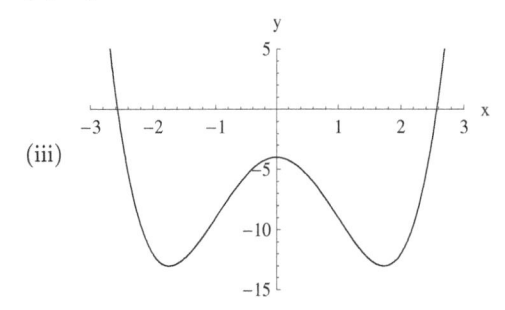

13. (i) g has three critical points $(-2, 18)$, $(0, 2)$, and $(2, 18)$.
    (ii) g has possible inflection points $(-2/\sqrt{3}, 10.9)$ and $(2/\sqrt{3}, 10.9)$.
    (iii)

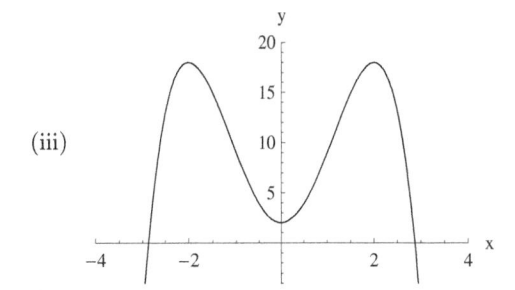

15. (i) h has critical points $(0, -2)$ and $(1.5, -3.69)$.
    (ii) h has possible inflection points $(0, -2)$ and $(1, -3)$.
    (iii)

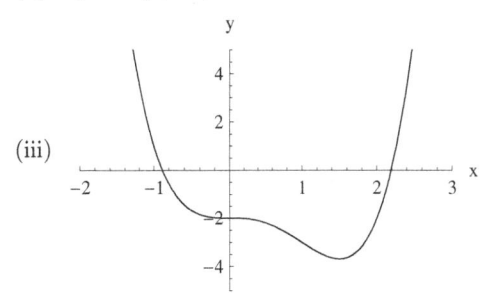

17. (i) f has a critical point $(1, -1)$.
    (ii) f has no inflection point.

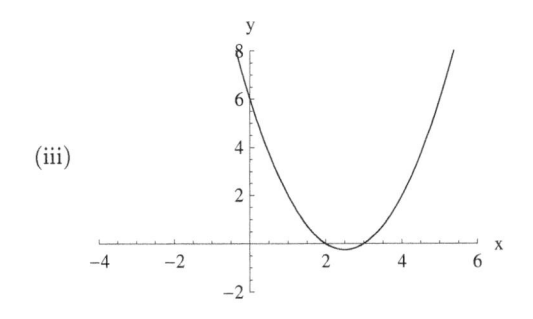

476   CHAPTER 6. FUNCTIONS OF TWO VARIABLES

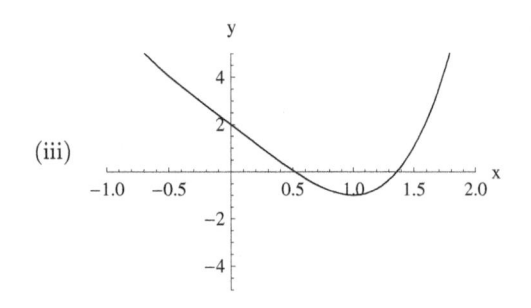

19. (i) s has critical points $(1, 2)$, $(0, 0)$, and $(-1, -2)$.
    (ii) s has possible inflection points $(-1/\sqrt{2}, -1.24)$, $(0, 0)$, and $(1/\sqrt{2}, 1.24)$.
    (iii) 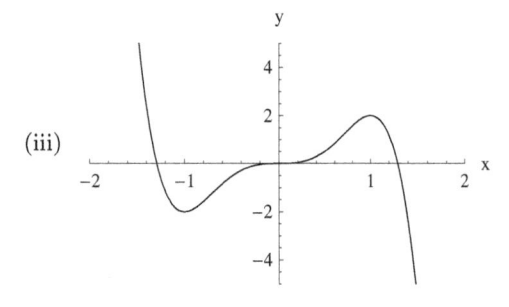 [graph]

21. (i) f has critical points $(0, 0)$ and $(2, 0.54)$.
    (ii) f has possible inflection points $(0.59, 0.19)$ and $(3.41, 0.38)$.
    (iii) 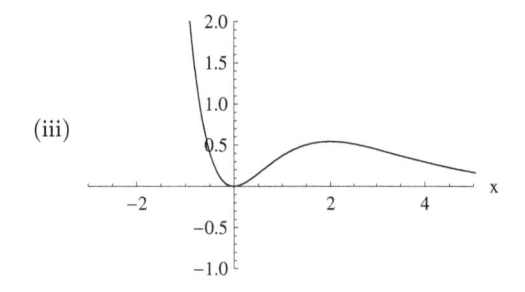 [graph]

23. (i) f has a critical point $(-0.5, -0.184)$.
    (ii) f has a possible inflection point $(-1.5, -0.1)$.
    (iii) See the graph in Example 3.5.5.

25. (i) g has a critical point $(-3, 2)$.
    (ii) g has no inflection point.
    (iii) 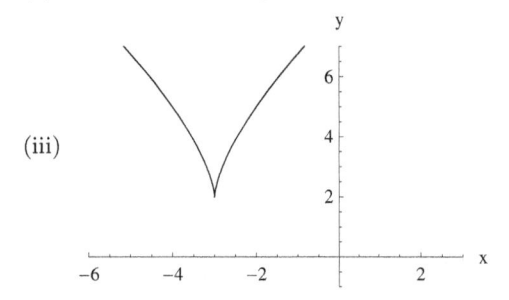 [graph]

27. (i) h has a critical point $(-3, 2)$.
    (ii) h has a possible inflection point $(-3, 2)$.
    (iii) 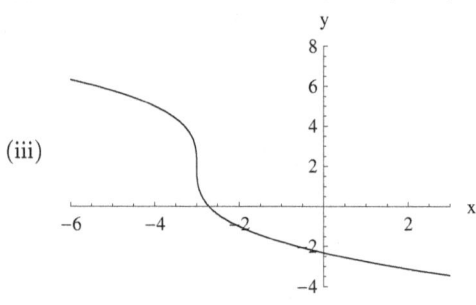 [graph]

29. (i) f has a relative minimum at $(2.5, -0.25)$
    f has no relative maximum
    (ii) f is increasing on $(2.5, \infty)$
    f is decreasing on $(-\infty, 2.5)$
    (iii) f has no inflection point
    (iv) f is concave up on $(-\infty, \infty)$
    (v) See the graph in solution of 1 (iii).

31. (i) g has a relative maximum at $(0, 4)$
    g has a relative minimum at $(2, 0)$
    (ii) g is increasing on $(-\infty, 0)$ and $(2, \infty)$
    g is decreasing on $(0, 2)$
    (iii) g has an inflection point at $(1, 2)$
    (iv) g is concave up on $(1, \infty)$
    g is concave down on $(-\infty, 1)$
    (v) See the graph in solution of 3 (iii).

33. (i) h has a relative maximum at $(1, 8\frac{1}{3})$
    h has a relative minimum at $(5, -2\frac{1}{3})$
    (ii) h is increasing on $(-\infty, 1)$ and $(5, \infty)$
    h is decreasing on $(1, 5)$
    (iii) h has an inflection point at $(3, 3)$
    (iv) h is concave up on $(3, \infty)$
    h is concave down on $(-\infty, 3)$
    (v) See the graph in solution of 5 (iii).

35. (i) f has a relative maximum at $(-2, 16)$
    f has a relative minimum at $(2, -16)$
    (ii) f is increasing on $(-\infty, -2)$ and $(2, \infty)$
    f is decreasing on $(-2, 2)$
    (iii) f has an inflection point at $(0, 0)$
    (iv) f is concave up on $(0, \infty)$
    f is concave down on $(-\infty, 0)$
    (v) See the graph in solution of 7 (iii).

37. (i) s has a relative maximum at $(2, 5)$
    s has a relative minimum at $(0, 1)$
    (ii) s is increasing on $(0, 2)$
    s is decreasing on $(-\infty, 0)$ and $(2, \infty)$
    (iii) s has an inflection point at $(1, 3)$
    (iv) s is concave up on $(-\infty, 1)$
    s is concave down on $(1, \infty)$
    (v) See the graph in solution of 9 (iii).

39. (i) f has a relative maximum at $(0, -4)$
   f has relative minima at $(-\sqrt{3}, -13)$
   and $(\sqrt{3}, -13)$
   (ii) f is increasing on $(-\sqrt{3}, 0)$ and $(\sqrt{3}, \infty)$
   f is decreasing on $(-\infty, -\sqrt{3})$ and $(0, \sqrt{3})$
   (iii) f has inflection points at $(-1, -9)$
   and $(1, -9)$
   (iv) f is concave up on $(-\infty, -1)$ and $(1, \infty)$
   f is concave down on $(-1, 1)$
   (v) See the graph in solution of 11 (iii).

41. (i) g has relative maxima at $(-2, 18)$
   and $(2, 18)$, a relative minimum at $(0, 2)$
   (ii) g is increasing on $(-\infty, -2)$ and $(0, 2)$
   g is decreasing on $(-2, 0)$ and $(2, \infty)$
   (iii) g has inflection points
   at $(-2/\sqrt{3}, 10.9)$ and $(2/\sqrt{3}, 10.9)$
   (iv) g is concave up on $(-2/\sqrt{3}, 2/\sqrt{3})$
   g is concave down on $(-\infty, -2/\sqrt{3})$
   and $(2/\sqrt{3}, \infty)$
   (v) See the graph in solution of 13 (iii).

43. (i) h has no maximum
   h has a relative minimum at $(1.5, -3.69)$
   (ii) h is increasing on $(1.5, \infty)$
   h is decreasing on $(-\infty, 1.5)$
   (iii) h has inflection points at
   $(0, -2)$ and $(1, -3)$
   (iv) h is concave up on $(-\infty, 0)$ and $(1, \infty)$
   h is concave down on $(0, 1)$
   (v) See the graph in solution of 15 (iii).

45. (i) f has no maximum
   f has a relative minimum at $(1, -1)$
   (ii) f is increasing on $(1, \infty)$
   f is decreasing on $(-\infty, 1)$
   (iii) f has no inflection point
   (iv) f is concave up on $(-\infty, \infty)$
   (v) See the graph in solution of 17 (iii).

47. (i) s has a relative maximum at $(1, 2)$
   s has a relative minimum at $(-1, -2)$
   (ii) s is increasing on $(-1, 1)$
   s is decreasing on $(-\infty, -1)$ and $(1, \infty)$
   (iii) s has inflection points at
   $(-1/\sqrt{2}, -1.24)$, $(0, 0)$, and $(1/\sqrt{2}, 1.24)$
   (iv) s is concave up on $(-\infty, -1/\sqrt{2})$
   and $(0, 1/\sqrt{2})$, s is concave down
   on $(-1/\sqrt{2}, 0)$ and $(1/\sqrt{2}, \infty)$
   (v) See the graph in solution of 19 (iii).

49. (i) f has a relative maximum at $(2, 0.54)$
   f has a relative minimum at $(0, 0)$
   (ii) f is increasing on $(0, 2)$
   f is decreasing on $(-\infty, 0)$ and $(2, \infty)$

   (iii) f has inflection points
   at $(0.59, 0.19)$ and $(3.41, 0.38)$
   (iv) f is concave up on $(-\infty, 0.59)$
   and $(3.41, \infty)$, f is concave down
   on $(0.59, 3.41)$
   (v) See the graph in solution of 21 (iii).

51. (i) g has a relative minimum at $(-3, 2)$
   g has no relative maximum
   (ii) g is increasing on $(-3, \infty)$
   g is decreasing on $(-\infty, -3)$
   (iii) g has no inflection point
   (iv) g is concave down on $(-\infty, -3)$
   and $(-3, \infty)$
   (v) See the graph in solution of 25 (iii).

53. (i) h has no relative extremum
   (ii) h is decreasing on $(-\infty, \infty)$
   (iii) h has an inflection point at $(-3, 2)$
   (iv) h is concave up on $(-3, \infty)$
   h is concave down on $(-\infty, -3)$
   (v) See the graph in solution of 27 (iii).

55. A plausible graph

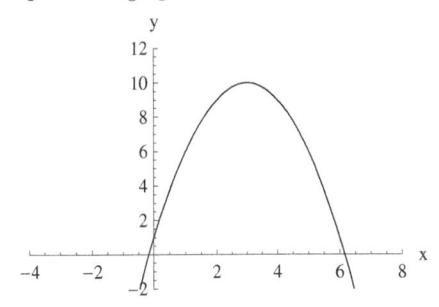

57. A plausible graph

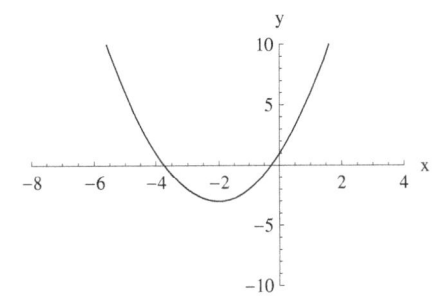

59. A plausible graph

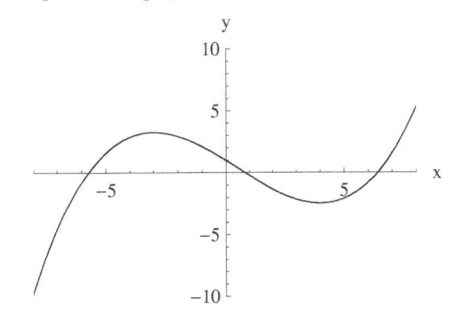

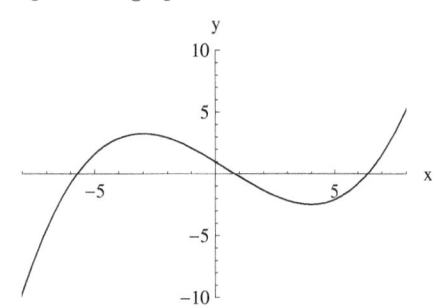

61. A plausible graph

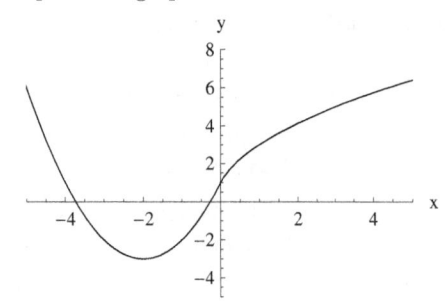

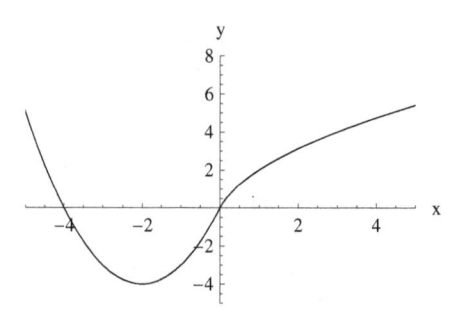

63. A plausible graph

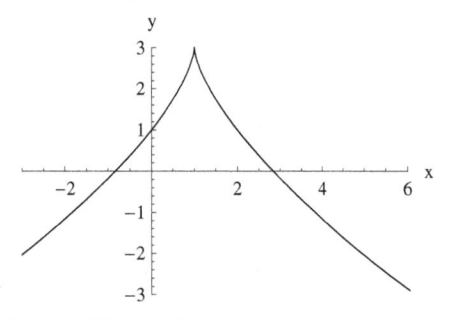

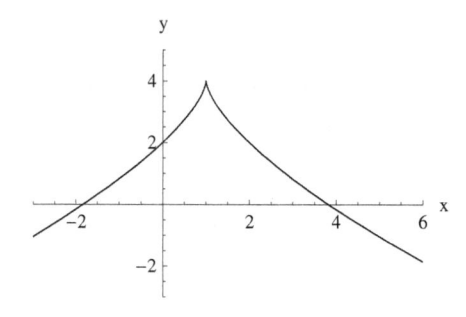

65. A plausible graph

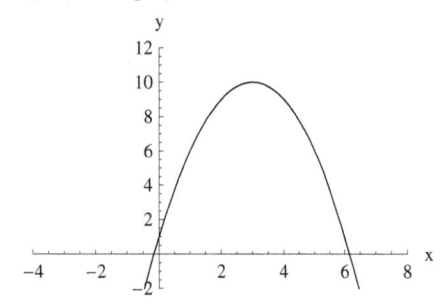

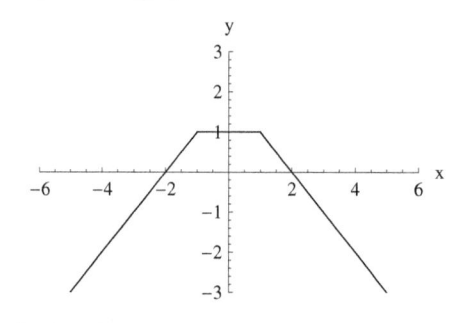

67. A plausible graph

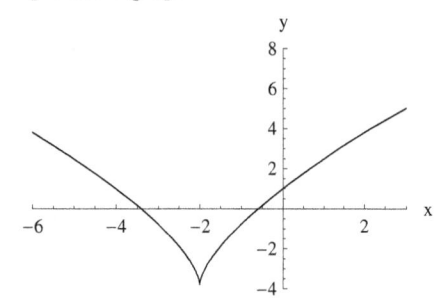

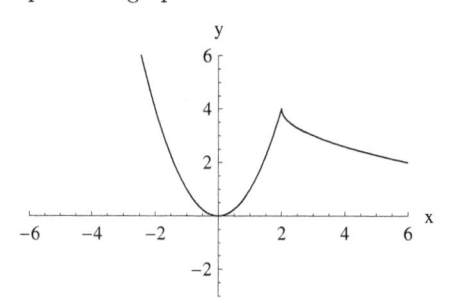

69. A plausible graph

71. A plausible graph

73. A plausible graph

75. A plausible graph

77. A plausible graph

79. A plausible graph

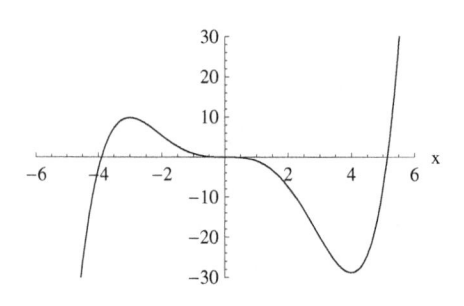

81. (a) $P(x) = 70x - 0.2x^2 - 1,500,\ 0 \leq x \leq 400$

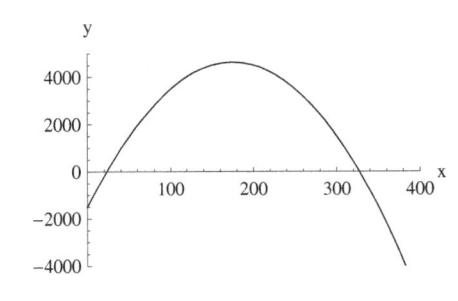

CHAPTER 6 TEST 479

(b) $x = 175$ pairs

(c) $MP(x) = 70 - 0.4x$, $0 \leq x \leq 400$

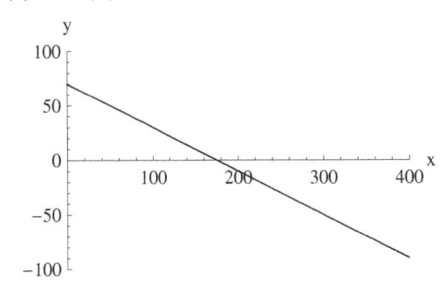

(d) $x = 175$ pairs

(e)

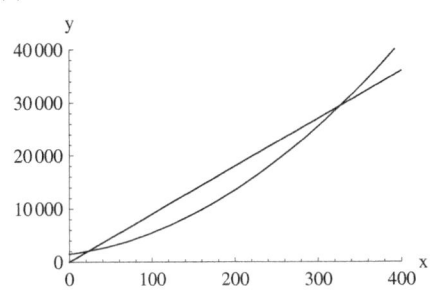

(f)

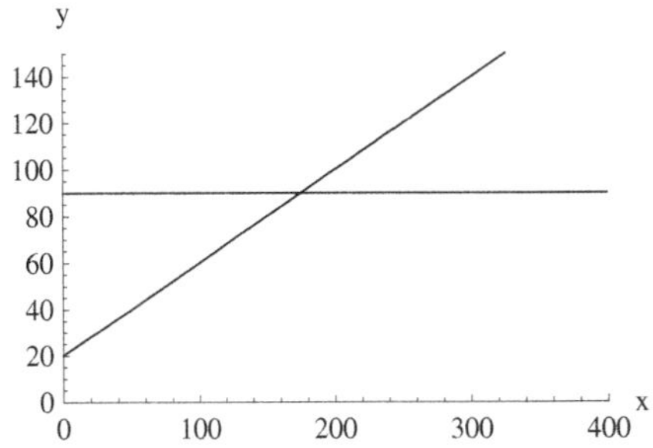

(g) $x = 175$ pairs

83. (a) $P(x) = 464x - 4.5x^2 - 4{,}000$, $0 \leq x \leq 100$

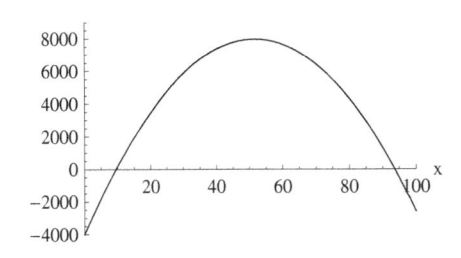

(b) $x \approx 52$ microwave ovens

(c) $MP(x) = 464 - 9x$, $0 \leq x \leq 100$

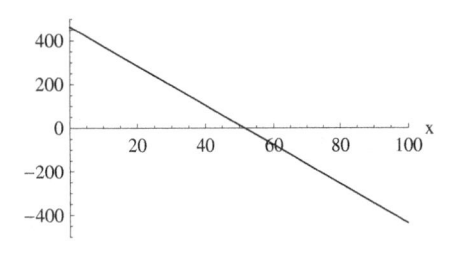

(d) $x \approx 52$ microwave ovens

(e)

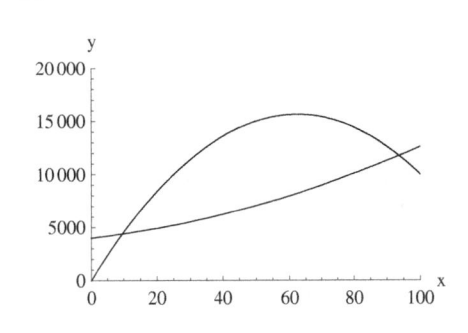

(f)

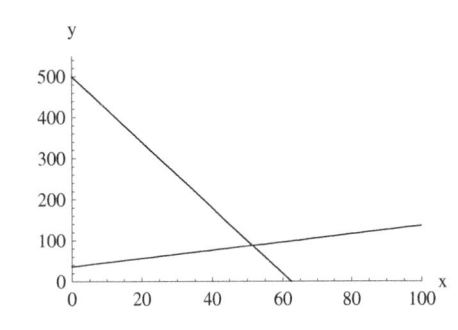

(g) $x \approx 52$ microwave ovens

85. $x_E \approx 62$, $p_E \approx 16$

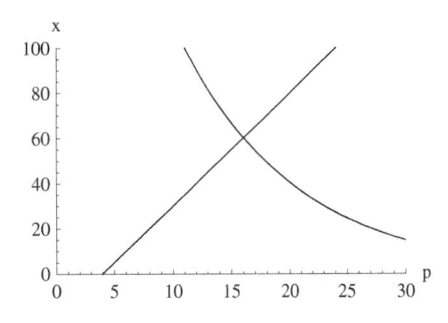

87. $x_E \approx 7$, $p_E \approx 9$

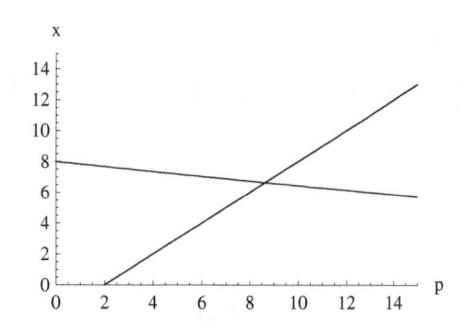

480   CHAPTER 6. FUNCTIONS OF TWO VARIABLES

Exercises 3.6   [Ref. Example]

1. $x = 2$ 3. $x = -4, x = 4$ [2]
5. $x = 1, x = 2$ 7. $x = 1$ [2]
9. There is no vertical asymptote
11. There is no vertical asymptote
13. $y = 2$ 15. $y = 0$ [3]
17. $y = 2$ 19. $y = 3$ [3]
21. There is no horizontal asymptote   [3]
23. $y = -1$ [3]
25. (a) Domain: $x \neq 0$. Vertical asymptote: $x = 0$. Horizontal asymptote: $y = 0$.

    (b) f has no critical point. f has no inflection point.

    (c)

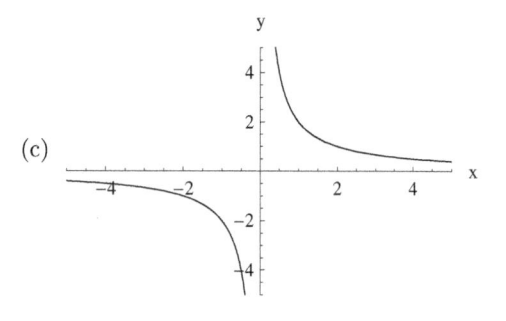

27. (a) Domain: $x \neq 2$. Vertical asymptote: $x = 2$. Horizontal asymptote: $y = 0$

    (b) g has no critical point. g has no inflection point.

    (c)

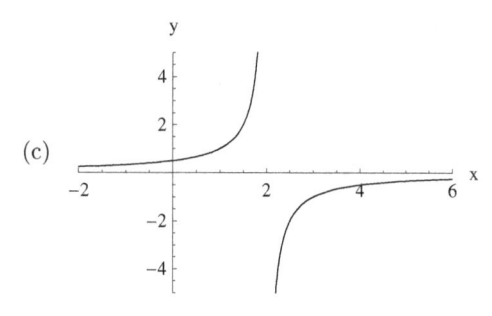

29. (a) Domain: $x \neq -1$. Vertical asymptote: $x = -1$. Horizontal asymptote: $y = 0$.

    (b) h has no critical point. h has no inflection point.

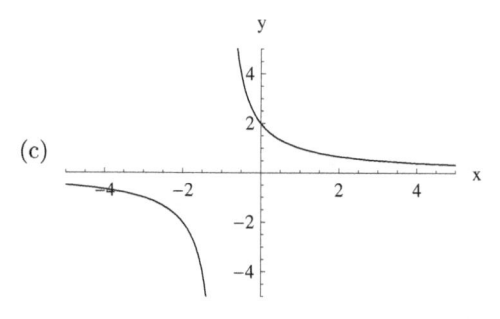

31. (a) Domain: $x \neq 0$. Vertical asymptote: $x = 0$. Horizontal asymptote: $y = 0$.

    (b) K has no critical point. K has no inflection point.

    (c)

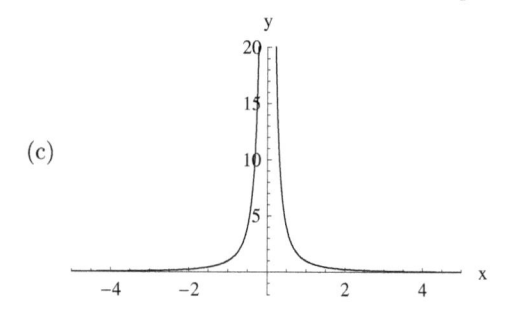

33. (a) Domain: $x \neq 0$. Vertical asymptote: $x = 0$. Horizontal asymptote: $y = 1$.

    (b) s has no critical point. s has no inflection point.

    (c)

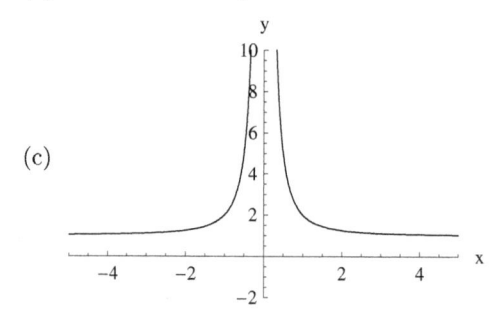

35. (a) Domain: $x \neq 0$. No intercept   [8]
    Vertical asymptote: $x = 0$. There is no horizontal asymptote. Slant asymptote: $y = x$.

    (c) f has critical points $(-\sqrt{2}, -2\sqrt{2})$ and $(\sqrt{2}, 2\sqrt{2})$. f has no inflection point.

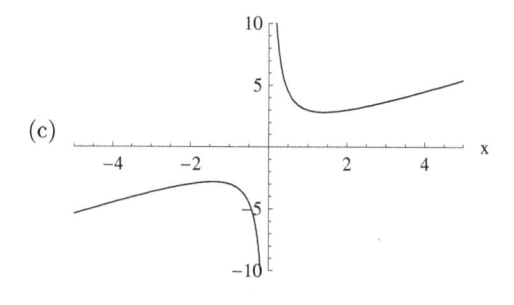

CHAPTER 6 TEST

37. (a) Domain: $x \neq 2$. Vertical asymptote: $x = 2$. Horizontal asymptote: $y = 2$.

(b) g has no critical point. g has no inflection point.

(c)

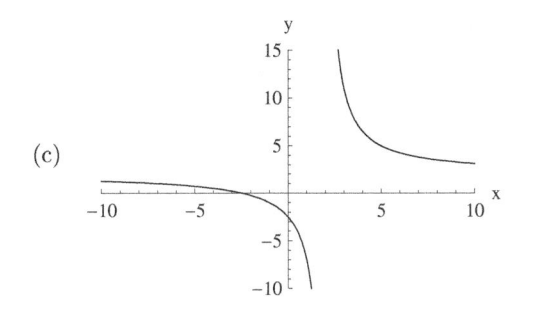

39. (a) Domain: $x \neq -2, 2$. [6]

Vertical asymptotes: $x = -2$, $x = 2$. Horizontal asymptote: $y = 0$.

(b) u has a critical point $(0, -1/2)$. u has no inflection point.

(c)

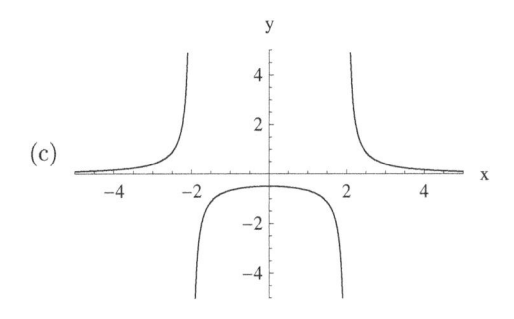

41. (a) Domain: $x \neq 0$. Vertical asymptote: $x = 0$. Horizontal asymptote: $y = 2$.

(b) h has no critical point. h has no inflection point.

(c)

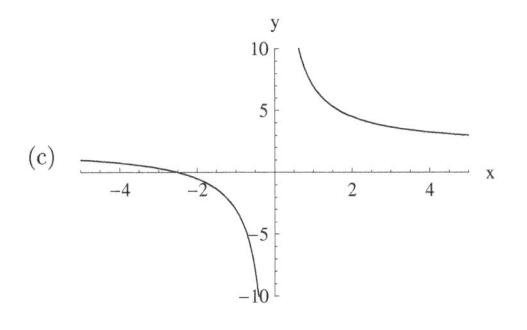

43. (a) Domain: $x \neq 0$. [8]

Vertical asymptote: $x = 0$. There is no horizontal asymptote. Slant asymptote: $y = x$.

(b) K has critical points $(-\sqrt{2}, -2\sqrt{2})$ and $(\sqrt{2}, 2\sqrt{2})$. K has no inflection point

(c)

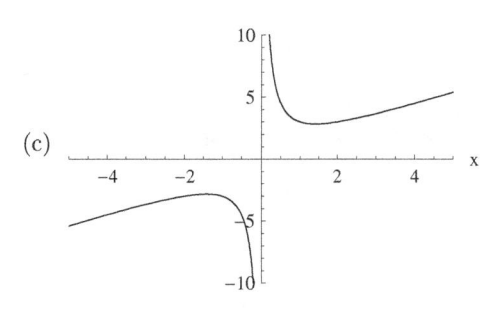

45. $u(x) = \dfrac{1}{x - 2}$, $x \neq -2$.

(a) Domain: $x \neq -2, 2$. Vertical asymptote: $x = 2$. Horizontal asymptote: $y = 0$

(b) u has no critical point. u has no inflection point.

(c)

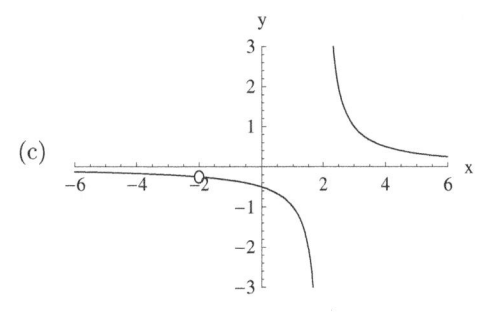

47. (a) Domain: $(-\infty, \infty)$. [7]
There is no vertical asymptote Horizontal asymptote: $y = 0$

(b) f has a critical number $(0, 2)$. f has possible inflection points $(-1/\sqrt{3}, 3/2)$ and $(1/\sqrt{3}, 3/2)$.

(c)

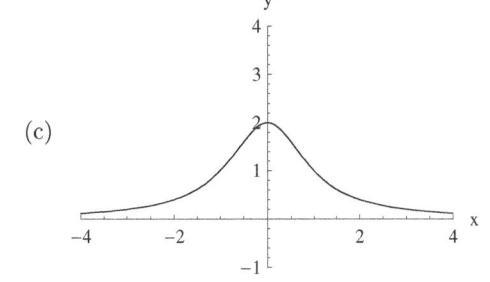

49. (a) Domain: $x \neq -1, 1$. [6]
Vertical asymptotes: $x = -1$, $x = 1$. Horizontal asymptote: $y = -1$

(b) g has a critical point $(0, 0)$. g has no inflection point.

(c)

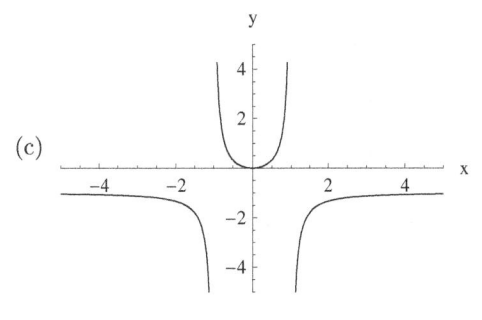

51. (a) Domain: $x \neq 2$. [8] Vertical asymptote: $x = 2$. There is no horizontal asymptote Slant asymptote: $y = x/2 + 1$

(b) h has critical points $(-0.24, -0.3)$ and $(4.24, 4.24)$. h has no inflection point

(c)

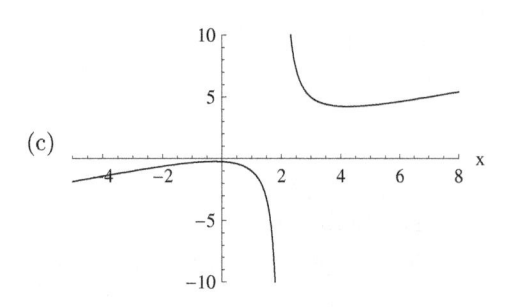

53. $s(x) = x - 2$, $x \neq -2$.

(a) Domain: $x \neq -2$. There is no asymptote

(b) s has no critical point. s has no inflection point.

(c)

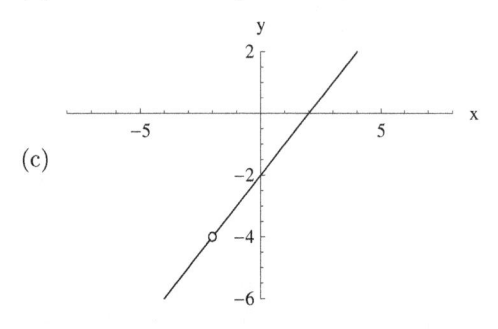

55. (a) Domain: $x \neq 0$. No intercept

(b) Vertical asymptote: $x = 0$
Horizontal asymptote: $y = 0$

(c) f has no relative extremum

(d) f is decreasing on $(-\infty, 0)$ and $(0, \infty)$

(e) f has no inflection point

(f) f is concave up on $(0, \infty)$
f is concave down on $(-\infty, 0)$

(g)

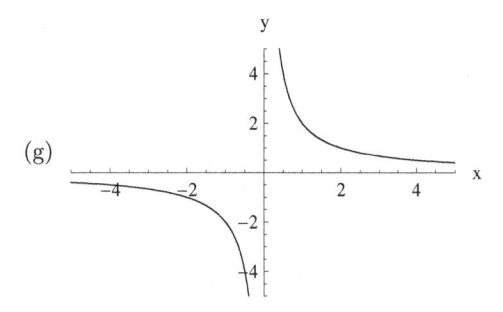

57. (a) Domain: $x \neq 2$. y-intercept $(0, 1/2)$

(b) Vertical asymptote: $x = 2$
Horizontal asymptote: $y = 0$

(c) g has no relative extremum

(d) g is increasing on $(-\infty, 2)$ and $(2, \infty)$

(e) g has no inflection point

(f) g is concave up on $(-\infty, 2)$
g is concave down on $(2, \infty)$

(g)

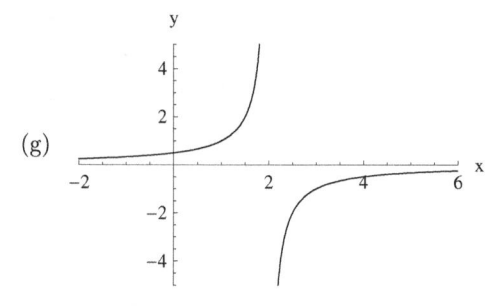

59. (a) Domain: $x \neq -1$. y-intercept $(0, 2)$

(b) Vertical asymptote: $x = -1$
Horizontal asymptote: $y = 0$

(c) h has no relative extremum

(d) h is decreasing on $(-\infty, -1)$ and $(-1, \infty)$

(e) h has no inflection point

(f) h is concave up on $(-1, \infty)$
h is concave down on $(-\infty, -1)$

(g)

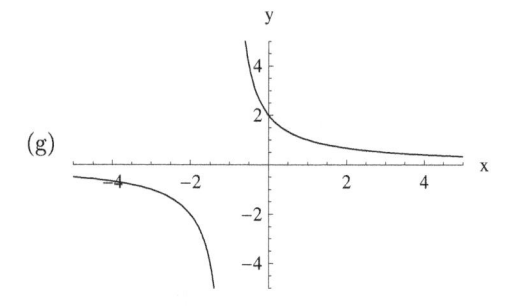

61. (a) Domain: $x \neq 0$. No intercept

(b) Vertical asymptote: $x = 0$
Horizontal asymptote: $y = 0$

(c) K has no relative extremum

(d) K is increasing on $(-\infty, 0)$
K is decreasing on $(0, \infty)$

(e) K has no inflection point

(f) K is concave up on $(-\infty, 0)$ and $(0, \infty)$

(g)

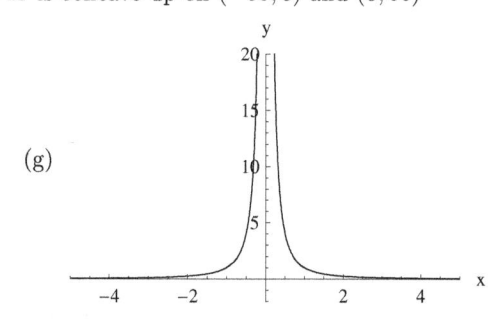

63. (a) Domain: $x \neq 0$. No intercept

(b) Vertical asymptote: $x = 0$
Horizontal asymptote: $y = 1$

(c) s has no relative extremum

(d) s is increasing on $(-\infty, 0)$
s is decreasing on $(0, \infty)$

(e) s has no inflection point

(f) s is concave up on $(-\infty, 0)$ and $(0, \infty)$

(g)

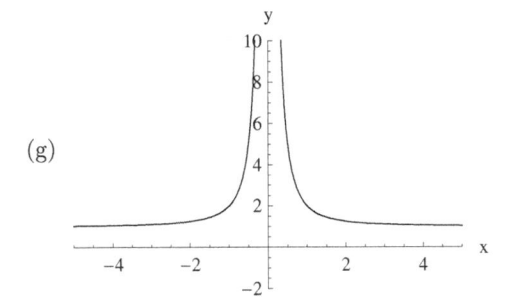

65. (a) Domain: $x \neq 0$. No intercept [11]

(b) Vertical asymptote: $x = 0$
There is no horizontal asymptote
Slant asymptote: $y = x$

(c) f has a relative maximum
at $(-\sqrt{2}, -2\sqrt{2})$, f has a
relative minimum at $(\sqrt{2}, 2\sqrt{2})$

(d) f is increasing on $(-\infty, -\sqrt{2})$
and $(\sqrt{2}, \infty)$, f is decreasing
on $(-\sqrt{2}, 0)$ and $(0, \sqrt{2})$

(e) f has no inflection point

(f) f is concave up on $(0, \infty)$
f is concave down on $(-\infty, 0)$

(g)

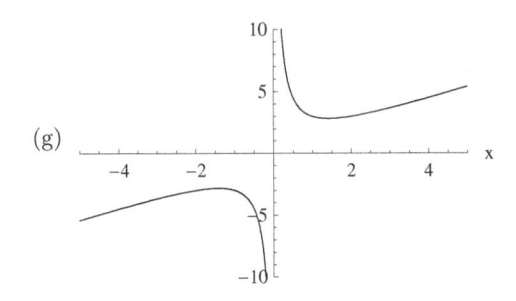

67. (a) Domain: $x \neq 2$.
Intercepts $(0, -2.5)$, $(-2.5, 0)$

(b) Vertical asymptote: $x = 2$
Horizontal asymptote: $y = 2$

(c) g has no relative extremum

(d) g is decreasing on $(-\infty, 2)$ and $(2, \infty)$

(e) g has no inflection point

(f) g is concave up on $(2, \infty)$
g is concave down on $(-\infty, 2)$

(g)

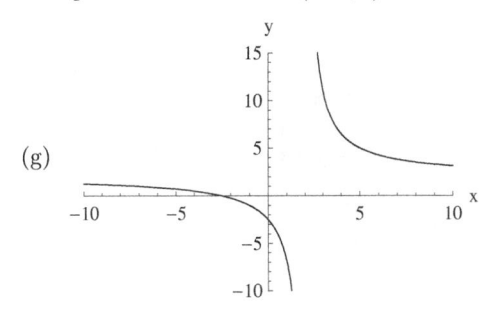

69. (a) Domain: $x \neq -2, 2$. [6]
y-intercept $(0, -1/2)$

(b) Vertical asymptotes: $x = -2$, $x = 2$
Horizontal asymptote: $y = 0$

(c) u has a relative maximum at $(0, -1/2)$
u has no relative minimum

(d) u is increasing on $(-\infty, -2)$ and $(-2, 0)$
u is decreasing on $(0, 2)$ and $(2, \infty)$

(e) u has no inflection point

(f) u is concave up on $(-\infty, -2)$ and $(2, \infty)$
u is concave down on $(-2, 2)$

(g)

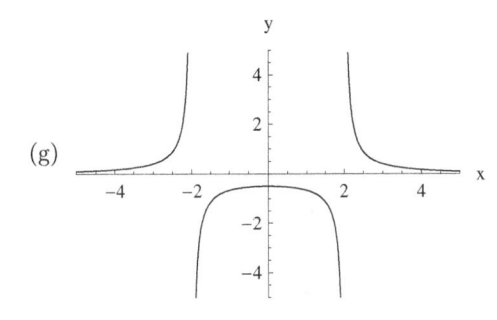

71. (a) Domain: $x \neq 0$. x-intercept $(-5/2, 0)$

(b) Vertical asymptote: $x = 0$
Horizontal asymptote: $y = 2$

(c) h has no relative extremum

(d) h is decreasing on $(-\infty, 0)$ and $(0, \infty)$

(e) h has no inflection point

(f) h is concave up on $(0, \infty)$
h is concave down on $(-\infty, 0)$

(g)

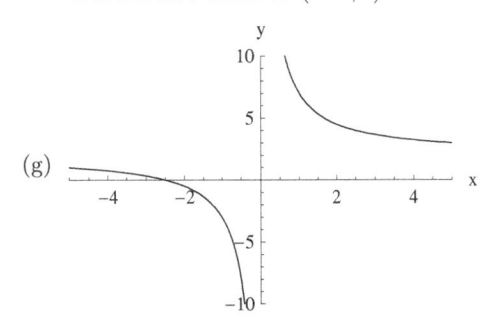

73. (a) Domain: $x \neq 0$. No intercept
   (b) Vertical asymptote: $x = 0$
   There is no horizontal asymptote
   Slant asymptote: $y = x$
   (c) K has a relative maximum at $(-\sqrt{2}, -2\sqrt{2})$, K has a relative minimum at $(\sqrt{2}, 2\sqrt{2})$
   (d) K is increasing on $(-\infty, -\sqrt{2})$ and $(\sqrt{2}, \infty)$, K is decreasing on $(-\sqrt{2}, 0)$ and $(0, \sqrt{2})$
   (e) K has no inflection point
   (f) K is concave up on $(0, \infty)$
   K is concave down on $(-\infty, 0)$
   (g) [graph]

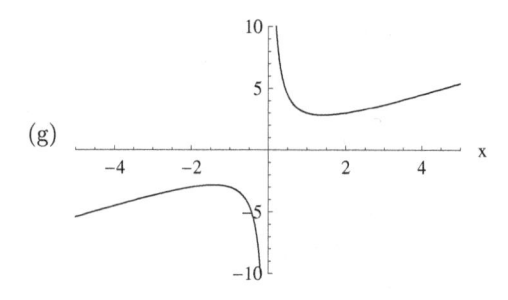

75. $s(x) = x - 2$, $x \neq -2$.
   (a) Domain: $x \neq -2$.
   Intercepts $(0, -2)$, $(2, 0)$
   (b) There is no asymptote
   (c) s has no relative extremum
   (d) s is increasing on $(-\infty, -2)$ and $(-2, \infty)$
   (e) s has no inflection point
   (f) s is neither concave up nor concave down
   (g) [graph]

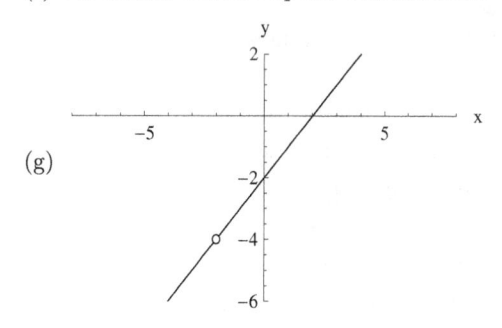

77. $u(x) = \dfrac{1}{x-2}$, $x \neq -2$.
   (a) Domain: $x \neq -2, 2$
   y-intercept $(0, -1/2)$
   (b) Vertical asymptote: $x = 2$
   Horizontal asymptote: $y = 0$
   (c) u has no relative extremum
   (d) u is decreasing on $(-\infty, -2)$, $(-2, 2)$, and $(2, \infty)$
   (e) u has no inflection point
   (f) u is concave up on $(2, \infty)$
   u is concave down on $(-\infty, -2)$ and $(-2, 2)$
   (g) [graph]

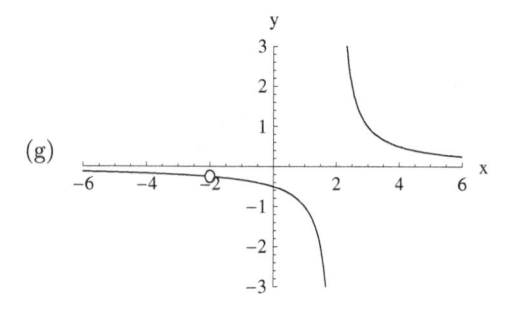

79. (a) Domain: $(-\infty, \infty)$. [9]
   y-intercept $(0, 2)$
   (b) There is no vertical asymptote
   Horizontal asymptote: $y = 0$
   (c) f has a relative maximum at $(0, 2)$
   (d) f is increasing on $(-\infty, 0)$
   f is decreasing on $(0, \infty)$
   (e) f has inflection points at $(-1/\sqrt{3}, 3/2)$ and $(1/\sqrt{3}, 3/2)$
   (f) f is concave up on $(-\infty, -1/\sqrt{3})$ and $(1/\sqrt{3}, \infty)$, f is concave down on $(-1/\sqrt{3}, 1/\sqrt{3})$
   (g) [graph]

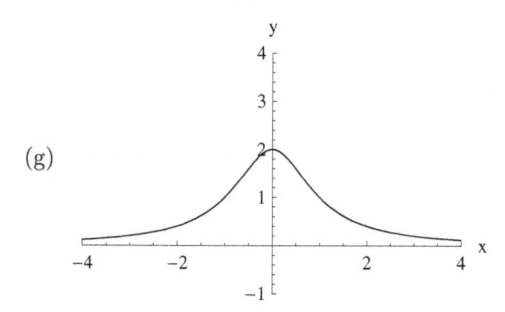

81. (a) Domain: $x \neq -1, 1$. [9]
   y-intercept $(0, 0)$
   (b) Vertical asymptotes: $x = -1$, $x = 1$
   Horizontal asymptote: $y = -1$
   (c) g has a relative minimum at $(0, 0)$
   (d) g is increasing on $(0, 1)$ and $(1, \infty)$
   g is decreasing on $(-\infty, -1)$ and $(-1, 0)$
   (e) g has no inflection point
   (f) g is concave up on $(-1, 1)$
   g is concave down on $(-\infty, -1)$ and $(1, \infty)$

CHAPTER 6 TEST   485

(g) [graph]

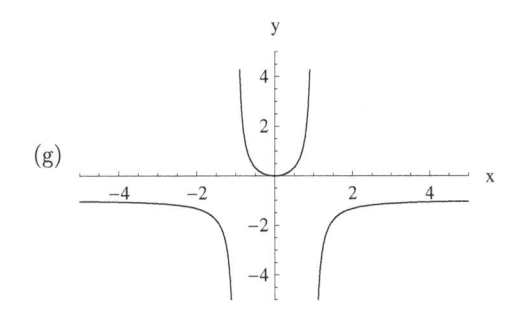

83. (a) Domain: $x \neq 2$. y-intercept $(0, -1/4)$

(b) Vertical asymptote: $x = 2$
There is no horizontal asymptote
Slant asymptote: $y = x/2 + 1$

(c) h has a relative maximum
at $(-0.24, -0.3)$, h has a relative minimum
at $(4.24, 4.24)$

(d) h is increasing on
$(-\infty, -0.24)$ and $(4.24, \infty)$
h is decreasing on $(-0.24, 2)$ and $(2, 4.24)$

(e) h has no inflection point

(f) h is concave up on $(2, \infty)$
h is concave down on $(-\infty, 2)$

(g) [graph]

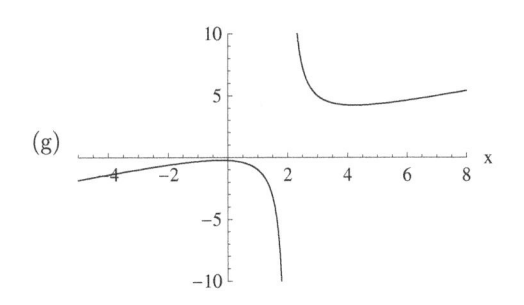

85. (a) $P(x) = 70x - 0.2x^2 - 1,500$, $0 \leq x \leq 400$

(b) $A_P(x) = 70 - 0.2x - 1,500/x$,
$1 \leq x \leq 400$

(c) $y = -0.2x + 70$

(d) [graph]

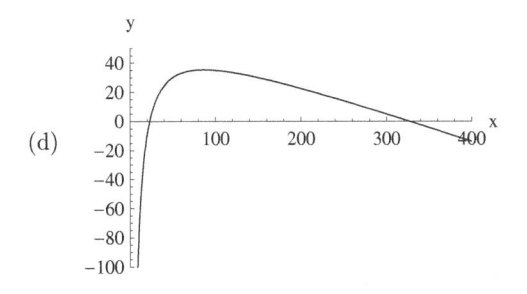

87. $p_E \approx 1.5$, $x_E \approx 4.4$

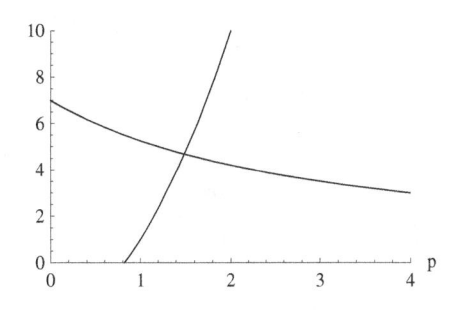

[graph]

89. $p_E = 20$, $x_E = 3$

[graph]

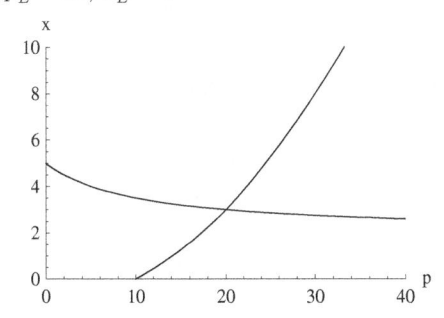

Exercises 3.7   [Ref. Example]

1. (a) abs. max = $g(2.5) = 2.5$,
abs. min = $g(-2.5) = -2.5$

(b) abs. max = $g(2) = 2.3$,
abs. min = $g(-2) = -2$

(c) abs. max = abs. min = 0.5 on $[-1, 1]$

(d) abs. max = 0.5 on $[-1, 1]$,
abs. min = $g(-2) = -2$

3. (a) No abs. max, abs. min = $f(-1.5) = -3$

(b) abs. max = $f(1.5) = 2$,
abs. min = $f(0) = -1$

(c) abs. max = $f(-3) = -1.3$, no abs. min

(d) abs. max = $f(0) = -1$,
abs. min = $f(-1.5) = -3$

5. abs. max = $8\frac{1}{3}$ at $x = 1$,   [1,2]
abs. min = $-2\frac{1}{3}$ at $x = 5$

7. abs. max = 0.54 at $x = 2$,   [1,2]
abs. min = 0 at $x = 0$

9. abs. max = 0.5 at $x = 2$,   [1,2]
abs. min = -0.5 at $x = -2$

11. abs. max = 6 at $x = 0$,   [1,2]
abs. min = -0.25 at $x = 2.5$

13. abs. max = 3 at $x = -1$,   [1,2]
abs. min = -3 at $x = 2$

15. abs. max = $6\frac{2}{3}$ at $x = 2$,   [1,2]
abs. min = $-2\frac{1}{3}$ at $x = 5$

17. abs. max = 29.6 at $x = -2$, [1,2]
    abs. min = 0 at $x = 0$

19. abs. max = 0.5 at $x = 2$, [1,2]
    abs. min = 0 at $x = 0$

21. abs. max = 10.98 at $x = 4$, [1,2]
    abs. min = 0 at $x = -3$

23. abs. max = 5 at $x = 1, 4$, [1,2]
    abs. min = 4 at $x = 2$

25. abs. max = 2 at $x = 0$, [1,2]
    abs. min = -79 at $x = -3, 3$

27. abs. max = 0.54 at $x = 2$, [1,2]
    abs. min = 0.37 at $x = 1$

29. abs. max = 0 at $x = 0$, [1,2]
    abs. min = -0.5 at $x = -2$

31. abs. max = 9.9 at $x = 3$, [1,2]
    abs. min = 3 at $x = -2$

33. abs. max = -4 at $x = -2$, [1,2]
    abs. min = -5 at $x = -1$

35. abs. max = $8\frac{1}{3}$ at $x = 1$, [5,6]
    no abs. min

37. abs. max = 0.54 at $x = 2$, [5,6]
    no abs. min

39. abs. max = 0.5 at $x = 2$, [5,6]
    no abs. min

41. No abs. max, [5,6]
    abs. min = -0.25 at $x = 2.5$

43. No abs. max, [5,6]
    abs. min = -7 at $x = 4$

45. No abs. max, no abs. min [5,6]

47. No abs. max, [5,6]
    abs. min = 0 at $x = 0$

49. abs. max = -0.4 at $x = -1$, [5,6]
    abs. min = -0.5 at $x = -2$

51. No abs. max, [5,6]
    abs. min = 0 at $x = -3$

53. abs. max = -4 at $x = -2$, [5,6]
    no abs. min

55. No abs. max, [5,6]
    abs. min = -3 at $x = 2$

57. $20,000 per month must be spent to achieve the maximum number of 2,012 television sets sold.

59. (a) $P(x) = 70x - 0.2x^2 - 1,500$, $0 \leq x \leq 400$

    (b) 175 pairs of shoes should be produced and sold each week;
        maximum profit is $4,625 per week.

61. (a) 63 microwave ovens should be sold each week; maximum revenue is $15,624 per week.

(b) $P(x) = 464x - 4.5x^2 - 4,000$,
    $0 \leq x \leq 100$

(c) 52 microwave ovens should be produced and sold each week;
    maximum profit is $11,960 per week.

Exercises 3.8 [Ref. Example]

1. 15 and 15

3. 11 and 12

5. 2.5 and -2.5

7. 10 and 10

9. (a)

    Area = $x \cdot y$

(b) $A(x) = 10x - x^2$

(c) $0 \leq x \leq 10$

(d) $x^* = 5$

(e) 5 inch $\times$ 5 inch

11. (a)

    Perimeter = $2x + 2y$

(b) $P(x) = 2x + 128/x$

(c) $0 < x < \infty$

(d) $x^* = 8$

(e) 8 inch $\times$ 8 inch

13. (a)

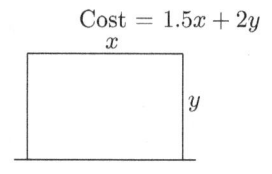

    Cost = $1.5x + 2y$

(b) $C(x) = 1.5x + 400/x$

(c) $0 < x < \infty$

(d) $x^* = 16.33$

(e) 16.33 ft $\times$ 12.25 ft
    with 16.33 ft along boundary

15. (a)

Cost = $40x + $20y + $30y

(fence diagram with sides y top and x right, Fence on bottom)

(b) $C(x) = 40x + 100{,}000/x$

(c) $0 < x < \infty$

(d) $x^* = 50$

(e) 50 yd × 40 yd
    with fence side = 40 yd

17. 38.73 yd × 25.82 yd
    with divider = 25.82 yd

19. 54.77 yd × 18.26 yd
    with 54.77 yd along the river

21. Base = 2.29 m × 2.29 m, height = 1.91 m

23. Base = 9.7 in, height = 12.94 in

25. Cut: 2 in. × 2 in. Maximum volume = 128 in^3

27. (a) $p = -0.2x + 148$, $0 \leq x \leq 740$ [3]

(b) $R(x) = -0.2x^2 + 148x$, $0 \leq x \leq 740$

(c) $74 per microwave oven

(d) $97 per microwave oven

29. (a) $p = -3x + 300$, $0 \leq x \leq 80$ [4]

(b) $P(x) = -3x^2 + 285x$, $0 \leq x \leq 80$

(c) $159 per room

31. (a) $p = -0.4x + 105$, $0 \leq x \leq 240$ [4]

(b) $R(x) = -0.4x^2 + 105x$, $0 \leq x \leq 240$

(c) $52.60 a ticket

33. 32 trees per acre

35. (a) $P(t) = 150{,}000{,}000$ [5]
    $-150{,}000{,}000 e^{-0.05t} - 12{,}000t$

(b) About 129 days

37. (a) $P(t) = 150{,}000{,}000$ [5]
    $-150{,}000{,}000 e^{-0.04t} - 10{,}000t$

(b) About 166 days

39. (a) $C(x) = 25x + 40{,}000/x + 16{,}000$, [6]
    $1 \leq x \leq 800$.

(b) Lot size = 40, place order 20 times

41. (a) $C(x) = 25x + 48{,}000/x + 16{,}000$, [7]
    $1 \leq x \leq 800$.

(b) Lot size = 45, place order 18 times

43. (a) $C(x) = 20x + 8{,}000/x + 2{,}000$, [6]
    $1 \leq x \leq 200$.

(b) Lot size = 20, place order 10 times

45. (a) $C(x) = 25x + 8{,}000/x + 2{,}000$, [7]
    $1 \leq x \leq 200$.

(b) Lot size = 17, place order 12 times

Exercises 3.9 [Ref. Example]

1. (a) 6% demand decrease

(b) 6% demand increase

3. (a) $E(p) = \dfrac{5p}{220 - 5p}$

(b) 0.29, inelastic (c) 22

5. (a) $E(p) = \dfrac{2p^2}{250 - p^2}$

(b) 1.33, elastic (c) 9.13

7. (a) $E(p) = \dfrac{p}{100 - 2p}$

(b) 0.75, inelastic (c) 33.33

9. (a) $E(p) = 1$

(b) 1, unitary (c) Any price

11. (a) $E(p) = \dfrac{p}{50}$

(b) 4, elastic (c) 50

13. (a) $E(p) = \dfrac{p}{20}$

(b) 2, elastic (c) 20

15. (a) $E(p) = \dfrac{2p + 2p^2}{30 - 2p - p^2}$

(b) 0.55, inelastic (c) 2.57

17. (a) $E(p) = \dfrac{2p}{2 + p}$

(b) 0.67, inelastic (c) 2

19. (a) $E(p) = \dfrac{\sqrt{p}}{10 - 2\sqrt{p}}$ [3]

(b) 0.33, inelastic

(c) Raise the price

21. (a) $E(p) = \dfrac{p}{30 - p}$ [3]

(b) 1.5, elastic. Lower the price

(c) $15 per blouse

23. (a) $x = -5p + 330$ [1,2]

(b) $R(p) = -5p^2 + 330p$

(c) $33 per barstool

(d) $E(p) = \dfrac{5p}{330 - 5p}$

(e) 10, elastic. Lower the price

(f) $33 per barstool

25. (a) $E(p) = \dfrac{p + 2p^2}{30 - p - p^2}$ [3]

(b) 0.71, inelastic

(c) 2.2, elastic

(d) $2.85 per bag

(e) 1903 bags

(f) decrease

27. (a) $E(p) = \dfrac{p}{(150-p)\ln(150-p)}$

(b) 0.51, inelastic. Increase

29. (a) $E(p) = 0.04p$

(b) 0.8, inelastic. Raise the price

(c) $25 per day-pack

31. (a) $E(p) = 0.8$ (b) No

(c) No   (d) No   (e) Yes

Chapter 3 Review Exercises

1. (a) When the price is set at $120 per skateboard, 8,000 skateboards are sold.

   (b) When the price is set at $120 per skateboard, the demand is decreasing at a rate of 80 skateboards sold/dollar per skateboard.

   (c) About 7,840 skateboards are sold.

2. (a) $\Delta y = 0.110903$, $dy = 0.11$

   (b) $f(1.01) \approx 6.11$

3. (a) $P(x) = -0.1x^2 + 102x - 1200$

   (b) $2,950, $8,400, $5,450

   (c) $32 per suit, $120 per suit, $88 per suit

   (d) $3,078, $8,880, $5,802

   (e) 510 suits

4. (a) f is increasing on $(-1, 1.5)$,
   f is decreasing on $(-4, -1)$ and $(1.5, 4)$

   (b) f' is positive on $(-1, 1.5)$,
   f' is negative on $(-4, -1)$ and $(1.5, 4)$,
   $f' = 0$ at $x = -1, 1.5$, f' exists on $(-4, 4)$

   (c) -1 and 1.5 (d) $(1.5, 2)$

   (e) $(-1, -2.5)$ (f) 2.5 (g) -2.5

   (g) f is concave up on $(-4, 0)$
   f is concave down on $(0, 4)$

5. (a) $(1, 8\tfrac{1}{3})$ and $(5, -2\tfrac{1}{3})$

   (b) f is increasing on $(-\infty, 1)$ and $(5, \infty)$,

   (c) f is decreasing on $(1, 5)$

   (d) f has a relative maximum at $(1, 8\tfrac{1}{3})$
   f has a relative minimum at $(5, -2\tfrac{1}{3})$

   (e) f has an inflection point at $(3, 3)$

   (f) f is concave up on $(3, \infty)$
   f is concave down on $(-\infty, 3)$

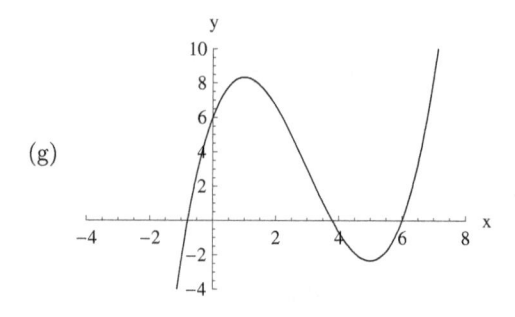

(g)

6. (a) $(0, -2)$ and $(1.5, -3.69)$

   (b) f is increasing on $(1.5, \infty)$

   (c) f is decreasing on $(-\infty, 1.5)$

   (d) f has no maximum
   h has a relative minimum at $(1.5, -3.69)$

   (e) f has inflection points at
   $(0, -2)$ and $(1, -3)$

   (f) f is concave up on $(-\infty, 0)$ and $(1, \infty)$
   f is concave down on $(0, 1)$

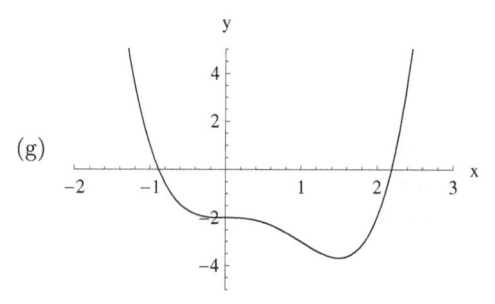

(g)

7. f has a relative minimum at $(-3, 2)$
   f has no relative maximum

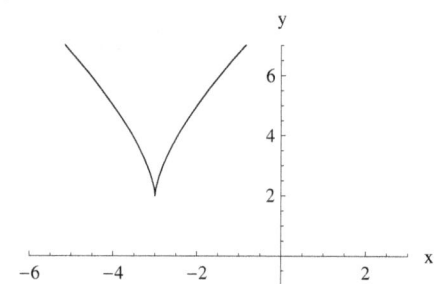

8. (a) Domain: $x \neq -2, 2$.
   y-intercept $(0, -1/2)$

   (b) Vertical asymptotes: $x = -2$, $x = 2$
   Horizontal asymptote: $y = 0$

   (c) f has a relative maximum at $(0, -1/2)$
   f has no relative minimum

   (d) f is increasing on $(-\infty, -2)$ and $(-2, 0)$
   f is decreasing on $(0, 2)$ and $(2, \infty)$

   (e) f has no inflection point

   (f) f is concave up on $(-\infty, -2)$ and $(2, \infty)$
   f is concave down on $(-2, 2)$

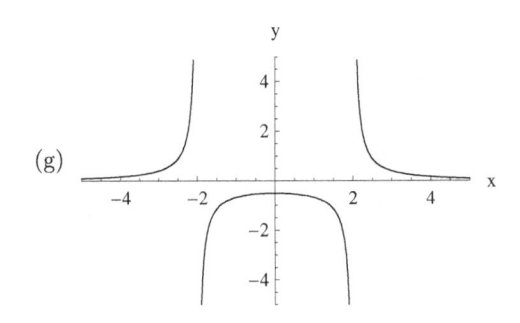

9. (a) Domain: $(-\infty, \infty)$
   y-intercept $(0, 0)$

   (b) f has no vertical asymptote
   Horizontal asymptote: $y = 4$

   (c) f has no relative maximum
   f has a relative minimum at $(0, 0)$

   (d) f is increasing on $(-\infty, 0)$
   f is decreasing on $(0, \infty)$

   (e) f has inflection points
   $(-1/\sqrt{3}, 1)$ and $(1/\sqrt{3}, 1)$

   (f) f is concave up on $(-1/\sqrt{3}, 1/\sqrt{3})$
   f is concave down on
   $(-\infty, -1/\sqrt{3})$ and $(1/\sqrt{3}, \infty)$

   (g)

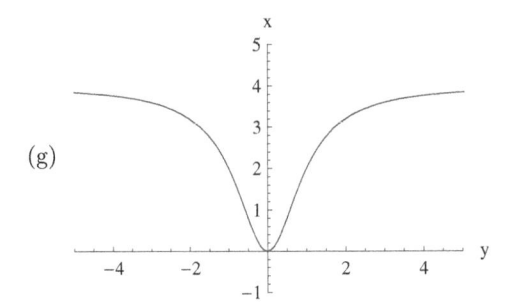

10. $g(x) = \dfrac{1}{x-2}, \; x \neq -2$.

    (a) Domain: $x \neq -2, 2$
    y-intercept $(0, -1/2)$

    (b) Vertical asymptote: $x = 2$
    Horizontal asymptote: $y = 0$

    (c) g has no relative extremum

    (d) g is decreasing on $(-\infty, -2)$,
    $(-2, 2)$, and $(2, \infty)$

    (e) g has no inflection point

    (f) g is concave up on $(2, \infty)$
    g is concave down on $(-\infty, -2)$ and $(-2, 2)$

    (g)

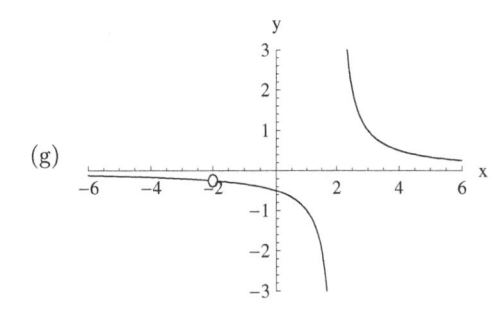

11. abs. max $= 6\frac{2}{3}$ at $x = 2$,
    abs. min $= -2\frac{1}{3}$ at $x = 5$

12. abs. max $= 0.54$ at $x = 2$,
    abs. min $= 0$ at $x = 0$

13. abs. max $= 0.5$ at $x = 2$,
    abs. min $= -0.5$ at $x = -2$

14. abs. max $= 0.54$ at $x = 2$, no abs. min

15. abs. max $= 0.5$ at $x = 2$, no abs. min

16. No abs. max, abs. min $= -3$ at $x = 2$

17. 4 and -4

18. (a)

    Cost $= 1.5x + 2y$

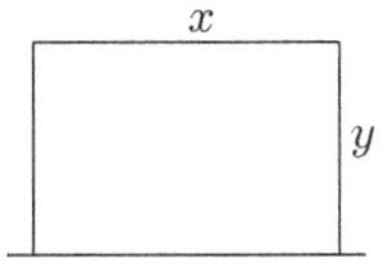

    (b) $C(x) = 1.5x + 486/x$

    (c) $0 < x < \infty$

    (d) $x^* = 18$

    (e) 18 ft × 13.5 ft
    with 18 ft along boundary

19. 44.72 yd × 33.54 yd
    with divider $= 33.54$ yd

20. (a) $C(x) = 25x + 40,000/x + 16,000$

    (b) Lot size $= 40$, re-order 20 times

21. (a) $E(p) = \dfrac{p}{30-p}$

    (b) 1.5, elastic. Lower the price

    (c) $15 per blouse

22. (a) $p = -0.05x + 70$, $0 \leq x \leq 1,000$

    (b) $R(x) = -0.05x^2 + 75x$,
    $0 \leq x \leq 1,000$

    (c) $32.50 a ticket

Chapter 4 Exercises Answers

Exercises 4.1 [Ref. Example]

1. (a) 7 (b) 6 (c) 3 (d) 14
3. (a) 7 (b) 3.25 (c) 1 (d) 11
5. 18   7. 19 [3]
9. (a) 245 miles (b) 122.5 miles [3]
   (c) 140 miles (d) 122.5 miles
11. (a) $950 (b) $1,250 [4]
    (c) $1,550 (d) $1,900
13. Total sales in 20 weeks are $200,000.
15. Total distance travelled in 20 minutes is 200 meters.
17. Total variable cost of producing 20 TV sets is $200.
19. Total profit from producing and selling 2,000 pairs of shoes is $200,000.
21. (a) [3]

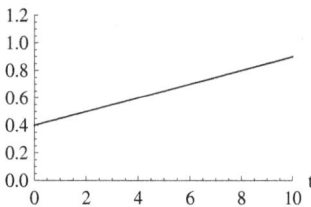

    (b) 6.5 miles
23. (a) [3]

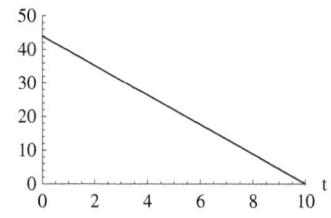

    (b) 220 feet
25. (a) [4]

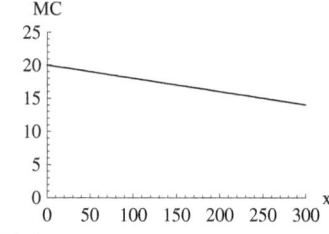

    (b) $5,275
27. (a) [4]

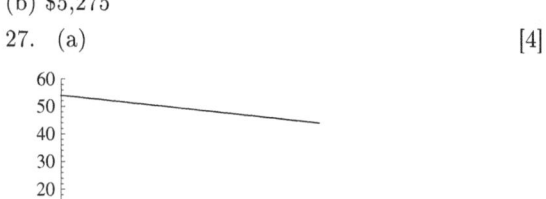

    (b) $6,350
29. (a) [4]

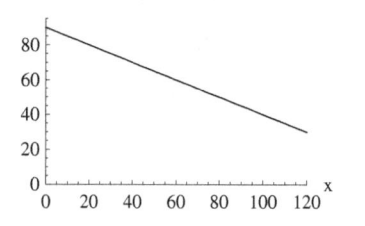

    (b) $6,500
31. (a) [3]

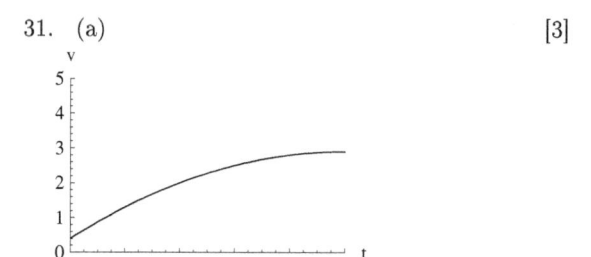

    (b) 18 miles (c) underestimates
33. (a) [3]

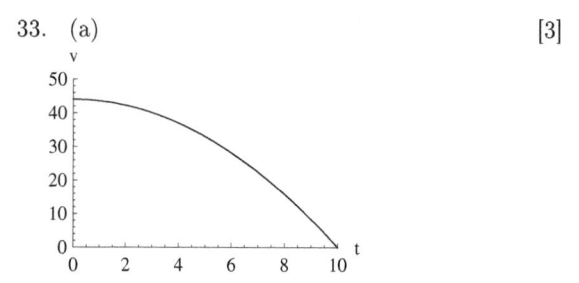

    (b) 343.75 ft
35. (a) [4]

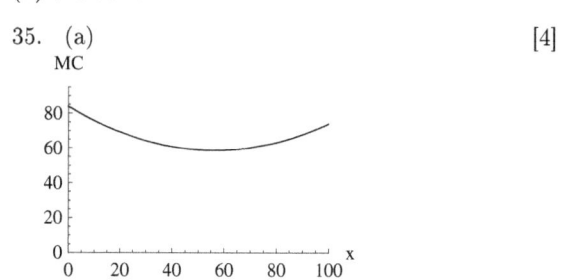

    (b) $6,256
37. (a) [4]

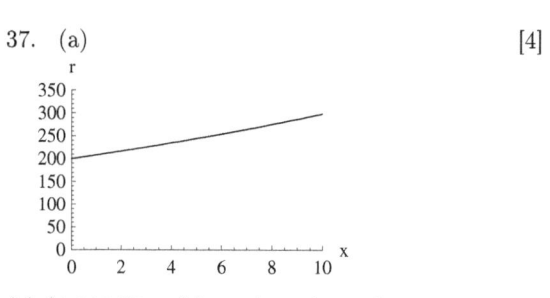

    (b) $2,362.07 (c) underestimated
39. (a) [4]

CHAPTER 6 TEST

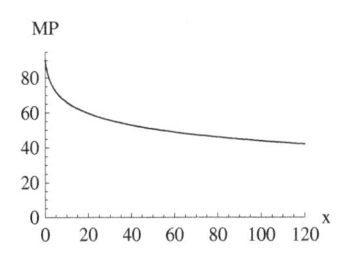

(b) $6,119.84

Exercises 4.2 [Ref. Example]

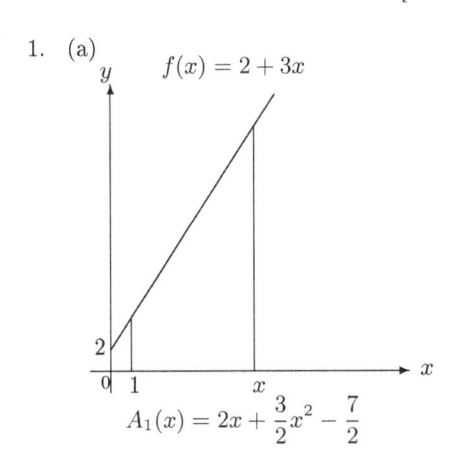

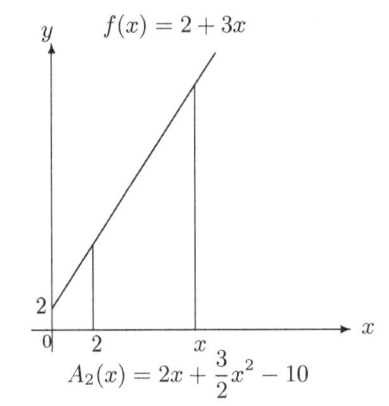

1. (a) [2]

$A_1(x) = 2x + \frac{3}{2}x^2 - \frac{7}{2}$

$A_2(x) = 2x + \frac{3}{2}x^2 - 10$

(b) $A_1'(x) = A_2'(x) = 2 + 3x$, both conform with Theorem 1.

3. $\frac{x^4}{4} + C$ 5. $3x + C$

7. $\frac{1}{2.4}x^{2.4} + C$ 9. $\frac{1}{1-\pi}x^{1-\pi} + C$

11. $\frac{1}{2}x^4 - 4x + C$ 13. $2\sqrt{x^3} + C$

15. $5\sqrt[5]{x^{12}} + C$ 17. $-\frac{15}{8}\sqrt[5]{x^8} + C$

19. $-\frac{2}{x} + C$ 21. $4\sqrt{x} + C$

23. $\ln|x| + C$ 25. $10x^{0.4} + C$

27. $-\frac{15}{2}\sqrt[5]{x^2} + C$ 29. $-\frac{1}{2x^2} + 4\ln|x| + C$

31. $2\sqrt{x} - \frac{9}{2}\sqrt[3]{x^2} + C$ 33. $\frac{-1}{(\pi-1)x^{\pi-1}} + C$

35. $\frac{8}{3x^{1.5}} + C$ 37. $e^x + C$

39. $\frac{2}{3}e^{3x} + C$ 41. $\frac{2}{5}e^{5x} + 5x + C$

43. $-300e^{-0.02x} + C$ 45. $\frac{4}{3}e^{x/2} + C$

47. $-3\ln|x| - \frac{1}{2}e^{-10x} - \frac{6}{5}\sqrt[3]{x^5} + C$

49. $6\sqrt{x} + 4\ln|x| + \frac{2}{3}e^{-3x} + C$

51. $f(x) = x^2 + 4x - 2$ [10]

53. $f(x) = \frac{1}{3}x^3 + 4x + 3$ [10]

55. $f(x) = \frac{2}{3}x^3 - \frac{1}{2}x^2 + 5x + 2$ [10]

57. $f(x) = \frac{1}{3}x^3 - 2x^2 + 3x + \frac{2}{3}$ [10]

59. $f(x) = \frac{2}{3}e^{3x} + 3\frac{1}{3}$ [10]

61. $f(x) = 6\sqrt{x} - 3$ [10]

63. $f(x) = 3\ln|x| + 2$ [10]

65. $s(t) = \frac{4}{3}t^3 + 4$ [11]

67. $s(t) = \frac{3}{2}t^2 + 2t + 5$ [11]

69. $s(t) = \frac{2}{3}t^3 + 3t + 10$ [11]

71. (a) $s(t) = 0.4t + 0.1t^2 - \frac{0.02}{3}t^3$

(b) 2.7 miles

73. (a) $s(t) = 44t - 2.2t^2$

(b) 220 feet

75. (a) $C(x) = 900 + 20x - 0.01x^2$
$0 \le x \le 300$

(b) $C(250) = \$5,275$

77. $C(x) = 1,200 + 84x - 0.45x^2 + 0.008x^3/3$,
$0 \le x \le 200$,
and $C(100) = \$7,766.67$

79. (a) $R(x) = 90x - 0.25x^2$
$0 \le x \le 120$

(b) Yes. No sale, no revenue.

(c) $R(100) = \$6,500$

81. (a) $V(t) = 5,000e^{0.04t} - 4,000$
$0 \le t \le 10$

(b) $V(10) = \$3,459,123$

83. $P(x) = 30x - 0.015x^{4/3} - 300$,
$0 \le x \le 100$,
and $P(60) = \$1,496.48$

85. $s(t) = 600 - 16t^2$ [12]
Hit the ground at 6.12 second.
$v(6.12) = -196$ ft/sec. The firm can use the existing containers.

87. $S(p) = 0.00006p^3 + 0.02p^2 + 0.6p - 39.42$, $p \geq 30.9$

Exercises 4.3 [Ref. Example]

1. (a) 10

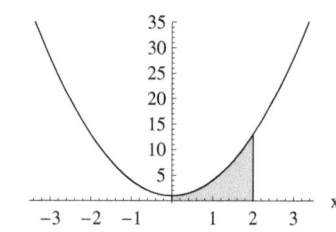

   (b) 28

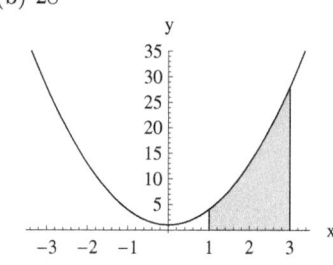

   (c) 10

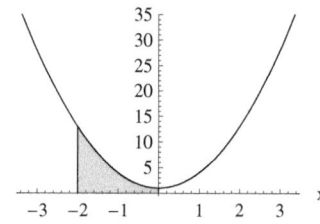

   (d) 32

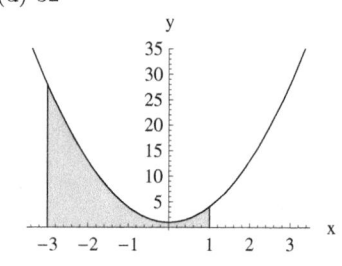

3. (a) 4.918

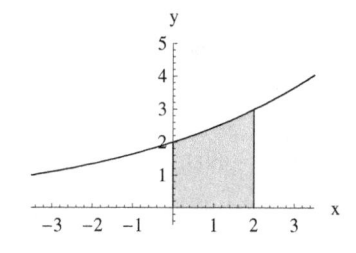

   (b) 6.007

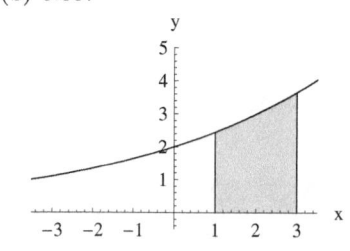

   (c) 3.297

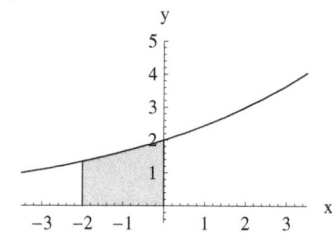

   (d) 6.726

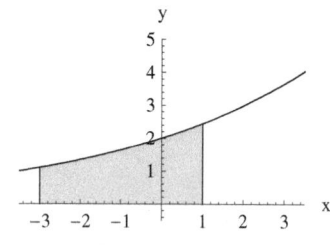

5. (a) $4\sqrt{2} - 1 \approx 4.657$

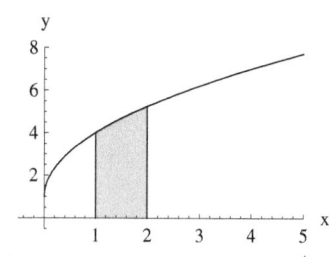

   (b) 17

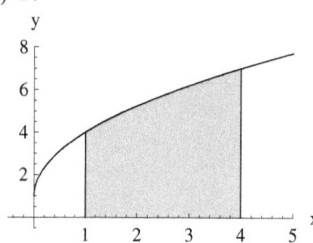

7. (a) $2\ln 2 + 1.5 \approx 2.89$

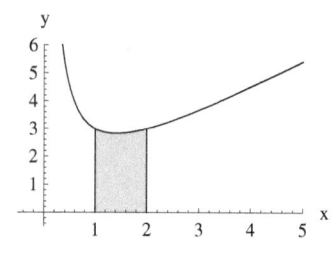

   (b) $4\ln 2 + 7.5 \approx 10.27$

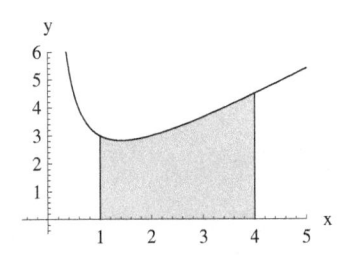

CHAPTER 6 TEST

9. 33 11. $\frac{8}{3}$

13. $\frac{4}{3}$ 15. 13.39

17. 0.8 19. 3

21. 4 23. $1.5e^2 + 0.5 \approx 11.58$

25. (a) $2\ln 3$

(b) $2\ln 3 =$ shaded area
$f(x)=2/x$

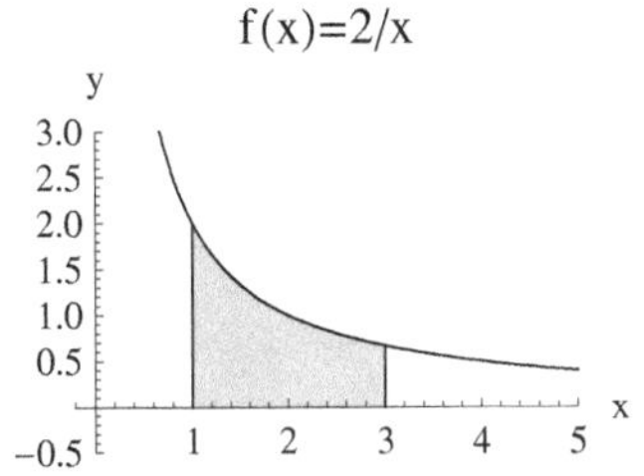

27. (a) -1.5

(b) $1.5 =$ shaded area
$f(x)=1-5/x^2$

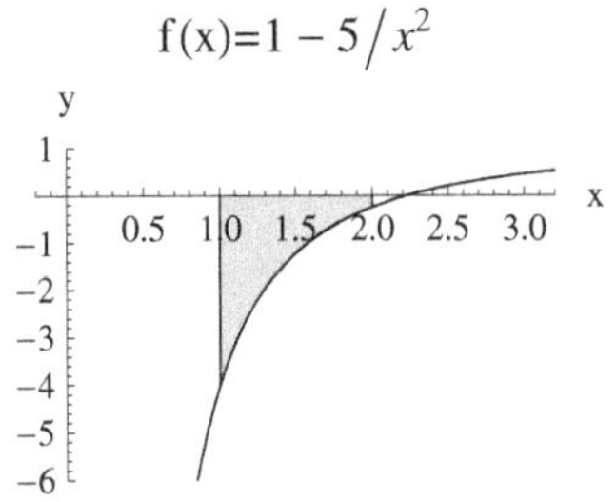

29. (a) 3

(b) $3 = A - B$
$f(x)=4-x^2$

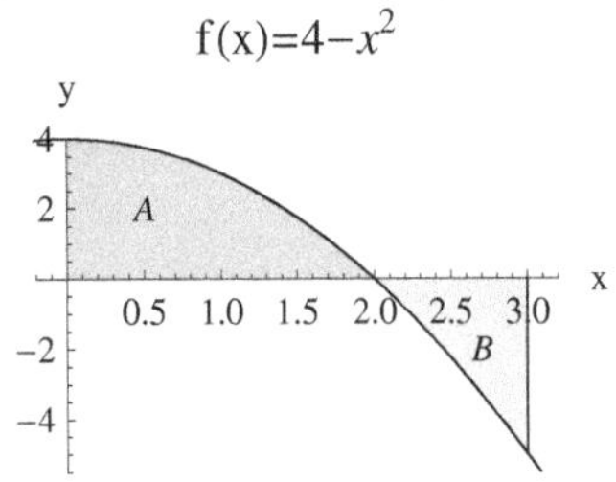

31. $6,566.67

33. 2.7 miles

35. 220 feet

37. $5,275

39. $6,500

41. $3,459,123

43. $1,496.48

45. 135 miles east of home [6]

Exercises 4.4 [Ref. Example]

1. (a) 18 [1]

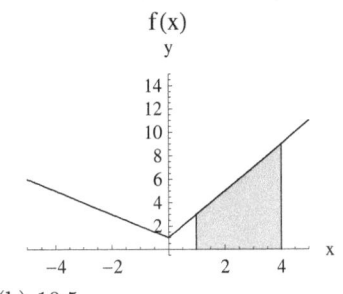

(b) 10.5

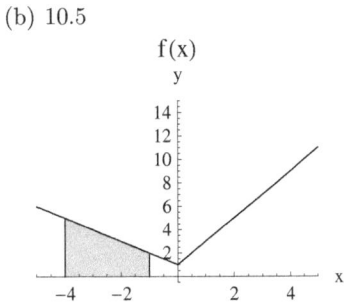

(c) 32

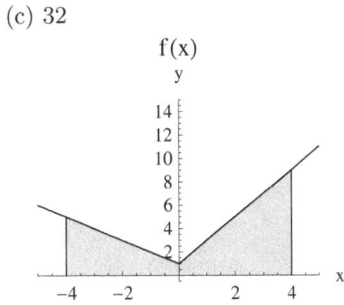

3. (a) 10.2 [1]

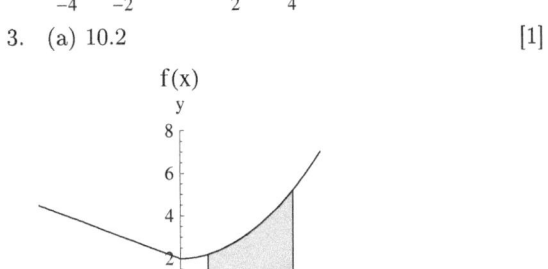

(b) 9.75

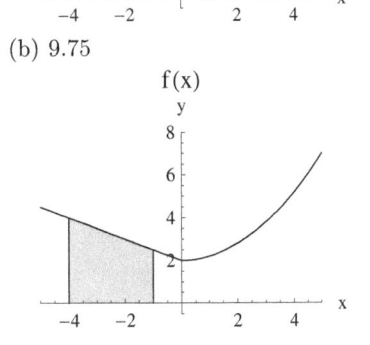

494 CHAPTER 6. FUNCTIONS OF TWO VARIABLES

(c) 24.27

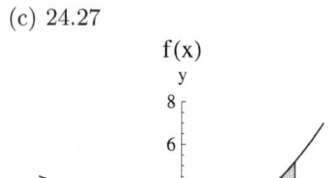

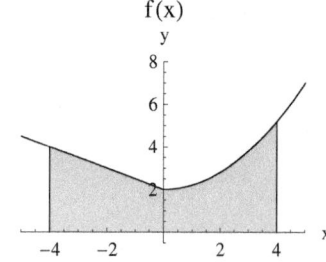

5. 19.83 [1]

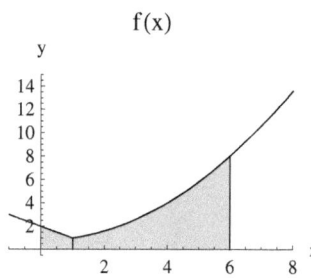

7. $77\frac{1}{3}$ [1]

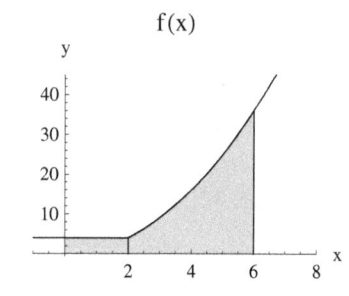

9. $-2, 2$ 11. $-1, 2$

13. 4 15. No intersect

17. $A = \int_{-1}^{1}[2 - (x + 2x^2 - x^3)]\,dx$
$+ \int_{1}^{2}[(x + 2x^2 - x^3) - 2]\,dx = 3.083$ [4]

19. $A = \int_{1}^{5}[(-x^2 + 6x + 2)$
$- (x^2 - 6x + 12)]\,dx = \frac{64}{3}$ [3]

21. $A = \int_{-2}^{2}[(x^2 + 3) - (x + 1)]\,dx = \frac{40}{3}$ [2]

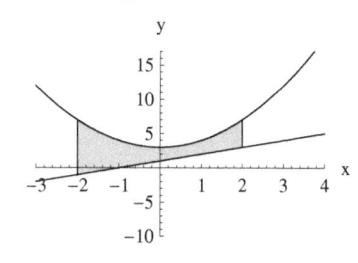

23. $A = \int_{-2}^{2/3}[(1 - x) - (2x - 1)]\,dx$
$+ \int_{2/3}^{3}[(2x - 1) - (1 - x)]\,dx = \frac{113}{6}$ [4]

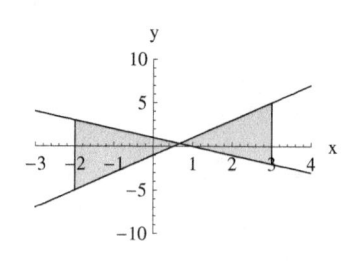

25. $A = \int_{1}^{4}[(x + 3) - (\sqrt{x} - 2)]\,dx = \frac{35}{6}$ [2]

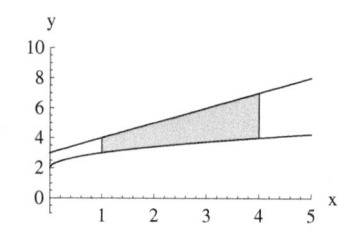

27. $A = \int_{1}^{4}[(\sqrt{x} + 2) - (x - 3)]\,dx = \frac{73}{6}$ [2]

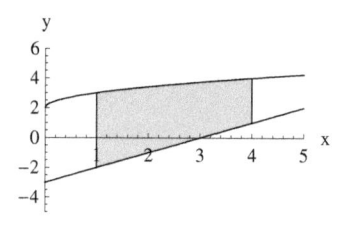

29. $A = \int_{1}^{2}\left(\frac{2}{x} - \frac{x^2}{4}\right)dx$
$+ \int_{2}^{4}\left(\frac{x^2}{4} - \frac{2}{x}\right)dx = \frac{49}{12}$ [4]

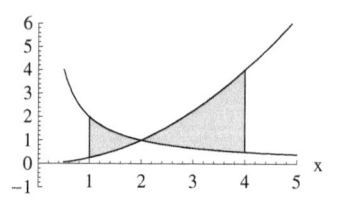

31. $A = \int_{0}^{1}(x - x^2)\,dx = \frac{1}{6}$ [3]

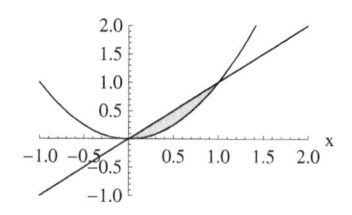

33. $A = \int_{-1}^{3}[(3 - x^2) - (-2x)]\,dx = \frac{32}{3}$ [3]

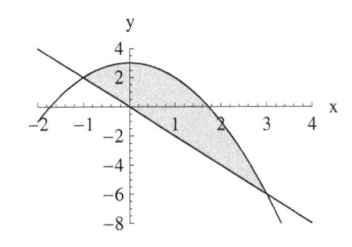

CHAPTER 6 TEST 495

35. $A = 2\int_0^2 (4x - x^3)\, dx = 8$ [4]

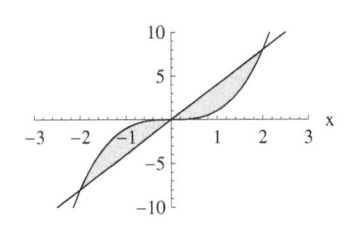

37. $A = \int_0^9 (3 - \sqrt{x})\, dx = 9$ [3]

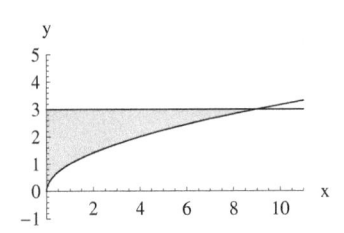

39. $A = \int_{-2}^2 [(4 - x^2) - (x^2 - 4)]\, dx = \dfrac{64}{3}$ [3]

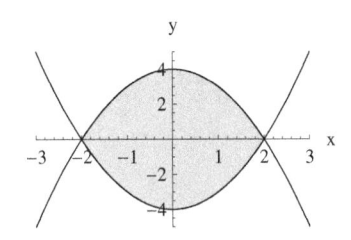

41. $A = \int_{-1}^2 [(x^2 + 1) - (x^2)]\, dx = 3$ [2]

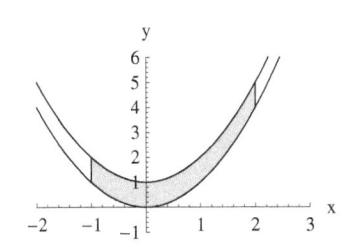

43. 3.19 [6]

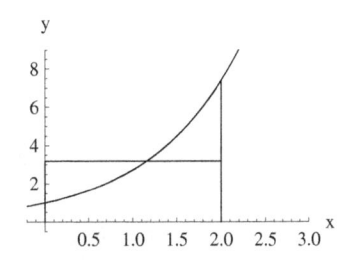

45. $\dfrac{8}{3}$ [6]

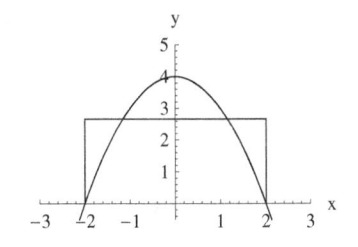

47. 5 [6]

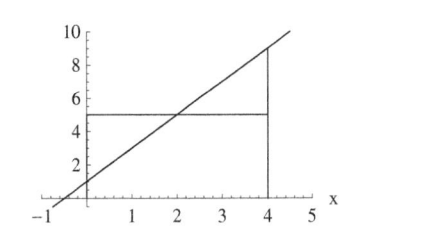

49. 1 [6]

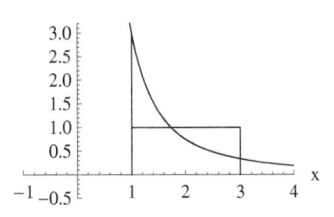

51. $\dfrac{7}{3}$ [6]

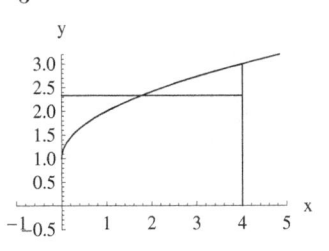

53. $S_A = 7.33$ miles, $S_B = 7.83$ miles. [7]
Car B is ahead by 0.5 mile

55. $68,410 [5]

57. (a) 80, 5   (b) 28 cases [8]

59. 54.69 mph,   13.67 miles [7]

61. (a) $5,153.05   (b) $61.84 [8]

63. (a) $215.63 (b) $2.59 [8]
65. $1,096.16 [8]
67. $1,267.76 [8]
69. $245,912 per year [8]

Exercises 4.5 [Ref. Example]

1. $\frac{2}{3}x^6 + \frac{3}{4}x^4 - \frac{5}{3}x^3 + C$ [1]
3. $\frac{6}{5}\sqrt{x^5} + 3x^2 - \frac{2}{3}\sqrt{x^3} - 2x + C$ [1]
5. $x^2 + \frac{4}{x} + C$ [2]
7. $-\frac{2}{3}e^{-3x} + C$
9. $\frac{1}{7}(x^2 + 1)^7 + C$ [3]
11. $\frac{1}{6}(2x^3 - 5)^6 + C$ [3]
13. $\frac{1}{6}(\sqrt{x^3} + 2)^4 + C$ [6]
15. $\frac{1}{6}e^{3x^2} + C$ [7]
17. $\ln(x^4 + 1) + C$ [5]
19. $\frac{1}{3}\ln|3x + 4| + C$
21. $\frac{1}{4}(2\sqrt{x} + 3)^4 + C$
23. $\frac{1}{2}(\ln(3x))^2 + C$ [8]
25. $-8\sqrt{4-x} + C$ [6]
27. $\frac{3}{4(2e^x + 1)^2} + C$
29. $\frac{1}{3}\sqrt{(x^2 + 4)^3} + C$
31. $\sqrt{e^{2x} + 1} + C$
33. $\frac{-1}{3(3x + 5)} + C$
35. $2(\ln x)^2 + C$
37. $-\frac{1}{2}e^{-2x+1} + C$
39. $6e^{x/3} + C$
41. $2e^{\sqrt{x}} + C$
43. $\ln|\ln x| + C$
45. $-e^{2/x} + C$
47. $\frac{1}{7}(x^3 + x^2 + x + 1)^7 + C$
49. $\frac{1}{6}(1 + \sqrt{x^3})^4 + \frac{1}{3}e^{3x^2} + C$
51. $\frac{3}{16}$ 53. 0 [9]
55. $\frac{4}{3}$ 57. $e - 1$ [9]

59. $\frac{\ln(10/7)}{3}$ 61. 1 [9]
63. $\frac{1}{3}$ 65. 1 [9]
67. $\frac{52}{3}$ 69. $\frac{1}{70}$ [9]
71. (a) $C(x) = 100 + 40\sqrt{x^2 + 100}$
    $0 \le x \le 100$
    (b) $C(60) = \$2,533.11$
73. (a) $V(t) = 3,500 - 2,500e^{-0.04t^2}$
    $0 \le t \le 10$
    (b) $V(10) = \$3,454,210$
75. (a) $1,249.24. This quantity represents the total profit from producing and selling the 21th suit to 60th suit.
    (b) $P(x) = 20\sqrt{2x^2 - 400} - 1,058.3$, $15 \le x \le 100$

Exercises 4.6 [Ref. Formula #]

1. $-\frac{2}{9}(3x + 1)e^{-3x} + C$ [4]
3. $x\ln(5x) - x + C$ [6]
5. $x^2(\frac{\ln(2x)}{2} - \frac{1}{4}) + C$ [8]
7. $(x^2 + 1)\ln(x^2 + 1) - (x^2 + 1) + C$ [6]
9. $\frac{4}{9} - \frac{x}{3} - \frac{4}{9}\ln|3x + 4| + C$ [13]
11. $-\frac{1}{2}\ln\left|\frac{2 + \sqrt{x^2 + 4}}{x}\right| + C$ [12]
13. $\frac{x}{2}\sqrt{4x^2 - 9} - \frac{9}{4}\ln|2x + \sqrt{4x^2 - 9}| + C$ [17]
15. $\ln|x + \sqrt{x^2 + 4}| + C$ [11]
17. $\frac{2}{3}\ln\left|\frac{x-3}{x+3}\right| + C$ [10]
19. $\frac{4}{9(3x + 4)} + \frac{1}{9}\ln|3x + 4| + C$ [14]
21. $\frac{1}{x^2}(-\frac{\ln(x)}{2} - \frac{1}{4}) + C$ [8]
23. $\frac{2}{5}\ln\left|\frac{x}{5 - 3x}\right| + C$ [15]
25. $-2\ln\left|\frac{2 + \sqrt{4 - x^2}}{x}\right| + C$ [12]
27. $\frac{1}{10}(6x - 2)(1 + 2x)^{3/2} + C$ [18]
29. $\frac{2}{7}x^2(4 + x)^{3/2} - \frac{32}{105}(3x - 8)(4 + x)^{3/2} + C$ [19]
31. $\frac{1}{6}(2x + 2)\sqrt{2x - 1} + C$ [20]

Chapter 4 Review Exercises

1. (a) $950   (b) $1,250
   (c) $1,550   (d) $1,900

CHAPTER 6 TEST

2. (a) Total profit in dollars from producing and selling 20 chairs

    (a) Total variable cost in dollars of producing 20,000 screwdrivers

3. (a)

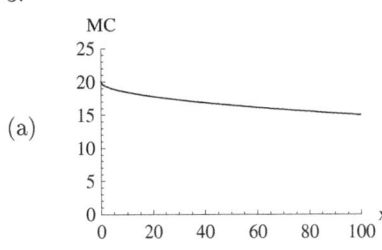

    MC graph with values from 0 to 25 on y-axis, 0 to 100 on x-axis

    (b) $1,925.13   (c) $1,866.67

4. $x^6 + \frac{8}{3}e^{3x} - 4x + C$

5. $-\frac{15}{2}\sqrt[5]{x^2} - \frac{1}{2x^2} + 4\ln|x| + C$

6. $\frac{4}{3}$

7. $2e^2 - 5$

8. $f(x) = \frac{3}{2}e^{2x} + \frac{1}{2}$

9. (a) $\frac{16}{3}$ (b) $\frac{5}{3}$ (c) $\frac{32}{3}$

10. (a) 3

    (b) $3 = A - B$

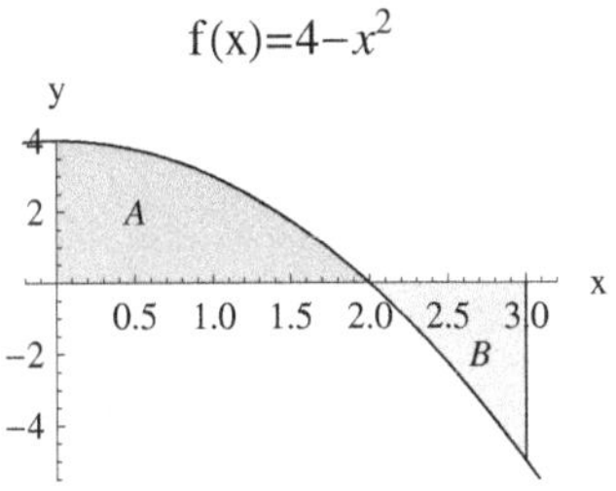

    $f(x) = 4 - x^2$

11. 2.7 miles

12. $\frac{64}{3}$

13. $68,410

14. $\frac{4}{5}x^{5/2} + \frac{4}{3}x^{3/2} + C$

15. $x^2 + \frac{4}{x} + C$

16. $\frac{1}{6}(2x^3 - 5)^6 + C$

17. $\frac{1}{6}e^{3x^2} + C$

18. $\ln(x^4 + 1) + C$

19. $\frac{3}{4(2e^x + 1)^2} + C$

20. $\frac{1}{6}(\sqrt{8} - 1) \approx 0.3$

21. $\frac{4}{3}$

22. $-\frac{2}{9}(3x + 1)e^{-3x} + C$

23. $x\ln(5x) - x + C$

24. $\frac{4}{375}(15x + 8)(5x - 4)^{3/2} + C$

25. $\frac{2}{3}x^6 + \frac{3}{4}x^4 - \frac{5}{3}x^3 + C$

26. $\frac{1}{42}(2x^3 + 7)^7 + C$

27. $\frac{4}{9} - \frac{x}{3} - \frac{4}{9}\ln|3x + 4| + C$

28. $2(\ln x)^2 + C$

29. $\frac{1}{x^2}(-\frac{\ln(x)}{2} - \frac{1}{4}) + C$

30. $-\frac{1}{16}(4x + 1)e^{-4x} + C$

31. $-\frac{1}{4}e^{-4x^2} + C$

32. $\frac{1}{3}\ln|3x + 4| + C$

33. (a) $5,153.05 (b) $61.84

Chapter 5 Exercises Answers

Exercises 5.1 [Ref. Example]

1. (a)   [1,2,7]

    $C.S. = \int_0^{x_0} D(x)\,dx - p_0 \times x_0$

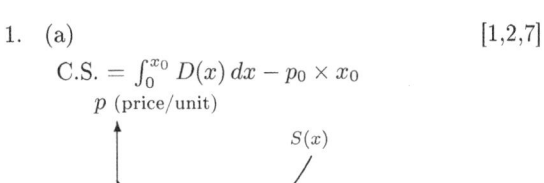

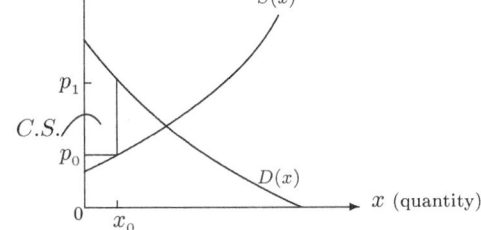

    p (price/unit) graph with S(x), D(x) curves, showing C.S., p_1, p_0, x_0

    (b)

    p (price/unit) graph with S(x), D(x) curves, showing C.S., p_1, p_0, x_1

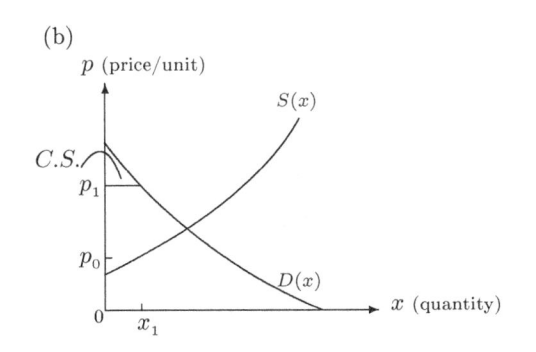

(c) P.S. $= p_0 \times x_0 - \int_0^{x_0} S(x)\, dx$

(d)

(e)

(f)

(g)

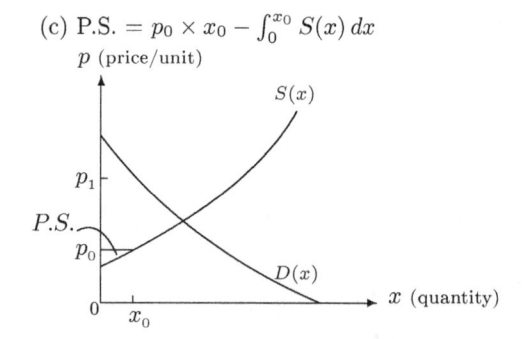

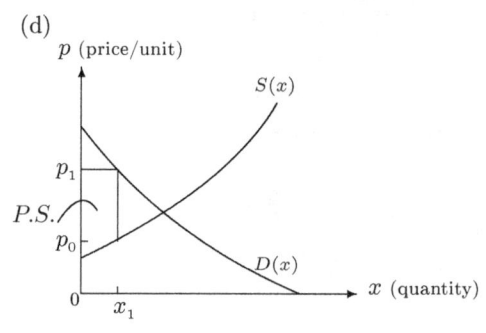

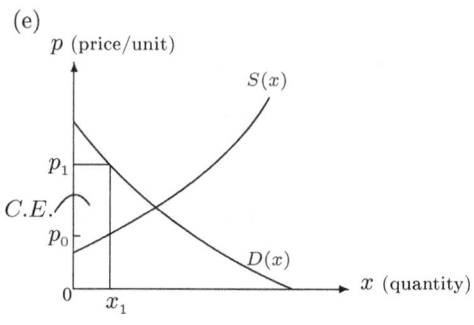

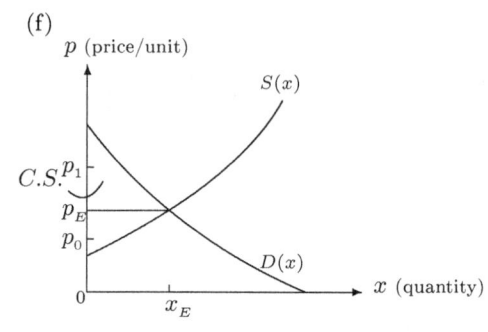

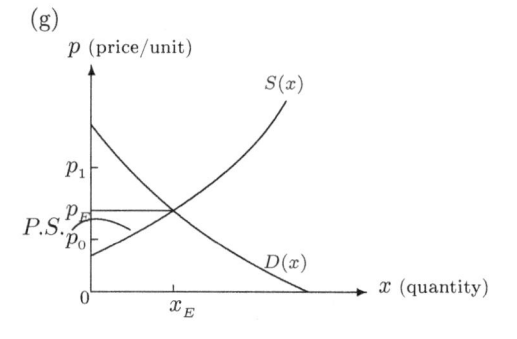

(h) At the equilibrium price

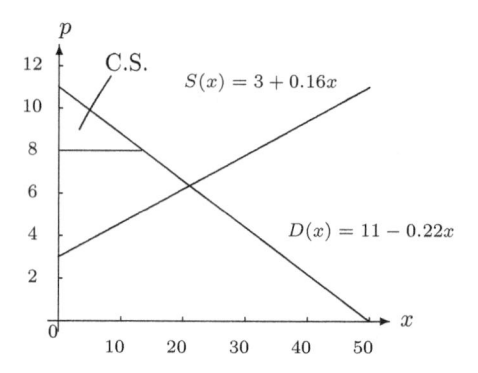

3. (a) [1,2,7]
C.S. $= \int_0^{13.64}(11 - 0.22x)\, dx$
$- 8 \times 13.64 = 20.45$

(b) P.S. $= 8 \times 13.64$
$- \int_0^{13.64}(3 + 0.16x)\, dx = 53.32$

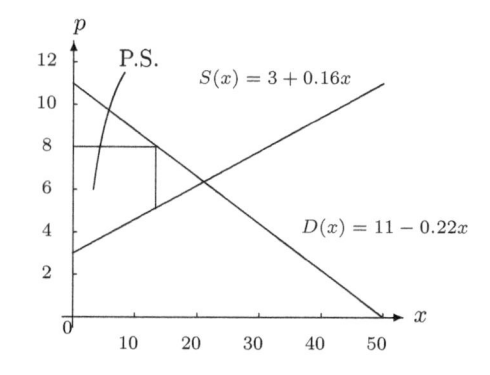

(c) C.S. $= \int_0^{21.05}(11 - 0.22x)\, dx$
$- 6.73 \times 21.05 = 48.72$

(d) P.S. $= 6.73 \times 21.05$
$- \int_0^{21.05}(3 + 0.16x)\, dx = 35.49$

5. 2,400   7. 16,000   [1]
9. 27   11. 18.67   [1]
13. 18,000   [1]
15. 8,909.91   17. 21,383.9   [1]
19. 30,000   21. 500   [2]
23. 5.814   25. 180   [2]
27. 27   29. 554.89   [2]
31. 4,194.53   [2]
33. (a) $(25, 245)$ [3]
(b) 937.5   (c) 1,562.5
35. (a) $(70, 76.5)$ [3]
(b) 612.5   (c) 1,873.5
37. (a) $(44, 142.56)$ [3]
(b) 2,271.57   (c) 3,407.36
39. (a) $(35, 397.5)$ [3]

(b) 2,858.33   (c) 8,575

41. (a) (3,3)   [3]

(b) 4.5   (c) 7.98

43. (a) (80,9)   [3]

(b) 576   (c) 234.67

45. (a) 52.8   (b) 1,548.8   [3]

47. (a) $D(x) = 50 - 0.1x$

(b) $x_E = 230$ (c) 2,645

49. (a) (1,200, $80)   (b) $36,000

(c) $72,000   (d) $108,000

51. (a) New $S(x) = 22.5 + 0.05x$ [4]

New $x_E \approx 1,183$, new $p_E \approx \$81.7$

(b) $34,987.2   (c) $70,033.6

(d) $105,021

(e) All three surpluses are reduced.

53. (a) New $D(x) = 185.19 - 0.0926x$ [5]

New $x_E \approx 1,158$, new $p_E \approx \$77.92$

(b) $33,549.6   (c) $62,129.1

(d) $95,678.7

(e) All three surpluses are reduced.

55. (a) Short-supply.   [6]

1,000 pairs of sunglasses will be sold

(b) $25,000   (c) $80,000

(d) $105,000

(e) Consumers' surplus is higher, but producers' and total surpluses are reduced.

57. (a) Over-supply.

1,100 pairs of sunglasses will be sold

(b) $46,750   (c) $60,500

(d) $107,250

(e) Producers' surplus is higher, but consumers' and total surpluses are reduced.

Exercises 5.2   [Ref. Example]

1. $68,731.3   3. $16,275   [2]

5. 19.1 years   [3]

7. 14.7 years

9. $2,853.13   11. $49.88   [3]

13. $695,102   [4]

15. $834,799

17. $25,284.8   19. $98,895   [5]

21. $18,252.3   [6]

23. Can apply for a loan of $24,538.2   [6]

Loan plus savings = $32,538, not enough for the car.

25. (a) $1,680.78   [7]

(b) 13.83 years   (c) $7,036.39

27. $2,851.96   [8]

29. (a) $667,761   (b) $308.90   [9]

31. $3,258.65

33. $662.38

35. (a) $34,193.2   (b) $248.82   (c) $677.75

Exercises 5.3   [Ref. Example]

1. Convergent, = 2   3. Divergent

5. Convergent, = 2.5   7. Divergent

9. Convergent, = 1   11. Divergent

13. Divergent   15. Divergent

17. Divergent   19. Divergent

21. Convergent, = 4   23. Convergent, = 1

25. Divergent   27. Divergent

29. 1/8   31. 5

33. (a) $V(t) = 6,000 - 5,000e^{-0.04t}$ (b) $6,000,000

35. $48,000   [3]

37. (a) $800,000   (b) $370   [4]

Exercises 5.4   [Ref. Example]

17. $k = 2$ 19. $k = 1/12$ [2]

21. $k = \dfrac{1}{3}$ 23. $k = \dfrac{1}{e^2 - 1}$ [2]

25. (a) 40%   (b) 80%   [2]

27. (a) 64%   (b) 64%   [2]

(b) 2.36 minutes   (c) 64%   (d) 64%

29. (a) $\dfrac{8}{21} = 0.38095$ [3]

(b) 79.6%   (c) 11.1%   (d) 244 light bulbs

31. (a) $f(t) = \dfrac{1}{120}e^{-t/120}, 0 \le t$ (b) 34%   [4]

Exercises 5.5   [Ref. Example]

1. (a) $\mu = 2.5$ (b) $\sigma = 0.866$ [1]

3. (a) $\mu = 4/3$ (b) $\sigma = \sqrt{2}/3 \approx 0.4714$ [1]

5. (a) $\mu = 2.25$ (b) $\sigma = 0.581$ [1]

7. (a) $\mu = (3\ln 3)/2 \approx 1.648$ (b) $\sigma = 0.533$ [1]

9. (a) $\mu = e - 1 \approx 1.718$ (b) $\sigma = 0.492$ [1]

11. (a) $\mu = 0$ (b) $\sigma = 1.549$ [1]

13. (a) $\mu = 5$ (b) $\sigma = 5$ [1]

15. (a) $\mu = 1.5$ (b) $\sigma = 0.866$ [1]

17. 0.4834   [3]

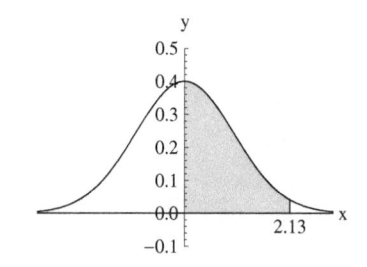

500   CHAPTER 6. FUNCTIONS OF TWO VARIABLES

19. 0.3508 [3]

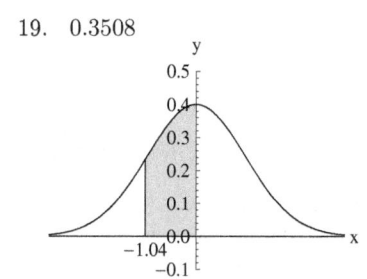

21. 0.7008 [3]

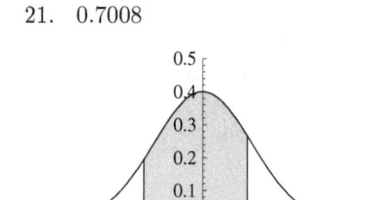

23. 0.3578 [3]

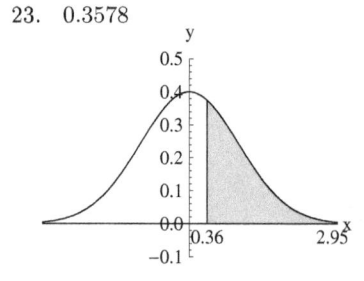

25. 0.3339 [3]

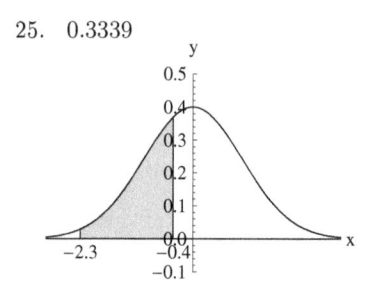

27. 0.8944 [3]

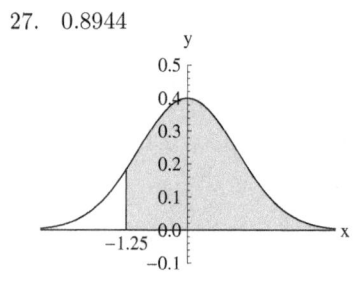

29. 0.5 [3]

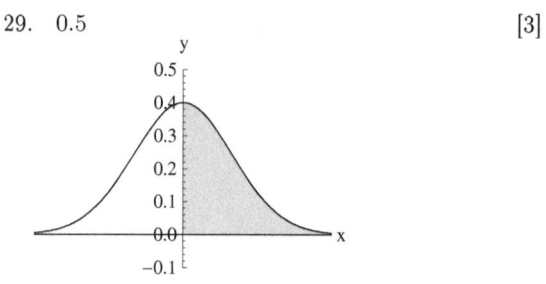

31. 0.0359 [3]

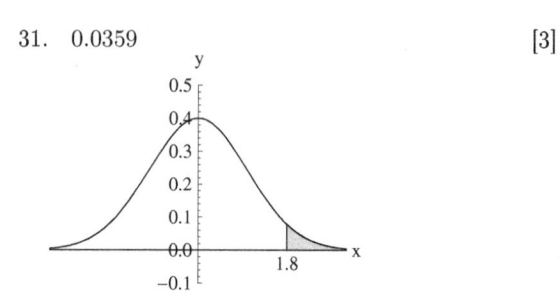

33. $a = -1$, $b = 0.5$, $p = 0.5328$ [4]

35. $a = 0.5$, $b = 1.5$, $p = 0.2417$ [4]

37. $a = -0.5$, $p = 0.6915$ [4]

39. (b) $\sigma = 1/k$

41. (a) 5 minutes [1]
    (b) 2.9 minutes   (c) 40%   (d) 80 %

43. (a) 6.67 minutes [1]
    (b) 2.36 minutes   (c) 64%   (d) 64 %

45. (a) $\dfrac{8}{21} = 0.38095$ [3]
    (b) 1,518 hours   (c) 363 hours

47. (a) $P(x \leq -2) = 0.0228 = 2.28\%$ [6]
    (b) $P(x > \dfrac{A - 20}{0.25}) = 0.023$, $A = 20.5$ oz

49. $P(x > \dfrac{22.5 - \mu}{0.3}) = 0.024$, [6]
    $\mu = 21.9$ oz

51. (a) $a = 0.4$, $p = 0.3446 = 34.46\%$ [5]
    (b) $b = -0.8$, $p = 0.2119 = 21.19\%$
    (c) $a = -2$, $b = -0.8$, $p = 0.1891 = 18.9\%$
    $600(0.1891) \approx 113$ light bulbs

Exercises 5.6   [Ref. Example]

1. $xy' + y - 2x$
   $= x(1 + 1/x^2) + (x - 1/x) - 2x$
   $= x + 1/x + x - 1/x - 2x = 0$ for all x

3. $y'' + y - 6y$
   $= 4e^{2x} + 2e^{2x} - 6e^{2x}$
   $= (4 + 2 - 6)e^{2x} = 0$ for all x

5. $y' + y^2$
   $= \dfrac{-1}{(x+1)^2} + \dfrac{1}{(x+1)^2} = 0$ for all x

CHAPTER 6 TEST

7. $y = \pm\sqrt{x^2 + C}$ 9. $y = Ce^{x^3}$ [2, 1]

11. $y = \pm\sqrt[4]{2x^2 + C}$ 13. $y = Cx$ [2]

15. $y = \pm\sqrt[4]{e^x + C}$ 17. $y = \pm\sqrt{6x + C}$ [2]

19. $P = Ce^{0.02t}$ 21. $y = -4 + Ce^{x^2/2}$ [1,3]

23. $y = Ce^{(x+2)^2/2}$ 25. $y = \sqrt{x^2 + 3}$ [3,2]

27. $y = 3e^{x^3}$ 29. $y = \sqrt[3]{9x + 1}$ [1]

31. $P = 100e^{0.02t}$ 33. $y = -4 + 9e^{x^2/2}$ [2,3]

35. $x = \dfrac{6}{p}$ 37. $x = 5e^{2/p}$ [6]

39. $x = 10(100 - p)$ [6]

41. (a) $P(t) = 20,000 - 10,000e^{0.04t}$

(b) $\$7,287.51$, $\$6,228.72$

(c) About 17.33 years

43. (a) $P(t) = 20,000 + 4,000e^{0.04t}$

(a) $\$25,085$, $\$25,508.5$

(c) Never

45. $\$20,000$

Chapter 5 Review Exercises

1. See solution to Exercise #1 in Section 5.1
2. (a) $(1,9)$ (b) $10/3$ (c) $5/3$
3. $\$68,731.3$ 4. 19.1 years
5. $\$2,853.13$ 6. $\$18,252.3$
7. $\$1,680.78$ 8. (a) $\$667,761$ (b) $\$308.90$
9. Convergent, $=2$ 10. Divergent
11. Convergent, $=6$ 12. Divergent
13. $1/8$ 14. $\$800,000$
15. $k = 1/(\displaystyle\int_2^4 \dfrac{1}{x^2}\, dx) = 4$
16. (a) 5 minutes (b) 2.87 minutes (c) 40% (d) 80%
17. (a) $f(t) = 3.5e^{-3.5t}$
    (b) 82.6%   (c) 3%
18. 0.4834   19. 0.4429
20. 0.9279   21. 0.3339
22. 0.1056
23. (a) $P(x \le -2) = 0.0228 = 2.28\%$
    (b) $P(x > \dfrac{A - 20}{0.25}) = 0.023$, $A = 20.5$ oz
24. (a) $a = 0.4$, $p = 0.3446 = 34.46\%$

(b) $b = -0.8$, $p = 0.2119 = 21.19\%$

(c) $a = -2$, $b = -0.8$, $p = 0.1891 = 18.9\%$

$800(0.1891) \approx 151$ light bulbs

25. $y = Ce^{x^3}$ 26. $y = \sqrt{x^2 + 3}$

27. $y = \pm\sqrt{C - \dfrac{1}{x}}$ 28. $y = \sqrt[3]{9x + 1}$

29. $y = -4 + Ce^{x^2/2}$ 30. $P = 100e^{0.02t}$

31. $x = \dfrac{6}{p}$ 32. $P(t) = 20,000(1 - \tfrac{1}{2}e^{0.04t})$

Chapter 6 Exercises Answers

Exercises 6.1   [Ref. Example]

1. $f_x(x,y) = 14x^6 y^3$ $f_y(x,y) = 6x^7 y^2$ [2]

3. $f_x(x,y) = \dfrac{3\sqrt{y}}{2\sqrt{x}}$ $f_y(x,y) = \dfrac{3\sqrt{x}}{2\sqrt{y}}$ [2]

5. $f_y(x,y) = 2e^{y^2}$ $f_y(x,y) = 4xye^{y^2}$ [2]

7. $f_x(x,y) = \dfrac{2x}{x^2 + y^2}$ $f_y(x,y) = \dfrac{2y}{x^2 + y^2}$ [2]

9. $f_x(x,y) = \dfrac{-4}{x^3 y}$ $f_y(x,y) = \dfrac{-2}{x^2 y^2}$ [2]

11. $f_x(x,y) = \dfrac{-1}{(x + 2y)^2}$ $f_y(x,y) = \dfrac{-2}{(x + 2y)^2}$ [2]

13. $\dfrac{\partial z}{\partial x} = 2xy + 12x^3 y^2$ [2]

$\dfrac{\partial z}{\partial y} = x^2 + 6x^4 y$

15. $\dfrac{\partial z}{\partial x} = 2y^2 e^{2xy^2}$ $\dfrac{\partial z}{\partial y} = 4xy e^{2xy^2}$ [2]

17. $\dfrac{\partial z}{\partial x} = e^x \ln xy + \dfrac{e^x}{x}$ $\dfrac{\partial z}{\partial y} = \dfrac{e^x}{y}$ [2]

19. $\dfrac{\partial z}{\partial x} = \dfrac{1}{y}$ $\dfrac{\partial z}{\partial y} = \dfrac{-x}{y^2}$ [2]

21. $\dfrac{\partial z}{\partial x} = 6x^2 e^y - 20xy + 6e^x$ [2]

$\dfrac{\partial z}{\partial y} = 2x^3 e^y - 10x^2 - 6y$

23. $\dfrac{\partial z}{\partial x} = \dfrac{3\sqrt{y}}{2\sqrt{x}} - 3ye^x - \dfrac{2y^2}{x^3}$ [2]

$\dfrac{\partial z}{\partial y} = \dfrac{3\sqrt{x}}{2\sqrt{y}} - 3e^x - \dfrac{2y}{x^2}$

25. $f_x(1,2) = 20$ $f_y(1,2) = -11$ [3]
27. $f_x(1,2) = 2/5$ $f_y(1,2) = 4/5$ [3]
29. $f_x(1,2) = 4e^2$ $f_y(1,2) = e^2$ [3]
31. $f_{xx}(x,y) = 24x^2 + 6xy^2$ [7]

$f_{xy}(x,y) = 6x^2 y$

$f_{yy}(x,y) = 2x^3 - 8$

33. $f_{xx}(x,y) = 4y^3 + e^{x+y}$ [7]
$f_{xy}(x,y) = 12xy^2 + e^{x+y}$
$f_{yy}(x,y) = 12x^2y + e^{x+y}$

35. $f_{xx}(x,y) = y^3 e^x - \dfrac{1}{x^2}$ [7]
$f_{xy}(x,y) = 3y^2 e^x$
$f_{yy}(x,y) = 6ye^x$

37. $f_{xx}(x,y) = \dfrac{-4}{(2x+3y)^2}$ [7]
$f_{xy}(x,y) = \dfrac{-6}{(2x+3y)^2}$
$f_{yy}(x,y) = \dfrac{-9}{(2x+3y)^2}$

39. $f_{xx}(x,y) = 6(x^2-y^2)^2 + 24x^2(x^2-y^2)$ [7]
$f_{xy}(x,y) = -24xy(x^2-y^2)$
$f_{yy}(x,y) = -6(x^2-y^2)^2 + 24y^2(x^2-y^2)$

41. $f_{xx}(x,y) = \dfrac{20x^3}{y^3}$ [7]
$f_{xy}(x,y) = \dfrac{-15x^4}{y^4}$
$f_{yy}(x,y) = \dfrac{12x^5}{y^5}$

43. $f_{xx}(x,y) = y^5 e^{xy}$ [7]
$f_{xy}(x,y) = (4y^3 + xy^4)e^{xy}$
$f_{yy}(x,y) = (6y + 6xy^2 + x^2y^3)e^{xy}$

45. $f_{xx}(x,y) = (2y + 4x^2y^2)e^{x^2y}$ [7]
$f_{xy}(x,y) = (2x + 2x^3y)e^{x^2y}$
$f_{yy}(x,y) = x^4 e^{x^2y}$

47. (a) $V_r(3,5) = 10\pi$ [4]
When $r = 3$ in and $h = 5$ in, if the radius increases by 1 in and the height remains constant at 5 in, then the volume of the cone increases by approximately 10π in^3.
(b) $V_h(3,5) = 3\pi$
When $r = 3$ in and $h = 5$ in, if the height increases by 1 in and the radius remains constant at 3 in, then the volume of the cone increases by approximately 3π in^3.

49. (a) $f_P(2{,}000, 3) = \$1.13$ per dollar
When \$2,000 is invested for 3 years, if the principal increases by \$1 and the time remains constant at 3 years, then the balance in the fund increases by approximately \$1.13.
(b) $f_t(2{,}000, 3) = \$90.03$ per year
When \$2,000 is invested for 3 years, if the time increases by 1 year and the principal remains constant at \$2,000, then the balance in the fund increases by approximately \$90.03.

51. (a) $f_t(5,3) = \$348.06$ per year
When \$10,000 is invested for 5 years at an annual interest rate of 3% compounded monthly, if the time increases by 1 year and the annual interest rate remains constant at 3%, then the balance in the fund increases by approximately \$348.06.
(b) $f_r(5,3) = \$579.36$ per percent
When \$10,000 is invested for 5 years at an annual interest rate of 3% compounded monthly, if the annual interest rate increases by 1% and the time remains constant at 5 years, then the balance in the fund increases by approximately \$579.36.

53. (a) $R_x(5,2) = 15$ [4]
When spending \$5,000 weekly on television advertising and \$2,000 weekly on radio advertising, if the weekly spending on television advertising increases by \$1,000 and the weekly spending on radio advertising remains constant at \$2,000, then the weekly revenue increases by approximately \$15,000.
(b) $R_y(5,2) = 6$
When spending \$5,000 weekly on television advertising and \$2,000 weekly on radio advertising, if the weekly spending on radio advertising increases by \$1,000 and the weekly spending on television advertising remains constant at \$5,000, then the weekly revenue increases by approximately \$6,000.

55. (a) $R(x,y) =$ [5]
$\quad 120x + 150y - 0.5x^2 - 0.8xy - 0.5y^2$
(b) $R_x(65,35) = 27$
When 65 basic models and 35 luxury models of the cellular phones are sold per week, if the sales of the basic model increase by 1 unit per week, and the sales of the luxury model remain constant at 35 units per week, then the total revenue increases by approximately \$27 per week.
(c) $R_y(65,35) = 63$
When 65 basic models and 35 luxury models of the cellular phones are sold per week, if the sales of the luxury model increase by 1 unit per week, and the sales of the basic model remain constant at 65 units per week, then the total revenue increases by approximately \$63 per week.

57. (a) $P(x,y) = 110x + 110y - 0.5x^2$
$\quad -0.8xy - 0.5y^2 - 1{,}000$
(b) $P_x(65,35) = 17$
When 65 basic models and 35 luxury models of the cellular phones are produced and sold per week, if the production and sales of the basic model increase by 1 unit per week, and the production and sales of the luxury model remain constant at 35 units per week, then the total profit increases by approximately \$17 per week.

(c) $P_y(65, 35) = 23$

When 65 basic models and 35 luxury models of the cellular phones are produced and sold per week, if the production and sales of the luxury model increase by 1 unit per week, and the production and sales of the basic model remain constant at 65 units per week, then the total profit increases by approximately $23 per week.

59. (a) $f(250, 120) = 3727.92$ [6]

When the company has a labor force of 250 labor hours per day and has $120,000 worth of equipment, the production is about 3,728 compact fluorescent light bulbs per day.

(b) $f_x(250, 120) \approx 8.95$.

When the company has a labor force of 250 labor hours per day and has $120,000 worth of equipment, if the labor hours increase by 1 hour per day to 251, and the value of equipment remains $120,000, then the production increases by approximately 9 compact fluorescent light bulbs per day.

(c) $f_y(250, 120) \approx 12.43$.

When the company has a labor force of 250 labor hours per day and has $120,000 worth of equipment, if the value of the equipment increases by one thousand dollars to $121,000, and the labor hours remain 250 per day, then the production increases by approximately 12 compact fluorescent light bulbs per day.

61. (a) $f(50, 10) = 4,554.6$ [6]

(b) $f_x(50, 10) \approx 59.21$.

When the company has a labor force of 50,000 labor hours per year and has $100,000 worth of equipment, if the labor hours increase by 1,000 hours per year to 51,000, and the value of equipment remains $100,000, then the production increases by approximately 59 refrigerators per year.

(c) $f_y(50, 10) \approx 159.41$.

When the company has a labor force of 50,000 labor hours per year and has $100,000 worth of equipment, if the value of the equipment increases by ten thousand dollars to $110,000, and the labor hours remain 50,000 per year, then the production increases by approximately 159 refrigerators per year.

Exercises 6.2 [Ref. Example]
1. $(-3, 5)$ 3. $(-10, 20)$

5. $(3, 2)$ 7. $(-8/3, 16/3)$

9. $(0, 2), (0, -2), (2, 2), (2, -2)$ [1]

11. Critical point $(-\frac{1}{3}, \frac{2}{3})$, relative minimum [3]

13. Critical point $(-6, 12)$, saddle point [3]

15. Critical point $(\frac{2}{3}, \frac{2}{3})$, relative maximum
    Critical point $(0, 0)$, saddle point

17. Critical point $(3, 2)$, relative maximum

19. Critical point $(2, 2)$, relative minimum
    Critical points $(2, -2)$ and $(0, 2)$, saddle points
    Critical point $(0, -2)$, relative maximum

21. Critical point $(1, -1)$, relative minimum
    Critical point $(0, 0)$, saddle point

23. Critical point $(1, 0)$, relative maximum
    Critical point $(1, 2)$, saddle point

25. Critical point $(-2, 1)$, saddle point

27. Critical point $(0, 0)$, relative minimum

29. Critical point $(0, 0)$, saddle point

31. (a) $30,000 on labor [5]
    $20,000 on equipment upgrade
    (b) Maximum profit $100,000

33. (a) $R(x, y) =$ [4,5]
    $600x - 0.4x^2 + 400y - 0.2y^2$
    (b) $P(x, y) = 480x - 0.4x^2 + 200y$
    $-0.2y^2 - 0.2xy - 60,000$
    (c) 329 units of Product A, 336 units of Product B should be produced and sold per week.

35. (a) $R(x, y) =$ [5]
    $200x - 1.5x^2 + 300y - 0.8y^2 - 0.4xy$
    (b) 43 units of the basic models and 177 units of the luxury models should be sold per week.

37. (a) $P(x, y) = 160x - 1.5x^2$ [5]
    $+220y - 0.8y^2 - 0.4xy - 1,000$
    (b) 36 units of the basic models and 130 units of the luxury models should be produced and sold per week.

39. (a) $P(x, y) = 65x - 0.08x^2$
    $+65y - 0.08y^2 - 0.08xy - 15$
    (b) Plant A: 27,083 units; Plant B: 27,083 units

Chapter 6 Review Exercises

1. 2
2. $4xe^y - \frac{1}{x}$
3. $2x^2e^y + 15y^2$
4. $4e^y + \frac{1}{x^2}$
5. $4xe^y$
6. $4xe^y$
7. $2x^2e^y + 30y$
8. $3ye^{xy} + 2xy^3$
9. $3xe^{xy} + 3x^2y^2$
10. $3y^2e^{xy} + 2y^3$
11. $3e^{xy} + 3xye^{xy} + 6xy^2$
12. $3e^{xy} + 3xye^{xy} + 6xy^2$
13. $3x^2e^{xy} + 6x^2y$

14. (a) $R(x,y) = 120x + 150y - 0.5x^2 - 0.8xy - 0.5y^2$
    (b) $R_x(65, 35) = 27$
    When 65 basic models and 35 luxury models of the cellular phones are sold per week, if the sales of the basic model increase by 1 unit per week, and the sales of the luxury model remain constant at 35 units per week, then the total revenue increases by approximately $27 per week.
    (c) $R_y(65, 35) = 63$
    When 65 basic models and 35 luxury models of the cellular phones are sold per week, if the sales of the luxury model increase by 1 unit per week, and the sales of the basic model remain constant at 65 units per week, then the total revenue increases by approximately $63 per week.

15. Critical point $(\frac{2}{3}, \frac{2}{3})$, relative maximum
    Critical point $(0,0)$, saddle point

16. Critical point $(3, 2)$, relative maximum

17. Critical point $(1, -1)$, relative minimum
    Critical point $(0, 0)$, saddle point

18. Critical point $(-2, 1)$, saddle point

19. (a) $30,000 on labor
    $20,000 on equipment upgrade
    (b) Maximum profit $100,000

20. (a) $R(x, y) = 600x - 0.4x^2 + 400y - 0.2y^2$
    (b) $P(x, y) = 480x - 0.4x^2$
    $+ 200y - 0.2y^2 - 0.2xy - 60,000$
    (c) 329 units of Product A, 336 units of Product B should be produced and sold per week.

21. $x = 9$, $y = 4$

22. $x = -5$, $y = 5$

23. $x = 8$, $y = 4$

Chapter 7
TABLES

Table 1: Integration Formulas

1. $\int x^n \, dx = \dfrac{1}{n+1} x^{n+1} + C, \; n \neq -1$

2. $\int \dfrac{1}{x} \, dx = \ln x + C, \; x > 0$

3. $\int e^{ax} \, dx = \dfrac{1}{a} e^{ax} + C$

4. $\int x e^{ax} \, dx = \dfrac{1}{a^2} \cdot e^{ax} (ax - 1) + C$

5. $\int x^n e^{ax} \, dx = \dfrac{x^n e^{ax}}{a} - \dfrac{n}{a} \int x^{n-1} e^{ax} \, dx + C$

6. $\int \ln x \, dx = x \ln x - x + C$

7. $\int (\ln x)^n \, dx = x(\ln x)^n - n \int (\ln x)^{n-1} \, dx + C, \; n > 1$

8. $\int x^n \ln x \, dx = x^{n+1} \left[\dfrac{\ln x}{n+1} - \dfrac{1}{(n+1)^2} \right] + C, \; n \neq -1$

9. $\int a^x \, dx = \dfrac{a^x}{\ln a} + C, \; a > 0, \; a \neq 1$

10. $\int \dfrac{1}{x^2 - a^2} \, dx = \dfrac{1}{2a} \ln \left| \dfrac{x-a}{x+a} \right| + C$

11. $\int \dfrac{1}{\sqrt{x^2 \pm a^2}} \, dx = \ln \left| x + \sqrt{x^2 \pm a^2} \right| + C$

12. $\int \dfrac{1}{x \sqrt{a^2 \pm x^2}} \, dx = -\dfrac{1}{a} \ln \left| \dfrac{a + \sqrt{a^2 \pm x^2}}{x} \right| + C$

13. $\int \dfrac{x}{a+bx} \, dx = \dfrac{a}{b^2} - \dfrac{x}{b} - \dfrac{a}{b^2} \ln |a + bx| + C$

14. $\int \dfrac{x}{(a+bx)^2} \, dx = \dfrac{a}{b^2(a+bx)} + \dfrac{1}{b^2} \ln |a + bx| + C$

15. $\int \dfrac{1}{x(a+bx)} \, dx = \dfrac{1}{a} \ln \left| \dfrac{x}{a+bx} \right| + C$

16. $\int \dfrac{1}{x(a+bx)^2} \, dx = \dfrac{1}{a(a+bx)} + \dfrac{1}{a^2} \ln \left| \dfrac{x}{a+bx} \right| + C$

17. $\int \sqrt{x^2 \pm a^2} \, dx = \dfrac{x}{2} \sqrt{x^2 \pm a^2} \pm \dfrac{a^2}{2} \ln \left| x + \sqrt{x^2 \pm a^2} \right| + C$

18. $\int x \sqrt{a+bx} \, dx = \dfrac{2}{15b^2} (3bx - 2a)(a+bx)^{3/2} + C$

19. $\int x^n \sqrt{a+bx} \, dx = \dfrac{2}{b(2n+3)} \left[x^n (a+bx)^{3/2} - na \int x^{n-1} \sqrt{a+bx} \, dx \right] + C$

20. $\int \dfrac{x}{\sqrt{a+bx}} \, dx = \dfrac{2}{3b^2} (bx - 2a) \sqrt{a+bx} + C$

Table 2: Areas of a Standard Normal Distribution

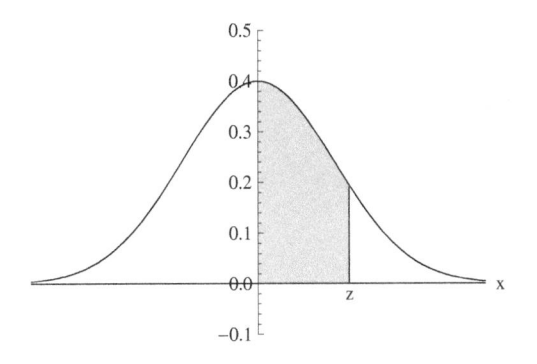

z-score	0.00	0.01	0.02	0.03	0.04	0.05	0.06	0.07	0.08	0.09
0.0	0.0000	0.0040	0.0080	0.0120	0.0160	0.0199	0.0239	0.0279	0.0319	0.0359
0.1	0.0398	0.0438	0.0478	0.0517	0.0557	0.0596	0.0636	0.0675	0.0714	0.0753
0.2	0.0793	0.0832	0.0871	0.0910	0.0948	0.0987	0.1026	0.1064	0.1103	0.1141
0.3	0.1179	0.1217	0.1255	0.1293	0.1331	0.1368	0.1406	0.1443	0.1480	0.1517
0.4	0.1554	0.1591	0.1628	0.1664	0.1700	0.1736	0.1772	0.1808	0.1844	0.1879
0.5	0.1915	0.1950	0.1985	0.2019	0.2054	0.2088	0.2123	0.2157	0.2190	0.2224
0.6	0.2257	0.2291	0.2324	0.2357	0.2389	0.2422	0.2454	0.2486	0.2517	0.2549
0.7	0.2580	0.2611	0.2642	0.2673	0.2704	0.2734	0.2764	0.2794	0.2823	0.2852
0.8	0.2881	0.2910	0.2939	0.2967	0.2995	0.3023	0.3051	0.3078	0.3106	0.3133
0.9	0.3159	0.3186	0.3212	0.3238	0.3264	0.3289	0.3315	0.3340	0.3365	0.3389
1.0	0.3413	0.3438	0.3461	0.3485	0.3508	0.3531	0.3554	0.3577	0.3599	0.3621
1.1	0.3643	0.3665	0.3686	0.3708	0.3729	0.3749	0.3770	0.3790	0.3810	0.3830
1.2	0.3849	0.3869	0.3888	0.3907	0.3925	0.3944	0.3962	0.3980	0.3997	0.4015
1.3	0.4032	0.4049	0.4066	0.4082	0.4099	0.4115	0.4131	0.4147	0.4162	0.4177
1.4	0.4192	0.4207	0.4222	0.4236	0.4251	0.4265	0.4279	0.4292	0.4306	0.4319
1.5	0.4332	0.4345	0.4357	0.4370	0.4382	0.4394	0.4406	0.4418	0.4429	0.4441
1.6	0.4452	0.4463	0.4474	0.4484	0.4495	0.4505	0.4515	0.4525	0.4535	0.4545
1.7	0.4554	0.4564	0.4573	0.4582	0.4591	0.4599	0.4608	0.4616	0.4625	0.4633
1.8	0.4641	0.4649	0.4656	0.4664	0.4671	0.4678	0.4686	0.4693	0.4699	0.4706
1.9	0.4713	0.4719	0.4726	0.4732	0.4738	0.4744	0.4750	0.4756	0.4761	0.4767
2.0	0.4772	0.4778	0.4783	0.4788	0.4793	0.4798	0.4803	0.4808	0.4812	0.4817
2.1	0.4821	0.4826	0.4830	0.4834	0.4838	0.4842	0.4846	0.4850	0.4854	0.4857
2.2	0.4861	0.4864	0.4868	0.4871	0.4875	0.4878	0.4881	0.4884	0.4887	0.4890
2.3	0.4893	0.4896	0.4898	0.4901	0.4904	0.4906	0.4909	0.4911	0.4913	0.4916
2.4	0.4918	0.4920	0.4922	0.4925	0.4927	0.4929	0.4931	0.4932	0.4934	0.4936
2.5	0.4938	0.4940	0.4941	0.4943	0.4945	0.4946	0.4948	0.4949	0.4951	0.4952
2.6	0.4953	0.4955	0.4956	0.4957	0.4959	0.4960	0.4961	0.4962	0.4963	0.4964
2.7	0.4965	0.4966	0.4967	0.4968	0.4969	0.4970	0.4971	0.4972	0.4973	0.4974
2.8	0.4974	0.4975	0.4976	0.4977	0.4977	0.4978	0.4979	0.4979	0.4980	0.4981
2.9	0.4981	0.4982	0.4982	0.4983	0.4984	0.4984	0.4985	0.4985	0.4986	0.4986
3.0	0.4987	0.4987	0.4987	0.4988	0.4988	0.4989	0.4989	0.4989	0.4990	0.4990

Index

Absolute and relative extrema
    Reality Check, 227
Absolute extrema, 181, 445
    Absolute extremum, 445
    Absolute maxima, 445
    Absolute maximum, 445
    Absolute maximum and minimum, 181
    Absolute maximum, minimum, extrema, 227
        on a closed interval, 228
        on other intervals, 230
        relative and, 227
        strategies for determining, 229, 231
    Absolute minima, 445
    Absolute minimum, 445
    Warning, 229
Absolute extrema and constant functions
    Reality Check, 227
Absolute Maximum, 263
Absolute Minimum, 263
Absolute-value function, 36
Accumulated present value
    of continuous money flow, 374
Accumulation, 274
Accuracy, 74
Alfred Marshall, 22
$\int$, 291
Antiderivative, 274
    General, 291
Antiderivatives, 288, 291
    antidifferentiating, 291
    indefinite integral, 291
    integrating, 291
    specific, 294
Antidifferentiating, 291
x approaches a, 74
Approximation
    linear, 157
Area, 274
Area Between Two Curves, 316
Area between two curves
    Warning, 316
Areas and integrals
    Area and antiderivatives, 290
    Area and definite integrals, 307
Asymptote, 212, 263

    horizontal, 214, 263
    horizontal asymptote, 82
    slant, 263
    vertical, 212, 263
    vertical asymptote, 82
Asymptotes
    slant or oblique, 216
Asymptotes and the graph of a function
    Reality Check, 215
average, 390
Average cost, revenue, profit, 169
Average rate of change, 94
    difference quotient, 97
    slope of secant line, 97
average rate of change, 145
Average value, 321
Average value of function, 321

Base of exponential function, 40
    change of bases, 52
Basic Integration Formulas, 293
Basic Integration Properties, 293
Boundary, 445
Break-even point, 19

Chain Rule, 133, 135, 136
    composite function, 133
    special cases of
        Extended Natural Exponential Rule, 130
        Extended Natural Logarithmic Rule, 140
        Extended Power Rule, 129
Change of Bases, 52
Clairaut's Theorem, 441
Close, 74
Common logarithmic, 49
Composition, 134
Compound interests, 40
    and the base e, 42
Concavity, 183, 262
    concave up and down, 183
    inflection points, 183
Constant function
    derivative of, 112
Constant Multiple Property, 116
Constructing a probability density function
    Reality Check, 388

Consumers' expenditure, 355
Consumers' surplus, 355
    at equilibrium point, 360
    effects of taxes and price control on,, 362
Continuity, 84, 86
    and limit, 88
    pencil test, 86
Continuity carelessness
    Warning, 89
Continuos money flows
    future value of,, 370
    present value of,, 374
Continuous money flows, 369
Continuous vs. discrete frequency
    Reality Check, 43
Convergent integrals, 382
Convergent or divergent?
    Reality Check, 382
Critical number, 260
Critical numbers and points, 190
Critical point, 446

Decreasing, 260
Decreasing functions, 9, 178
    first-derivatives and, 179
Definite and indefinite integrals
    Reality Check, 304
Definite integral, 274
Demand and supply
    Reality Check, 21
Demand function, 21
Derivative
    calculation by definition, 104
    calculation by rules
        Natural Exponential Rule, 115
        Natural Logarithmic Rule, 138
        Power Rule, 113
        Product Rule, 123
        Quotient Rule, 125
    constant multiple property, 116
    defined, 103
    equation of tangent line, 118
    higher order, 119
    instantaneous rate of change, 96
    nonexistence of, 107
    notations for, 111
    of $y = a^x$ and $y = \log_a x$, 141
    of constant function, 112
    of natural exponential function, 115
    of natural logarithmic function, 138
    partial, 436
    slopes of tangent lines, 118
    sum and difference property, 117
Derivative notation
    Warning, 112
Derivative of $A(x)$, 290
Difference quotient, 97
    derivative definition, 104
    differentiation using limit of, 104

Differential equations, 407
    general solution, 409
    separation of variables, 408
Differential equations and solutions
    Reality Check, 409
Differentials, 161
Differentiation Rules
    chain rule, 135
    Natural Exponential Rule, 115
    Natural Logarithm Rule, 138
    Power Rule, 113
    Product Rule, 123
    Quotient Rule, 125
Discrete vs. continuous models, 369
Distance vs. positive velocity
    Reality Check, 288
Divergent integrals, 382
Domain, 3
domain, 196
Doubling time, 54
    and growth rate, 54
    and Rule of 70, 54
Doubling Time and Growth Rate, 54

Economists switch axes
    Warning, 22
Effect of rate
    Reality Check, 46
Elasticity of demand, 250, 264
    elastic, inelastic, unitary, 253
    relationship to revenue, 254
Elasticity-Revenue Relationship, 254
Equilibrium point, 22
    equilibrium price p_E, 22
    equilibrium quantity x_E, 22
Estimating areas
    Reality Check, 281
Evaluating a limit by just plugging the number in
    is wrong
    Warning, 79
Existence of a Limit, 75
Existence of absolute extrema
    Reality Check, 228
Expected value, 395
Exponential decay model, 54
Exponential density function, 391
Exponential equations, 48
    and doubling time, 52
Exponential functions, 40
    and financial models, 42
    base of, 40
    change of bases, 52
    compound interests and the base e, 42
    defined, 40
    exponential decay models, 54
    exponential growth models, 54
    initial value of, 40
    present and future values, 43
    vs. power functions, 41

INDEX

Exponential growth model, 54
Exponential vs. power functions
    Reality Check, 41
Extended Natural Exponential Rule, 130
Extended Natural Logarithmic Rule, 140
Extended Power Rule, 129
Extrema
    absolute, 445
    relative, 445
Extreme Value Theorem, 228
Extremum
    absolute, 445
    relative, 445

First-Derivative Test, 261
First-Derivative Test for Relative Extrema, 193
First-Derivatives and Increasing, Decreasing, 179
Fixed cost, 16
Fixed costs and variable costs
    Reality Check, 16
Formulas for computing σ
    Reality Check, 395
Fraction
    size of
        Mnemonic, 79
Functions, 2
    absolute extrema, 181
    absolute maximum, minimum, extrema, 227
        on a closed interval, 228
        on other intervals, 230
        relative and, 227
        strategies for determining, 229, 231
    absolute-value, 36
    application to economics, 16
        cost, 16
        demand, 21
        profit, 19
        revenue, 18
        supply, 21
    concavity, 183
    decreasing, 9, 178
    defind, 3
    dependent variable, output, 2
    domain, range, 3
    evaluating, 3
    exponential, 40
    graphs of, 5
    increasing, 9, 178
    independent variable, input, 2
    linear, 10
    logarithmic, 49
    notations, 2
    of two variables, 434
    piecewise defined, 36, 89
    polynomial, 26
    power, 34
    rational, 32
    relative extrema, 182
Functions Having the Same Derivatives, 291

Fundamental Theorem of Calculus, 274, 305
Future value, 44
    of constant money flow, 371
    of a continuos money flow, 370

General antiderivative, 291
Graph
    sketching, 207, 222
        using a graphing device, 203, 218
Graphical inspection of limits
    Reality Check, 77
Graphical meaning of differentials
    Reality Check, 162
Graphing
    Reality Check, 5
Graphs of functions, 4
    evaluating, 6
    sketching, 5
    vertical line test, 6
Growth rate, 54

Horizontal asymptote, 82, 263
Horizontal asymptotes and non-rational functions
    Reality Check, 215
Horizontal asymptotes and rational functions
    Reality Check, 215
Horizontal intercepts, 30
Horizontal lines, 8
How to determine u and du, 329
How to perform the first-derivative test
    Reality Check, 193
How to perform the second-derivative test
    Reality Check, 195
How to pick the correct integration formula
    Warning, 335

Illegal notation for partial derivatives, 437
    Warning, 437
Illegal use of limit properties
    Warning, 85
Improper Integrals, 382
    convergent, 382
    divergent, 382
Incomplete information
    Reality Check, 279
Increasing, 260
Increasing functions, 9, 178
    first-derivatives and, 179
Indefinite
    integral, 274
Infinite (output) limit, 79
    definition, 80
∞ is not a number
    Warning, 80
Inflection point, 262
Inflection points, 183
Instantaneous rate of change, 96
    derivative, 103
instantaneous rate of change, 145

Integral, 274
    definite, 274
Integrals, 291
    definite integral, 303
        splitting property of, 314
    indefinite integral, 291
    integrand, 292
    sign, 292, 304
Integration, 274, 291
    antiderivatives, 291
    areas and integrals, 306
    basic formulas, 293, 328
    basic properties, 293
    by algebraic manipulation, 326
    by substitution, 328
    Fundamental Theorem of Calculus, 305
    integrable, 292
Intercepts
    vertical intercept, 10

Laws of operations with exponents and radicals, 34
Left-hand limit, 74
Left-rectangle approximation, 279
Leibniz notation, 111
Limit
    and continuity, 88
Limits, 73, 74
    Existence of a Limit theorem, 75
    graphical, 73
    infinite (output) limit, 79
    intuitive definition, 74
    left-hand limit, 74
    $\lim_{x \to a} f(x)$, 73
    numerical, 73
    one-sided limit, 74
    properties, 85
        illegal use of, 85
    right-hand limit, 74
    two-sided limit, 74
Linear approximation, 157
linear approximation, 10
Linear equation, 10
Linear functions, 10
    applications: linear models, 12
    defined, 10
    slope and, 10
Lines, 8
    horizontal, 8
    slope of, 9
    vertical, 8
Lines and concavity
    Reality Check, 183
List of Continuous Functions, 89
Logarithmic functions, 49
    change of bases, 52
    common, 49
    natural, 49

Marginal analysis, 260

Marginal cost, revenue, profit, 166
Marginal Test for Maximum Profit, 171
Marginal vs. actual
    Reality Check, 167
Marginal vs. average
    Reality Check, 170
Mathematical models, 12
    exponential decay models, 54
    exponential growth models, 54
    linear models, 12
Maxima
    absolute, 445
    relative, 445
Maximum, 260
    absolute, 445
    relative, 445
Mean, 395
mean, 390
Mean equals median in the uniform model
    Reality Check, 390
median, 390
Minima
    absolute, 445
    relative, 445
Minimum, 260
    absolute, 445
    relative, 445
Mnemonics
    Fraction
        size of, 79
    Pencil test, 86
    The Chain Train, 134

Natural Exponential Rule, 115
    Extended, 130
Natural logarithmic, 49
Natural Logarithmic Rule, 139
    Extended, 140
Newton's law of cooling, 55
No even roots of negative numbers
    Reality Check, 34
Nondifferentiable, 109
Normal distribution, 397
    standard, 398
Normal vs. standard normal
    Reality Check, 402
Numerical integration, 344

Oblique asymptote, 216
One-sided limit, 74
One-sided limits don't tell you about the value at
        the point
    Warning, 76
Optimization, 260
    Warning, 239
Optimization problems, 237
    steps for solving, 239

Parabola, 27

INDEX

Partial derivative, 436
    geometric meaning, 437
    notations, 436, 441
    second order, 440
Pencil test, 86
    Mnemonic, 86
Piecewise defined functions, 36, 89
Polynomial equations, 29
Polynomial functions, 26
    quadratic, 27
Power functions, 34
    vs. exponential functions, 41
Power Rule, 113, 114
    Extended, 129
Present and Future Values, 44
Present value, 44
    of constant money flow, 374
    of continuous money flow, 374
present value, 44
Present value vs. future value
    Reality Check, 44
Probability density functions, 387
Probability distributions and density functions
    exponential, 391
    mean (expected value) and standard deviation, 395
    normal distribution, 397
    uniform, 389
Process for calculating area
    Reality Check, 302
Producers' surplus, 358
    at equilibrium point, 360
    effects of taxes and price control on,, 362
Product Rule, 123
Profit and break-even point
    Reality Check, 19
Properties of limits, 85

Quadratic formula, 30
Quadratic functions, 27
    parabola, 27
    vertex, 27
Quotient Rule, 125

Radical notations, 34
Range, 3
Rate of change
    average, 94
    instantaneous, 96
Rational functions and equations, 32
Reality Checks
    Absolute and relative extrema, 227
    Absolute extrema and constant functions, 227
    Asymptotes and the graph of a function, 215
    Constructing a probability density function, 388
    Continuous vs. discrete frequency, 43
    Convergent or divergent?, 382
    Definite and indefinite integrals, 304

    Demand and supply, 21
    Differential equations and solutions, 409
    Distance vs. positive velocity, 288
    Effect of rate, 46
    Estimating areas, 281
    Existence of absolute extrema, 228
    Exponential vs. power functions, 41
    Fixed costs and variable costs, 16
    Formulas for computing σ, 395
    Graphical inspection of limits, 77
    Graphical meaning of differentials, 162
    Graphing, 5
    Horizontal asymptotes and non-rational functions, 215
    Horizontal asymptotes and rational functions, 215
    How to perform the first-derivative test, 193
    How to perform the second-derivative test, 195
    Incomplete information, 279
    Lines and concavity, 183
    Marginal vs. actual, 167
    Marginal vs. average, 170
    Mean equals median in the uniform model, 390
    No even roots of negative numbers, 34
    Normal vs. standard normal, 402
    Present value vs. future value, 44
    Process for calculating area, 302
    Profit and break-even point, 19
    Relative and absolute extrema, 181
    Relative and local, 182
    Relative extrema, 182
    Relative extrema and concavity, 183
    Relative extrema and constant functions, 182
    Revenue, 18
    Second derivatives and concavity, 183
    Simplifying constants of integration, 409
    Slant asymptotes and rational functions, 217
    The advantage of the second-derivative test, 195
    The integration formula for powers, 292
    The limit and sum expression of a definite integral, 303
    The middle variable u, 328
    The value of k in an exponential model, 391
    Total variable cost vs. marginal cost, 288
    Units in applications, 17
    Using a graphing device to determine absolute extrema, 234
    Vertical asymptotes and non-rational functions, 213
    Vertical asymptotes and rational functions, 213
    Vertical line test, 6
    Volume and surface area, 161
    Why the second-derivative test may fail, 195
Relative and absolute extrema
    Reality Check, 181
Relative and local

Reality Check, 182
Relative extrama, 182
Relative extrema, 445
    first-derivative test for, 193
    Reality Check, 182
    Relative extremum, 445
    Relative maxima, 445
    Relative maximum, 445
    Relative minima, 445
    Relative minimum, 445
    second-derivative test for, 195
Relative extrema and concavity
    Reality Check, 183
Relative extrema and constant functions
    Reality Check, 182
Relative Extrema and Critical Numbers, 191
Relative Extrema and Critical Points, 447
Relative Maximum, 261
Relative maximum and minimum, 181
Relative Minimum, 261
Revenue
    Reality Check, 18
Right-hand limit, 74

Saddle point, 447
Secant line, slope of, 95
Second derivatives and concavity
    Reality Check, 183
Second-Derivative Test, 261
Second-Derivative Test for Relative Extrema, 195
Second-Derivatives and Concavity, 184
Separation of variables, separable equation, 408
Significance of linear approximations, 161
Simplifying constants of integration
    Reality Check, 409
Slant asymptote, 216, 263
slant asymptote, 217
Slant asymptotes and rational functions
    Reality Check, 217
slope of the secant line, 145
slope of the tangent line, 145
Slope-intercept equation, 10
Slope-point equation, 10
Slopes, 8
    defined, 9
    negative, 9
    of linear functions, 9
    of secant line, 95
    of tangent line, 106
    positive, 9
Splitting Property of a Definite Integral, 314
Standard deviation, 395
Standard normal distribution, 398
Sum and Difference Property, 117
Supply function, 21
Surplus, 354
    consumers', 355
        at equilibrium point, 360
        effects of taxes and price control on,, 362
    producers', 358
        at equilibrium point, 360
        effects of taxes and price control on,, 362
Symbols
    $\int$, 291, 292
    Δx, 157
    Δy, 158
    $\frac{dy}{dx}$, 111
    $\frac{d^n}{dx^n}f(x)$, 120
    $\frac{d^2y}{dx^2}$, 119
    dx, 162
    dy, 162
    $f^{(n)}(x)$, 119
    $f'(x)$, 103
    $f''(x)$, 119
    $-\infty$, 79
    ∞, 80
    $\lim_{x \to a} f(x)$, 74
    $\lim_{x \to a^+} f(x)$, 74
    $\lim_{x \to a^-} f(x)$, 74
    y', 111

Tangent line, slope of, 106
The advantage of the second-derivative test
    Reality Check, 195
The Chain Train
    Mnemonic, 134
The integration formula for powers
    Reality Check, 292
The limit and sum expression of a definite integral
    Reality Check, 303
The middle variable u
    Reality Check, 328
The value of k in an exponential model
    Reality Check, 391
The D-Test, 447
Theorems
    Area Between Two Curves, 316
    Basic Integration Formulas, 293
    Basic Integration Properties, 293
    Chain Rule, 136
    Change of Bases, 52
    Clairaut's Theorem, 441
    Constant Multiple Property, 116
    Derivative of $A(x)$, 290
    Doubling Time and Growth Rate, 54
    Elasticity-Revenue Relationship, 254
    Existence of a Limit, 75
    Extended Natural Exponential Rule, 130
    Extended Natural Logarithmic Rule, 140
    Extended Power Rule, 129
    Extreme Value Theorem, 228
    First-Derivative Test for Relative Extrema, 193

INDEX

First-Derivatives and Increasing, Decreasing, 179
Functions Having the Same Derivatives, 291
Fundamental Theorem of Calculus, 274, 305
Laws of operations with exponents and radicals, 34
List of Continuous Functions, 89
Marginal Test for Maximum Profit, 171
Natural Exponential Rule, 115
Natural Logarithmic Rule, 139
Power Rule, 114
Present and Future Values, 44
Product Rule, 123
Properties of limits, 85
Quadratic formula, 30
Quotient Rule, 125
Relative Extrema and Critical Numbers, 191
Relative Extrema and Critical Points, 447
Second-Derivative Test for Relative Extrema, 195
Second-Derivatives and Concavity, 184
Splitting Property of a Definite Integral, 314
Sum and Difference Property, 117
The D-Test, 447
Unique critical number and absolute extrema, 231

Total cost function, 16
    fixed cost, 16
    variable cost, 16
Total profit function, 19
Total revenue function, 18
Total variable cost vs. marginal cost
    Reality Check, 288
Two-sided limit, 74

Uniform density function, 389
Uniform distribution, 390
Unique critical number and absolute extrema, 231
Units in applications
    Reality Check, 17
Using a graphing device to determine absolute extrema
    Reality Check, 234
Utility, 355

Variable cost, 16
Vertex, 27
Vertical asymptote, 82, 263
Vertical asymptotes
    Warning, 218
Vertical asymptotes and non-rational functions
    Reality Check, 213
Vertical asymptotes and rational functions
    Reality Check, 213
vertical intercept, 10
Vertical line test, 6
    Reality Check, 6
Vertical lines, 8
Volume and surface area
    Reality Check, 161

Warnings
    Absolute extrema, 229
    Area between two curves, 316
    Continuity carelessness, 89
    Derivative notation, 112
    Economists switch axes, 22
    Evaluating a limit by just plugging the number in is wrong, 79
    How to pick the correct integration formula, 335
    Illegal notation for partial derivatives, 437
    Illegal use of limit properties, 85
    ∞ is not a number, 80
    One-sided limits don't tell you about the value at the point, 76
    Optimization, 239
    Vertical asymptotes, 218
Why the second-derivative test may fail
    Reality Check, 195

x-intercepts, 30

y-intercept, 10